AF871460

Foundations of Colour Science

Foundations of Colour Science

From Colorimetry to Perception

Alexander D. Logvinenko
Glasgow Caledonian University
UK

Vladimir L. Levin[†]
Central Economics and Mathematics Institute of the Russian Academy of Sciences
Russia

WILEY

This edition first published 2023
© 2023 John Wiley & Sons Ltd

All rights reserved. No part of this publication may be reproduced, stored in a retrieval system, or transmitted, in any form or by any means, electronic, mechanical, photocopying, recording or otherwise, except as permitted by law. Advice on how to obtain permission to reuse material from this title is available at http://www.wiley.com/go/permissions.

The right of Alexander D. Logvinenko and Vladimir L. Levin to be identified as the authors of this work has been asserted in accordance with law.

Registered Offices
John Wiley & Sons, Inc., 111 River Street, Hoboken, NJ 07030, USA
John Wiley & Sons Ltd, The Atrium, Southern Gate, Chichester, West Sussex, PO19 8SQ, UK

Editorial Office
John Wiley & Sons Ltd, The Atrium, Southern Gate, Chichester, West Sussex, PO19 8SQ, UK

For details of our global editorial offices, customer services, and more information about Wiley products visit us at www.wiley.com.

Wiley also publishes its books in a variety of electronic formats and by print-on-demand. Some content that appears in standard print versions of this book may not be available in other formats.

Limit of Liability/Disclaimer of Warranty
In view of ongoing research, equipment modifications, changes in governmental regulations, and the constant flow of information relating to the use of experimental reagents, equipment, and devices, the reader is urged to review and evaluate the information provided in the package insert or instructions for each chemical, piece of equipment, reagent, or device for, among other things, any changes in the instructions or indication of usage and for added warnings and precautions. While the publisher and authors have used their best efforts in preparing this work, they make no representations or warranties with respect to the accuracy or completeness of the contents of this work and specifically disclaim all warranties, including without limitation any implied warranties of merchantability or fitness for a particular purpose. No warranty may be created or extended by sales representatives, written sales materials or promotional statements for this work. The fact that an organization, website, or product is referred to in this work as a citation and/or potential source of further information does not mean that the publisher and authors endorse the information or services the organization, website, or product may provide or recommendations it may make. This work is sold with the understanding that the publisher is not engaged in rendering professional services. The advice and strategies contained herein may not be suitable for your situation. You should consult with a specialist where appropriate. Further, readers should be aware that websites listed in this work may have changed or disappeared between when this work was written and when it is read. Neither the publisher nor authors shall be liable for any loss of profit or any other commercial damages, including but not limited to special, incidental, consequential, or other damages.

Library of Congress Cataloging-in-Publication Data

Names: Logvinenko, Alexander D., author. | Levin, Vladimir L., author.
Title: Foundations of colour science : from colorimetry to perception / Alexander D. Logvinenko, Glasgow Caledonian University, UK, Vladimir L. Levin, Central Economics and Mathematics Institute of the Russian Academy of Sciences, Russia.
Description: First edition. | Chichester, UK : Wiley, 2023. | Includes bibliographical references and index.
Identifiers: LCCN 2021060391 (print) | LCCN 2021060392 (ebook) | ISBN 9781119885917 (cloth) | ISBN 9781119885900 (adobe pdf) | ISBN 9781119885948 (epub)
Subjects: LCSH: Color. | Colorimetry. | Color vision.
Classification: LCC QC495 .L55 2022 (print) | LCC QC495 (ebook) | DDC 535.6–dc23/eng20220215
LC record available at https://lccn.loc.gov/2021060391
LC ebook record available at https://lccn.loc.gov/2021060392

Cover Design: Wiley

Set in 9.5/12.5pt STIXTwoText by Straive, Chennai, India

Contents

Preface *xi*

1 **Outline for Readers in a Hurry** *1*

Part I Light Colour *59*

2 **Colour Stimulus Space and Colour Mechanisms** *61*
2.1 Grassmann structures and Grassmann colour codes *64*
2.2 Continuous Grassmann structures and continuous Grassmann colour codes *70*

3 **Identification of Grassmann Structures Based on Metameric Matching** *73*
3.1 Colour matching functions *74*
3.2 Monochromatic primaries and colour matching functions in the trichromatic case ($n = 3$) *79*
3.3 Fundamental colour mechanisms in human colour vision *81*
3.3.1 König's approach to identification of the fundamental colour mechanisms *86*
3.3.2 Some estimates of the cone fundamentals used in colour research and applications *89*

4 **Colour-Signal Cone** *93*
4.1 Strong colour-signal-cone-boundary hypothesis *96*
4.2 Empirical status of the strong colour-signal-cone-boundary hypothesis *100*
4.3 Colour-signal-cone-boundary hypothesis *105*
4.4 The colour-signal cone of a 3-pigment Grassmann–Govardovskii structure *109*

5 **Colour Stimulus Manifold** *111*
5.1 Three-dimensional colour stimulus manifold *113*
5.2 Non-linear colour stimulus map. Colour stimulus transformation caused by the medium *116*
5.2.1 The colour stimulus shift caused by the medium variations *117*
5.2.2 Colour robustness to medium variations *119*
5.3 Causes of individual differences in trichromatic colour matching *120*
5.3.1 Effect of the photopigment peak sensitivity on the $qp\lambda$-coordinates *121*
5.3.2 Effect of the ocular media transmittance on $qp\lambda$-coordinates *125*
5.3.3 Trade-off between the ocular media spectral transmittance and the photopigment peak sensitivity in their effect on colour *128*

5.3.4 Dependence of the equivalent peak wavelength shift on light. Impossibility to overcome colour deficiency using a coloured filter *130*
5.3.5 Parametric identification of fundamental colour mechanisms *132*

6 Light Metamerism *135*
6.1 Metamer sets *136*
6.2 Colour mechanisms' transformations preserving light metamerism *139*
6.3 Light metamerism index *141*

7 Light Metamer Mismatching *143*
7.1 Metamer-mismatch regions *143*
7.2 Indices of light metamer mismatching *147*
7.3 Computing trichromatic metamer-mismatch regions *152*
7.3.1 Effect of the spectral positioning of photopigments on metamer mismatching *155*
7.3.2 Effect of the peak photopigment absorbance on metamer mismatching *158*
7.3.3 Metamer mismatching depending on the position in the chromaticity diagram *158*
7.3.4 Metamer mismatching induced by pre-receptoral filters *161*
7.3.5 Differences between cone fundamentals as revealed by metamer mismatching *163*
7.3.6 Metamer mismatching for the 10° colour matching functions of Stiles and Burch *167*
7.3.6.1 Metamer mismatching computation for the 10° colour matching functions of Stiles and Burch *168*
7.3.6.2 Metamer mismatching evaluated for the colour matching functions collected from same individuals *172*
7.3.6.3 Metamer mismatching and physical similarity between colour matching functions *174*
7.3.7 Metamer mismatching induced by neutral density filters *179*
7.3.8 Metamer mismatching produced by camera sensors *182*

8 Light-Colour Perception *187*
8.1 Achromatic scales and achromatic codes *191*
8.1.1 Ordinal brightness scales *192*
8.1.2 Grassmann brightness code. Luminance *196*
8.2 Hue, purity, and brightness fibre bundles. Cylindrical and psychophysical colour coordinates *202*
8.3 Colour transformation caused by media and metamer mismatching, as expressed in the psychophysical colour coordinates *208*
8.4 Light-colour perception in dichromats *210*
8.5 Chromatic structures *215*
8.5.1 Partial hue-matching *218*
8.5.2 Experiment on partial hue-matching *223*
8.5.3 Colour categories *226*
8.5.4 Chromatically ordered structures *229*
8.5.5 Chromatic scales and chromatic codes *230*
8.5.6 Hue, purity, and saturation in chromatic structures *232*
8.6 Light-colour manifold *234*
8.6.1 Hue cyclic order *235*
8.6.2 Light-colour manifold *238*

8.6.3 Circular Hering structures, their representation and experimental identification *239*
8.6.3.1 Opponent chromatic scales. Light-colour map. Perceptual light-colour coordinates. Light-colour gamut *240*
8.6.3.2 Hering–Grassmann structures. Opponent chromatic scales and codes *242*
8.6.3.3 Grassmann-opponent-chromatic-code hypothesis *243*
8.6.3.4 Empirical status of the Grassmann-opponent-chromatic-code hypothesis *244*
8.6.4 Light-colour manifold vs. colour stimulus manifold *248*

9 Typology of Light-Colour Perception. Inter-Individual Differences *253*

10 Colour Matching Structures and Matching Metamerism *263*
10.1 Colour matching structures *268*
10.2 Matching metamerism *276*

11 Identification of Grassmann Structures Induced by Colour Matching Structures *279*
11.1 Colour matching set, threshold set, and sensitivity function *280*
11.2 Regular and strongly regular tolerance extensions *282*
11.3 Identification of Grassmann structures induced by colour matching tolerance relations *285*
11.3.1 Identification of the linear colour mechanism space as a subspace in the linear span of a given set of linearly independent functionals *286*
11.3.2 Deriving the linear colour mechanism space from the colour matching set (the method of tangential hyperplane) *291*
11.3.3 Deriving the fundamental colour mechanisms from the colour matching set that they generate (the method of quadratic approximation) *294*

12 Identification of Indiscriminate Relations. Colour Detection and Discrimination *301*
12.1 Colour detection models *303*
12.1.1 Single-channel detection models *303*
12.1.2 Fundamental colour mechanisms revisited *306*
12.1.3 Multi-channel detection models *307*
12.2 Peak-detector model equivalent to a sublinear colour detection model *308*
12.2.1 Sublinear colour detection models *309*
12.2.2 Multi-channel sublinear models *310*
12.2.3 The most sensitive colour mechanisms *311*
12.3 Colour discrimination models *315*

13 In Search of Colour Mechanisms in the Eye and the Brain *317*
13.1 Do the cone photoreceptor responses encode the colour stimulus? *317*
13.1.1 Local non-linearity of the photoreceptor response *317*
13.1.2 Light adaptation in photoreceptors *319*
13.1.3 Spatial interaction between the cone photoreceptors *320*
13.1.4 Why the colour stimulus cannot be derived from the cone photoreceptor responses *320*

13.2 Do cone-opponent neural cells encode the opponent chromatic codes? 321
13.3 Transition to a different paradigm 326
13.3.1 From symmetric to asymmetric colour matching 326
13.3.2 From light stimulus to light stimulus array 328
13.3.3 On the notion of 'neural image' 330
13.4 Spatio-chromatic processing in the visual cortex 334
13.4.1 Estimating luminance pattern gradient using simple cortical cells 334
13.4.2 Directional gradient encoding with double-opponent cells 341
13.4.2.1 $M - L$-double-opponent cells 341
13.4.2.2 $S - (M + L)$-double-opponent cells 343
13.4.3 Difference in spatial sensitivity of $(M + L)$-, $(M - L)$-, and $S - (M + L)$-cells, and its implication for colour perception 344
13.4.4 Representation of the colour-signal surface in the form of its tangent bundle 345

Part II Object Colour 351

14 Object-Colour Solid 357
14.1 General properties of the object-colour solid 358
14.2 Optimal object stimuli 359
14.3 Elementary step functions as optimal object stimuli 361
14.4 Optimal object stimuli for trichromatic human observers 362
14.5 Condition for all step functions of degree $m < n$ to be optimal object stimuli 363

15 Trichromatic Regular Object-Colour Solid 365
15.1 Meridians of the trichromatic regular object-colour solid 365
15.2 Equator of the trichromatic object-colour solid and strictly optimal object stimuli 369

16 Object-Colour Stimulus Manifold 375
16.1 Object metamerism 375
16.2 Object atlas 378
16.3 Object-colour stimulus manifold. Illuminant-induced non-linear object-colour stimulus map 380
16.4 Trichromatic object-colour stimulus manifold 381
16.4.1 Trichromatic regular object-colour stimulus manifold and its spherical representation 381
16.4.2 Spherical representation of the trichromatic object-colour stimulus manifold and the object-colour stimulus gamut 385
16.4.3 Object-colour stimulus shift induced by the illuminant change 386

17 Object-Colour Perception in a Single-Illuminant Scene 389
17.1 Perceptual object-colour coordinates 394
17.2 Perceptual correlates of $c\Delta\lambda\bar{\lambda}$-coordinates 396
17.3 Effect of illumination on object colour in a single-illuminant scene: Object-colour shift induced by illumination 399
17.4 Object-colour perception by dichromats in a single-illuminant scene 401

18 Object Metamer Mismatching *409*
18.1 Metamer-mismatch regions *409*
18.2 Numerical evaluation of metamer-mismatch regions *413*
18.3 Indices of object metamer mismatching *415*
18.4 Object-metamerism-preserving transformations of colour mechanisms *418*

19 Object-Colour Perception in a Multiple-Illuminant Scene *421*
19.1 Object/light colour equivalence and its inseparability *425*
19.2 Object/light atlas *426*
19.3 Object/light colour stimulus manifold *427*
19.3.1 Asymmetric colour matching *427*
19.3.2 Material colour *430*
19.3.3 Lighting colour *431*
19.3.4 Object/light colour stimulus manifold. Material and lighting components of object/light colour stimulus manifold. Material- and lighting-colour coordinates *433*
19.4 Material colour shift induced by illumination change. Implication for the problem of 'colour constancy' *436*

20 Object-Colour Indeterminacy *439*
20.1 Trade-off between object and light components *439*
20.2 Trade-off between material and lighting colours *443*
20.2.1 Invariant relationship between lightness and lighting brightness *444*
20.2.2 Invariant relationship between lightness, lighting brightness and shading brightness *448*
20.2.3 Shading as a sensory basis of shape *449*
20.2.4 Invariant relationship between material-colour image and lighting-colour image in the chromatic domain *451*
20.3 Object-colour indeterminacy in variegated scenes. Impact of articulation *451*
20.4 Implication for measuring object colour *454*

21 On Perception in General: An Outline of an Alternative Approach *459*
21.1 What is colour for? *461*
21.2 The need for a new approach to perception. A linguistic metaphor *464*

22 Epilogue *473*

References *477*

Appendix A Some Auxiliary Facts from Functional Analysis *497*
A.1 Banach spaces of measures and functions, and stimulus spaces *497*
A.2 Convex analysis *499*

Appendix B Proofs *503*

Index *531*

Preface

The trigger for writing this book was, for me, the realisation that Grassmann's laws – the cornerstone of colorimetry (which is the basis of colour science) – seemed not to be quite correct. True, it has been reported from time to time before that Grassmann's laws get violated under some special circumstances such as high light intensity and/or at the far periphery in the retina. However, I realised that one of the most important aspects of Grassmann's laws – the transitivity law – just could not be possibly true for colour matching even under normal condition of viewing. Intransitivity of colour matching undermines the major achievement of colorimetry: the definition of colour as a class of equivalence (i.e., a class of visually indiscriminable lights) since in this case colour matching cannot be conceptualised as an equivalence relation. The problem, as I saw it, was not only that the whole grand construction of colorimetry, it turned out, was built on sand, but that, despite this, colour television, the very idea of which would have been impossible without colorimetry, still successfully worked.

In the end, resolving this apparent paradox proved to be rather easy (Logvinenko, 2006). However, in doing this I had been becoming more and more confident in the opinion that the very foundations of colour science should be reconsidered. Furthermore, it should be done on quite different grounds, involving rather different mathematics.

It so happened that thus far the basic mathematical tool of colour science has been linear algebra. There is good reason for this. 'Many modern ideas in linear algebra – especially the concept of linear functional and the resulting distinction between a linear space and its dual space – seem to be specially designed to clarify thinking about colour theory' (Krantz, 1975a, p. 283).

Really, lights can be mixed together. Formally, this translates into the statement that there is an algebraic operation on the set of lights. Although colour is a mental, not physical attribute of light, a similar algebraic operation can be defined on the set of colours, namely, colour mixing. As known, the colour of the mixture of two lights depends on the colour of each light rather than their individual spectral compositions. Grassmann was, perhaps, the first to realise that this implies there being an algebraic structure of colour, which is relatively independent of that of light (Grassmann, 1853).

It is the algebraic structure of colours that enables us to represent colours as vectors in the linear space. Also it allows us to use matrix algebra to perform colour computation, that has been widely used in colour science and its applications (e.g., colorimetry, colour imaging, computer graphics). Formally, the algebraic structure of colour is defined as that of the classes of visually equivalent (i.e., subjectively indistinguishable) lights. The vector representation of colours results from mapping the lights onto the classes of visual equivalence (the colour stimulus map). The vector representation of colour, empirically based on colour matching, puts colorimetry and its methods on par with the physical sciences, which has been acknowledged by the inclusion of colorimetry by Richard Feynman in his celebrated lectures on physics (Feynman et al., 1963, Ch. 35).

It should be said, however, that there are natural limitations for the application of linear algebra even in colorimetry, not to mention colour science in general. The finite dimensionality of linear spaces which linear algebra deals with requires the discretisation of continuous functions that naturally arise in colorimetry (e.g., the light energy distributions over the visible spectrum and colour matching functions), which is a problem in itself. Being free of this limitation, functional analysis (dealing with infinite-dimensional linear spaces) seems more appropriate for colour science purposes. We found that many problems in colorimetry are much easier to tackle from the infinite-dimensional standpoint.

The linear structure is not all that colour inherits from light. There is a natural physical measure of proximity between different lights. The fact that close lights have colours which look subjectively close to each other testifies that the light proximity is somehow retained in colour. It means that the colour stimulus map is, in a sense, continuous. While Grassmann stated the law of continuity for colour mixing more than 150 years ago, it has not been elaborated on to the same extent as his other laws of colour mixing. Moreover, the topological concepts underlying the notion of continuity still remain alien to colour science despite the fact that, as shown in this book, they prove to be rather effective for solving a number of problems in colour science.

Topological methods are particularly pertinent to colour science because the colour manifold itself, as well as some other important objects (e.g., the object-colour solid), is convex. All these objects inherit convexity from their physical prototypes. For example, all lights, as a whole, can be represented by a convex cone (in the infinite-dimensional vector space). As the colour stimulus map preserves convexity, the full gamut of colours can also be represented as a convex cone in the vector space, but with finite dimensions. Using the methods of convex analysis for infinite-dimensional topological vector spaces, we have solved some long-standing problems in colour science.

It should be mentioned that the achievements of colorimetry in colour measurement have been made at a cost of neglecting colour perception (i.e., the structure of colour appearance). Really, colorimetry can predict whether two lights will appear of same colour or different, but not what exactly they will look like. Although a number of colour appearance models have been put forward (Wyszecki, 1986; Fairchild, 2005), they are quite far from meeting the colorimetric standards of precision and accuracy. It is hardly surprising since colorimetry has proven to be, in fact, the science about photopigments in the human eye. The participation of a human observer in a colorimetric experiment is intentionally reduced to the role of a comparator, with only two outputs: same or different. The success of a colorimetric study depends on how well a human turns into, virtually, a technical device. Ideally, the influence of all factors (including perceptual factors), with the exception of photopigments, is supposed to be eliminated. The effectiveness of colour television suggests that we are pretty close to this ideal. One has, however, to admit that colorimetry has not quite reached the level of exact sciences, like physics (as evidenced by the aforementioned violation of Grassmann's laws). The reason is that, surprisingly, a human, alas, is unable to entirely simulate such a simple device as a comparator. The fact is that human responses, even such simple ones as binary 'yes/no' s, are inherently fuzzy, whereas colorimetry implies crisp responses, simply ignoring this fuzziness. It was interesting for us to see what would become of colorimetry after the revision of Grassmann's laws, taking into account the fuzziness of human responses.

In order to look into colour perception, we chose to completely break with colorimetry and start from scratch. Guided by Hering's idea of component hues, we attempted a new approach to colour perception based on the same theoretical framework as that used by us to expose the main results obtained in colorimetry. It allowed us to develop techniques to measure colour appearance with an accuracy meeting, we believe, the standards adopted in colorimetric measurements.

Although, human colour vision was our main concern, we hope many results presented in the book can be applicable to animals' colour vision as well as various technical devices equipped with artificial photosensors (e.g., cameras). Our intention was to outline mathematical foundations – the most general principles – which can be used when dealing with a particular system of colour vision based on the same principles as human colour vision. For example, a notion of metamerism is applicable to any live or artificial system of photosensors with different spectral characteristics. The mathematical problem of describing metameric lights (i.e., indistinguishable for these photosensors) naturally arises. It turns out that there is a general solution to this problem (an algorithm is presented in Chapter 6). In fact, much of the book is devoted to various mathematical problems arising in colour science, and their solving.

Going beyond the usual repertoire of mathematical methods used by colour scientists would not have been possible without the participation of a professional mathematician, such as was Vladimir Levin. He made proofs of almost all theorems, lemmas, and propositions in the book. Unfortunately, his untimely death in 2012 prevented him from taking part in the final stage of work on this book. In particular, the Outline (Chapter 1), Chapters 9, 13, 17, 19–21, and the Epilogue were written after he passed away.

It is also worth mentioning the following context, which played a not insignificant part in the genesis of the premise for this book. Since colour is the subject of various disciplines, and given the rapid progress in almost every one of them in recent years, there is an urgent need for a guide that presents the subject from a unified perspective and covers it as completely as possible. At the moment, a person's ideas about colour are very much dependent on which textbook they happen to read, recalling to mind the ancient Indian parable of the blind men and the elephant. This book attempts to at least partially realise this goal, by outlining our own vision of what the science of colour could be like. We, of course, were aware that such a task was too daunting for two. We also understood that we would inevitably have to intrude into adjacent areas in which we are not experts. That being said, such intrusions concerned only colour, not mathematics, so the responsibility for all inevitable inaccuracies and errors lies entirely with me. I shall be most grateful to those readers who discover such errors, if they do not see it as too troublesome to report them to the author. In general, any comments and criticism will be gratefully received if they aid the improvement of this book.

I am also grateful to Sara Hutchinson, for her generous and attentive contribution to the grammatical proofing of the text.

And last but not least, I would never have finished this book if it had not been for the constant support of my daughter Alisa, who has shown a keen interest in my work, always enthusiastically supported me in those rare moments when I lost spirit, and, unlike me, always believed that I would have enough stamina to finish the job.

Alexander D. Logvinenko
July 18, 2022
Glasgow

1

Outline for Readers in a Hurry

Colour is a peculiar, though fascinating, subject for a scientist to explore. A scientific study implies, in general terms, (i) outlining the subject matter; (ii) the experimental method whereby empirical data are collected; and (iii) the theoretical paradigm within which these data are interpreted.

When studying colour, difficulties begin from the very first step: it is not easy to describe colour as a subject for research, let alone give it a scientific definition.[1] Whilst most of us take for granted the colourfulness of our environment, it is worth remembering that the physical objects themselves are colourless. 'For the Rays to speak properly are not coloured' (Newton, 1730). As such, colour is the spawn of our mind. The problem here is not that colour belongs to the perceptual world, but that there is nothing in the physical world that the colour might reflect. Apparent size and apparent shape are also the attributes of the perceptual world; however, one can readily point out their physical counterparts.

Thus, the perception of colour cannot be put in the same category as the perception of size, shape, speed, and other physical variables, which are usually given attention in textbooks. Indeed, size and shape, as they appear to us, are usually regarded as direct representations, even reflections of the corresponding physical entities in our perception. So, even though we do not know how such representations are made possible, the existence of such perceptual attributes as apparent size and apparent shape seems natural.[2] On the contrary, perception of colour raises a problem as there is no colour in the world to which size and depth belong. Why does something that does not exist in the physical world nonetheless appear in our mind? To answer this question, which is in fact a philosophical one,[3] is not easy. At any rate, there is no commonly accepted answer as yet. We will present our views on this matter at the very end of this book (Section 21.1).

In such a situation, there is nothing left but to try to define colour as a mental fact without referring to a particular physical entity behind it (that is, not assuming any physical fact that it reflects). For example, colour is often defined as the perceptual attribute that we mean when using such words as red, green, yellow, brown, and the like. However, these words refer only to one, albeit very

1 An interesting list of the definitions of colour in various areas is given by Koenderink (2010, pp. 3–4).
2 Though, Gibson (1979) insisted that the situation with size and shape, as well as with depth, and in general, with space, is no better.
3 In philosophy it has been debated for a long time whether colour is a subjective or an objective phenomenon (e.g., Hilbert, 1987; Hardin, 1988; Hatfield, 2003).

Foundations of Colour Science: From Colorimetry to Perception, First Edition. Alexander D. Logvinenko and Vladimir L. Levin.
© 2023 John Wiley & Sons Ltd. Published 2023 by John Wiley & Sons Ltd.

important, attribute of colour, namely, *hue*.[4] And then, how to deal with such words as white and black? On the one hand, they do not seem to point out a hue; on the other, they undoubtedly belong to the colour names. Does it follow that white is a colour?[5] Also, are various shades of grey colours? While inclining to an affirmative answer, which is in line with an established tradition (Kaiser & Boynton, 1996), we have to admit that these, usually called *achromatic colours*, are somewhat different from so-called *chromatic colours*, which are referred to with words red, green, pink, etc.

Perhaps, all these questions arise because we are trying to go from colour names to colours, and not vice versa, which, in our opinion, would be more appropriate. It should be borne in mind that our verbal colour categories are much coarser than our colour experiences. Some obvious perceptual differences, pertinent to colour, cannot be rendered by colour names. For instance, we use the same colour name, say, red, to describe physically different things: (i) solid opaque objects (e.g., red apple), (ii) liquids (e.g., red wine), and (iii) self-luminous objects (e.g., red light). Is the redness, which we experience in all these three cases, same or different? On the one hand, they are different since one cannot find, for instance, a red light which will look absolutely identical to the red opaque object. On the other, some argue that, while appearing not quite same, they share the same quality of redness. However, we believe that the redness, that a red apple has, red wine has, and red light has, and which we can identify for all three objects, is a conceptual notion (an idea) rather than a perceptual experience. We do not perceive it. We can only think of it. Arguments in favour to this important conceptual distinction are the same as those Gibson (1979) was putting forward, arguing that space cannot be perceived, it can only be 'conceived' by means of abstract thinking. Specifically, Gibson claimed that it is the environment rather than the physical world that to be perceived. As a result, he considerably modified the concept of what, in his opinion, could be 'seen', including, for example, edibility, into it; but in his concept of 'seeable' no place had been left for space. In general, the issue of what we can 'see' turns out to be far from as trivial as it might seem to a layman and becomes decisive when investigating visual perception.

Because of certain historical reasons, understanding colour as abstract qualia (e.g., redness, greenness, orangeness, etc.), common to all the perceptual experiences, prevails in the science of colour. One reason is that when the science of colour only evolved as independent discipline within psychology the latter was dominated by the philosophical theory of sensations, colour being understood as an attribute of visual sensations (Boring, 1942). Perception was believed to be based on the sensations. Although the Gestalt psychologists (e.g., Koffka, 1935) and Gibson (1979) have convincingly showed that it is a percept (e.g., a red apple) rather than sensations that is immediately experienced, according to the major dogma of the psychology of sensations, a percept can be decomposed, by using introspection, into sensations underlying this percept, in a similar way as a molecule can be decomposed into atoms. Briefly, sensations were thought of as elements of perception, in the same way as atoms were considered as elements of matter. Sensations were believed to have some attributes, colour being one of them. Along with colour, extension and duration were also listed as unalienated attributes of visual sensations (Boring, 1942).

Within the psychology of sensations, an experimental paradigm has been formed for studying them. The way to experimentally study sensations has been paved by the belief that for each

---

4 Being a perceptual attribute, colour itself is commonly described in terms of its own attributes into which the introspective analysis decomposes it. Apart from hue (which is defined as above, using the colour names (Wyszecki & Stiles, 1982, p. 487), colour is usually characterised by its saturation and brightness, the former being understood as the pronouncedness (i.e., chromatic strength) of colour, the latter as the subjective magnitude of the visual sensation in general (Hunt, 1977; Kaiser & Boynton, 1996).

5 'Whiteness is like justice. Everyone is sure he knows what it is, but as soon as he tries to tell someone else what he knows, he gets all confused'. (MacAdam, 1955, p. 78)

attribute of a sensation there exists the *stimulus*, that is, a physical agent which causes and determines this attribute.[6] For example, the frequency of a sound wave is commonly accepted to be the stimulus for such an attribute of auditory sensations as pitch. Its amplitude (generally, sound pressure level) is thought of as the stimulus for loudness. So is light intensity for brightness (i.e., the subjective magnitude of a visual sensation). It is ascertaining the relationship between the stimulus intensity and the subjective magnitude of the corresponding sensation (known as the *psychophysical function*) that psychophysics began from[7] (Fechner, 1860).

Over time, the psychophysical approach has gone beyond a separate discipline and turned into a way of thought that dominates vision science. Colour science has developed within the psychophysical approach, adopting and elaborating psychophysical experimental methods. According to this approach, the main objective of vision research is to ascertain stimuli for the attributes of visual sensation and then to establish the corresponding psychophysical relationship between them, just as the psychophysical functions have been established. However, it did not take long to realise that the search for psychophysical stimuli was a much more complicated task than it seemed at first. Admittedly, in the most general sense, the stimulus for visual sensations is believed to be light. However, it turns out that establishing what characteristics of light can be taken as the stimulus for particular attributes of visual sensations is not easy. For example, apart from the intensity of a light patch its brightness is also affected, to a certain degree, by the size of the patch, its duration, and some others dimensions (Wyszecki & Stiles, 1982). It turns out that finding all physical parameters which determine brightness of light is rather difficult, if possible at all.[8] As a palliative, light intensity is considered as the core stimulus for brightness, the contribution of all other factors affecting brightness being believed to be relatively small.

To find what in the light can be considered as the stimulus for the other attributes of its colour, and especially for the hue, turned out to be even more difficult. An account of what is known on this issue will be preceded by a quick summary of the basic notions and facts concerning light which will be used throughout the book.[9]

Modern science holds that light is a form of *electromagnetic radiation* which propagates through space at a constant speed. Moreover, the physical theory states that light, as well as other forms of matter, has both a wave nature and a particle nature (*wave-particle duality of light*).

In the wave model of electromagnetic radiation, light is viewed as an *electromagnetic wave*, that is, oscillation of electric and magnetic fields. The electric and magnetic field components oscillate at right angles to each other, and to the direction of propagation. An important characteristic of an electromagnetic wave is *frequency*, that is, its rate of oscillation measured in hertz. The frequency of the electromagnetic wave, f, is related to its *wavelength*, λ (measured in metres), by the equation: $c = f\lambda$, where c is the speed of the wave. The electromagnetic waves of all possible wavelengths (in practice, from thousands of metres to 10^{-12} m) are referred to as the *electromagnetic spectrum*.

6 The notion of stimulus came from physiology and became the central notion in *behaviourism* (J. B. Watson, 1919). It is also the core concept in *psychophysics* (Fechner, 1860) and *colorimetry* (Schanda, 2007a).

7 Fechner argued that the sensation magnitude was related to the stimulus intensity through a logarithm function (referred to as *Fechner's* or *Psychophysical logarithmic law*). Fechner's law was challenged by Stevens who believed that brightness (as well as loudness and many other sensory dimensions) grew as a power function of light intensity (referred to as *Stevens's* or *Psychophysical power law*) rather than the logarithmic one (Stevens, 1961). Though, as far as brightness is concerned, this dispute has proven to be of pure theoretical nature because the value of the exponent relating brightness to light intensity was found to be approximately equal to 0.3. Both a logarithmic and a power function with this exponent provide nearly the same goodness of fit to data over quite a range.

8 No wonder, Stevens proclaimed that the problem of stimulus was the major issue in psychophysics (Stevens, 1957).

9 For more detailed introduction to optics from the colour vision standpoint see e.g., Riggs (1965), Wyszecki & Stiles (1982, Ch. 1).

The range of the electromagnetic spectrum within which the sun radiation hits a maximum (wavelengths between approximately 4×10^{-7} and 8×10^{-7} m) is called the *visible spectrum*. Light is said to be electromagnetic radiation in this range of wavelengths. The wavelength of light is commonly expressed in terms of nanometres (10^{-9} m).

Thc particle model of electromagnetic radiation states that the energy of electromagnetic waves (i.e., *radiant energy*) is emitted by a source (and absorbed by an object) in discrete packets of energy, or quanta. A quantum of light is often referred to as a *photon*. Within this point of view, light can be considered as a stream of photons. The energy of a photon, E, is proportional to its frequency, f, and inversely proportional to its wavelength, λ:

$$E = hf = \frac{hc}{\lambda}, \tag{1.1}$$

where h is *Planck's constant* and c is the speed of light in a vacuum.

Electromagnetic waves of a single wavelength are referred to as *monochromatic*. No real electromagnetic radiation can be, strictly speaking, monochromatic. Even such sources as lasers, which are designed to produce monochromatic radiation, do, in fact, emit radiant energy across a whole (though, quite narrow) band of the spectrum. Hence, real light is always a complex wave, that is, a superposition of electromagnetic waves of different wavelengths. An important characteristic of light is its *radiant spectrum* which describes how the radiant energy[10] is distributed with wavelength (i.e., across the spectrum).

Electromagnetic waves reveal themselves only when interacting with other objects. When any electromagnetic waves interact with (i.e., are absorbed by) an object, the energy of the waves is converted to heat. As to the eye, apart from warming, light can result in another, specific effect – isomerisation of the light-sensitive molecules in the retina. This triggers a complex (and not yet fully understood) process of converting a photon into electrical signals in the eye, called *visual phototransduction* (Schnapf & Baylor, 1987; M. E. Burns & Pugh, 2014).

Light absorption causes a change to the photopigment (light-sensitive molecules) contained in the outer segment of the photoreceptors.[11] The probability that a photon is caught by the photopigment is different for different wavelengths. Such a spectral selectivity (or spectral tuning) of the light absorption in the photopigment is characterised by the *photopigment absorption spectrum*, which indicates the amount of light absorbed depending on the wavelength.[12] However, having been captured, each photon contributes into the photoreceptor output signal regardless of the wavelength (the *univariance principle*) (Rushton, 1972). In other words, the process of transforming light absorption into electrical response is assumed to be wavelength-independent, the output electrical signal of a photoreceptor being proportional to the number of absorbed photons (the *quantum catch*).[13]

The probability of a photon capture by a photopigment is maximal for a particular wavelength (different for different photopigments), the photopigment absorption spectrum having an inverted

10 Radiant energy or other radiant entities such as radiant power, radiance, irradiance, etc.

11 The photochemical process in the photopigments is a great deal similar in all vertebrates and most of invertabrates (Bowmaker, 1991).

12 In Section 3.3, we introduce various indices that are commonly used to quantify the efficiency of a photoreceptor in absorbing radiant energy (e.g., absorptance, absorbance). Until then, the term 'absorption spectrum' will be used as a generic term for the spectral tuning of photoreceptors.

13 Admittedly, the phototransduction process is far more intricate. As discussed in Section 13.1.2, the magnitude of the photoreceptor electrical response depends on not only the quantum catch but also on the photoreceptor adaptation state (and some other factors).

U-shaped form.[14] Actually, the end points of the visible spectrum wavelength interval are defined as the wavelengths where the photon capture probability is negligibly small for all the photoreceptors involved. Monochromatic electromagnetic waves falling out of this range are not considered as stimuli for vision.

The relationship between the radiant spectrum of a light and its colour has been noticed a long time ago (for a historical review see e.g., Mollon, 2003). For example, it has been known since Newton's time that monochromatic lights differ in hue, each monochromatic light having its own hue. The rainbow allows everyone to see that literally with their own eyes. However, not every hue can be found in the rainbow. For instance, purple, lilac, violet are not there. Such hues are called *non-spectral*. However, what determines the hue of a light with the broadband spectrum had remained unclear until the nascence of *colorimetry* (Schanda, 2007a).

The major idea behind colorimetry is rather simple. Since the light that enters the eye is the only reality the visual system encounters, it seems a good idea to first examine how the colour of light depends on its radiant spectrum in the laboratory, where you can control all the characteristics of light with the desired accuracy. Over time, certain standards for conducting such experiments were formed (Wyszecki & Stiles, 1982), that will be referred to, for brevity, as the *colorimetric experimental paradigm*. In experiments of this sort, either a single spotlight is presented in isolation (in complete darkness) or light is projected directly to the eye using special optics (with no objects or medium between the light and the observer).[15] The rationale is to study disembodied light, that is, light as such. Behind this is an implicit belief that studying light as such will result in the knowledge of colour as such. Indeed, in this situation, the colour is usually perceived as being attributed to neither object nor medium. A term *unrelated colour* is used to distinguish this type of colour experience from a more natural one (referred to as *related colour*) when colours are perceived as being attributed to objects or media (Kaiser & Boynton, 1996, pp. 30–41). Alternatively, the terms *light colour*, and *object* (or *medium*) *colour* will be used in this book.

Significant advances have been made in studying light colours under the colorimetric experimental paradigm. First of all, it has been found that lights which are physically very different may be indistinguishable for humans. Such lights obviously *match* each other in colour. Furthermore, it has been discovered that lights evoking an equal quantum catch in the cone photoreceptors of the same type appear to make a match.[16] Thus, being indistinguishable for the brain, all the lights producing the same quantum catch match each other, and they have same colour.

Then, it has been established that for each light either it matches a mixture of some amount of daylight and a monochromatic light, or its mixture with some monochromatic light matches daylight. The wavelengths of these monochromatic lights are called *dominant* or *complementary*

---

14 For the vast majority of humans (those known as *trichromats*), their colour vision is mediated by three types of photoreceptors, called *cones*, which have maximal spectral sensitivity in either the short (*S-cones*), the middle (*M-cones*), or the long (*L-cones*) wavelength parts of the visible spectrum (Packer & Williams, 2003). Still, there are people equipped with cone photoreceptors of only two, or even only one, type (referred to as *dichromats* and *monochromats*, respectively).

15 Recent advances in adaptive optics allow us to overcome the optical aberrations of the eye and to form a retinal stimulus image with such high quality that was impossible until recently (Roorda, 2011).

16 Some caveats must be made at this point. The equal quantum catch does not always guarantee the colour match, especially outside the colorimetric paradigm. Even under the colorimetric paradigm, an important necessary condition is the cone photoreceptors being in an identical adaptation state (see Section 13.1.2).

wavelengths for the light, respectively (see Section 5.1). Dominant (respectively, complementary) wavelength is often called the stimulus correlate of hue.[17]

It must be emphasised, however, that colorimetry holds its explanatory capacity only within the colorimetric experimental paradigm. However, the latter implies so impoverished an environment (e.g., a dark laboratory room, presentation of a homogeneous light patch of fixed size and duration, etc.) that one cannot help questioning its relation to natural vision (i.e., vision in the natural environment). The justification of the colorimetric paradigm is rooted in the psychology of sensations the major dogma of which is that sensation as an element of perception is identical to elemental perception, the latter being understood as perception of an isolated stimulus. However, that this dogma does not withstand any criticism was clear already for Gestalt psychologists (e.g., Koffka, 1935).

Furthermore, colorimetry remains vulnerable even on its own territory. No doubt, it has been a great achievement for colour science that a theory of light colour has been created that can predict rather successfully the colour of lights viewed under some standardised conditions (i.e., within the colorimetric experimental paradigm) (Wyszecki & Stiles, 1982). However, apart from being very restrictive, these conditions are practically impossible to implement. Indeed, these conditions ideally imply studying light which is structured neither in space nor in time so as to prevent confounding these variables with the spectral variable (i.e., wavelength). However, it is very difficult to present absolutely homogeneous light. Furthermore, it turns out that light that is absolutely homogeneous in space and time always tends to be perceived as neutral (i.e., hueless, such as daylight) irrespective of its radiant spectrum. Whilst this goes against our everyday life experience, it has been confirmed in experiments with optical stabilisation of the image on the retina. The human visual system was found to rapidly cease responding to unchanging visual stimulation. Specifically, a stabilised image was found to fade within 3–5 seconds (Krauskopf, 1963; Yarbus, 1965/1967).

Therefore, to study how the colour of a patch of light depends on its radiant spectrum, the patch should be presented not in isolation but be accompanied by other lights. However, the colour of the patch may change dramatically in the presence of other patches of light (despite that all the patches are perceived unrelated to any objects). Colour induction in its two varieties – *chromatic simultaneous contrast* and *chromatic assimilation* – can serve as an example (Kaiser & Boynton, 1996, pp. 30–41). Colorimetry loses its predictive power in this situation. Although a number of models have been put forward to explain the effect of the spatial and/or temporal context on colour (e.g., Wyszecki, 1986; Chichilnisky & Wandell, 1995; Mausfeld, 1998), there is no general account as yet.

True, the effect of the spatial and/or temporal context on colour in principle does not undermine colorimetry but only shows its limitations. Really, the colorimetric experimental paradigm implies varying the radiant energy only across the spectrum, keeping the spatial and temporal parameters of the stimulus display constant. As a result, one always has to remember that the explanatory power is limited by this particular spatial and temporal configuration of the stimulus display. For the stimulus display with a different spatio-temporal configuration one will need, strictly speaking, a new colorimetry.[18]

This state of affairs is by no means satisfactory. It seems reasonable to expect that the visual environment for which the theory holds its explanatory power (let us call it the *admissible environment* for the theory) is considerably wider than the *test environment* (i.e., which was used to

---

17 Note that dominant (and/or complementary) wavelength is not a stimulus in the strict sense since, firstly, it is just a number, and secondly, it is not a physical characteristic of the light the hue of which is meant. Instead, it is a characteristic of the monochromatic light involved in matching the light in question. That is why textbooks cautiously use the less definitive term 'stimulus correlate', rather than 'stimulus'.

18 As we will see later, this is exactly what actually happens when different colorimetric standards (e.g., colour matching functions) are approved for visual displays of different size (Wyszecki & Stiles, 1982).

conduct the experiments on which the theory is based). Ideally, the admissible environment for the colour science should encompass the natural scenes. Colour induction indicates that the admissible environment for colorimetry is way far from that objective. An obvious reason for this is that the colorimetric description of light as a radiant spectrum is too restrictive and should be reconsidered. Really, since (i) spatial and temporal modulations are necessary for light to be perceived as coloured, and (ii) the spatial and/or temporal context affect the colour, one should describe light as the radiant energy distribution not only across the spectrum but also across the space and time. An attempt at such generalisation will be given at the end of Part I (Section 13.3.2).

Such a generalisation is a recognition of the fact that a theory of colour as such (i.e., regardless of all other dimensions, such as space and time) cannot be created; that it can only be created as part of the general theory of perception.[19] Besides, this generalisation is made in an attempt to broaden the admissible environment so as to make the theory accountable for more realistic conditions of observation (up to real scenes). However, in the natural conditions, the spatio-temporal modulation of light is produced by reflecting objects, transparent media, and other elements of the environment. Furthermore, under the natural viewing conditions, we see the colour of these objects and/or media rather than that of light. Therefore, we have to include in consideration object colours and transparent medium colours. After all, the ability to see the colour of objects (i.e., the related colours) is much more important from an ecological point of view than the colour of lights.[20]

At this point, we have to return to the issue already raised above. How do unrelated and related colours 'relate' to each other? Is the colour of an object the same as the colour of the light reflected from that object? As mentioned above, the major dogma of the psychology of sensations entails the positive answer. Although there is abundant evidence against this dogma (provided by Gestalt Psychologists and Gibson[21]) these questions in colour science remain without a clear answer.

Perhaps, one reason for the viability of this dogma is that under natural (daylight) illumination objects which look, say, red (tomatoes, apples, etc.) will reflect light with the dominant wavelength belonging to the long-wavelength end of the spectrum and those looking blue to the short-wavelength end of the spectrum. However, the light reflected from an object, generally, depends on both the object's reflectivity and the incident light. Therefore, the same light may be reflected by different objects depending on the incident light. For instance, a white object illuminated by a reddish light and a red object illuminated by a neutral source (daylight) may happen to reflect lights indistinguishable for the eye; however, the colour appearance of the objects will be rather different. Indeed, in the natural environment, we readily distinguish between a piece of chalk illuminated by red light and a red apple under daylight despite that the

19 In fact, one cannot perceive colour separately from space and/or time. The spatio-temporal distribution of colour constitutes the very sensory fabric of the visual image. This is what we find through direct introspection. This basic fact was acknowledged by Maxwell who wrote 'All vision is colour vision, for it is only by observing differences of colour that we distinguish the forms of objects. I include differences of brightness or shade among differences of colour'. (Maxwell, 1872/1970, p. 75) Further analytical breakdown of it leads to such abstractions as colour as such, space as such and the like. Assuming that the colour is processed regardless of space and/or time leads then to unsolvable problems such as the so-called binding problem (see, e.g., Revonsuo & Newman, 1999): 'How do these abstractions then combine into a holistic visual image?'

20 Curiously enough, James J. Gibson was convinced that 'light, properly speaking, can never be seen'. (Gibson, 1979, p. 54).

21 Gestalt Psychologists and Gibson dealt a severe blow to this dogma at the theoretical level. At the empirical level, the inconsistency of this dogma becomes self-evident in the following well-known observation. Look through a small hole in the concealing screen at the cornflower, which appears blue in the sunshine. However, what you see through the hole (i.e., the light reflected by the flower) might be dramatically different from blue. This observation proves that the colour of an object might dramatically differ from the colour of the light reflected by this object. Hence, the object colour cannot be reduced to the light colour.

quantum catches produced by the reflected lights might happen to be the same. Moreover, we will see both a difference in objects' material (yellow lemon vs. blue flowers) and a difference in objects' lighting (yellowish highlights vs. bluish shadows). Therefore, lighting and material spectral characteristics, which are hopelessly confounded in the radiant spectrum of the reflected light, 'miraculously' emerge again in the perceptual image as *lighting colour* and *material colour* (Tokunaga & Logvinenko, 2010c,b,a). It still remains a challenge for the visual science to ascertain how it becomes possible (Foster, 2003; Logvinenko, Funt, Mirzaei, & Tokunaga, 2015).

Another side of the same coin is that, according to immediate observation, material colour appears to remain relatively constant during illumination changes. Really, the white wall remains white, the red apple remains red despite significant changes in ambient illumination throughout the course of the day. Such a relative independence of the colour of an object of its illumination is called *colour constancy*. While colour constancy of objects has been extensively studied (Katz, 1911/1999; D. Jameson & Hurvich, 1989; Foster, 2011; Logvinenko, Funt, Mirzaei, & Tokunaga, 2015; Brainard & Radonjic, 2014), there is still no clear understanding how the visual system manages to achieve it.

So, objects reflecting lights of different (unrelated) colour might appear of the same (related) colour; and vice versa, objects appearing of different (related) colour might reflect lights of the same (unrelated) colour. Therefore, the colour of an object is not the same as the colour of the light reflected by this object.

Moreover, the major reason why the object colours cannot be reduced to unrelated colours is that the nomenclature of the object colours (i.e., the object-colour palette[22]) is richer than that of the unrelated colours (Logvinenko, 2015c). Indeed, an object can be brown or black. However, there is no brown or black light.[23] Nor is there light having, say, khaki colour. For this reason alone, one needs a separate theory of object colours, which does not reduce them to unrelated colours (Logvinenko, 2013b).

Strangely enough, the main obstacle in the way of creating a theory of object colours, as we see it, is colorimetry; to be more exact, its success in accounting for seeing light colours (under the colorimetric viewing conditions). Really, in studying object colours, colorimetry has turned into an obstacle, since all effort has been made to amend colorimetry so as to predict object colour perception. One can say that the history of studying object colours in colour science can be described as a series of attempts to derive the object colour from the quantum catch produced by the light reflected by this object. In Part II where we discuss object colours, we argue at length why such derivation is impossible. The major reason is that the psychophysical determination, which is, by and large, true for light colours (under the colorimetric experimental paradigm), fails for object colours. A change in the theoretical paradigm is needed. Our proposals on this issue will be presented at the end of Part II (Chapter 21).

Part I is devoted to light (unrelated) colours considered under the colorimetric experimental paradigm.[24] The first three chapters cover the range of topics traditionally included in textbooks and chapters on colorimetry (Ohta & Robertson, 2005; Schanda, 2007a; Brainard & Stockman, 2010; Koenderink, 2010). However, the level of consideration of the material is, perhaps, more formal and

---

22 'Colour palette' will be used throughout the book as a loose term referring to all the colours that a particular observer is able to experience.

23 Light can be dark but not black. Only an object can be black. Blackness is not just the absence of light. To see black in some place, one needs light, however not in that place, but nearby (Volbrecht & Kliegl, 1998). For example, a television screen, being dark grey when 'off', can appear black in certain areas (e.g., the 'frame' around the central picture) when 'on', not only due to the absence of any light shining out of those non-lit-up areas, but, importantly, by the contrast created with the lit-up areas, i.e., when areas adjacent to them are more brightly lit up.

24 So, throughout Part I the colour will mean 'light colour', unless otherwise indicated.

systematic than in the previously published mathematical treatises on this subject (Suppes et al., 1989; Dubois, 2009).

We begin (Chapter 2) addressing an important theoretical issue: how to formally represent light, considered as a stimulus, as far as the colorimetric experimental paradigm is concerned. Since in this paradigm the only characteristic of light that varies is its radiant spectrum, the issue reduces to the mathematical representation of radiant spectra. By axiomatising some physical properties of light (Krantz, 1975a), one can show that it is elements of some linear spaces that can be chosen to formally represent light as a stimulus (Example 2.1). The choice of the linear space (referred to as the *light stimulus space*) is important as it essentially predetermines the subsequent analysis. For example, if the radiant spectrum is specified in terms of functions (e.g., the spectral distribution function of radiant power) then monochromatic lights are left behind, since *generalised* (and not ordinary) *functions* (Kolmogorov & Fomin, 1999) are required to describe their radiant spectra in a similar way. This is why we choose to use the space of all Borel measures on the visible spectrum wavelength interval (see Section A.1) as the most general model of the light stimulus space. In fact, it is only a part of the light stimulus space taking the form of a convex cone (referred to as the *light stimulus cone*) that represents lights. Specifically, the cone of positive Borel measures serves as the light stimulus cone.

As pointed out above, being physically different, some lights might have the same colour, or, expressing the same thing somewhat differently, make a colour match. Evidently, such lights will appear to have the same hue, saturation and brightness. It is an important task to ascertain, in experiment, groups of lights matching each other in colour. However, if we instruct an observer to make a judgement concerning the identity or the difference of the light colours, some, especially inexperienced observers, might have problems with such an instruction.[25] To work around these problems, colour matching is often operationalised in terms of equivalent substitution (Wyszecki & Stiles, 1982).

Specifically, two lights are assumed to match each other if one light can substitute the other so that this substitution will remain unnoticed by the observer. The substitution of one light by another can be carried either in time or in space. For instance, one light spot can be repeatedly replaced with another (of the same size and shape). When the observer does not see any flicker (i.e., the light spot is perceived as steady one) the lights are considered as matching each other. Along with such a temporal substitution, its spatial analogue can also be used. In this case, the observer is presented two spatially homogeneous (i.e., unstructured in space) patches of lights next to each other (e.g., as in Fig. I.1). The lights are taken as matching each other when the border between them is invisible. This operationalisation of colour matching is widely used in colorimetry.

A distinctive feature of both the (temporal and spatial) substitution techniques is that these procedures do not involve any subjective judgements on how the light appears. In other words, colour matching operationalised this way falls to class A observations as defined by Brindley (1960). In a sense, a colour match based on the absence of flicker and/or the border in the bipartite field in Fig. I.1 is the extreme opposite of a colour match based on the identity of subjective appearance of the lights.[26] Certainly, there are other ways to operationalise colour matching (e.g., see Wyszecki & Stiles, 1982).

It must be said that the elimination of subjectivity in the equivalent substitution procedure occurs at the cost of the danger of losing the very subject matter. One might argue that what we are dealing

25 For instance, they might ask to specify what is meant by colour; or whether the difference in brightness should be taken as the difference in colour, and so on.

26 Although, when the lights are separated in space (and/or in time), the subjective identity might be the only option.

with in this procedure is the flicker or pattern perception rather than the perception of colour. True, in the case of the temporal substitution, the observer performance can be interpreted as a detection of temporal inhomogeneity (or temporal homogeneity) in the stimulus light spot. Likewise, the spatial substitution relies on detecting (or not detecting) the border in the bipartite field. Such a diversion into the field of spatial and/or temporal detection raises some methodological questions concerning the eligibility of such ways of operationalising colour matching which will be discussed at length in Chapters 10 and 13.

Being an empirical relation between lights, colour matching is known to obey some laws. Depending on how these laws are formulated, we obtain this or that conceptualisation of colour matching, which ultimately determines what sort of colour theory we will eventually come up with. Historically, the first attempt at such a conceptualisation is believed to have been made by Grassmann (1853). To honour his pioneering work, the colour matching laws are often called *Grassmann's laws*. Krantz formalised colour matching as an equivalence relation on the light stimulus cone compatible with the linear structure of the ambient light stimulus space (Krantz, 1975a). Following Krantz we introduce a notion of *metamerism* as an equivalence relation on the light stimulus cone satisfying Grassmann's laws; the light stimulus cone with metamerism satisfying Grassmann's laws being referred to as a *Grassmann structure* (Definition 2.2).

It should be pointed out that metamerism is a key theoretical notion in this book. It has been introduced as a theoretical binary relation formalising colour matching (i.e., an empirical relation between lights which reflects lack of distinction between them). However, as will be shown below, metamerism can be defined independently of colour matching. In this case, colour matching can be modelled as a relation between lights which might be different from metamerism. Such an alternative conceptualisation of colour matching will be considered in Chapter 10 where colour matching does not entail metamerism. When we do not distinguish between colour matching and metamerism (as in Chapters 2–7) we will use the term *metameric matching*.

Being an equivalence relation, metamerism determines a partition of the light stimulus cone into subsets of metamers, i.e., subsets of lights metameric to each other (referred to as *metamer sets*). Each light in a metamer set is supposed to evoke the same colour. In other words, as colour stimuli all the lights in a metamer set are equivalent to each other. Because of this a metamer set will be referred to as the *colour stimulus*, the family of the metamer sets being referred to as the *colour stimulus set*.[27] Since (according to one of Grassmann's laws) metamerism is translatory invariant with respect to summation of lights (i.e., for any three lights a, b, and c the lights a and b are metameric if and only if the light $a + c$ is metameric to the light $b + c$), one can define summation of colour stimuli (i.e., subsets of metamers) (Krantz, 1975a; Wyszecki & Stiles, 1982; Koenderink & Doorn, 2003). Furthermore, the colour stimulus set can be embedded in the finite-dimensional linear space (Theorem 2.1). The dimensionality of this space (referred to as the *colour stimulus space*) is predetermined by one of the axioms assumed for metamerism (the n-chromacy axiom in Definition 2.2). The colour stimulus set inherits the conical structure and convexity of the light stimulus cone. As the colour stimulus set makes a n-dimensional convex cone in the ambient colour stimulus space, it is referred to as the *colour stimulus cone*.

27 Admittedly, one could have called the set of subsets of metamers *the colour set* since its elements are in one-to-one correspondence with the set of colours by definition (see e.g., Koenderink, 2010). However, we will not use the term 'colour set' till Chapter 8 where the internal (perceptual) structure of the set of light colours is considered (Section 8.6.2). We believe that it is this structure that distinguishes the colour set from the family of metamer sets. Thus, the term 'colour stimulus set' emphasises here that we abstract away from this internal colour structure, considering the set of colours just as a formal set.

Such an embedding (of the colour stimulus set into the colour stimulus space) entails there existing such n linear functionals, $\varphi_1, \dots, \varphi_n$, on the light stimulus space $\mathbf{A}$ that metamerism $\sim$ between lights a and b takes place if and only if for every i and for each $a, b \in \mathbf{A}$ one has $a \sim b \Leftrightarrow \varphi_i(a) = \varphi_i(b)$ (Krantz, 1975a). Therefore, each colour stimulus (thus, each colour) can be represented by n real numbers (as a vector in the arithmetic linear space $\mathbf{R}^n$). These numbers can be considered as linear coordinates of the colour stimulus (thus, the corresponding colour) in $\mathbf{R}^n$. We will refer to these as the *colorimetric coordinates* of the colour stimulus.[28]

This representation will remain true even if each linear functional φ_i is followed by an arbitrary one-to-one function g_i. Let us call a superposition $g_i \circ \varphi_i$ a *colour mechanism*,[29] and linear functional φ_i its *linear part*. Then, one can say that that metamerism $\sim$ between lights a and b takes place if and only if the response of each of these colour mechanisms to the light a equals that to the light b. More formally, for each $a, b \in \mathbf{A}$, we have $a \sim b \Leftrightarrow g_i(\varphi_i(a)) = g_i(\varphi_i(b))$ $(i = 1, \dots, n)$. The n values each colour mechanism $g_i \circ \varphi_i$ takes on light a (i.e., $(g_1(\varphi_1(a)), \dots, g_n(\varphi_n(a)))$) will be referred to as the *colour signal* of this light. Therefore, the major objective of colorimetry – to predict which lights will appear the same colour – is formally fulfilled by introducing the notion of colour signal. The lights with the same colour signals, and only them, will have the same colour.

Chapter 3 deals with theoretical issues arising from the problem of experimental identification of Grassmann structures. Theoretically, finding n colour mechanisms is enough for a n-dimensional Grassmann structure identification. As the dimensionality of the Grassmann structure underlying colour vision of a particular individual is unknown in advance, one has to start from its identification (i.e., ascertaining the dimensionality of the colour stimulus space associated with it). It has been established that, for most human observers, there exist triplets of lights (*primaries*) such that either every light matches a mixture of these primaries, or a mixture of this light with one of the primaries matches a mixture of the other two primaries (the law of trichromatism). This entails that the colour stimulus set can be embedded in the three-dimensional linear space where it makes a three-dimensional convex cone (i.e., the colour stimulus cone is three-dimensional). Such human observers are called *trichromats*, and their vision *trichromatic*.[30] Yet, a small proportion of population (just over 2%) have so-called *dichromatic* vision. For such observers, the number of primaries is reduced to two, the colour stimulus cone being two-dimensional. They are called *dichromats*.

It should be borne in mind, however, that the law of trichromatism has been tested only for monochromatic lights. An intuitive justification for this was the belief (descending from E. Schrödinger) that, having been proved for monochromatic lights, the law of trichromatism can be extended to the set of all lights because of continuity. Although Grassmann claimed the law of continuity, it has never been tested. Moreover, it has never been clear how to test this law. We introduce a notion of *continuous Grassmann structure* (Definition 2.11) so as to accommodate

28 Along with the colorimetric coordinates, in Section 8.2 there will be introduced the cylindrical coordinates and the psychophysical colour coordinates, and also, in Section 8.6.3, the perceptual colour coordinates. The colorimetric and cylindrical coordinates are solely based on the metamerism. They represent only the metamerism, containing no information about the colour attributes at all. So these coordinates are suitable for artificial sensors as well. The psychophysical colour coordinates can apply only to human observers since they imply the existence of such colour attributes as hue, saturation, and brightness. Being, in principle, independent of metamerism, the perceptual colour coordinates reflect the perceptual structure of light colours.

29 It must be noted that the term 'colour mechanism', put into circulation by Stiles, is, maybe, one of the most popular in colour literature. Usually, the colour mechanism is thought of as some neural unit related to the perception of colour. Strangely enough, it is always left without a rigorous definition, appealling to our intuition. In this book, the colour mechanism is defined as a quite formal thing, namely, as a function on the light stimulus set, without any speculation concerning its possible implementation in the visual system.

30 As will be shown below, this definition of trichromatism is equivalent to the one given above (through the number of different cone photoreceptor types).

Grassmann's idea of continuity. We show that Grassmann's law of continuity amounts to the continuity of colour-matching functions for monochromatic lights (see Definition 3.1). We prove it is only for continuous Grassmann structures that, first, the law of trichromatism established for monochromatic lights can be extended to the set of all lights. Second, for continuous Grassmann structures, the linear part of a colour mechanism allows an integral representation as in Eq. (2.12) in Section 2.2, the integrand (i.e., function $\varphi(\lambda)$ under the sign of integral in Eq. (2.12)) being referred to as the *spectral sensitivity function* of the colour mechanism in question. Third, the spectral sensitivity function of any colour mechanism (in a n-dimensional Grassmann structure) can be represented as a linear combination of n colour matching functions.

There is a general belief that, for trichromatic human observers, amongst many possible colour mechanisms there might exist a particular triplet (or generally, triplets) implemented somehow in their visual system. In other words, the belief is that there are some units (mechanisms or processes) in the visual system which can be formally described (i.e., conceptualised) as colour mechanisms. Should this be the case then these units (mechanisms or processes) can be considered as the anatomo-physiological basis of human colour vision. Since their spectral sensitivity functions should be linearly related to the observer's colour matching functions, finding this linear transformation has been considered as an important problem for quite a long time. As noted by Maxwell, Helmholtz and König, combining the colour-matching data from trichromats with those from dichromats allows their identification up to a scaling factor (Section 3.3.1).

Specifically, consider linearly independent linear functionals, $\varphi_1, \dots, \varphi_n$, on the light stimulus space, $\mathbf{A}$. Define an equivalence relation $\sim$ on $\mathbf{A}$ as follows. For each $a, b \in \mathbf{A}$, $a \sim b \Leftrightarrow \varphi_i(a) = \varphi_i(b)$ for every i. Satisfying all Grassmann's laws, this equivalence relation induces a Grassmann structure, becoming its metamerism. Functionals $\varphi_1, \dots, \varphi_n$ will be, of course, colour mechanisms with respect to this Grassmann structure. To stress their generative role, these colour mechanisms will be referred to as the *fundamental* (for this particular Grassmann structure) *colour mechanisms*, and equivalence relation $\sim$ as the *fundamental metamerism* (generated by the fundamental colour mechanisms). Proposition 3.4 shows that, given the Grassmann structure generated by some unknown fundamental colour mechanisms $\varphi_1, \dots, \varphi_n$, under certain conditions, one can restore these mechanisms if we know (i) a set of n colour matching functions for this Grassmann structure, and (ii) $n - 1$ colour-matching functions for the Grassmann substructures generated by the subset $\varphi_1, \dots, \varphi_{k-1}, \varphi_{k+1}, \dots, \varphi_n$ for every $k = 1, \dots, n$.

Since the end of the 19th century, this approach has been used repeatedly to derive the hypothetical fundamental colour mechanisms, using the colour-matching data obtained at different times for different trichromatic and dichromatic human observers. After it becomes possible to directly measure the light absorption spectra in the cone photopigments (Dartnall et al., 1983), it has turned out that they are in fairly good agreement with the hypothetical fundamental colour mechanisms derived from the colour matching data.[31] Because of that the latter have been dubbed the cone fundamentals. In Section 3.3.1, we discuss the various versions of the cone fundamentals derived from the colour matching functions by various researchers for a few decades.

It is important to realise that there are a few assumptions implicitly implied by the notion of cone fundamentals. The most crucial are the following. First, colour matching is assumed to meet Grassmann's laws. Second, the cone photoreceptors of three types play the role of the fundamental colour mechanisms. Third, colour matching is assumed to be based entirely on the equality of the

31 Since the light reaching the photopigment is screened by the cornea, lens, macular pigment, and ocular media, for this agreement to take place the cone photopigment absorption spectrum should be measured relative to light at the cornea in order to take into account the spectral selectivity of all these pre-receptoral filters. Measured this way, the cone photopigment absorption spectrum is also called the *cone spectral sensitivity*.

cone responses. Upon closer examination, however, it turns out that none of these assumptions can withstand criticism.

In Chapter 10, we will show that the first assumption simply contradicts the obvious facts and consider a more realistic conceptualisation of colour matching. As for the second assumption, admittedly, the first stage of the visual phototransduction process, namely, photon absorption in cone photopigment (which will be referred to as *cone excitation*), can be described as a linear process (cf. the univariance principle above), so it can be conceptualised as a colour mechanism (Brindley, 1960; Suppes et al., 1989). More specifically, the cone excitation (i.e., the quantum catch) induced by a particular light can be described by an integral of the spectral power distribution function of this light times the cone spectral sensitivity. Therefore, it can be considered a linear functional on the light stimulus space. These functionals can, in principle, play the role of fundamental colour mechanisms. It is these colour mechanisms (considered as fundamental ones) that we will refer to as *cone fundamentals*. It does not follow, however, that the cone photoreceptor, as a device converting incident light into an electrical response, can serve as a colour mechanism. In the end, we will see that the cone photoreceptors as such do not implement any colour mechanisms. Detailed arguments in favour of such the conclusion will be given in Chapter 13. For this reason alone, the third assumption cannot stand, especially along with the first assumption.

Moreover, as shown in Chapter 10, the third assumption cannot be rescued even by replacing 'cone response' by 'cone excitation'. In fact, a model of colour matching asserting that a colour match is based on the equality of the responses of a set of some fundamental colour mechanisms (referred to as the colorimetric model of colour matching) should be recognised as oversimplified, whatever the fundamental colour mechanisms.

In Chapter 10, we will consider a more plausible generalisation of the colorimetric model of colour matching. Specifically, postulating a set of fundamental colour mechanisms (e.g., the cone fundamentals), we will assume that lights that evoke equal responses of all of these mechanisms match each other. Yet, we will not assume that the converse is also true. An important consequence of this is that the relationship between colour matching and the fundamental metamerism becomes more complicated than it has been thought so far. In particular, although the spectral sensitivities of the cone fundamentals might appear to be in a good agreement with the colour matching function, they are, strictly speaking, not related to each other by a linear transformation.[32] How to identify the fundamental colour mechanisms in such a case will be discussed in Chapter 11.

It should be said that our model of colour matching includes the colorimetric model as a particular (albeit somewhat degenerate) case. Because of its role in colorimetry, we will repeatedly resort to the colorimetric model of colour matching in Chapters 4–7, using the cone fundamentals with various spectral sensitivity functions. As is customary in literature on colour vision, the cone spectral sensitivities will be modelled as the product of the photopigment spectral absorptance and the spectral transmittance of the pre-receptoral filters (Judd & Wyszecki, 1975; Smith & Pokorny, 2003). In particular, the photopigment absorptance spectrum model by Govardovskii and colleagues (Govardovskii et al., 2000) will be used throughout the book to model the cone fundamentals (referred to as *Govardovskii's fundamental colour mechanisms*). Also, when dealing with particular problems arising in the course of the presentation, we will make use of a number of exemplars of the cone spectral sensitivities, derived by various researchers from the colour matching functions under the assumption that the latter are linearly related to the spectral cone sensitivities (despite that we do not share this common assumption).

---

32 This is the main difference between our definition of cone fundamentals and the traditional one. Although both are based on the cone spectral sensitivities, we do not expect the cone fundamentals to be a linear transformation of the colour matching functions.

When a Grassmann structure is defined through the fundamental colour mechanisms, a set of responses of these mechanisms to a light will be referred to as the *fundamental colour signal* in response to this light. The 3D space endowed with the fundamental colour mechanisms as coordinate functionals will be referred to as the *fundamental colour-signal space*,[33] the set of all colour signals being referred to as the *fundamental colour-signal cone*. In the case of linear fundamental colour mechanisms (as in colorimetry), the fundamental colour-signal space will be referred to as the colorimetric space. When three arbitrary (not necessarily fundamental) colour mechanisms are concerned, the terms *colour signal*, *colour-signal space*, and *colour-signal cone* will be used.

In Chapter 4, we explore the geometry of the colour-signal cone. An important role in our analysis is played by the so-called *spectral curve* – the set of the colour signals caused by all monochromatic lights in the visible spectrum. We prove (Theorem 4.1) that the colour-signal cone can be represented as the convex hull of the *spectral surface* (defined as the cone through the spectral curve). The spectral curve is shown not to have to entirely belong to the colour cone boundary, since its ends might come into the colour cone interior. The fragment of the spectrum for which the spectral curve lies on the colour cone boundary is called the *effective visible spectrum*. Actually, the colour-signal cone is shown to be a convex hull of the conical surface through the effective visible spectrum. The effective visible spectrum end points prove to be determined by the photopigment spectral absorption, being independent of the pre-receptoral filters. In Section 4.4, it is shown how these end points vary with the photopigment peak absorbance wavelength for Govardovskii's fundamental colour mechanisms.

In the case of a Grassmann structure for which the fundamental colour mechanisms are not defined, any set of linear colour mechanisms can be used to produce a colour-signal cone. Representing one and the same colour stimulus cone, all these colour-signal cones are related to each other by a linear transformation (Remark 4.1). When two different metamerisms are defined on the same light stimulus cone, the colour stimulus cones (thus, the colour-signal cones) induced by them are not linearly related (i.e., there is no linear transformation which maps one colour-signal cone onto the other) even if the dimensionalities of the colour stimulus cones are equal. Furthermore, there is no (even non-linear) one-to-one map between different colour stimulus cones (and the corresponding colour-signal cones) such that it puts in correspondence the colour signals produced by the same lights. The reason for this is that different metamerisms induce different metamer sets. Specifically, two lights belonging to the same metamer set in terms of one metamerism (thus, mapping to the same point in one colour stimulus cone) may happen to be not metameric in terms of another metamerism (thus, mapping to different points in another colour stimulus cone), and vice versa. This phenomenon, called *metamer mismatching* (Wyszecki & Stiles, 1982), is discussed at length in Chapter 7.

It follows that two human observers equipped with cone photoreceptors with different peak sensitivities induce colour stimulus cones (thus, the colour-signal cones) between which one cannot establish any meaningful one-to-one relationship. Moreover, this is the case even for human observers equipped with the same cone photoreceptors but with different pre-receptoral filters. In particular, there is no clear physical basis for establishing correspondence between the colour stimulus cones describing the colour vision of the same observer before and after putting on sunglasses. Still, wearing sunglasses we hardly notice any changes in the colour of the outer world. Hence, the visual system puts somehow in correspondence the colour signals obtained with sunglasses on and off. As a matter of fact, the colour palettes (with sunglasses on and off) are experienced as the same.

---

33 In the case of cone fundamentals, the fundamental colour-signal space is also called the cone excitation space (Smith & Pokorny, 1996).

To be more exact, wearing the sunglasses we experience, firstly, that all the colours are somewhat different (they appear darker, or tinged with some chromatic shade, depending on the tint of the glasses). Secondly, we realise that, in principle, the colour palette remains the same. At any rate, the hues we perceive can be verbally categorised with the same words (yellow, red, green, etc.) as before. As the colour signals evoking the same hues are different with sunglasses and without them, one can conjecture that the visual system establishes some correspondence between the two colour-signal cones (thus, the colour stimulus cones). It is important to describe this correspondence. In full this problem will be discussed in Chapter 9. To tackle it, appropriate mathematical tools are required.

One of these tools (suitable only for trichromatic vision) is described in Chapter 5. It is based on a rather straightforward idea. Given two different colour stimulus cones, consider such a subset of light stimuli that each metamer set (in either colour stimulus cone) contains exactly one of these light stimuli. Such a subset of light stimuli (referred to as the *light-colour atlas* in Logvinenko (2013b)) can be used to simultaneously mark the metamer sets in both the colour stimulus cones. As the light-colour atlas is in a one-to-one correspondence with both the colour stimulus cones, it represents both of them. A particular example of the light-colour atlas is given in Section 5.1.

The example rests on the fact, mentioned above, that for each light, either it is metameric to a mixture of a some amount of daylight and a monochromatic light, or its mixture with some monochromatic light is metameric to daylight. This fact allows us to pick in every metamer set a stimulus of the form: $p\sigma + q\sigma_0$, where p and q are non-negative numbers; σ is either a monochromatic light with some wavelength μ (denoted as δ_μ), or a mixture of the two monochromatic lights constituting the end points of the effective visible spectrum; and σ_0 is daylight (or any other fixed light mapping to the interior of the colour-signal cone). The set of all such mixtures is shown to make a light-colour atlas. Furthermore, we show that in the case of trichromatic colour vision, the set of these mixtures ($p\sigma + q\sigma_0$) is a three-dimensional smooth manifold with boundary diffeomorphic to a half-space. It is called the *colour stimulus manifold*. It represents the colour stimulus cones for all trichromatic observers. (It should be pointed out that, as observers might differ in effective visible spectrum, their light-colour atlases might be different. Nevertheless, these atlases are diffeomorphic to the same manifold with boundary, namely, the three-dimensional half-space.)

A closely related representation of stimulus lights can be obtained if the number p is allowed to be negative. In this case, $\sigma \sim p\delta_\mu + q\sigma_0$ means that either a mixture of lights $p\delta_\mu$ and $q\sigma_0$ is metameric to σ (when $p > 0$), or a mixture of lights $|p|\delta_\mu$ and σ is metameric to $q\sigma_0$ (when $p < 0$). In the first case, wavelength μ is nothing more than what in colorimetry is called *dominant wavelength* for light σ; in the second μ is the *complementary wavelength*.

Given light σ_0, the numbers p and q and the wavelength μ are uniquely determined for a light σ and can be considered as coordinates for it. When σ_0 is daylight, these coordinates lend themselves as a more appropriate representation of the light-colour appearance than colorimetric coordinates. Indeed, in this case, for $p > 0$, μ is the dominant wavelength, and the hue of light σ is rather close to the hue of the monochromatic light with wavelength μ. When μ is a complementary wavelength (i.e., $p < 0$), then the hues of light σ and monochromatic light with wavelength μ are nearly opponent to each other.[34]

Numbers p and q determine where the ray through point σ intersects the oriented line determined by the segment between σ_0 and δ_μ (considered as the points in the light stimulus cone). Specifically, ratio p/q describes the ratio with which this ray divides the line segment in question. When $p < 0$ the ratio is negative which means that the point of intersection precedes σ_0 and δ_μ

34 The definition of opponent hues is given in Section 8.6.3.

(provided that the order in the line is such that σ_0 precedes δ_μ). It proves that when σ_0 is hueless the ratio $p/(q+p)$ correlates with the saturation of the colour of σ. When $p = 0$ (thus, $p/(q+p) = 0$) light σ is metameric to $q\sigma_0$, that is, its saturation is zero. When $q = 0$ and $p > 0$ (thus, $p/(q+p) = 1$) light σ is metameric to $p\delta_\mu$, therefore its saturation is maximal for its hue. In general, the saturation of light σ is monotonically related to $|p/(q+p)|$. Since the cross-ratio (the particular case of which is p/q) is an affine invariant, $p/(q+p)$ remains the same in a colour-signal cone as well. In colorimetry ratio $p/(q+p)$ $(p > 0)$ is known as *colorimetric purity* of light σ (with respect to σ_0) (Wyszecki & Stiles, 1982).

Different points of the colour stimulus manifold represent different metamer sets (thus, colours). Hence, the initial stage of colour vision can be, formally, described as mapping the light stimulus cone onto the colour stimulus manifold. (This map is referred to as the *colour map*). Some observers may happen to perceive identically the colour stimulus manifold.[35] However, even for such observers, their colour maps might be different. Hence, the individual differences in colour vision might result from the differences in both the colour maps and the way they perceive the colour stimulus manifold.

In Section 5.3, we consider the case when the observers have the light stimulus cones which can be represented by the same light-colour atlas, and they perceive the lights in this atlas identically. In such a case, the individual differences can be derived from their colour maps. Specifically, the differences can be specified in terms of numbers p and q and dominant wavelength μ. Particularly, the individual differences arising from the differences in photopigment absorbance (Section 5.3.1) and spectral transmittance of the ocular media (Section 5.3.2) are discussed.

In Section 5.3.3, the trade-off between the ocular media spectral transmittance and the photopigment peak sensitivity is analysed. Specifically, given a small alteration of the ocular media spectral transmittance, we found the so-called equivalent peak wavelength shifts, that is, the shift of the photopigment peak sensitivity wavelengths for the cone photoreceptors that induces the same change of the p, q, and μ coordinates for a particular light. Taken with the opposite sign, these shifts will make up for the spectral transmittance alteration in question. In other words, applied together, these photopigment peak sensitivity shifts and the ocular media spectral transmittance alteration will leave the p, q, and μ coordinates of this light unchanged. Conversely, given a triplet of (small) shifts in photopigment peak sensitivity (between two observers), we evaluate the alteration of the ocular media spectral transmittance which will produce the same effect on colour matching. To be more exact, we evaluate such an alteration of the ocular media spectral transmittance that makes up for this photopigment peak sensitivity shift.

This allows to get insight into the individual differences caused by a photopigment peak sensitivity shift. For instance, one can compute the spectral transmittance of a filter which will cause the same changes in colour perception of a light as those induced by a particular photopigment peak sensitivity shift. Then wearing the glasses with this spectral transmittance, one can literally see with your own eyes how you would see the light colour with the photoreceptors differing from your own by the photopigment peak sensitivity shift in question.

As shown in Section 5.3.4, the equivalent peak wavelength shift essentially depends on the light for which it is evaluated. Because of this dependence, it is impossible to find a filter which will make up for a particular photopigment peak sensitivity shift for all the lights simultaneously. It follows,

---

35 Generally, even monochromatic lights might be perceived differently by different observers. For example, one observer may see a monochromatic light of a particular wavelength, say, 577 nm, as pure yellow, whereas the other will see it as yellow contaminated with some tinge of green or red.

in particular, that it is impossible to cure dichromacy and anomalous trichromacy by means of coloured filters.[36]

In Section 5.3.5, a new method of identification of the cone fundamentals based on metameric matching is proposed. Specifically, described is an algorithm of derivation of the unknown photopigment sensitivity peaks and the ocular media spectral transmittance from the results of colour matching.

In Chapter 6, we deal with a few problems concerning light metamerism. It should be noted that in vision literature, the term 'metamerism' has different meanings in different contexts (Fairman, 1991). Although the main meaning of this term is 'colour match' (i.e., visual indiscriminability) (as in Chapter 2), there is also a term 'observer metamerism' which has a rather different meaning. As pointed out above, when the fundamental colour mechanisms of one observer differ from those of the other, it may happen that two lights metameric for the first observer will not be metameric for the second, and vice versa. While in this case, observer metamerism is often said to take place, as said above, we will call this phenomenon, following Wyszecki & Stiles (1982), metamer mismatching in order to prevent a terminological confusion.

First of all, we seek for a solution of the following problem, referred to as *light metamerism* (Section 6.1): given a set of linear colour mechanisms (e.g., the cone fundamentals), describe the *metamer set* for a particular light (i.e., all the metamers of this light) (Wyszecki & Stiles, 1982; Cohen, 2001). Note that this is different from another problem (referred to as *object metamerism*): given a set of linear colour mechanisms and an illuminant, describe all the objects reflecting (under given illuminant) metameric lights (Wyszecki & Stiles, 1982). The solution to this problem will be given in Part II (Section 16.1). While a number of techniques for generating object metamers have been suggested (Takahama & Nayatani, 1972; Ohta, 1975; Ohta & Wyszecki, 1976a,b; Schmitt, 1976; Worthey, 1988; Trigt, 1994; Cohen, 2001), very few attempts to evaluate a metamer set for an arbitrary light have been made by far (Schmitt, 1976; Cohen & Kappauf, 1982, 1985).

It should be noted that all previous authors dealt with the problem of light metamerism (as well as object metamerism) substituting the spectral power distribution functions (respectively, the spectral reflectance functions) with a sequence of their values at a discrete set of wavelengths, treating the resulting samples as a finite-dimensional approximation of the original functions. Even if we leave aside the unavoidable discretisation errors and dependence on the sampling wavelengths, this discretisation-based approach is aimed at solving, strictly speaking, a different problem, rather than the full characterisation of the light metamer sets. Such an approach is unable, in principle, to produce all the metamers. In Section 6.1, we provide a systematic way to evaluate metamer sets. We prove that metamer sets are convex and compact, and they can be represented as a convex hull of their extreme points. We propose an algorithm to solve for the extreme points of metamer sets. Besides from its theoretical interest, this algorithm can serve as a practical method of generating metamers. As a working example, we show that, for trichromatic human colour vision, the set of lights metameric to a given light can be represented as the convex hull of triads of monochromatic lights which are uniquely determined by this given light.

It turns out that there is no metamerism for monochromatic lights falling within the effective visible spectrum (i.e., those mapping to the curved part of the colour-signal cone boundary). In other words, the metamer set for such a monochromatic light consists of only this monochromatic light itself. On the other hand, it is known that there are a lot of metamers for a neutral light (Wyszecki & Stiles, 1982). This led some researchers to raise the question of the *degree of metamerism*. For instance, some shared an intuitive view that the larger the Euclidean distance between two

36 Still, advertising of glasses, designed for colour blindness treatment, quite often appear on the Internet.

metamers, the bigger degree of metamerism between them. We introduce a quantitative measure of metamerism (the index of metamerism), defined as the diameter of a metamer set, which corresponds to such intuition.

In Section 6.2, we consider the problem of which transformations when applied to a given set of the fundamental colour mechanisms will not break metamerism, that is, it will not change the Grassmann structure determined by these mechanisms. This problem gains its significance in relation to such an important visual phenomenon as adaptation. As known, the appearance of a light might change, sometimes rather dramatically, during prolonged observing the light even if the light itself does not change. In colour literature, the term adaptation refers to both the very visual phenomenon, and 'the process by which the visual mechanism adjusts itself to the conditions under which the eyes are exposed to radiant energy' (Judd & Wyszecki, 1975, p. 352). While the physiological nature of the process of adaptation is far from being fully understood, it is well established that the photoreceptors in the human eye adjust their sensitivity in response to the general lighting conditions (Wyszecki & Stiles, 1982, pp. 429–451). We will use the term *light adaptation* to refer to this process[37] (e.g., on light adaptation in photoreceptors see Section 13.1.2).

The generally accepted view is that light adaptation does not break metamerism in humans (D. Jameson & Hurvich, 1972). For instance, Judd & Wyszecki (1975) state this as a colorimetric principle of persistence of the colour match. 'This principle says that, for the conditions under which Grassmann's laws of additive colour mixture hold, a metameric colour match is unaffected by the adaptive state of the observer's eye'. (Judd & Wyszecki, 1975, pp. 361–362) Yet, some authors claimed otherwise; namely, that lights which were metameric before adaptation can become not metameric afterwards. For example, it was reported that even adaptation to sunlight of moderately high levels might affect trichromatic colour matches, thus metamerism (Jordan & Mollon, 1997).

In this regard, it is of a great interest to find out what sort of transformation the photoreceptors, and more generally, the fundamental colour mechanisms must undergo in order to break metamerism. Obviously, any linear transformation of the fundamental colour mechanisms will retain the metamerism induced by them. The oldest model of light adaptation (known as the von Kries Coefficients Law (Kries, 1905)) meets this constraint. It turns out that all transformations preserving metamerism are exhausted by the linear ones. Hence, light adaptation preserving metamerism is confined to a linear transformation of the photoreceptors' sensitivities. This, of course, does not mean that the von Kries model exhausts the problem of light adaptation. The fact is that any even non-linear recombination of the photoreceptor outputs (but not the photoreceptor sensitivities) will preserve metamerism provided that it is of a one-to-one type. Thus, any post-receptoral (one-to-one) transformation of the cone excitations cannot disturb metamerism provided the receptor spectral sensitivities remain intact.

In Chapter 7, we address the problem of light metamer mismatching. As mentioned above, an observer might distinguish lights metameric for the other observers. We prove that the colour signals, produced by the first observer in response to the metamer set of the second observer, make up a convex compact set (*metamer mismatch region*) which can be represented as a convex hull of its extreme points. We propose an algorithm which can be used to practically evaluate the extreme points of a metamer mismatch region (Section 7.1). We also introduce an index of metamer mismatching as a normalised diameter of the metamer mismatch region (Section 7.2). This index can be considered as a natural generalisation of the observer metamerism index suggested by the

37 Though light adaptation is quite often used to refer to only one particular kind of adaptation caused by the transition from the dark to light (another term for it is *photopic adaptation*). *Dark adaptation* (or *scotopic adaptation*) is used for the reverse condition. Lastly, *chromatic adaptation* is usually associated with a change not only in light intensity but also (and mainly) in the spectral composition.

Commission Internationale de l'Eclairage (CIE) (Wyszecki & Stiles, 1982). However, it depends on the choice of the colour mechanisms. So, we propose another metamer mismatching index which is invariant of the choice of the colour mechanisms. In order to illustrate how this works in practice, the metamer mismatch regions and corresponding metamer mismatching indices have been evaluated for the various colour matching functions and cone fundamentals available in colour literature (Section 7.3).

Chapter 8 is devoted to what is referred to in colour literature as colour appearance. One usually resorts to this term when discussing the perceptual attributes of light colours such as hue, saturation, brightness, and some others (Kaiser & Boynton, 1996; Hunt, 1998). The necessity to bring these attributes into consideration results from the aforementioned fact that colorimetry does not predict what a light will look like. True, it predicts which two lights will look alike but not their appearance, that is, not how they are perceived. Hence, strictly speaking, colour in colorimetry is just a class of lights looking alike. However, we do not perceive colours as disjointed elements of some mosaic that have nothing in common with each other. On the contrary, in perception, light colours form a special perceptual structure. The fact that the light colours are experienced by us as possessing the perceptual attributes mentioned above is the manifestation of the existence of this perceptual structure. In this chapter, we go beyond colorimetry and analyse the perceptual structure of the light colours, that is, the light-colour perception.

Still, we begin with a generalisation of previous attempts to incorporate these perceptual attributes in the framework of colorimetry. In particular, much effort has been made to ascertain some functions on the colorimetric space which would predict hue, saturation, brightness, and the other perceptual attributes which are included in the list of colour appearance attributes (Hunt, 1977; Fairchild, 2005). For example, in almost all colour models there is a sort of hue descriptor which is to predict the hue of a light. However, the significance of theoretical constructions of this kind is undermined by there being no rigorous definition of hue. Hue prediction implies a possibility to measure hue in the quantitative terms. Figuratively speaking, hue scaling should precede hue predicting. Otherwise all attempts to predict hue will appear like little more than lightweight embellishments on the solid edifice of colorimetry.

This state of affairs seems puzzling because, even in the colorimetric paradigm, hue can be defined the same way (i.e., as rigorous) as colour: namely, as a class of equivalence on the set of lights. As has been known for a long time, lights can match in hue. The hue matching is a well-known procedure in colour science. There are just as many reasons to believe that hue matching is an equivalence relation as that colour matching is. So, hue is defined in Section 8.2 as a class of equivalence arising from hue matching. So is brightness: a class of equivalence produced by (heterochromatic) brightness matching (Section 8.1). As a result, the colour stimulus manifold (thus, the colour-signal cone) proves to be stratified by, on the one hand, the subsets of the same hue (referred to as the *hue fibres*), and on the other, the equi-bright subsets. Furthermore, the three-dimensional colour stimulus manifold can be decomposed into (i) a closed one-dimensional (hue) *fibre bundle* (topologically equivalent to a circle); and (ii) the neutral line (i.e., the subset of hueless colour stimuli). Likewise, the colour stimulus manifold can also be sliced into a family of equi-bright subsets. Every equi-bright subset can be represented as a closed area topologically equivalent to a disc. The traces of hue fibres in an equi-bright subset make a fan-like pattern of curvilinear radii (usually referred to as constant hue loci) diverging from the neutral point (i.e., the trace of the neutral line) (S. A. Burns et al., 1984; Hung & Berns, 1995; Pridmore, 2007). When moving along such a curvilinear radius towards the neutral point, saturation is known to decrease, vanishing at the neutral point.

It is generally accepted that two lights of different brightness can be ordered with respect to which is brighter. Starting from this self-evident fact, we develop a theoretical framework for measuring brightness in terms of achromatic scale (Section 8.1) in accordance with the classical axiomatic approach to ordinal and interval scaling (Krantz et al., 1971). Although various brightness scaling procedures are widely used by vision scientists (Judd & Wyszecki, 1975; Wyszecki & Stiles, 1982), its theoretical analysis in the context of the general measurement theory has never been properly done before.

A hue fibre can be ordered not only in relation to brightness, but also in relation to the strength of the hue, which is expressed by words such as saturation, chromatic purity, and the like. Departing from this order, the existence of which is self-evident, we define *purity* as the colour attribute with respect to which the order in question (referred to as the purity order) is carried out by observers. The axioms of the measurement theory are listed, which are sufficient for the existence of ordinal and interval scales for brightness and purity.

Parametrising hue with an angular variable, and employing the purity and brightness scales, the cylindrical coordinates are proposed in the colour stimulus manifold. These are called the *psychophysical colour coordinates*. As the colour stimulus manifold is a psychophysical entity, the coordinates are called psychophysical. However, since the purity and brightness scales are based on the corresponding colour attributes, the word ‘colour’ is also added. It should be mentioned, however, that representing hue by a formal angular variable does not reflect the perceptual structure of this colour attribute. This structure will be revealed in Section 8.5 by using a different (not colorimetric) approach. And then proper perceptual colour coordinates will be introduced in Section 8.6.3.

In Section 8.3, it is shown how to use the psychophysical colour coordinates to quantify the colour transformation caused by media (Section 5.2.1). It is argued that the psychophysical colour coordinates seem to be a natural tools to express the colour stimulus shift caused by the transition from one medium to the other. Also, it seems more natural to characterise metamer mismatching in terms of maximal shifts expressed in the psychophysical colour coordinates. Although the metamer mismatch volume (in the colour-signal space) provides some information concerning the magnitude of metamer mismatching, this information is of relative nature. It allows to compare the magnitudes of metamer mismatching, but it does not provide information about the absolute magnitudes. In contrast, the maximal shift in, say, brightness indicates the maximal difference in brightness that can be caused by metamer mismatching. The same is the case for hue and purity.

In Section 8.4, we address the old problem of what colours dichromats can see and offer a solution based on the approach outlined in Logvinenko (2014). The problem arises from that we cannot rely on the verbal reports by dichromats, because it is unclear how the colour names that they have learned are related to what they see. The approach we adopt is based on the so-called *reductionist hypothesis*. Specifically, we consider one and the same trichromat in two situations: (i) a normal state and (ii) a hypothetical situation when the outputs from the cone photoreceptors of one type are blocked (or nullified) for this trichromat. So, by dichromat in Section 8.4, we shall mean the trichromat whose photoreceptors of one type are taken out of the game. This allows us to assume that the light-colour palette of such the dichromat is some part of the trichromatic light-colour palette, the problem being to describe this part.

That the dichromatic light-colour palette is only a proper part of the trichromatic one follows from the fact that, because of the lack of a one type of the cone photoreceptors, dichromats do not discriminate (thus, confuse) some colour signals discriminated by the trichromats. Almost every colour signal is confused with a whole interval of colour signals (called a *dichromatic confusion interval*). However, it turns out that there exist exactly two rays in the colour-signal cone the points of which are not confused with any other colour signal. In accord with the reductionist hypothesis, we conjecture that the lights mapping to these rays will be identically perceived by both trichromats and dichromats. In other words, in the trichromatic chromaticity diagram, there are two points that represent the chromaticity of those lights that are perceived by the dichromat in the same way as the trichromat. Since the lights mapping to a ray are perceived approximately of the same hue, one can argue that we found two hues of the dichromatic light-colour palette. Furthermore, since the lights mapping to the two rays in question make the boundary of the dichromatic colour-signal cone, it follows that these two hues exhaust the whole dichromatic light hue palette provided the reductionist hypothesis holds true. Really, as we know, the trichromatic light hue palette is determined by the boundary colour signals. So the reductionist hypothesis dictates that it should be the case for the dichromatic light hue palette as well. Although this conclusion is in line with the prediction from, for example, Hurvich & Jameson's model (Hurvich & Jameson, 1957), there is abundant evidence for that the dichromats are able to perceive more than two hues. We will come back to this controversy in Section 17.4 when considering the hue palette of dichromats for object colours.

As stated above, representing the hues in the form of the fibre bundle topologically equivalent to a circle does not exhaust the hue perceptual structure that we find through introspective analysis. Really, hue is commonly believed to consist of a number of component (elemental) hues. Specifically, Hering (1920/1964) believed that the light colours can be phenomenologically described as unitary or binary combinations of the four component hues: yellow, blue, red, and green[38] (e.g., orange is a combination of red and yellow, purple is a combination of red and blue). Moreover, it has been known for a long time that human observers readily take on the task of judging how much of each component hue there is in a given colour. For example, an observer may say that a colour contains 70% red and 30% blue. So, a common practice is to quantify a colour in terms of the relative magnitude of the component hues in it (known as *hue scaling*) (Boynton & Gordon, 1965; Gordon & Abramov, 1988; Gordon et al., 1994). It is not clear, however, how meaningful and consistent such judgements are (a detailed discussion of this issue is given in Section 8.6.3). Moreover, the existing procedures of hue scaling assume the predetermined number of component hues (namely, yellow, blue, red, and green) which are experienced as unique (or elemental). As there is no formal definition of hue in general and unique hue in particular, it is usually left to the observers to decide what these mean. As the notions of hue, component hues and unique hues are of great importance for the colour theory, in Sections 8.5 and 8.6, we present a new approach within which the concept of hue, and the other related notions find their rigorous definition.

The notion of hue is the departing point for this approach (Logvinenko, 2012). We assume that each colour comprises a number of component hues which are understood as just some unknown colour features. The number of component hues is supposed to be arbitrary and unknown. It is shown that the component hues can be derived from the observer's judgements concerning whether the pairs of colours contain some component hues in common. More specifically, for a given colour sample, one can derive a family of the sample subsets such that each subset is the largest one comprising all the colours containing a particular component hue (referred to as the *chromaticity class*). For the practical implementation of this idea, a special experimental technique – the *partial hue matching*[39] – was developed (Logvinenko, 2012) (see Section 8.5.2). Evidently, the number of chromaticity classes indicates the number of component hues. A notion of *m-chromatic structure* is introduced as a light stimulus set with *m* chromaticity classes defined

38 Colours comprising just one component hue are also called 'unique' or 'elemental'.

39 Alternatively, this technique can be named the partial colour matching. A control experiment in which observers were instructed to decide whether two colours contain any shade of any colour in common brought about similar results. We consider this as a confirmation that, beyond the colour science community, the words hue and colour mean about the same thing.

on it (Section 8.5.1). (When the number of component hues is not specified the term *chromatic structure* will be used.) It plays a definitive role within the new approach (comparable to that played by the Grassmann structure in the previous chapters).

Since chromatic structure is a fairly broad and general concept, it is worth narrowing and concretising it, endowing this concept with some important properties that were revealed in experiments on partial hue matching (Logvinenko & Beattie, 2011; Logvinenko & Geithner, 2015).

When a light belongs only to one chromaticity class, its colour is called *unitary*. When a light is included in exactly two chromaticity classes, its colour is called *binary*. This is in line with the traditional terminology according to which unitary colours consist of just one component hue, and binary colours consist of exactly two component hues. Note, however, that our definition of unitary and binary colours is based on a behaviouristic rather than introspective criterion.

The results over all observers participating in the partial hue matching experiments gave rise to a single, special chromaticity class that does not overlap with any other chromaticity class. This means that, although one can formally treat this chromaticity class as manifesting some hue, this hue is very special since it does not combine with any other hue so as to make a binary colour. This chromaticity class, as well as the corresponding component hue, will be referred to as *improper*, all the other chromaticity classes, as well as component hues, being called *proper*. While there can, theoretically, exist many different improper component hues in a chromatic structure, we will consider in the book only chromatic structures with a single improper component hue. Moreover, the term m-chromatic structure will apply to chromatic structures with m proper and one improper component hues. The term component hue will usually mean 'proper component hue' unless the otherwise is said. The unitary colour corresponding to the only improper component hue will be called *neutral* (or *achromatic*).

As to the number of component hues, the majority of observers in experiments on partial hue matching reveal five chromaticity classes, indicating the existence of one improper and four proper (yellow, blue, red, and green) component hues, in line with Hering's conjecture. Nevertheless, it would be unwarranted to restrict our consideration to only four-chromatic structures. The issue of the number of component hues is more complicated than it might seem, and surprisingly, not as important, as we will see later.

Unlike the number of component hues, the following fact can be considered as firmly established in experiments on partial hue matching: there are only primary and binary light colours. No matter how many chromaticity classes were produced by a particular observer, no light was found to belong simultaneously to three chromaticity classes. In other words, no ternary colours were found. In the traditional terminology, this means that there was found no light possessing more than two component hues.

Also, an important property of human colour vision, noticed by Hering, is the fact that not every component hue can be combined with any other in a light colour. Two component hues making a binary colour for some light will be referred to as *concomitant* component hues. It turns out that each component hue has exactly two concomitants (i.e., two concomitant component hues). The importance of this fact (which will be referred to as the *two-concomitant hue condition*) is that it allows the component hues be arranged on a circle so that each component hue is between its two concomitants.

Another important property of human colour vision is that observers are able to order each chromaticity class with respect to the strength (subjective magnitude) of the component hue underlying this chromaticity class (Logvinenko & Beattie, 2006). Such an ability is conceptualised in terms of chromatic ordering. A chromatic structure, each proper chromaticity class of which is chromatically ordered, is referred to as a *chromatically ordered structure* provided it has some additional technical properties (Definition 8.20). In Section 8.5.5, we present a set of axioms for chromatic ordering sufficing the existence of the interval chromatic scale in each chromaticity class (i.e., for each component hue). Although empirical testing these axioms has never been carried out, they do not seem implausible. Nothing what we know of human colour vision testifies against them. At any rate, most of the hue scaling techniques used by colour scientists lose meaning if interval chromatic scales do not exist.

Ascertaining a chromatic scale value for particular light stimuli will be called *component hue scaling*. Although this term will be used in the book primarily as a theoretical concept, it also implies certain techniques for the practical determination of the chromatic scale values for concrete light stimuli. Admittedly, these techniques have yet to be properly developed in accordance with the standards adopted in the theory and practice of scaling[40] (Krantz et al., 1971).

As known from immediate introspection, for any component hue, there are lights which do not contain it. It seems natural to assume the corresponding chromatic scale value for these lights to be zero. We will also assume that for any component hue, there is a light on which the corresponding chromatic scale achieves its maximum. The colours of these lights will be referred to as the *full colours*. Assigning to this light one makes the interval chromatic scale uniquely determined with the range $[0, 1]$.

Taking a value on each chromatic scale, every light gets so-called *chromaticness* – a set of the chromatic scale values assigned to this light. The chromaticness of unitary colours has only one non-zero component. The chromaticness of binary colours has exactly two non-zero components. Lights with equal chromaticness are called *chromatically equivalent*. The set of all classes of chromatic equivalence is called the *chromaticity palette*. It inherits the chromatic orders from the chromatic structure. The chromaticity palette with the inherited chromatic orders makes the *chromaticity structure*.

Hue (in the chromaticity structure) is defined as the subset of lights with proportional chromaticnesses. An improper hue is defined as one for which all the chromatic scale values equal zero. Taking away the improper element from the set of all classes of chromatic equivalence, one gets what will be called the *proper chromaticity palette* and the *proper chromaticity structure*. The lights of the same hue can be ordered with respect to the coefficient of proportionality relating their chromaticnesses. If the proportionality coefficient is more than 1, the first light is said to be of higher purity than the second.

One more interesting and important feature of human perception is the following. For any pair of binary colours comprising the same component hues, h_i and h_j, if component hue h_i is more pronounced in the first colour than in the second, then component hue h_j is less pronounced in the first colour than in the second, and vice versa. (This feature will be referred to as a *concordance condition*.) This allows generalising the circular order, arising on the set of component hues from the two-concomitant hue condition, on the whole set of hues. This circular order determines a circular topology on the set of proper hues. It is noteworthy that this is the same topology no matter what the number of component hues is. This makes the issue of the number of component hues be of secondary importance.

In view of their importance in further consideration, chromatically ordered structures, meeting the two-concomitant hue condition and the concordance condition, are singled out to a special

40 The main difference between the component hue scaling and the hitherto used hue scaling procedures mentioned above is that the former implies presenting the whole chromaticity class and ordering it, whereas the latter usually suggests estimating a single stimulus in terms of a verbally described variable, such as 'redness' or 'greenness'. As a matter of fact, this is nothing else than the magnitude estimation – one of the scaling techniques proposed by S. Stevens (within his 'subjective psychophysics') which was deservedly subjected to serious criticism.

group, called *circular chromatic structures*. Their proper chromaticity palette can be thought as a (hue) fibre bundle over a circle, each (hue) fibre being topologically equivalent to $(0, 1]$. The chromaticity structure itself (i.e., the proper chromaticity palette plus the improper hue) can be geometrically represented as a disc with the improper hue in the centre and the full colours along the circumference of the disc.

Although all the lights in the improper (neutral) chromaticity class have the same (zero) chromaticness, some will be readily experienced as different. A perceptual attribute according to which the lights in the improper (neutral) chromaticity class differ from each other (and according to which they can be ordered) will be called *brightness*.[41] Observers can also order, with respect to brightness, defined this way, the lights in any other element of the chromaticity palette. Moreover, we believe that, for any element of the chromaticity palette, a brightness scale can be constructed. These brightness scales will be called local, since they are defined separately and independently for each element of the chromaticity palette.

An important question is whether these local brightness scales can be combined into a single brightness scale for the entire light stimulus set (as suggested in Section 8.1). The answer to this question depends on whether observers are capable to make consistent judgements such as the following. Is the brightness of, say, a yellow light larger than the brightness of, say, the blue light taken as a standard? Is the brightness of the yellow light twice as much as the brightness of the blue light? In the book, we consider the case when this is possible, and assume, in most cases, the existence of the global brightness scale (as in Section 8.1).

Let us define an equivalence relation on the light stimulus set (referred to as the *light-colour equivalence*) as follows. Two lights will be called light-colour equivalent if, first, they are chromatically equivalent, i.e., they belong to the same element of the chromaticity palette (thus, they have identical chromaticness); and second, they have the same brightness. Note that this definition does not imply the global brightness scale because the definition requires the equality of brightness within the same element of the chromaticity palette. *Light colour* is defined then as a class of the light-colour equivalence. So, two lights have the same light colour if equal are all the chromatic scale values and the brightness scale values for these lights.

In the case of a circular chromatic structure determined by continuous chromatic and brightness scales, the set of all light colours turns out to make a manifold with boundary of the same kind as the colour stimulus manifold (i.e., a three-dimensional smooth manifold with boundary diffeomorphic to a half-space). It will be referred to as the *light-colour manifold*. It should be noted that the dimensionality of the light-colour manifold does not depend on the number of component hues provided the chromatic structure is circular.

The three dimensionality of the light-colour manifold is intuitively clear if we recall that the chromaticity structure (of the circular chromatic structure) can be represented as a disc in $\mathbf{R}^2$. Therefore, adding to each point of the disc (representing some chromaticness), a set of points (along the additional third dimension) representing all possible brightness values associated with this particular chromaticness will make up a three-dimensional subset in $\mathbf{R}^3$.

An important particular case is a circular four-chromatic structure, which aroused so much interest thanks to Hering. It will be referred to as the *circular Hering structure*. In a circular Hering structure for each component hue, there is just one non-concomitant component hue. In the terminology of Hering, these two component hues make up an *opponent* pair.[42] The chromatic scales

---

41 Note that this is an alternative conceptualisation of brightness, which does not assume the observer to have an intuitive understanding of the words 'brightness', 'subjective magnitude', and the like.

42 Adopting this terminology, we will call two component hues *opponent* to each other when they are the only non-concomitant for each other.

corresponding to a component hue and its non-concomitant counterpart can be combined into a single coordinate function (see Eqs. (8.46) and (8.47)), resulting in two *opponent scales* in terms of which the chromaticness of every light can be expressed. Then, each light can be assigned a triplet comprising these two coordinate values and the brightness value (Section 8.6.3). We call this triplet the *perceptual light-colour coordinates* of the light, thereby emphasising that the triplet represents not only the class of the light-colour equivalence (thus, the light colour) but also the hue components, their magnitude, and brightness. The three-dimensional space, $\mathbf{R}^3$, endowed with the perceptual light-colour coordinates, will be called the *perceptual light-colour space*. A map of the light stimulus set into the perceptual light-colour space will be referred to as the *light-colour map*.[43] All the points in this space which are the coordinates of some light (i.e., the range of this map) make the *perceptual light-colour gamut*. It represents the light-colour manifold in the perceptual light-colour space. Its projection parallel to the brightness axis makes what we call the *chromaticness gamut*. When the chromatic scales are interval scales (which is most likely the case for human observers), the chromaticness gamut is unique for each observer, possibly varying between the observers.

It should be emphasised that the light-colour equivalence and all the other related notions (such as the light-colour manifold, the perceptual light-colour space, etc.) have been developed independently of the concept of metamerism and other colorimetric concepts. For example, the circular chromatic structures (e.g., circular Hering structures) are defined within the conceptual framework which is different from (and independent of) that of colorimetry. Yet, these two approaches do not exclude each other. Particularly, they both can be used for studying human colour vision. In this case, a question arises: how are the notions of these two different conceptual frameworks related? For instance, if we assume the fundamental colour mechanisms to exist in the visual system, then, how is the fundamental metamerism (determined by them) related to the light-colour equivalence? Or, since the chromatic scales are derived from the partial hue matching data independently from the classical colour matching data, what is the relationship between the light-colour equivalence and colour matching? Sure, it can be clarified only in experiment. Moreover, the answer to these questions depends on how colour matching is operationalised, namely, what operational criterion is used by observers to make judgement concerning matching. In particular, when colour matching is operationalised as subjective identity, it seems natural to expect the colour matching to coincide with the light-colour equivalence. At any rate, it is of great interest whether observers will accept as a match two lights with identical chromaticness and brightness (according to their own settings).

Also of great importance is the issue concerning the light-colour equivalence and metamerism. At the first glance, it seems natural to expect the light-colour equivalence to include the metamerism. Moreover, when the number of the fundamental colour mechanisms is three, these two equivalence relations are expected to be equal. However, if the number of the fundamental colour mechanisms is more than three, the light-colour equivalence might be coarser than the metamerism. For example, there is ample evidence that the rods participate in colour vision in the mesopic range of light intensity (Wyszecki & Stiles, 1982; Buck, 2014). The rod involvement should increase the number of the fundamental colour mechanisms up to four. Still, there is no evidence for experiencing additional hues during mesopic vision (Pokorny et al., 2006). Therefore, the dimensionality of the light-colour manifold might be different from that of the colour-signal space. They can coincide only when there are three independent fundamental colour mechanisms.

Admitting the possibility of dissociation between the fundamental metamerism and the light-colour equivalence leads to some conceptual collisions. For instance, when the number of fundamental colour mechanisms is more than three, it should be possible, in principle, that

43 It should not be confused with the notion of colour map (Definition 5.2).

some not metameric lights will match each other in colour. It follows that, on the one hand, such the lights should be experienced, at least in some situations, as different (as they produce different colour signals), but on the other hand, they do not differ in colour. At the first glance, it seems an impossible perceptual experience. Indeed, in what other phenomenological form can the difference between such the lights manifest itself, if not in the form of a difference in colour? Still, such a paradoxical experience was reported in a study of discrimination between chromatic sinusoidal gratings (Logvinenko, 1996b). For some gratings, when their contrast is such that the grating is barely visible, one can recognise a striped pattern but not its colour; and vice versa, for some other gratings, one can perceive the grating colour without recognising its frequency. Hence, if this is a paradox, then the paradox is conceptual, indicating the limitation of the colorimetric paradigm.[44]

Ascertaining a connection between the notions of the two conceptual frameworks gives an opportunity to reveal such features of light-colour perception that cannot be established by using either of them alone. Certainly, the primary role in this belongs to experimental studies. Unfortunately, there are not much data to that end. However, the role of a theoretical exploration in this direction should not be underestimated. For instance, if a Grassmann structure and a circular chromatic structure (e.g., circular Hering structure) are simultaneously defined on the light stimulus set, and the metamerism is included in the light-colour equivalence, then the chromatic and brightness scales become the colour codes (see Definition 2.8). When the circular chromatic structure is a circular Hering structure (referred to as *Hering–Grassmann structure*), metameric stimuli have identical perceptual light-colour coordinates. In other words, the metameric stimuli are interchangeable in calculating the perceptual light-colour coordinates. The light stimulus cone in the light-colour map can, therefore, be replaced by the colour-signal cone. The coordinate functions become the functions of colour signal (i.e., the functions of n variable, where n stands for the dimension of the Grassmann structure). We refer to these coordinate functions as the *opponent chromatic codes*. Although the existence of the opponent chromatic codes is usually taken for granted, we show that they exist only if the chromatic scales satisfy some conditions (Section 8.5.5).

The chromatic scales are not necessarily related to the colour matching functions or the fundamental colour mechanisms. Admittedly, such a relation is possible. For example, this is the case in a Hering–Grassmann structure where the chromatic scales are the colour codes. Moreover, it is only in the case of a Hering–Grassmann structure (where the opponent chromatic codes exist) that Hering's idea of the colour opponent mechanisms can be implemented (Section 8.6.3). Although Hering believed in the colour opponent mechanisms in the visual system, as we will see in Chapter 13, this issue requires some theoretical analysis before it can be put at the level of experimental investigation.

Interestingly, Hurvich & Jameson (1957) went even further, assuming, though rather unwarrantedly, that Hering's hypothetical colour opponent mechanisms were linear Grassmann codes, that is, they could be represented as a linear combination of the cone fundamentals. However, there is strong evidence against the linearity of the opponent chromatic codes (for more detail, see Section 8.6.3). Though, this does not undermine the role of the opponent chromatic code notion in the theoretical analysis of light-colour perception presented above.

Consider, for example, that the light-colour equivalence and colour matching coincide. If this is the case, we have to conclude that (since the light-colour equivalence is based on the equality of the opponent chromatic codes rather than the equality of the cone fundamental values) it is the

44 As mentioned above, when perceiving real objects lit by real light sources, one can perceive objects sending the metameric lights to the eye as completely different, e.g., yellow lemon lit by blue light vs. purple plum lit by yellow light.

opponent chromatic codes rather than the cone fundamentals that mediate colour matching. Of course, if the opponent chromatic codes are assumed to be linearly related to the cone fundamentals (as many authors believe), this distinction is a formal one. However, the linearity assumption fails, as stated above. So, it is important to point out that the issue concerning what colour codes underlie colour matching remains open. In Section 12.1.2, we will come back to this problem.

In Chapter 9, the problem of inter-individual differences – as far as light-colour perception is concerned – is considered. First of all, let us ask ourselves a seemingly simple question: what kind of people can be said to perceive colour the same way? The major problem here is not only to answer this question, but first to make it meaningful, that is, to specify more precisely what it is we are asking. Indeed, what might it mean for two observers 'to perceive colour the same way'?

From the colorimetric standpoint, two observers have the same colour vision if their colour matches are identical. That is, one observer accepts all the colour matches set by the other and vice versa. The colour vision of such observers can be conceptualised in terms of one and the same Grassmann structure (or to be more exact, two isomorphic Grassmann structures). However, the agreement between the observers on what lights look alike does not imply that these lights will appear identically for these observers. As follows from Chapter 8, the possibility to describe the colour vision of two observers, by using one and the same Grassmann structure, does not guarantee yet the identity of their light-colour perception. For example, they might have different numbers of component hues.

Besides, even a tiny difference in the peak sensitivity of the cone photoreceptors of at least one type for two observers will result in metamer mismatching for an infinite number of lights. Such observers cannot be, strictly speaking, described with isomorphic Grassmann structures. Therefore, from a colorimetric point of view, one has to admit that these observers have different colour vision. Moreover, as follows from Section 7.3.7, putting on a neutral filter in front of just one eye will result in a similar effect. There is no Grassmann structure that can simultaneously model the colour vision for both the eyes in this case. Yet, it would be against our immediate introspection to accept the hypothesis that an observer with a monocular neutral filter on one eye will experience different light-colour perception with that eye. More likely, the observers with different, from the colorimetric standpoint, colour vision might have the same light-colour perception.

Thus, if we are interested in inter-individual differences in the perception of colour, we must consider the above-mentioned question in terms of the colour perception conceptual framework developed in Chapter 8, rather than the colorimetric one.

The identity of the colour perception of two observers implies, by common sense, that they experience the same colours. However, is it possible to verify that two people experience the same colours, i.e., same colour qualia? Unfortunately, this hypothesis does not lend itself for scientific verification. Undoubtedly of interest to philosophers (e.g., Shoemaker, 1982), this issue goes beyond the scope of scientific cognition. Still, one can find out if the colours experienced by one observer form the same structure as those experienced by the other observer. The methods explored in Chapter 8 (the partial hue matching and the component hue scaling), allow to test whether these structures are same or different. Specifically, the identity of the chromatic structures is a necessary condition for two observers to have identical light-colour perception. However, it is not sufficient. It guarantees only the identity of their light-colour palettes. Still, the same light might be assigned different light colours. For the colour perception of light to be identical for two observers, one has to require that the identity map from the light stimulus set onto itself is an isomorphism of the chromatic structures of these observers. It is a rather strong requirement. Introducing a neutral filter will violate it. Thus, one might expect unavoidable differences between observers even with the same light-colour palettes.

The main attention in Chapter 9 is given to the analysis of the differences between the individuals whose light-colour perception can be modelled by Hering–Grassmann structures described in Section 8.6. It is for these individuals that the notion of the perceptual light-colour gamut is defined (Section 8.6.3). They can be differentiated by the shape of their perceptual light-colour gamuts. This means in particular that while having identical hue palettes, they might have different maximal purity for particular hues. Obviously, observers should have identical perceptual light-colour gamuts in order to be considered as having identical light-colour perception.

It is worth noting that the identity of the perceptual light-colour gamuts for two observers entails a natural one-to-one correspondence between their perceptual light-colour gamuts. This, in turn, induces a one-to-one correspondence between their colour-signal cones based on the equality of the perceptual light-colour coordinates. Thus, a problem that does not have a solution in colorimetry: what points in the colour-signal cones of two observers can be put in correspondence, finds, eventually, a natural solution: those which appear for the observers to be of the same colour (i.e., having the identical chromaticness and brightness). It must be stressed that the difficulty of this problem is not just to establish a correspondence (which can be done in various ways, some of which are given in Chapter 5), but rather to do it in a meaningful way, namely, on the basis of equality of colours. The latter is impossible without the perceptual light-colour coordinates derived from the component hue (as well as brightness) scaling.

As pointed out above, the identity of their perceptual light-colour gamuts does not guarantee that the observers will identically perceive all the lights. In particular, because of metamer mismatching some lights perceived identically by one observer will be of different colours for the others. For instance, a light appearing unique green for one observer might look blueish-green for the other. The difference in perceiving a particular light between two observers can be expressed as the difference between the perceptual light-colour coordinates that this light has in their perceptual light-colour spaces (referred to as the *light-colour shift*). For example, the light-colour shift in the example given above is that the unitary (green) colour becomes the binary colour gaining the second (blue) hue component.

Notably, the light-colour shift is a characteristic of this particular light rather than the light-colour itself. Another light, which also appears the same unique green for the first observer, might gain a different (say, yellow) second hue component (due to the metamer mismatching), or remain unitary green for the second observer. Hence, a change in light colour caused by the transition from one observer to another would be better described as a fuzzy, light-colour dispersion rather than a clear cut, light-colour shift.

In Hering–Grassmann structures, light-colour dispersion is nothing more than metamer mismatching expressed in terms of the perceptual light-colour coordinates. In particular, a metamer mismatch region evaluated in the perceptual light-colour coordinates is called the *light-colour dispersion region*. It is a region in the perceptual light-colour space that shows the possible light-colour variation for one observer that accompanies the perception of a particular light colour experienced by the other observer. The coordinate dimensions of the light-colour dispersion region can be used to quantify the uncertainty of the light-colour shift caused by metamer mismatching. Various measures are proposed to characterise the difference in seeing light colours between observers and its uncertainty arising from metamer mismatching.

Chapter 10 begins with a more profound discussion of the empirical status of Grassmann's laws for colour matching. As noted earlier, known deviations from Grassmann's laws under special observation conditions (e.g., for lights of high intensity or for peripheral viewing) do not undermine, in general opinion, the validity of Grassmann's laws for the standard colorimetric range of observation conditions (Brindley, 1960; Wyszecki & Stiles, 1982). However, it is not difficult to see that Grassmann's laws do not (and actually cannot), strictly speaking, hold true no matter what observation conditions. Curiously enough, in order to show this there is no need in a sophisticated experiment. For instance, to refute the transitivity law, the following *Gedankenexperiment* suffices.

Consider a series of different monochromatic lights with adjacent ones in the series having wavelengths close enough to match each other in colour. Clearly, the series can be made long enough that the first and last lights will not match each other, thus contradicting transitivity. Hence, as an empirical binary relation, colour matching cannot be, strictly speaking, transitive.

One could argue, though, that this intransitivity could be caused by 'incomplete' colour matching. However, a colour match between monochromatic lights with the wavelength difference as small as, say, 0.01 nm can hardly be classed as 'incomplete'. Alternatively, one could just ignore the intransitivity in question, believing that it is due to 'piling up of small errors' which 'must be eliminated by making matches with sufficient statistical precision' (Krantz, 1975a). In either case, we run into a problem of statistically testing the transitivity of human colour matching judgements. However, the methodological repertoire of modern statistics is not quite suitable for testing logical hypotheses – the hypothesis of transitivity in particular. Since pursuing this direction (e.g., Logvinenko et al., 2004) might lead us beyond the scope of the book, we choose instead to admit the intransitivity of colour matching and find out to what extent that might undermine the results obtained in the previous chapters.[45]

When colour matching (considered as a binary relation) is reflexive and symmetric but intransitive, it can be decomposed into an equivalence relation and a reflexive, symmetric, and intransitive binary relation defined on the classes of this equivalence. More specifically, given such an intransitive colour matching relation, define a relation assuming two lights being in this relation if and only if they match the same lights (Definition 10.1). It is shown in Chapter 10 that such a relation is an equivalence (referred to as the *matching equivalence*).

It is worth noting that the transitivity law occupies a special place among other Grassmann's laws. It is necessary for colour matching to obey Grassmann's law of additivity and proportionality. Specifically, it proves that if colour matching is reflexive and symmetric, and obeys Grassmann's additivity and proportionality laws, then it must be transitive (Chapter 10). Therefore, an intransitive colour matching cannot obey Grassmann's additivity and proportionality laws.

Still, nothing prevents the matching equivalence from meeting Grassmann's laws. Assuming the matching equivalence to satisfy Grassmann's additivity and proportionality laws, we introduce a new notion – *colour matching structure* (Definition 10.5) – which is a generalisation of Grassmann structure. The matching equivalence in a colour matching structure will be referred to as the *matching metamerism*.[46] A Grassmann structure determined by the matching metamerism will be referred to as the *induced* (by the colour matching structure) *Grassmann structure*.

A Grassmann structure can be easily converted into a colour matching structure by postulating a more complex relation between colour matching and metamerism (underlying the Grassmann structure) than their identity (as being assumed in colorimetry). Specifically, given a Grassmann

45 Section 16.1.1 in Suppes et al. (1989) entitled 'Three approaches to nontransitive data' exposes the issue in a broader context.

46 Matching metamerism should not be confused with metameric matching (as defined in Chapter 3) – the term widely used in colorimetry. Although both refer to an equivalence relation satisfying Grassmann's laws, the latter is used when colour matching itself is assumed to satisfy Grassmann's laws. In this case, metameric matching is thought of as a theoretical concept, and colour matching as its operational analog. When colour matching does not obey Grassmann's laws (e.g., when it is intransitive), it ceases being 'metameric matching'. Although being determined in terms of colour matching, matching metamerism differs from it. In contrast with metameric matching, matching metamerism is inaccessible for direct observation. So the problem of its identification is much more complicated than for metameric matching (the identification of which is considered in Chapter 3).

structure, let us define colour matching as follows. Two lights match each other when their colour signals are close enough to each other. The proximity between the colour signals resulting in a colour match (of the lights producing these colour signals) can be defined, for example, in terms of some metric (on the colour-signal space). Specifically, one can assume that two lights match each other if and only if the distance (in terms of this metric) between the colour signals induced by them does not exceed some threshold value ε. In this case, the equality of the colour signals for two lights implies their colour match. However, the colour match between lights does not imply the equality of the colour signals for these lights. In other words, for a light a to match light b, the former does not have to produce the same colour signal as (i.e., to be metameric to) the latter. Even if a induces a different colour signal, it will match b provided that this colour signal belongs to the ε-ball around the colour signal evoked by b.

Generalising this example, we come up with the following representation of colour matching. Given a Grassmann structure, for any point x in the colour-signal cone, define such a volume around it that any other colour signal inside this volume results in a match. This volume will be referred to as the *indiscriminable set of* x. The net of the indiscriminable sets around each colour signal specifies what is called in Chapter 10 the *indiscriminate relation*. Being reflexive, symmetric and intransitive this relation along with the metamerism (of the underlying Grassmann structure) determines an intransitive colour matching.

It should be mentioned that not always a Grassmann structure followed by an arbitrary indiscriminate relation produces a colour matching structure, the induced Grassmann structure of which coincides with the underlying Grassmann structure. Chapter 11 discusses at length what properties the indiscriminate relation should satisfy, so as to avoid this kind of pathology.

Experimental identification of a colour matching structure implies the identification of both the matching metamerism (to be more exact, the induced Grassmann structure) and the indiscriminate relation. For identification of the induced Grassmann structure, one needs to find some linear colour mechanisms which could represent the matching metamerism. All such linear colour mechanisms make a (dual) linear subspace (referred to as the *linear colour mechanism subspace*). In fact, the identification of a Grassmann structure reduces to the identification of its linear colour mechanism subspace. So any set of linear colour mechanisms which make a basis in the linear colour mechanism subspace will do. However, the search for such colour mechanisms is significantly hampered by the fact that the matching metamerism is not as directly observed as metameric matching.

Identification of the indiscriminate relation reduces to describing all the indiscriminable sets. When the indiscriminate relation is represented in terms of some proximity function (as in the example above), then this function (referred to as the *discrimination function* (Chapter 12)) is to be identified.

As discussed above, an important particular case is when the matching metamerism is determined by some fundamental colour mechanisms. That is, the model of colour matching consists of a number of colour mechanisms (playing the role of fundamental ones) followed by a discrimination function (determining indiscriminable sets). For identifying such a model, one needs to know not only the subspace determined by the fundamental colour mechanisms but also the fundamental colour mechanisms themselves.

That this colour matching model is more realistic than the colorimetric one (which, being based on the equality of the cone fundamental values, assumes singleton indiscriminable sets) follows from it being in line with colour discrimination models. Recall that the two notions – colour discrimination and colour matching – are intimately related to each other. Indeed, being presented with two lights, a human observer can either discriminate or fail to discriminate them. Hence, when colour matching is understood as the lack of colour discrimination,[47] any model of colour discrimination is at the same time a model of colour matching, and vice versa. Strangely enough, in spite of the complementarity of the notions of colour discrimination and colour matching, the models of colour discrimination and colour matching, available in colour vision literature, seem to be in an apparent contradiction. Really, all the modern colour discrimination models predict proper (i.e., larger than singletons) indiscriminable sets in the colour-signal space, thus implying intransitivity of colour matching, whereas the colour matching transitivity is the major dogma of colorimetry.[48]

Specifically, all the colour discrimination models proposed over the last few decades (Eskew et al., 1999; Smith & Pokorny, 2003; Eskew, 2008; Stockman & Brainard, 2010; T. G. Shepard et al., 2016) assume, as the first stage, the cone fundamentals, and, as the second stage, some way to combine their outputs so as to predict the colour discrimination data (e.g., the indiscriminable sets revealed in the empirical studies of colour discrimination).

Curiously, in colour discrimination studies, the main attention is usually paid to the second stage, taking for granted that the first stage consists of either the cone fundamentals themselves, or some linear transformation of them. The pioneer works in the field were driven by the hope to make up a more or less simple discrimination function which could successfully predict the indiscriminable sets in the fundamental colour-signal space. In particular, much effort has been made to find a metric which could serve as the discrimination function compatible with the colour discrimination data.[49] As pointed out above, when the indiscriminate relation is defined in terms of a metric, all the indiscriminable sets are the balls of ε-radius. (For an arbitrary indiscriminate relation the indiscriminable sets are some arbitrary neighbourhoods around the colour signals.) However, there is experimental evidence against using a metric as the discrimination function. Specifically, the indiscriminable sets were not always found to be symmetric with respect to the origin (Logvinenko, Tyurin, & Sawey, 2002). The asymmetry of, at least some, indiscriminable sets means that colour discrimination cannot be modelled by any metrics. Therefore, any attempt to derive colour metric from the colour discrimination data – the task that inspired many prominent scientists (Helmholtz, Schrödinger and many others) – is doomed to failure.

However, the main drawback of all the modern colour discrimination models is that they incorporate (at the first stage) the cone fundamentals as derived in colorimetry. In other words, these models of colour discrimination are based on the results obtained by using a different (colorimetric) model of colour discrimination which is, as argued above, incorrect (or, at best, can be considered only as the very rough approximation) and is incompatible with these models. In fact, the experimental identification of matching metamerism and/or the fundamental colour mechanisms should precede modelling colour discrimination. It is important to emphasise that, admitting the intransitivity colour matching, one cannot, strictly speaking, reduce the experimental identification of matching metamerism to evaluating the colour matching functions the way it is done in colorimetry. So, one needs an alternative approach to identifying the matching metamerism (and/or the fundamental colour mechanisms).

Such an approach is outlined in Chapter 11. Specifically, we present there a method which can be used to identify the matching metamerism in terms of its linear colour mechanism subspace.

---

47 In particular, the colorimetric bipartite display (Fig. I.1) implies such understanding.

48 It seems that the only resonable explanation for why colorimetry clings to the identification of metamerism and colour matching is that without this identification another capital dogma of colorimetry will collapse. Namely, in this case one will have to admit that the fundamental colour mechanisms cannot be represented as a linear combination of the colour matching functions.

49 In particular, MacAdam's data on the colour matching variability registered in the CIE chromaticity diagram which appeared to be ellipse-like (MacAdam, 1942), inspired many to seek in this direction.

When the matching metamerism is supposed to be determined by the fundamental colour mechanisms, another method, also presented in this chapter, can be used to identify these fundamental colour mechanisms. Notably, these methods have been developed basically without any specific assumption concerning the indiscriminate relation. To be more exact, the methods assume the indiscriminable sets to be only convex and smooth. It is important to refrain from making concrete assumptions about the indiscriminate relation when experimentally identifying the matching metamerism (and/or the fundamental colour mechanisms) because any such assumption might essentially contaminate the identification results.

For example, Stiles attempted to derive the fundamental colour mechanisms from colour discrimination data by using his two-colour threshold technique (Wyszecki & Stiles, 1982). However, as pointed out by Graham (1989) and others, this method is based on several assumptions concerning colour discrimination whose validity is in serious doubt. Not surprisingly, the results obtained with the two-colour threshold technique were hard to conciliate with the classical colorimetry. Even the number of the fundamental colour mechanisms revealed by this method (the so-called π mechanisms) significantly exceeded three (Wyszecki & Stiles, 1982).

Another example comes from spatial and temporal discrimination which have been studied maybe even more thoroughly and profoundly than colour discrimination.[50] A number of two-stage models have been put forward, the first stage comprising linear mechanisms in the spatial and/or temporal domain (Luce & Green, 1974; R. De Valois & De Valois, 1988; Kelly & Burbeck, 1984; Graham, 1989). While various terminology was used (mechanisms, analysers, filters, neurons, channels[51]), these spatio-temporal mechanisms were, usually, described similar to the colour mechanisms in this book. The spatio-temporal characteristics of these mechanisms were to be derived from the discrimination data. In spite of lots of effort, the spatio-temporal mechanisms remain a controversial concept still to be identified (Wilson & Wilkinson, 1997). The major obstacle was the trade-off between the mechanisms' characteristics and the way their outputs are combined. Specifically, the same discrimination data can be predicted by models with different mechanisms provided that the way their outputs are combined is adjusted properly (Graham, 1989).

At the first glance, such a trade-off makes impossible to derive the hypothetical mechanisms from the discrimination data. As far as colour vision is concerned, the trade-off in question casts serious doubts on the possibility to identify not only the fundamental colour mechanisms underlying colour matching and/or discrimination, but, in general, the matching metamerism itself (thus, the induced Grassmann structure). Nevertheless, addressing this problem in its entire generality, we show that the solution can be found. Although the presentation of this solution is made in terms of colour mechanisms, the results hold true if they are reformulated by replacing the wavelength with the spatial and/or temporal variable.

Specifically, in Chapter 11, the matching metamerism identification is posed as a mathematical problem, and two different approaches to its solution are outlined by using the methods of convex

---

50 It should be pointed out that the distinction into spatial and temporal, and colour discrimination is conventional and quite superficial. It is just the custom in the field to mark discrimination with the term referring to the varying stimulus parameter when the other parameters being constant. For example, colour discrimination usually indicates that the spatial and temporal characteristics of the stimulation are fixed. However, the fact of colour discrimination established, for example, in a bipartite field is at the same time the fact of detection of the vertical spatial border in the display, that is, detection of a spatial pattern (or discrimination between the presence of this pattern and its absence, i.e., the presence of the spatially homogeneous display).

51 It should be borne in mind that although the term 'channel' is rather often found in vision literature, its meaning is rarely determined strictly. Furthermore, depending on the context, different writers seem to put into this term different meanings. Specifically, the channel is thought of as either a linear mechanism (which can be formalised as a linear functional followed by a local non-linearity), or a family of linear mechanisms of the same kind. We will conceptualise this term in more detail in Chapters 12 and 13.

analysis. As there is every indication that the indiscriminable sets are convex (Logvinenko, Tyurin, & Sawey, 2002), it seems natural to assume that their preimages in the light stimulus space (referred to as the *colour matching sets*[52]) are also convex.[53] It is also safe to believe that the colour matching sets are smooth enough to allow the existence of the tangent hyperplane at most of its boundary points. In this case, one can use the *method of tangential hyperplane* for identifying the matching metamerism (Section 11.3.2). The idea of this method, as applied to the identification of the spatial channels, was outlined elsewhere (Logvinenko, 1996a). In relation to the identification of the matching metamerism it is, in short, the following. The hyperplane tangent to the colour matching set can be represented as a level set of some linear functional on the light stimulus space. It proves that the linear functionals determining the hyperplanes tangent to the colour matching set belong to the linear colour mechanism subspace determined by the matching metamerism. Therefore, the matching metamerism identification reduces to estimating a few hyperplanes tangent to the colour matching set at some points. The number of these points should not be fewer than the expected dimensionality of the matching metamerism (i.e., the induced Grassmann structure).

Empirically, an approximation of the hyperplane tangent to the colour matching set at a boundary point can be made by evaluating a fragment of the colour matching set boundary surface around this point. This surface is made up of the light stimuli which are equally discriminable from the colour matching set centre (i.e., from the light stimulus specifying this colour matching set). So the colour matching set boundary will be referred to as the *equi-discrimination surface.* Admittedly, an approximation of the equi-discrimination surface can be practically done only in a finite-dimensional subspace of the light stimulus space. This subspace must be chosen in advance. Such restricting to a finite-dimensional light stimulus subspace amounts to seeking for an approximation of the linear colour mechanism subspace in a finite-dimensional subspace of linear functionals. The dimensionality of this subspace should be big enough so as to guarantee either, at best, encompassing the linear colour mechanism subspace, or its more or less accurate approximation.

We start from considering a particular case when the matching metamerism is determined by some fundamental colour mechanisms belonging to a concrete family of linear functionals (on the light stimulus space) (Section 11.3.1), passing then to the general case when the linear colour mechanism subspace is assumed to belong to the ambient finite-dimensional subspace spanned over a set of some basis linear functionals (Section 11.3.2). At the first glance, the assumption that the variety of fundamental colour mechanisms is finite and predetermined might seem far too restrictive. However, if, as suggested above, the fundamental colour mechanisms are based on the cone spectral sensitivity functions, then such an assumption does not sound that implausible. Indeed, as known from molecular genetics of the cone photopigments, for a particular individual, there is a finite number of possible cone spectral sensitivity functions, which is determined by a finite number of possible wavelengths at which the photopigments have the maximal absorption (J. Neitz & Neitz, 2011).

Practically, estimation of the equi-discrimination surface around some of its point reduces to threshold measurements. In fact, the method of tangential hyperplane is a generalisation of the subthreshold summation technique widely used in attempts to identify the spatio-temporal channels (Kulikowski & King-Smith, 1973; Graham, 1989).

It is worth pointing out that even if the matching metamerism is determined by the fundamental colour mechanisms, it is the linear colour mechanism subspace rather than the fundamental colour

52 In Chapter 10, a colour matching set for a light a is defined as all the lights that match a.

53 Actually, the indiscriminable set convexity implies the convexity of the correspondingcolour matching set provided that some technical axiom holds true.

mechanisms themselves that can be identified with the method of tangential hyperplane. In other words, using the method of tangential hyperplane, the fundamental colour mechanisms can be evaluated only up to an arbitrary linear transformation. Though, this is the same degree of certainty as that provided by the traditional (colorimetric) approach based on colour matching functions.

In order to identify the fundamental colour mechanisms (determining the matching metamerism) as such, one has to resort to the method of quadratic approximation described in Section 11.3.3. It enables us to recover the fundamental colour mechanisms up to a scaling factor. The idea of this method as applied to identification of the spatio-temporal channels is described elsewhere (Logvinenko, 2003b). As reflected in its name, the method is based on local approximation of the equi-discrimination surface by a quadratic (elliptic) surface. To secure the existence of such approximation, we make some, rather weak assumptions about the form of the discrimination function (which are met by all current models of colour discrimination). Specifically, along with convexity, we also assume the indiscriminable sets to be smooth enough for the quadratic approximation to exist, though allowing them to be of any shape (unless it violates the convexity and smoothness). The method of quadratic approximation allows us to restore the fundamental colour mechanisms no matter of what exact shape the indiscriminable set is (except for the case when it is a Euclidean sphere).

Thus, despite the trade-off between the fundamental colour mechanisms and the discrimination function mentioned above, one can identify the fundamental colour mechanisms, thus, the matching metamerism, separately and independently from the indiscriminate relation (Chapters 11 and 12). This is quite possible because, as it turns out, even with the trade-off in question, we remain within the linear subspace spanned over the fundamental colour mechanisms. True, different sets of the colour mechanisms followed by different discrimination functions can be compatible with (i.e., can induce) the same colour matching sets. However, both the sets of such colour mechanisms will lie in the same linear colour mechanism subspace determined by these colour matching sets (thus, the same induced Grassmann structure). This allows, given a colour matching structure, considering the discrimination function as determined on the colour-signal space (determined by the induced Grassmann structure) rather than on the light stimulus space. Thus, the analysis of colour discrimination is transferred to the finite-dimensional colour-signal space, making it much simpler.

The study of colour discrimination is reduced to finding out what the indiscriminate relation is. From a geometrical point of view, one needs to describe the entire net of the indiscriminable sets (in the colour-signal space determined by the induced Grassmann structure). Analytically, this can be done in terms of the discrimination function. Experimental identification of the discrimination function appears an extremely difficult task for at least two reasons. Firstly, as mentioned above, there is every indication that the indiscriminable sets have different shapes. Secondly, one and the same indiscriminate relation (thus, the indiscriminable sets) can be represented by different discrimination functions.

The first difficulty can be circumvented by considering one indiscriminable set at a time. A particular case of discrimination when one of the two stimuli to be discriminated is fixed (and considered as the background) and the second varies (defining the corresponding indiscriminable set) is called *detection* in psychophysical literature. Respectively, when one of the two variables of the discrimination function is fixed, the resulting function of one variable is called the *detection function*.

In Chapter 12 using the conceptual repertoire of convex analysis, we develop a theoretical framework for detection/discrimination. We believe that this framework can be suitable for studying detection/discrimination not only in the chromatic, but also in the spatial and temporal domains. Surprisingly, there is a remarkable conceptual parallelism between these two fields, at the first glance very distant from each other. For instance, as is known from convex analysis, a convex body (which by our assumption is any indiscriminable set) can be uniquely represented by its *Minkowski function* (i.e., as its sublevel set). We propose using the Minkowski function of the indiscriminable set as the canonical detection function. (Any two detection functions inducing the same indiscriminable set are called *psychophysically equivalent*.) According to the psychophysical terminology, the Minkowski function turns out to be the sensitivity function, that is, the value that the Minkowski function takes on light stimulus a is the reciprocal to the threshold for a.

Another useful fact from convex analysis is that a Minkowski function can be represented in the form of the so-called support function. In psychophysical terms, this means that the sensitivity function admits a representation as an envelope of the linear colour mechanisms. In psychophysical literature, such a representation is referred to as the *peak detector model* (Graham, 1989). Therefore, for any detection model, there is a peak detector model psychophysically equivalent to it. Admittedly, there can be many peak detector models psychophysically equivalent to a given one. They all differ from each other in the composition of the linear colour mechanisms that form them. The set of linear colour mechanisms constituting a particular peak detector model is called its (colour) *preprocessor*. The preprocessor generated by the support function is maximal (for a given indiscriminable set). Since in the terminology of convex analysis this preprocessor is nothing more than the *polar* (see Section A.2) of the indiscriminable set, the corresponding model is referred to as the *polar peak detector model*.

Because of the uniqueness of the polar of a convex set, it makes sense to consider the polar peak detector model as canonical. For any detection model, there exists the only psychophysically equivalent polar peak detector model. The psychophysically equivalent polar peak detector models of two arbitrary detection models are equal if and only if these detection models are psychophysically equivalent. Therefore, the equality of the psychophysically equivalent polar peak detector models can be used as a criterion of the psychophysical equivalence of detection models. In this regard, an important problem is to find the psychophysically equivalent polar peak detector model for a given detection model. Much of the Chapter 12 is devoted to developing mathematical tools for solving this problem.

A peak detector model, with the cone fundamentals making up the preprocessor, can be thought of as the simplest plausible generalisation of the colorimetric model of colour matching[54] (thus, colour discrimination). This model predicts the indiscriminable set to have the form of a parallelepiped with its sides parallel to the coordinate planes in the colour-signal space (i.e., those defined by the cone fundamentals). The shape of the equi-discrimination surfaces revealed in the experiments reported thus far does resemble a parallelepiped (i.e., the data points can be interpreted as the empirical realisation of a parallelepiped). However, these empirical parallelepipeds have somewhat rounded rather than sharp vertices and edges. Such roundness is widely believed to reflect probability summation between the underlying colour mechanisms. In addition, their sides are not at all parallel to the coordinate planes. It follows that the cone fundamentals do not play the role of fundamental colour mechanisms in colour discrimination. The orientation of the empirical parallelepipeds is more consistent with a hypothesis that colour discrimination is mediated by the following colour mechanisms: $M + L$; $M - L$; and $S - (M + L)$, where S, M, and L stand for the cone fundamentals. The directions in the cone fundamental colour-signal space corresponding to these mechanisms have been dubbed as *cardinal* (Krauskopf et al., 1982). Further studies have shown that the issue is much more complicated. So, it is not surprising that the problem of the colour mechanisms involved in colour discrimination has been intensively debated in the last two decades. Although the issue remains open (Eskew, 2008;

54 Recall that being taken literally, the colometric model implies singleton indiscriminable sets.

Stockman & Brainard, 2010), one thing seems to be firmly established: these are not the cone fundamentals.

Interestingly, if colour match is thought of as simply the lack of colour discrimination, then one has to conclude that the role of the cone fundamentals in colour matching is the same as in colour discrimination. If the latter is not directly mediated by the cone fundamentals, then neither is the former. Therefore, either the major presumption of colorimetry is incorrect, or the relation between the concepts of colour matching and colour discrimination is more complicated.

As for the probability summation, in the spatio-temporal domain, its effect is commonly taken into account using the detection function based on the p-norm with $1 < p < \infty$ (the case of $p = \infty$ amounts to the peak detector model). For such a detection function, the polar preprocessor (for the psychophysically equivalent polar peak detector model) can be analytically evaluated (Section 12.2.1). It comprises an infinite number of colour mechanisms. This is an example of the aforementioned trade-off between the preprocessor and the detection function. The same indiscriminable set is compatible with, on the one hand, a finite number (three) of the colour mechanisms the outputs of which are combined via the p-norm-based ($p < \infty$) detection function; and on the other hand, an infinite number of the colour mechanisms followed by the maximum-output-based detection function. More generally, if the detection function is a norm on the fundamental colour-signal space (thus, the indiscriminable set is a ball with respect to this norm), then the polar preprocessor is shown to be a ball in the linear colour mechanism subspace with respect to the dual norm (Section 12.2.1).

It must be said, however, that for more complicated (than a norm) detection functions, it is highly unlikely to express explicitly the polar preprocessor for the psychophysically equivalent polar peak detector model. Still, it can be characterised implicitly by means of the notion of the *most sensitive colour mechanism* (see Definition 11.5). Specifically, for any light stimulus, a, in the polar preprocessor, there is a colour mechanism most sensitive to a. Moreover, if the indiscriminable set is smooth enough, such a colour mechanism is unique. In this case, the most sensitive colour mechanism is referred to as the *colour detector* (of a). A peak detector model comprising only the colour detectors (for all the light stimuli) will be the simplest one in the sense that its preprocessor is minimal. In Section 12.2.3, we present a technique enabling us to evaluate the most sensitive colour mechanism for an arbitrary light stimulus. Hence, although the minimal preprocessor as a whole cannot be expressed explicitly with a single formula, it can be thought of as given implicitly since its element most sensitive to an arbitrary light stimulus can be computed. In other words, having been parametrised with light stimuli, the minimal preprocessor is given parametrically so that for any parameter value its element can be evaluated.

Empirically, the evaluation of the colour detector for a light stimulus a amounts to determining a plane tangent at a to the indiscriminable set. This, in turn, reduces to an approximation of the equi-discrimination surface by a plane at point a. This task was already considered in Chapter 11 when describing the method of tangential hyperplane (Section 11.3.2).

As mentioned above, from what we know of the shape of indiscriminable sets follows that they do not lend themselves to more or less simple detection/discrimination models. It might happen, however, that the equi-discrimination surface can be partitioned into a few local fragments each of which is compatible with a relatively simple detection/discrimination model. Being different these models will be applicable only within their own part of the equi-discrimination surface. Each of these models is referred to as a colour channel and the whole set of these models as the multi-channel model of colour detection (respectively, discrimination). The idea of multiple channels in spatio-temporal detection and discrimination had been intensively debated in the 1970–1980s (Graham, 1989). In Section 12.2.2, we present a theory of multi-channel detection models based on a sort of algebra allowing, for example, to derive the polar preprocessor of the multi-channel model provided that the polar preprocessors of the channels constituting it are known. Also we show how the resulting polar preprocessor depends on the way, the outputs of the individual channels are combined with each other.

Chapter 12 concludes with discussing the models of colour discrimination known in literature on vision. This discussion is, perforce, brief, because only at the dawn of the colour science, genuine models with an explicit discrimination function were put forward, and even then only for wavelength discrimination. More recent studies revealed that the equi-discrimination surfaces even for a fixed background were too complicated to be specified by a single, more, or less apprehensible detection function. It is not surprising that the emphasis has shifted to the problem of colour mechanisms that mediate colour detection/discrimination, leaving in the shadow the detection/discrimination function (Eskew, 2008, 2009; Stockman & Brainard, 2010).

From a formal point of view, a colour discrimination model is just a family of colour detection models each of which corresponds to a particular background stimulus. Therefore, the study of colour discrimination can be formally thought of as finding out how the background stimulus affects the detection function. It should be borne in mind that the background stimulus plays a dual role in colour discrimination. First, the background stimulus is a necessary element of the paradigm of discrimination: a stimulus with respect to whom the test stimulus is supposed to be detected. Clearly, changing the background stimulus is most likely to affect the detection of the test stimulus. Second, a change of the background stimulus can affect the adaptation state of the neural units involved in the colour discrimination process; particularly, the cone photoreceptors, which are known to be subject to light adaptation (see Section 13.1.2). In order to separate these two background effects, one can use two background stimuli (e.g., as in Fig. 11.1): immediate and remote ones. The former can be a light spot, preferably of the same size as the test stimulus. The latter, surrounding the test and the immediate background stimulus, should be spatially large enough to make a major contribution to light adaptation. Keeping the large remote background constant, one can study the colour discrimination under a practically constant adaptation state.

Empirical estimation of the equidiscrimination surfaces around two different background stimuli allows (i) to ascertain the colour mechanisms mediating the colour discrimination and (ii) to find out whether these are the same or different for these two backgrounds. As mentioned above, revealing the colour mechanisms involved in colour discrimination has been considered by many authors as one of the major problems in the field (Eskew, 2008, 2009; Stockman & Brainard, 2010). Despite much effort, there is no consensus on this subject as yet. There is perhaps one thing, on which the researchers in this field do agree with each other. This is that neither the cone fundamentals nor the 'cardinal mechanisms' could directly underlie colour discrimination. In order to emphasise this fact, the term 'multiple, higher-order colour mechanisms' has been introduced (for a review see e.g., Eskew, 2009). Being unable to enter into details, we note in this regard only the following.

First, attempting to ascertain the spectral sensitivity of the hypothetical higher-order colour mechanisms underlying colour discrimination, the researchers employed various techniques (e.g., subthreshold summation, chromatic masking, chromatic adaptation, and many others (see the list in (Eskew, 2009)), not caring at all about the methodological validity and theoretical justification of these techniques. For example, the chromatic selectivity of the effect of the masking stimulus on the discrimination of the test stimulus is often, not critically, taken as a signature of the spectral sensitivity of the hypothetical higher-order colour mechanisms. However, without a clear understanding, and/or a detailed hypothesis about how the masking stimulus affect the discrimination of the test stimulus, using chromatic masking as a tool to estimate the spectral sensitivity of the hypothetical higher-order colour mechanisms seems even more controversial than using chromatic adaptation for this purpose. It must be admitted that Stiles' two-colour threshold technique, based on chromatic adaptation, was elaborated thoroughly enough to explicate the effect of the adapting stimulus on the spectral sensitivity of the hypothetical colour mechanisms. It is this elaboration that allowed later to understand what assumptions underlying the technique were wrong.

Second, if the hypothetical higher-order colour mechanisms do exist, and they contribute to colour detection and discrimination the way suggested in Section 11.3.3, then the method of quadratic approximation (Section 11.3.3) can be used to evaluate their spectral sensitivities.

Third, it should be remembered, however, that the availability of convincing psychophysical evidence in favour of the existence of those or other colour mechanisms is not sufficient yet for their existence in the eye and/or brain. As shown above, there exist psychophysically equivalent models containing different colour mechanisms, so knowing the physiology of the eye and the brain can help decide which of these potential colour mechanisms really exists in the form of neural units, if any. Thus, the issue of the anatomo-physiological substratum of the colour mechanisms, the plausibility of which is fully justified at the psychophysical level, deserves a separate investigation.

Part I ends with Chapter 13, which discusses the possible anatomical and physiological status of the various colour mechanisms that emerged in the previous chapters. Until now, they figured solely as psychophysical entities.

At the onset, it is worth asking: are there any units in the retina or brain at all that are consistent with the notion of colour mechanism (as defined within the theoretical framework of Grassmann structures)? Fortunately, this question is not difficult to answer. As generally agreed, the cone photopigment excitation can be described as a linear function of the light stimulus. Therefore, the cone photopigment of each type determines a colour mechanism.

After appropriate adjustments for the transmittance of the lens and ocular media, these mechanisms, based on the cone photopigment excitation, turn into the cone fundamentals, becoming a basis for various models of human colour vision. As a matter of fact, all recognised models of human colour vision contain as the first stage three cone fundamental colour mechanisms the outputs of which are recombined to feed into the subsequent hypothetical colour mechanisms (e.g., Hurvich & Jameson, 1957; Guth, 1991). The latter differ from model to model depending on what the model is designed for: to account for chromatic detection and/or discrimination, or colour appearance, or both. Striving for universality, some authors proposed models consisting of the three successive stages (R. L. De Valois & De Valois, 1993). Although these models are sometimes quite different, one feature in common unites them all: the subsequent stages operate with what is called in this book the fundamental colour signal. In other words, all the colour mechanisms in these models can be formally described as colour signal codes[55] (with respect to the cone fundamental metamerism, i.e., the one determined by the cone fundamentals). So are the opponent chromatic codes underlying the component hues (thus, the colour matching structure) introduced in Chapter 8, as well as the various colour mechanisms arising from the analysis of the chromatic detection and discrimination data. In fact, we assumed them to belong to the linear subspace spanned over the cone fundamental colour mechanisms. This is also taken for granted by current researchers when they use the cone excitation space to represent the chromatic detection and discrimination data (Eskew, 2008, 2009).

Taking into account the above, it would seem natural to expect that the output of the fundamental colour mechanisms (i.e., the quantum catch (isomerisation rate) in the cone photoreceptors) is to be transferred further along the visual pathways. However, curiously enough, this does not seem to happen. Moreover, it would be incorrect to state that (i) the cone photoreceptors implement the

55 The definition of colour signal code is given in Remark 2.9.

fundamental colour mechanisms and (ii) the cone photoreceptors' responses encode the fundamental colour signal. One can point out at least two reasons for this.

The first reason is that, because of light adaptation in it, a photoreceptor might respond differently to the same light stimulus. Specifically, it is known that the photoreceptor response to a light (i.e., the electrical signal transmitted further to the subsequent retinal layers) depends not only on this light but also on the state in which the photoreceptor is at that moment (Section 13.1) (Shapley & Enroth-Cugell, 1984). The photoreceptor adaptation state is determined by the whole prehistory of its stimulation. That is, the stimulus light affects both the photoreceptor adaptation state and photoreceptor response. The photoreceptor turns out to be a dynamic system for which the relationship between the input and the output is more complicated than a functional relationship (Marmarelis & Marmarelis, 1978). As a result, the same light might result in different photoreceptor responses. In other words, due to photoreceptor adaptation, there is a many-to-many correspondence between the cone excitation and the cone response. It follows that, first, the former cannot be uniquely derived from the latter. Second, the cone response cannot be considered as a function of the light stimulus, letting alone a colour mechanism. In literature on colour vision, this problem is usually circumvented by considering only stimulation conditions that leave the photoreceptor adaptation state constant, or simply assuming that the latter is constant or does not change significantly. This is a very restrictive assumption. However, without it, the relation between cone excitations and cone responses becomes enormously complicated, if not altogether intractable.

The second reason is that the notion of colour mechanism as given in Definition 2.9 is too abstract to be implemented by a photoreceptor, a neuron, and generally, by any material device. Really, apart from spectral characteristics, any such device has spatio-temporal characteristics as well. Therefore, for the problem of the colour mechanism implementation to be posed, one needs to make a notion of colour mechanism more concrete and realistic, extending its domain so as to encompass space and time. In turn, doing this requires a similar concretisation of the notion of light stimulus. In fact, one has to make the transition from the concept of light to that of illumination[56] which Gibson accomplished in his last book (Gibson, 1979). Specifically, the notion of *light stimulus array* introduced in Section 13.3.2 can be thought of as the formalisation of Gibsonian concept of ambient optic array. Respectively, a colour mechanism has been generalised there as a linear functional (perhaps followed by a local non-linearity) on the light stimulus arrays, the spectral sensitivity function of which is assumed to be independent of the spatial (and/or temporal[57]) variable. This spatio-chromatic separability is important because without it the very term 'colour' mechanism will make no sense because of multiplexing the chromatic and spatial information.

Now that the notion of light stimulus includes the spatial variable, we had better start considering the whole spatial array of the cone photoreceptors in the retina rather a single photoreceptor in isolation, as it has been the case so far. Really, firstly, as the light stimulus varies across the retina (which is emphasised by the notion of light stimulus array), it seems logical to mark explicitly for what spatial location the photoreceptor response is registered. Secondly, the cone photoreceptor response is determined by not only the quantum catch in this receptor (and its adaptation state) but also by the excitation of the photoreceptors in the vicinity nearby. As a matter of fact, there exists a spatial interaction between neighbouring photoreceptors due to the electrotonic coupling between them (Bloomfield & Völgyi, 2014), and because of the shunting effect on photoreceptors from the horizontal cells (Kramer, 2014). For this photoreceptor spatial interaction to be taken into account, one needs to consider the photoreceptor array as a whole, as a sort of a 'photoreceptor channel'.

---

56 The term 'illumination' will be extensively used in Part II, albeit in a more traditional manner, as is common in vision literature.
57 For brevity, the temporal variable is excluded from consideration in the book.

As pointed out above, although the term 'channel' is one of the most frequently used terms in vision literature, the meaning that some authors hold in it is often rather loosely defined. Quite important is how the channel's output is understood. First, it can be thought of as a scalar quantity that can be represented by a real number. A photoreceptor, a neuron are examples, as their responses can be described by a number (accompanied by a unit of measurement). Since such channels are usually represented by a linear functional on a set of some (stimulus) functions considered as the channel's input, we will refer to these as *channel functionals*. We believe that our new definition of colour mechanism can be considered as a formalisation of the term of channel functional.

Depending on what characteristic – chromatic, spatial, or temporal – is in focus, the terms 'colour channels', 'spatial channels', or 'temporal channels' are used in vision literature. However, from our point of view, each of these represents a particular aspect of a colour mechanism (according to our new definition). For example, when the spatial and temporal properties of the colour mechanism are neglected, it is termed as a 'colour channel'. When the term 'spatial channel' is engaged, the chromatic and temporal variations are usually ignored. Furthermore, in the case of 'spatial channels', their chromatic properties are implicitly assumed to be such that the channel's output is nothing more than luminance. Thus, the same colour mechanism (according to our terminology) can be classified, according to the traditional terminology, as a luminance channel and a spatial channel. In other words, all the spatial channels discussed in literature are the luminance (thus, colour) channels according to the present terminology. On the other hand, the colour opponent (red-green and yellow-blue) channels, as suggested by Hering and elaborated by Hurvich & Jameson (1957), can be taken as an example when the emphasis is mainly placed on their chromatic properties, the spatial properties being left unspecified.[58]

Alternatively, the channel's output is sometimes understood as an 'image', a sort of 'picture'. Such a channel can be thought of as a bundle of identical channel functionals. We will call such channels *channel operators*. An example is a spatial array of the photoreceptors (or neurons) of the same type. A separate photoreceptor (neuron) can be viewed as a channel functional. The spatial array of the responses of these photoreceptors (neurons) makes up the output of the channel operator (constituted by these photoreceptors (neurons)). It is sometimes called the 'photoreceptor (or neural) image' (Robson, 1980; Brainard, 2015).

Although the concept of a photoreceptor image seems natural, it is not very useful in our analysis. Really, because of the lack of a functional relationship between the excitation and response of the photoreceptor (due to the photoreceptor adaptation), there is no functional relationship between the photoreceptor image and the light-stimulus array either. In plain words, the colour of a light-stimulus array (whatever that means) cannot be derivied from the photoreceptor image. So, it is not surprising that there seems no evidence that the photoreceptor image is transmitted through the optic nerve to the brain. In other words, a photoreceptor image exists for us, but not for the brain.

In view of the importance of this, albeit negative, conclusion, let us dwell on it in more detail. Let us begin with a generalisation of the concept of fundamental colour signal, which is necessary due to incorporating a spatial variable into consideration. Specifically, given a colour signal code,[59] define the corresponding *colour signal code pattern* as the spatial distribution of the values of this colour signal code in response to the light stimulus array. For example, consider a set of cone fundamentals. Each cone fundamental induces the corresponding *cone fundamental pattern*: the S, M, and L patterns. The colour signal code pattern evaluated as the sum of the M and L patterns is naturally to be called the luminance ($M + L$) pattern. Likewise, one can formally introduce

58 True, in some studies, the spatial properties of these channels have also been explored, by using the same psychophysical techniques as in works on 'spatial channels' (e.g., Mullen, 1985).
59 The colour signal code is meant with respect to the cone fundamental metamerism here.

the $(L - M)$ pattern, the $(S - (M + L))$ pattern, and generally the $(k_S S + k_M M + k_L L)$ pattern for any coefficients k_S, k_M, and k_L. The triplet of the cone fundamental patterns (i.e., the S-, M-, and L-patterns) will be referred to as the *colour signal array*. In fact, the colour signal array can be thought of as a colorimetric transformation of the light-stimulus array.

In colorimetry the concepts 'cone-excitation triplet' and 'fundamental colour signal' have meant virtually almost the same thing. However, their generalising into the spatial domain brings essential difference. Let us call a spatial array of the excitations of the cone photoreceptors of the same type the *cone excitation pattern*. The fact is that the S (respectively, M, and L) cone excitation pattern would amount to the S (respectively, M and L) pattern only in the case of ideally perfect eye optics. However, an unavoidable optical blur induced by the real eye optics makes them different. For instance, the cone excitation pattern induced by a point light source is not point-like (when only one cone gets excited) but resembles (to the first approximation) the disc into which the point light source is blurred because of optical aberrations. For the same reason, the luminance distribution in the retina differs from the luminance $(M + L)$ pattern.

As the optical blur can be taken into account, the cone excitation pattern can be, in principle, described as the corresponding cone fundamental pattern, appropriately transformed. Really, the effect of the optical blur on the cone fundamental pattern can be described by using a non-degenerate linear transformation. Therefore, using the inverse transformation, one can restore the cone fundamental pattern from the cone excitation pattern.[60] This fact will be referred to as that the cone excitation pattern encodes the corresponding cone fundamental pattern.

However, during converting into a cone response pattern, the cone-excitation patter undergoes such a complex transformation that, in the end, the functional relationship between them is lost. Indeed, the cone response pattern (i.e., the cone photoreceptor image) differs from the cone excitation pattern because of the lateral interactions between photoreceptors even in the absence of the light adaptation in photoreceptors. True, when the light adaptation is approximately the same for all the cone types, the cone photoreceptor image can be thought of as the linear transformation (spatial filtering) of the cone excitation pattern. The character of the lateral interactions between photoreceptors indicates that this spatial filtering is a sort of smoothing, and it admits an inverse transformation. Hence, when the light adaptation in the photoreceptors can be neglected, the cone fundamental pattern is encoded by the corresponding cone photoreceptor image. However, because of the light adaptation in the cone photoreceptors, the cone photoreceptor images not only do not encode but also they simply cannot encode the colour-signal array at all.

There are two important implications of the fact that, by their design, the cone photoreceptors cannot record and, thus, transmit further the colour-signal array. First, as the cone photoreceptor images are not a neural implementation of the colour-signal array, there is no need to convey them further to the brain intact and secure. Indeed, paradoxically, it may seem, there is no evidence that the cone photoreceptor images are transmitted via the optic nerve. The signal processing in the retina is striking in its complexity and incomprehensibility, however, what is transmitted along the optic nerve is clearly not the encoded replica of the cone photoreceptor images.

60 It should be mentioned that apart from the optical blur, there are other factors contributing to transforming the cone fundamental pattern into the cone excitation pattern. One of these is the discretness of the photoreceptor mosaic. Because of it a so-called spatio-chromatic aliasing occurs (Brainard, 2015). As a result, two different cone fundamental patterns can induce the same cone excitation pattern. In this case, the inverse transformation in question does not exist. Although one could, in principle, take into account the spatio-chromatic aliasing, this would unnecessarily complicate the book, leading beyond its scope. So, we will leave the effect of the spatio-chromatic aliasing out of consideration.

If, contrary to these facts, we assumed that the cone response patterns are nevertheless transmitted somehow to the brain and underlie the colour computation, we would have to face insuperable difficulties in explaining some colour phenomena. In particular, the spatial smoothing (blurring) of the cone excitation pattern depends on the spatial density of the cone mosaic. The spatial distribution of different cones across the retina has been established to be different (Williams & Roorda, 1999). In particular, the *S*-cones are a few times sparser than the *M*- and *L*-cones. Therefore, the *S*-cone image will be more blurred than the *M*- and *L*-cones images. As a result, an achromatic edge should be seen with a chromatic fringe. However, this never happens. Therefore, the colour appearance of the achromatic edge is not in line with the superposition of the three cone images.

Still, the lack of an illusory colour fringe when observing an achromatic sharp edge can be accounted for if one assumes that it is the outputs of the luminance and colour opponent channels that determine the colour of the achromatic edge. Since the colour opponent channels will yield zero-response to such an edge, the poorer spatial resolution of the yellow-blue channel (to which the *S*-cones contribute) does not reveal itself when the achromatic edges are perceived. To put it the other way round, the colour appearance of the achromatic edge is in line with the superposition of the spatial distribution of the outputs from the luminance and the two colour opponent channels even if the latter yield the 'images' of poorer quality (of higher blur) than the luminance channel. However, when at least one side of the edge is coloured, the colour 'illusion' might be observed because of the *S*-cone sparsity (e.g., see Fig. 13.10 in Chapter 13). This 'illusion' cannot be accounted for by the simple difference in the cone image quality either (Logvinenko, 2001). However, the illusory colours in Fig. 13.10 can be explained by the poor spatial resolution of the yellow-blue channel in the same way as the lack of an illusory colour fringe for the achromatic edges (Logvinenko & Hutchinson, 2007).

This demonstration corroborates the fact, acknowledged by some previous authors (e.g., Brainard, 2015), that colour and space are intimately interwoven in perception. One cannot fully understand perception of colour as such, abstracting from spatial and temporal distribution of light. Admittedly, up to some point, this abstraction might prove useful (especially at the beginning). However, sooner or later one has to widen the context and to generalise the notion of colour stimulus including the spatial (as well as temporal) dimensions.

Second, since all the subsequent neurons in the retina, the lateral geniculate nucleus (LGN) and the visual cortex are, generally, fed by the photoreceptors' responses, none of them can serve as a colour mechanism. In Section 13.2, we argue in detail why, contrary to the general belief, the so-called colour opponent ganglion cells cannot be considered as the neural substratum of the colour opponent mechanisms. The main reason is that the spatio-chromatic sensitivity function of these cells cannot be factored into independent spatial and spectral sensitivity functions. Therefore, they cannot play the role of any colour mechanism because, as pointed out above, such cells will confound in their response the effect of the spectral and spatial variables. What is then the contribution of these ganglion cells to colour vision? What is their role, if any, in colour processing? In the end, it is the pattern of their activity that is sent to the visual cortex.

One idea is that the ganglion cells perform a sort of spatial (and temporal) differentiation. Although, since Ernst Mach's time, a lot has been stated on differentiation in vision, it is important to specify what is being differentiated and how, in order to prevent this claim from remaining just a metaphor. For instance, the neural image based on the ganglion non-colour opponent (achromatic) cells can be thought of as the result of smoothing and subsequent (spatial) differentiation of the luminance image (Young, 1987). Indeed, although the response range compression (due to the non-linearity) and the gain control prevent the photoreceptors from encoding the luminance pattern itself, they are quite suitable for its differentiation.[61]

However, as Marr (1982) rightly pointed out, it would be incorrect to claim that the colour opponent ganglion cells perform differentiation of any sort (see Section 13.2). Still, so-called double-opponent cells in the primary visual cortex (Shapley & Hawken, 2002) are quite suitable

61 Admittedly, it is numerical differentiation that is meant here.

for this role (Lindeberg, 2013). We hypothesise that what these cells are actually doing can be interpreted as (i) smoothing the colour signal code pattern determined by the spectral sensitivity of these cells, followed by (ii) computing the gradient of the resultant smoothed pattern at each its point. Since a neural network can only implement numerical differentiation, these two sequential operations can be combined into one, namely, numerical computing the gradient of the colour signal code pattern. (See Section 13.4.1 for the *directional gradient encoding hypothesis*.) The result is the gradient field of the colour signal code pattern in question.[62] In other words, the neural image produced by the responses of these cells renders not the colour signal code pattern but its gradient field. Still, the colour signal code pattern can be reconstructed from the neural image generated by these cells, albeit up to an arbitrary constant. Consequently, the loss of information accompanying gradient encoding is minimal, and the advantage is enormous.

Furthermore, among the various separate layers of the visual cortex, there are, generally speaking, as many as three to which the directional gradient encoding hypothesis is, in our opinion, applicable. Specifically, we argue that, apart from the layer comprising the already mentioned double-opponent cells, which presents the gradient field of the $(k_M M - k_L L)$ pattern, there are two more. One of them presents the gradient field of the $(k_S S - (k_M M + k_L L))$ pattern (here, as above, k_S, k_M, and k_L are positive real numbers) and the other contains the gradient field of the luminance $(M + L)$ pattern. Having simultaneously appropriate access to these three gradient fields, it is possible to reconstruct the colour-signal array up to an unknown affine transformation. The linear part of this transformation relates the three colour-signal codes (i.e., $M + L$, $k_M M - k_L L$, and $S - (k_M M + k_L L)$) with the cone fundamentals S, M, and L. The unknown additive constant of the transformation results from the differentiation.

Since the three colour-signal codes in question remain, by and large, fixed for an individual, the fact that they are unknown do not bring more uncertainty. In contrast to this, the unknown additive constant results in an ambiguity of the cortical representation of the colour-signal array. Note, however, that this ambiguity is of much lesser extent than that which would occur if the colour-signal array were derived directly from the cone responses point by point. As the cone response is ambiguous, in order to eliminate this ambiguity the visual system had to transmit to the brain not only the response of every cone photoreceptor but also information of its adaptation state. But that would have doubled the amount of information transmitted through the optic nerve. As a matter of fact, there has been found no evidence that this happens. In the case of the gradient encoding, the ambiguity of the whole cone photoreceptor image can be resolved by anchoring it with a single value (by assigning a certain value to the unknown additive constant). In other words, the gradient encoding eliminates the need to resolve the ambiguity of each cone response individually.

An important feature of the gradient encoding is its holistic nature; that is, it is a way to encode the colour-signal array as a whole rather than point by point. The response of a single double-opponent cell is as ambiguous as that of a single photoreceptor. Neither a single double-opponent cell nor a triplet of such cells can encode the colour at a particular location of the visual field. The gradient encoding is a way to 'encode colour' without colour mechanisms, moreover, even without colour codes.[63]

---

62 Recall that the colour signal code pattern is defined as a function of the two spatial variables, α_1 and α_2, the value of which at point (α_1^0, α_2^0) is the colour signal code value at (α_1^0, α_2^0). Of course, in order to be able to talk about a gradient, the corresponding assumption is made about the smoothness of this function.

63 It should be borne in mind that, while the expression 'to encode colour' is quite common in vision literature, in the present context it can be taken only as a sheer metaphor. Indeed, since colour is not present in the physical world, it cannot be, strictly speaking, encoded. More rigorous would be to rephrase this sentence as follows. 'The gradient encoding is a way to encode a cone excitation pattern in the absence of colour mechanisms …'.

This calls into question the widespread belief (or the 'linking proposition' (Teller, 1984)) that there are neurons (or groups of neurons) the functioning of which is accompanied by the experience of particular colours. Perhaps this idea is expressed most straightforwardly in the single-neuron doctrine, according to which there exist some neural units whose activity brings about some specific features of the perceptual image (Barlow, 1972). From a philosophical point of view, this doctrine is a particular case of *psycho-physiological parallelism*.[64] Although this doctrine is vulnerable to criticism[65] (e.g., Dennett, 1991), many vision scientists share this, or similar point of view. As one can see, the single-neuron doctrine is hardly compatible with the gradient encoding of the colour-signal array because of its holistic nature.

Furthermore, the way of encoding the colour-signal array accomplished by the visual system (i.e., the gradient encoding) casts serious doubts concerning the pertinency of the very psychophysical approach in colour science. Really, the keystone of psychophysics is the belief in a functional relation between stimuli and perceptual responses.[66] Psychophysical models are constructed so that knowledge of the stimulus allows, at least in principle, to predict the percept. The psychophysical paradigm implies strict determinism. The gradient encoding cannot secure such determinism. The relationship between the colour-signal array (the stimulus) and the gradient fields of the corresponding colour signal code patterns (underlying the neural response) is indeterministic due to the uncertainty brought by the additive (vector) constant.

Actually, the ambiguity of the photoreceptor response (particularly, due to light adaptation) is a well-recognised fact. It is rendered by such terminology as 'contrast coding' (Stockman & Brainard, 2015), 'contrast colours' (Whittle, 2003) and the like. It should be said, however, that the significance of this fact for colour science has not yet been fully realised. The very terminology, involving the notion of contrast, seems to us somewhat misleading. Contrast implies just two adjacent patches of light, usually, these are the test and the background. There is ample evidence for that it is contrast between the test and background that determines the colour appearance of the test (Chichilnisky & Wandell, 1995; Whittle, 2003) and its discriminability from the background (Eskew, 2009). If a photoreceptor, or a visual neuron encodes the test vs. background contrast, there is a temptation to believe that it is these photoreceptors or neurons that determine the test appearance and discriminability. And given that the background determines the photoreceptor (respectively, the neuron) adaptation state, one cannot help falling into illusion that we are on the strong ground of determinism.

It must be borne in mind, however, that such an account is restricted to only a two-field (or Mondrian-like[67]) display and is completely inapplicable to more general, variegated displays producing smooth colour-signal arrays where the colour signal gradually varies from point to point. For such displays, the notion 'contrast' loses its meaning. In a sense, the gradient field, as defined above, is a generalisation of the notion of contrast for colour signal code patterns. Besides, even given a test-background display, it is not clear under such an explanation what determines the colour of the background. Numerous demonstrations are known that show that when, due to the contextual effect, the colour of the background changes the colour of the test changes as well, even though the test/background contrast remains constant. For example, in achromatic scenes, the luminance

64 It must not be confused with *psycho-physical parallelism* enunciated by René Descartes as far back as the 17th century.
65 Suffice it to mention that this belief does not differ too much from a very old idea that has first been formulated, perhaps, by Johannes Peter Müller in 1835 in his *Law of specific nerve energies*.
66 Actually, it is a particular case of the behaviouristic stimulus response principle.
67 It was noticed that some of celebrated Piet Mondrian's paintings, containing only coloured rectangles defined by vertical and horizontal lines (e.g., Composition with Red, Blue, and Yellow, 1930) are strikingly similar to what is shown to observers in colour experiments.

contrast between the test and background seems to determine the brightness ratio rather than the individual brightness values for the test and background (Gilchrist, 2006).

We conclude Part I discussing two fundamental questions arising. Why is the cortical representation of the colour-signal array as ambiguous as described above? If it is as ambiguous as claimed above, why does this ambiguity not reveal itself in our immediate perceptual experience? Our everyday perceptual experience testifies in favour of stable and veridical perception which is in clear discord with the ambiguity in question.

It should be noted that it is only perception of real scenes that seems to be stable and veridical. As pointed out above, there are many laboratory demonstrations corroborating the perceptual ambiguity. Curiously enough, the answer to the first question becomes almost obvious in the case of real perception. Indeed, in this case even if the information capacity of the optic nerve were doubled, and the ambiguity of each cone response were resolved, this would not help too much because there is another ambiguity arising when perceiving real objects illuminated by real light sources. Consider, for example, an achromatic surface illuminated by daylight. As in this case only the gradient field of the luminance pattern is not zero, one can restrict consideration only to the luminance variable. Even if the luminance of the reflected light is known, it is not enough to restore the spectral reflectance of the surface. So, the luminance of the reflected light is ambiguous as a stimulus for achromatic object colour (Gilchrist, 2006). Our ability to see the colour of objects more or less independently of their illumination testifies for that this ambiguity is resolved somehow by the visual system. Notice that if this ambiguity is resolved at least for one point in the surface, then the reflectance of all the other points can be computed even if it is the gradient field of the luminance pattern rather than the luminance pattern itself that is known. Hence, from this point of view, there is no need in recording the luminance pattern as such, its gradient field will do. We will show in Chapter 21 that this is the case for all object colours not only achromatic ones.

Part II is devoted to object colours. Light is assumed now to be reflected by an opaque object before entering the observer's eye. Although undoubtedly the reflected light plays an essential role in perception of objects in general and their colour in particular, as emphasised above, the colour of an object is by no means the colour of the light reflected by this object. We start from the simplest case when there is one spatially homogeneous light source (*illuminant*) common for different reflecting objects. Each reflecting object is assumed to be spatially homogeneous and represented by its spectral reflectance function (referred to as an *object stimulus*). Even in a such single-illuminant scene one can see that, as pointed out above, the colour palette of object colours is rather different from that of light colours (see Chapter 17 for more detail). Specifically, the object-colour palette is richer than the light-colour palette. It is interesting since if one assumes that the object colour is derived somehow from the colour signal produced by the reflected light, then one would expect just the opposite because the set of colour signals that the reflecting objects can induce under a fixed illuminant is a great deal smaller than the colour-signal cone. Specifically, a set of the colour signals produced by all possible reflecting objects under a particular illumination (referred to as the *object-colour solid* (Wyszecki & Stiles, 1982, p. 179)) is shown to be a closed convex body inscribed into the colour-signal cone. The origin is the only common point between the object-colour solid and the colour-signal-cone boundary. It is referred to as the south pole (of the object-colour solid). It is the perfect absorber (the object reflecting no light at all) that maps to the south pole. The object-colour solid is symmetric with respect to its centre. The object reflecting all the light (at each wavelength) maps to the point symmetric to the south pole (referred to as the north pole). The interval connecting both the poles is called the achromatic interval. The objects mapping to the achromatic interval (under neutral illumination), as a rule, appear achromatic, that explains the name.

Despite the long history of studying (for a review see, e.g., Wyszecki & Stiles, 1982; Maximov, 1984; Koenderink, 2010), there are some unsolved fundamental issues concerning the object-colour solid. We address them in Chapter 14. Those objects which produce colour signals lying on the boundary of the object-colour solid, are called the *optimal object stimuli* (Wyszecki & Stiles, 1982, p. 179). They fully determine the object-colour solid. The first attempt to describe the optimal object stimuli was made by Schrödinger as early as 1920. He claimed that, first, the optimal object stimuli do not depend on the spectral sensitivity of the photoreceptors. Second, the optimal object stimuli can either reflect all the light (for some wavelengths), or none (for the rest of the wavelengths). Third, the number of transitions from 100% to 0% reflectivity is limited to no more than two across the visible spectrum wavelength interval (Schrödinger, 1920). That is, the optimal object stimuli have a form of so-called *step functions* (Definition 14.4) with the number of transition not more than two. While all these claims have been taken for granted since then (e.g., Wyszecki & Stiles, 1982, pp. 182–184), only the second one is correct and even then, only for human trichromatic observers.[68] As a matter of fact, Schrödinger mistakenly took for granted some property of the human colour matching functions that, in reality, is not met (Condition 14.3). We show in Chapter 14 that the optimal object stimuli are essentially determined by the photoreceptors' spectral sensitivity. Furthermore, we present a method to derive the optimal object stimuli from the spectral sensitivity of the fundamental colour mechanisms (Section 14.2). It turns out that the set of optimal object stimuli remains the same for any illuminant with everywhere positive spectral power distribution function. Moreover, placing a filter (with everywhere positive spectral transmittance function) in front of the photoreceptor pigments does not affect optimal object stimuli. Hence, when calculating optimal object stimuli, only the photopigment absorbance spectra could be taken into account.

Using this method, the optimal object stimuli were evaluated for a 3-*pigment Grassmann structure* (Definition 3.2). It was found that for trichromatic human colour vision based on such a structure, first, there were optimal object stimuli with spectral reflectance functions which have three and four transitions. Second, not all the step functions with one and/or two transitions were optimal object stimuli. We conclude Chapter 14 with a condition (the *regularity assumption* in Definition 14.5), which the fundamental colour mechanisms must meet so that the optimal object stimuli comprise *all* the spectral reflectances with the number of transitions not more than two, and only them (Proposition 14.3).

The object-colour solid for the fundamental colour mechanisms satisfying the regularity assumption has an interesting geometry. In particular, one can arrange the internal coordinates on its boundary surface in terms of the transition wavelengths of the optimal object stimuli. The object-colour solid for such fundamental colour mechanisms is easy to evaluate: one has to map all the two-transition spectral reflectances into the colour-signal space. When the fundamental colour mechanisms do not satisfy the regularity assumption, mapping all the two-transition spectral reflectances into the colour-signal space yields the volume falling inside the object-colour solid. Interestingly, all the previous authors (except for (Maximov, 1984)) computed just this volume instead of the proper object-colour solid. We refer to this volume (i.e., the image of the two-transition spectral reflectances) as a regular approximation of the object-colour solid or, in short, the *regular object-colour solid*. It turns out that for Govardovskii's cone fundamentals the regular object-colour solid can be a rather good approximation of the proper object-colour solid (Logvinenko, 2009b). Furthermore, the internal coordinates on the boundary of the regular

68 As follows from Theorem 14.2, the optimal object stimuli can, in principle, take a value between zero and one within a whole interval of wavelengths for non-human observers.

object-colour solid can be used to parametrise not only the regular object-colour solid but the proper object-colour solid too. So, we explore the geometry of the regular object-colour solid at length in Chapter 15.

At the end of Chapter 15, the following problem was also addressed. What spectral reflectances produce the colour signals most remote from the achromatic interval in the (not necessarily regular) colour solid? Intuitively, it seems naturally to expect such objects to have the maximal chromatic strength (purity). These spectral reflectances (rather misleadingly dubbed in colour literature as 'full colours', and referred to, by us, as strictly optimal object stimuli in Definition 15.2) map to the closed curve on the object-colour solid boundary named the equator (of the object-colour solid) (Section 15.2). The strictly optimal object stimuli are evaluated for Govardovskii's cone fundamentals. Although the strictly optimal object stimuli are different for different illuminants, they are found to be independent of illuminant if the fundamental colour mechanisms are subject to light adaptation, that is, if they adjust their responsivity so that their response to the perfect reflector remains constant.[69] In this case while the shape of the object-colour solid changes with illuminant, the position of the apex (the north pole) of the object-colour solid remains in the same position irrespective of illuminant.

In Chapter 16 we, following the well-established tradition, attempt to apply the same (actually, colorimetric) approach to studying object colours as that used to study light colours in Chapters 2–5. Specifically, in Section 16.1, we define object metamerism (Definition 16.1) the same way as light metamerism in Definition 2.2. Applying the same line of reasoning as in Chapter 2, we come up with such notions as object metamer set (Section 16.1), object-colour stimulus, the object-colour stimulus manifold (Section 16.3), and some others which formally are just reformulation of similar notions in Chapters 2–5 only in terms of spectral reflectance function (of an object) rather than spectral power distribution function (of a light). Although the principal limitation of this approach is revealed very quickly, a few important results on object colour, obtained within the colorimetric conceptual framework are noteworthy. These results are related to solving the similar problems as in Part I only for objects rather than lights (e.g., evaluating the object metamer sets and the metamer mismatch regions). It must be said that while posing the problems for objects is similar to that for light, their solution requires a somewhat different approach. For instance, while the problem of describing all the spectral reflectances metameric to a given one seems similar to that for metameric lights, its solution requires a different mathematical technique.

Specifically, two objects are called metameric if, being illuminated by the same light, they reflect metameric lights. All the mutually metameric objects (the *object metamer set*) map to the same point in the object-colour solid. They make what is referred to as the *object-colour stimulus*. We prove that the object metamer set for a given spectral reflectance is the convex hull of object stimuli, taking either 1 or 0, that produce the same colour signal as the given spectral reflectance (this follows from Theorem 16.1). Also, we show that there is no metamerism on the object-colour solid boundary (Theorem 16.2).

It is shown that the set of all the object-colour stimuli (i.e., the object metamer sets) can be treated as a topological manifold with boundary, homeomorphic to the unit ball. It will be called the *object-colour stimulus manifold*. Choosing one object stimulus in each object metamer set one can make a sample of object stimuli that bijectively represents the object-colour stimulus set. Such sample is referred to as an *object atlas*. Since in a single-illuminant scene metameric objects are commonly believed to match each other in colour (and objects inducing different colour signals

69 In particular, if the fundamental responses are normalised to unity.

differ in colour), an object atlas can also be thought of as a representation of all the object colours (in a single-illuminant scene).

An important example of an object atlas is given in Example 16.1. It comprises all the optimal object stimuli, x_{opt}, and some arbitrary object stimulus, x_0, completed with all weighted sums $(1-\alpha)x_0 + \alpha x_{opt}$ $(0 < \alpha < 1)$. Such an atlas is referred to as an *optimal object atlas*. This object atlas is full; that is, every object stimulus is metameric to some element of the atlas, which is called its *optimal metamer*. When the illumination alters, the classes of metameric object stimuli get reshuffled, but exactly one element of the atlas continues to remain in each class. Hence, an optimal object atlas can uniquely represent the object-colour stimulus manifold irrespective of illumination.

A particular case, with a flat reflectance mapping to the object-colour solid centre as x_0 (i.e., $x_0(\lambda) \equiv 0.5$), was used to represent a trichromatic object-colour stimulus manifold (Logvinenko, 2009b). As mentioned above, when the colour mechanisms satisfy the regularity assumption (Definition 14.5), the optimal object stimuli are exactly all the elementary step functions of degree $m \leq 2$ (Proposition 14.3). The elements of this atlas are, then, the spectral reflectances taking two values, c or $1-c$ $(0 \leq c \leq 1)$, with two transitions (at wavelengths λ_1 and λ_2) across the visible spectrum. As its members have a rectangular spectral profile, this atlas will be called the *rectangular object atlas*, its elements being called *rectangular metamers*.

Using a geographic coordinate system, each element of the rectangular object atlas (thus, each object-colour stimulus) can be geometrically represented as a radius vector so that its length equals c, the latitude and longitude being proportional to $\Delta\lambda = |\lambda_1 - \lambda_2|$ and $\bar{\lambda} = 0.5(\lambda_1 + \lambda_2)$, respectively. We refer to the former as *spectral bandwidth* and to the latter as *central wavelength* for the rectangular metamer, c being referred to as its *chromatic contrast*. The whole rectangular object atlas maps onto the unit ball, with the optimal object-colour stimuli mapping onto its boundary. Thus, a spherical representation of the trichromatic object-colour stimulus manifold is obtained. The numbers c, $\Delta\lambda$, and $\bar{\lambda}$ are treated as the coordinates on the rectangular object atlas considered as the object-colour stimulus manifold. We will refer to these as $c\Delta\lambda\bar{\lambda}$-coordinates. The equator (as well as any circle parallel to it) represents the visible spectrum. The meridian semicircles are loci of constant central wavelength. The north and south poles represent $x(\lambda) \equiv 1$ and $x(\lambda) \equiv 0$, respectively.

In general, when the regular object-colour solid is only a proper part of the genuine object-colour solid (as in the case of Govardovskii's cone fundamentals), the optimal object atlas does not coincide with the rectangular object atlas. Yet, the latter can be used to represent the object-colour stimulus manifold even in such a case provided that the colour mechanisms and illuminant meet some assumptions (Definitions 16.8 and 16.9). To do this, allow chromatic contrast, c, to take values greater than 1. True, when $c > 1$, the rectangular metamer is impossible to physically implement and cannot be treated as a spectral reflectance function at all. Nevertheless, treating such rectangular metamers as *imaginary*, one can construct what is called an extended rectangular atlas (that will include, along with the ordinary ones, also imaginary rectangular metamers) so that it unambiguously represents the object-colour stimulus manifold, as well as the object-colour solid.

Furthermore, using one and the same geographic coordinate system based on the $c\Delta\lambda\bar{\lambda}$-coordinates, one can simultaneously represent all the trichromatic object-colour stimulus manifolds (meeting the conditions in Definitions 16.8 and 16.9). Each of these object-colour stimulus manifolds will be represented as a volume (encompassing the unit ball in $\mathbf{R}^3$), which will be referred to as the *object-colour stimulus gamut*. The shape of these volumes will be determined by the cone fundamentals. All the regular cone fundamentals (that is, meeting the *regularity assumption* in Definition 14.5) bring about the spherical object-colour stimulus gamut.

As a rule, a change in illumination causes a rearrangement of object metamer sets. When this happens, an object stimulus can become metameric to a different optimal metamer. If this occurs, we will say that an *object-colour stimulus shift* induced by the illuminant change takes place. For those trichromatic object-colour stimulus manifolds that admit a representation based on the $c\Delta\lambda\bar{\lambda}$-coordinates, the object-colour stimulus shift can be decomposed into a shift in (i) the chromatic contrast, (ii) spectral bandwidth, and (iii) central wavelength. Each of these three component shifts is explicitly expressed as a function of the illuminant change provided that the latter is relatively small (Section 16.4.3).

In Chapter 17, a perceptual structure that the object colours form in a single-illuminant scene is analysed by applying the same approach as that used for light colours (Chapter 8). Specifically, the colour of an object is specified by the amounts of the component hues it contains. Using the partial hue matching technique (Section 8.5.1), six proper chromaticity classes have been revealed for object colours, which indicates that there are six corresponding component hues – black, white, red, green, yellow, and blue – corroborating the views of Leonardo da Vinci, Ewald Hering, and many others. Each of the component hues mingles with any other except one, creating binary and ternary object colours. As a result, six component hues fall into three pairs of opponent component hues. These are black and white, red and green, and yellow and blue. Also, the grey chromaticity class was found that does not intersect with any other chromaticity class. According to the theory of chromatically ordered structures (Section 8.5.4), which is as applicable for object colours as for light colours, the grey is a neutral component hue.

As in the case of light colours, the observer's ability to judge in which of the two elements of the chromaticity class the corresponding component hue is more pronounced is conceptualised in terms of the chromatic ordering (in this chromaticity class). As before (in Chapter 8), a scale representing the chromatic ordering is called a chromatic scale. The existence of chromatic ordering allows us to define a chromatically ordered structure and use the theory developed in Section 8.5.4. Specifically, it is shown that the six chromatic orderings (with the properties described above) determine the chromatically ordered structure (referred to the *spherical Hering structure*) that can be considered as a smooth manifold with boundary, topologically homeomorphic to a ball in $\mathbf{R}^3$. It must be stressed that this topological structure arises from the properties of the chromaticity classes established in experiment, without any reference to the topology of the object-colour stimulus manifold. Obtained independently, the sphericity of the chromatically ordered structure is in line with that of the object-colour stimulus manifold, which strengthens the credibility of both of these alternative approaches.

With there being three pairs of opponent hues, the six chromatic scales can be compressed into three, in a similar manner as in Chapter 8 for light colours (Section 8.6.3). Specifically, the red and green chromatic scales are combined into the opponent red-green scale, the yellow and blue chromatic scales into the opponent yellow-blue scale, and the white and black chromatic scales into the opponent white-black scale (Section 17.1). The latter is also referred to as the lightness scale. It does not follow, however, that we, as some authors using this term, believe in the existence of such a special object-colour attribute as lightness. It is just a terminological convention made solely for technical convenience. Note that, first, contrary to traditional terminology, our lightness, as well as the other two opponent scales, can take negative value, which should be interpreted as a value of blackness taken with the negative sign. Second, there are many object colours that contain neither whiteness nor blackness (for example, a unitary yellow, blue, red, and green). In our terminology, their lightness is zero. However, in all existing models, the only object with zero lightness is the perfect absorber, and all the rest are assigned non-zero lightness.

The three opponent chromatic scales can be thought of as coordinate functions, and the triplet of their values as the *perceptual object-colour coordinates*. The colour of an object can, then, be represented by a point in $\mathbf{R}^3$. The set of points in $\mathbf{R}^3$ that represent all object-colours is called the *object-colour gamut* (in a single-illuminant scene). Special attention is paid to the case (the plausibility of which is more than likely) when metameric objects, and only they, have the same perceptual object-colour coordinates. In this case, the opponent chromatic scales are colour codes, and the perceptual object-colour coordinates can be expressed as functions of the $c\Delta\lambda\bar{\lambda}$-coordinates. It follows that, in order to establish the perceptual object-colour coordinates of a particular object, there is no need in experiment with this object. The experiment with its rectangular metamer will suffice. Generally, if the perceptual object-colour coordinates for the full object atlas are known, the perceptual object-colour coordinates of the rest of objects can be calculated. An object atlas endowed with the perceptual object-colour coordinates is referred to as *colour-charted*.

As shown by Logvinenko & Beer (2012), objects lit with chromatic light give rise to the same spherical Hering structure as under neutral light. Although the chromaticity classes were found to alter with illuminant, the effect of the illuminant on the chromaticity classes was relatively moderate. The effect of the illuminant on the colour of an object can be assessed by using the perceptual object-colour coordinates. Specifically, it is proposed (Section 17.3) to employ the difference between the perceptual object-colour coordinates obtained under two different illuminations as a measure of the illuminant effect (referred to as the *object-colour shift induced by the illuminant change*). Since the perceptual object-colour coordinates can be considered as determined on the object atlas, in particular, the extended rectangular object atlas, the object-colour shift can be decomposed into the object-colour stimulus shift (induced by the illuminant change) and a change in colour charting (induced by the illuminant change). The former can be theoretically evaluated. For the latter to be determined, one has to conduct a special experiment. Namely, the perceptual object-colour coordinates of the extended rectangular object atlas should be evaluated for both the illuminants in question. In general, if we have a full object atlas that is colour-charted for two illuminants, then the object-colour shift, as induced by the change in illuminant from one to the other, can be predicted on theoretical grounds. It must be said, however, that a perceptual change in the colour of an object caused by an illumination change is much more involved, so we will return to this issue in more detail in Chapter 19.

In the end of Chapter 17, we address the problem of object-colour perception by dichromats (Section 17.4). The theoretical approach is based on the same reductionist hypothesis which we exercised when considering light colours in Section 8.4. Recall that a loss of one of the three cone photoreceptors will result in the confusion of some colour signals (namely, those belonging to the same confusion interval) in the colour-signal cone, and thus, in the object-colour solid. Therefore, the dichromats will confuse the corresponding object colours. In particular, all the optimal metamers (or the elements of any other object atlas) mapping to the same confusion interval will be perceived as having the same colour. Yet, as shown in Section 17.4, there are some points in the object-colour solid boundary which are not confused with any colour signal in the object-colour solid. For such points, the confusion interval to which they belong includes only the colour signals from outside of the object-colour solid. These points form a closed loop in the object-colour solid boundary (thus, in the object-colour stimulus gamut and/or the object-colour gamut), which will be referred to as the *dichromatic orbit*. We show that the optimal metamers forming the dichromatic orbit are exactly those that are the optimal stimuli for the dichromatic observer in question. Therefore, evaluating the dichromatic optimal object stimuli as described in Chapter 14 will bring the full specification of the dichromatic orbit. Since the set of the optimal object stimuli remains the same for all the everywhere positive illuminants, the dichromatic orbits do not depend on illumination, being fully determined by the cone photopigment absorbance spectra. The dichromatic orbits for protanopes, deuteranopes, and tritanopes were calculated by using Govardovskii's cone fundamentals and plotted in the $\Delta\lambda\bar{\lambda}$-coordinates (Fig. 17.9).

Clearly, the optimal metamers of the dichromatic orbit are not confused with any other optimal metamer. Hence, the object colours in the dichromatic orbit of the object-colour gamut should be perceived by the dichromats the same way as by the trichromats provided that the reductionist hypothesis holds true. In other words, the part of the trichromatic object-colour palette formed by these colours is included in the dichromatic object-colour palette. Furthermore, if the reductionist hypothesis is understood in the sense that the dichromatic object-colour palette is organised similarly to the trichromatic one, that is, the dichromatic hue palette is fully determined by the full object colours which are, in turn, determined by the optimal object stimuli, then the dichromatic orbit exhausts the dichromatic hue palette. In other words, for dichromats, each object colour (unless it is grey) has the same hue as one of the colours in the dichromatic orbit, differing only in purity.

If the perceptual object-colour coordinates were known for rectangular metamers, the dichromatic orbits for protanopes, deuteranopes, and tritanopes could be plotted in the object-colour gamut (by converting the $\Delta\lambda\bar{\lambda}$-coordinates into the perceptual object-colour coordinates). This would answer the long-standing question: what colours the dichromats perceive when viewing real objects lit by a real light? However, for the lack of a properly colour-charted object atlas, we tried to shed light on this issue by finding elements in the Munsell atlas that are closest to the dichromatic orbits. It follows from this sample of Munsell papers (Fig. 17.10) that the dichromatic object-colour palettes contain all the six component hues. The important feature of dichromatic colour perception is that those combinations of the four chromatic component hues (yellow, blue, red, and green), that dichromats are able to see, are accompanied by only one or two values of whiteness (and/or blackness). Because of that, the green object colours appear to be always very whitish for dichromats, whereas the red object colours mostly look quite blackish, as many researchers have pointed out earlier.

In Chapter 18, we return to the issue of metamer mismatching, now in the context of object colours. Since, likewise metamerism of lights, object metamerism is based on the equality of colour signals, a replacement of one observer by another leads to object metamer mismatching, formally, by the same reasons as light metamer mismatching. Moreover, object metamer mismatching can take place even in the case of a single observer due to a replacement of the illuminant. Really, objects metameric under one illumination may happen not to be metameric under another, scattering (under the second illuminant) into a whole set of colour signals called the *metamer mismatch region*. We prove that every metamer mismatch region forms a closed convex body in the object-colour solid (thus, confirming the conjecture of some previous authors (Maximov, 1984; Schmitt, 1976)). The shape and volume of metamer mismatch regions are different for different colour signals. For instance, we prove that the metamers mismatch region for an optimal object stimulus shrinks into a single point.

As mentioned above, the estimation of the boundary of the metamer mismatch bodies remains an unresolved theoretical problem. All previous attempts were, firstly, based on the same finite-dimensional approximation of the spectral reflectance functions (Wyszecki & Stiles, 1982, pp. 200–212), which has already been critisied above in connection with the evaluation of the light metamer sets. Secondly, even in the finite-dimensional case, the theoretical boundaries were not, actually, fully described. We present a general solution to this problem (Theorem 18.1) putting forward the method of evaluating the boundary of the metamer mismatch bodies based on convex analysis. Theoretical issues encountered when practically computing the metamer mismatch bodies are discussed in Section 18.2. In Section 18.3, a number of indices is proposed to quantify the degree of object metamer mismatching. In the end of Chapter 18 (Section 18.4), we show, firstly, that if the colour mechanisms are replaced by ones from the linear subspace spanned over these mechanisms (i.e., their linear combinations), there is no metamer mismatching. Secondly, it is linear transformations (of the colour mechanisms), and only they, that do not cause metamer mismatching. Thirdly, any change in the illuminant will lead to non-trivial object metamer mismatching, unless the change reduces to multiplying the spectral power distribution function of the illuminant by a positive number.

In Chapter 19, attempting to more successfully model the environment, we enrich the experimental paradigm (making it somewhat more realistic) by allowing for it to contain several light sources. Now consideration is given to various objects lit by various lights. The entire set of such object/light pairs, taken into consideration, is referred to as the *object/light stimulus set*. As one would expect, some physically different object/light pairs appear absolutely identical for an observer, that is, they make an absolute match. Just as colour matching for lights is conceptualised as metamerism of lights, colour matching for object/light pairs is formalised as an equivalence relation on the set of object/light pairs, called the *object/light colour equivalence* (Section 19.1). The latter is assumed to imply the metamerism of the reflected lights but not the other way round. Together with object metamer mismatching this assumption leads to the inseparability of the object/light colour equivalence, that is, the latter cannot be represented as a combination of an equivalence relation on the set of objects and another equivalence relation on the set of illuminants.

Picking up one object/light pair in each class of the object/light colour equivalence, we get an *object/light atlas* which can be considered as a generalisation of the hitherto defined the light and object atlases. Specifically, the object/light atlas is considered as the Cartesian product of an object atlas and a light atlas. The elements of the object atlas are referred to as the *object components*, and those of the light atlas the *light components* of the object/light atlas. The object/light atlas is used to represent the set of object/light colour equivalence classes, referred to as the *object/light colour stimulus set*. Each element of the latter represents an object colour (in the multi-illuminant scene).

It is worth emphasising that our notion of object colour is rather wider than usually assumed. Really, it does not take a colour scientist to notice that the same objects lit by different lights, as a rule, look different. According to our definition, this difference is the difference in the object colour. True, the colour difference produced by light is experienced as qualitatively different from that produced by objects. A piece of white paper under a red light does not look like a red paper under daylight, whatever the shade of its redness.

The most common interpretation of the colour appearance in the multiple light scenes, descending from Helmholtz, is that the perception of the colour of the objects is accompanied by the perception of the colour of the illuminants, both of which occur simultaneously and independently. That is, there exist the object colour perception and the illumination colour perception; and they are independent of each other providing us with two different percepts. However, such a view contradicts the inseparability of the object/light colour equivalence. Besides, if this view were correct, then the colour of an object would be constant despite the illumination changes. Nevertheless, numerous experiments undertaken to register colour constancy (a phenomenon that we supposedly experience daily) have shown that, strictly speaking, the constancy of the object colour never occurs. One might argue that the visual system manages to achieve only approximately object-colour constancy. However, the degree of colour constancy was found to be different for different objects and different illuminants. Moreover, whilst the colour constancy of objects has been studied extensively for more than a century, there is still no clear understanding of how it is in principle possible to get the colour of objects, at least approximately, constant.

A variety of different approaches to the colour constancy problem have been revolving around the following simple idea, also enunciated by Helmholtz. If the brain knew the illuminant change, it could evaluate the transformation which the colour signal should undergo as a result of this change. Some believe that by taking the inverse transformation, one can restore the 'true' colour signal, i.e., that produced under the previous illumination (e.g., Zaidi, 1998). However, the problem is that the illuminant-related transformation of the colour signal is such that there is no unique inverse transformation for it. Indeed, because of metamer mismatching, two objects which produce different colour signals under one illumination may produce the same colour signal under another, so the inverse image of the latter colour signal cannot be unique. This problem is usually either ignored or underestimated by visual scientists involved in the study, as well as in computation modelling of the object-colour constancy. In the latter case, there is a belief that while in theory the inverse illuminant-related transformation of the colour signal is not unique, in practice, especially, for natural illuminants with a broadband spectrum, this non-uniqueness is negligible, though no theoretical estimation of it has been made so far. The non-uniqueness of the illuminant-related transformation of the colour signal can be quantitatively specified in terms of the index of metamer mismatching. Specifically, the volumes of a metamer mismatch regions can be used to ascertain to what extent metamerism is a problem for the approach to the colour constancy problem based on the inverse illuminant-related transformation of the colour signal. Some preliminary estimations show that for chromatic lights the degree of metamer mismatching is so large that it completely undermines the Helmholtz idea (Logvinenko, Funt, Mirzaei, & Tokunaga, 2015).

We find the concept of 'colour constancy' inefficient and misleading because it puts a cart before the horse. In the days of Helmholtz, when appropriate quantitative measurements were not available, it might have been excusable to simply proclaim the colour constancy, as a plausible hypothesis. However, now, when the hundreds of experiments carried out prove otherwise, this idea has become a drag on the way of object colour studies. We should admit that object colour changes with illumination. Instead of studying how imperfect colour constancy is, we had better study systematically how object colour changes with illumination. However, this cannot be done without revealing the internal, perceptual structure of the set of object colours, which Section 19.3 is devoted to.

Critical attitude towards the notion of colour constancy itself should not prevent us from appreciating the following important fact established in experiments driven by the idea of colour constancy. Observers are capable of seeing certain resemblance between some objects lit by different lights. For example, when observing white paper under different illumination, one cannot help having a strong impression that, in a certain sense, the colour of the paper remains 'white' regardless of the lighting. No wonder that the observers do not consider it impossible to establish a 'match' between objects lit by different lights. Such a match between stimuli, which are not object/light colour equivalent, is usually called an *asymmetric colour match* in vision literature. Of course, it cannot be a genuine, absolute match because of the unavoidable perceptual difference in lighting. However, as proved by Logvinenko & Tokunaga (2011), such a match satisfies the axioms of an equivalence relation, likewise, a symmetric colour match. We define the notion of *material colour* (mentioned informally above) via this equivalence relation. Specifically, all the elements of an equivalence class, determined by the asymmetric object-colour matching, are assumed to have, by definition, an identical material colour. Also, for the sake of generality, we consider the case when a similar asymmetric colour match exists that allows us to give a formal definition of *lighting colour*. A few simple technical axioms are discussed, which ensure that for each element in an object/light atlas there is exactly one pair of material and lighting colours. These are called the *material* and *lighting components* of the object colour.

As the lighting colour cannot be thought of as the colour of the illuminant, there is no reason to expect that, when equally illuminated, the elements of an object atlas will have the same lighting colour. Quite probably, some variations in lighting colour can be also observed. Yet, there is every reason to believe that the range of the lighting colour variation in this situation will be rather small, if any. Hence, the set of object colours produced by the object atlas in a single-illuminant scene can be considered as a suitable approximation to the set of material colours (paired with the constant lighting colour). Recall that the experiments conducted with the illuminants of various chromaticity revealed the chromatically ordered structure of the same type: a spherical Hering structure (Chapter 17). It does not follow, however, that the chromaticity classes comprised exactly the same stimulus objects under every illuminant. Just to the contrary, the chromaticity classes varied depending on the illuminant, but the type of the chromatically ordered structure remained the same. Therefore, it seems reasonable to assume that the material colours make a spherical Hering structure that can be represented by the perceptual object-colour coordinates. In other words, for a given lighting colour, all the material colours can be represented by a volume in $\mathbf{R}^3$ which is topologically equivalent to a ball. We refer to this volume as the *material colour gamut* (associated with a particular lighting colour). Needless to say, the material colour gamuts for different lighting colours can differ significantly from each other.

It follows that when the illumination of an object changes, its material colour is highly likely to change too because, firstly, the object is most likely to become object/light colour equivalent to a different element in the object/light atlas; and secondly, the very material colour gamut is transformed by the illumination change. Thus, the perceptual object-colour coordinates in the object/light atlas are most likely to change as well. That is, even the constancy of material colour is hard to expect under the illumination change. Thus, an attempt to interpret such a situation with the use of the concept of 'constancy of colour' is doomed from the very beginning to failure.

As for lighting colours, in order to understand what kind of structure they form, one can begin with the variation of illumination, keeping the object on which the light falls, constant. Strictly speaking, one cannot take for granted that the material colour will remain constant as well. However, immediate introspection indicates that the material colour changes, caused only by changes in light, are rather small, if any. Since this experimental set-up can be considered as an implementation of the colorimetric paradigm, it seems quite justified to assume that, for a given material colour, the lighting colours (paired with it), likewise the light colours, constitute a circular Hering structure (Section 8.6.3) that can be represented by a volume in $\mathbf{R}^3$ which is topologically equivalent to a circular cylinder.[70] We refer to this volume as the *lighting colour gamut* (associated with a particular material colour). Admittedly, the lighting colour gamuts corresponding to different material colours can differ.

The set of object colours can be, therefore, considered as a family of material colour gamuts corresponding to various lighting colours. From the experiment, we know that small variations in the illumination (thus, in the lighting colour) lead to small changes in the object colour in a single-illuminant scene (thus, in the material colour). Consequently, it can be assumed that the material colour gamuts are smoothly 'glued together' to form a smooth manifold (referred to as the *object/light colour manifold*). Locally, that is, in a neighbourhood of a pair of particular lighting and material colours, this manifold can be thought of as a fragment of the Cartesian product of the

70 Recall that in Section 8.6.3 we consider two possible representations – in the form of a cone and a cylinder – for the light-colour manifold depending on whether the colour of the dark is included or not. Obviously, only the cylindrical representation is suitable for lighting colours, since material colours go out of existence in the dark.

spherical Hering structure (representing the material colours) and the circular Hering structure (representing the lighting colours). It does not follow, however, that the whole object/light colour manifold can be represented as a Cartesian product of the spherical Hering structure of material colours and the circular Hering structure of lighting colours. The global structure of the object/light colour manifold is a complicated problem that requires separate study.

In view of the complete lack of experimental data on this score, we accept the simplifying hypothesis that the object/light colour manifold can be represented not only locally, but also as a whole as a Cartesian product of the spherical Hering structure of material colours and the circular Hering structure of lighting colours. In this case, the object/light colour manifold will have the same topological structure as the object/light colour stimulus manifold. Recall that the latter is the Cartesian product of the object atlas and the light atlas. The object atlas is topologically equivalent to a ball in $\mathbf{R}^3$. So is the spherical Hering structure of material colours. The light atlas (not including the full dark) is topologically equivalent to a cylinder through a disc in $\mathbf{R}^3$. So is the circular Hering structure of lighting colours. Therefore, from a topological point of view, the object/light colour stimulus manifold and, due to our assumption, the object/light colour manifold too can be thought of as a cylinder through a five-dimensional ball in $\mathbf{R}^6$.

It should be borne in mind that a topological representation in general, and in the form of a cylinder through the five-dimensional ball in $\mathbf{R}^6$ in particular, is quite crude. Specifically, it does not reflect the differences in the material colour gamuts for different lighting colours, which are simply irrelevant from a topological point of view. A more accurate representation, taking into account these differences, can be obtained if we combine together all the material colour gamuts by putting in correspondence with every point of the light-colour gamut the corresponding material colour gamuts. This will lead to some irregular volume in $\mathbf{R}^6$.

We note, however, that in such a representation, the spherical Hering structures (for material colours), corresponding to different lighting colours, are represented independently, without assuming that they are related to each other. Nevertheless, there is reason to believe that perhaps such a relationship exists. In this regard, it is worth recalling that the chromaticity classes obtained using a single-illuminant display were found to be exactly the same as for a multi-illuminant display. For example, if for a single-illuminant display, a certain chromaticity class shrinks under the red illumination as compared to the neutral illumination, then when using both the red and neutral illuminants, the same shrinking takes place (Logvinenko & Beer, 2012). This means that the component hue behind this chromaticity class was identified by the observers as the same. This allows us to assume that all the spherical Hering structures registered for various illuminants (in the single-illuminant displays) are perceptually combined in a common structure. In particular, one can conjecture that the perceptual object-colour coordinates can be expressed in the same units for all the illuminants. True, this would be justified only if the observers are able to order the elements of the same chromaticity classes obtained for different illuminations. Although such an ability has not been tested as yet, it does not seem implausible. Indeed, when observing the same red paper under bright and dim lights, it seems self-evident that the redness of the bright paper is more pronounced than that of the dim one. Assume, for example, that reducing the illuminant intensity results in decreasing the chromaticness of each object by the same factor. In this case, the object-colour gamuts as obtained in a single-illuminant display will be identical (because of normalisation). However, for a multiple-illuminant display they will be different, albeit similar in shape. (Of course, in reality, one can expect that the object-colour gamut will change its shape, even if the light source varies only in intensity, not to mention the change in chromaticity.) For a more detailed description of the geometry of the object/light colour manifold, additional experiments are needed.

Chapter 20 is devoted to the problems arising from the fact that many different pairs of material and lighting colours may occur in response to the same colour signal (produced by the reflected light). Specifically, we examine the possible relationship between the material and lighting colours brought about by the same colour signal. First of all, we show that there is a trade-off between the object and light components of those elements (in the object/light atlas) that induce the same colour signal (Section 20.1). Specifically, it is shown that, given a particular colour signal, there is a whole three-dimensional subset in the object/light atlas that can induce this colour signal. Therefore, the colour signal in no way can be treated as a determinant (or stimulus) for the colour of an object. Still, the former imposes a strong constraint on the latter. Indeed, when the light component is fixed, then the object component is uniquely determined for the given colour signal, and vice versa. In Section 20.1, using the inverse function theorem, we calculate an explicit approximation to this functional relationship between the object and light components for some object/light atlases.

At this point, an analogy arises itself with the relationship between size and distance producing the same angular size. Considered as a proximal stimulus, the angular size does not allow to determine uniquely the size of the object either. Yet, the angular size uniquely specifies the size/distance ratio. Generally, the proximal stimulus specifies the relationship between some physical properties of the distal stimulus (i.e., the object). Recall that an element of the object/light atlas is a physical object (i.e., the distal stimulus from the psychophysical standpoint). Likewise, for a psychophysicist, a colour signal can be considered as the proximal stimulus. Therefore, what is established in Section 20.1 is that if such a proximal stimulus as a colour signal encodes anything at all, then it is a kind of reciprocal relationship between such two physical properties of the distal stimulus as its spectral reflectivity and the spectrum of its illumination.

Interestingly, there is a perceptual counterpart of the size/distance invariance for a given angular size. Specifically, the ratio of apparent size to apparent distance is known to hold, by and large, constant as long as the angular size is kept constant. This apparent size/distance invariance is manifested in the fact that any misperception of distance causes a corresponding misperception of size, and vice versa. In particular, as shown in the famous Ames demonstrations, if an object is perceived as deceptively further (respectively, closer) than it really is, it is perceived as larger (respectively, smaller). In Section 20.2, we provide some evidence for that there exists the invariant relationship of the kind between material and lighting colours too. For instance, lightness and lighting brightness appear to be locked in a reciprocal relationship provided that the retinal illuminance is constant (Section 20.2.1). Likewise, if the lighting colour is illusory shifted towards, say, yellow, the material colour of an object appears more bluish (Section 20.2.4).

Here, we bump into a serious problem. The reciprocal relationship between apparent size and apparent distance can be considered as a mirror reflection of its physical counterpart. In other words, it seems reasonable to assume that a certain relationship between the physical attributes of the distal stimulus can be built in perception as prior knowledge (a prior rule). However, as repeatedly argued before, there is no distal stimulus for the object colour. To put it the other way round, there is no physical counterpart for the invariant relationships revealed for material and lighting colours. What is the point, then, in binding the material and lighting colour by any relationship if it does not reflect any physical relationship? What purpose can this serve? In our opinion, the answer is hard to find until we get an answer to the following, more general question: what is, in general, colour for?

The key to answering this fundamental question, in our opinion, is the repeatedly reported effect of shape or, more generally, spatial layout on colour. Actually, it is an apparent rearrangement of the spatial layout of a scene rather than the shape of separate objects that can lead to a significant alteration of the colour in the scene. It must be noted that it is not about small but about very massive changes in the colour of the object depending on its position in the scene. For instance, Gilchrist (1977) reported that the lightness of a card could vary within a broad range depending on its apparent slant. Section 20.2.1 discusses more demonstrations of the considerable effect of apparent slant and/or depth on lightness. Since it is the apparent (rather than physical) shape (more generally, spatial layout) that affects the material colour, a question arises whether the material colour is locked with the apparent shape in an invariant relationship as it is locked with the lighting colour. The answer to this question can be found only in the experiment. However, its transfer to the level of experimental verification implies the existence of a sensory attribute representing the shape. But what can serve as a sensory fabric for the shape? In particular, what is the sensory fabric of the slant? What exactly do we mean by saying that we see some surface (e.g., a grey card) being slanted? Do we see something in particular, or just apprehend that it is slanted? Imagine we start rotating the card. Will, at this time, any sensory attribute begin changing in our perception? When this happens in a casual room lit by a normal light source, then it is easy to notice that if the card in its rotation deviates from the light source, it will look more and more shaded as it rotates. This is the so-called attached shadow. Although attached shadows look very similar to cast shadows, it has been found that they are different perceptual attributes. In particular, it was found that shading caused by a slant cannot be compensated for by an increase in illumination (Logvinenko, 2015a). To distinguish between these attributes, we use the terms 'shading brightness' and 'lighting brightness'. Furthermore, we argue that, as a perceptual attribute on its own, shading represents shape in our perception (Section 20.2.3). Admittedly, as shown by Gibson (1979), the ambient optic array contains another sources of information about spatial layout and shape (e.g., the gradient of optical texture), but in the present context it is important to emphasise the role of shading.

In Section 20.2.3 and later in the last chapter (Chapter 21), we describe several demonstrations testifying that lighting brightness can convert into shading brightness and vice versa. Moreover, these transformations are governed by relationships similar to those that link lighting brightness and lightness.

Thus, shading turns out to be, to the same extent, both a colour attribute and a spatial layout attribute. It is certainly such a colour attribute that does not evoke questions concerning its role. If shading represents shape (thus, playing a crucial role in representing spatial layout), then it is clear what it is for (see Section 21.1). Indeed, the ecological validity of apparent shape is self-evident. Locomotion is impossible without a representation of the spatial layout in one form or another. Hence, a veridical perception of shape (and more broadly, spatial layout) is vitally important for animals.

When the surface turns away from a light source, it sends light of lower luminance into the eyes. As shown elsewhere, with certain limitations on its spectral reflectivity, the shape of the surface can be derived from the luminance pattern provided that illumination is constant and homogeneous (Horn, 1989). Without such limitations (e.g., when spectral reflectivity can vary across the surface), the effect on luminance caused by shape will be confounded with that caused by spectral reflectivity. In order to successfully derive the surface shape from the luminance pattern, then, one has to unravel these two effects. Once such a disambiguation is done, and the surface shape is evaluated, it seems reasonable to utilise the effect produced by the spectral reflectivity variations somehow too, since it might contain some information that can be useful from an ecological point of view. Therefore, the contribution of both factors should be represented in our perception as two different perceptual attributes (i.e., shape and material colour). Note that in this context, the question of a more or less accurate assessment of the surface reflectivity as such is not worth it and is relegated to the background. The evaluation of shape takes precedence; reflectivity is estimated to the extent necessary to eliminate the ambiguity in question. In other words, the resulting material colour is a kind of by-product of shape perception.

When the illumination varies over the surface, it also contributes to the formation of the luminance pattern. In this case, the contribution of the three factors is subject to disentanglement: shape, surface reflectivity, and light spectrum. Accordingly, along with the shading and the material colour, a perceptual attribute such as the lighting colour emerges. Yet, like material colour, lighting colour has a subordinate role as compared to shading.

As we can see, having evolved from shape perception, colour perception is closely interlaced with shape perception, being in a sense just an aspect of the latter. The fact that colour perception is inextricably linked with shape perception explains why alteration of such (seemingly extraneous to colour) a depth cue as binocular disparity may dramatically affect object colour. In the pseudoscopic demonstrations described in Section 20.2.3, reversing the sign of binocular disparity results in an apparent depth reversal, thus changing the spatial layout, which in turn leads to a spectacular change in the object colour. Watching pseudoscopic demonstrations convinces us that it is hard to decide where shape perception ends and colour perception begins. Moreover, it is becoming more and more obvious that it is hardly a good idea first to divide the holistic perception into particular perceptions of certain physical characteristics (such as size, distance, depth, shape, and speed), and then investigate them separately, using a psychophysical approach.

As we argue, such an approach does not work. In any case, it does not work as far as colour is concerned because the psychophysical approach implies the existence of a stimulus, whereas it is impossible to indicate which physical entity could possibly serve as a stimulus for the colour of an object.

At the end (Chapter 21), we discuss the possibility of an alternative approach. Specifically, we discuss the role of the invariant relationships between the various perceptual attributes (such as material colour, lighting colour, and shading) on the one hand, and the attributes of the colour signal code pattern on the other. We regard them as a kind of 'perceptual equations' where perceptual attributes play the role of 'unknowns'. Although a single equation contains far too many unknowns for a unique solution, a system of such equations significantly constrains a possible solution. Natural scenes, rich in various features bring about a large number of perceptual equations, the indeterminacy of which is significantly reduced.

Being unable to be the only direct cause of the colour of an object, the colour signal code pattern, nevertheless, determine it through the perceptual equations which play the role of a kind of normative rules. One can draw an analogy with the grammatical and syntactic rules of the language. In Section 21.2, we argue that not only is this analogy appropriate, but the linguistic metaphor in general is quite fruitful for understanding the very essence of visual perception.

The book ends with two appendices. In Appendix A, we summarise the main properties of Banach spaces of measures and functions that are used as stimulus spaces (Section A.1). It also includes facts from the convex analysis that are necessary for understanding some theorems and their proofs (Section A.2). Appendix B contains proofs of all the mathematical results given in the main text.

Part I

Light Colour

In this part, we deal with light colours, that is, those perceived to be not attributed to any object or medium. As mentioned above, such colours are experienced under special experimental conditions when disembodied light is presented to the human eye. The major goal of using such conditions is to control with desirable precision the spectral, spatial, and temporal characteristics of light coming into the eye. Special apparatus (such as integrating spheres, colorimeters and the like) and special techniques (such as colour matching, colour discrimination, etc.) have been developed for this purpose. The details can be found in Wyszecki & Stiles (1982). Moreover, recent advances in adaptive optics (Miller & Roorda, 2010) make it possible to control not only the light entering the eye but also the quality of the image formed on the retina.

When studying light colours, very simple spatial distributions of light are usually used. In colour matching experiments, a bipartite field (presented against some spatially homogeneous background) is usually used as a stimulus display (Fig. I.1). Both halves of the field are spatially homogeneous, differing only in radiant spectrum. The background usually remains unchanged. The stimulus display in experiments on colour detection and discrimination can be even simpler consisting of only two (rather than three as in Fig. I.1) spatially homogeneous light patches: a relatively small simple geometrical shape (e.g., a disk or a square), referred to as a *target*, surrounding by a large spatially homogeneous *background*.

Figure I.1 Bipartite field, widely used in experiments on colour matching. Its two halves emit different lights. When the human observer cannot see the border dividing the bipartite field, it is taken as a colour match.

Foundations of Colour Science: From Colorimetry to Perception, First Edition. Alexander D. Logvinenko and Vladimir L. Levin.
© 2023 John Wiley & Sons Ltd. Published 2023 by John Wiley & Sons Ltd.

In the first chapters, we will assume that the media (between the observer and the stimulus display) are transparent enough to assume that their effect on the light entering the observer's eye is negligibly small. In Chapter 5 we will consider the effect of the medium spectral transmittance on the colour of light passing through the medium. In the concluding chapter of this part (Chapter 13), a more complicated stimulus display will be considered. Specifically, each of the two fields will be presented on their own background (Section 13.3.1).

2

Colour Stimulus Space and Colour Mechanisms

Lights can be mixed (i.e., physically superimposed) with each other. Such light mixing lends itself to be formalised as an algebraic operation of summation. Also, there is a straightforward physical meaning for multiplying a light by a positive real number, say, t: changing the radiant energy at each wavelength by the constant factor[1] t. We will assume that we deal with a set of lights which is large enough for these two operations to be possible for each pair of lights and each positive number. More specifically, we will assume that the lights we consider constitute what is defined below as a light stimulus cone (Suppes et al., 1989, p. 231).

Definition 2.1 *Suppose* $\mathbf{A}$ *is a non-empty set, and two operations,* $\oplus$ *(mixing):*

$$\mathbf{A} \times \mathbf{A} \ni (a, b) \mapsto a \oplus b \in \mathbf{A},$$

and $*$ *(intensity alteration):*

$$\mathbf{R}_+ \times \mathbf{A} \ni (t, a) \mapsto t * a \in \mathbf{A},$$

are defined[2]*. We say* $\mathbf{A}$ *is a light stimulus cone (and its elements are light stimuli) if two axioms are satisfied as follows:*

1. *For all* $a, b, c \in \mathbf{A}$, $(a \oplus b) \oplus c = a \oplus (b \oplus c)$ *(associativity),* $a \oplus b = b \oplus a$ *(commutativity), and* $a \oplus c = b \oplus c$ *implies* $a = b$ *(cancellation).*
2. *For all* $a, b \in \mathbf{A}$ *and* $t, u \in \mathbf{R}_+$,

   (i) $t * (a \oplus b) = (t * a) \oplus (t * b)$;
   (ii) $(t + u) * a = (t * a) \oplus (u * a)$;
   (iii) $t * (u * a) = (tu) * a$;
   (iv) $1 * a = a$.

---

1 We will say that lights a and ta *differ only in intensity* as we will need to distinguish this particular case from a more general one when two lights, a and b, have different radiant spectral distributions, that is, there is no $t > 0$ such that $a = tb$. Note that we do not define the word 'intensity' itself (as is done, for example, in the physical sciences, where it denotes time-averaged energy flux). In the vision sciences, there are a few measures which are usually interpreted as light intensity: radiant intensity and irradiance, to mention two. The former, measured in watts per steradian, is defined as 'the quotient of the radiant power emitted by the source in an infinitesimal element of solid angle containing the given direction, by the element of solid angle' (Wyszecki & Stiles, 1982, p. 690). The latter, measured in watts per meter squared, is defined as 'the quotient of the radiant power incident on an infinitesimal surface element containing the point, by the area of that surface element' (Wyszecki & Stiles, 1982, p. 690). Still, we will, if possible, refrain from using this word because of its ambiguity. In the rare cases when it will nevertheless appear, it should be treated simply as one of the words of the English language.

2 Here and below, $\mathbf{R}$ and $\mathbf{R}_+$ will be used to denote the set of real numbers and its nonnegative subset respectively.

Foundations of Colour Science: From Colorimetry to Perception, First Edition. Alexander D. Logvinenko and Vladimir L. Levin.
© 2023 John Wiley & Sons Ltd. Published 2023 by John Wiley & Sons Ltd.

By (ii) and (iv), for every $a \in \mathbf{A}$, $a = a \oplus (0 * a)$, and, by applying associativity, commutativity, and cancellation conditions from Axiom 1, we obtain $b \oplus a = b \oplus (a \oplus (0 * a)) = b \oplus ((0 * a) \oplus a) = (b \oplus (0 * a)) \oplus a \Rightarrow b = b \oplus (0 * a)$ whenever $a, b \in \mathbf{A}$. On the other hand, $b = b \oplus (0 * b)$, hence $b \oplus (0 * a) = b \oplus (0 * b)$, and the cancellation condition gives us $0 * a = 0 * b$, that is, there exists the unique element $\mathbf{0}$ in $\mathbf{A}$ such that, for all $a \in \mathbf{A}$, $0 * a = \mathbf{0}$ and $a \oplus \mathbf{0} = a$.

Now consider the set $\mathcal{E}$ of pairs $(a, b) \in \mathbf{A} \times \mathbf{A}$ with the equivalence relation[3] $\sim$ on it, as given by

$$(a, b) \sim (a', b') \Leftrightarrow \exists c, c' \in \mathbf{A} : (a + c, b + c) = (a' + c', b' + c').$$

Clearly $(a, b) \sim (\mathbf{0}, \mathbf{0})$ if and only if $a = b$ (this follows from the cancellation condition). Further, it follows from Axioms 1 and 2 that $\mathcal{E}$ is a vector space relative to operations

$$(a, b) + (a', b') = (a \oplus a', b \oplus b')$$

$$t(a, b) = \begin{cases} (t * a, t * b), & \text{if } t \geq 0, \\ ((-t) * b, (-t) * a), & \text{if } t < 0. \end{cases}$$

Then $\mathcal{E}_0 := \{(a, b) \in \mathcal{E} : (a, b) \sim (\mathbf{0}, \mathbf{0})\} = \{(a, a) : a \in \mathbf{A}\}$ is its vector subspace, and the set of the equivalence classes $E := \mathcal{E}/\sim$ is the factor space $\mathcal{E}/\mathcal{E}_0$.

Let us define a mapping $\varphi : \mathbf{A} \to E$, $\varphi(a) := \pi(a, \mathbf{0})$, where $\pi : \mathcal{E} \to E$ is the canonical mapping onto the factor space: $\pi(a, b) = [(a, b)]$ is the equivalence class of (a, b). Clearly, φ is a one-to-one mapping of $\mathbf{A}$ onto the convex cone $A := \varphi(\mathbf{A}) = \{[(a, \mathbf{0})] : a \in \mathbf{A}\}$ in E, $\varphi(a \oplus b) = \varphi(a) + \varphi(b)$, $\varphi(t * a) = t\varphi(a)$ whenever $a, b \in \mathbf{A}$, $t \in \mathbf{R}$. We have $A \cap (-A) = \{0\}$, and every element $[(a, b)] \in E$ is represented as $[(a, \mathbf{0})] + [(\mathbf{0}, b)] = [(a, \mathbf{0})] - [(b, \mathbf{0})] \in A - A$, that is, $E = A - A$. Moreover, it can be easily seen that if there is another vector space E', a convex cone A' in it such that $E' = A' - A'$, and a one-to-one mapping φ' of $\mathbf{A}$ onto A', then there exists a one-to-one linear transformation ψ of E onto E' such that $\psi(A) = A'$ and $\varphi' = \psi\varphi$. The mapping ψ acts as follows. Every element in A is of the form $\varphi(a)$ where $a \in \mathbf{A}$ is uniquely determined, and ψ is defined on A by the equality $\psi(\varphi(a)) := \varphi'(a)$ and then it is extended on $E = A - A$ in a natural way: $\psi(\varphi(a) - \varphi(b)) := \psi(\varphi(a)) - \psi(\varphi(b))$. Clearly, ψ is a one-to-one linear mapping of E onto E' and $\psi(A) = A'$.

Thus, the light stimulus cone $\mathbf{A}$ from Definition 2.1 is identified with a convex cone A in a vector space E, and we obtain the following embedding proposition (cf. Krantz, 1975a, Theorem 1 or Suppes et al., 1989, pp. 231–232).

Proposition 2.1 *If $\mathbf{A}$ satisfies Axioms 1 and 2 of Definition 2.1, then there exist a vector space E, a convex cone A in it such that $A \cap (-A) = \{0\}$ and $E = A - A$, and a one-to-one mapping φ of $\mathbf{A}$ onto A, such that $\varphi(a \oplus b) = \varphi(a) + \varphi(b)$ and $\varphi(t * a) = t\varphi(a)$ whenever $a, b \in \mathbf{A}$, $t \in \mathbf{R}_+$. Moreover, if a vector space E', a convex cone A' in it, and a mapping φ' of $\mathbf{A}$ onto A' have the same properties then there is a one-to-one linear transformation ψ of E onto E' such that $\psi(A) = A'$ and $\varphi' = \psi\varphi$.*

So, as a mathematical model of lights, we will use the convex cone A (referred to as the light stimulus cone) in some vector space E such that $E = A - A$ (this means that each $e \in E$ is represented as a difference of two elements from A, $e = a_1 - a_2$), and that A is line-free (this means that $-a \notin A$ whenever $0 \neq a \in A$). It follows that $A \cap (-A) = 0$. The null vector 0 will be referred to as the vertex of the cone A. Though the null vector 0 means the absence of light (i.e., the full dark), sometimes, it will be convenient to assume that $0 \in A$ formally considering it as an *improper light* (the *nil light*[4]).

---

3 An equivalence relation is a binary relation $\sim$, which is reflexive ($a \sim a$), symmetric ($a \sim b$ if and only if $b \sim a$), and transitive ($a \sim b$, $b \sim c$ implies $a \sim c$).

4 The colour of the nil light is quite often called 'black' (Wyszecki & Stiles, 1982, pp. 182–184). So is the colour of the object reflecting no light (the *perfect absorber*). As we intend to distinguish between object colours and light colours, we will refer to the colour of the nil light as *'full dark'*, leaving the term *'black'* for the colour of the perfect absorber.

The choice of the vector space E depends on what lights we wish to take into consideration. Such a representation of the light stimulus cone becomes rather straightforward for lights the radiant spectrum of which can be specified by a function which we will refer to as a radiant spectral density function. The physical meaning of the *radiant spectral density function* of a light is the radiant energy (or power) carried by the light per unit wavelength.[5] A radiant spectral density function is supposed to be (Lebesgue) integrable, the integral of which has the meaning of the radiant entity in question (i.e., energy, power, etc.) of the light. In this case, E is identified with the space $\mathbf{L}^1$ of integrable functions on the wavelength interval of the visible spectrum $\Lambda = [\lambda_{\text{min}}, \lambda_{\text{max}}]$, and the light stimulus cone A consists of non-negative functions in it.

If we take into consideration only lights with bounded (respectively, continuous) radiant spectral density function, E is the vector space of all bounded measurable (respectively, continuous) functions on Λ, and A is the cone of non-negative functions in it.

It should be kept in mind that monochromatic lights do not have radiant spectral density function because all their radiant energy is concentrated at just one wavelength. As we intend to take into consideration both light sources with a broadband radiant spectrum, and monochromatic sources, we must choose a vector space E that will be a suitable mathematical model for both. We believe that positive Borel measures on the real segment Λ are appropriate idealisations for A. Indeed, broadband lights can be specified by their radiant spectral density functions, while monochromatic lights can be represented by the delta functions. Both determine particular cases of positive Borel measures on the real segment Λ. So, in the most general case, we adopt a representation of light, considered as visual stimulus, as a positive Borel measure on Λ.

Yet, depending on the context, we may restrict our consideration to a narrower set of lights (e.g., only broadband or only monochromatic lights). It means that we will use different convex cones in different vector spaces as models of the light stimulus cone. The basic examples of light stimulus cones are as follows (for definitions of measure and function spaces $\mathbf{C}^*$, $\mathbf{L}^1$, $\mathbf{L}^\infty$, and positive cones in them, see Section A.1).

Example 2.1

(a) *$E = \mathbf{C}^*$ is the space of all Borel measures on Λ and $A = \mathbf{C}^*_+$ is the cone of positive Borel measures;*
(b) *$E = \mathbf{L}^1$ is the space of all absolutely continuous Borel measures on Λ identified with their densities and $A = \mathbf{L}^1_+$ is the cone of positive Borel measures in it;*
(c) *E is the space of absolutely continuous Borel measures on Λ whose densities are continuous functions (we will write $E = \mathbf{C}$ by identifying absolutely continuous measures with their densities) and $A = \mathbf{C}_+$ is the cone of absolutely continuous measures with non-negative continuous densities;*
(d) *E is the space of absolutely continuous Borel measures on Λ whose densities are bounded Lebesgue measurable functions (we will write $E = \mathbf{L}^\infty$ by identifying absolutely continuous measures with their densities) and $A = \mathbf{L}^\infty_+$ is the cone in E consisting of measures with non-negative densities.*
(e) *E is the space of purely atomic measures (= linear combinations of delta functions) on Λ and A is the cone of positive (non-negative) atomic measures (= non-negative combinations of delta functions).*

Note that in all cases $E = A - A$ and $A \cap (-A) = 0$.

---

5 The radiometric (and colorimetric) notions, closest to what is called here 'radiant spectral density function', are 'spectral distribution function', e.g., the spectral distribution function of radiant power; and 'relative spectral distribution function' (Wyszecki & Stiles, 1982, p. 691).

2.1 Grassmann structures and Grassmann colour codes

Lights appearing different in colour have different radiant spectra. The converse is not true – lights with different radiant spectra may look absolutely the same. When two (physically different) lights have identical appearance in colour one says that a colour match between them takes place (Wyszecki & Stiles, 1982). For example, although two semicircular patches of light in Fig. I.1 look very similar their radiant spectra might be very different. When they look completely identical (that is, when the vertical border between them disappears), we will say that these two (physically different) lights *match* each other in colour.

Colour matching can be considered as an empirical (binary) relation on the set of lights. Colour matching is obviously reflexive and symmetric. Moreover, there is a general belief that colour matching is also transitive (Wyszecki & Stiles, 1982). Although the transitivity of colour matching cannot be taken for granted (Logvinenko, 2006), it is a good departing point in our theoretical analysis. However, to be on the safe side, we will distinguish between empirical relations, such as colour matching, and their conceptualisations in the form of theoretical notions. In particular, following the well-established tradition in colour literature (Krantz, 1975a; Wyszecki & Stiles, 1982; Suppes et al., 1989), we will conceptualise colour matching as a theoretical binary relation which we will refer to as *metamerism*. In other words, introducing this binary relation (i.e., metamerism) we will assume (in this and a few subsequent chapters) that it is operationalised as colour matching. Thus, colour matching is conceptualised as metamerism, and metamerism is operationally reduced to colour matching.[6]

Metamerism will be assumed to be an equivalence relation on the light stimulus cone which is compatible with the operation summation and multiplication as defined in Definition 2.1. More specifically, metamerism will be assumed to define the so-called Grassmann structure on the set of light stimuli (Krantz, 1975a; Suppes et al., 1989).

Definition 2.2 *Given a light stimulus cone A, we say there is a* Grassmann structure *on it if an equivalence relation (*metamerism*) $\sim$ is defined on A which satisfies the following axioms (Grassmann's laws):*

1. *(Law of additivity) For any a, b, and c in A, $a \sim b$ if and only if $a + c \sim b + c$.*
2. *(Law of proportionality) For any a, b in A and any positive real number t, if $a \sim b$ then $ta \sim tb$.*
   *A Grassmann structure is called n-*dimensional*, if yet one axiom (Grassmann's law) is satisfied as follows:*
3. *(Law of n-chromacy) There exist $a_1, a_2, \dots, a_n$ in A such that for any positive t_i and u_i, $i = 1, \dots, n$, if*

$$\sum_{i=1}^{n} t_i a_i \sim \sum_{i=1}^{n} u_i a_i, \tag{2.1}$$

then $t_i = u_i$, $i = 1, \dots, n$, and n is the maximum integer satisfying that condition (referred to as dimensionality *of the* Grassmann structure*). The latter means that for any $m > n$ and for any $b_1, \dots, b_m$ in A, there exist positive numbers t_i and u_i, $i = 1, \dots, m$, such that $t_i \neq u_i$ for at least one i, and*

$$\sum_{i=1}^{m} t_i b_i \sim \sum_{i=1}^{m} u_i b_i. \tag{2.2}$$

6 In Chapter 10, colour matching will be alternatively conceptualised as intransitive binary relation.

Thus, speaking in a more formal way, an n-dimensional Grassmann structure is a pair $\langle A, \sim\rangle$, satisfying Axioms 1–3.

Remark 2.1 *Given an n-dimensional Grassmann structure $\langle A, \sim\rangle$, for any cone $A' \subset A$ the Grassmann structure $\langle A', \sim\rangle$ will be called a restriction of the Grassmann structure $\langle A, \sim\rangle$ onto A'.*

Definition 2.3 *Consider two Grassmann structures, $\langle A, \sim\rangle$ and $\langle A, \approx\rangle$, defined on the same light stimulus cone A. Then $\langle A, \approx\rangle$ will be called a* Grassmann substructure *of $\langle A, \sim\rangle$ if for all $a, b \in A$ $a \sim b$ implies $a \approx b$.*

Remark 2.2 *Given an n-dimensional Grassmann structure $\langle A, \sim\rangle$, the dimensionality of its any Grassmann substructure $\langle A, \approx\rangle$ is not larger than n.*

Definition 2.4 *Let $\langle A, \sim\rangle$ be an n-dimensional Grassmann structure. Any $m \leq n$ lights $a_1, \dots, a_m \in A$ satisfying the condition that for any positive t_i and u_i, $i = 1, \dots, m$,*

$$\sum_{i=1}^{m} t_i a_i \sim \sum_{i=1}^{m} u_i a_i,$$

implies $t_i = u_i$, $i = 1, \dots, m$, will be called colour independent. *Any n colour independent lights will be called* primary light stimuli *(or just* primaries*).*

Example 2.2 *Let us consider the cone $A = \mathbf{C}_+^*$ of positive Borel measures on Λ, and n functions $f_1, \dots, f_n$ on Λ such that for each of them there exists an integral $\int_\Lambda f_i(\lambda)\, a(d\lambda)$ for all $a \in A$. Then $\langle A, \sim\rangle$ is a Grassmann structure if metamerism $\sim$ is defined as follows. For every $a, b \in A$ $a \sim b$ if and only if*

$$\int_\Lambda f_i(\lambda)\, a(d\lambda) = \int_\Lambda f_i(\lambda)\, b(d\lambda) \tag{2.3}$$

for every $i = 1, \dots, n$. We shall see that the Grassmann structure $\langle A, \sim\rangle$ is n-dimensional if and only if the functions $f_1, \dots, f_n$ are linearly independent (see Proposition 2.3(I) below).

Given an n-dimensional Grassmann structure on the light stimulus cone A, the equivalence relation $\sim$ is extended with preserving Axioms 1 and 2 to the whole of $E = A - A$ as follows. For any elements $e, e' \in E$, we fix some of their representations $e = a_1 - a_2, e' = b_1 - b_2$, where $a_1, a_2, b_1, b_2 \in A$, and say that $e \sim e'$ if $a_1 + b_2 \sim b_1 + a_2$. Taking into account that $\sim$ is an equivalence relation on A, it follows easily from Axiom 1 of Grassmann structure that so extended relation does not depend on concrete representations of e and e' (i.e., it is well defined). Moreover, from Axioms 1 and 2 of Grassmann structure, it follows that the extended $\sim$ is an equivalence relation on E which satisfies these axioms, and from Axiom 3 we derive that for every $e \in E$ there exists the unique linear combination $e_0(e) = t_1 a_1 + \cdots + t_n a_n$ (some t_i may be zero or negative even if $e \in A$), such that $e \sim e_0(e)$. The uniqueness of the linear combination $e_0(e)$ implies, in turn, its linearity in e: $e_0(e + e') = e_0(e) + e_0(e')$ for any $e, e' \in E$ and $e_0(te) = te_0(e)$ for any $e \in E$ and $t \in \mathbf{R}$. Also, from Axiom 3 it follows that colour independent lights are linearly independent. (The converse is in general not true.)

We now consider a vector subspace in E (referred to as the *null-space of the metamerism* $\sim$),

$$E(\sim) := \{e = a - b : a, b \in A, a \sim b\} = \{e \in E : e \sim 0\} \tag{2.4}$$

(here and throughout the book, the symbol $:=$ stands for 'is equal by definition'), and turn to the quotient space $E/E(\sim)$. This is a n-dimensional vector space,[7] whose elements are equivalence classes of $\sim$. The algebraic operations in $E/E(\sim)$ are defined in a natural way: if $[e]$ stands for the equivalence class of $e \in E$, then $[e] + [e'] := [e + e']$ and $\alpha[e] := [\alpha e]$ whenever $e, e' \in E, \alpha \in \mathbf{R}$. The canonical map π of E onto $E/E(\sim)$, $\pi(e) = [e]$, is a linear operator, and the image of A, $\pi(A)$, is a convex cone in $E/E(\sim)$.

Definition 2.5 *Let $\langle A, \sim\rangle$ be an n-dimensional Grassmann structure. Then the linear space $E = A - A$ will be called the* light stimulus space. *Those elements in E which do not belong to A will be called* imaginary *light stimuli, the elements in E which do belong to A being referred to as* proper *light stimuli. The factor space $E/E(\sim)$, where $\sim$ is the equivalence relation on E induced by the metamerism, will be called the* colour stimulus space. *The canonical homomorphism $\pi : E \to E/E(\sim)$ will be referred to as the* colour stimulus map. *For every light stimulus $a \in A$, the element $\pi(a) = [a]$ in $E/E(\sim)$ is called* colour stimulus brought about by *a, and the cone of all colour stimuli*

$$\pi(A) = \{[a] : a \in A\} \tag{2.5}$$

is called the colour stimulus cone. *For the sake of generality, we shall also refer to the elements in $E/E(\sim)$ which do not belong to $\pi(A)$ as* imaginary colour stimuli.

Remark 2.3 *Following the terminology established in colour vision literature, we shall call the colours of monochromatic lights* spectral colours.

When using this terminology, one can say that two lights are metameric if and only if they produce equal colour stimuli: for any a, b in A, $a \sim b \Leftrightarrow [a] = [b]$.

Definition 2.6 *Given a light a, a light b is called its* metamer *if b is metameric to a ($b \sim a$). The set of all metamers of a particular light a will be referred to as its* metamer set *and denoted as $Met(\sim; a)$. That is,*

$$Met(\sim; a) := \{b \in A : b \sim a\} = \{b \in A : b \in [a]\}. \tag{2.6}$$

Definition 2.7 *We say a Grassmann structure $\langle A, \sim\rangle$ is* regular *if 0 is the only proper light stimulus belonging to the null-space of $\sim$:*

$$A \cap E(\sim) = \{0\}.$$

It is clear that Grassmann structures on cones A from Example 2.1 and the metamerism $\sim$ from Example 2.2 where every functions on Λ, $f_1, \dots, f_n$, is non-negative and their sum is strictly positive are regular.

Remark 2.4 *Note that the metamer set of a light a contains only proper light stimuli (i.e., those elements $e \in E$ that are lights themselves). In other words, imaginary light stimuli cannot be metamers (i.e., not included in $Met(\sim; a)$). Thus, while every element in $E(\sim)$ is equivalent (in the sense of the*

7 Since for every $e \in E$ there exists the unique linear combination $e_0(e) = t_1 a_1 + \cdots + t_n a_n$, such that $e \sim e_0(e)$, $E/E(\sim)$ is isomorphic to the subspace $Lin(a_1, \dots, a_n)$ in E spanned over $a_1, \dots, a_n$. Thus, for any n colour independent lights, the vector subspace of E spanned over them may be treated as an isomorphic representation of $E/E(\sim)$.

equivalence relation on E induced by the metamerism) to the zero light stimulus, 0, the metamer set of 0, Met(~; 0), is a proper subset in E(~) which becomes the singleton {0} in the case of regular Grassmann structure. In other words, according to our terminology, in a regular Grassmann structure there is no proper light stimulus metameric to the nil light and different from it.[8]

The next proposition (2.2) follows directly from the definitions.

Proposition 2.2 *Let $\langle A, \sim\rangle$ be an n-dimensional Grassmann structure, then m lights $b_1, \ldots, b_m \in A$ are colour independent if and only if the corresponding colour stimuli $[b_1], \ldots, [b_m]$ are linearly independent in $E/E(\sim)$.*

Given n colour independent lights $a_1, \ldots, a_n$, the set $\{e_0(a) : a \in A\}$, where $e_0(a)$ is the unique element of $Lin(a_1, \ldots, a_n)$, which is metameric to a, is a convex cone in E. Since $e_0(a) \in [a]$ whenever $a \in A$, $e_0(a)$ may be identified with the equivalence class $[a]$, hence this cone may be considered as a concrete representation in E of the colour stimulus cone, and any set of n colour independent lights generates such a representation.

An important theoretical consequence of Grassmann's laws is the next representational result (Krantz, 1975a).

Theorem 2.1 *(Krantz Theorem)* Given a *n-dimensional Grassmann structure on A, there exists a homomorphism of A onto a convex cone C in $\mathbf{R}^n$, that is, a map Φ from A onto C, such that for all a, b in A and any positive number t*

(i) $\Phi(a + b) = \Phi(a) + \Phi(b)$;
(ii) $\Phi(ta) = t\Phi(a)$;
(iii) $a \sim b$ if and only if $\Phi(a) = \Phi(b)$.

If there is another homomorphism Φ' of *A onto a convex cone C' in $\mathbf{R}^n$ satisfying (i)–(iii), then there exists a non-singular linear operator $T : \mathbf{R}^n \to \mathbf{R}^n$ such that for all a in A, $T\Phi(a) = \Phi'(a)$.*

Remark 2.5 *It follows from (iii) that $\Phi(a_1), \ldots, \Phi(a_n)$ are linearly independent elements of C; therefore, any cone C satisfying the statement of the Krantz Theorem has a non-empty interior, and $\mathbf{R}^n = C - C$. Any such a cone C may be considered as a concrete (coordinate) representation of the colour stimulus cone $\pi(A)$ in $\mathbf{R}^n$. It will be called the* colour-signal cone *(see Definition 4.1), and its elements* colour signals, *the homomorphism Φ being referred to as the* colour-signal map. *For the sake of terminological coherence, we will sometimes call $\Phi(E)$ the* colour-signal space.[9]

Remark 2.6 *(A sketch of the proof[10]) A cone C in $\mathbf{R}^n$ and a colour-signal map Φ from A onto C may be obtained as follows. Fix n linearly independent vectors $v_1, \ldots, v_n$ in $\mathbf{R}^n$ and let $\Phi(a_i) = v_i$, $i = 1, \ldots, n$, where $a_1, \ldots, a_n$ are the primaries. Let $E_n = Lin(a_1, \ldots, a_n)$ stand for the vector space spanned over $a_1, \ldots, a_n$. Now, for every $e \in E$ we take the unique element $e_0 = e_0(e) = \sum_{i=1}^{n} t_i a_i \in E_n$,*

8 In colour literature, $E(\sim)$ is sometimes referred to as *metameric black* (Cohen & Kappauf, 1982; Wyszecki & Stiles, 1982). We will refrain from using this terminology because, firstly, as mentioned above, according to our terminology, unlike objects, lights do not appear black. Secondly, the set of lights metameric to the nil light ('metameric full dark') contains only the nil light itself (i.e., it is not $E(\sim)$).

9 From the mathematical standpoint $\Phi(E)$ is, of course, nothing more than $\boldsymbol{R}^n$. However, different colour-signal maps may have different psychophysical meaning, so the psychophysical interpretation of the axes in different colour-signal spaces may also be different. The rationale for this terminology will be clear in Section 3.3.

10 This proof differs from that given by Krantz (1975a).

such that $e \sim e_0$. The formula $\Phi(e) := \sum_{i=1}^{n} t_i v_i$ determines the unique linear extension of Φ to E satisfying (iii). Here, the linearity of Φ is a direct consequence of the linearity of the mapping $e_0(\cdot) : E \to E_n$. (Since Φ is linear, (i) and (ii) are satisfied as well.) Then $C := \Phi(A)$ is a convex cone in $\mathbf{R}^n$ and the restriction of Φ to A is a homomorphism of A onto C. Moreover, if C' is another convex cone and Φ' is a homomorphism of A onto C' satisfying (i)–(iii), then $T\Phi = \Phi'$ where the non-singular linear operator T is uniquely determined by the equalities: $Tv_i = \Phi'(a_i)$, $i = 1, \dots, n$.

Remark 2.7 *Being a linear operator from $E = A - A$ onto $\mathbf{R}^n$, Φ generates n linearly independent linear functionals on E, $\varphi_1, \dots, \varphi_n$, so that $\Phi(e) = (\varphi_1(e), \dots, \varphi_n(e))$ whenever $e \in E$ or, briefly, $\Phi = (\varphi_1, \dots, \varphi_n)$. Here, $\varphi_i(e) := t_i$ for $e \in E$. Thus, any set of primaries $a_1, \dots, a_n$ and of n linearly independent vectors $v_1, \dots, v_n$ in $\mathbf{R}^n$ induces a set of linearly independent functionals $\varphi_1, \dots, \varphi_n$ representing a colour-signal map Φ.*

Remark 2.8 *Given a light stimulus cone A,* any *linear map $\Phi = (\varphi_1, \dots, \varphi_n)$, generated by n linear functionals on $E = A - A$, determines an equivalence relation $\sim_\Phi$ on A as follows:*

$$a \sim_\Phi b \Leftrightarrow \varphi_i(a) = \varphi_i(b), \quad i = 1, \dots, n. \tag{2.7}$$

Since Φ is a linear operator, $\sim_\Phi$ determines on A a Grassmann structure.

Proposition 2.3

I. *The Grassmann structure determined on A by $\sim_\Phi$ is n-dimensional if and only if functionals $\varphi_1, \dots, \varphi_n$ are linearly independent. Moreover, this Grassmann structure is regular if and only if $a \in A$, $a \neq 0$, implies $\Phi(a) \neq 0$.*
II. *Linear operators $\Phi = (\varphi_1, \dots, \varphi_n) : E \to \mathbf{R}^n$ and $\Phi' = (\varphi'_1, \dots, \varphi'_m) : E \to \mathbf{R}^m$ determine on A the same equivalence relation ($\sim_\Phi = \sim_{\Phi'}$) hence the same Grassmann structure if and only if the vector spaces spanned over the functionals $\varphi_1, \dots, \varphi_n$ and $\varphi'_1, \dots, \varphi'_m$ coincide: $Lin(\varphi_1, \dots, \varphi_n) = Lin(\varphi'_1, \dots, \varphi'_m)$.*

To prove Proposition 2.3 (and Theorem 3.1), we shall need the following general lemma.[11]

Lemma 2.1 *Given an abstract set A, which contains at least n points, and n real-valued functions on it, $\varphi_1, \dots, \varphi_n$, the functions are linearly independent if and only if there exist n points in A, $a_1, \dots, a_n$, such that $\det(\varphi_i(a_j)) \neq 0$. (Here $\det(\varphi_i(a_j))$ stands for the determinant of the matrix $(\varphi_i(a_j))$.)*

Corollary 2.1 *Let $\varphi_1, \dots, \varphi_n$ be real-valued functions on A. Given $m < n$ points $a_1, \dots, a_m$ in A such that the n-dimensional vectors $(\varphi_1(a_j), \dots, \varphi_n(a_j))$, $j = 1, \dots, m$, are linearly independent, one can add $n - m$ points $a_{m+1}, \dots, a_n \in A$ such that the enlarged system of n vectors $(\varphi_1(a_j), \dots, \varphi_n(a_j))$, $j = 1, \dots, n$, will remain linearly independent.*

Note that, as follows from statement (iii) of the Krantz Theorem, the operator Φ subserves not only colour matching but colour discrimination as well, that is, $\Phi(a) \neq \Phi(b)$ entails colour discrimination of lights a and b (i.e., $a \sim b$ is not valid). Real-valued functions subserving colour discrimination (but not necessarily colour matching) are usually called colour codes (Krantz, 1975a).

11 This lemma seems to be known, however we cannot find an appropriate reference.

Definition 2.8 *A real-valued function φ on A, such that if $\varphi(a) \neq \varphi(b)$, then a is discriminable from b, i.e., $a \sim b$ is false, is called a* colour code. *In other words, φ is a colour code if and only if $a \sim b$ implies $\varphi(a) = \varphi(b)$ whenever $a, b \in A$. A colour code φ is called* non-trivial *if there is at least one pair $a, b \in A$ such that $\varphi(a) \neq \varphi(b)$.*

Remark 2.9 *Any real-valued function f on the colour-signal cone $\Phi(A)$ determines a colour code $\varphi = f \circ \Phi$ on A. The function f will be, sometimes, referred to as a* colour-signal code.

We will call a colour code *linear* if it can be represented as a linear functional. Linear colour codes play an important role in colour theory since, by the Krantz Theorem, any n-dimensional Grassmann structure on A is completely determined by a homomorphism Φ, that is, by n (linearly independent, hence non-trivial) linear colour codes $\varphi_1, \dots, \varphi_n$ where $\Phi(a) = (\varphi_1(a), \dots, \varphi_n(a))$ for all $a \in A$. Moreover, according to Proposition 2.3, any n linearly independent linear functionals $\varphi_1, \dots, \varphi_n$ on A define an n-dimensional Grassmann structure with the metamerism $\sim$ determined as $a \sim b$ if and only if $\varphi_i(a) = \varphi_i(b)$ for every $i = 1, \dots, n$.

Given a light stimulus cone A with an n-dimensional Grassmann structure on it, any non-trivial linear colour code defines on A a one-dimensional Grassmann structure (hence a one-dimensional Grassmann substructure of the original n-dimensional Grassmann structure on A), but if a one-dimensional Grassmann structure is defined by some colour code, that code need not be linear. For example, let φ be a non-trivial linear colour code, and g a one-to-one real function on $\mathbf{R}$. Then a colour code $\phi = g \circ \varphi$ defines a one-dimensional Grassmann structure. However, as the following proposition shows, this example exhausts all the colour codes defining one-dimensional Grassmann structure.

Before formulating the proposition, it will be convenient to introduce a definition of the Grassmann colour code (thereby formalising a colour mechanism term widely used in colour literature).

Definition 2.9 *The colour code ϕ on A is called the* Grassmann colour code[12] or the colour mechanism *if the equivalence relation $\sim_\phi$ defined as $a \sim_\phi b$ if and only if $\phi(a) = \phi(b)$ induces a one-dimensional Grassmann structure on A.*

The next result follows straightforwardly from the Krantz Theorem.

Proposition 2.4 *(Krantz, 1975a). A colour code ϕ on A is a Grassmann colour code (i.e., colour mechanism) if and only if $\phi = g \circ \varphi$, where φ is a non-trivial linear functional on $E = A - A$ and g is an arbitrary one-to-one map $\mathbf{R} \to \mathbf{R}$ (g and φ will be referred to as non-linear and linear parts of the Grassmann colour code ϕ respectively[13]).*

Clearly, a Grassmann colour code defines on A the same one-dimensional Grassmann structure as its linear part. It follows that an n-dimensional Grassmann structure on A is completely specified by n Grassmann colour codes whose linear parts are linearly independent.

Note that, as follows from the Krantz Theorem, any linear Grassmann colour code is a linear function on $\Phi(A)$ that is, a linear combination of n coordinate functionals $\varphi_1, \dots, \varphi_n$.

---

12 For simplicity, we will sometimes omit word 'colour' and write just the *Grassmann code*.

13 Of course, they are determined up to a positive multiplier: alike (g, φ), a pair $(\alpha g, \alpha^{-1}\varphi)$ where $\alpha > 0$ can also be taken as a non-linear and linear parts of ϕ.

Proposition 2.5 *Let $\phi_1, \dots, \phi_n$ be Grassmann colour codes on a light stimulus cone A, and $\varphi_1, \dots, \varphi_n$ their linear parts (i.e., $\varphi_1, \dots, \varphi_n$ are linear functionals on A, and there are n one-to-one real functions $g_1, \dots, g_n$ such that $\phi_i = g_i \circ \varphi_i$). Let us define an equivalence relation $\sim$ on A:*

$$\forall \sigma_1, \sigma_2 \in A,\ \sigma_1 \sim \sigma_2 \Leftrightarrow \phi_i(\sigma_1) = \phi_i(\sigma_2), \quad i = 1, \dots, n. \tag{2.8}$$

Then $\langle A, \sim\rangle$ is the same Grassmann structure as that determined on A by the linear functionals $\varphi_1, \dots, \varphi_n$ (see Remark 2.8).

Indeed, since for each i, $g_i(x) = g_i(y) \Leftrightarrow x = y$, we have

$$\phi_i(\sigma_1) = \phi_i(\sigma_2) \Leftrightarrow \varphi_i(\sigma_1) = \varphi_i(\sigma_2). \tag{2.9}$$

Therefore (2.8) defines the same equivalence relation as the linear functionals $\varphi_1, \dots, \varphi_n$. As Grassmann's laws hold true for the equivalence relation determined by the linear functionals $\varphi_1, \dots, \varphi_n$ (Remark 2.8), it follows from (2.9) that they are satisfied for the equivalence relation (2.8) as well.

Thus, we have proved that if $\langle A, \sim\rangle$ is an n-dimensional Grassmann structure, $\Phi : A \to \mathbf{R}^n$ is a colour-signal map (as defined in Remark 2.5), and G is a one-to-one map $\mathbf{R}^n_+ \to \mathbf{R}^n_+$, induced by the functions g_i (as in Proposition 2.5) then $F = G \circ \Phi$ induces the same Grassmann structure. More precisely, let us rewrite (2.8) as

$$\forall \sigma_1, \sigma_2 \in A,\ \sigma_1 \approx \sigma_2 \Leftrightarrow F(\sigma_1) = F(\sigma_2). \tag{2.10}$$

Then, $\langle A, \approx\rangle$ is the same Grassmann structure as that determined on A by the colour-signal map Φ, that is, $\langle A, \approx\rangle$ is identical to $\langle A, \sim\rangle$.

Interestingly, this remains true even if G is an arbitrary one-to-one map $\mathbf{R}^n_+ \to \mathbf{R}^n_+$, that is, such that $G(x) = G(y) \Leftrightarrow x = y$ for any $x, y \in \mathbf{R}^n_+$. Indeed, in this case, likewise in Eq. (2.9), $F(\sigma_1) = F(\sigma_2) \Leftrightarrow \Phi(\sigma_1) = \Phi(\sigma_2)$. Therefore, Grassmann's laws will hold true for the equivalence relation $\approx$, defined by Eq. (2.10), even when G is an arbitrary one-to-one map $\mathbf{R}^n_+ \to \mathbf{R}^n_+$.

We will use the terms 'colour-signal map' and 'colour signals' to refer not only to Φ (in the Krantz Theorem) but also to $F = G \circ \Phi$, and their values respectively. The reason is that both provide one-to-one representations of the colour stimuli (statement (iii) in the Krantz Theorem holds truefor both Φ and F). Evidently, the representation using F is non-linear (since statements (i) and (ii) in the Krantz Theorem fail for F). To emphasise this we will call Φ in the Krantz Theorem the linear colour-signal map, and its values will called 'additive colour signals'.[14] Alternatively, F will be referred to as the non-linear colour-signal map, and its values will called 'non-additive colour signals'. Given a non-linear colour-signal map $F = G \circ \Phi$, we will call Φ the linear part of F.

It should be noted that whereas $\Phi(A)$ is a cone in $\mathbf{R}^n$, $F(A)$ does not have to be one. We will call $F(A)$ the *colour-signal gamut*, using the notion of a colour-signal cone only when the colour-signal map is linear. Also, we will refrain from using the notion of colour-signal space when the colour-signal map is non-linear. An example of a non-linear colour-signal map is given in Example 3.1 in Chapter 3 (Section 3.1).

2.2 Continuous Grassmann structures and continuous Grassmann colour codes

The fundamental issue in colour science is to identify (i.e., to establish practically) a Grassmann structure which underlies colour vision of a human observer, if any exists. As follows from the

14 Additive colour signals will be also referred to as *colorimetric coordinates*.

Krantz Theorem, an arbitrary n-dimensional Grassmann structure on $\mathbf{C}_+^*$ can be represented by n linear Grassmann colour codes (linear functionals on $\mathbf{C}^*$) $\varphi_1, \dots, \varphi_n$ inducing the corresponding colour-signal map Φ, $\Phi(\sigma) = (\varphi_1(\sigma), \dots, \varphi_n(\sigma))$. Therefore, identification of a Grassmann structure can be reduced to identification of the linear colour codes underlying some colour-signal map. In turn, this problem can be split into a problem of representation of linear functionals on $\mathbf{C}^*$ and an identification problem as such. For example, if a linear functional φ on $\mathbf{C}^*$ can be represented as

$$\varphi(\sigma) = \langle \varphi, \sigma \rangle := \int_\Lambda \varphi(\lambda)\, \sigma(d\lambda), \sigma \in \mathbf{C}^*(\Lambda), \varphi(\lambda) \in \mathbf{C}, \tag{2.11}$$

then its identification can be reduced to the identification of a continuous function $\varphi(\lambda)$.

However, not every linear functional φ on $\mathbf{C}^*$ has an integral representation (2.11). Therefore, in general, the linear Grassmann colour codes implementing an arbitrary n-dimensional Grassmann structure on $\mathbf{C}_+^*$ can have no integral representation (2.11).

Still, in human colour vision a *principle of continuity* applies. Already in Grassmann (1853) we can find a statement of this principle. Sometimes it is formulated as another law of Grassman (Judd & Wyszecki, 1975, p. 45). Specifically, close colour stimuli σ and σ' should produce close colour signals $\Phi(\sigma)$ and $\Phi(\sigma')$. In particular, monochromatic lights with close wavelengths produce close colour sensations. This leads us to conjecture that linear colour mechanisms implementing the colour-signal map Φ should be in a sense continuous. Thus, it is worth considering only those integral functionals (2.11) which are generated by continuous functions $\varphi(\lambda)$. Discontinuity of $\varphi(\lambda)$ entails the discontinuity of the functional (2.11). A simplest example is given by integral functionals of type (2.11) with φ being discontinuous bounded Borel functions. (Such linear functionals on $\mathbf{C}^*$ are norm-continuous but not weak* continuous. Moreover, there are linear functionals on $\mathbf{C}^*$, which are not even norm-continuous.)

A notion of continuity implies some topology on the light stimulus space $\mathbf{C}^*$. It is natural to consider $\mathbf{C}^*$ endowed with the weak* topology on it and to equip its vector subspaces E from Examples 2.1b, 2.1c, 2.1d, 2.1e with traces (restrictions to E) of that topology. Note (see Section A.1), that every linear functional φ on $\mathbf{C}^*$ which is weak* continuous, i.e., continuous relative the weak* topology, is identified with a function $\varphi \in \mathbf{C}$ as follows[15]:

$$\varphi(\sigma) = \langle \varphi, \sigma \rangle := \int_\Lambda \varphi(\lambda)\, \sigma(d\lambda), \sigma \in \mathbf{C}^*(\Lambda). \tag{2.12}$$

Also, one can consider stronger topologies in Examples 2.1b, 2.1c, and 2.1d when identifying the corresponding E with the Banach spaces $\mathbf{L}^1 = \mathbf{L}^1(\Lambda)$, $\mathbf{C} = \mathbf{C}(\Lambda)$, and $\mathbf{L}^\infty = \mathbf{L}^\infty(\Lambda)$ respectively. (Concerning these function spaces, see Section A.1.)

A question arises, what condition (axiom) is to be added to Grassmann's laws for the operator $\Phi = (\varphi_1, \dots, \varphi_n)$ in Krantz Theorem to be continuous with respect to an appropriate (weak* or norm) topology. This question is of great importance because arbitrary (discontinuous) linear functionals φ_i on concrete function spaces are rather obscure (non-constructive) objects which are practically impossible to be identified. For an accurate mathematical formulation of the problem we need the following definition.

Definition 2.10 *Let A be a convex cone in a locally convex space $E = A - A$, and $A \cap (-A) = \{0\}$. We say a Grassmann structure on A is* continuous *if the following axiom is satisfied:*

4. (Law of closedness) The vector subspace $E(\sim)$ as defined in (2.4) is closed in E.

---

15 Sometimes, however, we may wish to distinguish the two mathematical objects: the weak* continuous linear functional φ on $\mathbf{C}^*$ and the continuous function $\varphi(\lambda)$ on Λ such that (2.12) holds true. In this case, we will refer to $\varphi(\lambda)$ as the *spectral sensitivity function (or, more generally, kernel)* of the functional φ.

Proposition 2.6 *Suppose $E = A - A$ is a Hausdorff locally convex space with a light stimulus cone A in it. Given an n-dimensional Grassmann structure on A, the following assertions are equivalent:*

(i) *the Grassmann structure is continuous, that is, Axiom 4 is satisfied;*
(ii) *every homomorphism (linear operator) $\Phi = (\varphi_1, \dots, \varphi_n)$ representing the structure (that is, satisfying the statement of the Krantz Theorem) is continuous.*
Furthermore, if E is $\mathbf{C}^*$ with the weak topology in it and $A = \mathbf{C}^*_+$, then these assertions are equivalent to the following one:*
(iii) *$E(\sim)$ is weak* sequentially continuous in $\mathbf{C}^*$, that is, if a sequence $(\sigma_k) \subset E(\sim)$ converges weakly* to a measure $\sigma \in \mathbf{C}^*$, then $\sigma \in E(\sim)$.*
(An equivalent formulation of (iii) is as follows: if $\sigma_{1k}, \sigma_{2k} \in A$, $\sigma_{1k} \sim \sigma_{2k}$ $(k = 1, 2, \dots)$, $\sigma_1, \sigma_2 \in A$, and, for all $\varphi \in \mathbf{C}$,

$$\lim_{k\to\infty} \int_\Lambda \varphi(\lambda)\,(\sigma_{1k} - \sigma_{2k})(d\lambda) = \int_\Lambda \varphi(\lambda)\,(\sigma_1 - \sigma_2)(d\lambda),$$

then $\sigma_1 \sim \sigma_2$.)

We conclude this chapter with the following definition.

Definition 2.11 *A Grassmann structure on $\mathbf{C}^*_+$ is called* weak*continuous *if it is continuous with respect to the weak* topology in $\mathbf{C}^*$.*

3

Identification of Grassmann Structures Based on Metameric Matching

The theory of Grassmann structures outlined above can be seen as the first step towards the creation of a theoretical framework for human colour vision. Colour vision of each individual can be specified in terms of Grassmann structure. As there are unavoidable inter-individual differences in colour vision (Webster, 2015), which can be so dramatic that some people are classed as colour anomalous (see Section 3.3), the colour vision of individuals should be characterised by their own Grassmann structures. Such an important task – specifying the particular Grassmann structure that characterises an individual's colour vision – will be called the *identification of the Grassmann structure.*

Identifying a Grassmann structure is a relatively straightforward task when colour matching is understood as an operationalisation of metamerism (Krantz, 1975a; Wyszecki & Stiles, 1982; Suppes et al., 1989). In this case, human colour matching judgements are assumed to satisfy the axioms of equivalence relation, and to obey Grassmann's laws (Definition 2.2). It should be noted, however, that under some stimulus conditions, such as very intense light or very far periphery of the field of view, colour matching judgements do not follow Grassmann's laws (Brindley, 1960; Krantz, 1975a). Moreover, as we will see in Chapter 10, the empirical status of the axiom of equivalence relation and Grassmann's laws for colour matching is, in general, far more involved than it is usually thought. Therefore, it is only a hypothesis (referred to as the *metameric matching hypothesis*) that human colour matching obeys the axiom of equivalence relation and Grassmann's laws. Being commonly accepted in colour science (although see MacLeod, 2010), this hypothesis makes no distinction between colour matching and metamerism. In order to stress this, we will call colour matching satisfying Grassmann's laws *metameric matching*. In this chapter, we will consider identification of Grassmann structures based on metameric matching.[1]

An important aspect of identifying a Grassmann structure is the determination of its dimensionality. It should be pointed out that it is not easy to establish empirically the dimensionality of the Grassmann structure underlying colour matching judgements made by a particular human observer. In other words, it is not easy to establish how many primary light stimuli are needed to match any light. An obvious problem here is that since it is impossible to test in experiment all the light stimuli, there is always a possibility that there exists a light stimulus (among those which were not tested) which cannot be matched by a given set of primaries.

The issue becomes more tractable for weak* continuous Grassmann structures (Definition 2.11). In the next Section (3.1) we also show how to practically derive from colour matching data a colour-signal map (see Remark 2.5) representing a weak* continuous Grassmann structure. After a

1 In Chapter 11, we will consider a more complex case when colour matching does not obey all of Grassmann's laws.

Foundations of Colour Science: From Colorimetry to Perception, First Edition. Alexander D. Logvinenko and Vladimir L. Levin.
© 2023 John Wiley & Sons Ltd. Published 2023 by John Wiley & Sons Ltd.

colour-signal map representing the Grassmann structures for that particular observer is specified, one can calculate whether two lights are metameric or not, without making a colour matching experiment. In other words, a colour-signal map identified in a colour matching experiment servers an instrument of measuring colour stimuli. A corresponding part of colour science is called *colorimetry* (Ohta & Robertson, 2005; Schanda, 2007a; Brainard & Stockman, 2010). Note that colorimetry is not about measuring (specifying) colour as such; it is about measuring colour stimulus as defined in Definition 2.5.[2]

3.1 Colour matching functions

The following result shows that identifying a weak* continuous Grassmann structure $\langle \mathbf{C}_+^*, \sim \rangle$ can be reduced to identifying the Grassmann substructure $\langle A, \sim \rangle$ where $A \subset \mathbf{C}_+^*$ is the cone of non-negative combinations of delta functions (Example 2.1e).

Proposition 3.1 *Let $\langle \mathbf{C}_+^*, \sim \rangle$ be a weak* continuous m-dimensional Grassmann structure. Suppose there are $n \leq m$ different points $\lambda_1, \dots, \lambda_n \in \Lambda$ such that, for every $\lambda \in \Lambda$, there exists the unique linear combination $t_1\delta_{\lambda_1} + \cdots + t_n\delta_{\lambda_n}$ metameric to δ_λ,*

$$\delta_\lambda \sim t_1(\lambda)\delta_{\lambda_1} + \cdots + t_n(\lambda)\delta_{\lambda_n}. \tag{3.1}$$

If $t_1 = t_1(\lambda), \dots, t_n = t_n(\lambda)$ are continuous functions of λ, then $m = n$ (hence the Grassmann structure $\langle \mathbf{C}_+^*, \sim \rangle$ is n-dimensional), and for any $\sigma_1, \sigma_2 \in \mathbf{C}_+^*$,

$$\sigma_1 \sim \sigma_2 \Leftrightarrow \langle t_i, \sigma_1 \rangle = \langle t_i, \sigma_2 \rangle, \quad i = 1, \dots, n. \tag{3.2}$$

It follows, first, that the dimensionality of the Grassmann structure $\langle \mathbf{C}_+^*, \sim \rangle$ equals that of the Grassmann substructure of monochromatic lights. Second, given a set of monochromatic lights $\delta_{\lambda_1}, \dots, \delta_{\lambda_n}$, satisfying the conditions of Proposition 3.1, a particular set of colour mechanisms can be derived from the colour matching data obtained only for monochromatic lights by evaluating the functions $t_1(\lambda), \dots, t_n(\lambda)$. These functions will be referred to as the colour matching functions (see Definition 3.1).

In more detail, it is clear that monochromatic lights of the unit intensity[3] $\delta_{\lambda_1}, \dots, \delta_{\lambda_n}$ are colour independent (hence, are primaries in n-dimensional Grassmann structure $\langle \mathbf{C}_+^*, \sim \rangle$) if and only if they satisfy the supposition of Proposition 3.1, that is, for every monochromatic light δ_λ there exist real numbers depending on λ, $t_1 = t_1(\lambda), \dots, t_n = t_n(\lambda)$, such that

$$\delta_\lambda \sim t_1(\lambda)\delta_{\lambda_1} + \cdots + t_n(\lambda)\delta_{\lambda_n}, \tag{3.3}$$

and $t_1(\lambda)\delta_{\lambda_1} + \cdots + t_n(\lambda)\delta_{\lambda_n}$ is the unique linear combination of $\delta_{\lambda_1}, \dots, \delta_{\lambda_n}$ metameric to δ_λ. Such monochromatic lights will be referred to as *monochromatic primaries*.

Non-negative coefficients in (3.3) are treated as the intensities of the monochromatic lights in the mixture. If, for a given λ, some coefficients $t_i(\lambda)$ are negative, then (3.3) should be understood in the sense that, for $J_+(\lambda) = \{i : t_i(\lambda) > 0\}$ and $J_-(\lambda) = \{i : t_i(\lambda) < 0\}$, the lights $\delta_\lambda + \sum_{i \in J_-(\lambda)} |t_i(\lambda)| \delta_{\lambda_i}$ and $\sum_{i \in J_+(\lambda)} t_i(\lambda)\delta_{\lambda_i}$ are metameric:

$$\delta_\lambda + \sum_{i \in J_-(\lambda)} |t_i(\lambda)|\delta_{\lambda_i} \sim \sum_{i \in J_+(\lambda)} t_i(\lambda)\delta_{\lambda_i}. \tag{3.4}$$

2 An excellent review of colorimetry from the measurement theory point of view is given by Krantz in Suppes et al. (1989). Especially pertinent is a section entitled 'Four ways to misunderstand colour measurement' (Suppes et al., 1989, pp. 250–252).

3 By a monochromatic light with the wavelength $\lambda \in \Lambda$ and intensity $\alpha > 0$ is meant here $\alpha\delta_\lambda$ where δ_λ stands for the Dirac measure (delta function) at λ; see Appendix A.

Definition 3.1 *Given monochromatic primaries $\delta_{\lambda_1}, \dots, \delta_{\lambda_n}$, the (uniquely determined) coefficients in (3.3), $t_1(\lambda), \dots, t_n(\lambda)$, are called* colour matching functions.

The following theorem highlights the role of colour matching functions in the identification of weak* continuous Grassmann structures.

Theorem 3.1 *Let $\langle \mathbf{C}_+^*, \sim \rangle$ be an n-dimensional weak* continuous Grassmann structure. Suppose there are n different points $\lambda_1, \dots, \lambda_n \in \Lambda$ such that, for every $\lambda \in \Lambda$, there exists the unique linear combination $t_1\delta_{\lambda_1} + \cdots + t_n\delta_{\lambda_n}$ metameric to δ_λ. The following assertions are then equivalent:*

(i) *$t_1 = t_1(\lambda), \dots, t_n = t_n(\lambda)$ are continuous functions of λ;*
(ii) *there exist linear colour mechanisms $\mathfrak{f}_{t_1}, \dots, \mathfrak{f}_{t_n}$ such that, for every $\lambda \in \Lambda$,*

$$\mathfrak{f}_{t_i}(\delta_\lambda) = t_i(\lambda), \quad i = 1, \dots, n,$$

and for any $\sigma_1, \sigma_2 \in \mathbf{C}_+^*$,

$$\sigma_1 \sim \sigma_2 \Leftrightarrow \mathfrak{f}_{t_i}(\sigma_1) = \mathfrak{f}_{t_i}(\sigma_2), \quad i = 1, \dots, n. \tag{3.5}$$

It follows from Theorem 3.1 along with the Krantz Theorem and Proposition 2.3 that any linear colour-signal map $\Phi = (\varphi_1, \dots, \varphi_n)$ representing metamerism $\sim$ is connected with $\Phi_* = (t_1, \dots, t_n)$ by a non-singular linear operator T, that is, $\Phi(\sigma) = T\Phi_*(\sigma)$ for all $\sigma \in \mathbf{C}_+^*$, and $Lin(\varphi_1, \dots, \varphi_n) = Lin(t_1, \dots, t_n)$.

Thus, the identification of a weak* continuous Grassmann structure is reduced to that of colour matching functions. More specifically, one has, first, to ascertain its dimensionality, that is, the number of monochromatic primaries; and second, to evaluate colour matching functions with respect to some set of monochromatic primaries.

It was found that the dimensionality of a Grassmann structure which describes colour matching for human observers is, firstly, different for different observers; and secondly, it depends on the condition of observation (Wright, 1946; Wyszecki & Stiles, 1982). For example, for all observers it is one-dimensional when the illumination intensity is very low (i.e., under night, or so-called, *scotopic* vision). In other words, any two lights can be matched by adjusting their intensities if they do not exceed the scotopic level. The reason for this is that photoreceptors of only one type (*rods*) function during night vision (Rieke, 2008).

When the illumination intensity exceeds the scotopic level (turning into the *photopic* range), for most human observers one needs three monochromatic primaries to match any spectral light under most conditions of observation (Wyszecki & Stiles, 1982). In other words, metameric matching for these observers in these conditions satisfies Axiom 3 in Definition 2.2 with $n = 3$. Such observers, thus their colour vision, are said to be *trichromatic*. The physiological basis of trichromatism is generally believed to there being three types of photoreceptors – *cones* (Brindley, 1960) (though see a critical review by D. I. A. MacLeod (2010)). Although the existence of the cone photoreceptors of three types in the retina of trichromats have been proved relatively recently (Williams & Roorda, 1999), this idea has been claimed by a number of scientists of the 18–19th centuries: Lomonosov (1757), Palmer (1777), and Young (1802) to name a few.

A small population of human observers need only two monochromatic primaries to match any other monochromatic light (Wright, 1946; Wyszecki & Stiles, 1982). Their metameric matching is described by a two-dimensional Grassmann structure. They are called *dichromats*, and their colour vision *dichromatic*. Finally, there are some people (called *monochromats*) who are capable to match any monochromatic light by adjusting the intensity of only one monochromatic primary even in the photopic range of light intensities. Although these people can hardly enjoy ‘colour vision’ in

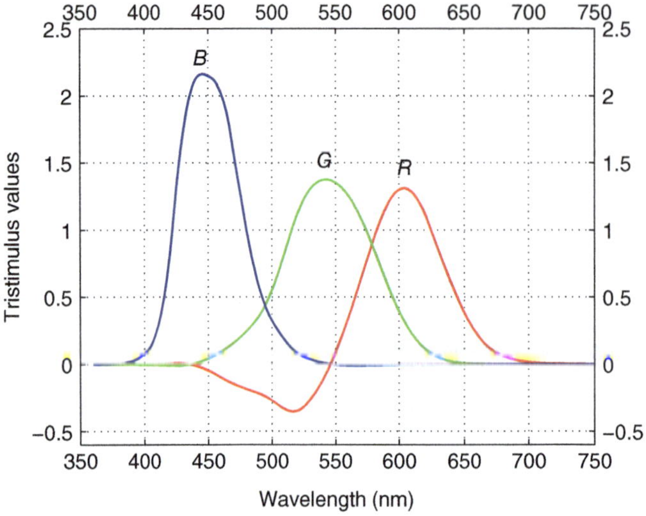

Figure 3.1 Colour matching functions of the CIE 1931 Standard Colorimetric Observer with respect to the three monochromatic primaries of wavelengths $\lambda_B = 435.8$, $\lambda_G = 546.1$, and $\lambda_R = 700$ nm. The primaries radiant powers were in ratios: 1.0 : 1.4 : 72.1. The vertical axis is named in accordance with the tradition in colorimetry to call colorimetric coordinates 'tristimulus values' (e.g., Wyszecki & Stiles, 1982, p. 120).

the strict sense of the term, for the sake of generality, and following literary convention, we will say that they have *monochromatic* colour vision. For a long time it was believed that dichromatic colour vision is caused by the lack of one of three types of cone photoreceptors (Wright, 1946; Wyszecki & Stiles, 1982).[4] Respectively, the retina of a monochromat is believed to be equipped with only one (of three) type of cone photoreceptors. We will discuss colour vision of dichromats and monochromats as compared to trichromats in more detail below (Section 3.3).

Colour matching functions have been measured for trichromatic and dichromatic human observers by a number of researchers, and the average results can be found in all standard textbooks on colour vision (Wright, 1946; Wyszecki & Stiles, 1982; Smith & Pokorny, 2003; Brainard & Stockman, 2010) and also at the CVRL database (http://www.cvrl.org). Figure 3.1 presents the colour matching functions accepted by the Commission Internationale de l'Eclairage (CIE) as standards under the name *CIE 1931 colour matching functions* (Wyszecki & Stiles, 1982; Smith & Pokorny, 2003). Though there is no human being with such colour matching characteristics, the CIE has coined a notion of *CIE 1931 Standard Colorimetric Observer*. As can be seen, there is every reason to believe that the CIE 1931 colour matching functions are continuous. Therefore, they can be taken as the spectral sensitivity functions of the linear colour mechanisms determining the Grassmann structure for the CIE 1931 Standard Colorimetric Observer.

The CIE 1931 colour matching functions were also converted via a linear transformation into another set of colour matching functions (Fig. 3.2) also adopted by the CIE as internationally accepted standards (Wyszecki & Stiles, 1982; Smith & Pokorny, 2003). The primaries (written as X, Y, and Z) with respect to which the colour matching functions in Fig. 3.2 were calculated, are imaginary ones (Definition 2.5). They were chosen such that, inter alia, (i) all three colour matching functions are non-negative; and (ii) the second colour matching function (i.e., with respect to the primary Y) is equal to the *spectral luminous efficiency function* (see Section 8.1.2).

4 More recent views on the nature of dichromatism will be discussed in Section 3.3.

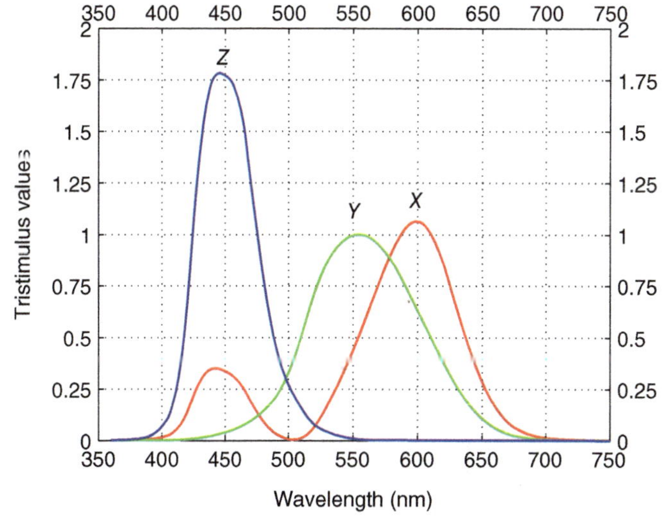

Figure 3.2 Colour matching functions of the CIE 1931 Standard Colorimetric Observer with respect to the imaginary primaries *X*, *Y*, and *Z*.

It should be borne in mind that colour matching functions in human observers depend on the size of the light stimuli and their location on the retina. One of the reasons for this is that the optical density of the macular pigment and the cone photopigment vary with retinal eccentricity. Unsurprisingly, a difference was found between the colour matching functions of Wright & Guild, who employed targets subtended 2 angular degrees, and Stiles & Burch's colour matching functions (see Fig. 5.2), obtained for a 10° viewing field (Stiles & Burch, 1959). Figure 3.3 presents the *XYZ* colour matching functions that define the so-called *CIE 1964 Complementary Standard Colorimetric Observer* (Wyszecki & Stiles, 1982, p. 142). They are derived from the colour matching functions of Stiles & Burch (1959) and Speranskaya (1959). A quantitative estimation of the

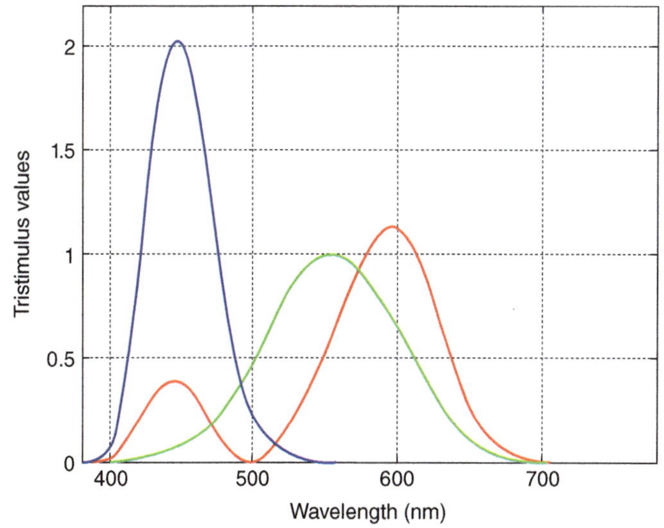

Figure 3.3 Colour matching functions of the CIE 1964 Complementary Standard Colorimetric Observer with respect to the imaginary primaries *X*, *Y*, and *Z*.

difference between the CIE 1931 Standard Colorimetric Observer and the CIE 1964 Complementary Standard Colorimetric Observer can be seen in Table 7.1.

The effect of stimulus size on colour matching can reveal itself even more dramatically. For example, the colour vision of trichromats becomes dichromatic when a stimulus light is projected only onto the central area of the retina, which is no more than 20 arc minutes (Williams et al., 1981). This is due to the fact that this part of the retina lacks cone photoreceptors of one type (with the peak sensitivity in the short-wave length part of the spectrum) (Curcio et al., 1991). As an alternative example of the dependence of the dimensionality of colour vision on the size of light stimuli, it can be mentioned that the colour matches of dichromats are reported to become trichromatic for large light stimuli (Smith & Pokorny, 1977). Although its nature is not completely understood, it is believed that the large-field-trichromatism of dichromats is mainly caused by the spatial heterogeneity of the distribution of photoreceptors in the retina. Particularly, many believe that it can be accounted for by that the rod photoreceptors (which are absent in the centre of the retina) get involved in colour matching when the light stimulus size become large enough (so-called the *rod intrusion hypothesis* (Wyszecki & Stiles, 1982)).

It should be said, however, that the CIE 1931 colour matching functions are not the result of direct matches. On the contrary, they were reconstructed from the relative measurements obtained by Wright (1928–29) and Guild (1931) from trichromatic observers (Wright, 1996). Expressed in relative units, these data were then averaged, smoothed, and adjusted to accommodate some additional constraints that were not really pertinent to colour matching. In particular, the CIE 1931 colour matching functions have been reconstructed from Wright & Guild's data so that the spectral luminous efficiency function can be expressed as a linear combination of them. Over the past nearly a 100 years, the CIE 1931 colour matching function have been repeatedly corrected (Judd, 1951; Vos, 1978) and subjected to serious criticism (e.g., Stockman, 2019). Eventually, a new set of colour matching functions (Fig. 3.4) was approved as a CIE standard (CIE, 2015) (referred to as the *CIE 2015 XYZ 2° colour matching functions*). These are based on the work of Stockman & Sharpe (2000), who used data from Stiles & Burch (1959), adapting them to the 2° condition (see Section 3.3 for more information).

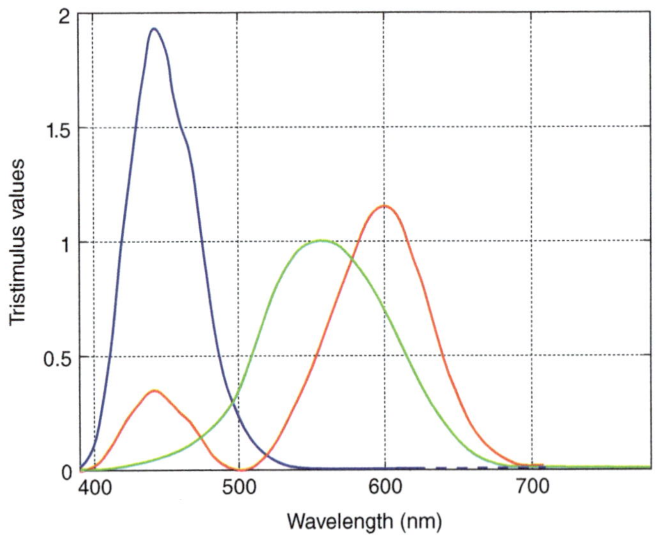

Figure 3.4 CIE 2015 *XYZ* 2° colour matching functions.

As we can see, the colour matches of a real human observer cannot be modelled by any single Grassmann structure. Still, this becomes possible when the observation conditions are specified and fixed. Following the established tradition, we will use the notion of a colorimetric observer, meaning simply either a Grassmann structure, or a set of colour matching functions that define it. Depending on conditions of observation, a real observer (i.e., colour matches made by her/him) can be modelled by various colorimetric observers.

Example 3.1 *Consider the CIE 1931 colour matching functions followed by the following non-linear transformation:*

$$L(X, Y, Z) = 116 f(Y/Y_0) - 16, \qquad (3.6)$$
$$a(X, Y, Z) = 500\left(f(X/X_0)\right) - f(Y/Y_0),$$
$$b(X, Y, Z) = 200\left(f(Y/Y_0) - f(Z/Z_0)\right),$$

where

$$f(t) = \begin{cases} \sqrt[3]{t}, & \text{if } t > \delta^3 \ (\delta = 6/29), \\ \frac{t}{3\delta^2} + \frac{4}{29}, & \text{otherwise.} \end{cases}$$

*Equation (3.6) defines a non-linear colour-signal map that preserves metamerism caused by the CIE 1931 colour matching functions. This is known as the CIE 1976 (L*a*b*)-space (or CIELAB colour space) (Wyszecki & Stiles, 1982, pp. 166–168). However, as noted before, we reserve the term "colour-signal space" for linear colour-signal maps only.*

3.2 Monochromatic primaries and colour matching functions in the trichromatic case ($n = 3$)

Let $\langle A, \sim \rangle$ be a regular n-dimensional Grassmann structure on the cone A from Example 2.1e. Given three monochromatic lights of the unit intensity, δ_{λ_1}, δ_{λ_2}, and δ_{λ_3}, we formulate two conditions, (I) and (II), as follows.

(I). (a) None of the lights $\delta_{\lambda_1}, \delta_{\lambda_2}, \delta_{\lambda_3}$ can be matched in colour by a mixture (positive linear combination) of two others: the equivalence

$$\delta_{\lambda_i} \sim t_j \delta_{\lambda_j} + t_k \delta_{\lambda_k} \qquad (3.7)$$

fails to be true whenever $i, j, k \in \{1, 2, 3\}, i \neq j, i \neq k, t_j, t_k \in \mathbf{R}_+$.

(b) For every monochromatic light δ_λ there exist real numbers depending on λ, $t_1 = t_1(\lambda)$, $t_2 = t_2(\lambda)$, and $t_3 = t_3(\lambda)$, such that

$$\delta_\lambda \sim t_1(\lambda)\delta_{\lambda_1} + t_2(\lambda)\delta_{\lambda_2} + t_3(\lambda)\delta_{\lambda_3}. \qquad (3.8)$$

(II). For every monochromatic light δ_λ, (3.8) holds true where: (a) at most one of the coefficients t_1, t_2, t_3 can be negative, and (b) if one of them equals zero then two others are non-negative. Here, as usually, non-negative coefficients are treated as the intensities of the lights in the corresponding mixture, and if, for a given λ, one of coefficients $t_i(\lambda)$, say $t_1(\lambda)$, is negative, then (3.8) should be understood in the sense that the lights $\delta_\lambda + |t_1(\lambda)|\delta_{\lambda_1}$ and $t_2(\lambda)\delta_{\lambda_2} + t_3(\lambda)\delta_{\lambda_3}$ are metameric:

$$\delta_\lambda + |t_1(\lambda)|\delta_{\lambda_1} \sim t_2(\lambda)\delta_{\lambda_2} + t_3(\lambda)\delta_{\lambda_3}.$$

It is empirically established that there exist triads of monochromatic lights $\delta_{\lambda_1}, \delta_{\lambda_2}, \delta_{\lambda_3}$ satisfying condition (I) (Wyszecki & Stiles, 1982; Smith & Pokorny, 2003). There is every indication that, for every such a triad, condition (II) is satisfied as well (e.g., see Fig. 3.1 and Table I(3.3.3) in

Wyszecki & Stiles 1982, pp. 750–751). Still, in the next Chapter (4) we shall see that, in general, condition (II) might hold for some colour independent monochromatic triads and fail for others (see Proposition 4.3 and Fig. 4.6).

It follows from (I) that, for every $\lambda \in \Lambda$, $t_1(\lambda)\delta_{\lambda_1} + t_2(\lambda)\delta_{\lambda_2} + t_3(\lambda)\delta_{\lambda_3}$ is the unique element in $Lin(\delta_{\lambda_1}, \delta_{\lambda_2}, \delta_{\lambda_3})$ which is metameric to δ_λ. (Indeed, the existence of two such elements would imply $u_1\delta_{\lambda_1} + u_2\delta_{\lambda_2} + u_3\delta_{\lambda_3} \sim 0$ for some $(u_1, u_2, u_3) \neq (0,0,0)$, and as the Grassmann structure $\langle A, \sim\rangle$ is regular, there are both positive and negative numbers among u_1, u_2, and u_3, which leads to (3.7) in contradiction with (I).) A similar argument shows that condition (Ia), along with regularity of $\langle A, \sim\rangle$, implies colour independence of $\delta_{\lambda_1}, \delta_{\lambda_2}, \delta_{\lambda_3}$. Thus, if there is a triad $\delta_{\lambda_1}, \delta_{\lambda_2}, \delta_{\lambda_3}$ satisfying (I), then the Grassmann structure $\langle A, \sim\rangle$ is three-dimensional, δ_{λ_i}, $i = 1, 2, 3$, are primaries for it, and the coefficients $t_i(\lambda)$ are the corresponding colour matching functions.

It is generally taken for granted that condition (II) is satisfied not only for monochromatic but for any arbitrary light as well (Wyszecki & Stiles, 1982). In other words, there is a general belief that an extended version of (II) holds true as follows:

(III). For every light σ there exist real numbers depending on σ, $t_1(\sigma)$, $t_2(\sigma)$, and $t_3(\sigma)$, such that

$$\sigma \sim t_1(\sigma)\delta_{\lambda_1} + t_2(\sigma)\delta_{\lambda_2} + t_3(\sigma)\delta_{\lambda_3}, \tag{3.9}$$

and conditions (II a) and (II b) are satisfied.

It follows from (I) and (III) that the *colour* of every light σ either is a mixture of primary colours,[5] $[\sigma] = t_1(\sigma)[\delta_{\lambda_1}] + t_2(\sigma)[\delta_{\lambda_2}] + t_3(\sigma)[\delta_{\lambda_3}]$, where all the coefficients $t_i(\sigma)$ are uniquely determined (uniqueness follows from (I)) and non-negative or, being mixed with one of them (say, the colour $[\delta_{\lambda_1}]$ for definiteness), equals a mixture of two other primary colours, i.e., $[\sigma] + |t_1(\sigma)|[\delta_{\lambda_1}] = t_2(\sigma)[\delta_{\lambda_2}] + t_3(\sigma)[\delta_{\lambda_3}]$, where $t_i(\sigma)$ are uniquely determined, exactly one of them ($t_1(\sigma)$) is negative, and two others ($t_2(\sigma)$ and $t_3(\sigma)$) are positive. Saying briefly (and somewhat figuratively), (III) along with (I) implies that each color can be obtained either by adding no more than three monochromatic lights, or by adding the difference of two monochromatic lights to a monochromatic light.

Although, one cannot exclude, strictly speaking, that there might exist a light which cannot be matched by a linear combination of a triad of monochromatic primaries $(\delta_{\lambda_1}, \delta_{\lambda_2}, \delta_{\lambda_3})$, it is generally thought that there is no need in experimental proofs (which is, in fact, impossible because all lights have to be examined) since this can be proved on the theoretical ground. For instance, Schrödinger believed that 'since each light can be regarded with sufficient approximation as a mixture of n pure spectral lights, it follows that *the pure spectral colours and binary mixtures exhaust completely the manifold of colours*' (Schrödinger, 1920/1970, p. 136). However, as follows from Proposition 2.6 and Theorem 3.1, this cannot be proved without additional assumptions on the Grassmann structure.

Moreover, condition (III) generally cannot be derived from (II) even for the Grassmann structure $\langle A, \sim\rangle$ on the cone A from Example 2.1e. Indeed, if $\delta_\lambda \sim t_1\delta_{\lambda_1} + t_2\delta_{\lambda_2} + t_3\delta_{\lambda_3}$, where $t_1 < 0$, $t_2 > 0$, $t_3 > 0$, and $\delta_{\lambda'} \sim t_1'\delta_{\lambda_1} + t_2'\delta_{\lambda_2} + t_3'\delta_{\lambda_3}$, where $t_2' < 0$, $t_1' > 0$, $t_3' > 0$, then $\delta_\lambda + \delta_{\lambda'} \sim \sum_{i=1}^{3}(t_i + t_i')\delta_{\lambda_i}$; however, it can occur that $t_1 + t_1' < 0$ and $t_2 + t_2' < 0$, in what case (III) fails.

To overcome that difficulty, we will formulate conditions (II′) and (III′) strengthening (II) and (III). But first note a simple formal consequence of (II) that strengthens (IIa) as follows.

Proposition 3.2 *Suppose that (II) holds true for any triad of monochromatic primaries. If, for given primaries $\delta_{\lambda_1}, \delta_{\lambda_2}, \delta_{\lambda_3}$ and a monochromatic light δ_λ, all coefficients $t_i(\lambda)$ in (3.8) are non-zero, then one of them is necessarily negative.*

---

5 Note that for brevity we use here the terminology that is common for colorimetry, identifying colour with colour stimulus, and colour mixing with the algebraic addition of colour stimuli.

Indeed, if $t_i(\lambda) > 0$, $i = 1, 2, 3$, then the monochromatic lights $\delta_{\lambda_1}, \delta_{\lambda_2}, \delta_\lambda$ are colour independent, since, if $u_1\delta_{\lambda_1} + u_2\delta_{\lambda_2} + u\delta_\lambda \sim 0$ then, as δ_{λ_1} and δ_{λ_2} are colour independent, $u \neq 0$ and $(u_1 + ut_1(\lambda))\delta_{\lambda_1} + (u_2 + ut_2(\lambda))\delta_{\lambda_2} + ut_3(\lambda)\delta_{\lambda_3} \sim 0$; therefore, $u_1 + ut_1(\lambda) = 0$, $u_2 + ut_2(\lambda) = 0$, and $ut_3(\lambda) = 0$, but this is impossible because $u \neq 0$ and $t_3(\lambda) > 0$. Hence $\delta_{\lambda_1}, \delta_{\lambda_2}$, and δ_λ may be taken as new primaries. Rewriting (3.8) as

$$\delta_{\lambda_3} \sim -\frac{t_1(\lambda)}{t_3(\lambda)}\delta_{\lambda_1} - \frac{t_2(\lambda)}{t_3(\lambda)}\delta_{\lambda_2} + \frac{1}{t_3(\lambda)}\delta_\lambda \tag{3.10}$$

we see that (II) fails for new primaries since two coefficients in (3.10) are negative.

We now formulate conditions (II′) and (III′) strengthening (II) and (III) respectively.

(II′). For every monochromatic light δ_λ there exist real numbers depending on λ, $t_1 = t_1(\lambda)$, $t_2 = t_2(\lambda)$, and $t_3 = t_3(\lambda)$, such that the statement of (II) holds true and, in addition, if one of the coefficients in (3.8) is negative then its absolute value cannot exceed any of other coefficients: $|t_i(\lambda)| \leq t_j(\lambda)$ whenever $i, j \in \{1, 2, 3\}$, $i \neq j$, $t_i(\lambda) < 0$.

Taking into account (II a) and (II b), condition (II′) may be restated as follows: for every $\lambda \in \Lambda$,

$$t_i(\lambda) + t_j(\lambda) \geq 0 \text{ whenever } i, j \in \{1, 2, 3\},\ i \neq j. \tag{3.11}$$

(III′). For every light σ there exist real numbers depending on it, $t_1(\sigma), t_2(\sigma), t_3(\sigma)$, such that (3.9) holds true and

$$t_i(\sigma) + t_j(\sigma) \geq 0 \text{ whenever } i, j \in \{1, 2, 3\},\ i \neq j. \tag{3.12}$$

Condition (3.12) means that all the coefficients $t_i(\sigma)$ in (3.9) are non-negative[6] or one of them is negative, two others are positive, and the absolute value of the negative coefficient does not exceed each of two positive coefficients.

Proposition 3.3 *If $(A, \sim)$ is a regular Grassmann structure on the cone A from Example 2.1 e, then (II′) implies (III′).*

Thus, if we adopt condition (II′) for a given triad of monochromatic primaries $\delta_{\lambda_1}, \delta_{\lambda_2}, \delta_{\lambda_3}$, and consider the colour stimulus cone A from Example 2.1e, then any light $\sigma \in A$ can be matched by a linear combination of $\delta_{\lambda_1}, \delta_{\lambda_2}, \delta_{\lambda_3}$, and coefficients of this combination satisfy (3.12) and prove to be linear colour mechanisms generating on A the original Grassmann structure.

3.3 Fundamental colour mechanisms in human colour vision

Let us consider a device equipped with n photosensors with linearly independent continuous spectral sensitivity functions $f_1(\lambda), \ldots, f_n(\lambda)$. Such a device naturally induces a continuous n-dimensional Grassmann structure if metamerism is defined as in Example 2.2. Any of the colour matching functions will be a linear combination of functions $f_1(\lambda), \ldots, f_n(\lambda)$. When dealing with that device, it seems natural to distinguish the colour mechanisms generated by functions $f_1(\lambda), \ldots, f_n(\lambda)$ among the other colour mechanisms representing the Grassmann structure in question. They will be referred to as the *fundamental colour mechanisms.*[7]

---

6 When the light σ is not monochromatic, all three coefficients may be strictly positive.

7 As *any* set of linearly independent linear colour mechanisms $\varphi_1, \ldots, \varphi_n \in Lin(f_1, \ldots, f_n)$ define the same metamerism, any of these can be, generally, used to specify this metamerism. Thus, the fundamental colour mechanisms is a characteristic of a given device rather than the Grassmann structure induced by this device. (See Section 12.1.2.)

The idea that some fundamental colour mechanisms exist, in one form or another, in the human visual system, is, perhaps, the only one which is shared by practically all colour scientists. The hypothesis of the existence of fundamental colour mechanisms (referred to as the *fundamental-colour-mechanisms hypothesis*) has become the cornerstone of colour science after it was enunciated by a few scientists independently in 18–19th century (Mollon, 2003). Debates have only been about their neural implementation and their location in the visual pathways.

More specifically, a view, commonly accepted in colour vision literature, is that colour match is mediated by some neural mechanisms, that is, there are such neurons that different lights match each other when the responses of those neurons are equal. 'In a colour match, two fields of different spectral radiation appear identical. The match represents a unique neural state in which neural signals generated by the fields are identical.' (Smith & Pokorny, 2003, p. 104) This will be referred further the *colorimetric model of colour matching*.

Note that it follows from the colorimetric model of colour matching that colour matching should have the properties of an equivalence relation (see the metameric matching hypothesis above). Although, as can be seen in this and subsequent chapters, the adoption of this hypothesis makes it possible to obtain a few remarkable results, generally speaking, it should be considered only as the first, rough approximation. As will be discussed in Chapter 10, colour matching, considered as an empirical relation, is, strictly speaking, not transitive. Because of this we will also consider a weakened form of this hypothesis (referred to as the *weak fundamental-colour-mechanisms hypothesis*): if the responses of the fundamental colour mechanisms to two lights are equal then these lights match each other (but the converse is not assumed).

As any neural unit in the visual pathways (no matter if it plays role of a fundamental colour mechanism or not) is eventually driven by the cone photoreceptors, it seems logical to suppose that it is the cone photoreceptors that serve as the fundamental colour mechanisms in human colour vision (Brindley, 1960; Stockman & Sharpe, 1999). Respectively, the weak fundamental-colour-mechanisms hypothesis can be specified as follows: If the responses of the cone photoreceptors to two lights are equal then these lights match each other. However, as we argue in Chapter 13, the cone photoreceptors are more complicated units than colour mechanisms. Still, the first stage of the process of forming of their electrical response (the cone photopigment excitation) can be modelled as a linear process, and thus, can be considered as a colour mechanism. So, we adopt the weak fundamental-colour-mechanisms hypothesis in the following form. If the quantum catch in all the cone photoreceptors produced by two lights are equal then these lights match (referred to as the *weak cone-photopigment-quantum-catch hypothesis*). If the converse is also true, we say that the *cone-photopigment-quantum-catch hypothesis* takes place.

The cone-photopigment-quantum-catch hypothesis has an implication which can be readily tested. If it were true, then the Grassmann structures, induced by the colour matching functions and the photopigment absorption spectra of the cone photoreceptors, must be identical. In particular, the photopigment absorption spectra could be represented as the linear transformation of the colour matching functions. For the last few decades a lot of effort has been made to prove the latter (Stockman & Sharpe, 1999). However, direct test of the two Grassmann structures in question reveals essential differences between them (Section 7.3.5). True, the result of such comparison depends on in what form the data on the photopigment absorption spectra are used, and how they are measured.

Generally, photopigment absorption spectra can be directly measured using microspectrophotometric techniques (Bowmaker et al., 1980; Dartnall et al., 1983). Under conditions in which the Beer–Lambert law is applicable, the absorbed part, F_a, of the incident radiant flux, F_i, is the following function of pigment concentration, c, light path length, l, and the wavelength-dependent function, $d(\lambda)$, characterising the spectral selectivity of absorption in this pigment:

$$F_a = F_i \left(1 - 10^{-cld(\lambda)}\right). \tag{3.13}$$

The fraction absorbed (referred to as the *absorptance*) is then:

$$\frac{F_a}{F_i} = \left(1 - 10^{-cld(\lambda)}\right). \tag{3.14}$$

The exponent in (3.14) is usually called *absorbance*,[8] or when considered as a function of wavelength, spectral absorbance. Sometimes, it is also called the optical density of the photopigment, although this terminology is now considered outdated. Grouping the concentration c and the length path l into a single factor

$$D = cl \tag{3.15}$$

and assuming it constant, we get that the spectral selectivity of absorptance (i.e., the *absorptance spectrum*) is determined by the shape of the function $d(\lambda)$ which we will call the *absorptivity spectrum* of the photopigment.[9] (There are other names for it, for example, the extinction coefficient.)

As noticed by Dartnall (1953), the absorbance spectra measured from different photopigments tend to fit the same curve (template) when plotted in terms of frequency (i.e., the reciprocal of wavelength), differing only in the wavelength of maximum of absorbance ($\lambda^{\max}$). This observation was elaborated in the subsequent studies (Lamb, 1999). Specifically, the photoreceptor absorbance spectra obtained in different experiments using different methods remarkably overlap when plotted on the same graph in terms of $\lambda^{\max}/\lambda$ (Lamb, 1995). A few template curves providing a good description of the common shape have been proposed. In particular, Govardovskii and colleagues Govardovskii et al. (2000) put forward the following mathematical expression for the absorptivity spectrum of visual photopigment using as a single parameter the absorbance peak wavelength $\lambda^{\max}$:

$$\begin{aligned} G\left(\lambda^{\max}/\lambda\right) &= a_0/(\exp(a_1(c_1 - \nu)) + \exp(a_6(a_7 - \nu)) + \exp(a_8(a_9 - \nu)) + a_{10}) \\ &\quad + b_0 \exp(-((c_2 - \lambda)/c_3)^2), \end{aligned} \tag{3.16}$$

where $\nu = \lambda^{\max}/\lambda$,

$$\begin{aligned} c_1 &= a_1^{-1}\left(a_2 + a_3 \exp\left(-\left(\left(\lambda^{\max} - a_4\right)^2\right)/a_5\right)\right), \\ c_2 &= b_1 + b_2\lambda^{\max}, \\ c_3 &= b_3 + b_4\lambda^{\max}, \end{aligned}$$

the remaining parameter values being in Table 3.1.

Given the photopigment absorptivity spectrum of a photoreceptor $G(\lambda^{\max}/\lambda)$, its absorptance spectrum can be expressed then as

$$g\left(\lambda; \lambda^{\max}\right) = 1 - 10^{-DG(\lambda^{\max}/\lambda)}, \tag{3.17}$$

where D is the peak photopigment absorbance (or the peak photopigment optical density according to the obsolete terminology). (As indicated in Eq. (3.15), it is proportional to both the concentration of the pigment and the length of the light path in the outer segment of the photoreceptor.) Formula (3.17) will be referred to as *Govardovskii's template*. Since, to the authors knowledge, this is the

8 Thus, absorbance is the decadic logarithm of incident to transmitted radiant power.

9 When normalised by their respective maxima, the absorptivity and absorbance spectra have the same shape, provided that D is not large. Note, however, that the absorptivity spectrum is an inherent property of the photopigment, whereas the absorbance spectra depends on the pigment concentration and the light path.

Table 3.1 Parameter values used by Govardovskii et al. (2000) in Eq. (3.16).

a_0	a_1	a_2	a_3	a_4	a_5	a_6	a_7
1	69.7	61.3	3.2	300	11940	28	0.922
a_8	a_9	a_{10}	b_0	b_1	b_2	b_3	b_4
−14.9	1.104	0.674	0.26	189	0.315	−40.5	0.195

Source: Based on Govardovskii et al. (2000).

most successful analytical approximation of the absorptance spectrum of visual pigments, which is currently available, it will be used later in the book.

There are also electrophysiological techniques for measuring the absorption spectrum of photoreceptors. These are based on measuring an electrical signal into which the photoreceptor converts absorbed light. Specifically, the intensity of monochromatic light needed to evoke electrical responses of a certain fixed rate, as a function of its wavelength, is called the *action spectrum* of the photoreceptor. It was found that the absorptance spectra of single isolated photoreceptors in the primate retina, measured using the microspectrophotometric method, are in excellent agreement with the action spectra derived from single-cell recording using a suction electrode technique (Schnapf & Schneewels, 1999). Consequently, the action and absorptance spectra are believed to represent two alternative forms of spectral sensitivity, obtained by different (i.e., electrophysiological and microspectrophotometric) methods.[10] So, depending on the context, we will sometimes use the general term *spectral sensitivity* (of a photoreceptor) to refer to both.

The template suggested by Govardovskii and colleagues provides a good fit to absorbance spectra recorded from photoreceptors of 39 types (amphibians, reptiles, and fishes) ranging from 357 to 620 nm, and also to action spectra from mammals (including primates) obtained in other studies (Govardovskii et al., 2000).

Figure 3.5 displays the spectral sensitivity of the cone photopigments with the peak absorbance wavelength λ^{max} at 430, 530, and 560 nm, D being equal to 0.3. These values are, by and large, in line with the average cone spectral sensitivities recorded from the macaque and human retina (Schnapf & Schneewels, 1999).

Definition 3.2 *We say that* $\langle \mathbf{C}^*_+, \sim \rangle$ *is a* n-pigment Grassmann structure *if* $\sim$ *is defined as in Example 2.2 with* $f_i(\lambda)$ *in (2.3) substituted by* $g_i(\lambda; \mu_i)$ *in (3.17) with* $\mu_i = \lambda^{max}$, *and* $G(\lambda^{max}/\lambda)$ *as in (3.16). It will be denoted by* $\langle \mathbf{C}^*_+, \sim, \mu_1, \ldots, \mu_n \rangle$.

There is the whole variety of different photoreceptor types (with different peak absorbance wavelengths λ^{max}) present in the eyes of different animals (Bowmaker, 1991). Some marine fish have only one type of photoreceptors (rods) with the peak sensitivity, λ^{max}, at about 480 nm (Bowmaker, 1991). They cannot, obviously, possess colour vision in the traditional sense since it requires photoreceptors of at least two distinct types with overlapping spectral sensitivity curves and operation within the same range of light intensities. Most mammals have rods with maximal absorbance between 490 and 500 nm, and two types of cones: with λ^{max} in the short-wavelength (between 420 and 450 nm) and long-wavelength (between 500 and 570 nm) range of the spectrum

10 The presence of oil droplets in photoreceptors brings about a difference between action and absorption spectra because they act as optical prefilters with their own spectral density.

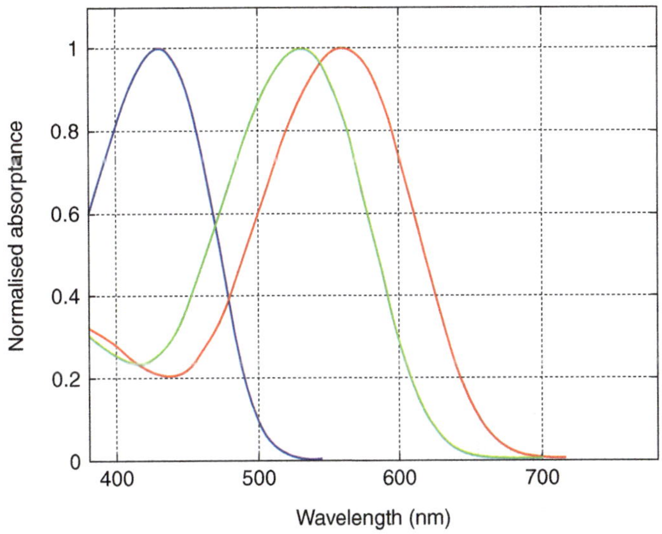

Figure 3.5 Scaled to unity peak, photopigment absorptance spectra, calculated using the Govardovskii model (Eq. (3.17)) with the peak absorbance wavelengths chosen at 430 nm (blue line), 530 nm (green line), and 560 nm (red line) and the peak photopigment absorbance D (Eq. (3.15)) set equal to 0.3. Source: Based on Govardovskii et al. (2000).

(Schnapf & Schneewels, 1999; Bowmaker, 1991). Old World primates (e.g., man, great apes, and the macaque monkey) are the only mammals having three types of cone photoreceptors (Bowmaker, 1991). Specifically, they possess so-called S-, M-, and L-cones with λ^{max} around 430, 530, and 560 nm, respectively. Fish, reptiles, and birds have cone photoreceptors of four types. For instance, some birds may have four cones with λ^{max} from approximately 360 to 570 nm. The most diverse photoreceptor equipment has been found in manatis shrimps. They have 16 types of photoreceptors (Marshall & Oberwinkler, 1999).

The nomenclature of photopigments in the human retina and the positions of their peak sensitivity were found to be genetically controlled at the molecular level (J. Neitz & Neitz, 2011). The M- and L-photopigments were found to be encoded by the genes carried on the X-chromosome, the autosomal gene encoding the S-photopigment. A lot of variability in the M- and L-photopigments due to genetic mutations and rearrangements has been found by studying the molecular genetics basis of human colour vision (J. Neitz & Neitz, 2011). Specifically, the whole families of the M- and L-photopigments variants were found to exist. The number and range of the M-like photopigments are found to be smaller than those of the L-like photopigments. The peak sensitivity (λ^{max}) of the M-like photopigments was established to vary around 530 nm; that of the L-like photopigments around 560 nm. The shift between the shortest λ^{max} and the longest λ^{max} for the L-like photopigments was found to be 27–30 nm, that for the M-like photopigments being slightly less (J. Neitz & Neitz, 2011).

The number of cone photopigments in a human individual varies from none (so-called *rod monochromats* – a very rare case of colour blindness) to three and even more. People with only rods and the S-cones are called *blue-cone monochromats*. Those who, apart from rods and the S-cones, have photoreceptors with one of the M-like photopigments are called *protanopes*. Individuals with rods, the S-cones, and photoreceptors with one of the L-like photopigments are called *deuteranopes*. About 1% of white men and 0.025% women are either protanopes or deuteranopes. People with rods, and one of the M- and one of the L-like photopigments are called *tritanopes*. Protanopes, deuteranopes, and tritanopes are referred to as *dichromats*. Those who possess cone photoreceptors with exactly three different photopigments are called *trichromats*. If all three photopigments are of different types (i.e., the *S*-, *M*-, and *L*-like) such trichromats are referred to as normal. When there are the *S*- and two *M*-like photopigments trichromats are called *protanomalous*. Trichromats with the *S*- and two *L*-like photopigments are called *deuteranomalous*.

Molecular genetic studies show that most people have more than two genes encoding both *M*-like and *L*-like photopigments. Although the phenotype-genotype interrelations are not straightforward, the existence of multiple photopigment genes led some authors to suggest that there exist individuals with more than three types of cone photopigment (M. Neitz et al., 1993). This suggestion is supported by microspectrophotometric data. These people should have been expected to have at least *tetrachromatic* colour vision (Jordan et al., 2010). However, tetrachromacy was confirmed only at *mesopic* light levels (i.e., when both rods and cones operate) that was usually accounted for by rod-intrusion. Such an apparent contradiction indicates that either the cone-photopigment-quantum-catch hypothesis or the metameric matching hypothesis, or both, are not valid.

It should be noticed that one can hardly expect the colour matching functions to be a linear transformation of the cone spectral sensitivities obtained from the isolated cone photoreceptors because the latter do not include the effect of so-called pre-receptoral filtering in the eye (produced by the ocular media, the lens and the macular pigment). Therefore, for the (weak) cone-photopigment-quantum-catch hypothesis to become more plausible, the cone spectral sensitivities should be measured so as to take into account the effect of these pre-receptoral filters (e.g., at the cornea). Let us call the fundamental colour mechanisms determined by such cone spectral sensitivities the cone fundamental colour mechanisms, and the cone spectral sensitivities themselves the *cone fundamentals*. Then, the cone-photopigment-quantum-catch hypothesis can be restated as follows: two lights match each other if and only if they cause equal cone fundamental responses[11] (the *cone-fundamentals hypothesis*). (Respectively can be stated a *weak cone-fundamentals hypothesis*: two lights causing equal cone fundamental responses match each other.[12])

In particular, the Govardovskii's template (Fig. 3.5) times the spectral transmittance of the macular pigment (Fig. 3.6) and the lens (Fig. 3.7) will be referred to as *Govardovskii's cone fundamentals*:

$$f(\lambda;\lambda^{\max}) = t(\lambda)g(\lambda;\lambda^{\max}), \tag{3.18}$$

where $t(\lambda)$ is the transmittance spectrum of the lens and the macular pigment; and $g(\lambda;\lambda^{\max})$ is the photopigment absorptance spectrum (Eq. (3.17)). Figures 3.8 and 3.9 show three (*S*-, *M*-, and *L*-) Govardovskii's cone fundamentals.

3.3.1 König's approach to identification of the fundamental colour mechanisms

The cone-fundamentals hypothesis implies that the colour matching functions can be linearly transformed into the cone photopigment absorptance spectra corrected for the spectral selectivity of the pre-receptoral filters. Since Clerk Maxwell (1855), colour scientists have tried to ascertain this transformation (the existence of which, though, they took for granted). However, even though the cone-fundamentals hypothesis is correct, the linear transformation related the cone fundamentals

11 If $f_i(\lambda)$ is a cone fundamental then its response to a light a is by definition an integral (2.3) in Example 2.2.

12 It will be shown in Chapter 10 that while the cone fundamental hypothesis fails, the weak cone fundamental hypothesis remains a rather plausible assumption for the human colour vision.

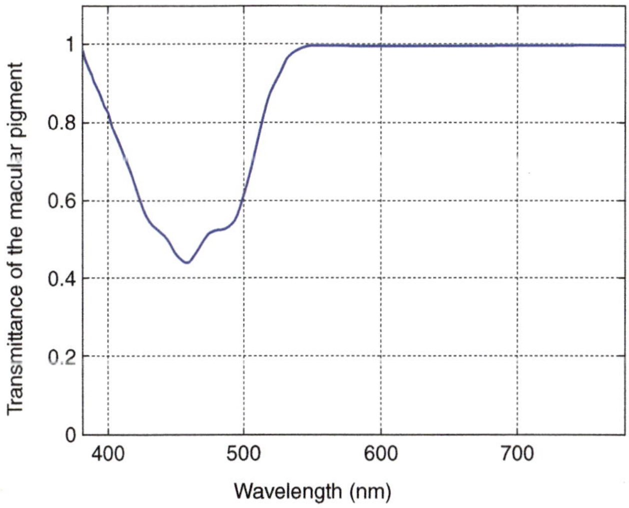

Figure 3.6 Transmittance of the macular pigment of the human eye as a function of wavelength. When plotting the graph, we used the macular pigment optical density spectrum for a 2° field, downloaded from the CVRL database. Source: Data from CVRL database.

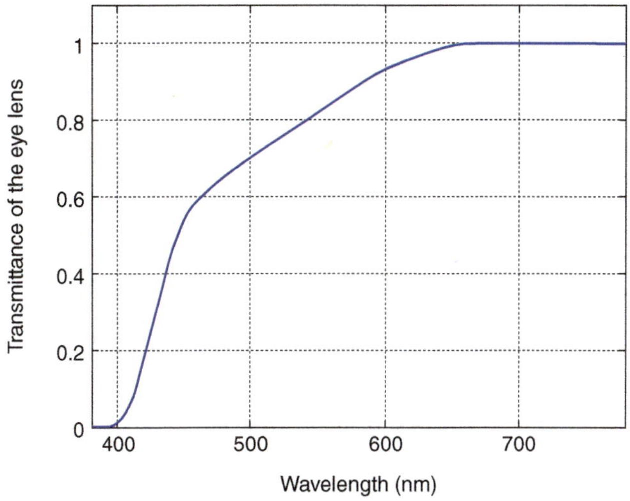

Figure 3.7 Transmittance of the young human eye lens relative to that at a wavelength of 700 nm. When plotting the graph, we used the lens optical density spectrum, downloaded from the CVRL database. Source: Data from CVRL database.

and the colour matching functions cannot be derived from the colour matching data of trichromats only.

Maxwell, and later König – a student of Helmholtz – suggested that this transformation could be evaluated if the set of the colour matching functions of trichromats was completed with that of dichromats (Maxwell, 1855; König & Dieterici, 1892). This approach is usually associated with König's name. It should be borne in mind that this approach assumes that, first, all trichromats produce the same set of colour matching functions (i.e., the individual variations in colour matching are neglected); and second, any two lights matching each other for trichromats will be a colour

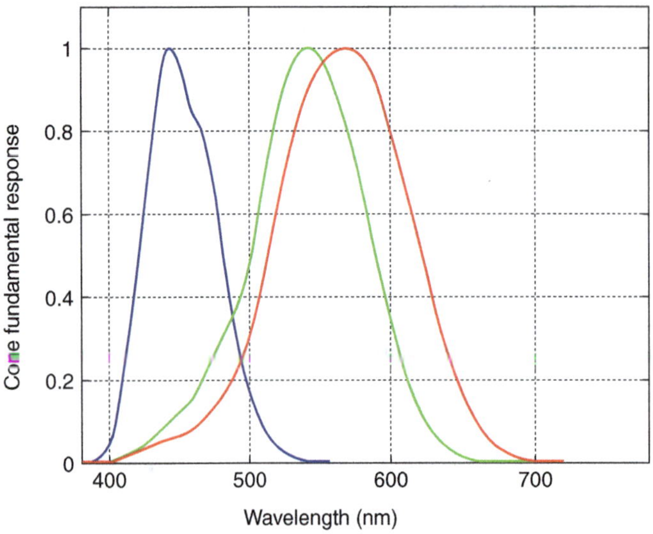

Figure 3.8 Normalised cone spectral sensitivities (cone fundamentals) based on the spectral photopigment absorptances depicted in Fig. 3.5.

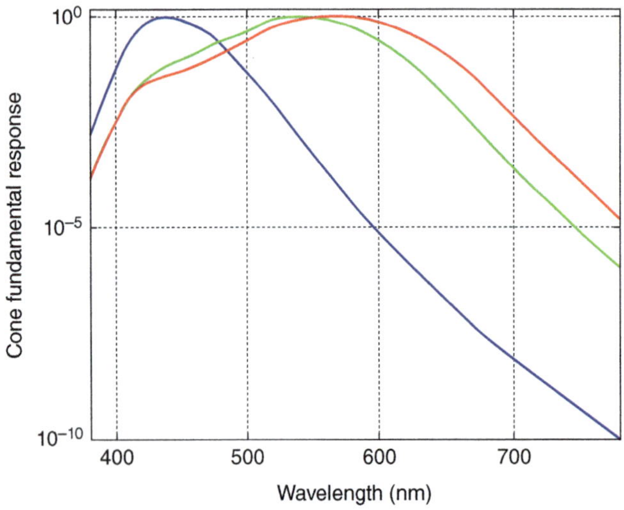

Figure 3.9 Govardovskii's cone fundamentals in semi-logarithmic coordinates.

match for dichromats too.[13] The latter will be the case if dichromatism is, as generally believed, a reduced form of trichromatism, that is, dichromats possess all but one of the fundamental colour mechanisms of trichromats.

Proposition below shows that considering additional (dichromatic) matching relations one can indeed infer about the fundamental colour mechanisms. More specifically, this proposition shows that a representation of a n-dimensional Grassmann structure together with n lower-dimensional (to be more exact, $(n-1)$-dimensional) Grassmann substructures $\langle A, \sim_i \rangle$ (Definition 2.3) turns out

13 This is, of course, an idealisation because of the unavoidable inter-individual differences between observers.

to be unique up to scaling factors provided that these Grassmann substructures satisfy the following condition (referred to as the *independence condition*).

Definition 3.3 *Suppose $\langle A, \sim\rangle$ is an n-dimensional Grassmann structure, and $\langle A, \sim_1\rangle, \ldots, \langle A, \sim_k\rangle$, $k \leq n$, are its $(n-1)$-dimensional Grassmann substructures. Each $\sim_i$ defines the $(n-1)$-dimensional vector space $E/E(\sim_i)$. Moreover, the mapping $\pi(e) \mapsto \pi_i(e)$ is a well-defined linear operator ψ_i of $E/E(\sim)$ onto $E/E(\sim_i)$, and $L_i := \psi_i^{-1}(0) = \pi(E(\sim_i))$ is a straight line (one-dimensional linear subspace) in the n-dimensional colour stimulus space $E/E(\sim)$. Let $\mathbf{a}_i$ be an arbitrary non-zero vector in L_i. If $\mathbf{a}_1, \ldots, \mathbf{a}_k$ are linear independent (this property does not depend on a particular choice of elements $0 \neq \mathbf{a}_i \in L_i$, $i = 1, \ldots, k$), we say that the Grassmann substructures $\langle A, \sim_1\rangle, \ldots, \langle A, \sim_k\rangle$ satisfy the* independence condition.

Proposition 3.4 *Let $\langle A, \sim\rangle$ be an n-dimensional Grassmann structure, and $\langle A, \sim_1\rangle, \ldots, \langle A, \sim_n\rangle$ its $(n-1)$-dimensional Grassmann substructures satisfying independence condition (Definition 3.3). Then there exist real-valued functions $\varphi_1, \ldots, \varphi_n$ on A such that $\Phi = (\varphi_1, \ldots, \varphi_n)$ is a homomorphism of $\langle A, \sim\rangle$ into $\mathbf{R}^n$ (i.e., a map $A \to \mathbf{R}^n$ satisfying conditions (i)–(iii) of the Krantz Theorem); and for each i, $\Phi_i = (\varphi_1, \ldots, \varphi_{i-1}, \varphi_{i+1}, \ldots, \varphi_n)$ is a homomorphism of $\langle A, \sim_i\rangle$ into $\mathbf{R}^{n-1}$. The functions $\varphi_1, \ldots, \varphi_n$ are unique up to scaling factors, that is, if there are another functions $\varphi'_1, \ldots, \varphi'_n$ on A with such properties then $\varphi'_i = \alpha_i\varphi_i$, $\alpha_i \neq 0$, $i = 1, \ldots, n$.*

The König approach implies, then, that, first, there exist three linear colour mechanisms (cone fundamentals), φ_S, φ_M, and φ_L; second, trichromatic colour matching is defined as in Example 2.2:

$$a \sim b \Leftrightarrow \varphi_S(a) = \varphi_S(b), \varphi_M(a) = \varphi_M(b), \varphi_L(a) = \varphi_L(b), \tag{3.19}$$

and third, there are exactly three types of dichromatism, thus three types of dichromatic metamerisms (written as $\sim_S$, $\sim_M$, and $\sim_L$) defined as follows

$$\begin{aligned} &a \sim_S b \Leftrightarrow \varphi_M(a) = \varphi_M(b), \varphi_L(a) = \varphi_L(b), \\ &a \sim_M b \Leftrightarrow \varphi_S(a) = \varphi_S(b), \varphi_L(a) = \varphi_L(b), \\ &a \sim_L b \Leftrightarrow \varphi_S(a) = \varphi_S(b), \varphi_M(a) = \varphi_M(b). \end{aligned} \tag{3.20}$$

Each of these dichromatic metamerisms defines a Grassmann substructure of $\langle A, \sim\rangle$. Proposition 3.4 entails that φ_S, φ_M, and φ_L are the only (up to scaling factors) linear colour mechanisms which bring conjoint representation of the trichromatic and dichromatic metamerisms (3.19) and (3.20).

Let us choose three non-zero vectors $\mathbf{a}_S$, $\mathbf{a}_M$, and $\mathbf{a}_L$ in the straight lines $\pi(E(\sim_S))$, $\pi(E(\sim_M))$, and $\pi(E(\sim_L))$ respectively (cf. Definition 3.3). Suppose $f_1(\lambda)$, $f_2(\lambda)$, and $f_3(\lambda)$ are the trichromatic colour matching functions evaluated with respect to primaries $a_1 = \delta_{\lambda_1}$, $a_2 = \delta_{\lambda_2}$, and $a_3 = \delta_{\lambda_3}$. Since a_1, a_2, a_3 are primaries, there are unique vectors $a_S, a_M, a_L \in Lin(a_1, a_2, a_3)$ satisfying $\pi(a_S) = \mathbf{a}_S$, $\pi(a_M) = \mathbf{a}_M$, and $\pi(a_L) = \mathbf{a}_L$. The vectors a_S, a_M, and a_L are colour independent because of linear independence of their π-images $\mathbf{a}_S$, $\mathbf{a}_M$, and $\mathbf{a}_L$. In colour literature, a_S, a_M, and a_L are called *missing primaries*.

3.3.2 Some estimates of the cone fundamentals used in colour research and applications

As Stockman rightly stated, 'until recently, the estimates by Smith & Pokorny (1975) have been widely used in science and vision research as a *de facto* standard' (Stockman, 2019, p. 90).

Figures 3.10 and 3.11 present the cone fundamentals derived by Smith & Pokorny (1975) from the CIE 1931 colour matching functions amended by Judd[14] (1951) and Vos (1978) (Fig. 3.12) using König's approach.[15] However, as mentioned above, these colour matching functions are based

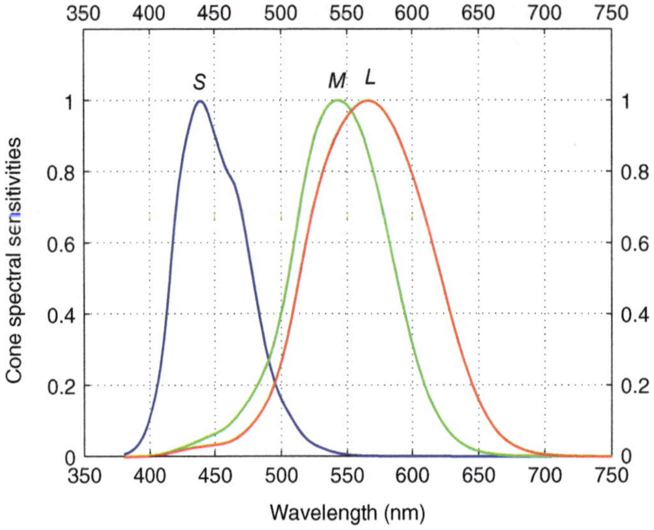

Figure 3.10 Smith and Pokorny's cone fundamentals normalised by their peak sensitivity. *S*, *M*, and *L* stand for the cone fundamentals with the peak sensitivity in the short-, middle-, and long-wave regions of the visible spectrum, respectively. Source: Based on Smith and Pokorny (1975).

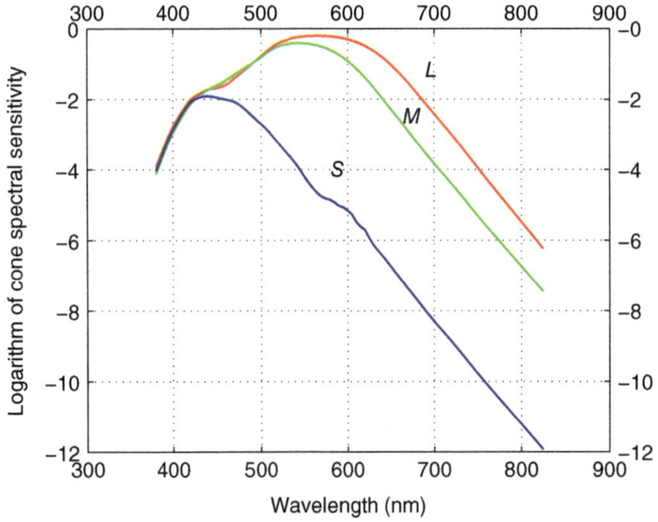

Figure 3.11 Logarithm of the relative spectral sensitivity of the cone fundamentals derived by Smith & Pokorny (1975). Source: Based on Smith and Pokorny (1975).

14 Judd modified the CIE 1931 colour matching functions so as to be in line with his new measurements of the luminous efficiency function (Wyszecki & Stiles, 1982, pp. 330–331).
15 They are in good correspondence with the cone fundamentals derived via an alternative approach not resorting to the dichromatic data (Logvinenko, 1998).

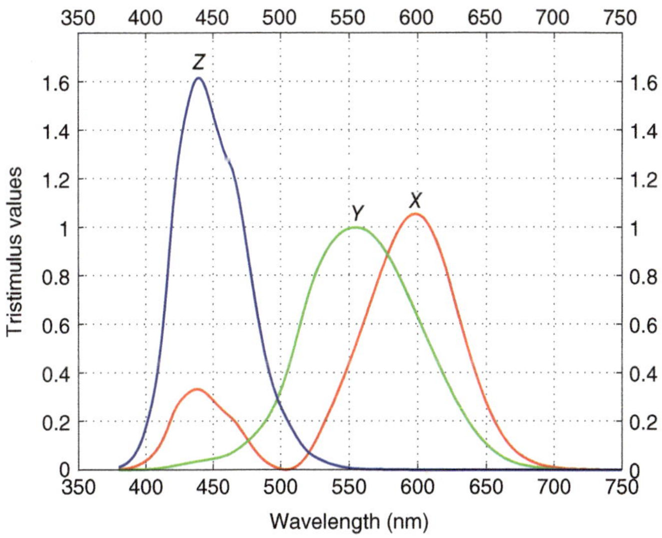

Figure 3.12 Colour matching functions of the CIE 1931 Standard Colorimetric Observer with respect to the imaginary primaries *X*, *Y*, and *Z*, as amended by Judd (1951) and Vos (1978). Source: Based on Judd (1951) and Vos (1978).

on colour matching data that have undergone many corrections, extrapolation, and smoothing. Evidently, being the most reliable and recognised colour matching data available so far, the 10° colour matching functions of Stiles & Burch (1959) seem to be more suitable for deriving cone fundamentals (Fig. 5.2). Stockman & Sharpe (2000) used them to derive the cone fundamentals not only for a 10° but also for a 2° viewing angle (Fig. 3.13). To do this, they made some adjustments based on their assumptions how the peak macular pigment density and the peak photopigment optical density change with angular size. Then, their approach was to find a linear transformation

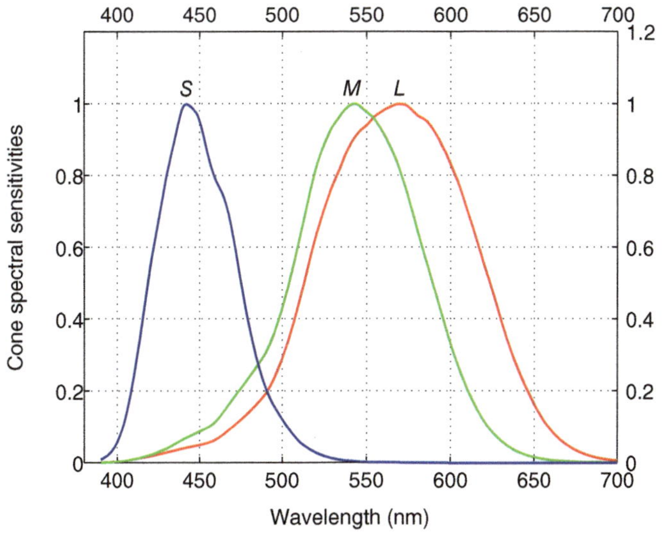

Figure 3.13 Stockman & Sharp 2° cone fundamentals based on the Stiles & Burch 10° colour matching functions adjusted to 2°.

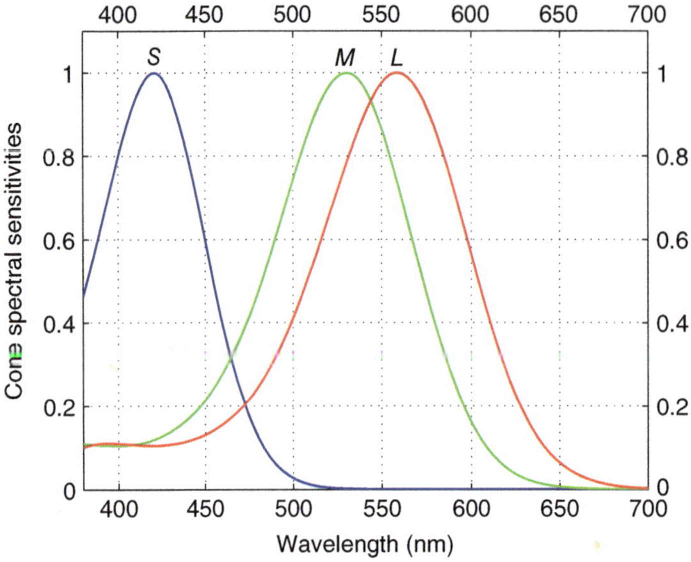

Figure 3.14 Spectral sensitivity of three types of human cones as evaluated by Stockman & Sharpe (2000). *S*-, *M*-, and *L*-cones reach maximum sensitivity at 421, 530, and 559 nm, respectively. Sensitivity curves are normalised to their maximum. Source: Based on Stockman and Sharpe (2000).

of the colour matching functions which best fits to the spectral sensitivity curves obtained from blue-cone monochromats, deuteranopes, and protanopes. Their cone fundamentals prove to be in good agreement with the photopigment absorptance spectrum obtained by direct methods (Stockman & Sharpe, 1999). In 2006, the cone fundamentals of Stockman & Sharpe (2000) were endorsed by the CIE CIE (2006). The cone fundamentals in Fig. 3.13 will be referred to as the *CIE 2006 2° LMS cone fundamentals*. Notably, the CIE 2015 *XYZ* 2° colour matching functions (Fig. 3.4) are just the linear transformation of the CIE 2006 2° LMS cone fundamentals. Figure 3.14 shows the cone photopigment absorptance spectra calculated from the cone fundamentals as evaluated by Stockman & Sharpe (2000).

It will be shown in Chapter 7 (Section 7.3.5) that, amongst all the sets of cone fundamentals considered above, Stockman & Sharpe's and Govardovskii's cone fundamentals are closest to each other. However, the advantage of the latter is its analytical expression given by Eqs. (3.17) and (3.16). So, we will use Govardovskii's cone fundamentals (Fig. 3.8) to represent the human trichromatic and/or dichromatic colour vision.

Definition 3.4 *Given* $f_1(\lambda;\mu_1),\ldots,f_n(\lambda;\mu_n)$ *where* $f_i(\lambda;\mu_i)$ *is Govardovskii's cone fundamental (i.e., Govardovskii's template with* $\lambda_i^{\max}=\mu_i$ *(3.16) times the spectral transmittance of the eye lens (Fig. 3.7) times the spectral transmittance of the macular pigment (Fig. 3.6)), we say that* $\langle \mathbf{C}_+^*, \sim\rangle$ *is a n-*pigment Grassmann–Govardovskii structure *if* $\sim$ *is defined as in Example 2.2 with* $f_i(\lambda)$ *in (2.3) substituted by* $f_i(\lambda;\mu_i)$. *It will be denoted as* $G\langle \mathbf{C}_+^*, \sim, \mu_1,\ldots,\mu_n\rangle$.

4

Colour-Signal Cone

In this and the subsequent chapters, we will assume n fundamental colour mechanisms to exist whose linear parts are specified by their spectral sensitivity functions, $\varphi_1(\lambda), \dots, \varphi_n(\lambda)$. We shall treat $\varphi_1(\lambda), \dots, \varphi_n(\lambda)$ as either cone fundamentals or the spectral sensitivity functions of some artificial photosensors, which are non-negative and continuous.[1] The latter means that they can be formally represented as elements of the Banach space $\mathbf{C} = \mathbf{C}(\Lambda)$. We shall adopt $\mathbf{C}_+^*$ as the light stimulus cone (Example 2.1a). Let $\mathsf{f}_\varphi(\sigma)$ stand for the response of a colour mechanism with the spectral sensitivity function $\varphi(\lambda)$ of its linear part to a light stimulus σ. Proposition 2.4 implies that $\mathsf{f}_\varphi(\sigma)$ can be represented as a linear functional $\langle\varphi, \sigma\rangle$ followed by a local nonlinearity g:

$$\mathsf{f}_\varphi(\sigma) = g\left(\langle\varphi, \sigma\rangle\right), \tag{4.1}$$

where g is a one-to-one map $\mathbf{R}_+ \to \mathbf{R}_+$, and

$$\langle\varphi, \sigma\rangle := \int_\Lambda \varphi(\lambda)\,\sigma(d\lambda), \ \ \varphi \in \mathbf{C}_+(\Lambda), \ \ \sigma \in \mathbf{C}_+^*(\Lambda). \tag{4.2}$$

We shall also assume that the linear parts of the colour mechanisms in question are linearly independent and cover the whole range of the visible spectrum wavelength interval. The latter means

$$\varphi_1(\lambda) + \cdots + \varphi_n(\lambda) > 0, \quad \lambda \in \Lambda, \tag{4.3}$$

which implies regularity of the Grassmann structure $\langle \mathbf{C}_+^*, \sim\rangle$ (Definition 2.7).

The map $\mathsf{F} : \mathbf{C}_+^*(\Lambda) \to \mathbf{R}^n$, $\mathsf{F}(\sigma) = (\mathsf{f}_{\varphi_1}(\sigma), \dots, \mathsf{f}_{\varphi_n}(\sigma))$, can be factored as $\mathsf{F} = G \circ \Phi$, where Φ is the linear colour-signal map (Remark 2.5), $\Phi(\sigma) = \left(\langle\varphi_1, \sigma\rangle, \dots, \left\langle\varphi_n, \sigma\right\rangle\right)$, and G is a one-to-one map $\mathbf{R}_+^n \to \mathbf{R}_+^n$, induced by the functions g_i. While in this chapter we consider only a linear case (i.e., $\mathsf{F} = \Phi$), the main results obtained in this and the following chapters will hold true for $\mathsf{F} = G \circ \Phi$. Since in the linear case a colour mechanism is completely determined by its spectral sensitivity function, we shall identify f_φ with the function φ unless it is otherwise said.

Since the colour-signal map Φ is linear, the set of all of the colour signals, $\Phi\left(\mathbf{C}_+^*\right)$, is a convex cone in $\mathbf{R}^n$.

Definition 4.1 *Given a linear colour-signal map Φ, the image of the light stimulus cone, $\Phi\left(\mathbf{C}_+^*\right)$, is called the* colour-signal cone.

1 Since linear colour mechanisms with discontinuous spectral sensitivity functions may be of interest when constructing artificial (technical) colour vision systems, we will notice those results which are free of the continuity assumption.

Foundations of Colour Science: From Colorimetry to Perception, First Edition. Alexander D. Logvinenko and Vladimir L. Levin.
© 2023 John Wiley & Sons Ltd. Published 2023 by John Wiley & Sons Ltd.

Remark 4.1 *The colour-signal map Φ induces an isomorphism (a one-to-one linear operator) Φ_0 of the colour stimulus cone $\pi(\mathbf{C}_+^*)$ onto the colour-signal cone $\Phi(\mathbf{C}_+^*)$, so that $\Phi = \Phi_0 \circ \pi$ where $\pi(\sigma) = [\sigma]$ is the colour stimulus (i.e., the class of metamerism) produced by a light stimulus $\sigma \in \mathbf{C}_+^*$. As already mentioned (Remark 2.5), the colour-signal cone may be considered as a concrete representation of the colour stimulus cone. Given another colour-signal map Φ' generating the same n-dimensional Grassmann structure on the light stimulus cone $\mathbf{C}_+^*$, the corresponding colour-signal cone $\Phi'(\mathbf{C}_+^*)$ is a different concrete representation of the colour stimulus cone, and the latter representation is connected with the former one via a nonsingular linear operator $T : \mathbf{R}^n \to \mathbf{R}^n$:*

$$\Phi'(\mathbf{C}_+^*) = T(\Phi(\mathbf{C}_+^*)).$$

Let P denote the set of probability measures on Λ. As is well-known (e.g., see Holmes, 1975), the set of extreme points[2] of P, $Ext\,(P)$, is exactly the set $\Delta = \{\delta_\lambda : \lambda \in \Lambda\}$ where δ_λ stands for the Dirac measure (delta function) at λ, $\langle \varphi, \delta_\lambda \rangle := \varphi(\lambda)$ for all $\varphi \in \mathbf{C}$. When λ varies from λ_{min} up to λ_{max}, $\Phi(\delta_\lambda) = (\varphi_1(\lambda), \ldots, \varphi_n(\lambda))$ runs a continuous curve in $\mathbf{R}^n$, $\Phi(\Delta)$, which will be referred to as the *spectral colour-signal curve* (or simply the *spectral curve*).

Figure 4.1 shows the spectral colour-signal curve plotted by using Govardovskii's cone fundamentals (Fig. 3.8). The spectral colour-signal curve based on Stockman & Sharpe's cone fundamentals (Fig. 3.13) is depicted in Fig. 4.2. According to the definition (Remark 2.5), various colour-signal maps, Φ, determine various colour-signal spaces, $\Phi(E)$. Still, to be in line with the terminology established in colour vision literature, we will refer to the colour-signal space based on any set of cone fundamentals as the *SML cone excitation space* (Smith & Pokorny, 1996; Kaiser & Boynton, 1996). For brevity, we denote the coordinates in the corresponding *SML* colour-signal space as S, M,

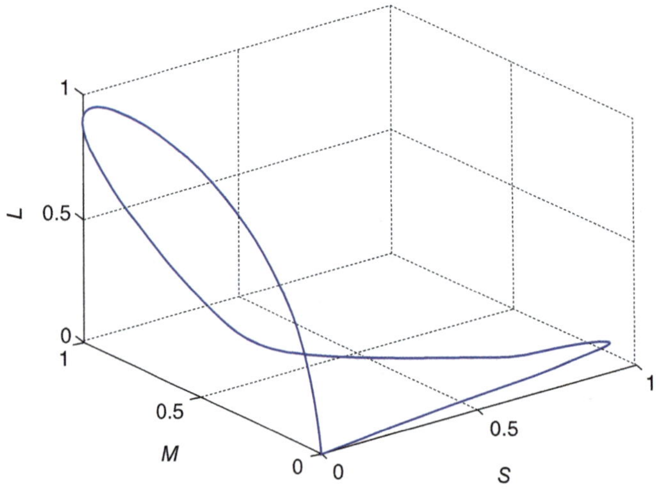

Figure 4.1 Spectral colour-signal curve in the *SML* cone excitation space evaluated for the Govardovskii's fundamentals depicted in Fig. 3.8. Notations S, M, and L are used for the response of the short, middle, and long wavelength maximally sensitive fundamentals, respectively.

---

2 Recall (e.g., see Holmes, 1975) that a point σ of a convex set P is called its extreme point if $\sigma = \frac{1}{2}(\sigma_1 + \sigma_2)$, $\sigma_1, \sigma_2 \in P$ implies $\sigma = \sigma_1 = \sigma_2$.

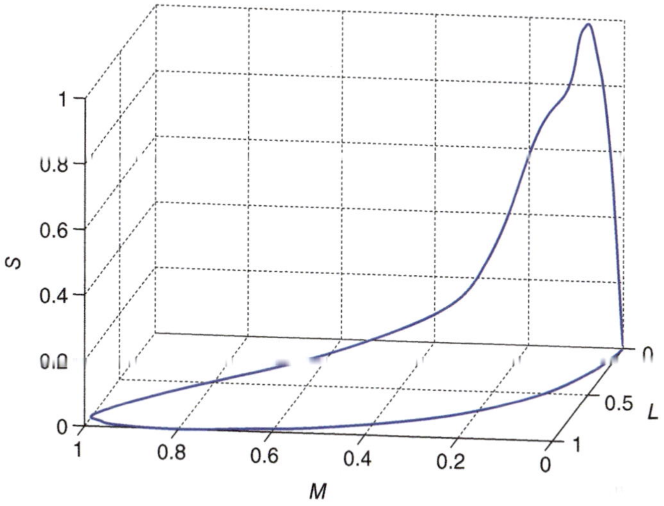

Figure 4.2 Spectral colour-signal curve in the *SML* cone excitation space evaluated for the cone fundamentals represented in Fig. 3.13. Notations *S*, *M*, and *L* are used for the response of the short, middle, and long wavelength maximally sensitive fundamentals, respectively.

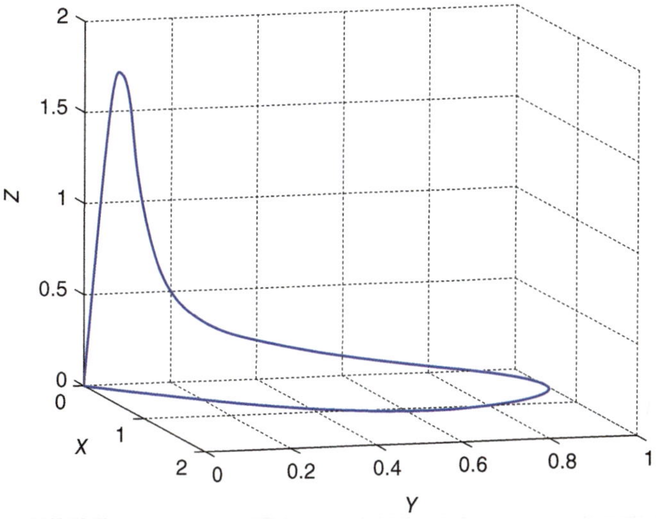

Figure 4.3 The colour-signal curve in the CIE 1931 colour-signal space. Notations *X*, *Y*, and *Z* are used for the corresponding tristimulus values (as in Fig. 3.12).

and L (assuming the fundamentals with the peak sensitivity in the short (S), middle (M), and long (L) wavelength regions of the visible spectrum).

Figure 4.3 shows the spectral colour-signal curve plotted for the CIE 1931 colorimetric observer. The colour-signal space induced by the colour-signal map determined by the CIE 1931 colour matching functions will be referred to as the CIE 1931 colour-signal space.

Theorem 4.1 *The Φ-image of P, $\Phi(P)$, is the convex hull of $\Phi(\Delta)$.*

In other words, $\Phi(P)$ is the convex hull (closing is not needed) of the colour signals produced by all monochromatic (spectral) lights of the unit intensity.[3] Since every non-zero positive measure lies on the ray in $\mathbf{C}_{+}^{*}$ passing through some probability measure, the colour-signal cone $\Phi(\mathbf{C}_{+}^{*})$ is the convex hull of the conical surface generated by the rays passing through points of the spectral colour-signal curve.

Remark 4.2 *The colour-signal cone $\Phi(\mathbf{C}_{+}^{*})$ is closed. Indeed, if $z^n \in \Phi(\mathbf{C}_{+}^{*})$ and $z^n \to z \neq 0$, then we can assume with no loss of generality that $z^n \neq 0$ for every n. Then $z^n = \alpha_n z_0^n$, where $\alpha_n > 0, z_0^n \in \Phi(P)$. Now, as $\Phi(P)$ is compact and $0 \notin \Phi(P)$, we can assume, by passing if needed to a subsequence, that $z_0^n \to z_0 \in \Phi(P)$ and $\alpha_n = \frac{\|z^n\|}{\|z_0^n\|} \to \alpha = \frac{\|z\|}{\|z_0\|} > 0$. Therefore, $z = \lim_{n\to\infty} \alpha_n z_0^n = \alpha z_0 \in \Phi(\mathbf{C}_{+}^{*})$, and the closedness of $\Phi(\mathbf{C}_{+}^{*})$ is thus established.*

We summarize the above reasoning in the following consequence of Theorem 4.1.

Corollary 4.1 *Given a weak* continuous linear colour-signal map Φ, the colour-signal cone $\Phi(\mathbf{C}_{+}^{*})$ is a closed convex cone in $\mathbf{R}^n$ that coincides with the convex hull of the conical surface generated by the rays passing through points of the spectral colour-signal curve.*

Proposition 4.1 *The interior of the colour-signal cone $\Phi(\mathbf{C}_{+}^{*})$ is nonempty.*

Indeed, otherwise the cone is contained in some hyperspace $H_0 = \{z \in \mathbf{R}^n : \langle c, z\rangle = 0\}$ where $c = (c_1, \dots, c_n) \neq 0$, which implies

$$\sum_{j=1}^{n} c_j \varphi_j(\lambda) = \langle c, \Phi(\delta_\lambda)\rangle = 0$$

for all $\lambda \in \Lambda$, in contradiction with linear independence of $\varphi_1, \dots, \varphi_n$.

4.1 Strong colour-signal-cone-boundary hypothesis

Consider a conical surface generated by the rays passing through points of the spectral colour-signal curve $\Phi(\Delta)$, which will be referred to as the *spectral colour-signal conical surface* (Fig. 4.4). The following question arises: Is this surface fully contained in the boundary of the colour-signal cone? Note that if the answer is affirmative for a given Φ (in this case, we shall say that Φ satisfies the *strong colour-signal-cone-boundary hypothesis*[4]), then it is also such for any Φ' consistent with the same Grassmann structure as Φ. (Indeed, since $\Phi' = T \circ \Phi$ where T is a linear one-to-one map of $\mathbf{R}^n$ onto itself, $\Phi'(\sigma)$ is a boundary point of $\Phi'(\mathbf{C}_{+}^{*})$ if and only if $\Phi(\sigma)$ is a boundary point of $\Phi(\mathbf{C}_{+}^{*})$.)

Suppose $H = \{z : \langle c, z\rangle = c_0\}$ (where $z = (z_1, \dots, z_n)$, $c = (c_1, \dots, c_n)$, $\langle c, z\rangle = c_1 z_1 + \dots + c_n z_n$) is a hyperplane in $\mathbf{R}^n$ such that $0 \notin H$, H has common points with the colour-signal cone, and no ray of the colour-signal cone $\Phi(\mathbf{C}_{+}^{*})$ is parallel to H. The class of all such hyperplanes is denoted as $\mathcal{H}(\Phi)$. If $H \in \mathcal{H}(\Phi)$ then its intersection with the colour-signal cone, $H \cap \Phi(\mathbf{C}_{+}^{*})$, is a compact convex set in H comprising convex mixtures of (at most n) points $g_H(\Phi(\delta_\lambda))$ of the continuous curve $g_H(\Phi(\Delta))$, where a (continuous) map g_H of $\Phi(\mathbf{C}_{+}^{*}) \setminus \{0\}$ onto H is defined by $g_H(z) = \frac{c_0}{\langle c,z\rangle} z$. (Note that $\langle c, z\rangle \neq 0$ since the ray passing through z is not parallel to H).

---

3 See the corresponding footnote on 'intensity' at the beginning of Chapter 2 (p. 61).
4 It is a particular case of a more general colour-signal-cone-boundary hypothesis; see Definition 4.2.

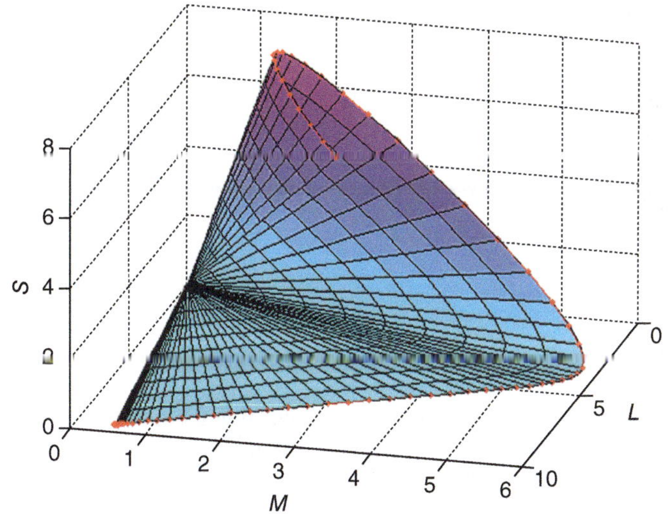

Figure 4.4 Spectral colour-signal conical surface plotted for Govardovskii's fundamentals (Fig. 3.8).

It follows that the strong colour-signal-cone-boundary hypothesis is true if and only if, for any $\lambda_1, \dots, \lambda_n, \lambda \in \Lambda$, $g_H(\Phi(\delta_\lambda))$ does not get into the interior (in H) of the polyhedron with vertexes $g_H(\Phi(\delta_{\lambda_j}))$, $j = 1, \dots, n$.

Fix a hyperplane $H \in \mathcal{H}(\Phi)$ and take a unique $c = (c_1, \dots, c_n) \in \mathbf{R}^n$ such that

$$H = \{z \in \mathbf{R}^n : \langle c, z \rangle = 1\}.$$

Then $\langle c, z \rangle > 0$ for all $z \in \Phi(\mathbf{C}_+^*) \setminus \{0\}$. (Indeed, if $z' \in H \cap \Phi(\mathbf{C}_+^*)$, then $\langle c, z' \rangle = 1$, and if we assume $\langle c, z'' \rangle \leq 0$ for some $z'' \in \Phi(\mathbf{C}_+^*)$, then either $\langle c, z'' \rangle = 0$ or $z := z'' - \langle c, z'' \rangle z' = z'' + |\langle c, z'' \rangle| z' \in \Phi(\mathbf{C}_+^*) \setminus \{0\}$ and $\langle c, z \rangle = 0$. In both cases, we obtain a ray passing respectively through z'' or through z and parallel to H in contradiction with $H \in \mathcal{H}(\Phi)$.)

Suppose now that $z \in \Phi(\mathbf{C}_+^*) \setminus \{0\}$, $z^j \in \Phi(\mathbf{C}_+^*) \setminus \{0\}$, $j = 1, \dots, n$, and $z = \sum_1^n \alpha_j z^j$. Then $\langle c, z^j \rangle > 0$ and $\frac{z^j}{\langle c, z^j \rangle} \in H$. In such a case,

$$\alpha := \sum_{j=1}^{n} \alpha_j \langle c, z^j \rangle = \langle c, z \rangle > 0$$

and

$$g_H(z) = \frac{z}{\alpha} = \sum_{j=1}^{n} \beta_j \frac{z^j}{\langle c, z^j \rangle} = \sum_{j=1}^{n} \beta_j g_H(z^j),$$

where $\beta_j := \frac{\alpha_j \langle c, z^j \rangle}{\alpha}$, $j = 1, \dots, n$. Note that $\beta_1 + \cdots + \beta_n = 1$; therefore, $g_H(z)$ lies in the convex polyhedron in H with vertexes $g_H(z^j)$ if and only if all α_j are non-negative. Moreover, $g_H(z)$ lies in the interior of that polyhedron if and only if all α_j are strictly positive, in which case z gets into the interior of the colour-signal cone $\Phi(\mathbf{C}_+^*)$. It can occur that $g_H(z)$ is a boundary point of the polyhedron but z gets into the interior of $\Phi(\mathbf{C}_+^*)$. Observe, in that connection, that since $H \cap \Phi(\mathbf{C}_+^*)$ comprises convex combinations of at most n points $g_H(\Phi(\delta_\lambda))$, for every interior point z of $\Phi(\mathbf{C}_+^*)$ there exist λ_j, $j = 1, \dots, n$, such that $\Phi(\delta_{\lambda_j})$ are linearly independent and $g_H(z)$ is an interior point of the polyhedron with vertexes $g_H(\Phi(\delta_{\lambda_j}))$, $j = 1, \dots, n$.

For $z^j = \Phi(\delta_{\lambda_j})$ the above reasoning may be summarized as follows.

Proposition 4.2 *Given a light $\sigma_0 \in \mathbf{C}_+^* \setminus \{0\}$ such that $\Phi(\sigma_0) = \alpha_1\Phi(\delta_{\lambda_1}) + \cdots + \alpha_n\Phi(\delta_{\lambda_n})$, three statements hold then true as follows.*

(a) *If $\alpha_j \geq 0, j = 1, \ldots, n$, and at least two of α_j are non-zero, then, for any $H \in \mathcal{H}(\Phi)$, $g_H(\Phi(\sigma_0))$ cannot be an extreme point of $H \cap \Phi(\mathbf{C}_+^*)$.*

(b) *If $\alpha_j > 0, j = 1, \ldots, n$, then $\Phi(\sigma_0)$ cannot be a boundary point of the colour-signal cone $\Phi(\mathbf{C}_+^*)$.*

(c) *If $\Phi(\sigma_0)$ is an interior point of $\Phi(\mathbf{C}_+^*)$, then there exist $\lambda_j, j = 1, \ldots, n$, such that $\Phi(\delta_{\lambda_1}), \ldots, \Phi(\delta_{\lambda_n})$ are linearly independent and $\alpha_j > 0, j = 1, \ldots, n$.*

Let us take a segment $I(\delta_{\lambda_{\min}}, \delta_{\lambda_{\max}}) = \{\alpha\delta_{\lambda_{\min}} + (1-\alpha)\delta_{\lambda_{\max}} \;:\; 0 \leq \alpha \leq 1\}$ (where $\lambda_{\min}$ and $\lambda_{\max}$ are the endpoints of Λ), add it to Δ, denote $\Delta' = \Delta \cup I(\delta_{\lambda_{\min}}, \delta_{\lambda_{\max}})$, and consider in $\mathbf{R}^n$ the closed curve $\Phi(\Delta')$.

If $n = 3$, then the closed curve $g_H(\Phi(\Delta'))$ determines in the two-dimensional plane H a closed domain $D = D(H)$, whose boundary it is, and the strong colour-signal-cone-boundary hypothesis means that the mentioned domain is convex and coincides with $H \cap \Phi(\mathbf{C}_+^*)$, hence the boundary of the colour-signal cone is exactly the conical surface generated by the rays passing through points of $g_H(\Phi(\Delta'))$. Moreover, any $H \in \mathcal{H}(\Phi)$ can be used to test the strong colour-signal-cone-boundary hypothesis: if so determined closed domain $D(H)$ in a particular plane H is convex, then a similar domain in any other $H \in \mathcal{H}(\Phi)$ is convex as well.

Note that, given a triad of primaries $\delta_{\lambda_1}, \delta_{\lambda_2}, \delta_{\lambda_3}$, the equality $\Phi(\delta_\lambda) = \alpha_1\Phi(\delta_{\lambda_1}) + \alpha_2\Phi(\delta_{\lambda_2}) + \alpha_3\Phi(\delta_{\lambda_3})$ means that $\alpha_j = t_j(\lambda)$, $j = 1, 2, 3$, and

$$\delta_\lambda \sim t_1(\lambda)\delta_{\lambda_1} + t_2(\lambda)\delta_{\lambda_2} + t_3(\lambda)\delta_{\lambda_3},$$

where t_1, t_2, t_3 stand for the colour matching functions determined by the primaries under consideration. Furthermore, if for a given λ all $t_j(\lambda)$ are strictly positive, then $\delta_{\lambda_1}, \delta_{\lambda_2}$, and δ_λ may be taken as new primaries, and one gets a representation

$$\delta_{\lambda_3} \sim t_1'(\lambda_3)\delta_{\lambda_1} + t_2'(\lambda_3)\delta_{\lambda_2} + t'(\lambda_3)\delta_\lambda$$

with new colour matching functions $t_1'(\lambda_3) = -\frac{t_1(\lambda)}{t_3(\lambda)}$, $t_2'(\lambda_3) = -\frac{t_2(\lambda)}{t_3(\lambda)}$, $t'(\lambda_3) = \frac{1}{t_3(\lambda)}$, and exactly two of them, t_1' and t_2', take negative values at λ_3 (cf. reasoning following Proposition 3.2). Taking into account this observation and Proposition 4.2, (c), we get the next result.

Proposition 4.3 *For $n = 3$ the following assertions are equivalent:*

(i) *the strong colour-signal-cone-boundary hypothesis fails;*

(ii) *there exists $H \in \mathcal{H}(\Phi)$ such that the domain $D(H)$ is not convex;*

(iii) *for every $H \in \mathcal{H}(\Phi)$, the domain $D(H)$ is not convex;*

(iv) *there exist a triad of primaries $\delta_{\lambda_1}, \delta_{\lambda_2}, \delta_{\lambda_3}$ and a monochromatic light δ_λ such that*

$$\delta_\lambda \sim t_1(\lambda)\delta_{\lambda_1} + t_2(\lambda)\delta_{\lambda_2} + t_3(\lambda)\delta_{\lambda_3}, \tag{4.4}$$

and the values at λ of all three colour matching functions, t_1, t_2, and t_3, are strictly positive;

(v) *there exist a triad of primaries $\delta_{\lambda_1}, \delta_{\lambda_2}, \delta_{\lambda_3}$ and a monochromatic light δ_λ such that (4.4) holds true and the values at λ of exactly two colour matching functions are strictly negative.*

Note that if all coefficients $t_1(\lambda), t_2(\lambda), t_3(\lambda)$ in (4.4) are non-zero and exactly two of them, say $t_1(\lambda)$ and $t_2(\lambda)$, are negative, then $\delta_{\lambda_1}, \delta_{\lambda_2}$, and δ_λ may be taken as new primaries, and rewriting (4.4) as

$$\delta_{\lambda_3} \sim t_1'(\lambda_3)\delta_{\lambda_1} + t_2'(\lambda_3)\delta_{\lambda_2} + t'(\lambda_3)\delta_\lambda,$$

where $t_1'(\lambda_3) = \frac{|t_1(\lambda)|}{t_3(\lambda)}$, $t_2'(\lambda_3) = \frac{|t_2(\lambda)|}{t_3(\lambda)}$, $t'(\lambda_3) = \frac{1}{t_3(\lambda)}$, yields a representation with three positive coefficients. Taking into account this observation, the next result is a direct formal consequence of Proposition 4.3.

Corollary 4.2 *For $n = 3$ the following assertions are equivalent:*

(i) *the strong colour-signal-cone-boundary hypothesis holds true;*
(ii) *the domain $D(H)$ is convex whenever $H \in \mathcal{H}(\Phi)$;*
(iii) *the domain $D(H)$ is convex for some $H \in \mathcal{H}(\Phi)$;*
(iv) *for each triad of primaries $\delta_{\lambda_1}, \delta_{\lambda_2}, \delta_{\lambda_3}$ the following condition is satisfied:*
For every monochromatic light δ_{λ}, (4.4) holds true where: (i) at most one of the coefficients t_1, t_2, t_3 can be negative, and (ii) if one of them equals zero then two others are non-negative. If for a given λ all three coefficients, $t_1(\lambda)$, $t_2(\lambda)$, and $t_3(\lambda)$, are non-zero, then exactly one of them is negative.

Here is an example where $D(H)$ is explicitly evaluated and the strong colour-signal-cone-boundary hypothesis is satisfied.

Example 4.1 *Let $n = 3$, $\Phi = (\varphi_1, \varphi_2, \varphi_3)$, $\varphi_i(\lambda) = p(\lambda)e^{-(\lambda-\lambda_i)^2}$ $(i = 1, 2, 3)$, where p is a positive continuous function on Λ, $\lambda_{\min} < \lambda_1 < \lambda_2 < \lambda_3 < \lambda_{\max}$ (see Fig. 4.5). Clearly, the plane $H = \{z = (z_1, z_2, z_3) : z_3 = 1\}$ belongs to the class $\mathcal{H}(\Phi)$. We have $g_H(z) = (\frac{z_1}{z_3}, \frac{z_2}{z_3}, 1)$ whenever $z \in \Phi(\mathbf{C}_+^*) \setminus \{0\}$; therefore,*

$$g_H(\Phi(\Delta)) = \left\{ \left(\frac{\varphi_1(\lambda)}{\varphi_3(\lambda)}, \frac{\varphi_2(\lambda)}{\varphi_3(\lambda)}, 1 \right) : \lambda \in \Lambda \right\}$$
$$= \left\{ \left(e^{-2\lambda(\lambda_3-\lambda_1)+\lambda_3^2-\lambda_1^2}, e^{-2\lambda(\lambda_3-\lambda_2)+\lambda_3^2-\lambda_2^2}, 1 \right) : \lambda \in \Lambda \right\},$$

and by identifying H with the plane $x = z_1$, $y = z_2$, we get a parameter representation of the curve $g_H(\Phi(\Delta))$ as

$$x = e^{-2\lambda(\lambda_3-\lambda_1)+\lambda_3^2-\lambda_1^2}, \quad y = e^{-2\lambda(\lambda_3-\lambda_2)+\lambda_3^2-\lambda_2^2}, \tag{4.5}$$

where the parameter λ varies from $\lambda_{\min}$ to $\lambda_{\max}$. We derive from (4.5) that

$$y = e^{\alpha \ln x + (1-\alpha)\beta}, \tag{4.6}$$

where

$$\alpha = \frac{\lambda_3 - \lambda_2}{\lambda_3 - \lambda_1}, \quad \beta = \lambda_3^2 - \lambda_1^2, \tag{4.7}$$

x runs from $x_{\min} = e^{-2\lambda_{\max}(\lambda_3-\lambda_1)+\lambda_3^2-\lambda_1^2}$ to $x_{\max} = e^{-2\lambda_{\min}(\lambda_3-\lambda_1)+\lambda_3^2-\lambda_1^2}$. Clearly, $0 < \alpha < 1$, $\beta > 0$. A simple calculation using (4.6), (4.7) yields

$$\frac{d^2y}{dx^2} = \frac{(\alpha^2 - \alpha)e^{\alpha \ln x + (1-\alpha)\beta}}{x^2} < 0,$$

therefore, function ((4.6)) on $[x_{\min}, x_{\max}]$ is strictly concave. Now, since the curve $g_H(\Phi(\Delta))$ is the graph of this function, and the domain D comprises points between it and the segment $g_H(I(\delta_{\lambda_{\min}}, \delta_{\lambda_{\max}}))$ joining the points $(x_{\min}, y(x_{\min}))$ and $(x_{\max}, y(x_{\max}))$ (i.e., $D = \{(x, y) : x = x_t, y_t \le y \le y(x_t), 0 \le t \le 1\}$, where $x_t := (1-t)x_{\min} + tx_{\max}$, $y_t := (1-t)y(x_{\min}) + ty(x_{\max})$), we see that D is convex, and the strong colour-signal-cone-boundary hypothesis is thus satisfied.

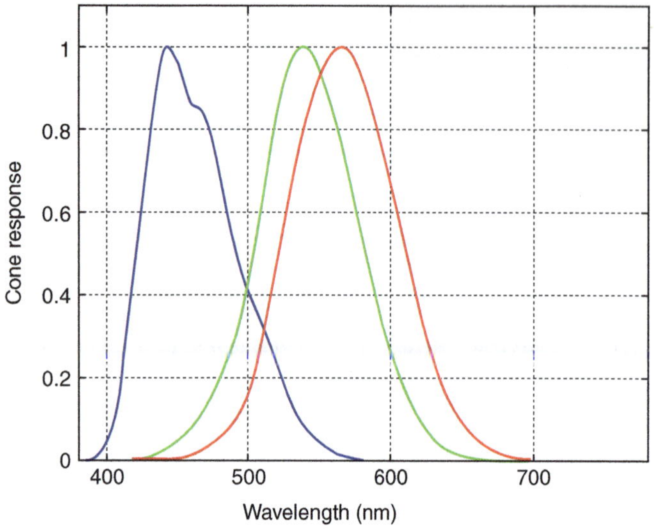

Figure 4.5 Gaussian cone fundamentals. They are obtained by replacing $g(\lambda; \lambda^{\max})$ in Eq. (3.18) by $\exp\left(-\left(\left(\lambda - \lambda_i\right)/80\right)^2\right)$ with $\lambda_i = 430$ nm (blue line), 530 nm (green line), and 560 nm (red line).

Remark 4.3 *Although the functions φ_i, $i = 1, 2, 3$, from Example 4.1 could be used as plausible photoreceptor models (compare Fig. 4.5 with Fig. 3.8 or Fig. 3.10) (D. I. A. MacLeod & Golz, 2003), their form is however too specific, and a little change of it may disturb convexity of the corresponding D thus violate the strong colour-signal-cone-boundary hypothesis. Indeed, as shown below, for Govardovskii's fundamentals – the most physiologically plausible model of the human cone spectral sensitivities – such a violation takes place. It is noticeable, though not large. To evaluate it, the Lebesgue measure of the set*

$$\{\lambda \in \Lambda : \Phi(\delta_\lambda) \text{ is an interior point of the colour-signal cone } \Phi(\mathbf{C}_+^*)\}$$

may be used.

4.2 Empirical status of the strong colour-signal-cone-boundary hypothesis

Given $H \in \mathcal{H}(\Phi)$, we consider a linear operator $Q : \mathbf{R}^3 \to \mathbf{R}^2$ such that its restriction to H is a one-to-one map of H *onto* the whole of $\mathbf{R}^2$. Then the Q-image of $D(H)$, $Q(D(H))$, is a closed domain in $\mathbf{R}^2$, which can be considered as a representation of $D(H)$. Since the domain $Q(D(H))$ is convex if and only if $D(H)$ is such, we derive from Corollary 4.2 that the strong colour-signal-cone-boundary hypothesis holds true if and only if the set $Q(D(H))$ is convex for any (for all) given $H \in \mathcal{H}(\Phi)$ and Q.

Given a 3-pigment Grassmann–Govardovskii structure (Definition 3.4), consider $H = \{(x, y, z) : x + y + z = 1\}$. Using the conventional notation, S, M, and L, for the coordinates in the corresponding SML colour-signal space, we have $g_H(S, M, L) = (S/(S + M + L), M/(S + M + L), L/(S + M + L))$. Let us consider $Q(x, y, z) = (z, y)$. In Fig. 4.6. the Q-image of $g_H(\Phi(\Delta))$ is plotted as a black line. The Q-image of the segment $g_H(I(\delta_{\lambda_{\min}}, \delta_{\lambda_{\max}}))$ is plotted as a purple line. The closed curve made by the black and purple lines in Fig. 4.6 represents $Q\left(g_H(\Phi(\Delta'))\right) = Q(g_H(\Phi(\Delta \cup I(\delta_{\lambda_{\min}}, \delta_{\lambda_{\max}}))))$. The area confined by this curve is the Q-image of $D(H)$. Therefore, as one can see in Fig. 4.6,

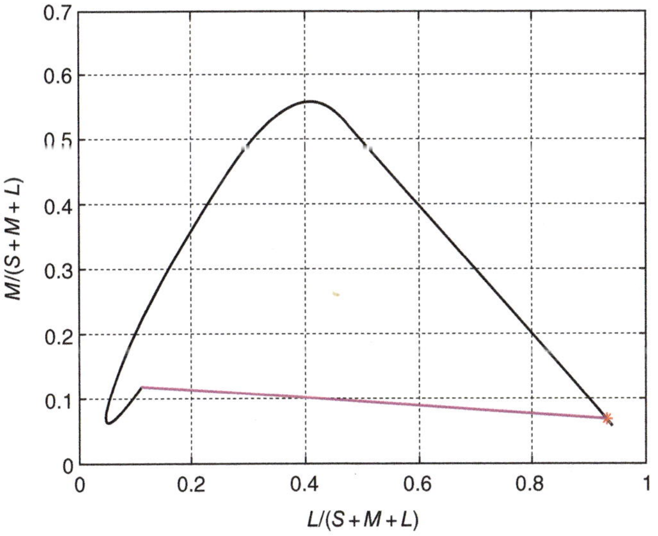

Figure 4.6 Spectral locus, i.e., the Q-image of $g_H(\Phi(\Delta))$ (black line), plotted for Govardovskii's fundamentals (Fig. 3.8). The purple line is the Q-representation of $g_H(\Phi(I(\delta_{\lambda_{min}}, \delta_{\lambda_{max}})))$, i.e., a line segment connecting the ends of the spectral locus.

the domain $Q(D(H))$ is not convex. Thus, it follows from Proposition 4.3 that the strong colour-signal-cone-boundary hypothesis fails for the 3-pigment Grassmann–Govardovskii structure (Definition 3.4).

We will refer to the Q-image of the plane H as the *chromaticity plane.*[5] The Q-image of the intersection of H with the colour-signal cone, $Q\big(H \cap \Phi(\mathbf{C}_+^*)\big)$, will be referred to as the *chromaticity gamut.*[6] The Q-image of the curve $g_H(\Phi(\Delta))$ will be referred to as the *spectral locus.* An obvious manifestation of the violation of the strong colour-signal-cone-boundary hypothesis is that the spectral locus clearly bends into the chromaticity gamut, forming an apparent 'beak' at its short-wavelength end (Fig. 4.7). Furthermore, while being disguised, there is also a beak at the opposite (long-wavelength) end of the spectral locus. It is hard to see because the long-wavelength part of the spectral locus seems to overlap itself. Note that the point corresponding to the long-wavelength end of the visible spectrum wavelength interval, $\lambda_{max} = 780\,\text{nm}$, (see the red asterisk in Figs 4.6 and 4.7) does not coincide with the apparent end of the curve $Q\big(g_H(\Phi(\Delta))\big)$. The latter corresponds to the wavelength of approximately 700 nm. This is the point at which the spectral locus makes a U-turn and starts moving backward. As a result, a line segment connecting the ends of the spectral locus (i.e., the Q-representation of $g_H(\Phi(I(\delta_{\lambda_{min}}, \delta_{\lambda_{max}})))$; it is usually referred to as the *purple interval*) in Figs 4.6 and 4.7 turns out to be inside the chromaticity gamut.

Self-overlapping of the long-wavelength end of the spectral locus results from the non-monotonicity of the L/M cone response ratio with wavelength (Fig. 4.8). As can be seen in Fig. 4.8,

5 The rationale for such terminology is that the lights differing from each other only by intensity are represented in $Q(H)$ by the same point. It is taken for granted that such lights appear different mainly in brightness, their hue and saturation being approximately the same. As it is hue and saturation that the term chromaticity refers to, the lights of the same chromaticity are represented in $Q(H)$ approximately by the same point (see more on chromaticity and the other colour appearance terminology in Chapter 8).

6 The chromaticity plane with the chromaticity gamut in it will be referred to as the *chromaticity diagram.* Such terminology is by and large in line with that established in colour vision literature (e.g., Kaiser & Boynton, 1996; Wyszecki & Stiles, 1982).

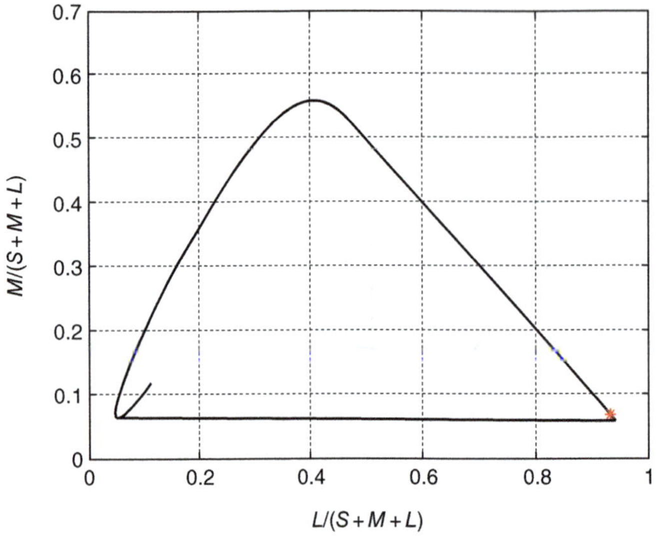

Figure 4.7 Chromaticity gamut (diagram) for Govardovskii's fundamentals (Fig. 3.8), i.e., convex closure of the spectral locus (the black line in Fig. 4.6).

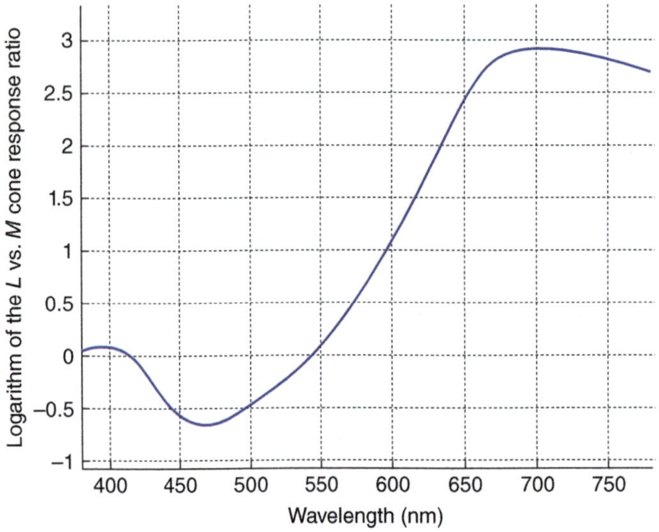

Figure 4.8 *L*- and *M*-cone response ratio as a function of wavelength, as calculated from the cone sensitivities based on Govardovskii's model of the spectral photopigment absorbances depicted in Fig. 3.5.

this ratio reaches a maximum at wavelength of 700 nm. For a given monochromatic light, its location in the chromaticity gamut in Fig. 4.7 is determined by the ratio of the *L*- and *M*-cone responses to it (rather than their absolute values). This leads to an overlap in the spectral locus occurring with the point of return being very near the wavelength at which the L/M cone response ratio curve shown in Fig. 4.8 reaches its maximum.

Interestingly, when the spectral locus is plotted by using the cone fundamentals of Stockman & Sharpe (2000) (Fig. 4.9) the long-wavelength overlap is present (see the red asterisk in Fig. 4.9) but the short-wavelength 'beak' is much less noticeable. Moreover, it is nearly inconspicuous in

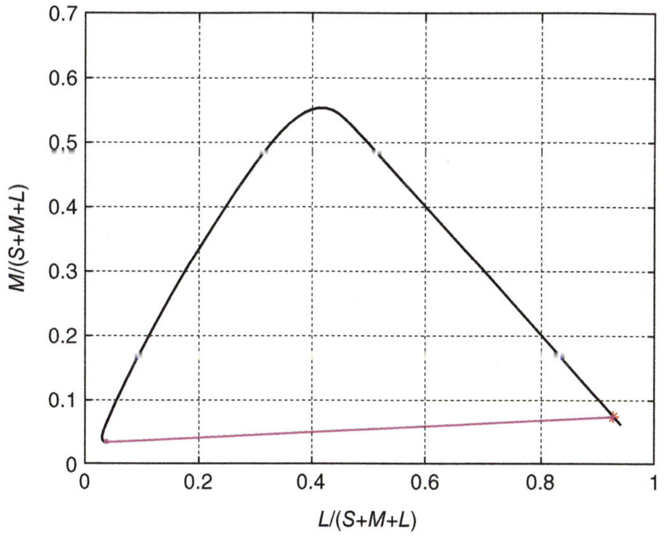

Figure 4.9 Spectral locus (black line) as derived from the Stockman & Sharpe (2000) cone fundamentals. The purple line is the line segment connecting the ends of the spectral locus.

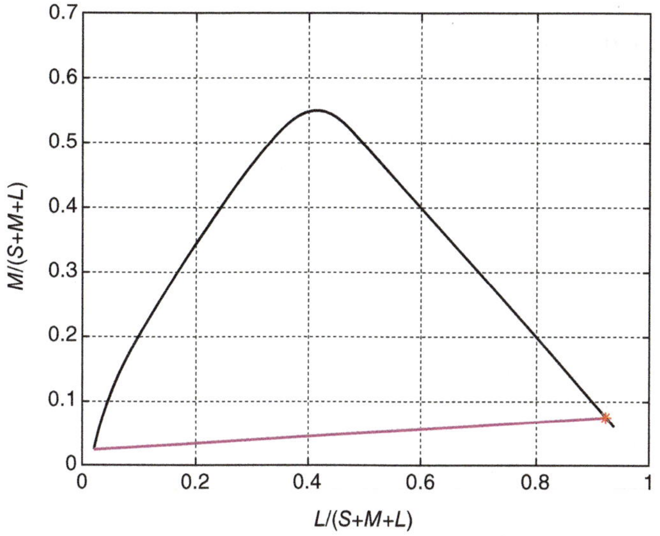

Figure 4.10 Spectral locus (black line) and purple interval (purple line) as derived from the Smith and Pokorny cone fundamentals (Smith & Pokorny, 1975), scaled to unity peak. Source: Based on Smith and Pokorny (1975).

Fig. 4.10, which is based on the spectral sensitivity functions derived from the CIE 1931 colour matching functions as amended by Judd and Vos (Smith & Pokorny, 1975). This is hardly surprising if one remembers that the original colour matching data on which the CIE 1931 colour matching functions were based were not collected for monochromatic lights of wavelengths shorter than 400 nm. Moreover, as mentioned earlier, the original colour matching data have undergone substantial amendment, extrapolation, interpolation and smoothing to become the CIE 1931 colour

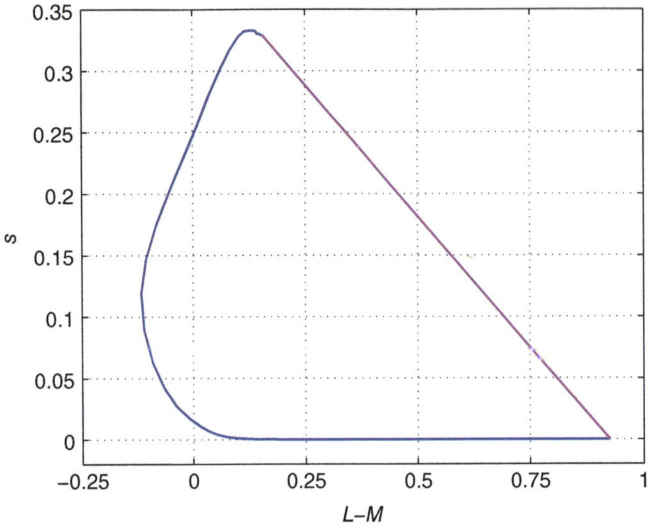

Figure 4.11 MacLeod–Boynton chromaticity diagram. The abscissa is $\frac{L-M}{S+M+L}$, where S, M, and L are the values of the Smith and Pokorny cone fundamentals. The ordinate is $\frac{S}{S+M+L}$. The blue curved line represents the Q-image of $g_H(\Phi(\Delta))$, the purple straight line is the Q-representation of $g_H(\Phi(I(\delta_{\lambda_{\min}}, \delta_{\lambda_{\max}})))$.

matching functions in their current form.[7] Not surprisingly, these functions turn out to be rather different from Govardovskii's cone fundamentals based on the cone photopigment absorptance spectra.

Different choice of linear operator Q brings about different chromaticity diagrams. In Fig. 4.11, the MacLeod–Boynton chromaticity diagram (D. I. A. MacLeod & Boynton, 1979; Boynton, 1996) based on the Smith and Pokorny cone fundamentals (Fig. 3.10) is presented. Here, $H = \{(x, y, z) : x + y + z = 1\}$ (hence, $g_H(S, M, L) = (S/(S + M + L), M/(S + M + L), L/(S + M + L)))$, and $Q(x, y, z) = (z - y, x)$. Once again, the domain $Q(D(H))$ seems to be convex, erroneously implying the validity of the strong colour-signal-cone-boundary hypothesis. This impression is even more compelling in Fig. 4.12 presenting the CIE 1931 chromaticity diagram (Wyszecki & Stiles, 1982), in which $H = \{(x, y, z) : x + y + z = 1\}$ (hence, $g_H(X, Y, Z) = (X/(X + Y + Z), Y/(X + Y + Z), Z/(X + Y + Z)))$, and $Q(x, y, z) = (x, y)$. As one can see in Fig. 4.12, the short-wavelength end 'beak' practically disappears.[8]

All these discrepancies between the spectral locus forms on different chromaticity diagrams should not prevent us from drawing the right conclusion. Indeed, as follows from Proposition 4.3, in order to show that the strong colour-signal-cone-boundary hypothesis fails it suffices to find at

7 As Wright explains, 'The C.I.E. Colorimetry Committee recently in their wisdom have been looking at the old 1931 observer and have been smoothing the data to obtain more consistent calculations with computers. This also involved some extrapolation and, in smoothing, they have added some additional decimal places. When I look at the revised table of the $\bar{x}, \bar{y}, \bar{z}$, functions, I am rather surprised to say the least. You see, I know how inaccurate the actual measurements really were. (Laughter) Guild did not take any observations below 400 nm and neither did I, and neither did Gibson and Tyndall on the V_λ curve, and yet at a wavelength of 362 nm, for example, we find a value $\bar{y}$ of.000004929604!' (Wright, 1996, pp. 541–542).

8 True, the choice of the linear operator Q can affect the pronouncedness of the 'beaks', but cannot lead to their complete disappearance.

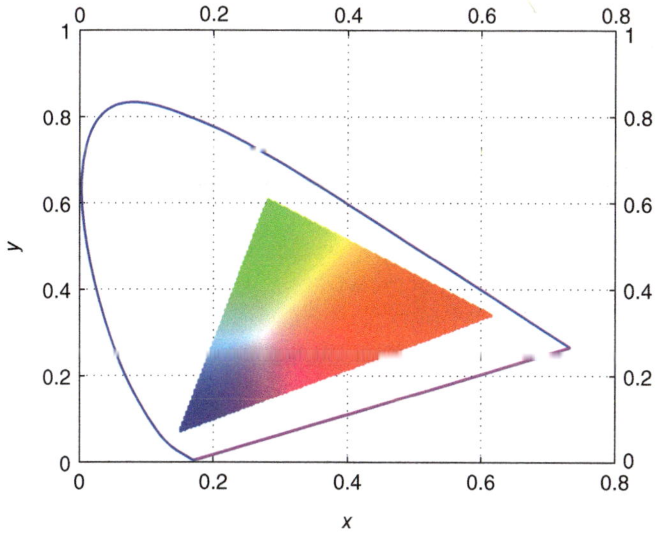

Figure 4.12 The CIE 1931 chromaticity diagram. The abscissa $x = \frac{X}{X+Y+Z}$, where X, Y, and Z are the tristimulus values. The ordinate $y = \frac{Y}{X+Y+Z}$. The coloured triangle inside represents those chromaticities that are within the colour gamut of the monitor of one of our laboratory computers. The rest of chromaticities cannot be rendered by the monitor. The blue curved line stands for the Q-image of $g_H(\Phi(\Delta))$, the purple straight line for the Q-image of $g_H(\Phi(I(\delta_{\lambda_{min}}, \delta_{\lambda_{max}})))$.

least one $H \in \mathcal{H}(\Phi)$ such that the domain $D(H)$ is clearly not convex.[9] For instance, the choise of H and Q leading to the chromaticity diagram in Figs 4.6 and 4.7 clearly testifies against the strong colour-signal-cone-boundary hypothesis. Note also that, being not affected by the pre-receptoral filters, the spectral locus in this figure depends only on the photopigment absorptance spectra. A direct calculation shows that the domain $D(H)$ remains not convex when the peak sensitivity of all the three cone photopigments varies within the range revealed in the genetic studies (J. Neitz & Neitz, 2011). Hence, we conclude that the strong colour-signal-cone-boundary hypothesis does not hold true for human trichromats in general.

4.3 Colour-signal-cone-boundary hypothesis

Note that, as can be seen in Fig. 4.6, the hypothesis would seem to hold true if the wavelength interval of the visible spectrum $[\lambda_{min}, \lambda_{max}]$ is shortened by replacing λ_{min} with some larger value λ'_{min}, and λ_{max} with some smaller value λ'_{max}. Indeed, it looks as if the spectral colour-signal curve belongs to the colour-signal cone boundary except for two short fragments at the short and long wavelength ends, which correspond to the segments $[\lambda_{min}, \lambda'_{min}]$ and $[\lambda'_{max}, \lambda_{max}]$. These fragments (i.e., $\Phi(I(\delta_{\lambda_{min}}, \delta_{\lambda'_{min}}))$ and $\Phi(I(\delta_{\lambda'_{max}}, \delta_{\lambda_{max}}))$) come into the colour-signal cone interior. This justifies the following generalisation of the strong colour-signal-cone-boundary hypothesis.

Definition 4.2 *We say the* colour-signal-cone-boundary hypothesis *holds true if there are* λ'_{min} *and* λ'_{max} *such that* $\lambda_{min} \leq \lambda'_{min} < \lambda'_{max} \leq \lambda_{max}$, *a fragment of the spectral colour-signal curve* $\{\Phi(\delta_\lambda)$:

9 As stated above, in this case it is not convex for all other H in $\mathcal{H}(\Phi)$, even if it is not quite visible on the chromaticity diagram.

$\lambda'_{\min} \le \lambda \le \lambda'_{\max}\}$ belongs to the colour-signal cone boundary, and if $\lambda \in [\lambda_{\min}, \lambda'_{\min}) \cup (\lambda'_{\max}, \lambda_{\max}]$ and $\Phi(\delta_\lambda)$ belongs to the colour-signal cone boundary, then there is t, $0 < t < 1$, such that $\Phi(\delta_\lambda) = t\Phi(\delta_{\lambda'_{\min}}) + (1-t)\Phi(\delta_{\lambda'_{\max}})$.

Thus, if the colour-signal-cone-boundary hypothesis holds true then, for every $\lambda \in \Lambda \setminus \Lambda'$, where Λ' as in Eq. (4.8), $\Phi(\delta_\lambda)$ is an interior point of the colour-signal cone or belongs to the segment $I(\Phi(\delta_{\lambda'_{\min}}), \Phi(\delta_{\lambda'_{\max}}))$.

An equivalent formulation of the colour-signal-cone-boundary hypothesis says that the boundary of the colour-signal cone is exactly the conical envelope of the set $\Phi(\Delta(\Lambda')) \cup I(\Phi(\delta_{\lambda'_{\min}}), \Phi(\delta_{\lambda'_{\max}}))$, where

$$\Lambda' = \{\lambda : \lambda'_{\min} \le \lambda \le \lambda'_{\max}\}, \tag{4.8}$$

$$\Delta(\Lambda') = \{\delta_\lambda : \lambda \in \Lambda'\}. \tag{4.9}$$

Let us consider the close curve

$$\Delta'(\Lambda') = \Delta(\Lambda') \cup I(\delta_{\lambda'_{\min}}, \delta_{\lambda'_{\max}}). \tag{4.10}$$

One more equivalent formulating the colour-signal-cone-boundary hypothesis is that, for some (for every) two-dimensional plane $H \in \mathcal{H}(\Phi)$, the closed domain in it, $D'(H)$, limited by the close curve $g_H(\Phi(\Delta'(\Lambda')))$, is exactly $H \cap \Phi(\mathbf{C}^*_+)$. (Therefore, the colour-signal-cone-boundary hypothesis holds true if and only if $D'(H)$ coincides with the convex hull of $g_H(\Phi(\Delta(\Lambda')))$; cf. Corollary 4.2.)

Also, as in the case of the strong hypothesis, one can consider a linear operator $Q : \mathbf{R}^3 \to \mathbf{R}^2$ such that its restriction to H is a one-to-one map of H onto the whole of $\mathbf{R}^2$, and since the domain $Q(D'(H))$ is convex if and only if $D'(H)$ is such, we conclude that the colour-signal-cone-boundary hypothesis holds true if and only if the set $Q(D'(H))$ is convex for any (for all) given $H \in \mathcal{H}(\Phi)$ and Q.

Note also that if the colour-signal-cone-boundary hypothesis is satisfied for a given Φ then it is satisfied for any other Φ' consistent with the considered Grassmann structure. The validity of the colour-signal-cone-boundary hypothesis can be treated thus as a *convexity* property of the corresponding Grassmann structure rather than concrete colour-signal cone.

We now consider some kind of a stronger *strict convexity* property. Suppose that the colour-signal-cone-boundary hypothesis holds true. Then the conical envelope of the set $I(\Phi(\delta_{\lambda'_{\min}}), \Phi(\delta_{\lambda'_{\max}}))$ is a plane fragment of the colour-signal cone boundary. A question arises, whether the boundary of $\Phi(\mathbf{C}^*_+)$ can contain another plane fragment. Taking into account Definition 4.2, this question can be equivalently reformulated as follows: Are there λ_1, λ_2, $\lambda'_{\min} < \lambda_1 < \lambda_2 < \lambda'_{\max}$, such that the whole linear segment $I(\Phi(\delta_{\lambda_1}), \Phi(\delta_{\lambda_2}))$ is contained in the boundary of $\Phi(\mathbf{C}^*_+)$, hence coincides with the fragment $\{\Phi(\delta_\lambda) : \lambda \in [\lambda_1, \lambda_2]\}$ of the spectral colour-signal curve? (In such a case, the conical envelope of that segment proves to be a flat fragment of the colour-signal cone boundary.) One more equivalent formulation of the above question is as follows: Given a plane $H \in \mathcal{H}(\Phi)$ and λ_1, λ_2 as above, does a fragment of the planar curve $\{g_H(\Phi(\delta_\lambda)) : \lambda_1 \le \lambda \le \lambda_2\}$ coincide with the linear segment $I(g_H(\Phi(\delta_{\lambda_1})), g_H(\Phi(\delta_{\lambda_2})))$ joining $g_H(\Phi(\delta_{\lambda_1}))$ and $g_H(\Phi(\delta_{\lambda_2}))$? The next proposition (4.4) answers this question.

Proposition 4.4 *Assume the colour-signal-cone-boundary hypothesis holds. Then, given $H \in \mathcal{H}(\Phi)$ and $\lambda'_{\min} < \lambda_1 < \lambda_2 < \lambda'_{\max}$, the following assertions are equivalent:*

(i) *the g_H-image of a spectral colour-signal curve fragment between λ_1 and λ_2 is a linear segment: $\{g_H(\Phi(\delta_\lambda)) : \lambda_1 \le \lambda \le \lambda_2\} = I(g_H(\Phi(\delta_{\lambda_1})), g_H(\Phi(\delta_{\lambda_2})))$;*

(ii) the fragment of the conical surface $\{t\Phi(\delta_\lambda) : t \geq 0, \lambda_1 \leq \lambda \leq \lambda_2\}$ *is flat;*
(iii) there exists at least one $\lambda' \in \Lambda'$ *such that* $\delta_{\lambda'} \sim t_1\delta_{\lambda_1} + t_2\delta_{\lambda_2}$ *where* $t_1 > 0, t_2 > 0$.

Definition 4.3 *We say the Grassmann structure* $\langle \mathbf{C}_+^*, \sim \rangle$ *has the* strict convexity property *if the colour-signal-cone-boundary hypothesis holds true but assertion (iii) of Proposition 4.4 fails.*

Remark 4.4 *Proposition 4.4 remains true when the map* g_H *in (i) is replaced by* $Q \circ g_H$ *where* $Q : \mathbf{R}^3 \to \mathbf{R}^2$ *is a linear operator such that its restriction to* H *is a one-to-one map of* H *onto* $\mathbf{R}^2$.

The next result is a direct consequence of Proposition 4.4.

Corollary 4.3 *The Grassmann structure* $\langle \mathbf{C}_+^*, \sim \rangle$ *has the strict convexity property if and only if the colour-signal-cone-boundary hypothesis holds true but no monochromatic light with the wavelength* $\lambda \in \Lambda'$ *can be metameric to a sum of two other monochromatic lights with wavelengths* $\lambda_1 \in \Lambda'$ *and* $\lambda_2 \in \Lambda'$.[10]

In the following example, we impose on the cone fundamentals some condition that entails the fulfillment of the assertions in Proposition 4.4.

Example 4.2 *Suppose there are* $\lambda_1, \lambda_2 \in \Lambda'$, $\lambda_1 < \lambda_2$, *such that* $\varphi_3(\lambda) = 0$ *whenever* $\lambda \geq \lambda_1$ *and*

$$\frac{\varphi_1(\lambda_1)}{\varphi_1(\lambda_1) + \varphi_2(\lambda_1)} \leq \frac{\varphi_1(\lambda)}{\varphi_1(\lambda) + \varphi_2(\lambda)} \leq \frac{\varphi_1(\lambda_2)}{\varphi_1(\lambda_2) + \varphi_2(\lambda_2)} \tag{4.11}$$

whenever $\lambda_1 \leq \lambda \leq \lambda_2$. *We take* $H = \{z = (z_1, z_2, z_3) : z_1 + z_2 + z_3 = 1\}$, *hence*

$$g_H(z) = \left(\frac{z_1}{z_1 + z_2 + z_3}, \frac{z_2}{z_1 + z_2 + z_3}, \frac{z_3}{z_1 + z_2 + z_3} \right)$$

whenever $z \in \mathbf{R}_+^3 \setminus \{0\}$, *and consider* $Q : \mathbf{R}^3 \to \mathbf{R}^2$, $Q(z_1, z_2, z_3) = (z_1, z_3)$. *Then for every* λ, $\lambda_1 \leq \lambda \leq \lambda_2$ *one has*

$$Q(g_H(\Phi(\delta_\lambda))) = \left(\frac{\varphi_1(\lambda)}{\varphi_1(\lambda) + \varphi_2(\lambda)}, 0 \right),$$

and by taking into account (4.11), we see that

$$\{Q(g_H(\Phi(\delta_\lambda))) : \lambda_1 \leq \lambda \leq \lambda_2\} = I(Q(g_H(\Phi(\delta_{\lambda_1}))), Q(g_H(\Phi(\delta_{\lambda_2})))).$$

Now, we derive from here, along with Remark 4.4, that all the assertions of Proposition 4.4 hold true.

Remark 4.5 *It can be of interest to compare Example 4.2 with some chromaticity diagrams as Fig. 4.11 based on Smith and Pokorny's cone fundamentals (see Fig. 3.10) or Fig. 4.6 based on Govardovskii's templates (see Figs 3.5 and 3.8). Though, as follows from (3.16), Govardovskii's template* $g_i(\lambda; \mu_i)$ *actually vanishes at no* λ*; therefore, it can occur that a fragment of the corresponding curve in the right-hand part of Fig. 4.6 only seems rectilinear but in fact is not such.*[11] *In such a case, the strict convexity property can be satisfied.*

10 The latter means that no non-negative solution to the following system of three equations in two unknowns x_1, x_2 can exist: $\varphi_i(\lambda) = x_1\varphi_i(\lambda_1) + x_2\varphi_i(\lambda_2)$, $i = 1, 2, 3$.
11 In that connection, see also Fig. 3.10 and a rectilinear fragment of the blue curved line in the bottom of Fig. 4.11.

Remark 4.6 *Suppose the colour-signal-cone-boundary hypothesis holds true; then the colour-signal cone boundary,* $bdry(\Phi(\mathbf{C}_+^*))$*, is a conical surface generated by rays passing through points of* $\Phi(\Delta'(\Lambda'))$*. We are interested whether this surface can be smooth at all its points except the origin. (Obviously, it cannot be smooth at the origin.)*

Clearly the set $S := bdry(\Phi(\mathbf{C}_+^*)) \setminus \{0\}$ *is smooth if and only if its intersection with every plane* $H \in \mathcal{H}(\Phi)$ *is a smooth close curve. Recall that any such* H *has a representation*

$$H = \{z : \langle c, z\rangle = 1\},$$

where $c = (c_1, c_2, c_3) \in \mathbf{R}_+^3 \setminus \{0\}$ *and the intersection* $H \cap S$ *is the* g_H*-image of the close curve* $\Phi(\Delta'(\Lambda'))$*, i.e., it is the close curve* $t \mapsto z(t) = (z_1(t), z_2(t), z_3(t))$*,* $\lambda'_{\min} \le t \le \lambda'_{\max} + 1$*, where*

$$z_i(t) = A_i(t) := \frac{\varphi_i(\lambda)}{c_1\varphi_1(\lambda) + c_2\varphi_2(\lambda) + c_3\varphi_3(\lambda)}$$

for $t = \lambda \in [\lambda'_{\min}, \lambda'_{\max}]$*, and*

$$z_i(t) = B_i(t) := \frac{(1 + \lambda'_{\max} - t)\varphi_i(\lambda'_{\max}) + (t - \lambda'_{\max})\varphi_i(\lambda'_{\min})}{(1 + \lambda'_{\max} - t)\sum_1^3 c_j\varphi_j(\lambda'_{\max}) + (t - \lambda'_{\max})\sum_1^3 c_j\varphi_j(\lambda'_{\min})}$$

for $t \in [\lambda'_{\max}, \lambda'_{\max} + 1]$*. Clearly* $A_i(\lambda'_{\min}) = B_i(\lambda'_{\max} + 1)$ *and* $A_i(\lambda'_{\max}) = B_i(\lambda'_{\max})$*,* $i = 1, 2, 3$*, hence* $z(t)$ *actually runs a close curve. This curve is smooth if and only if*

$$\frac{dA_i(\lambda'_{\min})}{dt} = \frac{dB_i(\lambda'_{\max} + 1)}{dt} \quad \text{and} \quad \frac{dA_i(\lambda'_{\max})}{dt} = \frac{dB_i(\lambda'_{\max})}{dt} \tag{4.12}$$

for all $i = 1, 2, 3$*. A simple calculation shows that equalities (4.12) may be equivalently rewritten as*

$$\begin{aligned}&\varphi'_i(\lambda'_{\min})\sum_{j=1}^3 c_j\varphi_j(\lambda'_{\min}) - \varphi_i(\lambda'_{\min})\sum_{j=1}^3 c_j\varphi'_j(\lambda'_{\min})\\ &= (\varphi_i(\lambda'_{\min}) - \varphi_i(\lambda'_{\max}))\sum_{j=1}^3 c_j\varphi_j(\lambda'_{\min})\\ &\quad - \varphi_i(\lambda'_{\min})\sum_{j=1}^3 c_j(\varphi_j(\lambda'_{\min}) - \varphi_j(\lambda'_{\max}))\end{aligned}$$

and

$$\begin{aligned}&\varphi'_i(\lambda'_{\max})\sum_{j=1}^3 c_j\varphi_j(\lambda'_{\max}) - \varphi_i(\lambda'_{\max})\sum_{j=1}^3 c_j\varphi'_j(\lambda'_{\max})\\ &= (\varphi_i(\lambda'_{\min}) - \varphi_i(\lambda'_{\max}))\sum_{j=1}^3 c_j\varphi_j(\lambda'_{\max})\\ &\quad - \varphi_i(\lambda'_{\max})\sum_{j=1}^3 c_j(\varphi_j(\lambda'_{\min}) - \varphi_j(\lambda'_{\max})),\end{aligned}$$

and as these equalities must be satisfied whenever $c = (c_1, c_2, c_3) \in \mathbf{R}_+^3 \setminus \{0\}$*, we obtain*

$$\varphi'_i(\lambda'_{\min})\varphi_j(\lambda'_{\min}) - \varphi_i(\lambda'_{\min})\varphi'_j(\lambda'_{\min}) = \varphi_i(\lambda'_{\min})\varphi_j(\lambda'_{\max}) - \varphi_i(\lambda'_{\max})\varphi_j(\lambda'_{\min}) \tag{4.13}$$

and

$$\varphi'_i(\lambda'_{\max})\varphi_j(\lambda'_{\max}) - \varphi_i(\lambda'_{\max})\varphi'_j(\lambda'_{\max}) = \varphi_i(\lambda'_{\min})\varphi_j(\lambda'_{\max}) - \varphi_i(\lambda'_{\max})\varphi_j(\lambda'_{\min}). \tag{4.14}$$

Thus, the surface S *is smooth if and only if conditions (4.13) and (4.14) are satisfied whenever* $i, j \in \{1, 2, 3\}$*.*

4.4 The colour-signal cone of a 3-pigment Grassmann–Govardovskii structure

As shown above (Fig. 4.6), a 3-pigment Grassmann–Govardovskii structure does not meet the strong colour-signal-cone-boundary hypothesis. Yet, as follows from the shape of the spectral locus in Fig. 4.6, there is every indication that it might meet the colour-signal-cone-boundary hypothesis (Definition 4.2). Specifically, there appear to be wavelengths $\lambda'_{\min}$ and $\lambda'_{\max}$ such that (i) a line through the spectral locus points corresponding to these wavelengths intersects the spectral locus at these two points; and (ii) the spectral locus lies fully on one side of this line. If this is the case then this line is supportive to the chromaticity gamut. Therefore, if there exist such wavelengths $\lambda'_{\min}$ and $\lambda'_{\max}$ that the line through the points corresponding to these wavelengths is supportive to the chromaticity gamut, then the colour-signal-cone-boundary hypothesis is satisfied; and it is an important task to ascertain these wavelengths $\lambda'_{\min}$ and $\lambda'_{\max}$. Following the terminology put forward elsewhere (Logvinenko, 2015b), the wavelength interval determined by $\lambda'_{\min}$ and $\lambda'_{\max}$ (i.e., Λ' in Eq. (4.8)) will be referred to as the *effective visible spectrum interval.*

The smoothness of the spectral locus entails that the desired supportive line, if any, is tangent to the spectral locus at two points. It follows that if there are $\lambda'_{\min}$ and $\lambda'_{\max}$ securing the colour-signal-cone-boundary hypothesis then along with condition (4.12) the line tangent to the spectral locus at the point corresponding to $\lambda'_{\min}$ is at the same time the line tangent to the spectral locus at the point corresponding to $\lambda'_{\max}$. It turns out that there is just one such line of double tangency.

More specifically, consider the spectral locus $c(\lambda) = (m(\lambda), l(\lambda))$, where m and l are given by

$$m(\lambda) = M(\lambda) / (S(\lambda) + M(\lambda) + L(\lambda)),$$
$$l(\lambda) = L(\lambda) / (S(\lambda) + M(\lambda) + L(\lambda)).$$

As before, $S(\lambda)$, $M(\lambda)$, and $L(\lambda)$ denote here the spectral sensitivity functions of the short (S), middle (M), and long (L) wavelength sensitive cones. Evidently, $m(\lambda)$ and $l(\lambda)$ do not depend on the ocular media. It follows that instead of using the spectral sensitivity functions in Eq. (3.18) one can use the spectral absorptance functions in Eq. (3.17) to evaluate them. As the Govardovskii model involves only smooth functions, the spectral curve as well as the spectral locus are smooth too.

The equation for the tangent line to the spectral locus at a point $c(\lambda_0)$ is given as

$$\frac{dm}{d\lambda}(\lambda_0)\,(l - l(\lambda_0)) = \frac{dl}{d\lambda}(\lambda_0)\,(m - m(\lambda_0)). \tag{4.15}$$

If a line is tangent to the spectral locus at two points $c(\lambda_1)$ and $c(\lambda_2)$ then it must satisfy the following two equations:

$$\frac{dm}{d\lambda}(\lambda_1)\,(l(\lambda_1) - l(\lambda_2)) = \frac{dl}{d\lambda}(\lambda_1)\,(m(\lambda_1) - m(\lambda_2)), \tag{4.16}$$
$$\frac{dm}{d\lambda}(\lambda_2)\,(l(\lambda_1) - l(\lambda_2)) = \frac{dl}{d\lambda}(\lambda_2)\,(m(\lambda_1) - m(\lambda_2)).$$

As a matter of fact, for the Govardovskii model, there is only one (unordered) pair of wavelengths (λ_1, λ_2) satisfying simultaneous equations (4.16). The particular value of λ_1 and λ_2 depends on the model parameters' values. In particular, given the parameters as in Eq. (3.16), λ_1 and λ_2 were evaluated to be 420 and 701 nm for the peak absorbance wavelengths of 430, 530, and 560 nm.

To make sure that the spectral locus lies fully on one side of the tangent line determined by Eq. (4.16), note that, given a point $c(\lambda^*)$ on the spectral locus, the sign of the determinant

$$\det\begin{pmatrix} m(\lambda^*) - m(\lambda_0) & l(\lambda^*) - l(\lambda_0) \\ \frac{dm}{d\lambda}(\lambda_0) & \frac{dl}{d\lambda}(\lambda_0) \end{pmatrix} \tag{4.17}$$

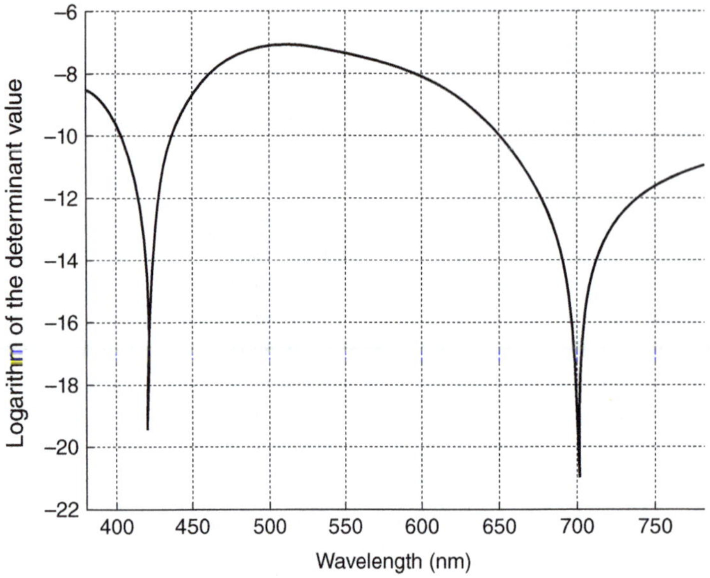

Figure 4.13 Logarithm of the determinant (4.17) as a function of wavelength λ^*.

indicates in which of the two half-planes (into which the tangent line (4.15) divides the chromaticity plane) $c(\lambda^*)$ lies. Direct calculations show that when $\lambda_0 = 420$ (or 701) nm the determinant (4.17) does not change sign when λ^* runs over the visible spectrum wavelength interval $[\lambda_{\min}, \lambda_{\max}]$. Figure 4.13 shows that it reaches zero at exactly two points: 420 and 701 nm. This means that the spectral locus remains on one side of the tangent line (4.15), that is, the line (4.15) does not intersect the spectral locus apart from the points of tangency at 420 and 701 nm.

Admittedly, the wavelengths $\lambda'_{\min}$ and $\lambda'_{\max}$ take different values depending on the photopigment peak absorbance wavelength values (i.e., $\lambda^{\max}$). For $\lambda_M^{\max} = 530 \pm 10$ nm and $\lambda_L^{\max} = 560 \pm 10$ nm, $\lambda'_{\min}$ varies between 415 and 424 nm, and $\lambda'_{\max}$ between 688 and 714 nm (Logvinenko, 2015b). The effect of $\lambda_S^{\max}$ on $\lambda'_{\min}$ was found to be less pronounced, being practically insignificant for $\lambda'_{\max}$ (Logvinenko, 2015b).

It should be noted that the model parameters presented in Table 3.1 were chosen by Govardovskii and colleagues so as to fit the data obtained from many different animals. However, the data obtained from monkey only are approximated much better if parameter b_0 in Eq. (3.16) is set equal to 0.15 (Govardovskii, personal communication). In this case, for the peak absorbance wavelengths of 430, 530, and 560 nm, $\lambda'_{\min}$ gets smaller: 413 nm, while $\lambda'_{\max}$ remains practically the same: 701 nm. Likewise, for the peak absorbance wavelengths of 430, 520, and 550 nm, $\lambda'_{\min}$ and $\lambda'_{\max}$ were found to be 408 and 688 nm. Respectively, for the peak absorbance wavelengths of 430, 540, and 570 nm, $\lambda'_{\min}$ and $\lambda'_{\max}$ were found to be 417 and 714 nm (Logvinenko, 2015b). The effect of the peak photopigment absorbance D (see Eq. (3.15)) was found to be quite small (Logvinenko, 2015b).

When the effective visible spectrum is shorter than the visible spectrum, the *purple interval* will refer to the straight segment connecting points $c(\lambda'_{\min})$ and $c(\lambda'_{\max})$, rather than the one that connects the ends of the spectral locus. In the past, there has always been some uncertainty about the purple interval because it has been defined as the interval connecting the wavelengths $\lambda_{\min}$ and $\lambda_{\max}$ (Wyszecki & Stiles, 1982) and the choice of $\lambda_{\min}$ and $\lambda_{\max}$ is, as mentioned above, rather arbitrary. In contrast, the proposed determination of the purple interval based on the effective ends of the visible spectrum wavelength interval is entirely unambiguous for the given photopigment spectral absorptivities.

5

Colour Stimulus Manifold

So far it has been assumed that the medium is ideally transparent and does not affect the light entering the eye. In this chapter, we abandon the assumption of the medium ideal transparency, assuming that the medium spectral transmittance is specified by a function $m(\lambda)$ ($0 \leq m(\lambda) \leq 1$ for $\lambda \in \Lambda$). The effect of the medium spectral transmittance can be taken into account the same way as it is commonly done for the ocular media transmittance, viz. incorporating it into the spectral sensitivity of the colour mechanisms. Specifically, the light metamerism produced by linear colour mechanisms (e.g., those induced by the cone photopigments) $\varphi_1(\lambda), \dots, \varphi_n(\lambda)$ under assumption that, before being absorbed by the photoreceptors, light passes a transparent medium with the spectral transmittance $m(\lambda)$ amounts to that induced by new linear colour mechanisms $\psi_1(\lambda), \dots, \psi_n(\lambda)$, such that

$$\psi_i(\lambda) = \varphi_i(\lambda)\, m(\lambda) \quad (i = 1, \dots, n). \tag{5.1}$$

Note that $m(\lambda)$ in these equations may be equally interpreted as the spectral reflectance of the target surface reflecting incident light (or a combined effect of both the surface and the medium). In this chapter, we will consider only continuous medium spectral transmittances. In Part II, we will drop off this restriction considering the effect of surface spectral reflectance which are not necessarily continuous.

In what follows, $E = \mathbf{C}^*(\Lambda)$ and $A = \mathbf{C}^*_+(\Lambda)$. Grassmann structures $\langle A, \sim\rangle$ and $\langle A, \approx\rangle$ induced by linear colour mechanisms $\varphi_1, \dots, \varphi_n$ and $\psi_1, \dots, \psi_n$ coincide (i.e., $\sim=\approx$) if and only if $Lin(\varphi_1, \dots, \varphi_n) = Lin(\psi_1, \dots, \psi_n)$ (Proposition 2.3, II). It follows from Proposition 18.4 proved in Part II that given (5.1), the equality $\langle A, \sim\rangle = \langle A, \approx\rangle$ is possible only when $m(\lambda) = const$, that is, when the medium is spectrally non-selective. Therefore, when $m(\lambda) \neq const$, there is no linear isomorphism T between the colour stimulus spaces $E/E(\sim)$ and $E/E(\approx)$ such that the colour stimulus cone $\pi_\sim(A)$ is mapped onto the colour stimulus cone $\pi_\approx(A)$, that is, $T\pi_\sim(A) = \pi_\approx(A)$.

Nevertheless, we will show that when $\langle A, \sim\rangle$ and $\langle A, \approx\rangle$ are rather general weak* continuous n-dimensional Grassmann structures (in particular, when $m(\lambda) \neq const$, or even when ψ_i is not related to φ_i by means of (5.1) at all), there is a homeomorphism between the colour stimulus cones $\pi_\sim(A)$ and $\pi_\approx(A)$. As noted in Remark 4.1, a colour-signal map $\Phi = (\varphi_1, \dots, \varphi_n)$ induces a non-singular linear operator $\Phi_0 : E/E(\sim) \to \mathbf{R}^n$ which maps the colour stimulus cone $\pi_\sim(A)$ onto the colour-signal cone $\Phi(A)$ so that $\Phi = \Phi_0 \circ \pi_\sim$ where $\pi_\sim$ is the colour stimulus map with respect to the Grassmann structure $\langle A, \sim\rangle$ (see Definition 2.5). Therefore, Φ_0 is a bijection of $\pi_\sim(A)$ onto $\Phi(A)$. Both Φ_0 and Φ_0^{-1} are continuous (as finite-dimensional linear operators). Thus, Φ_0 is a homeomorphism of $\pi_\sim(A)$ onto $\Phi(A)$. Similarly, given $\pi_\approx$ to be the colour stimulus map with respect to the Grassmann structure $\langle A, \approx\rangle$, the colour-signal map $\Psi = (\psi_1, \dots, \psi_n)$ induces a non-singular linear operator $\Psi_0 : E/E(\approx) \to \mathbf{R}^n$ which maps the colour stimulus cone $\pi_\approx(A)$ onto the colour-signal

Foundations of Colour Science: From Colorimetry to Perception, First Edition. Alexander D. Logvinenko and Vladimir L. Levin.
© 2023 John Wiley & Sons Ltd. Published 2023 by John Wiley & Sons Ltd.

cone $\Psi(A)$ so that $\Psi = \Psi_0 \circ \pi_{\approx}$ and Ψ_0 is a homeomorphism between $\pi_{\approx}(A)$ and $\Psi(A)$. As any two closed convex cones with non-empty interiors in $\mathbf{R}^n$ are homeomorphic, there is a homeomorphism between the colour-signal cones $\Phi(A)$ and $\Psi(A)$, and each of them is homeomorphic to the n-half-space $\mathbf{H}^n = \{(z_1, \dots, z_n) : z_n \geq 0\}$.

The homeomorphisms $\Phi_0 : \pi_{\sim}(A) \to \Phi(A)$ and $f_{\sim} : \Phi(A) \to \mathbf{H}^n = \{(z_1, \dots, z_n) \in \mathbf{R}^n : z_n \geq 0\}$ entail a homeomorphism $h_{\sim} = f_{\sim} \circ \Phi_0$ of the colour stimulus cone $\pi_{\sim}(A)$ onto the n-half-space $\mathbf{H}^n$. For the same reason, there is a homeomorphism $h_{\approx} = f_{\approx} \circ \Psi_0$ of $\pi_{\approx}(A)$ onto $\mathbf{H}^n$. (Here, $\Psi_0 : \pi_{\approx}(A) \to \Psi(A)$ and $f_{\approx} : \Psi(A) \to \mathbf{H}^n$.) Hence, the colour stimulus cones $\pi_{\sim}(A)$ and $\pi_{\approx}(A)$ are homeomorphic, and the map $h_{\approx}^{-1} \circ h_{\sim}$ provides a homeomorphism between them.

Note, however, that the homeomorphism $h_{\approx}^{-1} \circ h_{\sim}$ is rather arbitrary. Furthermore, while the signal (thus, stimulus) cone boundaries map onto each other, the spectral colour-signal curve $\Phi(\Delta)$ does not map to $\Psi(\Delta)$ because, generally, $h_{\sim} \circ \pi_{\sim}(\Delta) \neq h_{\approx} \circ \pi_{\approx}(\Delta)$.

Though, assuming that both the spectral colour-signal curves $\Phi(\Delta)$ and $\Psi(\Delta)$ are continuous curves without self-intersections, one can readily establish a homeomorphism between them, $\Phi(\delta_\lambda) \to \Psi(\delta_\lambda)$ for each $\lambda \in \Lambda$, which can be extended to the conical surface over the spectral colour-signal curve. However, there is no natural extension of this homeomorphism to the entire signal (thus, stimulus) cone.

At first glance, it seems natural to seek for a correspondence between those points in $\Phi(A)$ and $\Psi(A)$ that are images of the same lights. Specifically, let us put in correspondence to a point $z \in \Phi(A)$ the Ψ-image of the inverse Φ-image of z, that is, $\Psi(\Phi^{-1}(A))$. The resulting correspondence $\gamma : \Phi(A) \to \Psi(A)$,

$$\gamma(z) := \bigcup_{\sigma \in \Phi^{-1}(z)} \Psi(\sigma), \quad z \in \Phi(A), \tag{5.2}$$

proves to be a multi-valued map.

Note that restricting γ in (5.2) onto $\Phi(\Delta(\Lambda'))$ turns it into one-to-one mapping provided that the Grassmann structure $\langle \mathbf{C}_+^*(\Lambda), \sim \rangle$ has the strict convexity property (see Definition 4.3) and there is no metamerism for monochromatic lights, i.e.,

$$\alpha\delta_\lambda \sim \alpha'\delta_{\lambda'} \Rightarrow \lambda = \lambda' \text{ and } \alpha = \alpha'. \tag{5.3}$$

Indeed, as will be shown in Chapter 6 (Theorem 6.2), in this case no other lights can be metameric to monochromatic lights, that is, $\Phi^{-1}(\Phi(\delta_\lambda)) = \delta_\lambda$ whenever $\lambda \in \Lambda'$. Therefore, γ will map $\Phi(\delta_\lambda)$ to $\Psi(\delta_\lambda)$.

More generally, one can turn γ in (5.2) into a one-to-one correspondence if one picks up one light stimulus in each class of $\sim$ metamerism and in each class of $\approx$ metamerism. To be more exact, let us assume that there exists a light stimulus subset $\mathcal{A} \subset A$ such that $\Delta'(\Lambda') = \Delta(\Lambda') \cup I(\delta_{\lambda'_{\min}}, \delta_{\lambda'_{\max}}) \subset \mathcal{A}$ and there is exactly one element of $\mathcal{A}$ in each metameric class of $\pi_{\sim}(A)$ and in each metameric class of $\pi_{\approx}(A)$. Therefore, there exist bijections $b_{\sim} = \Phi_0 \circ \pi_{\sim|\mathcal{A}} : \mathcal{A} \to \Phi(A)$ and $b_{\approx} = \Psi_0 \circ \pi_{\approx|\mathcal{A}} : \mathcal{A} \to \Psi(A)$. A one-to-one map $b_{\approx} \circ b_{\sim}^{-1}$ is a bijection $\Phi(A) \to \Psi(A)$. Note that $b_{\approx} \circ b_{\sim}^{-1} = \Psi_{|\mathcal{A}} \circ \Phi_{|\mathcal{A}}^{-1}$, that is, $b_{\approx} \circ b_{\sim}^{-1}$ results from restricting Φ and Ψ onto $\mathcal{A}$ in formula (5.2) for γ. Admittedly, of interest are only those $\mathcal{A}$ for which the bijection $b_{\approx} \circ b_{\sim}^{-1}$ is a homeomorphism.

Let us consider a particular case when $\approx$ is a metamerism induced by a change in the spectral transmittance of the medium. In this case, each function $m(\lambda)$ induces a different metamerism $\approx = \sim_m$. If the original metamerism $\sim$ on $A = \mathbf{C}_+^*$ is induced by n linearly independent non-negative continuous functions $\varphi_1, \dots, \varphi_n$, then $\langle A, \sim \rangle$ is a weak* continuous Grassmann structure, and if, in addition, $m \in \mathbf{C}_+(\Lambda)$, then the functions $\psi_1, \dots, \psi_n$ given by (5.1) belong to $\mathbf{C}_+(\Lambda)$ as well, hence the induced Grassmann structure $\langle A, \sim_m \rangle$ is also weak* continuous. Let $\mathcal{M}$ denote the set of all $m \in \mathbf{C}_+(\Lambda)$ for which the functions $\psi_1, \dots, \psi_n$ given by (5.1) are linearly independent. In particular, this holds when m is strictly positive on Λ. Of great interest is whether it is possible to single out a subset $\mathcal{A} \subset A$ such that there is exactly one element of $\mathcal{A}$ in each metameric class in each $\pi_{\sim_m}(A)$, and $b_{\sim_m} \circ b_{\sim}^{-1}$ is a homeomorphism for each $m \in \mathcal{M}$. Of course, this question makes sense for $m \in \mathcal{M}$ only since for $m \notin \mathcal{M}$ the Grassmann structure $\langle A, \sim_m \rangle$ ceases to be n-dimensional (see Proposition 2.3, I). If such $\mathcal{A}$ exists, then all $\pi_{\sim_m}(A)$ are 'naturally' homeomorphic to each other.

Problem 5.1 *Let $\langle A, \sim \rangle$ and $\langle A, \approx \rangle = \langle A, \sim_m \rangle$ be n-dimensional weak* continuous Grassmann structures on the light stimulus cone $A = \mathbf{C}_+^*(\Lambda)$. Suppose they are induced by linear colour mechanisms with spectral sensitivities $\varphi_1(\lambda), \dots, \varphi_n(\lambda) \in \mathbf{C}_+(\Lambda)$ and $\psi_1(\lambda), \dots, \psi_n(\lambda) \in \mathbf{C}_+(\Lambda)$ related to each other as in (5.1). Is there such a light stimulus subset $\mathcal{A}$ in A that there exist bijections $b_\sim : \mathcal{A} \to \Phi(A)$ and $b_{\sim_m} : \mathcal{A} \to \Psi(A)$, such that $b_{\sim_m} \circ b_\sim^{-1}$ is a homeomorphism for every $m \in \mathcal{M}$?*

We will show below that such $\mathcal{A}$ exists for some three-dimensional weak* continuous Grassmann structures.

5.1 Three-dimensional colour stimulus manifold

Consider a three-dimensional weak* continuous Grassmann structure $\langle A, \sim \rangle$ on the light stimulus cone $A = \mathbf{C}_+^*(\Lambda)$ ($\Lambda = [\lambda_{\min}, \lambda_{\max}]$) determined by three linear colour mechanisms with spectral sensitivities $\varphi_1(\lambda), \varphi_2(\lambda), \varphi_3(\lambda) \in \mathbf{C}_+(\Lambda)$.

Definition 5.1 *Suppose the colour-signal-cone-boundary hypothesis holds true (see Definition 4.2). Given a light $\sigma_0 \in A$ such that $\Phi(\sigma_0)$ is an interior point of the colour-signal cone $\Phi(A)$, we say a Grassmann structure $\langle A, \sim \rangle$ is* standard with respect to σ_0 *if: (i) there is no metamerism for monochromatic lights, i.e., (5.3) holds true, (ii) for every $\lambda_1, \lambda_2 \in \Lambda'$, $\lambda_1 \neq \lambda_2$, there are no $t_1 > 0, t_2 > 0$ such that $\sigma_0 \sim t_1\delta_{\lambda_1} - t_2\delta_{\lambda_2}$, and (iii) for every $\lambda \in \Lambda' \setminus \{\lambda'_{\min}, \lambda'_{\max}\}$, there are no $t_1 > 0, t_2 > 0, t_3 > 0$ such that $\sigma_0 \sim t_1\delta_\lambda - t_2\delta_{\lambda'_{\min}} - t_3\delta_{\lambda'_{\max}}$.*

All these conditions look reasonable and not restrictive.[1]

We consider a subset in A:

$$\mathcal{A} = \mathcal{A}_{\sigma_0} := \{p\sigma' + q\sigma_0 \, : \, p \geq 0, q \geq 0, \sigma' \in \Delta'(\Lambda')\} \tag{5.4}$$

with the induced weak* topology in it. Recall (see Chapter 4) that

$$\Delta'(\Lambda') = \Delta(\Lambda') \cup I(\delta_{\lambda'_{\min}}, \delta_{\lambda'_{\max}}),$$

where $I(\delta_{\lambda'_{\min}}, \delta_{\lambda'_{\max}}) = \{\alpha\delta_{\lambda'_{\min}} + (1-\alpha)\delta_{\lambda'_{\max}} \, : 0 \leq \alpha \leq 1\}$.

Theorem 5.1 *Let $\langle A, \sim \rangle$ be a three-dimensional standard with respect to σ_0 and weak* continuous Grassmann structure on the light stimulus cone $A = \mathbf{C}_+^*(\Lambda)$, and $\Phi = (\varphi_1, \varphi_2, \varphi_3)$ a colour-signal map consistent with it. Suppose that: (i) the colour-signal-cone-boundary hypothesis holds true; (ii) all φ_i are continuous functions on Λ, and $\varphi_1(\lambda) + \varphi_2(\lambda) + \varphi_3(\lambda) > 0$ whenever $\lambda \in \Lambda$. Then $\Phi_{|\mathcal{A}}$ is a homeomorphism of $\mathcal{A}$ onto the colour-signal cone $\Phi(A)$.*

Note now that the colour-signal cone $\Phi(A) = \Phi(\mathbf{C}_+^*)$, being a closed convex cone with non-empty interior in $\mathbf{R}^3$ (see Corollary 4.1 and Proposition 4.1), is homeomorphic to the

1 One can compare (ii) and (iii) with condition (III) from Section 3.2.

half-space $\mathbf{H}^3 = \{(x, y, z) \in \mathbf{R}^3 : z \geq 0\}$. Let ϕ be a homeomorphism of $\Phi(A)$ onto $\mathbf{H}^3$. Then, it may be considered as a sole chart that determines in $\Phi(A)$ the structure of C^∞-differentiable three-dimensional ∂-manifold[2] diffeomorphic to $\mathbf{H}^3$. Since the colour-signal-cone-boundary hypothesis holds true, its boundary[3] is

$$\partial\Phi(A) = \phi^{-1}(\partial\mathbf{H}^3) = \{p\Phi(\sigma') : p \geq 0, \sigma' \in \Delta'\}.$$

Furthermore, if $\mathcal{A}$ is given by (5.4) then, under the hypotheses of Theorem 5.1, the map $\phi \circ \Phi_{|\mathcal{A}}$ is a homeomorphism of $\mathcal{A}$ onto $\mathbf{H}^3$. It may be considered as a sole chart defining in $\mathcal{A}$ the structure of C^∞-differentiable three-dimensional ∂-manifold diffeomorphic to $\mathbf{H}^3$. Its boundary[4] is then

$$\partial\mathcal{A} = \Phi_{|\mathcal{A}}^{-1}(\phi^{-1}(\partial\mathbf{H}^3)) = \{p\sigma' : p \geq 0, \sigma' \in \Delta'\}. \tag{5.5}$$

Thus, we have established a remarkable (but not very complicated) fact: *the topological space $\mathcal{A}$, which is an infinite-dimensional subset in $\mathbf{C}^*$ with the induced weak* topology in it, proves to be a three-dimensional C^∞-differentiable ∂-manifold, and the map $\Phi_{|\mathcal{A}}$ proves to be the corresponding diffeomorphism of $\mathcal{A}$ onto $\Phi(A)$.*

Note that all such $\mathcal{A}$ as given by (5.4) with various σ_0 prove to be diffeomorphic to each other.

Let us introduce some terminology that will be used throughout. We will call $\mathcal{A}$ the *colour stimulus manifold*. We will refer to the fragment of $\Delta'(\Lambda')$ comprising monochromatic lights (i.e., $\Delta(\Lambda')$) as the *spectral (or monochromatic) light stimulus curve*. $\Delta'(\Lambda')$ itself will be referred as the *spectral light stimulus closure*, $I(\delta_{\lambda'_{\min}}, \delta_{\lambda'_{\max}})$ being referred to as the *complementary light stimulus segment*. A cone over $\Delta'(\Lambda')$ (i.e., $con\Delta'(\Lambda') = \{p\sigma' : p \geq 0, \sigma' \in \Delta'\}$) will be called the *spectral light stimulus conical closure*, and the part of $con\Delta'(\Lambda')$, corresponding to $\Delta(\Lambda')$, the *spectral light stimulus conical surface*. Respectively, $\Phi\left(\Delta'(\Lambda')\right)$ will be referred as the *spectral colour-signal closure*; a cone over $\Phi\left(\Delta'(\Lambda')\right)$ the *spectral colour-signal conical closure*; $\Phi\left(\Delta(\Lambda')\right)$ the *spectral colour-signal curve*; and $\Phi\left(I(\delta_{\lambda'_{\min}}, \delta_{\lambda'_{\max}})\right)$ the *complementary colour-signal segment*.

Consider now another colour-signal map $\Psi = (\psi_1, \psi_2, \psi_3)$ defining a weak* continuous Grassmann structure $\langle A, \approx\rangle$ and satisfying the hypotheses of Theorem 5.1. Then all stated above for Φ, $\langle A, \sim\rangle$, and $\Phi\left(\mathbf{C}_+^*\right)$ holds true for Ψ, $\langle A, \approx\rangle$, and $\Psi\left(\mathbf{C}_+^*\right)$ as well. In particular, the colour-signal cone $\Psi\left(\mathbf{C}_+^*\right)$ is a three-dimensional C^∞-differentiable ∂-manifold diffeomorphic to the corresponding $\mathcal{A} = \mathcal{A}(\approx)$ similar to given by (5.4). Moreover, if $\langle A, \sim\rangle$ and $\langle A, \approx\rangle$ generate the same Λ' (this holds, in particular, when the strong colour-signal-cone-boundary hypothesis holds true for both the Grassmann structures), then the colour stimulus manifold $\mathcal{A}$ is obviously one and the same for both Grassmann structures.[5]

In the general case, the corresponding manifolds $\mathcal{A}$ can be different. If, for example, $\lambda'_{\min}(\sim) > \lambda'_{\min}(\approx)$ then, for every $p > 0$, $q \geq 0$, $p\delta_{\lambda'_{\min}(\approx)} + q\sigma_0$ belongs to $\mathcal{A}(\approx)$ but not to $\mathcal{A}(\sim)$, and generally $\Phi(\delta_{\lambda'_{\min}(\approx)})$ proves to be an interior point of the colour-signal cone $\Phi(A)$.

Assume that Φ and Ψ satisfy the hypotheses of Theorem 5.1 with the same σ_0 and the same Λ' and are related as in (5.1). When $m(\lambda)$ is interpreted as a ratio of the spectral transmittances

2 The corresponding information about differentiable ∂-manifolds (in other terminology, manifolds with boundary) may be found, for instance, in Hirsch (1976).
3 Clearly, its boundary in terms of manifold theory is exactly the topological boundary of the cone $\Phi(A)$ in $\mathbf{R}^3$.
4 Here the word 'boundary' is understood as the boundary in terms of manifold theory. Topologically, $\mathcal{A}$ as a subset of the light stimulus cone A endowed with the weak* topology has no interior points.
5 It can be shown that if: (i) (5.1) holds, where $m(\lambda) > 0$ for all $\lambda \in \Lambda$, (ii) $\Lambda'(\sim) \subseteq \Lambda'(\approx)$ or $\Lambda'(\approx) \subseteq \Lambda'(\sim)$, and (iii) both the Grassmann structures satisfy the hypotheses of Theorem 5.1 with the same σ_0, then $\Lambda'(\sim) = \Lambda'(\approx)$ hence the colour stimulus manifold $\mathcal{A}$ remains the same as well: $\mathcal{A}_{\sigma_0}(\sim) = \mathcal{A}_{\sigma_0}(\approx)$.
Furthermore, as follows from the calculations in Section 4.4, for a three-pigment Grassmann–Govardovskii structure, its effective visible spectrum interval (i.e., Λ') is not affected by the spectral transmittance of the medium provided that $m(\lambda) > 0$.

of two media[6], it is natural to think of the colour-signal cone $\Psi(\mathbf{C}^*_+)$ as a transformation of the colour-signal cone $\Phi(\mathbf{C}^*_+)$ induced by the change of the medium. As both $\Phi_{|\mathcal{A}}$ and $\Psi_{|\mathcal{A}}$ are diffeomorphisms of class C^∞, the composition $\Psi_{|\mathcal{A}} \circ \Phi^{-1}_{|\mathcal{A}}$ mapping $\Psi(\mathbf{C}^*_+)$ onto $\Phi(\mathbf{C}^*_+)$ is a diffeomorphism of class C^∞, as well.

Let $\Phi_0 : \pi_\sim(E) \to \Phi(E) = \mathbf{R}^3$ be the canonical linear isomorphism between two three-dimensional vector spaces ($\Phi_0([e]) = \Phi(e)\ \forall e \in E$; see Remark 4.1). The diffeomorphism $\Phi_{|\mathcal{A}}$ between $\mathcal{A}$ and $\Phi(A)$ implies then a diffeomorphism $\Phi_0^{-1} \circ \Phi_{|\mathcal{A}}$ between $\mathcal{A}$ and $\pi_\sim(A)$, which puts in correspondence to any element in $\mathcal{A}$ its class of metamerism. Both $\pi_\sim(A)$ and $\pi_\approx(A)$ can be considered then as the same three-dimensional C^∞-differentiable ∂-manifold, the concrete representation of which is $\mathcal{A}$. The diffeomorphism between $\mathcal{A}$ and the colour stimulus cone $\pi(A)$ enables us to identify $\sigma \subset \mathcal{A}$ with $\pi(\sigma) = [\sigma] \subset \pi(\mathcal{A})$ and justifies using the term 'colour stimulus manifold' for $\mathcal{A}$ and the term 'colour stimuli' for elements of $\mathcal{A}$. Specifically, $\mathcal{A}$ can be thought of as a concrete (non-linear) representation of the colour stimulus cone. Given $\sigma \in \mathcal{A}$, the parameters q, p, and λ, which determine the metamer of σ in $\mathcal{A}$ (i.e., $\sigma \sim p\delta_\lambda + q\sigma_0$), can be considered as a sort of coordinates of σ in $\mathcal{A}$. They will be referred to as *$qp\lambda$-coordinates*.

Given two light stimuli, σ_0 and σ, such that $\Phi(\sigma_0) \neq 0$ and $\Phi(\sigma) \neq 0$, we consider a plane $H(\Phi(\sigma), \Phi(\sigma_0)) := \{t_1\Phi(\sigma) + t_2\Phi(\sigma_0) : t_1, t_2 \in \mathbf{R}\}$ in the colour-signal space. If $\Phi(\sigma_0)$ is an interior point of the colour-signal cone, this plane intersects $\Phi(\Delta')$ at least at two points, and when the colour-signal-cone-boundary hypothesis holds true, that intersection comprises exactly two points. If, moreover, one of these two points lies in $\Phi(I(\delta_{\lambda'_{\min}}, \delta_{\lambda'_{\max}})) = I(\Phi(\delta_{\lambda'_{\min}}), \Phi(\delta_{\lambda'_{\max}}))$, then the plane $H(\Phi(\sigma), \Phi(\sigma_0))$ intersects the fragment $\Phi(\Delta(\Lambda'))$ of the spectral colour-signal curve at the sole point $\Phi(\delta_{\lambda^*})$. In general a point $\Phi(\delta_{\lambda^*}) \in \Phi(\Delta(\Lambda'))$ lies in the plane $H(\Phi(\sigma), \Phi(\sigma_0))$ if and only if $\lambda = \lambda^*$ satisfies the equation:

$$\begin{vmatrix} \varphi_1(\lambda) & \varphi_2(\lambda) & \varphi_3(\lambda) \\ z_1^0 & z_2^0 & z_3^0 \\ z_1^1 & z_2^1 & z_3^1 \end{vmatrix} = 0, \tag{5.6}$$

where $(z_1^0, z_2^0, z_3^0) = \Phi(\sigma_0)$ and $(z_1^1, z_2^1, z_3^1) = \Phi(\sigma)$. The vectors $\Phi(\sigma_0)$, $\Phi(\sigma)$, and $\Phi(\delta_{\lambda^*}) = (\varphi_1(\lambda^*), \varphi_2(\lambda^*), \varphi_3(\lambda^*))$ are coplanar, that is, there are real numbers k_1, k_2, and k_3 such that $k_1\Phi(\sigma_0) + k_2\Phi(\sigma) + k_3\Phi(\delta_{\lambda^*}) = 0$. Furthermore, if $\Phi(\sigma_0)$ is an interior point of the colour-signal cone and the colour-signal-cone-boundary hypothesis holds true then $k_2 \neq 0$; hence, there is a representation:

$$\Phi(\sigma) = q\Phi(\sigma_0) + p\Phi(\delta_{\lambda^*}). \tag{5.7}$$

In such a case, there exist at least one and at most two solutions to (5.6), and if the hypotheses of Theorem 5.1 are satisfied, then for at most one of them, representation (5.7) holds with $q \geq 0$, $p > 0$. The corresponding λ^* is then called the *dominant wavelength* of the light stimulus σ with respect to the light stimulus σ_0. When there is a non-dominant solution $\lambda = \lambda_1^*$ to (5.6), then in the corresponding representation (5.7),

$$\Phi(\sigma) = q_1\Phi(\sigma_0) + p_1\Phi(\delta_{\lambda_1^*}), \tag{5.8}$$

6 More specifically, it is assumed here that we deal with the same set of colour mechanisms operating under two different media, say, $m_1(\lambda)$ and $m_2(\lambda)$, and $m(\lambda) = m_2(\lambda)/m_1(\lambda)$.

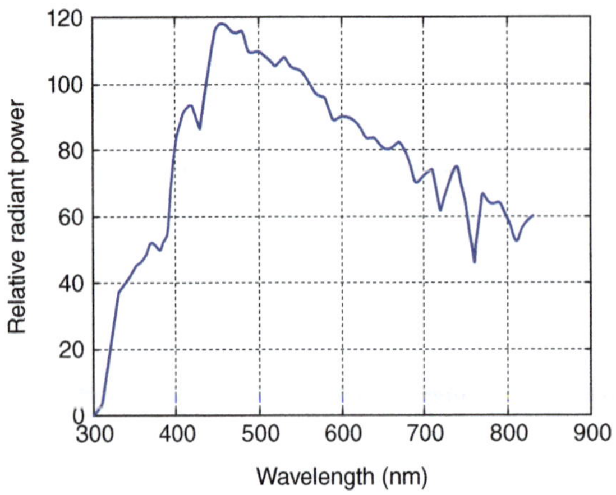

Figure 5.1 Radiant spectral density function of CIE illuminant D65.

one has $q_1 > 0, p_1 < 0$[7] (in this case, the corresponding λ_1^* is called the *complementary wavelength* of the light stimulus σ with respect to the light stimulus σ_0).

The rationale for such terminology is that two lights, σ_1 and σ_2, are often called *complementary* with respect to the light σ_0 if their mixture is metameric to σ_0 $\left(\text{thus, } \Phi(\sigma_1) + \Phi(\sigma_2) = \Phi(\sigma_0)\right)$.[8] In this terminology, if λ_1^* is the complementary wavelength of σ with respect to σ_0, then the light σ and the monochromatic light $\delta_{\lambda_1^*}$ are complementary with respect to σ_0.

For the human colour vision, there is a natural choice of σ_0 – daylight. Not only because the sun is the major source of light on earth but also because daylight looks neutral (i.e., achromatic) under natural conditions of observation. In this case, (5.7) and (5.8) entail that for every light σ either it is metameric to a mixture of the neutral light $(q\sigma_0)$ with the monochromatic light $p\delta_{\lambda^*}$ of the dominant wavelength λ^* (see (5.7) where $p \geq 0$), or the neutral light $(q_1\sigma_0)$ matches a mixture of the given light σ with the monochromatic light $|p_1|\,\delta_{\lambda_1^*}$ of the complementary wavelength λ_1^* (see (5.8) where $p_1 < 0$). It seems natural to call p and q (in representation (5.7)) the *chromatic* and *achromatic components* (of $qp\lambda$-coordinates). Unless it says otherwise, we will further assume that σ_0 is daylight. Figure 5.1 represents a radiant spectral density function (in terms of an arbitrary unit) which the Commission Internationale de l'Eclairage (CIE) recommends as so-called *standard illuminant D65*. It represents a phase of natural daylight (Wyszecki & Stiles, 1982, p. 144).

5.2 Non-linear colour stimulus map. Colour stimulus transformation caused by the medium

The colour stimulus map $\pi_\sim : A \to \pi_\sim(A)$ induces a map $A \to A_{\sigma_0}$.

---

7 Since $\Phi(\sigma) \neq 0$ and $\varphi_1(\lambda) + \varphi_2(\lambda) + \varphi_3(\lambda) > 0$ whenever $\lambda \in \Lambda$, no solution to (5.6) can exist, for which in representation (5.8) $q_1 = 0, p_1 \leq 0$. Since the Grassmann structure $\langle A, \sim \rangle$ is standard with respect to σ_0, the case where $q_1 < 0, p_1 > 0$ is also impossible. Indeed, if there are two solutions to (5.6), hence one of them, λ^* is dominant, then $q - q_1 > 0$ and $(q - q_1)\sigma_0 \sim p_1\delta_{\lambda_1^*} - p\delta_{\lambda^*}$ is in contradiction with condition (ii) of Definition 5.1. Similarly, when λ_1^* is the only solution to (5.6), there exist $p > 0, q > 0$, and $0 < \alpha < 1$ such that $(q - q_1)\sigma_0 \sim p_1\delta_{\lambda_1^*} - p\alpha\delta_{\lambda'_{\min}} - p(1-\alpha)\delta_{\lambda'_{\max}}$ is in contradiction with condition (iii) of Definition 5.1.

8 Note that this is the definition of complementarity for lights not for colours. The complementarity of light colours will be defined in Section 8.6.3.

Definition 5.2 *Let $\langle A, \sim \rangle$ be a three-dimensional standard with respect to σ_0 weak* continuous Grassmann structure on the light stimulus cone $A = \mathbf{C}_+^*(\Lambda)$, and $\mathcal{A}_{\sigma_0}$ the colour stimulus manifold. A map $\kappa_\sim = \pi_{\sim|\mathcal{A}}^{-1} \circ \pi_\sim : A \to \mathcal{A}_{\sigma_0}$ will be referred to as a* non-linear colour stimulus map.

Remark 5.1 *If the Grassmann structure $\langle A, \sim \rangle$ is induced by linear colour mechanisms $\varphi_1, \varphi_2, \varphi_3$, and $\Phi = (\varphi_1, \varphi_2, \varphi_3)$ is the corresponding colour-signal map then*

$$\kappa_\sim = \Phi_{|\mathcal{A}}^{-1} \circ \Phi, \tag{5.9}$$

and $\kappa_\sim$ is continuous.

Indeed, $\Phi = \Phi_0 \circ \pi_\sim$ hence $\Phi_{|\mathcal{A}}^{-1} \circ \Phi = \pi_{\sim|\mathcal{A}}^{-1} \circ \Phi_0^{-1} \circ \Phi_0 \circ \pi_\sim = \pi_{\sim|\mathcal{A}}^{-1} \circ \pi_\sim = \kappa_\sim$, and as Φ is continuous and $\Phi_{|\mathcal{A}}$, by Theorem 5.1, is a homeomorphism, the continuity of $\kappa_\sim$ follows.

When the medium transmittance changes, the new metamerism $\approx$ induces a new colour stimulus map $\pi_\approx : A \to \pi_\approx(A)$ (and a new colour-signal map $\Psi = (\psi_1, \psi_2, \psi_3)$), thus a new non-linear colour stimulus map $\kappa_\approx : A \to \mathcal{A}_{\sigma_0}(\approx)$. It follows that, as in many experiments demonstrated, changes in the medium may produce no colour match (in terms of $\approx$) between hitherto metameric (in terms of $\sim$) lights. In other words, some lights may shift from one metameric class to the other due to medium changes. A notion of non-linear colour stimulus map allows us to describe quantitatively such a shift, not only when it is caused by a change in the medium transmittance but also in the more general case, namely, when the colour stimulus manifold remains the same: $\mathcal{A}_{\sigma_0}(\sim) = \mathcal{A}_{\sigma_0}(\approx)$.

In what follows, we suppose that both Grassmann structures, $\langle A, \sim \rangle$ and $\langle A, \approx \rangle$, satisfy the hypotheses of Theorem 5.1 with the same σ_0 and the same Λ', hence the same colour stimulus manifold $\mathcal{A}$. If a light stimulus a is such that $\kappa_\sim(a) = \kappa_\approx(a)$, we say that the colour of a is *psychophysically invariant* with respect to the medium transmittance alteration. In such a case, we also say that the light stimulus a remains psychophysically invariant (in colour). This occur if and only if a belongs to the following subset

$$: \pi_\sim^{-1}\left([a]_\sim\right) \cap \pi_\approx^{-1}\left(\left[\kappa_\sim(a)\right]_\approx\right). \tag{5.10}$$

Note that this subset is not empty as it contains $\kappa_\sim(a)$.

5.2.1 The colour stimulus shift caused by the medium variations

When $\kappa_\sim(a) \neq \kappa_\approx(a)$ (i.e., the $qp\lambda$-coordinates of a are different), we say that the colour of a undergoes a *colour stimulus shift induced by the medium transmittance change.* (For the sake of generality, one can treat psychophysical invariance as a particular case of psychophysical shift, namely, a zero shift.) The medium-induced psychophysical shift means that the light stimulus in question shifts from one metameric class to the other, thereby changing its colour. Particularly, the light stimulus a which was in the same metamer class as the light stimulus $\kappa_\sim(a)$ under one medium shifts to the metamer class containing the light stimulus $\kappa_\approx(a)$ under the other medium.

Consider a light stimulus a and its class of metamerism $[a]_\sim$. The metamers of a disperses into the following subset in the colour stimulus manifold due to the medium transmittance change (i.e., from $\sim$ to $\approx$):

$$C(\sim, \approx; a) := \kappa_\approx(\{b \in A : b \sim a\}). \tag{5.11}$$

The subset $C(\sim, \approx; a)$ in $\mathcal{A}_{\sigma_0}$ fully describes to what extent the medium transmittance change can alter the colour (which a has under the original medium).

Lemma 5.1 *Let $\langle \mathbf{C}_+^*, \sim \rangle$ be a weak* continuous n-dimensional Grassmann structure induced by linear colour mechanisms with continuous spectral sensitivity functions $\varphi_1, \dots, \varphi_n$. Suppose (4.3) holds true, i.e.,*

$$\sum_{i=1}^{n} \varphi_i(\lambda) > 0 \quad \forall \lambda \in \Lambda, \tag{5.12}$$

then $\{b \in \mathbf{C}_+^* : b \sim a\}$ is a weak compact convex subset in $\mathbf{C}_+^*$ whenever $a \in \mathbf{C}_+^*$.*

Remark 5.2 *Condition (5.12) cannot be omitted. Indeed, if it fails then, for some $\lambda_0 \in \Lambda$, $\varphi_1(\lambda_0) = \dots = \varphi_n(\lambda_0) = 0$. In such a case, the set $\{b \in \mathbf{C}_+^*(\Lambda) : b \sim a\} = \{b \in \mathbf{C}_+^*(\Lambda) : \varphi_i(b) = \varphi_i(a), i - 1, \dots, n\}$ contains a measure $a + N\delta_{\lambda_0}$ whenever $N > 0$. Since $\left\| a + N\delta_{\lambda_0} \right\| = \|a\| + N > N$, this set is not norm bounded, thus is not weak* compact.*

The next result is a direct consequence of Lemma 5.1 taking into account hypothesis (ii) of Theorem 5.1 and the continuity of $\kappa_\approx$ (see Remark 5.1).

Corollary 5.1 *The set (5.11) is weak* compact.*

Quantitatively, the medium-induced psychophysical shift can be expressed in many different ways (cf. Section 8.3). Here, we begin from a rather abstract estimate in terms of the difference $\kappa_\approx(a) - \kappa_\sim(a)$ as:

$$Sh(a) := |\langle \varphi_0, \kappa_\approx(a) - \kappa_\sim(a) \rangle|, \tag{5.13}$$

where φ_0 is a weak* continuous linear functional on $\mathbf{C}^*$ determined by (and identified with) the continuous function $\varphi_0(\lambda) = \lambda$, $\lambda \in \Lambda$. Note that the quantity expressed by Eq. (5.13) is measured in the same units as wavelength λ (e.g., nanometers). When both $\kappa_\sim(a)$ and $\kappa_\approx(a)$ are probability measures, $Sh(a)$ is the absolute value of the wavelength mean value difference under the transfer from $\kappa_\sim(a)$ to $\kappa_\approx(a)$. Let $\kappa_\sim(a) = p_\sim(a)\sigma_1' + q_\sim(a)\sigma_0$ and $\kappa_\approx(a) = p_\approx(a)\sigma_2' + q_\approx(a)\sigma_0$ where $\sigma_1', \sigma_2' \in \Delta'(\Lambda')$, $p_\sim(a), q_\sim(a), p_\approx(a), q_\approx(a) \in \mathbf{R}_+$. Clearly, $Sh(a)$, as given by (5.13), equals

$$\left| p_\approx(a) \int_\Lambda \lambda \, \sigma_2'(d\lambda) - p_\sim(a) \int_\Lambda \lambda \, \sigma_1'(d\lambda) + (q_\approx(a) - q_\sim(a)) \int_\Lambda \lambda \, \sigma_0(d\lambda) \right|.$$

Notice that if $\sigma_1' = \sigma_2'$ and $q_\sim(a) = q_\approx(a)$, one has

$$Sh(a) = \langle \varphi_0, \sigma_1' \rangle |p_\approx(a) - p_\sim(a)|,$$

and if $\sigma_1' = \sigma_2'$ and $p_\sim(a) = p_\approx(a)$ then

$$Sh(a) = \langle \varphi_0, \sigma_0 \rangle |q_\approx(a) - q_\sim(a)|.$$

From Lemma 5.1, it follows that the *maximal colour stimulus shift* induced by the medium transmittance change (i.e., from $\sim$ to $\approx$):

$$Sh(\sim, \approx; a) = \max_{b \sim a} Sh(b) = \max_{b \sim a} |\langle \varphi_0, \kappa_\approx(b) \rangle - \langle \varphi_0, \kappa_\sim(a) \rangle| \tag{5.14}$$

is well defined, i.e., maximum in (5.14) is attained at some b.

One can also consider the *relative maximal colour stimulus shift*, which is expressed in relative units as:

$$\delta Sh(\sim, \approx; a) := \frac{\max_{b \sim a} Sh(b)}{\langle \varphi_0, \kappa_\sim(a) \rangle} = \frac{\max_{b \sim a} |\langle \varphi_0, \kappa_\approx(b) - \kappa_\sim(a) \rangle|}{\langle \varphi_0, \kappa_\sim(a) \rangle}. \tag{5.15}$$

It must be admitted that, being an abstract integral characteristic, the measure of colour stimulus shift (5.13) does not lend itself to a simple intuitive interpretation. It is easier to interpret the colour stimulus shift components: $q_{\approx}(a) - q_{\sim}(b)$, $p_{\approx}(a) - p_{\sim}(b)$, and the shift in terms of the dominant (complementary) wavelength.

By analogy with (5.14), one can introduce the following indices characterising the maximal shift with respect to the corresponding component:

$$\Delta p(\sim, \approx; a) := \max_{b\sim a} |p_{\approx}(b) - p_{\sim}(b)| = \max_{b\sim a} |p_{\approx}(b) - p_{\sim}(a)| \tag{5.16}$$

and

$$\Delta q(\sim, \approx; a) := \max_{b\sim a} |q_{\approx}(b) - q_{\sim}(b)| = \max_{b\sim a} |q_{\approx}(b) - q_{\sim}(a)|; \tag{5.17}$$

and by analogy with (5.15), introduce similar relative indices:

$$\delta p(\sim, \approx; a) := \frac{\max_{b\sim a}|p_{\approx}(b) - p_{\sim}(a)|}{p_{\sim}(a)} \tag{5.18}$$

and

$$\delta q(\sim, \approx; a) := \frac{\max_{b\sim a}|q_{\approx}(b) - q_{\sim}(a)|}{q_{\sim}(a)}. \tag{5.19}$$

When both $\kappa_{\sim}(a)$ and $\kappa_{\approx}(a)$ have either dominant wavelengths, say, $\mu_{\sim}$ and $\mu_{\approx}$, or complementary wavelengths, $\mu'_{\sim}$ and $\mu'_{\approx}$, one can consider either a shift in dominant wavelength – $|\mu_{\sim} - \mu_{\approx}|$ – or in complementary wavelength – $|\mu'_{\sim} - \mu'_{\approx}|$. However, when $\kappa_{\sim}(a)$ have a dominant wavelength, and $\kappa_{\approx}(a)$ a complementary wavelength (or the other way round), the difference $|\mu_{\sim} - \mu'_{\approx}|$ (or $|\mu'_{\sim} - \mu_{\approx}|$) does not make much sense. This problem can be overcome if the cylindrical coordinates, introduced in Section 8.2, are used to express the colour stimulus shift (Section 8.3).

5.2.2 Colour robustness to medium variations

An important characteristic of light in the present context is how robust it is with respect to medium transmittance change. For example, assume that one has to choose between two light sources (say for traffic lights) which are of the same colour (i.e., metameric), say, yellow, under clear atmosphere. However, in fog, the lights can change colour. For example, they can acquire a greenish tint in hazy air. It is natural to prefer the light source which becomes less greenish, that is, with a smaller psychophysical colour shift in dominant wavelength. We will say that this light source is more *robust* to atmosphere change. Such robustness to medium change can be quantified with the following index of *light robustness to medium transmittance change*:

$$rSh(\sim, \approx; a) := 1 - \frac{Sh(a)}{\max_{b\sim a} Sh(b)}. \tag{5.20}$$

Similarly, one can define robustness indices $rp(\sim, \approx; a)$ and $rq(\sim, \approx; a)$.

These indices of light robustness range from 0 to 1. Lights psychophysically invariant relative the medium transmittance change have the maximal robustness. The lights with maximal psychophysical shift have zero robustness.

When choosing light sources for traffic lights, one can notice that in the fog not only the yellow light source becomes greenish but the green light source also becomes yellowish. As a result, the yellow and green lights may lack the required distinctness in the foggy atmosphere. Moreover, as the psychophysical colour shift varies with light, the distinctness of the yellow and green traffic lights in fog will vary with light source. It is desirable to investigate the theoretical limits of such variability.

More specifically, we wish to establish which colours become metameric due to the medium transmittance change and which always remain distinct (not metameric). It turns out that it is more convenient to study a relation which is the opposite to the distinctness of colours.

Definition 5.3 *Let $\langle A, \sim\rangle$ and $\langle A, \approx\rangle$ be three-dimensional weak* continuous Grassmann structures with common colour stimulus manifold $\mathcal{A}$. Let $\kappa_\sim$ and $\kappa_\approx$ be the non-linear colour stimulus maps. Define a binary relation τ on $\mathcal{A}$:*

$$\forall c', c'' \in \mathcal{A}, \quad c'\tau c'' \Leftrightarrow \kappa_\approx\left(\kappa_\sim^{-1}(c')\right) \cap \kappa_\approx\left(\kappa_\sim^{-1}(c'')\right) \neq \emptyset. \tag{5.21}$$

We will say that the colours c' and c'' are confusable *(with respect to replacing $\sim$ with $\approx$) if and only if $c'\tau c''$. Otherwise (i.e., when $c'\tau c''$ fails), they will be called* well-separated *(with respect to replacing $\sim$ with $\approx$).*

When the Grassmann structures $\langle A, \sim\rangle$ and $\langle A, \approx\rangle$ describe the colour vision of the same observer under two different atmospheres, colours are confusable if there exist two lights having these colours under one atmosphere and such that they become metameric under the other atmosphere.

The relation τ is reflexive (since $\kappa_\approx\left(\kappa_\sim^{-1}(c)\right) \neq \emptyset$ for each c in $\mathcal{A}$) and symmetric. A reflexive and symmetric binary relation τ on a set A is called a *tolerance*, and a pair $\langle A, \tau\rangle$ a *tolerance space* (Zeeman, 1962; Roberts, 1973). We will deal with tolerance relations and their properties in Chapters 8 and 10. It is known that for each tolerance space $\langle A, \tau\rangle$, there is a family of its subsets $A_i (i \in \mathcal{I})$ such $\cup_{i\in\mathcal{I}} A_i = A$, and for each A_i(i) all its elements are tolerant to each other; and (ii) for every a in A_i there exists $a' \notin A_i$ such that a' is not tolerant to a. Such subsets are referred to as *tolerance classes*. Note that when a tolerance is transitive then the tolerance classes turn into the equivalence classes.

We will call the tolerance classes of the tolerance (5.21) the *colour confusion classes* (with respect to replacing $\sim$ with $\approx$). They are maximal subsets of the colour stimulus manifold such that all their colours are confusable with each other.

Problem 5.2 *Describe all the colour confusion classes for the colour stimulus manifold $\mathcal{A}$ (Definition 5.3).*

5.3 Causes of individual differences in trichromatic colour matching

A pair of lights looking indistinguishable for one observer may look rather different for others. Colour matching functions obtained for different observers differ. A good insight into the cross-individual differences in colour matching can be obtained from the 10° colour matching functions of Stiles & Burch (1955, 1959). They measured colour matching functions for 49 trichromatic observers (Fig. 5.2). Stiles and Burch's dataset is generally believed to be the most reliable and representative collection of trichromatic colour matching functions obtained by far.[9]

9 The Stiles & Burch data are available online at http://www.cvrl.org.

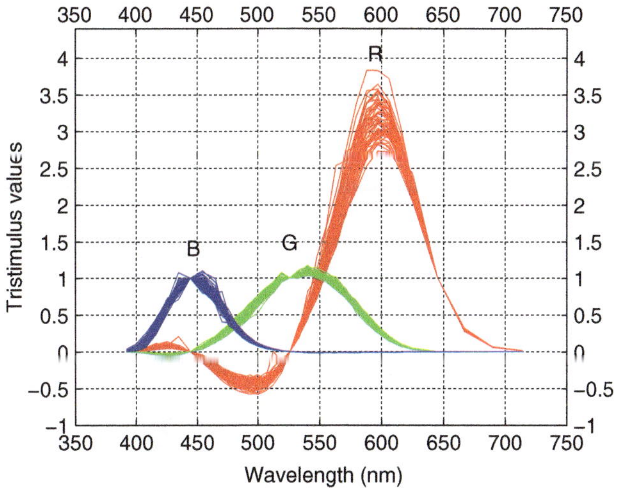

Figure 5.2 Individual colour matching functions of 49 observers employed by Stiles & Burch (1959) with respect to monochromatic primary stimuli of equal energy, B at $\lambda_B = 444.4$ nm, G at $\lambda_G = 525.3$ nm, and R at $\lambda_R = 645.2$ nm.

These individual differences could be caused by many factors the main of which are believed to be (i) the individual variations in the peak sensitivity ($\lambda^{\max}$) of the photopigments; and (ii) the variation in the spectral transmittance of the macular pigment and the lens (Webster & MacLeod, 1988). This leads to the fact that different three-dimensional Grassmann structures are required to model the colour vision of different trichromatic individuals. However, when one can use the same colour stimulus manifold $\mathcal{A}$ (Definition 5.3), all individual variations will be qualified in terms of the non-linear colour stimulus map.

We split the problem of inter-individual differences in the human trichromatic colour vision into the following two – differences in perceiving the lights comprising the colour stimulus manifold, and differences in the non-linear colour stimulus map between trichromatic individuals. The former are the differences in colour appearance of the monochromatic lights, daylight, and their mixtures. We will discuss this in Chapter 8. The latter can be analysed using the approach described above (Section 5.2).

5.3.1 Effect of the photopigment peak sensitivity on the $qp\lambda$-coordinates

The $qp\lambda$-coordinates obviously vary from observer to observer. In particular, they depend on the peak sensitivity of the photopigments. Generally, the relationship between the $qp\lambda$-coordinates and the photopigment peak sensitivities is quite complex and difficult for analytical expression in an explicit form. However, when the photopigment peak sensitivities vary within narrow limits, this relationship becomes rather transparent and allows a reasonable linear approximation. Figures 5.3 and 5.4 illustrate this point for the dominant wavelength, when the variation of the peak sensitivity of the L- and M- cones do not exceed that caused by genetic mutations.

In this section, we analyse the effect of the photopigment peak sensitivity of Govardovskii's cone fundamentals on the $qp\lambda$-coordinates. Let $f_1(\lambda; \mu_1^0)$, $f_2(\lambda; \mu_2^0)$, $f_3(\lambda; \mu_3^0)$ be Govardovskii's cone fundamentals, and $\mathcal{G}_0 = G\langle \mathbf{C}_+^*, \sim, \mu_1^0, \mu_2^0, \mu_3^0 \rangle$ the Grassmann–Govardovskii structure determined by them (Definition 3.4). Also we will consider a close Grassmann–Govardovskii structure $\mathcal{G} = G\langle \mathbf{C}_+^*, \approx, \mu_1, \mu_2, \mu_3 \rangle$ determined by the altered Govardovskii's cone fundamentals

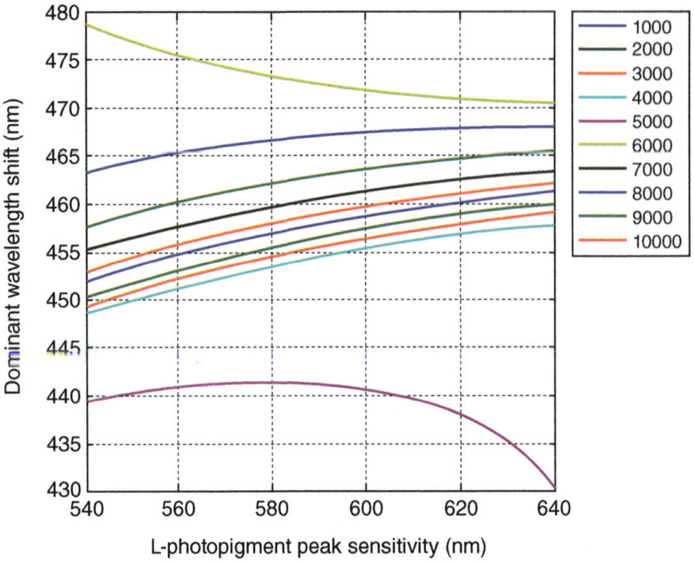

Figure 5.3 Dominant wavelength shift induced by a change in the L-photopigment peak sensitivity for the blackbody radiator of various temperatures (see Section 5.3.4 and Eq. (5.54)).

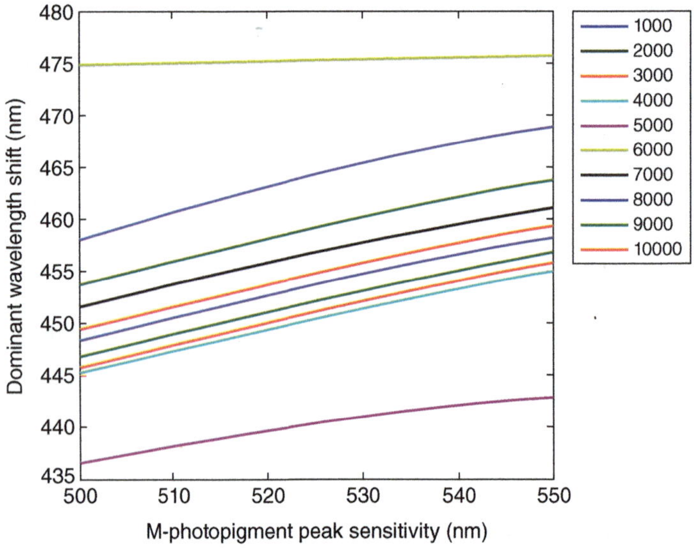

Figure 5.4 Dominant wavelength shift induced by a change in the *M*-photopigment peak sensitivity for the blackbody radiator of various temperatures.

$f_1(\lambda;\mu_1)$, $f_2(\lambda;\mu_2)$, and $f_3(\lambda;\mu_3)$. We will assume that $f_1(\lambda;\mu_1)$, $f_2(\lambda;\mu_2)$, and $f_3(\lambda;\mu_3)$ are continuously differentiable on Λ.[10] Given a light stimulus $\sigma_0 \in \mathbf{C}_+^*$, we suppose that both Grassmann–Govardovskii structures $\mathcal{G}_0$ and $\mathcal{G}$ are standard with respect to it and satisfy the hypotheses of Theorem 5.1. Also, we will suppose that the colour stimulus manifold $\mathcal{A}_{\sigma_0}$ is one and

10 In fact, we have to assume only that the spectral transmittance of the eye lens and the macular pigment are continuously differentiable. The Govardovskii templates (Eq. (3.17)) are smooth by definition.

the same for both $\mathcal{G}_0$ and $\mathcal{G}$[11]. Let $\kappa_{\sim}$ and $\kappa_{\approx}$ be the non-linear colour stimulus maps corresponding to $\mathcal{G}_0$ and $\mathcal{G}$, and σ a light stimulus with the dominant wavelengths λ_0 and λ with respect to $\mathcal{G}_0$ and $\mathcal{G}$, respectively. This means $\kappa_{\sim}(\sigma) = p_0\delta_{\lambda_0} + q_0\sigma_0$, thus,

$$\langle f_i(\cdot;\mu_i^0), p_0\delta_{\lambda_0} + q_0\sigma_0\rangle = \langle f_i(\cdot;\mu_i^0), \sigma\rangle, \quad i = 1,2,3, \tag{5.22}$$

and $\kappa_{\approx}(\sigma) = p\delta_\lambda + q\sigma_0$, thus,

$$\langle f_i(\cdot;\mu_i), p\delta_\lambda + q\sigma_0\rangle = \langle f_i(\cdot;\mu_i), \sigma\rangle, \quad i = 1,2,3. \tag{5.23}$$

Let us designate $x := (p, q, \lambda)^T$, $a := \left(p_0, q_0, \lambda_0\right)^T$, $y := \left(\mu_1, \mu_2, \mu_3\right)^T$, $b := \left(\mu_1^0, \mu_2^0, \mu_3^0\right)^T$ (it will be convenient to consider these vectors as column-vectors), and

$$F_i(x, y) := pf_i\left(\lambda;\mu_i\right) + \langle f_i(\cdot;\mu_i), q\sigma_0 - \sigma\rangle, \quad i = 1,2,3. \tag{5.24}$$

It follows from (5.23) that $F_i(x,y) = 0$ for $i = 1,2,3$; or designating $F(x,y) := \left(F_1(x,y), F_2(x,y), F_3(x,y)\right)$,

$$F(x,y) = 0.$$

Similarly,

$$F(a,b) = 0, \tag{5.25}$$

where a and b are connected via (5.22).

Equation (5.25) means that when metamerism $\approx$ is determined by Govardovskii's cone fundamentals $f_1\left(\lambda;\mu_1\right)$, $f_2\left(\lambda;\mu_2\right)$, and $f_3\left(\lambda;\mu_3\right)$, the $qp\lambda$-coordinates of a are equal to p, q, and λ. This condition determines which pairs of x and y are related by (5.25). (Obviously, the equality (5.25) can fail when x and y are chosen independently from some neighbourhoods of a and b, respectively, that is, when $p\delta_\lambda + q\sigma_0$ is an arbitrary stimulus from the colour stimulus manifold $\mathcal{A}_{\sigma_0}$, not necessarily $\kappa_{\approx}(\sigma)$). The dependence of x on y (or y on x) is, actually, a functional relationship, which can be derived from Eq. (5.25) using the implicit function theorem.

Abbreviating the Jacobi matrix $\left(J_F\right)_{|(a,b)}$ of the vector function F at (a,b) to $\left(\frac{\partial F}{\partial x}(a,b), \frac{\partial F}{\partial y}(a,b)\right)$, where

$$\frac{\partial F}{\partial x}(a,b) = \begin{pmatrix} \frac{\partial F_1}{\partial p}(a,b) & \frac{\partial F_1}{\partial q}(a,b) & \frac{\partial F_1}{\partial \lambda}(a,b) \\ \frac{\partial F_2}{\partial p}(a,b) & \frac{\partial F_2}{\partial q}(a,b) & \frac{\partial F_2}{\partial \lambda}(a,b) \\ \frac{\partial F_3}{\partial p}(a,b) & \frac{\partial F_3}{\partial q}(a,b) & \frac{\partial F_3}{\partial \lambda}(a,b) \end{pmatrix}, \tag{5.26}$$

and

$$\frac{\partial F}{\partial y}(a,b) = \begin{pmatrix} \frac{\partial F_1}{\partial \mu_1}(a,b) & \frac{\partial F_1}{\partial \mu_2}(a,b) & \frac{\partial F_1}{\partial \mu_3}(a,b) \\ \frac{\partial F_2}{\partial \mu_1}(a,b) & \frac{\partial F_2}{\partial \mu_2}(a,b) & \frac{\partial F_2}{\partial \mu_3}(a,b) \\ \frac{\partial F_3}{\partial \mu_1}(a,b) & \frac{\partial F_3}{\partial \mu_2}(a,b) & \frac{\partial F_3}{\partial \mu_3}(a,b) \end{pmatrix}, \tag{5.27}$$

let us evaluate the partial derivatives of F.

11 Admittedly, $\mathcal{G}_0$ and $\mathcal{G}$ might happen to have different effective visible spectrums. Still, firstly, as shown in Section 4.4, the difference between the effective visible spectrum ends caused by the difference in photopigment peak sensitivity (λ^{max}) is quite small. Secondly, reparametrising wavelength (as in Logvinenko, 2009b) one can avoid this problem. However, the ultimate solution to this problem would be to use the perceptual light colour coordinates (8.46–8.47).

We have for $i = 1, 2, 3$

$$\frac{\partial F_i}{\partial p}(a,b) = f_i(\lambda_0; \mu_i^0),$$
$$\frac{\partial F_i}{\partial q}(a,b) = \langle f_i(\cdot; \mu_i^0), \sigma_0\rangle,$$
$$\frac{\partial F_i}{\partial \lambda}(a,b) = p_0 \frac{\partial f_i}{\partial \lambda}\left(\lambda_0; \mu_i^0\right).$$

Recall that

$$f_i(\lambda; \mu) = f\left(\frac{\mu_i}{\lambda}\right),$$

where f is given by (3.16). Therefore,

$$\frac{\partial f_i}{\partial \lambda}\left(\lambda_0; \mu_i^0\right) = -\frac{\mu_i^0}{\lambda_0^2} f'\left(\frac{\mu_i^0}{\lambda_0}\right), \tag{5.28}$$

and

$$\frac{\partial f_i}{\partial \mu}\left(\lambda_0; \mu_i^0\right) = \frac{1}{\lambda_0} f'\left(\frac{\mu_i^0}{\lambda_0}\right). \tag{5.29}$$

Thus, we have

$$\frac{\partial F}{\partial x}(a,b) = \begin{pmatrix} f\left(\mu_1^0/\lambda_0\right) & \left\langle f\left(\mu_1^0/\cdot\right), \sigma_0\right\rangle & -(p_0\mu_1^0/\lambda_0^2) f'(\mu_1^0/\lambda_0) \\ f\left(\mu_2^0/\lambda_0\right) & \left\langle f\left(\mu_2^0/\cdot\right), \sigma_0\right\rangle & -(p_0\mu_2^0/\lambda_0^2) f'(\mu_2^0/\lambda_0) \\ f\left(\mu_3^0/\lambda_0\right) & \left\langle f\left(\mu_3^0/\cdot\right), \sigma_0\right\rangle & -(p_0\mu_3^0/\lambda_0^2) f'(\mu_3^0/\lambda_0) \end{pmatrix}. \tag{5.30}$$

Also, taking into account (5.29), we have

$$\frac{\partial F_i}{\partial \mu_i}(a,b) = \frac{p_0}{\lambda_0} f'\left(\frac{\mu_i^0}{\lambda_0}\right) + \frac{1}{\lambda_0}\left\langle f'\left(\frac{\mu_i^0}{\cdot}\right), q\sigma_0 - \sigma\right\rangle, \tag{5.31}$$

and for $i \neq j$, $\frac{\partial F_i}{\partial \mu_j}(a,b) = 0$, that is, the matrix (5.27) is diagonal.

If the matrix $\frac{\partial F}{\partial x}(a,b)$ is invertible (i.e., $\det\left(\frac{\partial F}{\partial x}(a,b)\right) \neq 0$), the function F implicitly defines locally a function $x = G(y)$ such that $a = G(b)$ and $F(G(y), y) \equiv 0$. In this case, differentiating the equation $F(x, y) = 0$ at (a, b) yields

$$\frac{\partial F}{\partial x}(a,b)\frac{\partial G}{\partial y}(b) + \frac{\partial F}{\partial y}(a,b) = 0,$$

(here $\frac{\partial G}{\partial y}$ stands for the matrix $\left(\partial G_i/\partial \mu_j\right)$) which entails

$$\frac{\partial G}{\partial y}(b) = -\left(\frac{\partial F}{\partial x}(a,b)\right)^{-1}\frac{\partial F}{\partial y}(a,b). \tag{5.32}$$

And finally, we have

$$dx = \frac{\partial G}{\partial y}(b)\, dy, \tag{5.33}$$

where $dx = (dq, dp, d\lambda)^T$ and $dy = \left(d\mu_1, d\mu_2, d\mu_3\right)^T$.

Equation (5.33) enables us to evaluate the shift in dominant wavelength ($d\lambda$) as well as the achromatic (dq) and chromatic (dp) shifts induced by alterations of the photopigment sensitivity peaks $d\mu_1, d\mu_2$, and $d\mu_3$.

The variations in the photopigment sensitivity peaks for trichromatic vision have been established by genetic studies (Section 3.3). There are some limits within which each of the three photopigments can vary (J. Neitz & Neitz, 2011). They are relatively small as compared to the values of the peak sensitivities, so an application of the formula (5.33) seems to be justified. This allows estimating the limits of colour matching variations (in terms of $qp\lambda$-coordinates) for any stimulus σ caused by interindividual differences in the photopigment sensitivity peaks, which can be observed for a population of trichromatic observers.

If the matrix $\frac{\partial F}{\partial y}(a, b)$ is also invertible, one can solve the inverse problem, that is, given shifts in dominant wavelength ($d\lambda$) and in achromatic (dq) and chromatic (dp) dimensions, to calculate the alteration in the photopigment sensitivity peaks $d\mu_1, d\mu_2$, and $d\mu_3$ which caused by the shift $(dq, dp, d\lambda)$. Indeed, differentiating the equation $F(x, y) = 0$ we get

$$\frac{\partial F}{\partial x}(a, b)\, dx + \frac{\partial F}{\partial y}(a, b)\, dy = 0. \tag{5.34}$$

If $\det\left(\frac{\partial F}{\partial y}(a, b)\right) \neq 0$, this equation can be resolved with respect to dy:

$$dy = -\left(\frac{\partial F}{\partial y}(a, b)\right)^{-1} \frac{\partial F}{\partial x}(a, b)\, dx. \tag{5.35}$$

We see that when both the partial derivatives $\frac{\partial F}{\partial x}(a, b)$ and $\frac{\partial F}{\partial y}(a, b)$ are invertible, the variables x and y are related to each other with a functional relationship in some neighbourhood of (a, b). Moreover, there is a bijective differentiable map $g : y \mapsto x$ and the inverse differentiable map $g^{-1} : x \mapsto y$, the Jacobi matrices of which are given in (5.32) and (5.35).

It follows that the match $\left(q\sigma_0 + p\delta_\lambda\right)$ for the stimulus σ (which can be determined in experiment) can be used to uniquely predict the photopigment sensitivity peaks $\left(\mu_1, \mu_2, \mu_3\right)$. It should be borne in mind, however, that this is the case only if the spectral transmittance of the ocular media (the eye lens and the macular pigment) are not subject to variability. When the spectral transmittance of the individual ocular media essentially deviates from that incorporated into the Govardovskii's cone fundamentals (Figs 3.7 and 3.6), such a deviation will contribute to the variability of the match. The effect of the ocular media transmittance on the colour match is studied in Section 5.3.2.

5.3.2 Effect of the ocular media transmittance on $qp\lambda$-coordinates

Let $g_1\left(\lambda; \mu_1\right)$, $g_2\left(\lambda; \mu_2\right)$, and $g_3\left(\lambda; \mu_3\right)$ be three Govardovskii's templates with fixed photopigment sensitivity peaks μ_1, μ_2, and μ_3. As the photopigment peak sensitivity will remain the same, we shall omit μ's abbreviating $g_1\left(\lambda; \mu_1\right)$, $g_2\left(\lambda; \mu_2\right)$, $g_3\left(\lambda; \mu_3\right)$ to $g_1(\lambda)$, $g_2(\lambda)$, $g_3(\lambda)$. We will assume that the spectral transmittance of the ocular media can be described by a function $t\left(\lambda, \xi_1, \dots, \xi_m\right)$ which depends on a set of parameters $\xi_1, \dots, \xi_m$. For instance, it can be represented as a linear combination of some continuously differentiable basis functions $h_1(\lambda), \dots, h_m(\lambda)$:

$$t\left(\lambda, \xi_1, \dots, \xi_m\right) = \sum_{j=1}^{m} \xi_j h_j(\lambda). \tag{5.36}$$

Let us choose a set of parameters $\left(\xi_1^0, \dots, \xi_m^0\right)$ and consider a Grassmann structure defined by the fundamentals: $g_i(\lambda)\, t\left(\lambda, \xi_1^0, \dots, \xi_m^0\right), i = 1, 2, 3$. Also, we will consider Grassmann structures defined by the altered fundamentals $g_i(\lambda)\, t\left(\lambda, \xi_1, \dots, \xi_m\right), i = 1, 2, 3$, assuming that all the structures under consideration have one and the same colour stimulus manifold $\mathcal{A}_{\sigma_0}$ and satisfy all the hypotheses of Theorem 5.1. Following the same line of reasoning as in Section 5.3.1, let us assume that there is a match between a light stimulus σ and a mixture $p_0\delta_{\lambda_0} + q_0\sigma_0$, that is,

$$\begin{aligned}\langle g_i(\cdot)t(\cdot,\xi_1^0,\dots,\xi_m^0),\sigma\rangle &= p_0 g_i(\lambda_0)t(\lambda_0,\xi_1^0,\dots,\xi_m^0)\\ &\quad+\langle g_i(\cdot)t(\cdot,\xi_1^0,\dots,\xi_m^0),q_0\sigma_0\rangle,\quad i=1,2,3.\end{aligned} \tag{5.37}$$

Using similar notations $x := (p,q,\lambda)^T$, $a := (p_0,q_0,\lambda_0)^T$, $y := (\xi_1,\dots,\xi_m)^T$, $b := (\xi_1^0,\dots,\xi_m^0)^T$, define a vector function:

$$F(x,y) := \left(F_1(x,y),F_2(x,y),F_3(x,y)\right),$$

where

$$F_i(x,y) := pg_i(\lambda)\,t(\lambda,y) + \langle g_i(\cdot)t(\cdot,y), q\sigma_0 - \sigma\rangle,\quad i=1,2,3.$$

Equation (5.37) implies that (a,b) satisfies the equation $F(x,y)=0$. Differentiating this equation, we get an equation similar to (5.34):

$$\frac{\partial F}{\partial x}(a,b)\,dx + \frac{\partial F}{\partial y}(a,b)\,dy = 0,$$

which can be resolved with respect to dx provided that $\det\left(\frac{\partial F}{\partial x}(a,b)\right)\neq 0$:

$$dx = -\left(\frac{\partial F}{\partial x}(a,b)\right)^{-1}\frac{\partial F}{\partial y}(a,b)\,dy. \tag{5.38}$$

The partial derivatives are

$$\begin{aligned}\frac{\partial F_i}{\partial p}(a,b) &= g_i(\lambda_0)\,t(\lambda_0,b),\\ \frac{\partial F_i}{\partial q}(a,b) &= \langle g_i(\cdot)t(\cdot,b),\sigma_0\rangle,\\ \frac{\partial F_i}{\partial \lambda}(a,b) &= p_0\left[g_i'(\lambda_0)\,t(\lambda_0,b) + g_i(\lambda_0)\frac{\partial t}{\partial\lambda}(\lambda_0,b)\right],\\ \frac{\partial F_i}{\partial \xi_j}(a,b) &= p_0 g_i(\lambda_0)\frac{\partial t}{\partial \xi_j}(\lambda_0,b) + \left\langle g_i(\cdot)\frac{\partial t(\cdot,b)}{\partial \xi_j}, q\sigma_0-\sigma\right\rangle.\end{aligned}$$

When the spectral transmittance t is given as in (5.36) its partial derivatives are

$$\frac{\partial t}{\partial \lambda}(\lambda,y) = \sum_{j=1}^m \xi_j h_j'(\lambda),\ \text{and}\ \frac{\partial t}{\partial \xi_j}(\lambda,y) = h_j(\lambda).$$

Therefore, we get

$$\frac{\partial F_i}{\partial \lambda}(a,b) = p_0\left[g_i'(\lambda_0)\sum_{j=1}^m \xi_j^0 h_j(\lambda_0) + g_i(\lambda_0)\sum_{j=1}^m \xi_j^0 h_j'(\lambda_0)\right], \tag{5.39}$$

and

$$\frac{\partial F_i}{\partial \xi_j}(a,b) = p_0 g_i(\lambda_0)\,h_j(\lambda_0) + \langle g_i h_j, q_0\sigma_0 - \sigma\rangle. \tag{5.40}$$

The above approach can be extended to the non-parametric (infinite-dimensional) case where the spectral transmittance of the ocular media is described by a continuously differentiable function $t_0(\lambda)$. Considering such a function as an element of the Banach space $\mathbf{C}^1 = \mathbf{C}^1(\Lambda)$ and following the same line of reasoning as in the finite-dimensional case, we assume that there is a match between a light stimulus σ and a mixture $p_0\delta_{\lambda_0} + q_0\sigma_0$, that is,

$$\langle g_i t_0, \sigma\rangle = p_0 g_i(\lambda_0)t_0(\lambda_0) + \langle g_i t_0, q_0\sigma_0\rangle, \quad i = 1, 2, 3.$$

Let us define a vector function $F(x,t) := (F_1(x,t), F_2(x,t), F_3(x,t))$ where $x = (p, q, \lambda)^T$, $t \in \mathbf{C}^1$,

$$F_i(x,t) := pg_i(\lambda)t(\lambda) + \langle g_i t, q\sigma_0 - \sigma\rangle, \quad i = 1, 2, 3.$$

Let $a = (p_0, q_0, \lambda_0)$. One has $F(a, t_0) = 0$, and there is an open set U in $\mathbf{R}^3 \times \mathbf{C}^1$ such that $(a, t_0) \in U$ and $F(x, t+h) - F(x,t) = L_t(x)h \ \forall (x,t) \in U, h \in \mathbf{C}^1$, where $L_t(x)$ is a continuous linear operator $\mathbf{C}^1 \to \mathbf{R}^3$ given by:

$$L_t(x)h = \left(pg_i(\lambda)h(\lambda) + \langle g_i h, q\sigma_0 - \sigma\rangle\right)_{i=1,2,3}. \tag{5.41}$$

It follows that F is Frechêt differentiable on U, and its partial Frechêt derivative with respect to x is determined by the matrix:

$$\frac{\partial F}{\partial x}(x,t) = \begin{pmatrix} \frac{\partial F_1}{\partial p}(x,t) & \frac{\partial F_1}{\partial q}(x,t) & \frac{\partial F_1}{\partial \lambda}(x,t) \\ \frac{\partial F_2}{\partial p}(x,t) & \frac{\partial F_2}{\partial q}(x,t) & \frac{\partial F_2}{\partial \lambda}(x,t) \\ \frac{\partial F_3}{\partial p}(x,t) & \frac{\partial F_3}{\partial q}(x,t) & \frac{\partial F_3}{\partial \lambda}(x,t) \end{pmatrix}, \tag{5.42}$$

and the partial Frechêt derivative with respect to t is determined by the linear operator (5.41),

$$\frac{\partial F}{\partial t}(x,t) = L_t(x). \tag{5.43}$$

The partial derivatives in the matrix (5.42) are as follows:

$$\frac{\partial F_i}{\partial p}(x,t) = g_i(\lambda)\,t(\lambda),$$
$$\frac{\partial F_i}{\partial q}(x,t) = \langle g_i t, \sigma_0\rangle,$$
$$\frac{\partial F_i}{\partial \lambda}(x,t) = p\left[\frac{\partial g_i}{\partial \lambda}(\lambda)\,t(\lambda) + g_i(\lambda)\,\frac{\partial t}{\partial \lambda}(\lambda)\right].$$

Thus, the partial Frechêt derivatives (5.41) and (5.42) exist whenever $(x,t) \in U$ and they are continuous in (x,t). Moreover, if the matrix $\frac{\partial F}{\partial x}(a, t_0)$ is invertible (i.e., $det\left(\frac{\partial F}{\partial x}(a,t_0)\right) \neq 0$) then, by the infinite-dimensional version of the implicit function theorem (see, e.g., Kolmogorov & Fomin, 1999), there exist $\varepsilon > 0$, $\delta > 0$, and a map $x = G(t)$ defined for $\|t - t_0\| < \delta$ such that: (i) G is differentiable at $t = t_0$ and satisfies $F(t, G(t)) = 0$ and $\|G(t) - a\| < \varepsilon$ whenever $\|t - t_0\|_{\mathbf{C}^1} < \delta$, (ii) if $\|t - t_0\|_{\mathbf{C}^1} < \delta$, $\|x - a\| < \varepsilon$, and $F(x,t) = 0$ then necessarily $x = G(t)$. Furthermore, in such a case one has

$$\frac{\partial G}{\partial t}(t_0) = -\left(\frac{\partial F}{\partial x}(a, t_0)\right)^{-1}\frac{\partial F}{\partial t}(a, t_0), \tag{5.44}$$

which, taking into account (5.43), can be rewritten as:

$$\frac{\partial G}{\partial t}(t_0) = -\left(\frac{\partial F}{\partial x}(a, t_0)\right)^{-1} L_{t_0}(a). \tag{5.45}$$

Now, by applying (5.45), we can evaluate the colour stimulus shift $dx = (dp, dq, d\lambda)^T$ caused by a small ocular media transmittance shift (from t_0 to $t = t_0 + h$). Up to an infinitesimal $o(\|h\|_{C^1})$, it equals

$$dx = \frac{\partial G}{\partial t}(t_0)h, \tag{5.46}$$

hence

$$dx = -\left(\frac{\partial F}{\partial x}(a, t_0)\right)^{-1} L_{t_0}(a)h,$$

and by (5.41),

$$L_{t_0}(a)h = \begin{pmatrix} p_0 g_1(\lambda_0)h(\lambda_0) & + & \langle g_1 h, q_0\sigma_0 - \sigma\rangle \\ p_0 g_2(\lambda_0)h(\lambda_0) & + & \langle g_2 h, q_0\sigma_0 - \sigma\rangle \\ p_0 g_3(\lambda_0)h(\lambda_0) & + & \langle g_3 h, q_0\sigma_0 - \sigma\rangle \end{pmatrix}.$$

5.3.3 Trade-off between the ocular media spectral transmittance and the photopigment peak sensitivity in their effect on colour

We see that an alteration of both the ocular media spectral transmittance and the photopigment peak sensitivity affect colour. In this section, we show that the effect produced by the ocular media spectral transmittance can be expressed in terms of the equivalent effect produced by the photopigment peak sensitivity.

Consider a Grassmann structure[12] defined by the fundamentals

$$g_i(\lambda; \mu_i)\, t(\lambda, \xi_1, \dots, \xi_m), \quad i = 1, 2, 3,$$

where $g_i(\lambda; \mu_i)$ is Govardovskii's template. Let us assume that for two sets of parameters, $a = (\mu_1^0, \mu_2^0, \mu_3^0)$ and $b = (\xi_1^0, \dots, \xi_m^0)$, there is a match between a light stimulus σ and a mixture $p_0\delta_{\lambda_0} + q_0\sigma_0$, that is

$$\langle g_i(\cdot; \mu_i^0)t(\cdot, \xi_1^0, \dots, \xi_m^0), p_0\delta_{\lambda_0} + q_0\sigma_0\rangle = \langle g_i(\cdot; \mu_i^0)t(\cdot, \xi_1^0, \dots, \xi_m^0), \sigma\rangle, \quad i = 1, 2, 3.$$

Given $x = (\mu_1, \mu_2, \mu_3)^T$ and $y = (\xi_1, \dots, \xi_m)^T$, define a vector function $F(x, y) := (F_1(x, y), F_2(x, y), F_3(x, y))$ where

$$F_i(x, y) := p_0 g_i(\lambda_0; \mu_i)\, t(\lambda_0, y) + \langle g_i(\cdot; \mu_i)t(\cdot, y), q_0\sigma_0 - \sigma\rangle, \quad i = 1, 2, 3. \tag{5.47}$$

The partial derivatives of $F_i(x, y)$ are

$$\frac{\partial F_i}{\partial \mu_i}(a, b) = p_0 \frac{\partial g_i}{\partial \mu_i}(\lambda_0; \mu_i^0)\, t(\lambda_0, b) + \left\langle \frac{\partial g_i}{\partial \mu_i}(\cdot; \mu_i^0)t(\cdot, b), q_0\sigma_0 - \sigma \right\rangle \tag{5.48}$$

and

$$\frac{\partial F_i}{\partial \xi_j}(a, b) = p_0 g_i(\lambda_0; \mu_i^0)\, \frac{\partial t}{\partial \xi_j}(\lambda_0, b) + \left\langle g_i(\cdot; \mu_i^0)\frac{\partial t}{\partial \xi_j}(\lambda_0, b), q_0\sigma_0 - \sigma \right\rangle. \tag{5.49}$$

When t is as in (5.36), we have

$$\frac{\partial F_i}{\partial \xi_j}(a, b) = p_0 g_i(\lambda_0; \mu_i^0)\, h_j(\lambda_0) + \langle g_i(\cdot; \mu_i^0)h_j(\cdot), q_0\sigma_0 - \sigma\rangle. \tag{5.50}$$

---

12 In this and the next subsections, all the Grassmann structures under consideration are supposed to satisfy hypotheses of Theorem 5.1 and to have one and the same colour stimulus manifold $\mathcal{A}_{\sigma_0}$.

If $\det\left(\frac{\partial F}{\partial x}(a,b)\right) \neq 0$ the equation:

$$\frac{\partial F}{\partial x}(a,b)\,dx + \frac{\partial F}{\partial y}(a,b)\,dy = 0,$$

can be resolved relative to dx;

$$dx = -\left(\frac{\partial F}{\partial x}(a,b)\right)^{-1}\frac{\partial F}{\partial y}(a,b)\,dy. \tag{5.51}$$

As one can see, the effect of the perturbation of the ocular media spectral transmittance t caused by $dy = \left(d\xi_1,\dots,d\xi_m\right)^T$ on the colour of σ is equivalent (at the first approximation) to that of the perturbation of the photopigment sensitivity peaks by $dx = \left(d\mu_1, d\mu_2, d\mu_3\right)^T$ given by

$$dx = \left(\frac{\partial F}{\partial x}(a,b)\right)^{-1}\frac{\partial F}{\partial y}(a,b)\,dy. \tag{5.52}$$

To put it the other way round, the effect of a perturbation of the ocular media spectral transmittance t by $dy = \left(d\xi_1,\dots,d\xi_m\right)^T$ on the colour of σ can be compensated by the perturbation of the photopigment sensitivity peaks by $dx = \left(d\mu_1, d\mu_2, d\mu_3\right)^T$ given by (5.51). The latter will be referred to as the *equivalent (filter-induced) peak wavelength shift*.

Remark 5.3 *In analogy to the argument at the end of the previous section, the above reasoning can be extended to the non-parametric (infinite-dimensional) case where the spectral transmittance of the ocular media can be described by an arbitrary continuously differentiable function $t(\lambda)$ (hence $y = (\xi_1,\dots,\xi_m)$ and $b = (\xi_1^0,\dots,\xi_m^0)$ are replaced with t and t_0, respectively). Given $a = (\mu_1^0,\mu_2^0,\mu_3^0)^T$ and $t_0 \in \mathbf{C}^1$, we suppose that there is a match between a light stimulus σ and a mixture $p_0\delta_{\lambda_0} + q_0\sigma_0$, that is,*

$$\langle g_i(\cdot;\mu_i^0)t_0(\cdot), p_0\delta_{\lambda_0} + q_0\sigma_0\rangle = \langle g_i(\cdot;\mu_i^0)t_0(\cdot),\sigma\rangle, \quad i=1,2,3,$$

and consider a vector function $F(x,t) = (F_1(x,t), F_2(x,t), F_3(x,t))$ where

$$F_i(x,t) := p_0 g_i\left(\lambda_0;\mu_i\right) t(\lambda_0) + \langle g_i(\cdot;\mu_i)t(\cdot), q_0\sigma_0 - \sigma\rangle, \quad i=1,2,3.$$

Clearly, it is Frechêt differentiable on some open neighbourhood of (a,t_0), and if its partial Frechêt derivative with respect to x at (a,t_0),

$$\frac{\partial F}{\partial x}(a,t_0) = \begin{pmatrix} \frac{\partial F_1}{\partial \mu_1}\left(a,t_0\right) & \frac{\partial F_1}{\partial \mu_2}\left(a,t_0\right) & \frac{\partial F_1}{\partial \mu_3}\left(a,t_0\right) \\ \frac{\partial F_2}{\partial \mu_1}\left(a,t_0\right) & \frac{\partial F_2}{\partial \mu_2}\left(a,t_0\right) & \frac{\partial F_2}{\partial \mu_3}\left(a,t_0\right) \\ \frac{\partial F_3}{\partial \mu_1}\left(a,t_0\right) & \frac{\partial F_3}{\partial \mu_2}\left(a,t_0\right) & \frac{\partial F_3}{\partial \mu_3}\left(a,t_0\right) \end{pmatrix},$$

is invertible (i.e., $\det(\frac{\partial F}{\partial x}(a,t_0)) \neq 0$) then one has

$$dx = -\left(\frac{\partial F}{\partial x}\left(a,t_0\right)\right)^{-1}\frac{\partial F}{\partial t}\left(a,t_0\right)h, \tag{5.53}$$

where the partial Frechêt derivative with respect to t is the linear operator $L_{t_0}(a) : \mathbf{C}^1 \to \mathbf{R}^3$,

$$L_{t_0}(a)h = (p_0 g_i(\lambda_0;\mu_i^0)h(\lambda_0) + \langle g_i(\cdot;\mu_i^0)h(\cdot), q_0\sigma_0 - \sigma\rangle)_{i=1,2,3},$$

and the effect of the perturbation of the ocular media spectral transmittance t_0 by h on the colour of σ is then equivalent (at the first approximation) to that of the perturbation of the photopigment sensitivity peaks by $dx = \left(d\mu_1, d\mu_2, d\mu_3\right)^T$ given by: (5.53).

5.3.4 Dependence of the equivalent peak wavelength shift on light. Impossibility to overcome colour deficiency using a coloured filter

It is important to emphasise that the trade-off between the perturbations $(d\xi_1, \ldots, d\xi_m)^T$ and $(d\mu_1, d\mu_2, d\mu_3)^T$ as described by Eqs. (5.52) and (5.53) depends on light stimuli σ. In other words, the equivalent peak wavelength shift induced by the same filter will be different for different light stimuli σ.

To illustrate this dependence, consider an observer equipped with three photopigments the spectral sensitivities of which are given by (3.17) with the peak absorbance wavelength $\lambda^{\max}$ at 430, 530, and 560 nm, D being equal to 0.3 (Fig. 3.5). Let us denote this observer O_N. Assume observer O_N (with no pre-receptoral filter) sees the light emitted by the blackbody radiator of colour temperature $T = 50\ 000$ K through a neutral filter, the spectral transmittance of which is 50% at each wavelength (i.e., $t_0(\lambda) \equiv 0.5$).

A blackbody radiator is an ideal object which emits light the radiant spectral density function of which is a function of its temperature:

$$p(\lambda; T) = 8\pi hc\lambda^{-5}\left(\exp\left(\frac{hc}{kT\lambda}\right) - 1\right)^{-1}, \tag{5.54}$$

where c is the speed of light, h is Planck's constant, k is Boltzmann's constant, and T is the blackbody temperature taken on the Kelvin scale (Fig. 5.5). The blackbody radiator with $T = 5800 \div 6500$ K roughly approximates the Sun. The light emitted by many incandescent light bulbs is fairly well approximated by the ideal blackbody radiator with $T = 2800 \div 3300$ K. Some other light sources such as certain fluorescent or high-intensity discharge lamps are also rather close to a blackbody radiator. In fact, Eq. (5.54) defines a one-parameter family of illuminants, T being the parameter. The colour appearance of the blackbody radiation varies with T from low-saturated red through yellowish to low-saturated blue. For this reason, T is usually called *colour temperature*.

Choose the equal energy light as σ_0 (specifically, a light the spectral power distribution function of which takes 1 at each wavelength). Then for observer O_N, $p(\lambda; 50\ 000)$ is metameric to a mixture

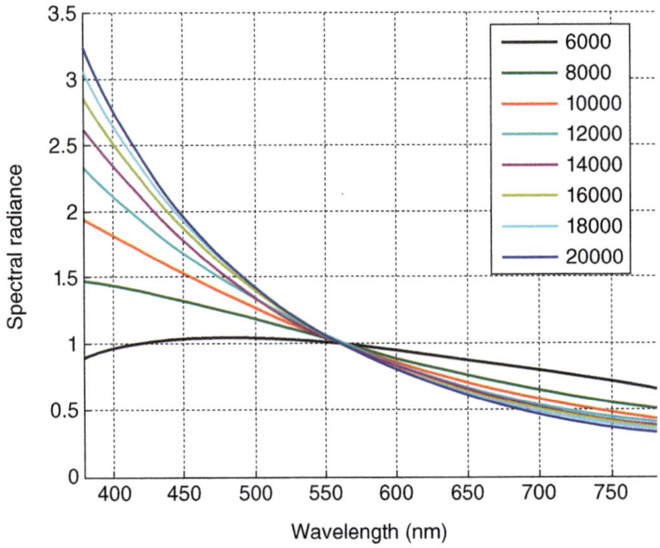

Figure 5.5 Spectral power distribution for the blackbody radiators of various colour temperatures ($T = 6000$–$20\ 000$).

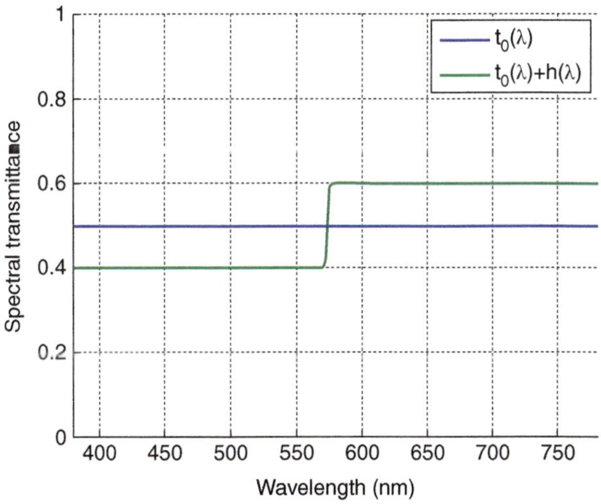

Figure 5.6 The profiles of $t_0(\lambda)$ and $t_0(\lambda)+h(\lambda)$. The parameters of the normal cumulative distribution function (CDF) determining $h(\lambda)$ are such that it is practically a step function with transition wavelength at 574 nm.

$p_0\delta_{\lambda_0}+q_0\sigma_0$ with $p_0=1.12$, $q_0=153.36$, and $\lambda_0=450.2$ nm. Let us evaluate the equivalent peak wavelength shift as defined by Eq. (5.53) using the cumulative distribution function of the normal distribution (with mean 574 and standard deviation 1) shifted down by 0.5 as $h(\lambda)$. When the amplitude of $h(\lambda)$ as in Fig. 5.6, the equivalent peak wavelength shift induced by $h(\lambda)$ is $d\mu_1=0$, $d\mu_2=7.11$ nm, and $d\mu_3=-23.03$ nm. Such an equivalent peak wavelength shift makes observer O_N a protanomalous trichromat (practically, a dichromat – protanope) when watching $p(\lambda; 50\ 000)$ through filter $t_0(\lambda)+h(\lambda)$.

Conversely, filter $t_0(\lambda)+h(\lambda)$ will enable a protanomalous observer with the photopigment peak absorbance wavelengths $\mu_1=430$ nm, $\mu_2=537.11$ nm, and $\mu_3=536.97$ nm (referred to as O_P) to see the light $p(\lambda; 50\ 000)$ practically the same way as normal observer O_N. Specifically, for O_P light $p(\lambda; 50\ 000)$ will have the dominant wavelength of 449.70 nm, q_0 and p_0 being 1.11 and 153.39, respectively. Therefore, for both the observer O_N (through the neutral filter $t_0(\lambda)$) and the observer O_P (through the filter $t_0(\lambda)+h(\lambda)$), light $p(\lambda; 50\ 000)$ will have almost the same dominant wavelength and colorimetric purity.

Imagine a unilateral protanope having one eye similar to that of observer O_N and the other to observer O_P. Then placing the neutral filter $t_0(\lambda)$ in front of the first eye and the coloured filter $t_0(\lambda)+h(\lambda)$ in front of the second is most likely to result in the identical (namely, normal) perception of the light $p(\lambda; 50\ 000)$ by both eyes. Does this mean that using the coloured filter $t_0(\lambda)+h(\lambda)$ one can overcome the protanomaly of the second eye?

Unfortunately, the answer is negative. The fact is that the equivalent peak wavelength shift produced by filter $h(\lambda)$ strongly depends on the light for which it was calculated. Figure 5.7 presents the equivalent peak wavelength shift evaluated for the same observer O_N and filter $h(\lambda)$ using black-body radiators of various colour temperatures.

As one can see, the equivalent peak wavelength shift for both the M- and L-cones varies considerably with colour temperature, though in opposite directions. The peak wavelength of the M-cones monotonically increases when colour temperature rises, whereas that of the L-cones decreases.

As a result, the equivalent peak wavelength difference between the M- and L-cones $(\mu_3-\mu_2)$ dramatically drops with colour temperature. It proves that for $T < 5770$ K, $\mu_3-\mu_2 > \mu_3^0-\mu_2^0 = 30$ nm.

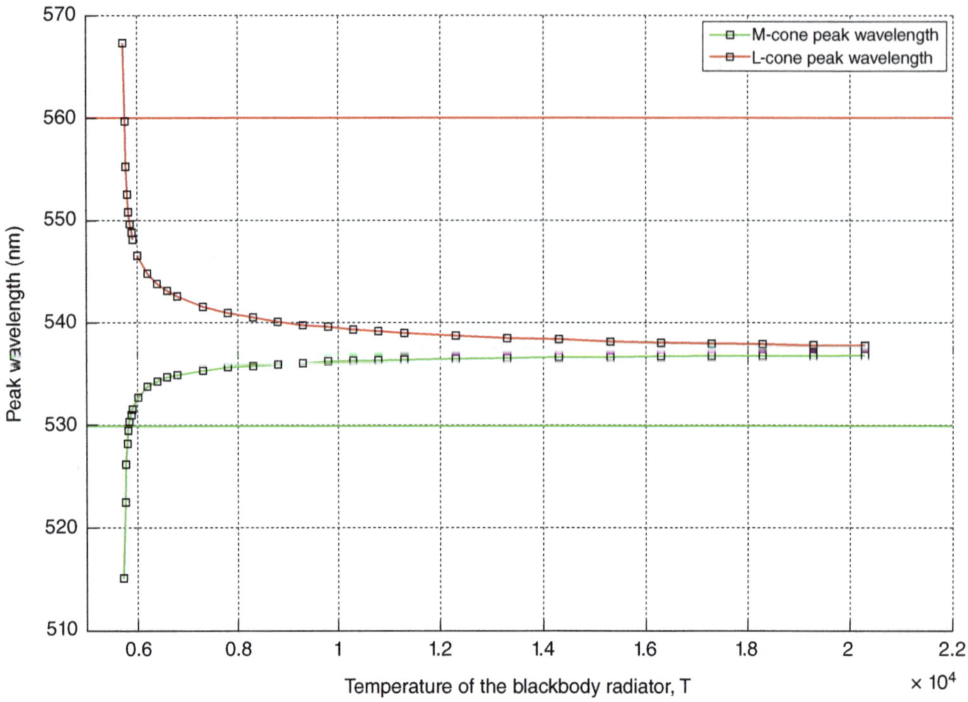

Figure 5.7 Equivalent peak wavelength shift as a function of the blackbody colour temperature.

So, depending on T, the same filter can produce an effect which is equivalent to moving the peak wavelengths both away from each other and towards one another.

5.3.5 Parametric identification of fundamental colour mechanisms

Consider trichromatic observers whose colour matching can be modelled by a three-chromatic Grassmann–Govardovskii structure determined by Govardovskii's fundamentals $f_i(\lambda; \mu_i) = g_i(\lambda; \mu_i)\, t(\lambda, \xi_1, \dots, \xi_m)$ $(i = 1, 2, 3)$, where $g_i(\lambda; \mu_i)$ is Govardovskii's template (3.16) and $t(\lambda, \xi_1, \dots, \xi_m)$ is the ocular media spectral transmittance. Assume that while unknown the parameters μ_1, μ_2, μ_3, and $\xi_1, \dots, \xi_m$ moderately deviate from some standard known values of these parameters which characterise on average the population of all the human trichromatic observers: $\mu_1^0, \mu_2^0, \mu_3^0$, and $\xi_1^0, \dots, \xi_m^0$, that is, the differences $\Delta\mu_i = \mu_i - \mu_i^0$ are small enough. Designating

$$y = \left(\mu_1, \mu_2, \mu_3, \xi_1, \dots, \xi_m\right)^T,$$
$$y^0 = \left(\mu_1^0, \mu_2^0, \mu_3^0, \xi_1^0, \dots, \xi_m^0\right)^T,$$
$$\Delta y = \left(\Delta\mu_1, \Delta\mu_2, \Delta\mu_3, \Delta\xi_1, \dots, \Delta\xi_m\right)^T,$$

we have $y = y^0 + \Delta y$.

We will refer to the Grassmann–Govardovskii structure determined by the fundamentals $g_i(\lambda; \mu_i^0)\, t(\lambda, \xi_1^0, \dots, \xi_m^0)$ as the standard observer, and that determined by the fundamentals $g_i(\lambda; \mu_i)\, t(\lambda, \xi_1, \dots, \xi_m)$ as the deviate observer. Let $p_0\delta_{\lambda_0} + q_0\sigma_0$ be a match for a light stimulus σ evaluated for the standard observer, and $p\delta_\lambda + q\sigma_0$ for the deviate observer. Designating $x := (p, q, \lambda)^T$, $x^0 := \left(p_0, q_0, \lambda_0\right)^T$, one can express a mismatch (for the deviated observer as compared to the standard observer) as a column vector $\Delta x := (\Delta p, \Delta q, \Delta\lambda)^T = x - x^0$. We will assume that one can determine the deviant observer match in experiment, so one can evaluate the mismatch Δx.

Define a vector function $F(x,y) := \left(F_1(x,y), F_2(x,y), F_3(x,y)\right)$, where

$$F_i(x,y) := pg_i\left(\lambda;\mu_i\right) t\left(\lambda,\xi_1,\ldots,\xi_m\right) + \langle g_i(\cdot;\mu_i)t(\cdot,\xi_1,\ldots,\xi_m), q\sigma_0 - \sigma\rangle, \tag{5.55}$$

$i = 1,2,3$. Note that $F\left(x^0,y^0\right) = 0$. Differentiating this equation at $\left(x^0,y^0\right)$ yields

$$\frac{\partial F}{\partial x}\left(x^0,y^0\right) dx + \frac{\partial F}{\partial y}\left(x^0,y^0\right) dy = 0. \tag{5.56}$$

The partial derivatives are given by:

$$\frac{\partial F_i}{\partial p}\left(x^0,y^0\right) = g_i\left(\lambda_0;\mu_i^0\right) t\left(\lambda_0,\xi_1^0,\ldots,\xi_m^0\right), \tag{5.57}$$

$$\frac{\partial F_i}{\partial q}\left(x^0,y^0\right) = \left\langle g_i\left(\cdot;\mu_i^0\right) t\left(\cdot,\xi_1^0,\ldots,\xi_m^0\right), \sigma_0\right\rangle, \tag{5.58}$$

$$\frac{\partial F_i}{\partial \lambda}\left(x^0,y^0\right) = p_0\left[t\left(\lambda_0,\xi_1^0,\ldots,\xi_m^0\right)\frac{\partial g_i}{\partial \lambda}\left(\lambda_0;\mu_i^0\right) + g_i\left(\lambda_0;\mu_i^0\right)\frac{\partial t}{\partial \lambda}\left(\lambda_0,\xi_1^0,\ldots,\xi_m^0\right)\right], \tag{5.59}$$

$$\begin{aligned}\frac{\partial F_i}{\partial \mu_i}\left(x^0,y^0\right) &= p_0\frac{\partial g_i}{\partial \mu_i}\left(\lambda_0;\mu_i^0\right) t\left(\lambda_0,\xi_1^0,\ldots,\xi_m^0\right)\\ &+\left\langle \frac{\partial g_i}{\partial \mu_i}\left(\cdot;\mu_i^0\right) t\left(\cdot,\xi_1^0,\ldots,\xi_m^0\right), q_0\sigma_0 - \sigma\right\rangle,\end{aligned} \tag{5.60}$$

$$\begin{aligned}\frac{\partial F_i}{\partial \xi_j}\left(x^0,y^0\right) &= p_0 g_i\left(\lambda_0;\mu_i^0\right)\frac{\partial t}{\partial \xi_j}\left(\lambda_0,\xi_1^0,\ldots,\xi_m^0\right)\\ &+\left\langle g_i\left(\cdot;\mu_i^0\right)\frac{\partial t}{\partial \xi_j}\left(\cdot,\xi_1^0,\ldots,\xi_m^0\right), q_0\sigma_0 - \sigma\right\rangle.\end{aligned} \tag{5.61}$$

If t is as in (5.36) the last equation takes a form:

$$\frac{\partial F_i}{\partial \xi_i}\left(x^0,y^0\right) = p_0 g_i\left(\lambda_0;\mu_i^0\right) h_j\left(\lambda_0\right) + \langle g_i(\cdot;\mu_i^0)h_j(\cdot), q_0\sigma_0 - \sigma\rangle. \tag{5.62}$$

If $\det\left(\frac{\partial F}{\partial x}\left(x^0,y^0\right)\right) \neq 0$ (5.56) entails

$$dx = -\left(\frac{\partial F}{\partial x}\left(x^0,y^0\right)\right)^{-1}\frac{\partial F}{\partial y}\left(x^0,y^0\right) dy. \tag{5.63}$$

The assumption made about the deviant observer can be specified now as that it justifies a replacement of the differentials dx and dy in (5.63) with the differences Δx and Δy, respectively:

$$\Delta x = -\left(\frac{\partial F}{\partial x}\left(x^0,y^0\right)\right)^{-1}\frac{\partial F}{\partial y}\left(x^0,y^0\right) \Delta y. \tag{5.64}$$

One can consider (5.64) as a system of linear equation with respect to Δy. Admittedly, as the number of the unknowns, $m+3$, is more than the number of equations, 3, it cannot be solved. Note, however, that one can determine a mismatch for another light stimulus. More specifically, choose M light stimuli $\sigma_1,\ldots,\sigma_M$. Let $x_k = \left(p_k, q_k, \lambda_k\right)^T$, and $\Delta x_k = \left(\Delta p_k, \Delta q_k, \Delta\lambda_k\right)^T$ be a match and a mismatch for σ_k $(k = 1,\ldots,M)$ respectively. Let D_k denote the matrix $-\left(\frac{\partial F}{\partial x}\left(x^0,y^0\right)\right)^{-1}\frac{\partial F}{\partial y}\left(x^0,y^0\right)$ as evaluated for $\sigma = \sigma_k$. Now consider $3M$ linear equations with $3+m$ unknowns $\Delta y = \left(\Delta\mu_1, \Delta\mu_2, \Delta\mu_3, \Delta\xi_1,\ldots,\Delta\xi_m\right)^T$,

$$\begin{aligned}\Delta x_1 &= D_1\Delta y,\\ &\cdots\\ \Delta x_M &= D_M\Delta y.\end{aligned} \tag{5.65}$$

Generally, this system has no exact solution. It should be said that even if the system (5.65) has a unique exact solution, it would be a better strategy to find a least squares solution to the system because the mismatches Δx_k found by observation are unavoidably contaminated by measurement error and noise. More specifically, let us make as many matches as to ensure $3M > m + 3$. Rewriting (5.65) in a matrix form:

$$A\Delta y = \overline{\Delta x}, \tag{5.66}$$

where

$$A := \begin{pmatrix} D_1 \\ \dots \\ D_M \end{pmatrix} \quad \text{and} \quad \overline{\Delta x} = \begin{pmatrix} \Delta x_1 \\ \dots \\ \Delta x_M \end{pmatrix},$$

let us find a vector $\overline{\Delta y}$ that minimises the Euclidean norm $\|A\overline{\Delta y} - \overline{\Delta x}\|$. As known, there is such a unique vector, and it is given by the pseudo-inverse matrix $A^{\#}$ of A. As known (for more on pseudo-inverse matrices, see Section 11.3.3), if the rows of A are linearly independent, then AA^T is invertible, and the pseudo-inverse matrix $A^{\#}$ can be expressed as:

$$A^{\#} = A^T(AA^T)^{-1}. \tag{5.67}$$

6

Light Metamerism

In this chapter, we will look into a problem of light metamerism. We shall assume an n-dimensional continuous Grassmann structure defined, as in Example 2.2, in terms of n fundamental colour mechanisms with spectral sensitivity functions $\varphi_1, \ldots, \varphi_n$ on Λ. Recall that two lights σ and σ' are called metameric with respect to these colour mechanisms if

$$\int_\Lambda \varphi_i(\lambda)\,\sigma(d\lambda) = \int_\Lambda \varphi_i(\lambda)\,\sigma'(d\lambda), \quad i = 1, \ldots, n. \tag{6.1}$$

Remark 6.1 *If lights σ and σ' are metameric, then the colour signals $\mathsf{F}(\sigma) = \left(g_1\left(\langle\varphi_1, \sigma\rangle\right), \ldots, g_n\left\langle\varphi_n, \sigma\right\rangle\right)$ and $\mathsf{F}(\sigma') = \left(g_1\left(\langle\varphi_1, \sigma'\rangle\right), \ldots, g_n\left\langle\varphi_n, \sigma'\right\rangle\right)$ are, obviously, equal. Moreover, $\mathsf{F}(\sigma) = \mathsf{F}(\sigma')$ even when $\mathsf{F} = G \circ \Phi$ where G is an arbitrary one-to-one map $\mathbf{R}^n_+ \to \mathbf{R}^n_+$. As stated earlier, the converse is also true. Namely, given $G : \mathbf{R}^n \to \mathbf{R}^n$ is one-to-one, and $\mathsf{F} = G \circ \Phi$, it follows from $\mathsf{F}(\sigma) = \mathsf{F}(\sigma')$ that $\left\langle\varphi_i, \sigma\right\rangle = \left\langle\varphi_i, \sigma'\right\rangle$ for each $i = 1, \ldots, n$. Therefore, metamerism is fully determined by the linear part of the colour mechanisms, that is, by the spectral sensitivity functions $\varphi_1, \ldots, \varphi_n$.*

A task of theoretical and practical importance is to describe all lights metameric to a particular light (the metamer set). This task is known as that of generating of metamers (Wyszecki & Stiles, 1982). The hitherto suggested algorithms of generating of metamers (Wyszecki & Stiles, 1982; Cohen, 2001) are based on the following obvious observation. All the metamers of a light σ_0 lie in the affine subspace passing through the point σ_0 parallel to a linear subspace

$$\ker(\Phi) := \{\sigma \in \mathbf{C}^* : \Phi(\sigma) = 0\}$$

(written as $\sigma_0 + \ker(\Phi)$). However, apart from proper light stimuli, $\sigma_0 + \ker(\Phi)$ contains imaginary ones (Definition 2.5). Hence, an important issue is to describe all the proper light stimuli metameric to an arbitrary light.

We give a general solution to this problem departing from the fact that the metamer set is convex. It turns out that positivity of the spectral sensitivity functions $\varphi_1, \ldots, \varphi_n$ on Λ implies existence of extreme points for it (Lemma 5.1). Next, we show how to evaluate these extreme points (Theorem 6.1). The metamer set is then represented as a convex hull of its extreme points (Corollary 6.1).

It must be said that while some previous studies made use of convex analysis when describing the metamer sets, they were restricted to discrete approximations of all the functions involved. Specifically, in these studies the visible interval of the spectrum, $\Lambda = [\lambda_{\min}, \lambda_{\max}]$, was divided into $N - 1$ equal sampling intervals $[\lambda_i, \lambda_{i+1}]$, $i = 1, \ldots, N - 1$, where $\lambda_i = \lambda_{\min} + \left(\lambda_{\max} - \lambda_{\min}\right)(i - 1)/(N - 1)$, $i = 1, \ldots, N$. The radiant spectral density functions of the lights and the spectral sensitivity functions of the fundamental colour mechanisms were then represented by N-dimensional vectors

Foundations of Colour Science: From Colorimetry to Perception, First Edition. Alexander D. Logvinenko and Vladimir L. Levin.
© 2023 John Wiley & Sons Ltd. Published 2023 by John Wiley & Sons Ltd.

(which could be interpreted as non-negative linear combinations of responses to monochromatic lights with wavelengths $\lambda_i, i = 1, \ldots, N$). In other words, in the previous studies the problem of describing the metamer set was posed only for the light stimulus space E from Example 2.1d. In this case, it is reduced to that of solving a system of n linear equations with respect to N unknowns.[1] The disadvantages of the finite-dimensional approach are obvious. For instance, it is not clear whether the solution of such a system will approach the metamer set when N goes to infinity. In other words, it is not clear if it is possible in principle to characterise the whole metamer set adopting the finite-dimensional approach.

Alternatively, we will completely describe the metamer sets in general (infinite-dimensional) case, where the lights are represented by arbitrary positive Borel measures on Λ. Practically, Theorem 6.1 provides the means to evaluate a finite-dimensional approximation (of an arbitrary high dimension) to the metamer set. The difference between traditional approaches and ours is that we give an approximation to the exact solution derived analytically, instead of an approximate solution in the approximating (finite-dimensional) light stimulus space.

6.1 Metamer sets

As follows from Remark 6.1, one can restrict our consideration to only linear colour mechanisms.

Definition 6.1 *Given $\varphi_1, \ldots, \varphi_n \in \mathbf{C}_+$ and a light $\sigma_0 \in \mathbf{C}_+^*$, we denote by $Met(\varphi_1, \ldots, \varphi_n; \sigma_0)$ the set of lights $\sigma \in \mathbf{C}_+^*$ that are metameric to σ_0 with respect to the colour mechanisms with spectral sensitivity functions $\varphi_1, \ldots, \varphi_n$ (the metamer set), that is,*

$$Met(\varphi_1, \ldots, \varphi_n; \sigma_0) := \{\sigma \in \mathbf{C}_+^* : \langle \varphi_i, \sigma \rangle = \langle \varphi_i, \sigma_0 \rangle,\ i = 1, \ldots, n\}. \tag{6.2}$$

Thus, $Met(\varphi_1, \ldots, \varphi_n; \sigma_0) = \mathbf{C}_+^* \cap (\sigma_0 + \ker(\Phi))$.

Remark 6.2 *It is easy to see that*

$$Met(\varphi_1, \ldots, \varphi_n; \sigma_0) = \bigcap_{i=1}^{n} Met(\varphi_i; \sigma_0). \tag{6.3}$$

Our objective is to describe $Met(\varphi_1, \ldots, \varphi_n; \sigma_0)$ for an arbitrary light σ_0. To this end, we shall use the following equivalent reformulation of Lemma 5.1.

Lemma 6.1 *Suppose*

$$\sum_{1}^{n} \varphi_i(\lambda) > 0 \text{ for every } \lambda \in \Lambda, \tag{6.4}$$

then $Met(\varphi_1, \ldots, \varphi_n; \sigma_0)$ is a weak* compact convex subset in $\mathbf{C}_+^*$.

It follows from (6.4) that $\langle \varphi_i, \sigma_0 \rangle > 0$ for at least one $i \in \overline{1, n}$. Note that the latter, in turn, entails $0 \notin Met(\varphi_1, \ldots, \varphi_n; \sigma_0)$. This means that the nil light is not metameric to any other (proper) light. Therefore, the metameric full dark, $Met(\varphi_1, \ldots, \varphi_n; 0)$, consists of the nil light only.[2]

---

1 For trichromatic human colour vision ($n = 3$) the number of sampling points N usually equals 31 (that corresponds to the 10 nm sampling interval).

2 As mentioned above, in colour vision literature the term 'metameric black' is used to refer to $\ker(\Phi)$. Note that the difference between this and our terminology is not only that, in accord with the terminology in Chapter 2, we use the term 'metameric full dark' instead of 'metameric black', but also that we use metameric full dark to refer to $\mathbf{C}_+^* \cap \ker(\Phi)$ rather than $\ker(\Phi)$. In other words, according to our definition, metamer sets contain only proper lights and/or the nil light.

By the Krein–Milman theorem (e.g., see Holmes, 1975), every weak* compact convex set in a dual Banach space is the weak* closed convex hull of its extreme points. Thus, knowing the extreme points of $Met(\varphi_1, \dots, \varphi_n; \sigma_0)$ enables us to obtain the whole metamer set.

Theorem 6.1 *Suppose condition (6.4) is satisfied.*

I. *If a light σ is an extreme point of $Met(\varphi_1, \dots, \varphi_n; \sigma_0)$, then a representation*

$$\sigma = \sum_{i=1}^{n} \alpha_i \delta_{\lambda_i} \tag{6.5}$$

holds, where $\alpha_i \geq 0$, $i = 1, \dots, n$, and

$$\sum_{i=1}^{n} \alpha_i \varphi_k(\lambda_i) = \langle \varphi_k, \sigma_0 \rangle, \quad k = 1, \dots, n. \tag{6.6}$$

II. *If (6.5) and (6.6) are satisfied and*

$$\det(\varphi_k(\lambda_i)) \neq 0, \tag{6.7}$$

then σ is an extreme point of $Met(\varphi_1, \dots, \varphi_n; \sigma_0)$.

Remark 6.3 *As follows from Lemma 2.1, condition (6.7) means that the functions $\varphi_1, \dots, \varphi_n$ are linearly independent on the subset $\{\lambda_1, \dots, \lambda_n\}$ in Λ.*

The general representation of $Met(\varphi_1, \dots, \varphi_n; \sigma_0)$ follows immediately from this and the Krein–Milman theorems along with Proposition 1.2 in Phelps (1966).

Corollary 6.1 *Consider a subset M of $\mathbf{R}^n \times \Lambda^n$ consisting of all sets of numbers $\alpha_1, \dots, \alpha_n, \lambda_1, \dots, \lambda_n$ satisfying (6.6), that is,*

$$M := \left\{ (\alpha_1, \dots, \alpha_n, \lambda_1, \dots, \lambda_n) : \sum_{i=1}^{n} \alpha_i \varphi_k(\lambda_i) = \langle \varphi_k, \sigma_0 \rangle,\ k = 1, \dots, n. \right\}.$$

For each $m = (\alpha_1, \dots, \alpha_n, \lambda_1, \dots, \lambda_n) \in M$ formula (6.5) defines a measure

$$\sigma_m := \sum_{i=1}^{n} \alpha_i \delta_{\lambda_i}.$$

Then, the metamer set $Met(\varphi_1, \dots, \varphi_n; \sigma_0)$ comprises the measures which can be represented as

$$\sigma = \int_M \sigma_m \mu(dm), \tag{6.8}$$

where μ runs over the set of probability measures on M.

Given n linearly independent functions $\varphi_1, \dots, \varphi_n \in \mathbf{C}_+$ satisfying (6.4), we consider the colour-signal map $\Phi = (\varphi_1, \dots, \varphi_n)$ determined by them. Recall (see Section 2.3) that $\mathcal{H}(\Phi)$ denotes the class of hyperplanes H in $\mathbf{R}^n$ such that $0 \notin H$, H has common points with the colour-signal cone $\Phi(\mathbf{C}_+^*)$, and no ray of the colour-signal cone is parallel to H.

Theorem 6.2 *Let us assume that two monochromatic lights, $\alpha\delta_\lambda$ and $\alpha'\delta_{\lambda'}$, are metameric if and only if $\lambda = \lambda'$ and $\alpha = \alpha'$. If $H \in \mathcal{H}(\Phi)$ and $\Phi(\sigma_0) \in Ext(H \cap \Phi(\mathbf{C}_+^*))$[3], then $Met(\varphi_1, \dots, \varphi_n; \sigma_0) = \{\sigma_0\}$.*

---

3 In such a case, $\sigma_0 = \alpha\delta_\lambda$ where $\alpha > 0$ and $\Phi(\sigma_0) \in Ext(H \cap \Phi(\mathbf{C}_+^*))$ for any hyperplane $H \in \mathcal{H}(\Phi)$ passing through σ_0.

It follows that if the conditions of Theorem 6.2 are satisfied with σ_0 being a monochromatic light, $\sigma_0 = \alpha\delta_\lambda$, then there is no metamerism for this light. The spectral colour (Remark 2.3) induced by such a monochromatic light will be called *simple spectral colour*. A simple spectral colour can be induced by only a single light stimulus (a monochromatic light of a particular wavelength). A compound (i.e., not monochromatic) light cannot have simple spectral colour. Therefore, a spectral colour cannot be a mixture (i.e., arithmetic sum) of any other colours.

There is every indication that, for human trichromatic observers, the first condition of Theorem 6.2 (two monochromatic lights, $\alpha\delta_\lambda$ and $\alpha'\delta_{\lambda'}$, are metameric if and only if $\lambda = \lambda'$ and $\alpha = \alpha'$) is satisfied for every λ and λ'. While, as pointed out in Chapter 10, two monochromatic lights, δ_λ and $\delta_{\lambda'}$ with close λ and λ' may happen to be indiscriminable, it follows from Fig. 3.1 that the colour signals they bring about will be different.

The second condition ($\Phi(\sigma_0) \in Ext(H \cap \Phi(\mathbf{C}_+^*))$ for $H \in \mathcal{H}(\Phi)$) is not satisfied for every monochromatic light $\sigma_0 = \alpha\delta_\lambda$ with $\Phi(\sigma_0) \in H$ even when the strong colour-signal-cone-boundary hypothesis holds true. True, the strong colour-signal-cone-boundary hypothesis entails that the spectral colour-signal curve $\Phi(\Delta)$ is contained in the colour-signal cone boundary. Yet, it is not enough. It is necessary to require that every point in the spectral colour-signal curve is not only a boundary point but also an extreme one, that is, $g_H(\Phi(\Delta)) \subseteq Ext(H \cap \Phi(\mathbf{C}_+^*))$. If this requirement is fulfilled, we shall say that the *strict colour-signal-cone-boundary hypothesis* takes place. It ensures the absence of metamerism for monochromatic lights.

Remark 6.4 *Note that the strict colour-signal-cone-boundary hypothesis can hold true only for Grassmann structures the dimensionality of which $n \geq 3$.*

The strict colour-signal-cone-boundary hypothesis implies, in particular, that the spectral colour-signal curve $g_H(\Phi(\Delta))$ cannot contain linear regions, hence the Grassmann structure $\langle \mathbf{C}_+^*, \sim \rangle$ has the strict convexity property (see Chapter 4). At first glance, the long wavelength parts of the spectral colour-signal curves in Figs 4.6, 4.9–4.12 look straight enough to challenge the strict colour-signal-cone-boundary hypothesis. However, this apparent straightness is caused by that the *S*-cone fundamental response to a monochromatic light δ_λ when $\lambda > 550$ nm is very small (e.g., Fig. 3.13). It is known that trichromatic human observers can set a match between a monochromatic yellow and a mixture of monochromatic green and red (Wyszecki & Stiles, 1982). Yet, it does not mean that the difference between the colour signals produced by these monochromatic yellow and mixture of monochromatic green and red is zero. It only means that the difference is so small that it falls within the indiscriminable set (Definition 10.9). Moreover, Fig. 3.11 supports our assumption (Eq. (6.4)) that all the fundamental colour mechanisms (including the *S*-cone fundamental) produce the non-zero response within the whole interval Λ.

As we know, instead of the strong colour-signal-cone-boundary hypothesis, human trichromatic observers obey the colour-signal-cone-boundary hypothesis. It follows that only the fragment of the spectral colour-signal curve, $\Phi(\Delta(\Lambda'))$ (see Eq. (4.9)), is contained in the colour-signal cone boundary. Moreover, the smoothness of $g_H(\Phi(\Delta(\Lambda')))$ in Fig. 4.6 allows to assume that $g_H(\Phi(\Delta(\Lambda'))) \subseteq Ext(H \cap \Phi(\mathbf{C}_+^*))$, that is, every point in $g_H(\Phi(\Delta(\Lambda')))$ is an extreme point. Therefore, human trichromatic observers are most likely to exhibit no metamerism for all the monochromatic lights δ_λ in $\Delta(\Lambda')$ (i.e., such that $\lambda'_{min} < \lambda < \lambda'_{max}$).

6.2 Colour mechanisms' transformations preserving light metamerism

Recall that $Met(\varphi_1, \dots, \varphi_n; \sigma_0)$ is completely determined by the vector subspace $Lin(\varphi_1, \dots, \varphi_n) := \{a_1\varphi_1 + \dots + a_n\varphi_n \, : \, a_i \in \mathbf{R}, \ i = 1, \dots, n\} \subset \mathbf{C}$ spanned over $\varphi_1, \dots, \varphi_n$, rather than by the mechanisms $\varphi_1, \dots, \varphi_n$ themselves. In other words, if a mechanism ψ is a linear combination of $\varphi_1, \dots, \varphi_n$, and a light σ is metameric to σ_0 with respect to $\varphi_1, \dots, \varphi_n$, then $\langle \psi, \sigma \rangle = \langle \psi, \sigma_0 \rangle$. Therefore, any mechanism in $Lin(\varphi_1, \dots, \varphi_n)$ does not disturb the metamerism for σ_0 as determined by $\varphi_1, \dots, \varphi_n$. However, there may be mechanisms preserving metamerism for σ_0 as defined by $\varphi_1, \dots, \varphi_n$ which do not belong to $Lin(\varphi_1, \dots, \varphi_n)$.

Definition 6.2 *Given mechanisms $\varphi_1, \dots, \varphi_n \in \mathbf{C}_+$ and a light $\sigma \in \mathbf{C}_+^*$, we say that a mechanism $\psi \in \mathbf{C}_+$* preserves the set of metamers *of σ_0 with respect to $\varphi_1, \dots, \varphi_n$ if*

$$Met(\varphi_1, \dots, \varphi_n; \sigma_0) \subseteq Met(\psi; \sigma_0). \tag{6.9}$$

The following lemma shows how a mechanism ψ, preserving metamerism with respect to $\varphi_1, \dots, \varphi_n$, is related to $\varphi_1, \dots, \varphi_n$. This relationship is described in terms of annihilators of subsets in $\mathbf{C}$ and $\mathbf{C}^*$. The annihilator of a set $L \subset \mathbf{C}$ is defined as the following set in $\mathbf{C}^*$:

$$L^0 := \{\mu \in \mathbf{C}^* \, : \, \langle \varphi, \mu \rangle = 0 \ \text{ for all } \varphi \in L\}. \tag{6.10}$$

Likewise, the annihilator of a set $H \subset \mathbf{C}^*$ is the following set in $\mathbf{C}$:

$$H^0 := \{\varphi \in \mathbf{C} \, : \, \langle \varphi, \mu \rangle = 0 \ \text{ for all } \mu \in H\}. \tag{6.11}$$

It follows at once from these definitions that L^0 is always a weak* closed vector subspace in $\mathbf{C}^*$, and H^0 is a closed vector subspace in $\mathbf{C}$.

Lemma 6.2 *The following statements are equivalent:*

(i) $Met(\varphi_1, \dots, \varphi_n; \sigma_0) \subseteq Met(\psi; \sigma_0)$;
(ii) $\psi \in H_{\sigma_0}^0$, *where*

$$H_{\sigma_0} = \{\mu \in L^0 \, : \, \mu = -\sigma_0 + \sigma, \ \sigma \in \mathbf{C}_+^*\}$$

and $L = Lin(\varphi_1, \dots, \varphi_n)$.

Since $L^{00} = L$ (this holds true for any closed vector subspace) and $H_{\sigma_0} \subset L^0$, one has $L = L^{00} \subseteq H_{\sigma_0}^0$. In some cases, the latter inclusion may prove to be strict. When this happens, adding a new mechanism $\psi \in H_{\sigma_0}^0 \backslash L$ to the set $\{\varphi_1, \dots, \varphi_n\}$ cannot change the set of metamers of σ_0. The following proposition presents the most extreme case of such a situation.

Proposition 6.1 *Let $\chi_\Lambda \in L = Lin(\varphi_1, \dots, \varphi_n)$ where $\chi_\Lambda(\lambda) = 1$ for all $\lambda \in \Lambda$. Fix $\lambda_0 \in \Lambda$ and suppose that there exists a mechanism $\varphi \in L$ satisfying $\varphi(\lambda_0) > \varphi(\lambda)$ for all $\lambda \in \Lambda, \lambda \neq \lambda_0$. Then $Met\left(\varphi_1, \dots, \varphi_n; \delta_{\lambda_0}\right) = \left\{\delta_{\lambda_0}\right\}$ and $H_{\lambda_0}^0$ is the whole space $\mathbf{C}$. Therefore,*

$$Met\left(\varphi_1, \dots, \varphi_n; \delta_{\lambda_0}\right) \subseteq Met\left(\psi; \delta_{\lambda_0}\right) \tag{6.12}$$

for all $\psi \in \mathbf{C}_+$, *that implies*

$$Met\left(\varphi_1, \ldots, \varphi_n; \delta_{\lambda_0}\right) = Met\left(\chi_\Lambda, \varphi, \psi_3, \ldots, \psi_n; \delta_{\lambda_0}\right)$$

for all $\psi_3, \ldots, \psi_n \in \mathbf{C}_+$.

Example 6.1 *Consider a set of three mechanisms given by* $\varphi_i(\lambda) = b_i - (\lambda - \lambda_i)^2$, *where* λ_1, λ_2, *and* λ_3 *are three different points of* Λ, *and*

$$b_i > \max\left\{(\lambda_{\min} - \lambda_i)^2, (\lambda_{\max} - \lambda_i)^2\right\}, \quad i = 1, 2, 3.$$

The inequalities secure that φ_i *are positive on* Λ.

Also suppose that a system of linear equations

$$\sum_1^3 x_i = 0, \quad \sum_1^3 \lambda_i x_i = 0, \quad \sum_1^3 \left(b_i - \lambda_i^2\right) x_i = 1$$

has a solution. Then $\sum_1^3 \varphi_i(\lambda) x_i \equiv 1$, *and by Proposition 6.1 inclusion (6.12) holds for* λ_0 *being any of the points* λ_1, λ_2, *and* λ_3.

As we can see, for some particular light stimulus σ the set of mechanisms preserving metamers of σ (with respect to $\varphi_1, \ldots, \varphi_n$), $H_\sigma^0 \cap \mathbf{C}_+$, can be much broader than $Lin(\varphi_1, \ldots, \varphi_n) \cap \mathbf{C}_+$ (e.g., the whole cone $\mathbf{C}_+$). However, of great theoretical interest is a maximal set of mechanisms which preserve metamerism (with respect to $\varphi_1, \ldots, \varphi_n$) for *all* the colour stimuli (i.e., for all $\sigma \in \mathbf{C}_+^*$), that is, the set $\cap_{\sigma \in \mathbf{C}_+^*} H_\sigma^0 \cap \mathbf{C}_+$.

Definition 6.3 *The set* $\cap_{\sigma \in \mathbf{C}_+^*} H_\sigma^0 \cap \mathbf{C}_+$ *will be called the* set of mechanisms preserving metamerism *(for all colour stimuli).*

The set of mechanisms preserving metamerism shows how the original colour mechanisms can be transformed without breaking the metamerism relation. It obviously contains $Lin(\varphi_1, \ldots, \varphi_n)$. Indeed, if $\psi_1, \ldots, \psi_n \in Lin(\varphi_1, \ldots, \varphi_n)$, and each of the two systems of mechanisms, $\varphi_1, \ldots, \varphi_n$ and $\psi_1, \ldots, \psi_n$, is linearly independent, then $Met(\varphi_1, \ldots, \varphi_n; \sigma_0) = Met(\psi_1, \ldots, \psi_n; \sigma_0)$. The following proposition, which is a particular case of Proposition 2.3,II,[4] shows that, in fact, $\cap_{\sigma \in \mathbf{C}_+^*} H_\sigma^0 \cap \mathbf{C}_+ = Lin(\varphi_1, \ldots, \varphi_n) \cap \mathbf{C}_+$, that is, only linear transformations can preserve metamerism for *all* colour stimuli.

Proposition 6.2 *Let* $\varphi_1, \ldots, \varphi_n$ *and* $\psi_1, \ldots, \psi_m$ *be two systems of continuous functions. The following statements are equivalent:*

(i) $Lin(\varphi_1, \ldots, \varphi_n) = Lin(\psi_1, \ldots, \psi_m)$;
(ii) $Met(\varphi_1, \ldots, \varphi_n; \sigma_0) = Met(\psi_1, \ldots, \psi_m; \sigma_0)$ *for every* $\sigma_0 \in \mathbf{C}_+^*$.

If one, thus both, of these equivalent statements holds true, and both systems, $\varphi_1, \ldots, \varphi_n$ *and* $\psi_1, \ldots, \psi_m$, *are linearly independent, then* $m = n$ *and there exists a unique linear operator* $T : Lin(\varphi_1, \ldots, \varphi_n) \to Lin(\psi_1, \ldots, \psi_n)$ *such that* $T\varphi_i = \psi_i$, $i = 1, \ldots, n$.

---

4 An independent proof is given in Appendix B.

It follows from Remark 6.1 that any invertible transformation G of the colour signal will preserve metamerism as well. Therefore, a linear transformation T, as described by Proposition 6.2, followed by an invertible transformation G will not break metamerism. And vice versa, if some transformation of the colour mechanisms does not violate metamerism then it should be possible to describe it as a linear transformation of the linear colour mechanisms, followed by an invertible transformation of additive colour signal. In particular, if, as many authors assume (e.g., Judd & Wyszecki, 1975), light adaptation retains metamerism, it can be modelled as a composition of a linear transformation and an invertible transformation (of additive colour signals).

6.3 Light metamerism index

As we know (Theorem 6.2), there may be no metamer for some lights. The metamer set for such a light consists of one element – the light itself. Generally, the metamer sets for different lights can vary significantly. Of particular interest is the question of the extent to which metameric lights can differ from each other. It seems natural to evaluate the difference between light stimuli in terms of some norm defined in the light stimulus space. The maximal difference in norm between the metameric lights can serve as a quantitative measure of the diversity in the metamer set.

Specifically, we suggest to quantify the degree of metamerism within a metamer set $Met(\varphi_1, \dots, \varphi_n; \sigma_0)$ induced by a light σ_0 as follows:

$$\frac{\sup\{\|\sigma\| : \sigma \in Met(\varphi_1, \dots, \varphi_n; \sigma_0)\} - \inf\{\|\sigma\| : \sigma \in Met(\varphi_1, \dots, \varphi_n; \sigma_0)\}}{\|\sigma_0\|}, \tag{6.13}$$

where $\|\sigma\|$ stands for the standard (dual) norm in $\mathbf{C}^*$ (see Section A.1). This ratio will then be called the *light metamerism index*.

7

Light Metamer Mismatching

As discussed in Section 5.3, lights which are metameric for one observer may be not metameric for the other observer, and vice versa. That is, lights metameric for one observer may appear different for the other. In other words, the lights mapping into a single point of the colour-signal space for one observer will map into a whole set of points for the other observer. This set will be called *metamer-mismatch region*.[1] In Section 7.1, we show how to evaluate these regions.

7.1 Metamer-mismatch regions

Consider the light stimulus cone $A = \mathbf{C}_+^*$ with two weak* closed n-dimensional Grassmann structures on it determined by two equivalence relations (metamerisms): an original one, $\sim$, induced by a given linear colour-signal map $\Phi = (\varphi_1, \dots, \varphi_n)$, and an altered one, $\approx$, induced by a new (changed) linear colour-signal map $\Psi = (\psi_1, \dots, \psi_n)$. Recall that the metamer set (relative to some metamerism relation) of a light σ is defined to be the set of all lights that are metameric to σ (Definition 2.6). Given a light $\sigma_0 \in A$, we denote its metamer set (relative to $\sim$) as

$$Met(\sim; \sigma_0) = \{\sigma \in \mathbf{C}_+^* : \sigma \sim \sigma_0\}. \tag{7.1}$$

Definition 7.1 *We define the* metamer-mismatch region of σ_0 *to be the subset in* $\mathbf{R}^n$

$$M(\Psi, \Phi; \sigma_0) := \Psi(Met(\sim; \sigma_0)). \tag{7.2}$$

Also, we shall consider its Ψ*-preimage in* $\mathbf{C}_+^*$

$$Msm(\approx, \sim; \sigma_0) := \Psi^{-1}(M(\Psi, \Phi; \sigma_0)) = \bigcup_{\sigma' \in Met(\sim;\sigma_0)} Met(\approx; \sigma'). \tag{7.3}$$

Thus, one has

$$Met(\sim; \sigma_0) = \Phi^{-1}(\Phi(\sigma_0)) = Met(\varphi_1, \dots, \varphi_n; \sigma_0), \tag{7.4}$$

$$M(\Psi, \Phi; \sigma_0) = \Psi(Met(\varphi_1, \dots, \varphi_n; \sigma_0)), \tag{7.5}$$

and

$$Msm(\approx, \sim; \sigma_0) = \Psi^{-1}\left(\Psi(Met(\varphi_1, \dots, \varphi_n; \sigma_0)\right). \tag{7.6}$$

1 It must be noticed that this terminology, quite common in colour vision literature (Wyszecki & Stiles, 1982) implies no distinction between matching and metamerism. While, as mentioned above, this is generally not the case, we will use this terminology in this chapter since here we continue to adhere to the metameric matching hypothesis.

Foundations of Colour Science: From Colorimetry to Perception, First Edition. Alexander D. Logvinenko and Vladimir L. Levin.
© 2023 John Wiley & Sons Ltd. Published 2023 by John Wiley & Sons Ltd.

Note that, as follows from (7.3), the Ψ-preimage of the metamer-mismatch region, $Msm(\approx, \sim; \sigma_0)$, is completely determined by two metamerisms, $\sim$ and $\approx$, and does not depend on concrete colour-signal maps Φ and Ψ. Since both Grassmann structures are assumed to be weak* closed, all φ_i, ψ_i are necessarily weak* continuous as linear functionals on $\mathbf{C}^*$ hence continuous as functions on Λ (see Proposition 2.6). Also, we suppose that all these functions are non-negative and both collections of n functions, $(\varphi_1, \ldots, \varphi_n)$ and $(\psi_1, \ldots, \psi_n)$, satisfy (6.4); that is,

$$\sum_{i=1}^{n} \varphi_i(\lambda) > 0, \quad \lambda \in \Lambda, \tag{7.7}$$

and

$$\sum_{i=1}^{n} \psi_i(\lambda) > 0, \quad \lambda \in \Lambda. \tag{7.8}$$

(In such a case, (7.7) and (7.8) are satisfied for any other collections of n non-negative functions representing $\sim$ and $\approx$ respectively.)

Proposition 7.1 *The metamer-mismatch region, $M(\Psi, \Phi; \sigma_0)$, is a convex compact subset in $\mathbf{R}^n$, and its Ψ-preimage, $Msm(\approx, \sim; \sigma_0)$, is a convex and weak* compact subset in $\mathbf{C}^*$.*

By the Caratheodory theorem (e.g., see Holmes, 1975), any point of the metamer-mismatch region $M(\Psi, \Phi; \sigma_0)$ is then represented as a convex combination of at most $n + 1$ its extreme points, and by the Krein–Milman theorem (e.g., see Holmes, 1975), the set $Msm(\approx, \sim; \sigma_0)$ is the weak* closed convex hull of its extreme points. In Theorem 7.1, we get necessary conditions for given points $z \in M(\Psi, \Phi; \sigma_0)$ and $\sigma \in Msm(\approx, \sim; \sigma_0)$ to be extreme points of the corresponding sets.

Theorem 7.1

I. *If a colour signal $z \in \Psi(\mathbf{C}^*_+)$ is an extreme point of the metamer-mismatch region $M(\Psi, \Phi; \sigma_0)$, then there exists a light σ such that $\Psi(\sigma) = z$ and σ is an extreme point of $Msm(\approx, \sim; \sigma_0)$.*
II. *If a light σ is an extreme point of the set $Msm(\approx, \sim; \sigma_0)$, then there exists a light $\sigma' \in Met(\sim; \sigma_0)$ such that:*
   (a) *σ' is an extreme point of $M(\sigma, \sigma_0) := \{\sigma' \in \mathbf{C}^*_+ : \sigma' \sim \sigma_0, \sigma' \approx \sigma\}$;*
   (b) *σ is an extreme point of $Met(\approx; \sigma')$.*

Furthermore, there exist points $\lambda_1, \ldots, \lambda_n, \lambda_1^1, \ldots, \lambda_n^1, \lambda_1^2, \ldots, \lambda_n^2 \in \Lambda$ and non-negative numbers $\alpha_1, \ldots, \alpha_n, \alpha_1^1, \ldots, \alpha_n^1, \alpha_1^2, \ldots, \alpha_n^2$ such that:

$$\sigma = \sum_{i=1}^{n} \alpha_i \delta_{\lambda_i}, \tag{7.9}$$

$$\sigma' = \sum_{i=1}^{n} \left(\alpha_i^1 \delta_{\lambda_i^1} + \alpha_i^2 \delta_{\lambda_i^2} \right), \tag{7.10}$$

$$\sum_{i=1}^{n} \left(\alpha_i^1 \varphi_k(\lambda_i^1) + \alpha_i^2 \varphi_k(\lambda_i^2) \right) = \langle \varphi_k, \sigma_0 \rangle, \quad k = 1, \ldots, n, \tag{7.11}$$

and

$$\sum_{i=1}^{n} \left(\alpha_i^1 \psi_k(\lambda_i^1) + \alpha_i^2 \psi_k(\lambda_i^2) \right) = \langle \psi_k, \sigma \rangle, \quad k = 1, \ldots, n. \tag{7.12}$$

Given $2n$ points $\lambda_1^1, \dots, \lambda_n^1, \lambda_1^2, \dots, \lambda_n^2 \in \Lambda$, we consider a subset $Z = Z(\lambda_1^1, \dots, \lambda_n^1, \lambda_1^2, \dots, \lambda_n^2)$ in $\mathbf{R}^n$ that consists of vectors $z = (z_1, \dots, z_n)$ with components

$$z_k = \sum_{i=1}^{n} \left(\alpha_i^1 \psi_k(\lambda_i^1) + \alpha_i^2 \psi_k(\lambda_i^2)\right), \quad k = 1, \dots, n, \tag{7.13}$$

where α_i^1, α_i^2 are non-negative numbers satisfying (7.11). Clearly, Z is a compact convex polyhedron. Every $z \in Z$ is represented as $z = \Psi(\sigma')$ where σ' is given by (7.10), and as (7.11) holds true, $\sigma' \in Met(\sim; \sigma_0)$. Thus, $Z \subseteq M(\Psi, \Phi; \sigma_0)$; therefore, if $z \in Z$ is an extreme point of $M(\Psi, \Phi; \sigma_0)$, then it is an extreme point of Z, too.

The next result is then a direct consequence of Theorem 7.1.

Corollary 7.1 *If $z = (z_1, \dots, z_n)$ is an extreme point of $M(\Psi, \Phi; \sigma_0)$, then there exist points $\lambda_1^1, \dots, \lambda_n^1, \lambda_1^2, \dots, \lambda_n^2 \in \Lambda$ such that z is an extreme point of $Z(\lambda_1^1, \dots, \lambda_n^1, \lambda_1^2, \dots, \lambda_n^2)$.*

Taking into account the Caratheodory theorem, we get a representation of the metamer-mismatch region as follows.

Corollary 7.2 *For every $z = (z_1, \dots, z_n) \in M(\Psi, \Phi; \sigma_0)$ there are $n+1$ non-negative numbers $\alpha_1, \dots, \alpha_{n+1}$, $\alpha_1 + \dots + \alpha_{n+1} = 1$, and $n+1$ collections*

$$(\alpha_1^{1j}, \dots, \alpha_n^{1j}, \alpha_1^{2j}, \dots, \alpha_n^{2j}, \lambda_1^{1j}, \dots, \lambda_n^{1j}, \lambda_1^{2j}, \dots, \lambda_n^{2j}) \in \mathbf{R}_+^{2n} \times \Lambda^{2n},$$

$j = 1, \dots, n+1$, such that each vector

$$z^j = \left(\sum_{i=1}^{n} \left(\alpha_i^{1j} \psi_1(\lambda_i^{1j}) + \alpha_i^{2j} \psi_1(\lambda_i^{2j})\right), \dots, \sum_{i=1}^{n} \left(\alpha_i^{1j} \psi_n(\lambda_i^{1j}) + \alpha_i^{2j} \psi_n(\lambda_i^{2j})\right) \right)$$

is an extreme point of $Z(\lambda_1^{1j}, \dots, \lambda_n^{1j}, \lambda_1^{2j}, \dots, \lambda_n^{2j})$, and

$$z_k = \sum_{j=1}^{n+1} \alpha_j z_k^j = \sum_{j=1}^{n+1} \sum_{i=1}^{n} \alpha_j \left(\alpha_i^{1j} \psi_k(\lambda_i^{1j}) + \alpha_i^{2j} \psi_k(\lambda_i^{2j})\right), \quad k = 1, \dots, n. \tag{7.14}$$

Remark 7.1 *Since $Z = Z(\lambda_1^1, \dots, \lambda_n^1, \lambda_1^2, \dots, \lambda_n^2)$ is a convex polyhedron, the set of its extreme points, Ext(Z), is finite, and each of them can be explicitly expressed from (7.11).*

Let us consider an illustrative example.

Example 7.1 *Suppose $n = 1$; then the metamer-mismatch region is a closed segment in $\mathbf{R}^1$:*

$$M(\psi, \varphi; \sigma_0) = \left[\min_{\sigma \in Met(\varphi;\sigma_0)} \langle \psi, \sigma \rangle, \max_{\sigma \in Met(\varphi;\sigma_0)} \langle \psi, \sigma \rangle \right],$$

and $Ext(M(\psi, \varphi; \sigma_0))$ consists of two its end points. Given any $\lambda^1, \lambda^2 \in \Lambda$, every z from $Z(\lambda^1, \lambda^2)$ is of the form $z = \alpha^1 \psi(\lambda^1) + \alpha^2 \psi(\lambda^2)$, where α^1, α^2 are non-negative numbers satisfying $\alpha^1 \varphi(\lambda^1) + \alpha^2 \varphi(\lambda^2) = \langle \varphi, \sigma_0 \rangle$. It follows that $Z(\lambda^1, \lambda^2)$ is a segment in $\mathbf{R}^1$ joining the points $\langle \varphi, \sigma_0 \rangle \frac{\psi(\lambda^1)}{\varphi(\lambda^1)}$ and $\langle \varphi, \sigma_0 \rangle \frac{\psi(\lambda^2)}{\varphi(\lambda^2)}$, which are its extreme points (if $\psi(\lambda^1)/\varphi(\lambda^1) = \psi(\lambda^2)/\varphi(\lambda^2)$, this segment degenerates into a point). Furthermore, since any $z \in Z$ is represented as $z = \langle \psi, \alpha^1 \delta_{\lambda^1} + \alpha^2 \delta_{\lambda^2} \rangle$, where $\alpha^1, \alpha^2 \in \mathbf{R}_+$ and $\alpha^1 \delta_{\lambda^1} + \alpha^2 \delta_{\lambda^2} \in Met(\varphi; \sigma_0)$, one has

$$\min_{\sigma \in Met(\varphi;\sigma_0)} \langle \psi, \sigma \rangle \leq \langle \psi, \alpha^1 \delta_{\lambda^1} + \alpha^2 \delta_{\lambda^2} \rangle \leq \max_{\sigma \in Met(\varphi;\sigma_0)} \langle \psi, \sigma \rangle; \tag{7.15}$$

therefore, $Z(\lambda^1, \lambda^2)$ is contained in the metamer-mismatch region.

Note now that, since the set $Met(\varphi; \sigma_0)$ *is convex and weak* compact, the minimum and the maximum in (7.15) are attained respectively at measures* σ_1 *and* σ_2*, which are its extreme points. Then, by Theorem 6.1, there are* $\lambda^1_*, \lambda^2_* \in \Lambda$ *such that* $\sigma_1 = \frac{\langle\varphi,\sigma_0\rangle}{\varphi(\lambda^1_*)}\delta_{\lambda^1_*}$, $\sigma_2 = \frac{\langle\varphi,\sigma_0\rangle}{\varphi(\lambda^2_*)}\delta_{\lambda^2_*}$, *hence*

$$\min_{\sigma\in Met(\varphi;\sigma_0)} \langle\psi,\sigma\rangle = \langle\varphi,\sigma_0\rangle\frac{\psi(\lambda^1_*)}{\varphi(\lambda^1_*)}, \qquad \max_{\sigma\in Met(\varphi;\sigma_0)} \langle\psi,\sigma\rangle = \langle\varphi,\sigma_0\rangle\frac{\psi(\lambda^2_*)}{\varphi(\lambda^2_*)}.$$

It follows that the left end point of $M(\psi,\varphi;\sigma_0)$ *coincides with the left end point of* $Z(\lambda^1_*, \lambda^2)$ *whenever* $\lambda^2 \geq \lambda^1_*$*, and the right end point of* $M(\psi,\varphi;\sigma_0)$ *coincides with the right end point of* $Z(\lambda^1, \lambda^2_*)$ *whenever* $\lambda^1] \leq \lambda^2_*$*. This implies*

$$\frac{\psi(\lambda^1_*)}{\varphi(\lambda^1_*)} = \min_{\lambda\in\Lambda}\frac{\psi(\lambda)}{\varphi(\lambda)}, \qquad \frac{\psi(\lambda^2_*)}{\varphi(\lambda^2_*)} = \max_{\lambda\in\Lambda}\frac{\psi(\lambda)}{\varphi(\lambda)}$$

and $M(\psi,\varphi;\sigma_0) = Z(\lambda^1_*, \lambda^2_*)$.

The metamer-mismatch regions for the two human M-cones (Fig. 3.14) differing in the pick sensitivity by 2 nm are presented in Figs 7.1 and 7.2.

The next result on a representation of $Msm(\approx, \sim; \sigma_0)$ follows immediately from Theorem 7.1 if one takes into account the Krein–Milman theorem along with Proposition 1.2 in Phelps (1966).

Consider a subset Ω in $\mathbf{R}^{3n} \times \Lambda^{3n}$ that comprises all collections of non-negative numbers and points in Λ

$$\omega = (\alpha_1, \dots, \alpha_n, \alpha^1_1, \dots, \alpha^1_n, \alpha^2_1, \dots, \alpha^2_n, \lambda_1, \dots, \lambda_n, \lambda^1_1, \dots, \lambda^1_n, \lambda^2_1, \dots, \lambda^2_n) \tag{7.16}$$

satisfying (7.11) and the equations:

$$\sum_{i=1}^{n} (\alpha^1_i\psi_k(\lambda^1_i) + \alpha^2_i\psi_k(\lambda^2_i)) = \sum_{i=1}^{n}\alpha_i\psi_k(\lambda_i), \quad k = 1, \dots, n.$$

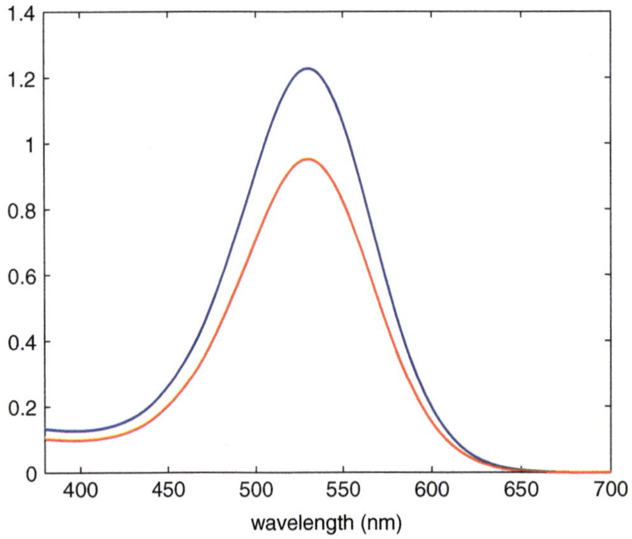

Figure 7.1 Metamer-mismatch region (between the blue and red lines) evaluated for the middle wavelength maximally sensitive human cone (the curve M in Fig. 3.14) considered as $\varphi(\lambda)$, and $\psi(\lambda) = \varphi(\lambda - \Delta\lambda)$ where $\Delta\lambda$ equals 2 nm. Specifically, the metamer-mismatch region for a monochromatic light δ_{λ_0} is the vertical interval between the blue and red lines at $\lambda = \lambda_0$. To be more exact, the blue line is $\max_{\lambda\in\Lambda}\frac{\psi(\lambda)}{\varphi(\lambda)}\varphi(\lambda)$, the red line being $\min_{\lambda\in\Lambda}\frac{\psi(\lambda)}{\varphi(\lambda)}\varphi(\lambda)$.

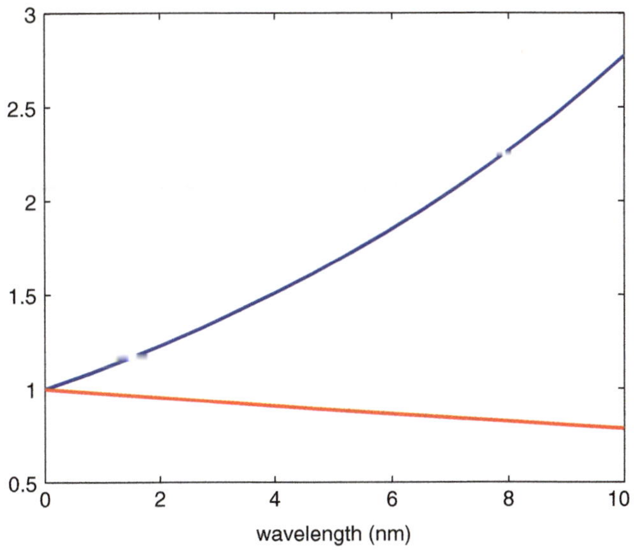

Figure 7.2 The graph of $\max_{\lambda \in \Lambda} \frac{\psi(\lambda)}{\varphi(\lambda)}$ (the blue line) and $\min_{\lambda \in \Lambda} \frac{\psi(\lambda)}{\varphi(\lambda)}$ (the red line) as a function of $\Delta\lambda$ (for more detail see Fig. 7.1) for $\lambda = 530$ nm.

Let M denote a subset in $\mathbf{R}^n \times \Lambda^n$ which is the natural projection of Ω onto coordinates $(\alpha_1, \dots, \alpha_n, \lambda_1, \dots, \lambda_n)$, $M = Pr_{\alpha,\lambda}(\Omega)$, where, for ω given by (7.16), $Pr_{\alpha,\lambda}(\omega) = (\alpha_1, \dots, \alpha_n, \lambda_1, \dots, \lambda_n)$. Taking into account (7.7) and (7.8) it is easy to see that M is a compact convex set in $\mathbf{R}^n \times \Lambda^n \subset \mathbf{R}^{2n}$.

Corollary 7.3 *For each*

$$m = \left(\alpha_1, \dots, \alpha_n, \lambda_1, \dots, \lambda_n\right) \in M$$

formula (7.9) defines a measure on Λ,

$$\sigma_m := \sum_{i=1}^{n} \alpha_i \delta_{\lambda_i},$$

and $Msm(\approx, \sim; \sigma_0)$ comprises the measures on Λ which can be represented as

$$\sigma = \int_M \sigma_m \mu\,(dm),$$

where μ runs over the set of probability measures on M.

7.2 Indices of light metamer mismatching

In this section, we will introduce several indices to quantify the degree of metamer mismatching. As a matter of fact, apart from its theoretical significance, metamer mismatching is of great practical importance. For example, metamer mismatching is a big problem in digital photography where it may be a cause of an essential difference between the appearance of the captured scene and its photographic picture. Manufacturers of cameras attempt to reduce metamer mismatching by making the spectral sensitivity of the camera's sensors as close as possible to the spectral sensitivity of human photoreceptors. However, because of manufacturing limitations and some other technical reasons, it is impossible to completely eliminate metamer mismatching for digital cameras. Hence, one needs a quantitative measure of metamer mismatching.

When two lights, metameric for one observer, look different in colour for the other, evidently, the larger the colour difference between these two lights for the second observer, the larger metamer mismatching for them. Establishing a reasonable quantitative measure of the similarity and/or dissimilarity between colours is then the first step when quantifying metamer mismatching. Although various colour difference formulae can be found in colour literature (Wyszecki & Stiles, 1982), they are all, in fact, variations on the theme of Euclidean distance in the colour-signal space. Admittedly, these formulae represent the difference between colour signals rather than colours. However, for lack of a better alternative we will accept, in this section, Euclidean distance between the colour signals as a measure of the difference in colour.[2]

Let $\Phi = (\varphi_1, \dots, \varphi_n)$ and $\Psi = (\psi_1, \dots, \psi_n)$ be the colour-signal maps of the two observers. Let us fix a light σ_0. Without loss of generality one can assume that $\Psi(\sigma_0) = \Phi(\sigma_0)$. (Indeed, this can be achieved by rescaling the functions $\varphi_1, \dots, \varphi_n$ preserving metamerism.) Let us evaluate the Euclidean distance between the colour signals $z_0 = \Psi(\sigma_0)$ and $z = \Psi(\sigma)$, where σ is a light metameric to σ_0 with respect to $\varphi_1, \dots, \varphi_n$, that is, $\sigma \in Met\left(\varphi_1, \dots, \varphi_n; \sigma_0\right)$. Since $\sigma \in Met(\varphi_1, \dots, \varphi_n; \sigma_0)$ means $\Phi(\sigma) = \Phi(\sigma_0)$, one has

$$||z - z_0|| = ||\Psi(\sigma) - \Psi(\sigma_0)|| = ||\Psi(\sigma) - \Phi(\sigma)|| = \sqrt{\sum_{i=1}^{n} \left(\langle \psi_i, \sigma\rangle - \langle \varphi_i, \sigma\rangle\right)^2},$$

and as $|\langle \psi_i, \sigma\rangle - \langle \varphi_i, \sigma\rangle| \le ||\sigma|| \, ||\psi_i - \varphi_i||_{\mathbf{C}}$, an estimate holds as follows:

$$||z - z_0|| \le ||\sigma|| \sqrt{\sum_{i=1}^{n} ||\psi_i - \varphi_i||^2_{\mathbf{C}}}. \tag{7.17}$$

One can observe that the closer the functions $\psi_1, \dots, \psi_n$ to $\varphi_1, \dots, \varphi_n$ in $\mathbf{C}$ with respect to the uniform norm $|| \cdot ||_{\mathbf{C}}$, the closer the point $z = \Psi(\sigma)$ to $z_0 = \Psi(\sigma_0)$ in $\mathbf{R}^n$.

A number of metamer-mismatch indices proposed so far for digital cameras are based on a simple idea that the closer the sensors' sensitivities are to the human cones' sensitivities the smaller the metamer-mismatch index should be (e.g., Neugebauer, 1956; Vora & Trussell, 1993). The estimate (7.17) justifies this intuitive idea. Specifically, assessing the goodness of two cameras – with sensors $\psi_1^1, \dots, \psi_n^1$ and $\psi_1^2, \dots, \psi_n^2$ – in comparison with human vision – with fundamental colour mechanisms $\varphi_1, \dots, \varphi_n$ – one must conclude that the camera for which $\sum_{i=1}^{n} ||\psi_i^j - \varphi_i||^2_{\mathbf{C}}, j = 1, 2;$ is smaller, will perform more favourably to human vision.

It should be kept in mind, however, that the idea to characterise metamer mismatching in general, that is, for all lights, is somewhat misleading since metamer mismatching is quite different for different lights, depending on σ_0. For example, if σ_0 has a simple spectral colour then $\Phi^{-1}\left(\Phi\left(\sigma_0\right)\right) = \sigma_0$. Thus, there will be no metamer mismatching for such σ_0. Therefore, any successful index of metamer mismatching should be local, that is, it should take into account the dependence of metamer mismatching on σ_0.

Note that $||z - z_0||$ in (7.17) does not exceed the diameter of the metamer-mismatch region $M(\Psi, \Phi; \sigma_0)$ which is defined as the biggest Euclidean distance between two points in $M(\Psi, \Phi; \sigma_0)$.

---

2 In our opinion, it would be more appropriate to define a measure of the similarity, and/or dissimilarity between colours in the light colour manifold (see Section 8.6) rather than in the colour stimulus manifold; not to mention the colour signal space. Such a measure must be validated in the experiment to make sure that it is, at least monotonously, related to the judgements of observers. At the end of Section 10.1 we show why the Euclidean metric is not suitable for this purpose.

Therefore, the diameter of the metamer-mismatch region $M(\Psi, \Phi; \sigma_0)$ can be taken as a measure of metamer mismatching. This measure can be used to compare between either two colour-signal maps (Ψ and Ψ') provided that σ_0 and Φ are fixed[3]; or between two lights (σ_0 and σ_0') provided that Φ and Ψ are fixed.

However, an obvious disadvantage of this measure is that it increases in direct proportion with light intensity even when the spectral composition of the light remains the same (i.e., $M(\Psi, \Phi; k\sigma_0) = kM(\Psi, \Phi; \sigma_0)$). One can make up for this shortcoming normalising the diameter of the metamer-mismatch region $M(\Psi, \Phi; \sigma_0)$ by the Euclidean length of the vector $\Psi(\sigma_0)$.

Definition 7.2 *Given a light σ_0, a colour-signal map $\Phi = (\varphi_1, \dots, \varphi_n)$, and an alterated colour-signal map Ψ, we define the* one-dimensional metamer-mismatch index *as a quotient*

$$I_1(\sigma_0; \Psi) := \frac{diam\Psi(Met(\varphi_1, \dots, \varphi_n; \sigma_0))}{||\Psi(\sigma_0)||}, \tag{7.18}$$

where $diam\Psi(Met(\varphi_1, \dots, \varphi_n; \sigma_0))$ stands for the diameter (the maximum Euclidean distance between two points) of $\Psi(Met(\varphi_1, \dots, \varphi_n; \sigma_0))$, $|| \cdot ||$ is the Euclidean norm in $\mathbf{R}^n$.

Remark 7.2 *Clearly $\Psi(Met(\varphi_1, \dots, \varphi_n; \sigma_0))$ is a compact convex set in $\mathbf{R}^n$; therefore its diameter equals the maximum distance between its extreme points.*

Remark 7.3 *The following estimation from below holds:*

$$I_1(\sigma_0; \Psi) \geq \frac{diam\Psi(Met(\varphi_1, \dots, \varphi_n; \sigma_0))}{||\Psi|| ||\sigma_0||},$$

where $||\Psi|| = \max_{\sigma \in \mathbf{C}^*} \frac{||\Psi(\sigma)||}{||\sigma||}$ is the norm of the linear operator $\Psi : \mathbf{C}^* \to \mathbf{R}^n$.

The following simple result gives an estimation of I_1 from above, which is coarse enough but might be useful in some applications.

Proposition 7.2 *Given σ_0, Φ, and Ψ, an estimate holds true as follows:*

$$I_1(\sigma_0; \Psi) \leq \frac{2\max\{||\sigma|| : \sigma \in Met(\varphi_1, \dots, \varphi_n; \sigma_0)\}}{||\Psi(\sigma_0)||} \sqrt{\sum_{i=1}^{n} ||\psi_i - \varphi_i||_{\mathbf{C}}^2}. \tag{7.19}$$

Here, the term $\sqrt{\sum_{i=1}^n ||\psi_i - \varphi_i||_{\mathbf{C}}^2}$ reflects the similarity between Φ and Ψ, and the factor $2\max\{||\sigma|| : \sigma \in Met(\varphi_1, \dots, \varphi_n; \sigma_0)\}/||\Psi(\sigma_0)||$ contains information related to σ_0.

The index (7.18) has two advantages: firstly, it allows a numerical estimate for specific σ_0, Φ, and Ψ; and secondly, it is scale invariant: $I_1(\alpha\sigma_0; \Psi) = I_1(\sigma_0; \Psi)$ and $I_1(\sigma_0; \alpha\Psi) = I_1(\sigma_0; \Psi)$ whenever $\alpha > 0$. A substantial disadvantage of the one-dimensional metamer-mismatch index is that it depends on the particular colour-signal map Ψ rather than the metamerism relation represented by this map. Namely, given two maps, Ψ and Ψ', representing the same metamerism relation ($\sim_\Psi = \sim_{\Psi'}$), the equality $I_1(\sigma_0; \Psi) = I_1(\sigma_0; \Psi')$, in general, does not hold. That is, being a characteristic of

3 It is intuitively clear that one of the two cameras (with the colour maps Ψ and Ψ') will be better with respect to the human colour vision (represented by the colour map Φ) if it disperses $\Phi^{-1}(\Phi(\sigma_0))$ into a smaller region in the corresponding (Ψ or Ψ') colour signal space.

a colour-signal map Ψ, the index $I_1(\sigma_0;\Psi)$, strictly speaking, cannot be regarded as an index of the metamer mismatching as such.

Now we introduce another index that does not have this drawback. To this end, consider two equivalence relations (metamerisms) on the light stimulus cone $\mathbf{C}_+^*$, an original metamerism $\sim$ and an altered metamerism $\approx$. We have thus two Grassmann structures on $\mathbf{C}_+^*$. Both structures are supposed to be n-dimensional and weak* closed. Given a light stimulus σ_0, its metamers (relative to $\sim$) form the set

$$Met(\sim;\sigma_0) := \{\sigma \in \mathbf{C}_+^* : \sigma \sim \sigma_0\}. \tag{7.20}$$

Further, we associate every $\sigma \in Met(\sim;\sigma_0)$ with its metamers relative to $\approx$:

$$Met(\approx;\sigma) := \{\sigma' \in \mathbf{C}_+^* : \sigma' \approx \sigma\}, \tag{7.21}$$

consider the set (see (7.3))

$$Msm(\approx,\sim;\sigma_0) = \bigcup_{\sigma \in Met(\sim;\sigma_0)} Met(\approx;\sigma), \tag{7.22}$$

and take its conic hull

$$con\ Msm(\approx,\sim;\sigma_0) := \{\alpha\sigma : \sigma \in Msm(\approx,\sim;\sigma_0), \alpha \in \mathbf{R}_+\}. \tag{7.23}$$

Given a colour-signal map Ψ representing the altered metamerism $\approx$, we define a norm $\|\cdot\|_\Psi$ in $\mathbf{R}^n$,

$$\|z\|_\Psi := \inf\{\|\sigma\| : \sigma \in \mathbf{C}^*, \Psi(\sigma) = z\}, \tag{7.24}$$

where $\|\sigma\|$ stands for the standard (dual) norm in $\mathbf{C}^*$ (see Section A.1).

Definition 7.3 *Given a light σ_0 and a colour-signal map Ψ representing the metamerism $\approx$, we define the n-*dimensional metamer-mismatch index as

$$I_n(\sigma_0;\Psi) := \frac{Vol_n\,(\Psi(con\ Msm(\approx,\sim;\sigma_0)) \cap B_\Psi)}{Vol_n\,(\Psi(\mathbf{C}_+^*) \cap B_\Psi)}, \tag{7.25}$$

where Vol_n stands for the standard n-dimensional volume in $\mathbf{R}^n$, and

$$B_\Psi := \{z \in \mathbf{R}^n : \|z\|_\Psi \le 1\}$$

is the unit ball in $(\mathbf{R}^n, \|\cdot\|_\Psi)$.

Though the index I_n is defined in terms of a concrete colour-signal map, Ψ, we shall see that, in fact, it does not depend on Ψ and is completely determined by $\approx$.

Let us consider a vector subspace $E(\approx) := \{\sigma \in \mathbf{C}^* : \sigma \approx 0\}$ (cf. (2.4)) in $\mathbf{C}^*$ and pass to the factor space $V := \mathbf{C}^*/E(\approx)$. Since $E(\approx)$ is a closed (even weak* closed) subspace in $\mathbf{C}^*$, the factor space V equipped with the standard factor norm

$$\|v\|_V := \inf\{\|\sigma\| : \sigma \in \mathbf{C}^*, \pi(\sigma) = v\},\ v \in V,$$

where $\pi(\sigma) = [\sigma]$ is the $\approx$-equivalence class of σ, is a Banach space. Note that Ψ is factored as $\Psi = \Psi_V \circ \pi$, where Ψ_V is a linear isometry of V onto $(\mathbf{R}^n, \|\cdot\|_\Psi)$. That is, $\Psi_V : V \to \mathbf{R}^n$ is a linear one-to-one map, and for all $v \in V$, $\|v\|_V = \|\Psi_V(v)\|_\Psi$ (the latter is a direct consequence of (7.24)).

If now Ψ' is another colour-signal map representing $\approx$, then the map $J_{\Psi\Psi'} := \Psi'_V \circ \Psi_V^{-1}$ proves to be a linear isometry of $(\mathbf{R}^n, \|\cdot\|_\Psi)$ onto $(\mathbf{R}^n, \|\cdot\|_{\Psi'})$, hence

$$B_{\Psi'} = J_{\Psi\Psi'}(B_\Psi). \tag{7.26}$$

Similarly, the colour-signal cone $\Psi'(\mathbf{C}_+^*)$ is the $J_{\Psi\Psi'}$-image of the colour-signal cone $\Psi(\mathbf{C}_+^*)$,

$$\Psi'(\mathbf{C}_+^*) = J_{\Psi\Psi'}(\Psi(\mathbf{C}_+^*)), \tag{7.27}$$

and

$$\Psi'(con\ Msm(\approx, \sim; \sigma_0)) = J_{\Psi\Psi'}(\Psi(con\ Msm(\approx, \sim; \sigma_0))). \tag{7.28}$$

Now, applying (7.25)–(7.28) yields

$$I_n(\sigma_0; \Psi') := \frac{Vol_n(J_{\Psi\Psi'}(\Psi(con\ Msm(\approx, \sim; \sigma_0)) \cap B_\Psi))}{Vol_n(J_{\Psi\Psi'}(\Psi(\mathbf{C}_+^*) \cap B_\Psi))}. \tag{7.29}$$

Since $J_{\Psi\Psi'}$ is a non-singular linear operator on $\mathbf{R}^n$, one has for any Borel set $B \subset \mathbf{R}^n$, $Vol_n(J_{\Psi\Psi'}(B)) = |\det(J_{\Psi\Psi'})| Vol_n(B)$, where $(J_{\Psi\Psi'})$ denotes the matrix of the operator $J_{\Psi\Psi'}$. Now, from (7.29) we infer that $I_n(\sigma_0; \Psi') = I_n(\sigma_0; \Psi)$; hence formula (7.25) determines a genuine metamer-mismatch index, which is fully determined by the metamerism $\approx$ and does not depend on its concrete representation Ψ.

We summarise the foregoing in the following statement.

Proposition 7.3 *Given two colour-signal maps, Ψ and Ψ', representing the altered metamerism $\approx$, the equality holds true as follows:*

$$I_n(\sigma_0; \Psi) = I_n(\sigma_0; \Psi'); \tag{7.30}$$

therefore, an invariant metamer-mismatch index

$$I(\sigma_0; \approx) := I_n(\sigma_0; \Psi) \tag{7.31}$$

is well defined.

Remark 7.4 *To evaluate numerically index (7.25), one has to calculate norms (7.24) of various $z \in Met(\varphi_1, \dots, \varphi_n; \sigma_0)$, what may be a fairly difficult task. Notice in that connection that if one replaces the ball B_Ψ in (7.25) with the Euclidean ball, so changed index ceases to be invariant. To overcome the above difficulty, we define now yet one index of metamer mismatching.*

Definition 7.4 *Given a light σ_0 and a colour-signal map Ψ representing the metamerism $\approx$, we define the* alternative n-dimensional metamer-mismatch index

$$I_n'(\sigma_0; \Psi) := \frac{Vol_n(M(\Psi, \Phi; \sigma_0))}{Vol_n(\Psi(P))}, \tag{7.32}$$

where $M(\Psi, \Phi; \sigma_0) = \Psi(Met(\varphi_1, \dots, \varphi_n; \sigma_0))$ is the metamer-mismatch region for σ_0, P is the set of probability measures on Λ, and $\Psi(P)$ is the convex hull of spectrum loci $\Psi(\Delta)$ (see Theorem 4.1).

If now Ψ' is a different linear colour-signal map representing $\approx$, then there exists a nonsingular linear operator $T : \mathbf{R}^n \to \mathbf{R}^n$ such that $\Psi' = T \circ \Psi$. Therefore $M(\Psi', \Phi; \sigma_0) = T(M(\Psi, \Phi; \sigma_0))$ and $\Psi'(P) = T(\Psi(P))$, and since a ratio of volumes in $\mathbf{R}^n$ is an affine invariant, alternative index (7.32) is invariant: $I_n'(\sigma_0; \Psi') = I_n'(\sigma_0; \Psi)$. Thus, index (7.32) is fully determined by the metamerism $\approx$ and does not depend on its concrete representation Ψ.

Remark 7.5 *Note that if the metamer-mismatch region, $M(\Psi, \Phi; \sigma_0)$, is not of full dimension, then both indices, (7.25) and (7.32) equal* 0.

7.3 Computing trichromatic metamer-mismatch regions

When $n > 1$ exact computation of metamer-mismatch regions (as in Example 7.1) is extremely difficult. In this section, we will calculate reasonable approximate estimates of the metamer-mismatch regions for some cone fundamentals considered in Section 3.3. Specifically, consider two sets of spectral sensitivity functions $s_1(\lambda), s_2(\lambda), s_3(\lambda)$; and $s_1'(\lambda), s_2'(\lambda), s_3'(\lambda)$, determining two colour-signal spaces Φ and Ψ respectively. Given a light σ_0, let us calculate Ψ-image of some extreme points of $\Phi^{-1}(\Phi(\sigma_0))$. They belong to the metamer-mismatch region $M(\Psi, \Phi; \sigma_0)$. Furthermore, some of them will also be extreme points of $M(\Psi, \Phi; \sigma_0)$. Therefore, the Ψ-image of a sample of extreme points of $\Phi^{-1}(\Phi(\sigma_0))$ can be considered as a reasonable approximation of the metamer mismatch region $M(\Psi, \Phi; \sigma_0)$. The larger the sample, the better the approximation.

More specifically, the metamer-mismatch region associated with some point z_0 (in the first colour-signal space) induced by light σ_0 was evaluated as follows. A triplet of wavelengths $\lambda_{i_1}, \lambda_{i_2}, \lambda_{i_3}$ was randomly chosen in the visible spectrum wavelength interval $[\lambda_{\min}, \lambda_{\max}]$. If the monochromatic lights corresponding to these wavelengths satisfied the conditions of Theorem 6.1 with some coefficients $k_{i_1} \geq 0, k_{i_2} \geq 0, k_{i_3} \geq 0$, that is,

$$
\begin{aligned}
k_{i_1} s_1(\lambda_{i_1}) + k_{i_2} s_1(\lambda_{i_2}) + k_{i_3} s_1(\lambda_{i_3}) &= \varphi_1(\sigma_0), \\
k_{i_1} s_2(\lambda_{i_1}) + k_{i_2} s_2(\lambda_{i_2}) + k_{i_3} s_2(\lambda_{i_3}) &= \varphi_2(\sigma_0), \\
k_{i_1} s_3(\lambda_{i_1}) + k_{i_2} s_3(\lambda_{i_2}) + k_{i_3} s_3(\lambda_{i_3}) &= \varphi_3(\sigma_0),
\end{aligned}
\tag{7.33}
$$

and

$$
\det \begin{pmatrix} s_1(\lambda_{i_1}) & s_1(\lambda_{i_2}) & s_1(\lambda_{i_3}) \\ s_2(\lambda_{i_1}) & s_2(\lambda_{i_2}) & s_2(\lambda_{i_3}) \\ s_3(\lambda_{i_1}) & s_3(\lambda_{i_2}) & s_3(\lambda_{i_3}) \end{pmatrix} \neq 0; \tag{7.34}
$$

then the Ψ-colour signal

$$
\begin{aligned}
&\Psi\left(k_{i_1}\delta(\lambda - \lambda_{i_1}) + k_{i_2}\delta(\lambda - \lambda_{i_2}) + k_{i_3}\delta(\lambda - \lambda_{i_3})\right) \\
&= \begin{pmatrix} k_{i_1} s_1'(\lambda_{i_1}) + k_{i_2} s_2'(\lambda_{i_2}) + k_{i_3} s_3'(\lambda_{i_3}) \\ k_{i_1} s_1'(\lambda_{i_1}) + k_{i_2} s_2'(\lambda_{i_2}) + k_{i_3} s_3'(\lambda_{i_3}) \\ k_{i_1} s_1'(\lambda_{i_1}) + k_{i_2} s_2'(\lambda_{i_2}) + k_{i_3} s_3'(\lambda_{i_3}) \end{pmatrix}
\end{aligned}
\tag{7.35}
$$

was calculated.[4] Point (7.35) belongs to the metamer-mismatch region $M(\Psi, \Phi; \sigma_0)$. Therefore, taking convex hull of a set of such points one can bring an approximation to the metamer-mismatch region $M(\Psi, \Phi; \sigma_0)$. (To remind that this is a closed convex set in the three-dimensional colour-signal space, it will be referred to here as the *metamer-mismatch volume*.)

First, the metamer-mismatch volumes were evaluated for a shift in the spectral peak sensitivity of the 3-*pigment Grassmann structure* (Definition 3.2). Specifically, the metamer-mismatch

4 Given determinant (7.34) is not zero, the coefficients were obtained by resolving Eq. (7.33) with respect to k_{i_1}, k_{i_2}, and k_{i_3}. Inequality (7.34) guaranteed the uniqueness of the solution. If this solution was negative (i.e., at least one of the coefficients was negative), it was dropped, and the next random sample of wavelengths was made.

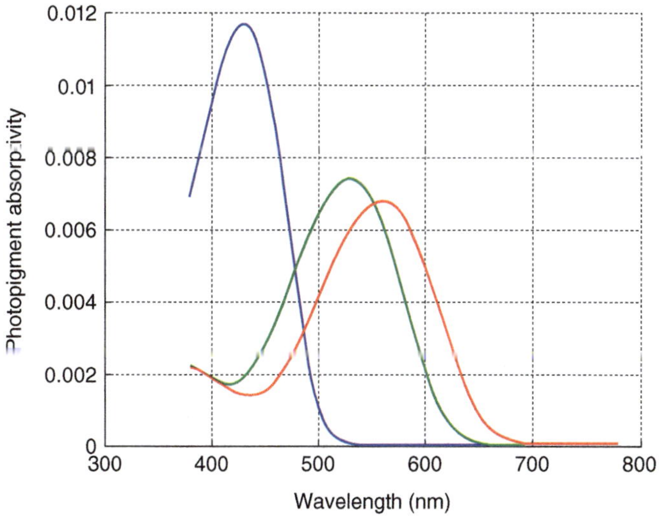

Figure 7.3 Cone photopigment absorptance curves, same as in Fig. 3.5 (with peak sensitivity λ^{max} at 430, 530, and 560 nm), scaled to unity area.

volumes were calculated for two sets of the cone photopigments (Fig. 7.3) with the following peak absorbance wavelengths. The S-cone peak absorbance wavelength λ_S^{max} was set up 429 and 431 nm for the first and second observer[5] respectively. The λ_M^{max} was set up 530 and 540 nm; and the L-cone peak absorbance wavelength λ_L^{max} was set up 550 and 560 nm respectively. Figure 7.4 presents the metamer-mismatch volume[6] induced by these two observers for the equal energy light (i.e., the spectral power of which is uniformly distributed across the visible spectrum).

Specifically, in this figure depicted are the colour signals of 10 000 of the metameric (for the first observer) mixtures of (randomly sampled) monochromatic triplets in the colour-signal space of the second observer. The convex hull of this colour-signal cluster (Fig. 7.5) is taken as an estimate of the metamer-mismatch volume. The one-dimensional (Eq. (7.18)) and n-dimensional (Eq. (7.32)) metamer-mismatch indices are 0.2850 and 0.0005 respectively. (For brevity, these indices will further on be referred to as *linear* (Eq. (7.18)) and *volumetric* (Eq. (7.32)).) For convenience, we rewrite Eq. (7.32) as

$$I_V\left(\mathbf{z}_0; \Phi, \Psi\right) = \frac{vol\left(M\left(\mathbf{z}_0; \Phi, \Psi\right)\right)}{vol(conv(C\,(\lambda)))}, \tag{7.36}$$

where $\mathbf{z}_0$ is a point in the colour-signal space $\Phi(\Lambda)$, $vol\left(M\left(\mathbf{z}_0; \Phi, \Psi\right)\right)$ is the volume of the metamer-mismatch volume $M\left(\mathbf{z}_0; \Phi, \Psi\right) = \Psi\left(\Phi^{-1}\left(\mathbf{z}_0\right)\right)$, and $vol(conv(C\,(\lambda)))$ is the volume of the convex hull of the spectral curve in the colour-signal space $\Psi\,(\Lambda)$. Index (7.36) will be referred to as *VMMI*. Also, for the sake of uniformity, we slightly change the notation for the linear index: $I_L(\sigma_0; \Phi, \Psi)$ instead of what is in Eq. (7.18).

5 By observer in the present context we mean a triplet of the spectral sensitivity functions defining the corresponding colour signal space. Notations Φ and Ψ will be used for the first and second (reference) observer respectively.
6 For brevity we call the cluster of points in this figure the metamer mismatch volume despite that the cluster is only its numerical approximation.

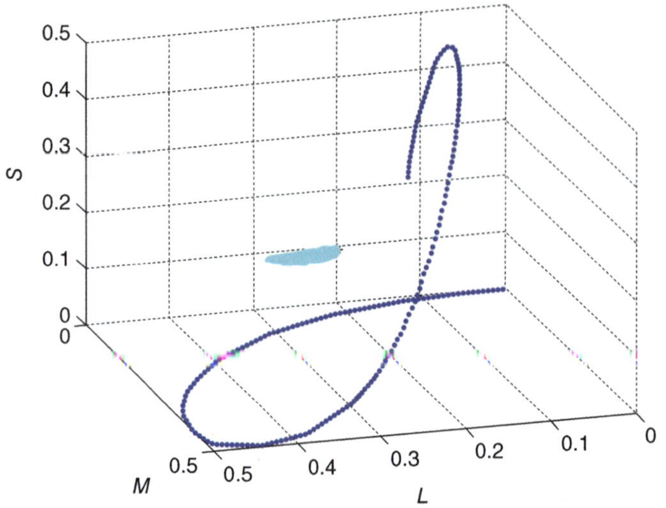

Figure 7.4 S-, M-, and L-cone photopigment excitations are along the axes. The spectral curve (blue) and the metamer-mismatch volume (cyan) are shown.

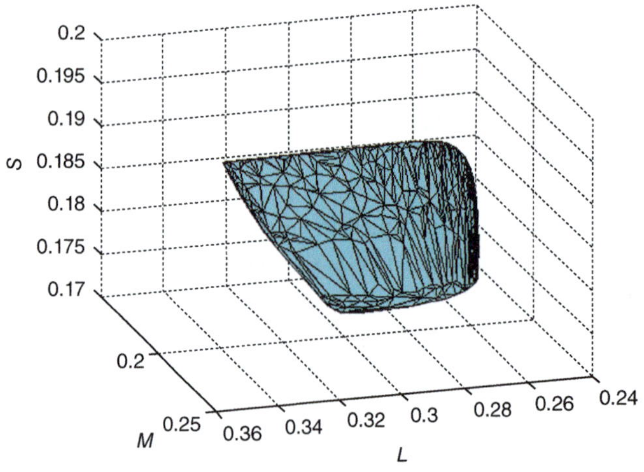

Figure 7.5 Convex hull of the cluster of the metamer-mismatch points depicted in Fig. 7.4.

As can be seen in Fig. 7.5, due to metamer mismatching, $\mathbf{z}_0$ scatters unequally along various coordinate axes. To demonstrate this difference linear metamer-mismatch indices for each of the three colour-signal components (S, M, and L) were computed. Specifically, S (respectively, M, and L) metamer-mismatch index was calculated as the maximal distance along the S (respectively, M, and L) coordinate axis between all the pairs in the metamer-mismatch volume. S, M, and L *coordinate* metamer-mismatch indices (denoted as *S-MMI*, *M-MMI*, and *L-MMI* respectively) were found to be 0.1209, 0.2164, and 0.3716 respectively.

The difference in the metamer mismatching range along the S and L axes is clearly observed in Fig. 7.6. As the cluster of points in Fig. 7.6 (referred to as the *chromaticity metamer-mismatch area*) shows the spread of only chromaticity induced by metamer mismatching, it is worthwhile to

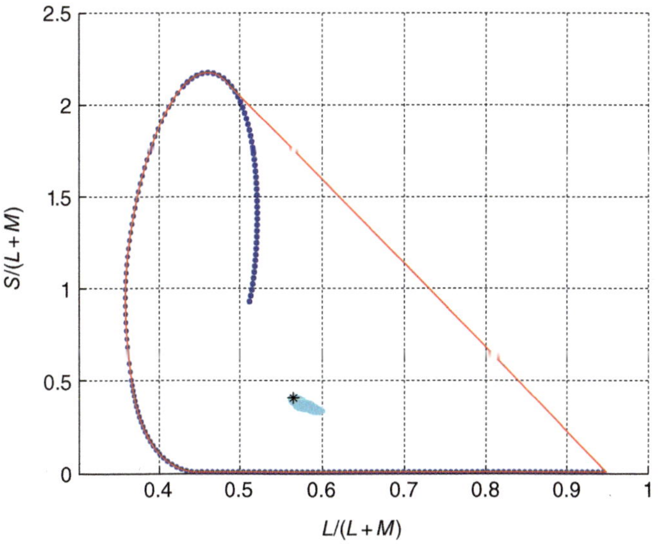

Figure 7.6 Chromaticity metamer-mismatch area for an equal energy light (marked with the black asterisk) induced by a shift in the peak photopigment absorption wavelengths. *S*, *M*, and *L* denote the excitation of the respective cone photopigments. Blue curve is the spectral locus, i.e., it represents the monochromatic lights. The cyan dots represent the chromaticity metamer-mismatch area. The red line outlines the convex hull of the spectral locus.

quantitatively characterise it with the following quotient:

$$I_C(\mathbf{z}_0;\Phi,\Psi) = \frac{area\left(M_A\left(\mathbf{z}_0;\Phi,\Psi\right)\right)}{area(conv(\bar{c}\,(\lambda)))}, \tag{7.37}$$

where $area\left(M_A\left(\mathbf{z}_0;\Phi,\Psi\right)\right)$ is the area of the projection of the metamer-mismatch volume on the chromaticity plane, and $area\,(conv\,(\bar{c}\,(\lambda)))$ is the area of the convex hull of the spectral locus $\bar{c}\,(\lambda)$. This quotient will be referred to as the *chromaticity metamer-mismatch index* (denoted as *CMMI*). It equals 0.0016 for the chromaticity metamer-mismatch area in Fig. 7.6.

It should be borne in mind, however, that the chromaticity metamer-mismatch index, generally, varies with choice of the chromaticity plane. For example, the chromaticity metamer-mismatch area in Fig. 7.7 is somewhat larger than in Fig. 7.6, the chromaticity metamer-mismatch index being 0.0023. (Note that here the spectral locus has a different shape than in Fig. 4.6 despited that it is based on cone photopigments with the same peak absorbance wavelengths. The reason is that the curves in Fig. 3.5 are normalised with respect to their peaks, whereas those in Fig. 7.3 with respect to their areas.)

Dispersing along the luminance axis can also be quantified with an index similar to the *S*, *M*, and *L* metamer-mismatch indices defined above. Specifically, it can be evaluated as the maximal distance along the luminance axis between all the pairs in the metamer-mismatch volume (referred to as LumMMI). Luminance metamer-mismatch index for the metamer-mismatch volume in Fig. 7.4 was found to be 0.3038.

7.3.1 Effect of the spectral positioning of photopigments on metamer mismatching

In this section, we explore how the amount of metamer mismatching depends on the peak absorbance shift for each cone type. Specifically, we ascertain the effect of variations in $\lambda_M^{\max}$ and

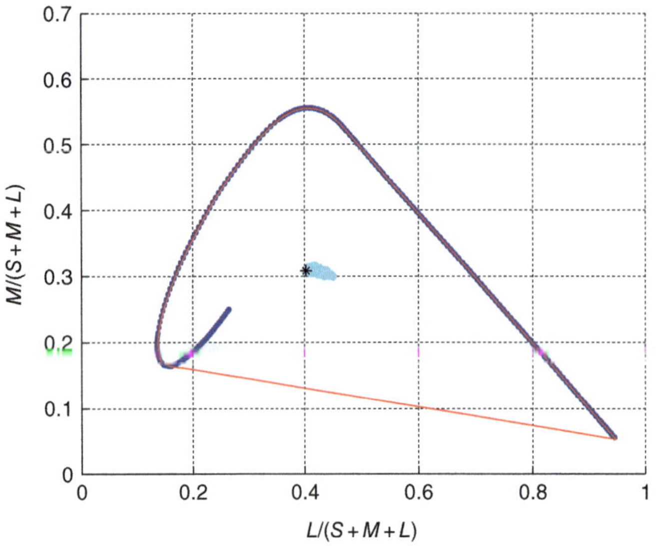

Figure 7.7 The same chromaticity metamer-mismatch area as in Fig. 7.6, only plotted in a different chromaticity plane. Notations are the same as in Fig. 7.6.

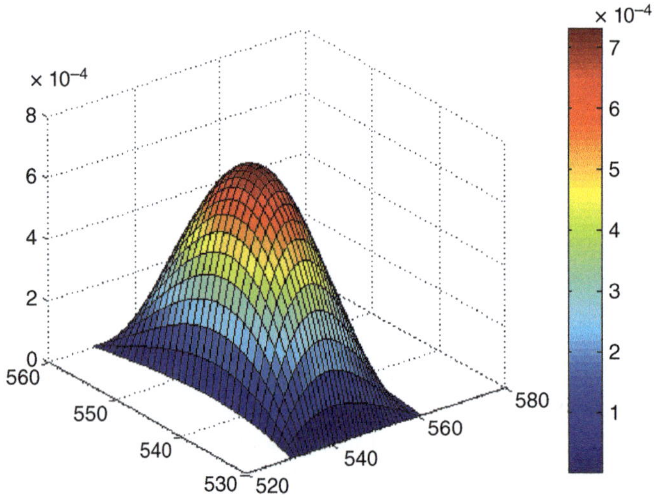

Figure 7.8 Volumetric metamer-mismatch index as a function of λ_M^{max} and λ_L^{max}.

λ_L^{max} on metamer mismatching by calculating the volumetric metamer-mismatch index (7.32) and the chromaticity metamer-mismatch index (7.37), and keeping λ_S^{max} constant at 429 and 431 nm for the first and second (reference) observer respectively. The range for varying λ_M^{max} and λ_L^{max} was chosen so as to simulate the variations which could occur due to genetic mutation and recombination (J. Neitz & Neitz, 2011). Specifically, the peak absorbance wavelengths λ_M^{max} and λ_L^{max} were set 530 and 560 nm for the reference observer. For the first observer λ_M^{max} varied between 530.5 to 560.5 nm, and λ_L^{max} from 530.25 to 560.25 nm with 1 nm step for both. Such samples for λ_M^{max} and λ_L^{max} were chosen to prevent degenerating the metamer-mismatch volumes to two-dimensional areas, that occurs when the peak absorbance wavelengths (for the first and second observers) coincide. In this case, volumetric metamer-mismatch index (7.32) degenerates to zero becoming an inappropriate measure of metamer mismatching.[7] Yet, chromaticity metamer-mismatch index (7.37) holds valid even for flat (two-dimensional) metamer-mismatch regions.

Not surprisingly, the graphs in Figs 7.8 and 7.9 are very different. Specifically, reaching its maximum (7.2736e-04) at $\lambda_M^{max} = 543.5$ nm and $\lambda_L^{max} = 543.25$ nm, volumetric metamer-mismatch index gradually decreases towards the periphery practically vanishing at the domain boundary (Fig. 7.8). In other words, the graph in Fig. 7.8 has its single peak at λ_M^{max} and λ_L^{max} which maximally distant from the λ_M^{max} and λ_L^{max} values set for the reference observer (i.e., 530 and 560 nm respectively). To the contrary, chromaticity metamer-mismatch index has its maximal value (0.0154) at $\lambda_M^{max} = 530.5$ nm and $\lambda_L^{max} = 530.25$ nm, and the second maximum (0.0103) at $\lambda_M^{max} = 560.5$ nm and $\lambda_L^{max} = 560.25$ nm. Hence, metamer mismatching (in terms of chromaticity metamer-mismatch index) is maximised when, firstly, both M- and L-photopigments have minimally separated peak absorbance wavelengths; and secondly, they both are maximally close to either the M- or L-photopigment of the reference observer.[8] Though, the first condition is not as significant since keeping the peak absorbance wavelength separation at the same value (0.25 nm) and shifting the location of λ_M^{max} and λ_L^{max} results in a significant reduction of chromaticity metamer mismatching. Indeed, in this case chromaticity metamer-mismatch index reaches its minimal value (0.0043) at $\lambda_M^{max} = 550.5$ nm and $\lambda_L^{max} = 550.25$ nm. On the other hand, the variation along the other diagonal of the domain ($\lambda_M^{max} = \lambda_L^{max}$) does not seem to be considerable.

All this is in line with the fact described elsewhere (Logvinenko, 2013a) that any variation in spectral tuning of the M-, and L-cone pigments such that $\lambda_M^{max} + \lambda_L^{max}$ remains constant leaves the spectral profile of the sum $p(\lambda; \lambda_L^{max}) + p(\lambda; \lambda_M^{max})$ effectively constant in shape, and the difference $p(\lambda; \lambda_L^{max}) - p(\lambda; \lambda_M^{max})$ same up to a multiplicative parameter. It follows that a symmetrical shift of the M-, and L-cone pigment curves relative to each other (i.e., such that $\lambda_M^{max} + \lambda_L^{max}$ remains

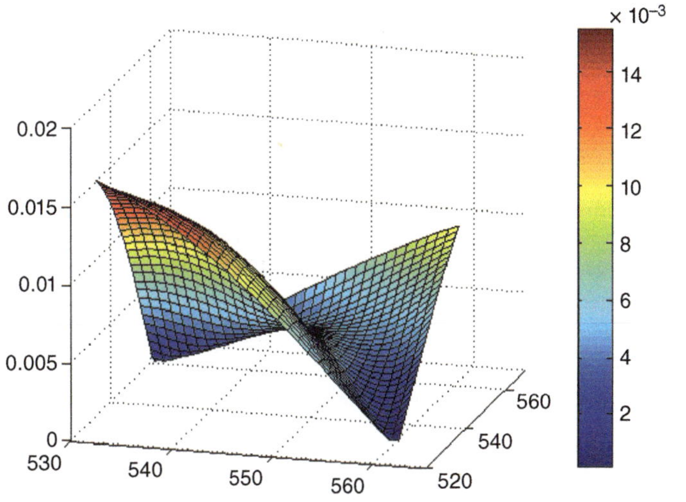

Figure 7.9 Chromaticity metamer-mismatch index as a function of λ_M^{max} and λ_L^{max}.

7 Because of this λ_S^{max} is chosen slightly different but not equal for the first and second observers despite that the main concern is the effect of variation of the other two peak absorbance wavelengths (λ_M^{max} and λ_L^{max}).

8 These two conditions are met for anomalous trichromats when instead of two well separated (M- and L-) types of cones there are two different types of either M-cones – protanomalous trichromacy – or L-cones – deuteranomalous trichromacy (J. Neitz & Neitz, 2011).

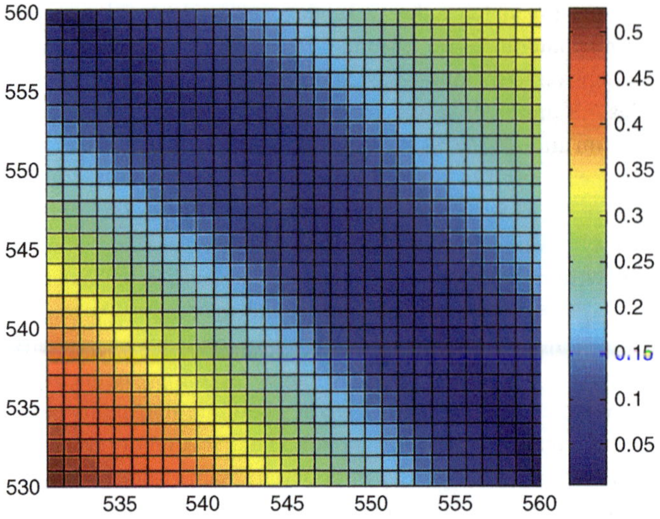

Figure 7.10 Chromaticity metamer-mismatch index as a function of λ_M^{max} and λ_L^{max} for tritanopic observers.

constant) effectively amounts to a linear transformation of these curves. Therefore, such a shift will practically produce no metamer mismatching. To be more exact, such a shift will not change the amount of metamer mismatching. That this is the case can be seen in Fig. 7.10 where chromaticity metamer-mismatch index is presented for a tritanopic observer (i.e., one equipped only with *M*- and *L*-cones) for the same sample of λ_M^{max} and λ_L^{max} values as in Fig. 7.9. The amount of chromaticity metamer mismatching clearly remains constant along the directions $\lambda_M^{max} + \lambda_L^{max} = const$, varying only along the orthogonal directions (i.e., where $\lambda_M^{max} - \lambda_L^{max} = const$). It follows that the amount of chromaticity metamer mismatching is determined by spectral positioning of the sum $p\left(\lambda; \lambda_L^{max}\right) + p\left(\lambda; \lambda_M^{max}\right)$ being independent of the amount of peak separation (i.e., $p\left(\lambda; \lambda_L^{max}\right) - p\left(\lambda; \lambda_M^{max}\right)$). Of course, including *S*-cone pigment somewhat smears the pattern in Fig. 7.9 but the general tendency of this sort is clearly present there.

7.3.2 Effect of the peak photopigment absorbance on metamer mismatching

Metamer mismatching can occur even across observers with identical peak absorbance wavelengths due to the difference in peak photopigment absorbance *D* (3.17). Figure 7.11 displays the chromaticity metamer-mismatch area for an equal energy light obtained for the observers with peak absorbance wavelengths $\lambda_S^{max} = 430\,\text{nm}$, $\lambda_M^{max} = 530\,\text{nm}$, and $\lambda_L^{max} = 560\,\text{nm}$, and peak photopigment absorbances $D = 0.3$ and $D = 0.6$. The metamer-mismatch indices are as follows: $VMMI = 0.0019$, $CMMI = 0.0037$, $S\text{-}MMI = 0.1801$, $M\text{-}MMI = 0.1130$, $L\text{-}MMI = 0.1908$, $\text{LumMMI} = 0.1103$.

7.3.3 Metamer mismatching depending on the position in the chromaticity diagram

Figure 7.12 displays the series of chromaticity metamer-mismatch areas obtained for various mixtures of the two monochromatic lights with wavelengths $\lambda_1 = 460\,\text{nm}$ and $\lambda_2 = 596\,\text{nm}$ for two sets of the cone photopigments with peak absorbance wavelengths at 429 and 431 nm; 529 and 531 nm; and 559 and 561 nm:

$$\sigma_0(\lambda) = a\delta\left(\lambda - \lambda_1\right) + (1 - a)\,\delta\left(\lambda - \lambda_2\right); \qquad 1 < a < 0. \tag{7.38}$$

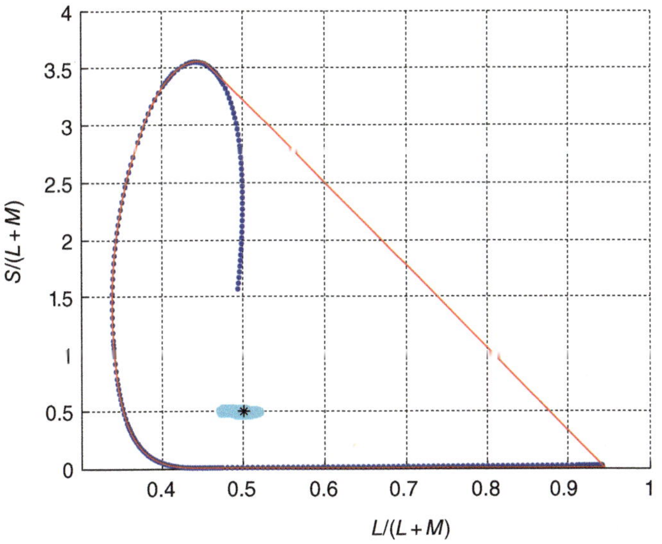

Figure 7.11 Chromaticity metamer-mismatch area for an equal energy light (marked with the black asterisk) induced by the difference in peak photopigment absorbance *D*.

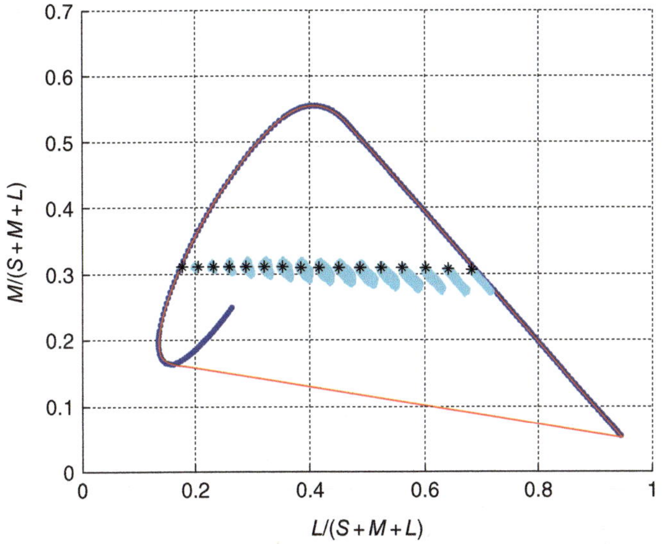

Figure 7.12 Series of chromaticity metamer-mismatch areas plotted in the same chromaticity plane as in Fig. 7.7. Notations are the same as in Fig. 7.6.

The colour signals of these mixtures fall in the straight interval connecting the spectral locus points corresponding to λ_1 and λ_2. Note that the chromaticity metamer-mismatch area appears to decrease when the point in the colour-signal cone (for the first observer) approaches the colour-signal cone boundary. This is hardly a surprise since there is no metamerism for monochromatic lights mapping to the colour-signal cone boundary (Theorem 6.2). This observation is corroborated by Figs 7.13–7.16 where the metamer-mismatch indices are plotted as a function of weighting coefficient a in (7.38).

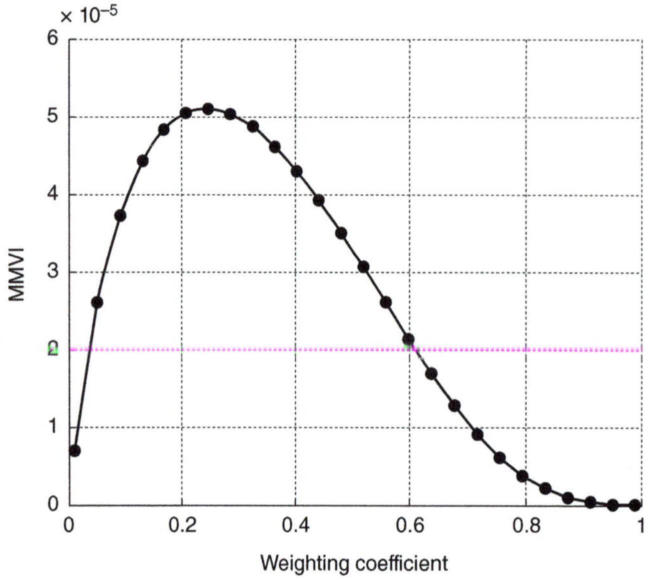

Figure 7.13 Volumetric metamer-mismatch index, $I_V(\mathbf{z}_0; \Phi, \Psi)$, computed for various values of parameter a in Eq. (7.38).

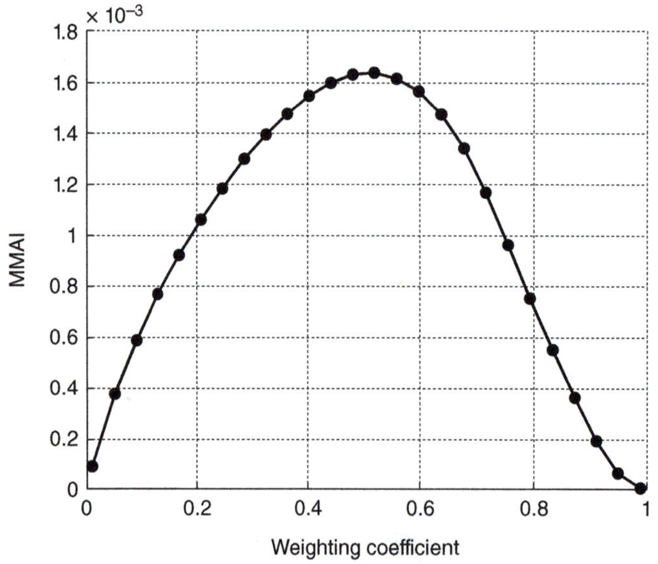

Figure 7.14 Chromaticity metamer-mismatch index, $I_C(\mathbf{z}_0; \Phi, \Psi)$, computed for the same light and for the same values of parameter a as in Fig. 7.13.

Remarkably, all the metamer-mismatch indices have an inverted U-shape form reaching their maximum at some intermediate value of a. Although this form is not quite pronounced in Figs 7.15 and 7.16, a close inspection of the left end of these curves shows a decline of the linear and luminance metamer-mismatch indices similar to that in the right end (Fig. 7.17).

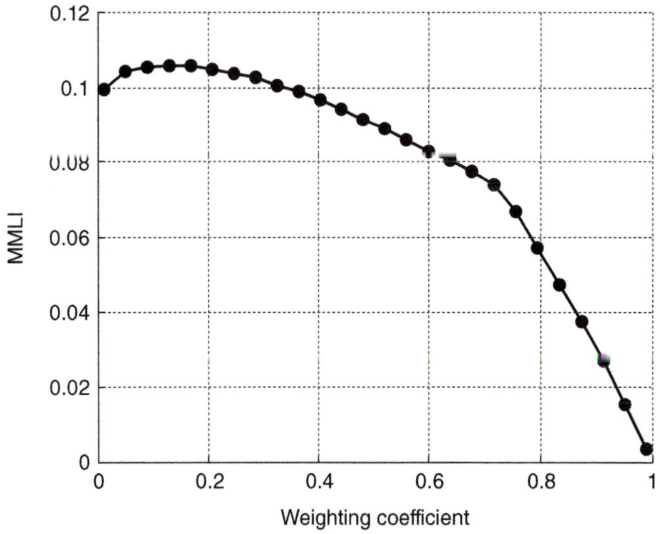

Figure 7.15 Linear metamer-mismatch index (Eq. (7.18)) computed for the same light and for the same values of parameter a as in Fig. 7.13.

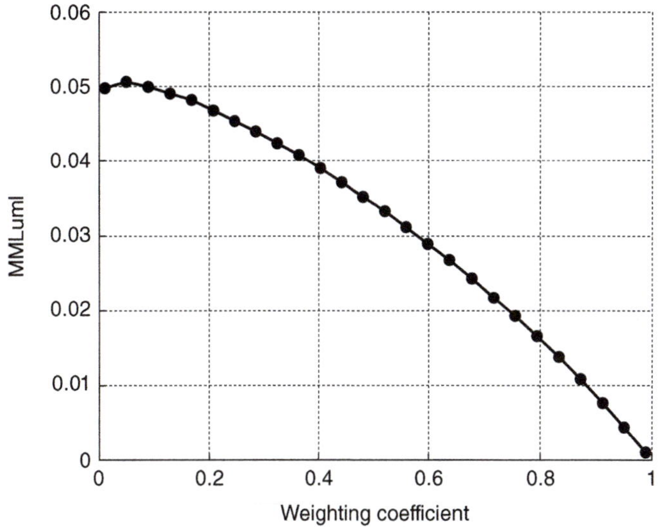

Figure 7.16 Luminance metamer-mismatch index computed for the same light and for the same values of parameter a as in Fig. 7.13.

7.3.4 Metamer mismatching induced by pre-receptoral filters

Apart from the peak-absorbance-wavelength and peak-photopigment-absorbance variations, the difference in spectral transmittance of the ocular media (e.g., the lens and macular pigment) also can result in the inter-individual differences in colour matching (Smith & Pokorny, 2003). Introducing the ocular media itself does not affect significantly the metamer mismatching. Figure 7.18 presents the metamer-mismatch volume for the spectral sensitivities (Fig. 7.19) based on the same

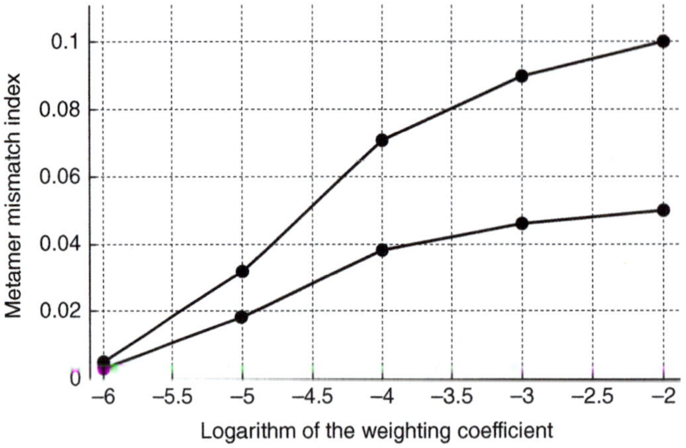

Figure 7.17 Linear (squares) and luminance (diamonds) metamer-mismatch indices for small values of parameter a the logarithm of which is plotted along the horizontal axis.

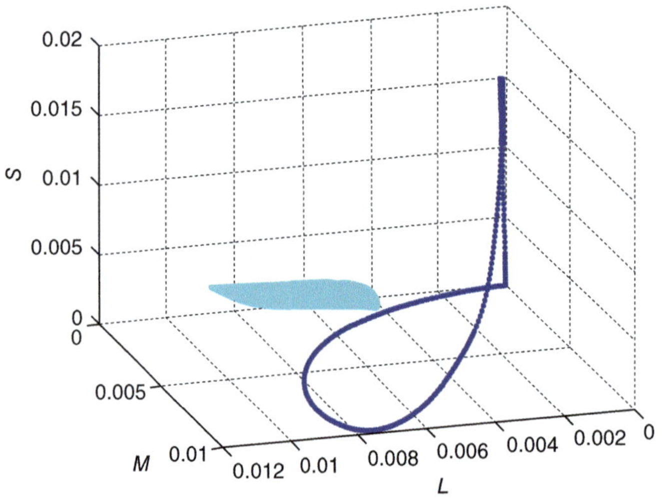

Figure 7.18 Metamer-mismatch volume for the equal energy light produced by shifting the peak sensitivity wavelengths of the cone spectral response functions (Fig. 7.19) from $\lambda_S^{max} = 429$ nm, $\lambda_M^{max} = 525$ nm, $\lambda_L^{max} = 555$ nm to $\lambda_S^{max} = 431$ nm, $\lambda_M^{max} = 535$ nm, $\lambda_L^{max} = 565$ nm.

cone photopigments as those for which the metamer-mismatch volume in Fig. 7.4 was computed. Specifically, the same photopigment spectral absorptance functions were multiplied by the average spectral transmittance of the lens and macular pigment (Wyszecki & Stiles, 1982, p. 587). Linear and volumetric metamer-mismatch indices for the metamer-mismatch volume in Fig. 7.18 are 0.4128 and 0.0070 respectively (also see Table 7.1). These values are rather close to those found for the photopigment spectral absorptance functions only.

In order to ascertain the contribution of the lens and macular pigment into metamer mismatching as such, the metamer-mismatch volume was evaluated for the equal energy light for the same photopigment spectral absorptance functions (with peak absorbance wavelengths 431, 535, and 565 nm) before and after multiplying by the lens and macular pigment spectral

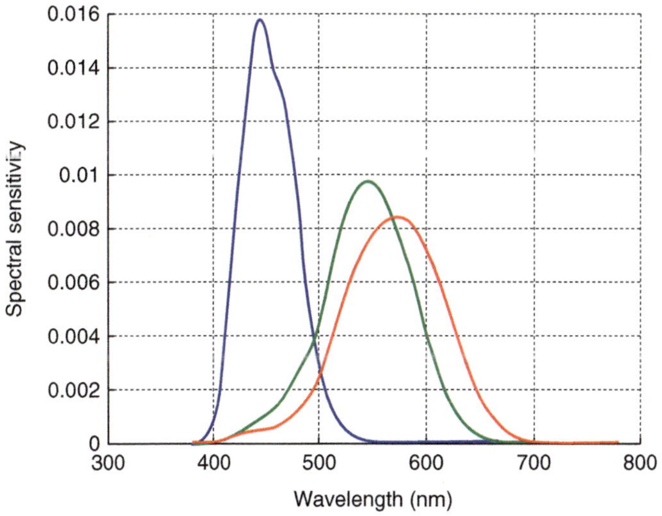

Figure 7.19 Cone spectral sensitivity functions based on the Govardovskii photopigment absorption model normalised to unity area.

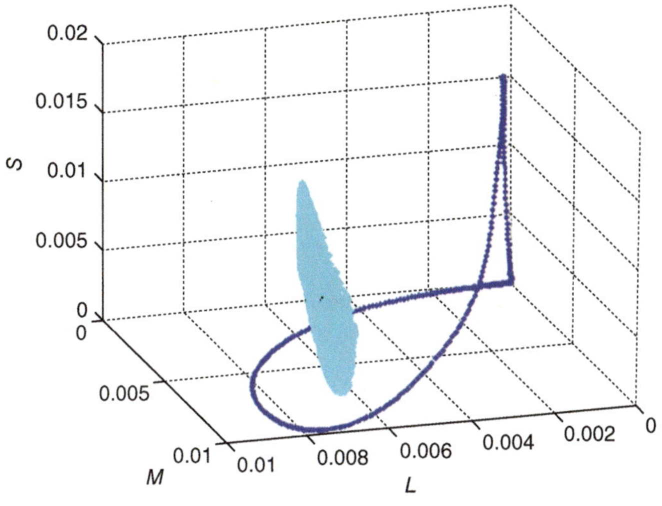

Figure 7.20 Metamer-mismatch volume induced by taking into account of the spectral transmittance of the lens and macular pigment (see text). Notations are the same as in Fig. 7.6.

transmittance (Figs 7.20 and 7.21). Linear and volumetric metamer-mismatch indices were found to be 1.0945 and 0.0515 which is significantly larger than those induced by the photopigment peak absorbance shift (Fig. 7.18). Hence, metamer mismatching caused by the ocular media is massive and considerably larger than that induced by the difference in cone peak absorbance variation.

7.3.5 Differences between cone fundamentals as revealed by metamer mismatching

Metamer mismatching can be used to quantify the difference between various cone fundamentals. Specifically, the metamer-mismatch volume $M(\Psi, \Phi; \sigma_0)$ for light (7.38) with $\lambda_1 = 480\,\text{nm}$,

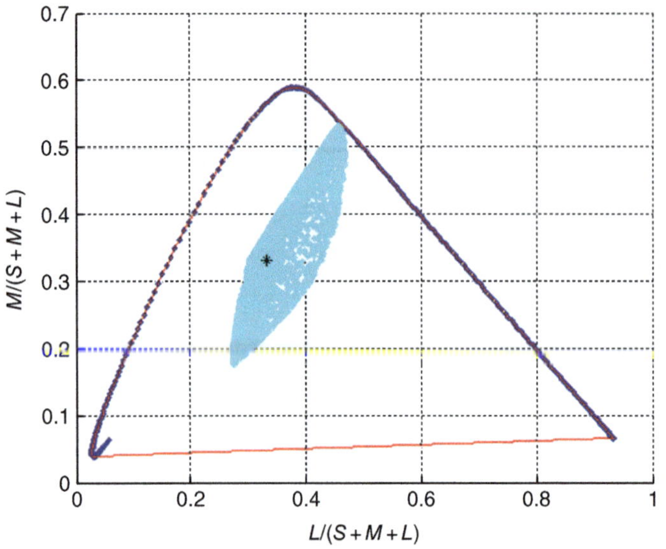

Figure 7.21 Chromaticity metamer-mismatch area induced by taking into account of the spectral transmittance of the lens and macular pigment (see text). Notations are the same as in Fig. 7.6.

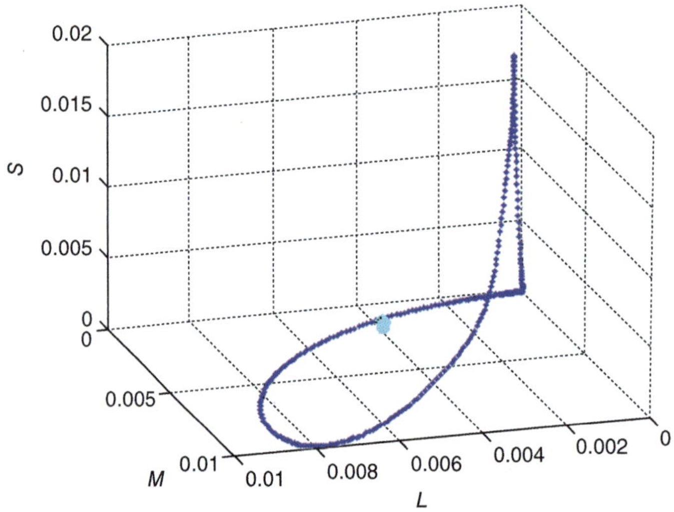

Figure 7.22 Metamer-mismatch volume for a mixture (in equal proportions) of two monochromatic lights with wavelengths 480 and 580 nm obtained for the CIE 2006 cone fundamentals and the cone sensitivities depicted in Fig. 7.19 in terms of which the colour signals are plotted in this figure.

$\lambda_2 = 580$ nm, and $a = 0.5$ was evaluated for the cone fundamentals proposed at different times by different authors (Section 3.3). Figures 7.22 and 7.23 present the results for the CIE (2006) 2° *LMS* cone fundamentals and the cone sensitivity functions based on the photopigment template put forward by Govardovskii et al. (2000) depicted in Fig. 7.19 (the reference observer). As one can notice, the metamer mismatching turns out to be very small. It must be said that the peak absorbance wavelengths (for the reference observer) were adjusted so as to minimise the metamer mismatching.

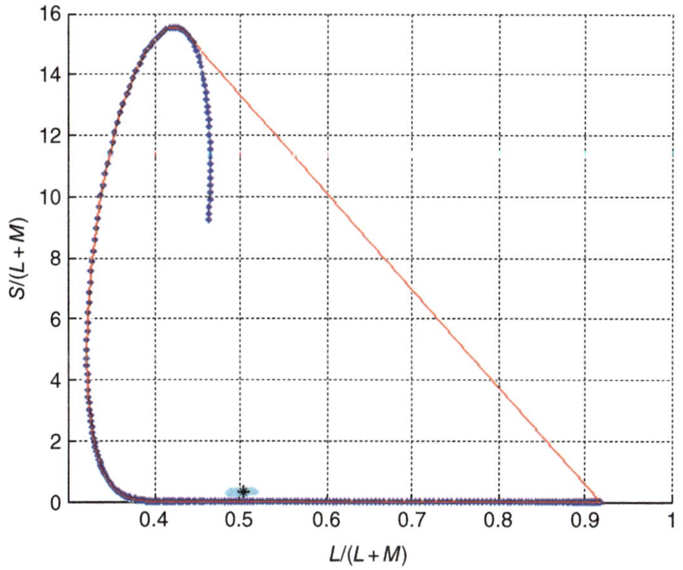

Figure 7.23 Chromaticity metamer-mismatch area for a mixture (in equal proportions) of two monochromatic lights with wavelengths 480 and 580 nm obtained for the cone sensitivities depicted in Fig. 7.19 and the CIE 2006 cone fundamentals.

The smallest metamer mismatching (see the metamer-mismatch indices in Table 7.1) was found under $\lambda_S^{\max} = 425\,\text{nm}$, $\lambda_M^{\max} = 533\,\text{nm}$, and $\lambda_L^{\max} = 560\,\text{nm}$.

Metamer mismatching induced by the CIE 2006 2° *LMS* cone fundamentals and Smith & Pokorny's cone fundamentals (Smith & Pokorny, 1975) can be seen in Figs 7.24 and 7.25 where the metamer-mismatch volume and the corresponding chromaticity metamer-mismatch area are presented (see Table 7.1 as well).

Table 7.1 Metamer-mismatch indices computed for various pairs of observers (denoted as Φ and Ψ).

Φ Ψ	GCS_1 GCS_2	CIE 2006 GCS	CIE 2006 SP 1975	CIE 2006 CIE 1931	CIE 1964 CIE 1931
$I_L(\sigma_0; \Phi, \Psi)$	0.4128	0.0927	0.1800	0.3897	4.0553
$I_V(z_0; \Phi, \Psi)$	0.0070	0.0002	0.0051	0.0018	0.1498
CMMI	0.0037	0.0022	0.0029	0.0185	0.1099
S-MMI (ZMMI)	0.2136	0.1483	1.8595	0.6485	8.1090
M-MMI (YMMI)	0.1756	0.1367	0.1373	0.1466	0.3547
L-MMI (XMMI)	0.6610	0.0603	0.1884	0.3858	1.6029
LumMMI	0.2463	0.0818	0.1713	0.1466	0.3547

Ψ served the reference observer. GSC stands for the cone sensitivity functions based on Govardovskii's template (3.17) with $\lambda_S^{\max} = 425\,\text{nm}$, $\lambda_M^{\max} = 533\,\text{nm}$, $\lambda_L^{\max} = 560\,\text{nm}$; GCS_1 with $\lambda_S^{\max} = 429\,\text{nm}$, $\lambda_M^{\max} = 525\,\text{nm}$, $\lambda_L^{\max} = 555\,\text{nm}$; and GCS_2 with $\lambda_S^{\max} = 431\,\text{nm}$, $\lambda_M^{\max} = 535\,\text{nm}$, $\lambda_L^{\max} = 565\,\text{nm}$. SP 1975 denotes Smith & Pokorny's fundamentals. So, in the first column are the metamer-mismatch indices for the metamer-mismatch volume in Fig. 7.18. Coordinate metamer-mismatch indices for the last two pairs (comprising the CIE Standard Colorimetric Observers) were computed in terms of the tristimulus values X, Y, and Z.

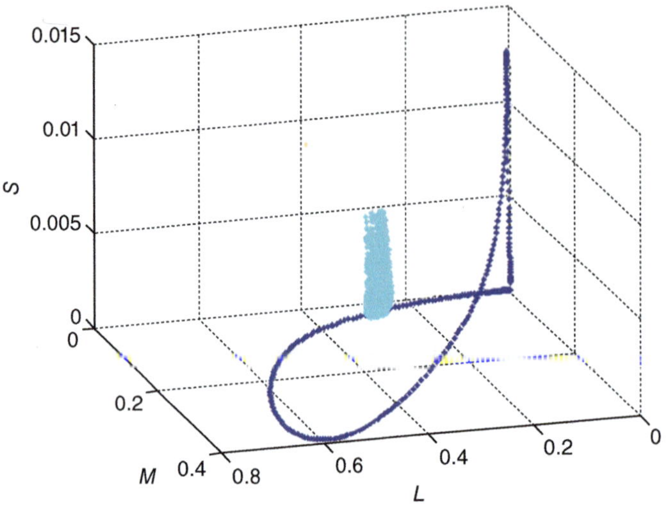

Figure 7.24 Metamer-mismatch volume produced for the CIE 2006 2° LMS cone fundamentals and Smith and Pokorny's cone fundamentals.

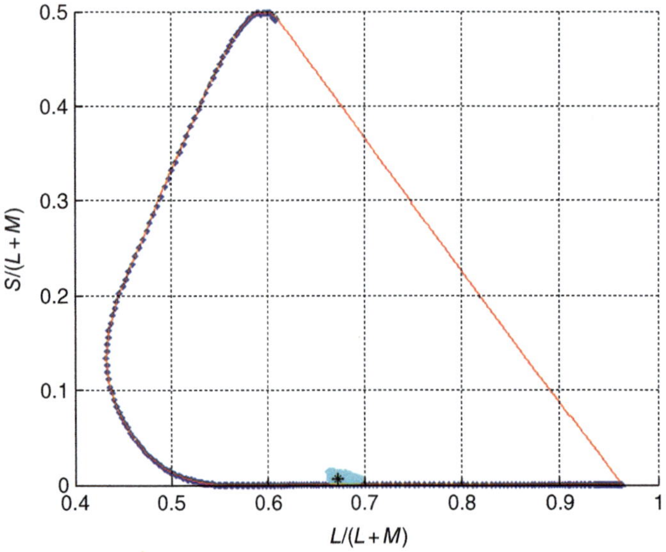

Figure 7.25 Chromaticity metamer-mismatch area produced for the CIE 2006 2° LMS cone fundamentals and Smith and Pokorny's cone fundamentals.

Metamer mismatching induced by the CIE (2006) 2° *LMS* cone fundamentals and CIE 1931 cone fundamentals can be seen in Fig. 7.26 where the chromaticity metamer-mismatch area is presented.

Figure 7.27 displays the chromaticity metamer-mismatch area induced by the CIE 1964 and 1931 Standard Colorimetric Observers. It is certainly the largest one. This is no surprise since the colour matching data underlying these observers were collected for different fields of view (10° and 2° respectively) that resulted in the difference in the ocular media spectral transmittance.

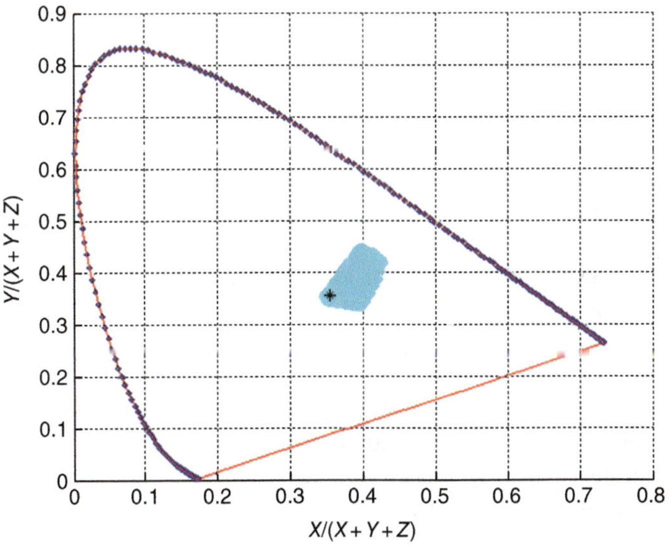

Figure 7.26 Chromaticity metamer-mismatch area obtained for the CIE 2006 2° LMS cone fundamentals and CIE 1931 colorimetric observers.

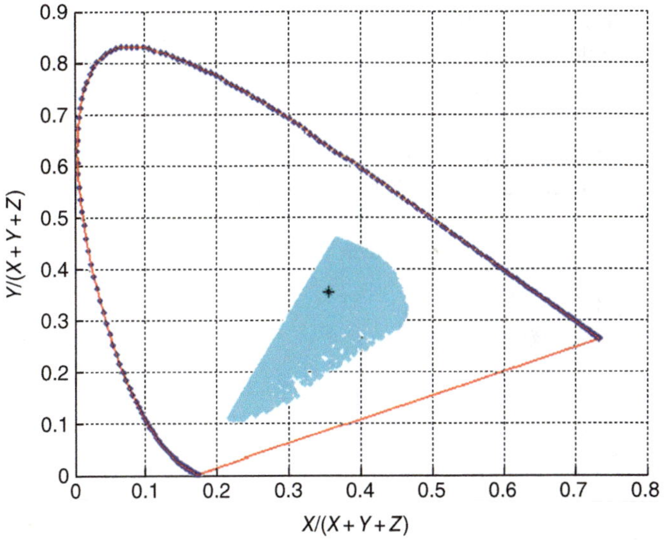

Figure 7.27 Chromaticity metamer-mismatch area obtained for the CIE 1964 and the 1931 colorimetric observers.

7.3.6 Metamer mismatching for the 10° colour matching functions of Stiles and Burch

The colour matching functions collected by Stiles & Burch from 49 individuals[9] (Stiles & Burch, 1959) lend themselves for metamer mismatch analysis. The significance of these data is that, firstly,

9 There are, actually, 53 sets of colour matching functions. However, the measurements were conducted twice for four of 49 individuals. Therefore, according to our terminology Stiles & Burch data provide us with 53 observers representing 49 individuals.

as pointed out above (Section 5.3), these are believed to be the most reliable colour matching data collected with high precision; and secondly, this dataset allows looking into the extent of metamer mismatching induced by the difference between the real trichromatic individuals. Figure 5.2 presents the whole collections of the 53 RGB colour matching functions obtained by Stiles & Burch.

7.3.6.1 Metamer mismatching computation for the 10° colour matching functions of Stiles and Burch

Metamer-mismatch volumes and indices were computed for all 2756(= 53 × 52) ordered pairs of observers (i.e., the colour matching function sets depicted in Fig. 5.2) for light σ_0 (7.38) with $\lambda_1 = 480\,\text{nm}$, $\lambda_2 = 580\,\text{nm}$, and $a = 0.5$. For each pair of the colour matching function sets metamer-mismatch volumes and metamer-mismatch indices were computed twice. More specifically, given observers Φ and Ψ, metamer-mismatch volumes $M(\Psi, \Phi; \sigma_0)$ and $M(\Phi, \Psi; \sigma_0)$ (as well as corresponding metamer-mismatch indices) were computed.

It should be mentioned that computing metamer-mismatch volumes for experimentally obtained colour matching functions becomes complicated because the colour matching functions values are contaminated by measurement errors and other noise. Indeed, in this case the entries of the matrix in (7.33) are noisy numbers. Solving such a system is a typical ill-posed problem (Tikhonov & Arsenin, 1977). Colour matching at the ends of the visible spectrum are known to be particularly subject to significant variability. Because of this the end points of the colour matching functions (i.e., 392 and 714 nm) were excluded from computation. Specifically, 5000 random triad samples were made from 400 to 690 nm. It should be stressed that the sampling was not only from the wavelengths where the colour matches were made but also from intermediate wavelength values by using spline interpolation.

Figures 7.28–7.30 represent all the volumetric, chromaticity and linear metamer-mismatch indices obtained. They are rather well correlated (Figs 7.31 and 7.32). Note that only 6 (of 2756)

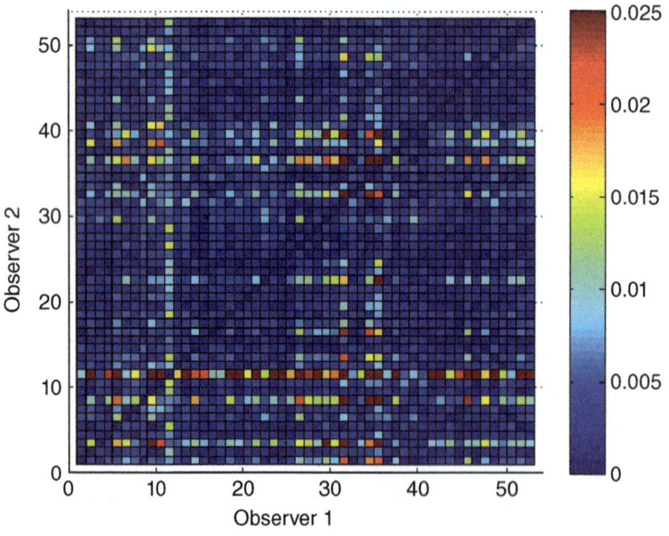

Figure 7.28 Volumetric metamer-mismatch indices for Stiles & Burch's observers. Colour of the *i*, *j* entry encodes the volumetric metamer-mismatch index for the *i*th and *j*th set of colour matching functions.

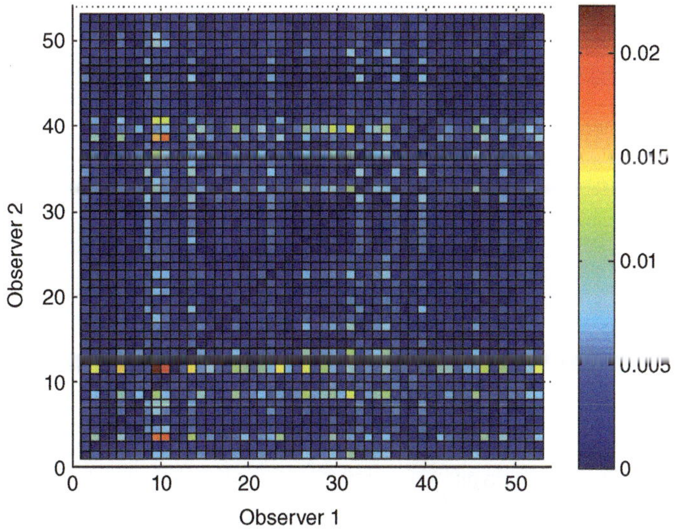

Figure 7.29 Chromaticity metamer-mismatch indices for Stiles & Burch's observers. Colour of the *i*, *j* entry encodes the chromaticity metamer-mismatch index for the *i*th and *j*th set of colour matching functions.

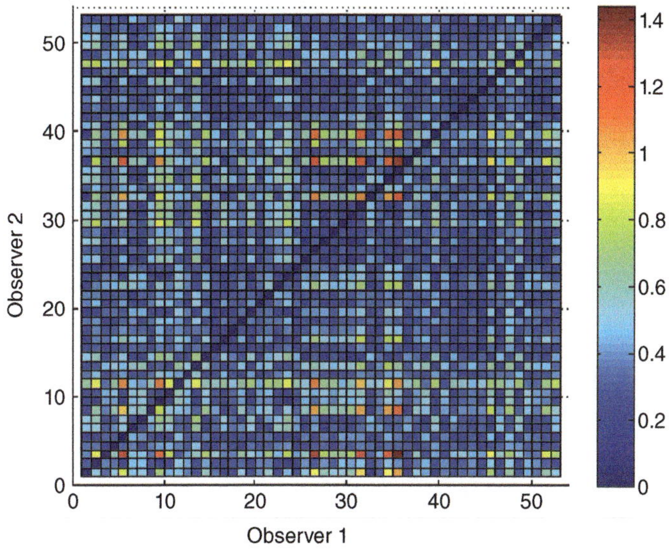

Figure 7.30 Linear metamer-mismatch indices for Stiles & Burch's observers. Colour of the *i*, *j* entry encodes the linear metamer-mismatch index for the *i*th and *j*th set of colour matching functions.

values of volumetric metamer-mismatch index exceeds 0.05. (The full range of metamer-mismatch indices as well as the mean and median values can be found in Table 7.2.) Notably, all these outliers come from Observer #11 (served as the first observer Φ). Moreover, the number of indices exceeding 0.025 is very small: 38. Once again, 32 of these indices were obtained for the pairs including Observer #11 (as the first observer Φ). They were excluded when plotting Fig. 7.28 and the histogram of volumetric metamer-mismatch indices (Fig. 7.33) (i.e., the volumetric

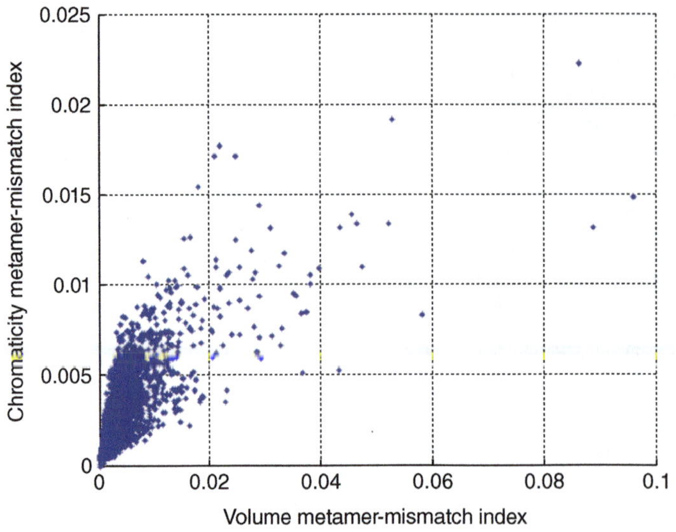

Figure 7.31 Volumetric vs. chromaticity metamer-mismatch indices for Stiles & Burch's observers.

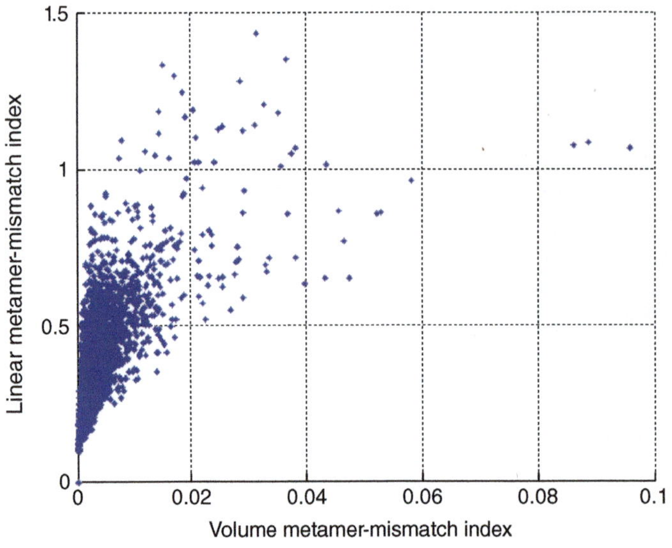

Figure 7.32 Volumetric vs. linear metamer-mismatch indices for Stiles & Burch's observers.

metamer-mismatch indices were capped with 0.025). The histogram of chromaticity and linear metamer-mismatch indices are shown in Figs 7.34 and 7.35.

Metamer-mismatching indices were found to be quite stable with respect to repetition of wavelength random sampling. Figure 7.36 presents a scatter plot of the volumetric metamer-mismatch indices obtained for two independent random samplings. The results are quite robust.

Figures 7.37 and 7.38 present histograms of row means for Figs 7.28 and 7.29 respectively. More specifically, for each observer Fig. 7.37 displays the mean of those volumetric metamer-mismatch indices in which this observer served as Φ. As one can see, Observer #11 is clearly an outlier.

Table 7.2 Minimal, maximal, mean, and median metamer-mismatch indices for 2756 pairs of Stiles & Burch's observers.

Metamer-mismatch index	Minimum	Mean	Median	Maximum
VMMI	1.26×10^{-4}	41×10^{-4}	24×10^{-4}	0.0960
CMMI	1.23×10^{-4}	28×10^{-4}	21×10^{-4}	0.0222
LMMI	0.0976	0.3818	0.3468	1.4347

VMMI, CMMI, and LMMI stand for the volumetric, chromaticity, and linear metamer-mismatch indices respectively.

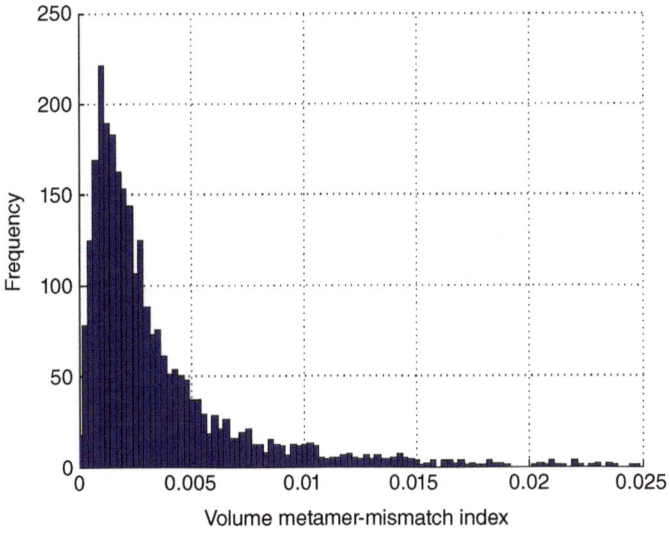

Figure 7.33 Histogram of volumetric metamer-mismatch indices for Stiles & Burch's observers. The horizontal axis is truncated at 0.025, cutting off approximately 2% of the indices (exceeding 0.025).

As can be seen in Fig. 7.28, the matrix of volumetric metamer-mismatch indices is not symmetrical. This is because, given light σ_0, $I_V(\Psi(\sigma_0); \Phi, \Psi) \neq I_V(\Phi(\sigma_0); \Psi, \Phi)$; that is, volumetric metamer-mismatch index depends on which of the two observers, Ψ or Φ, is taken as the reference one. The asymmetry observed in Figs 7.29 and 7.30 indicates that analogous inequalities take place for other metamer-mismatch indices, namely $I_C(\Psi(\sigma_0); \Phi, \Psi) \neq I_C(\Phi(\sigma_0); \Psi, \Phi)$ and $I_L(\Psi(\sigma_0); \Phi, \Psi) \neq I_L(\Phi(\sigma_0); \Psi, \Phi)$.

In order to quantify the volumetric metamer-mismatch index asymmetry, the following ratio was evaluated:

$$I_{V\,Asym} = \frac{\left| I_V(\Psi(\sigma_0); \Phi, \Psi) - I_V(\Phi(\sigma_0); \Psi, \Phi) \right|}{I_V(\Psi(\sigma_0); \Phi, \Psi) + I_V(\Phi(\sigma_0); \Psi, \Phi)}. \tag{7.39}$$

A similar index of asymmetry, $I_{C\,Asym}$, was computed for chromaticity metamer-mismatch indices replacing in (7.39) I_V with I_C. Figures 7.39 and 7.40 display the asymmetry indices computed for matrices in Figs 7.28 and 7.29. The corresponding histograms are presented in Figs 7.41 and 7.42. The mean $I_{V\,Asym}$ and $I_{C\,Asym}$ indices of asymmetry are equal to 0.2652 and 0.2096 respectively.

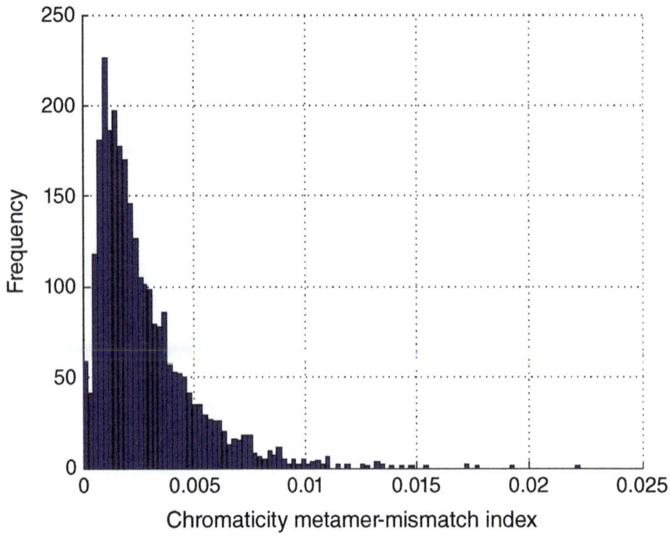

Figure 7.34 Histogram of chromaticity metamer-mismatch indices for Stiles & Burch's observers.

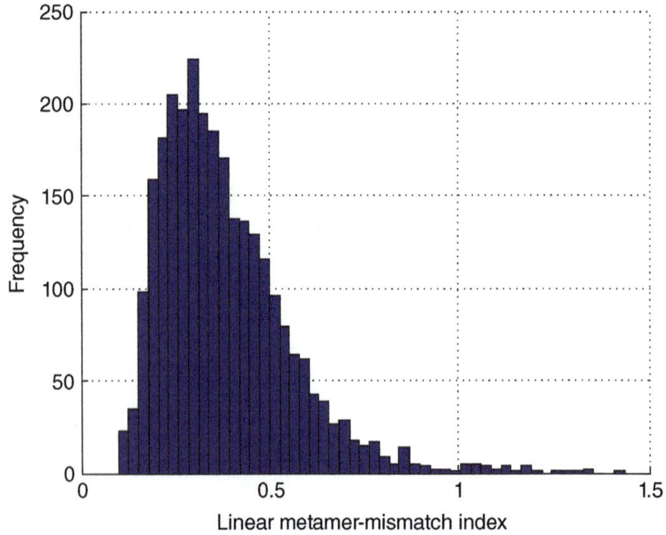

Figure 7.35 Histogram of linear metamer-mismatch indices for Stiles & Burch's observers.

7.3.6.2 Metamer mismatching evaluated for the colour matching functions collected from same individuals

As mentioned above, there are four pairs of the colour matching function sets, each of which was collected from the same individual. The metamer-mismatch volumes and metamer-mismatch indices for the colour matching function sets belonging to the same individual are of great theoretical interest because in this case the metamer mismatching will be due exclusively to the measurement errors and natural human observers' inability to reproduce their own colour matching

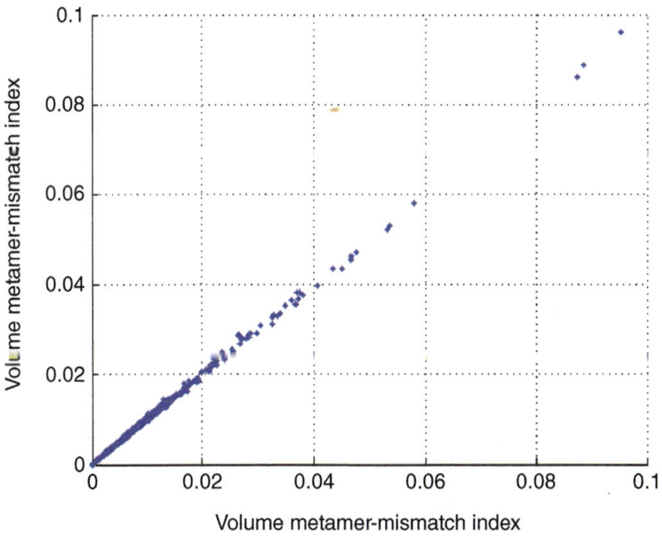

Figure 7.36 Volumetric metamer mismatch index was computed twice for each pair of 53 Stiles & Burch's observers. The abscissa is the index obtained for the pair in the first run, the ordinate in the second.

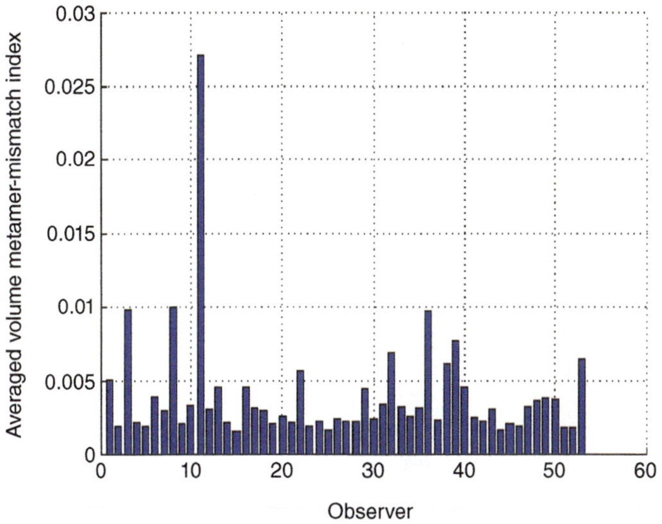

Figure 7.37 Average volumetric metamer-mismatch indices for Stiles & Burch's observers.

settings. Therefore, a metamer-mismatch index obtained for such eight (ordered) pairs of the colour matching function sets (Table 7.3) can be used as a gauge.

The mean values in Table 7.3 can be considered as the natural low bound on the metamer-mismatch indices. Indeed, it is hardly to expect the smaller metamer mismatching than these. Still, four (of 2226) observer pairs produced the volumetric metamer-mismatch index smaller than the smallest (1.713×10^{-4}) in Table 7.3. Perhaps these extremely low indices are due to the influence of noise and measurement errors.

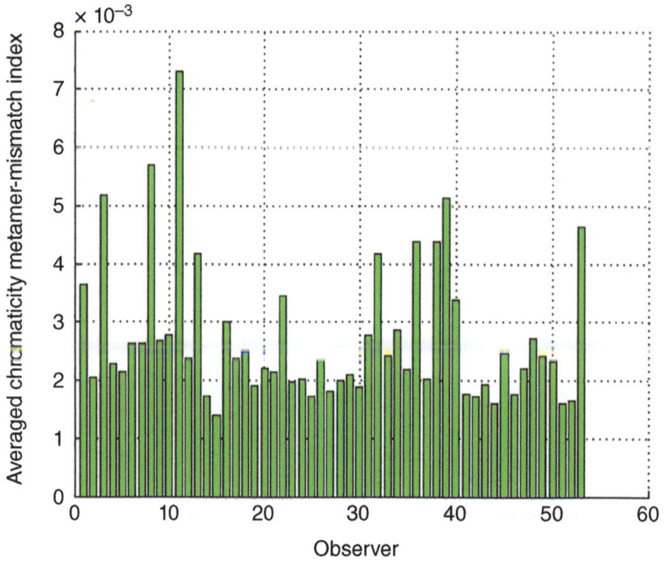

Figure 7.38 Average chromaticity metamer-mismatch indices for Stiles & Burch's observers.

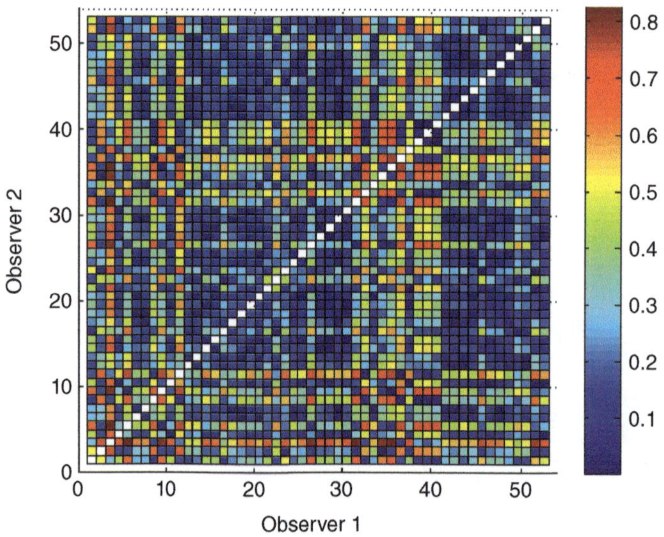

Figure 7.39 Asymmetry indices for volumetric metamer-mismatch indices for Stiles & Burch's observers. The diagonal entries are marked in white to remind that the asymmetry index (7.39) is not defined when $\Psi = \Phi$.

7.3.6.3 Metamer mismatching and physical similarity between colour matching functions

It seems natural to expect that the larger the difference between the colour matching functions, the larger the metamer mismatching. Still, if this were the case then metamer mismatching would be symmetrical. However, as we saw above it is not so. To look into this issue, various similarity

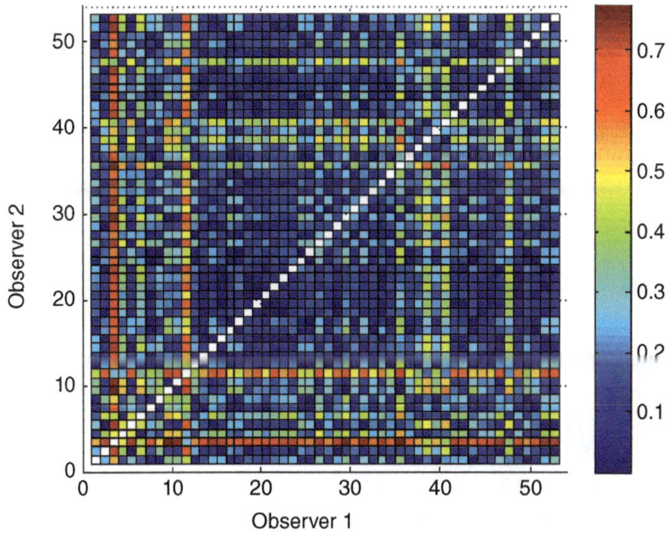

Figure 7.40 Asymmetry indices for chromaticity metamer-mismatch indices.

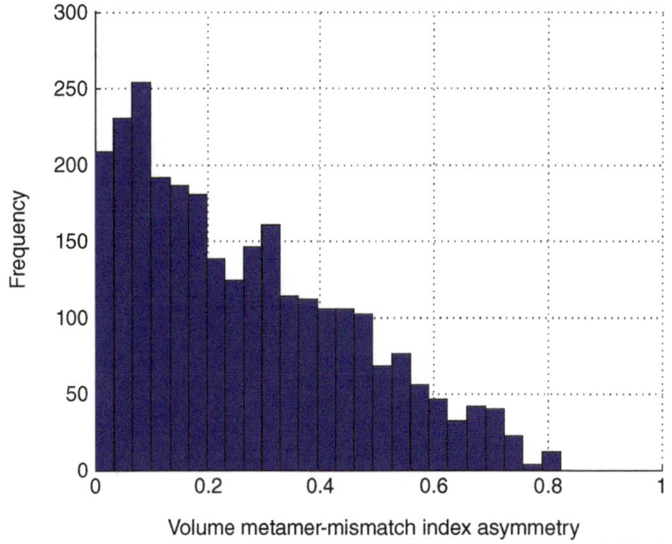

Figure 7.41 Histogram of indices of asymmetry, $I_{V\ Asym}$, for volumetric metamer-mismatch indices.

measures for the colour matching functions were employed to ascertain if they are in any correlation with the difference between the metamer-mismatch indices.

Firstly, considering the colour matching functions as the three 31-component vectors, the angle between the subspaces, spanned by these three vectors, for one observer and for another observer were computed. Observers #15 and #46 were found to make a pair with minimal angle between the corresponding subspaces (Fig. 7.43). The volumetric metamer-mismatch index for this pair is 1.767×10^{-4}. Only five more pairs have the smaller index. The maximal angle was observed

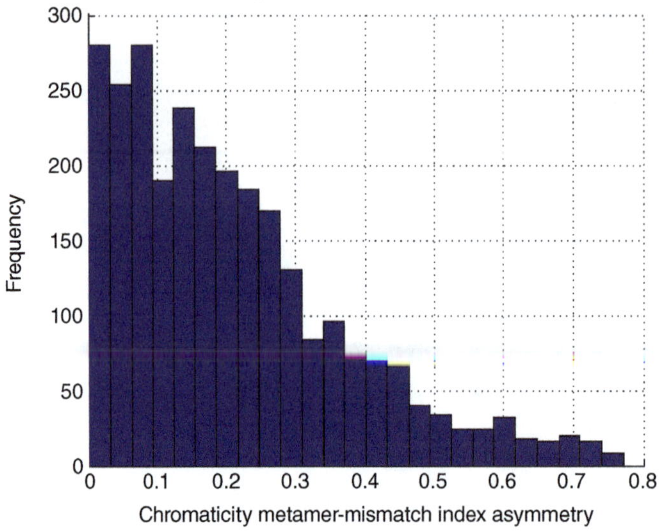

Figure 7.42 Histogram of indices of asymmetry, $I_{C\,Asym}$, for chromaticity metamer-mismatch indicies.

Table 7.3 Metamer-mismatch indices for those pairs of the colour matching function sets which were produced by the same individuals.

Pair of observers	VMMI	CMMI	LMMI
Obs #1 vs. Obs #32	0.0002	0.0005	0.105
Obs #32 vs. Obs #1	0.0002	0.0004	0.121
Obs #2 vs. Obs #52	0.0004	0.0009	0.209
Obs #52 vs. Obs #2	0.0006	0.0008	0.231
Obs #4 vs. Obs #45	0.0016	0.0014	0.288
Obs #45 vs. Obs #4	0.0013	0.0008	0.289
Obs #24 vs. Obs #44	0.0002	0.0004	0.126
Obs #44 vs. Obs #24	0.0002	0.0003	0.138
Mean	5.92×10^{-4}	6.75×10^{-4}	0.188

VMMI, CMMI, and LMMI stand for the volumetric, chromaticity, and luminance metamer-mismatch indices respectively.

between Observers #8 and #35 (Fig. 7.44) for whom the volumetric metamer-mismatch index is 0.0254. Only 37 (1.34%) pairs exceed this value. In Fig. 7.45, this angle (in radian) is plotted along the horizontal axis, the volumetric metamer-mismatch index for the corresponding observer being plotted along the vertical axis. While the cloud of points converge to a common origin in the left bottom part of the scatter plot, indicating that the observers with small angles exhibit, by and large, little metamer mismatching, the correlation between these two variables seems rather weak.

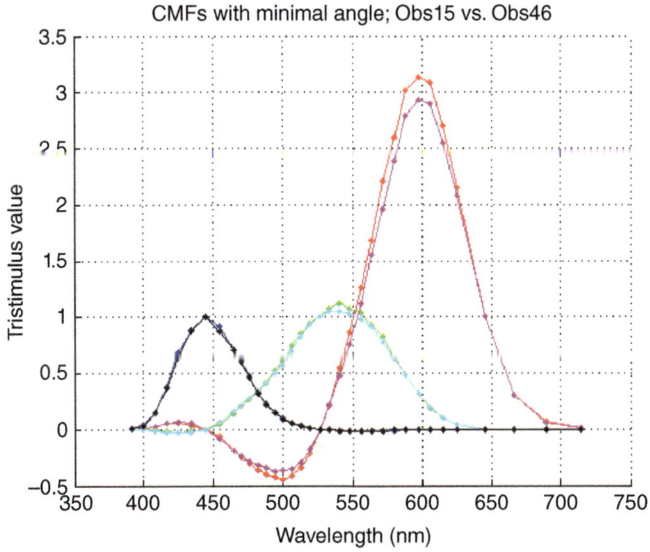

Figure 7.43 Two sets of colour matching functions, making the minimal angle between the three-dimensional subspaces spanned by them.

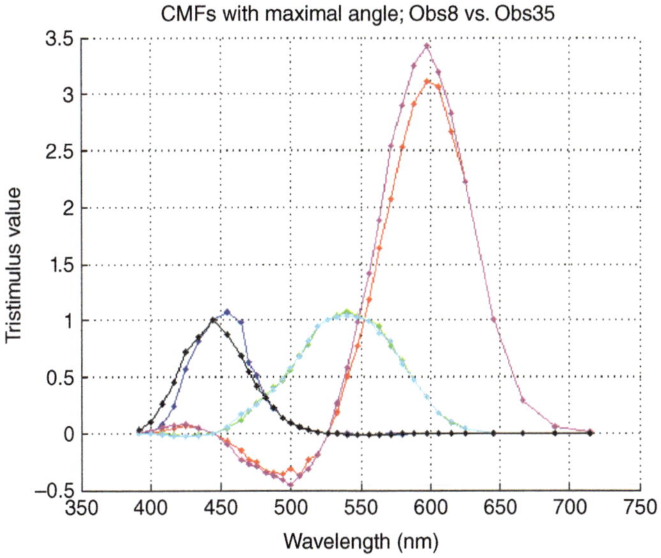

Figure 7.44 Two sets of colour matching functions, making the maximal angle between the three-dimensional subspaces spanned by them.

The other measures of proximity of colour matching functions turned out to be even less correlated to metamer-mismatch indices. In Figs 7.46 and 7.47 presented are the colour matching functions with smallest and largest inter-observer Euclidian difference. Volumetric metamer-mismatch index for Observers #16 and #19 (6.437×10^{-4}) is far from the minimal:

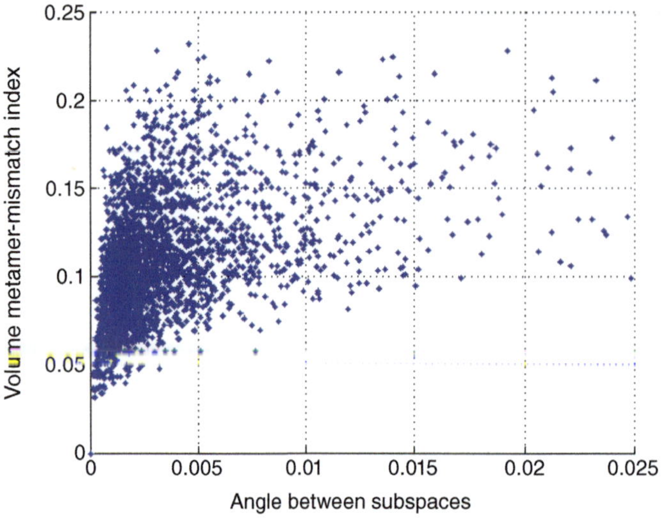

Figure 7.45 The abscissa of each point is the angle between the three-dimensional subspaces spanned by two triplets of the colour matching functions. The ordinate is the volume metamer-mismatch index for this pair of the colour matching triplets.

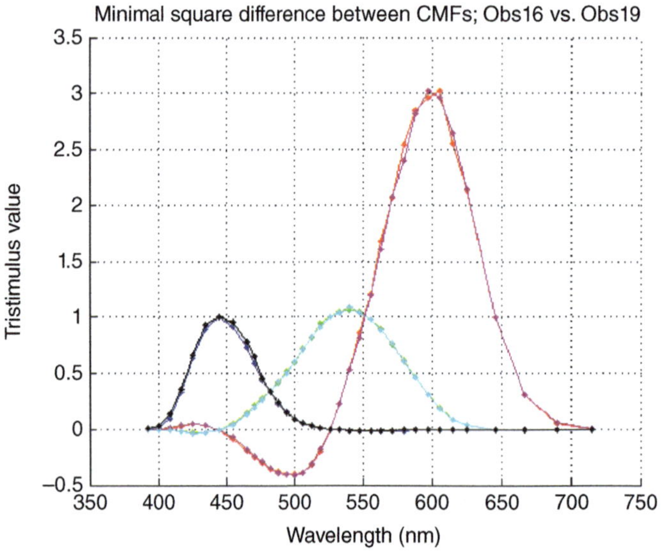

Figure 7.46 Colour matching functions with a minimal area difference between them.

there are 170 observer pairs with the smaller index. Likewise, 505 observer pairs produced larger volumetric metamer-mismatch index than Observers #13 and #47 (0.0055).

Thus, we see that metamer mismatching cannot be directly predicted from the similarity of the colour matching functions. It is important to keep this in mind when designing artificial sensors for such devices as cameras. Admittedly, when the sensor spectral sensitivity functions are in the subspace spanned by the human spectral sensitivity functions then there will be no metamer

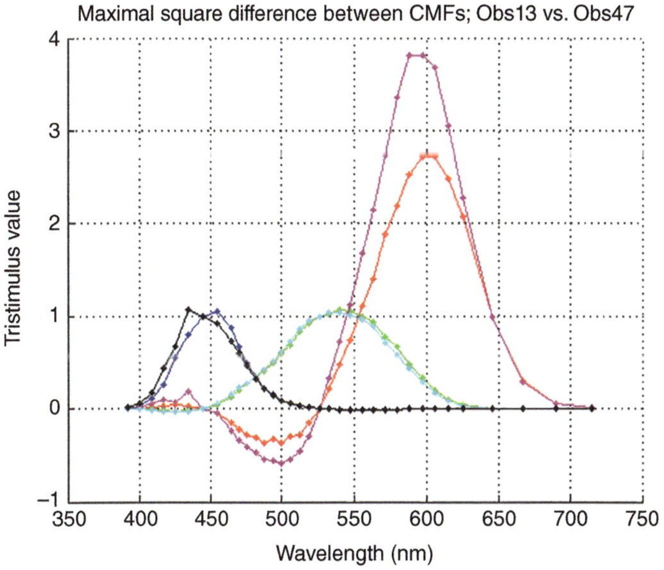

Figure 7.47 Colour matching functions with a maximal area difference between them.

mismatching (Proposition 6.2). However, when the subspaces spanned by the sensor spectral sensitivity functions and the human spectral sensitivity functions are different, there seems to be no physical measure of proximity between these subspaces which could be used to predict, more or less accurately, the degree of metamer mismatching. As one can see in Section 7.3.8, metamer mismatching produced by the difference between the sensor spectral sensitivity functions and the human spectral sensitivity functions can be surprisingly large. In fact, it is the amount of metamer mismatching (as expressed by metamer-mismatch indices introduced above) that can serve as such measure.

7.3.7 Metamer mismatching induced by neutral density filters

Neutral density filters are supposed to have equal spectral transmittance within the visible spectrum so as not to affect the radiant spectral density function of lights except for evenly attenuating their intensity. However, in reality their spectral transmittance curves are only approximately flat as it can be seen in Fig. 7.48. Therefore, such filters are not quite neutral. They will certainly alter the colour of some lights. Evaluating metamer mismatching produced by these filters will show the theoretical limits of such colour alteration.

Specifically, evaluated were the metamer-mismatch volumes for light (7.38) with $\lambda_1 = 480\,\text{nm}$, $\lambda_2 = 580\,\text{nm}$, and $a = 0.5$ induced by the filters in Fig. 7.48 for the CIE 1931 observer. Figure 7.49 and Table 7.4 display the corresponding chromaticity metamer-mismatch areas and indices. Interestingly, metamer mismatching of neutral density filters was found to decrease with their average transmittance. The smallest metamer-mismatch area and indices were found for filter ND70 (average transmittance 70%), the largest for filter ND10 (average transmittance 10%).

Figure 7.50 and Table 7.5 show the chromaticity metamer-mismatch areas and metamer-mismatch indices evaluated for the same light and the same filters using the photopigment

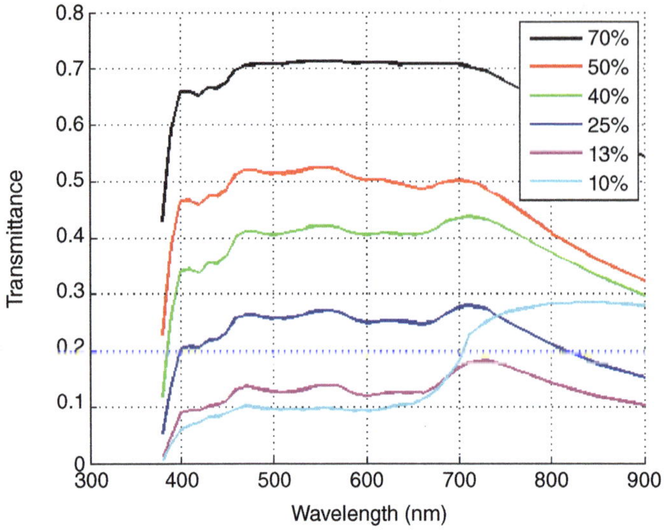

Figure 7.48 Spectral transmittance curves of a few neutral density filters from the Hoya Corporation website. Average transmittance at 400 ÷ 700 nm is given in the legend. Source: Data from Hoya Corporation website.

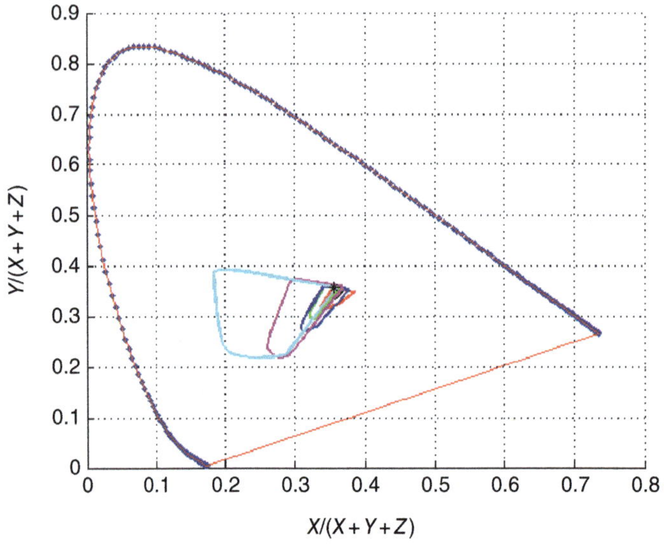

Figure 7.49 Boundary contours of the chromaticity metamer-mismatch areas induced by the neutral density filters in Fig. 7.48 for the CIE 1931 Standard Colorimetric Observer. The contour colours are the same as in Fig. 7.48. For example, the transmittance curve of the filter ND10 and the chromaticity metamer-mismatch area for the ND10 are both plotted in cyan, those for the ND13 in magenta, and so on.

spectral absorptance functions (3.17) with peak absorbance wavelengths $\lambda_S^{\max} = 425\,\text{nm}$, $\lambda_M^{\max} = 530\,\text{nm}$, and $\lambda_L^{\max} = 560\,\text{nm}$. Note that the chromaticity metamer-mismatch index induced by introducing neutral density filters ND10, ND13, and ND25 is larger than that computed for the chromaticity metamer-mismatch area depicted in Fig. 7.6 (0.0016). This means that these neutral

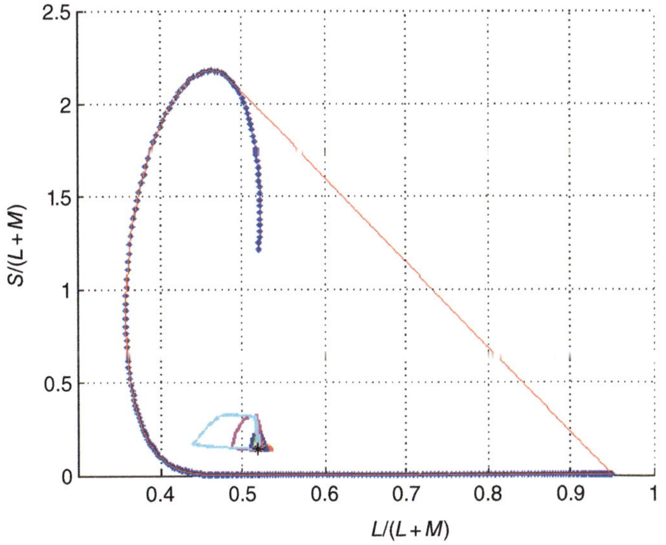

Figure 7.50 Chromaticity metamer-mismatch areas induced by the neutral density filters shown in Fig. 7.48, for the cone photopigment absorptance curves depicted in Fig. 7.3. The contour colours are the same as in Fig. 7.49.

Table 7.4 Indices of metamer mismatching induced by the filters shown in Fig. 7.48 for the CIE 1931 Standard Colorimetric Observer.

	ND10	ND13	ND25	ND40	ND50	ND70
VMMI	0.0047	0.0069	0.0012	$2.4e-04$	$1.7e-04$	$1.7e-05$
CMMI	0.0646	0.0267	0.0084	0.0035	0.0033	$9.7e-04$
ZMMI	1.7440	1.6697	0.8567	0.5668	0.3324	0.1910
YMMI	0.2635	0.1703	0.1030	0.0552	0.0791	0.0374
XMMI	0.9343	0.5917	0.2919	0.1599	0.1931	0.0994
LumMMI	0.2635	0.1703	0.1030	0.0552	0.0791	0.0374

The corresponding chromaticities metamer-mismatch areas are depicted in Fig. 7.48. ND10 denotes the filter with average transmittance of 10%, ND13 with average transmittance of 13%, and so on. Other notations are the same as in Table 7.1.

density filters alter the chromaticity of lights metameric (without the filter) to light (7.38) to a greater extent than the 10 nm shift in the peak absorbance wavelength.

Metamer mismatching produced by neutral density filter ND10 was also computed using the 10° colour matching functions of Stiles & Burch (1959). Figures 7.51 and 7.52 presents the histograms of volumetric and chromaticity metamer-mismatch indices respectively. These values seem to be comparable with those produced by the inter-individual differences between colour matching functions. For example, Fig. 7.53 shows that for Observer #1 only 8 (of 52) observers (i.e., 15%) induce chromaticity metamer-mismatch indices larger than neutral density filter ND10. This proportion was evaluated for all 53 observers (Fig. 7.54). As can be seen in Fig. 7.54, for 44 observers this

Table 7.5 Metamer-mismatch indices corresponding to the chromaticity metamer-mismatch areas depicted in Fig. 7.50.

	ND10	ND13	ND25	ND40	ND50	ND70
VMMI	0.0047	0.0053	$9.1e-04$	$2.2e-04$	$1.4e-04$	$1.8e-05$
CMMI	0.0158	0.0071	0.0021	$8.0e-04$	$8.0e-04$	$2.2e-04$
S-MMI	1.7485	1.6739	0.8534	0.5621	0.3361	0.1908
M-MMI	0.2458	0.2658	0.1380	0.0808	0.0514	0.0253
L-MMI	0.4990	0.3360	0.1506	0.0939	0.1005	0.0476
LumMMI	0.3770	0.2886	0.1414	0.0866	0.0710	0.0346

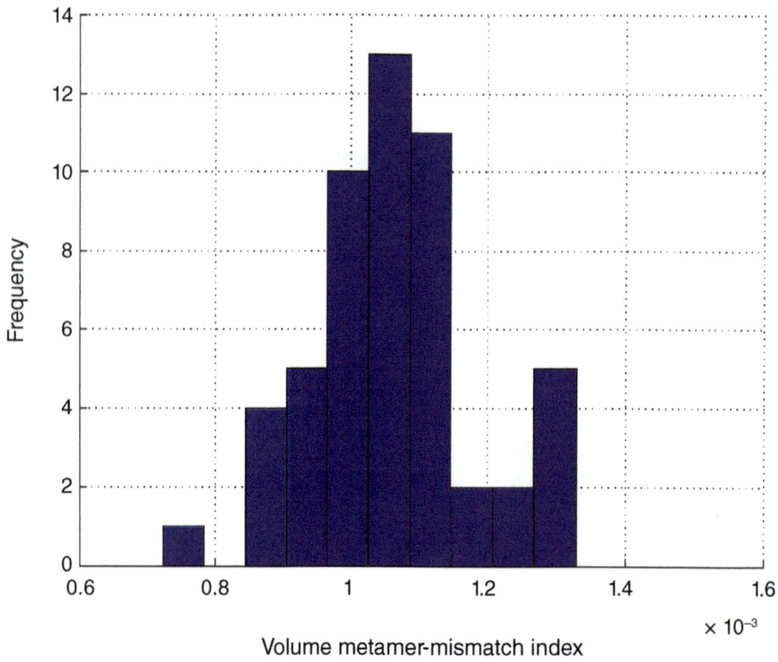

Figure 7.51 Volumetric metamer-mismatch indices produced by the neutral density filter ND10 for 53 Stiles & Burch's observers.

proportion is less 20%. For each of these 44 observers their differences in seeing lights metameric to light (7.38) as compared to more than 80% of the other observers do not exceed the difference induced by introducing neutral density filter ND10.

7.3.8 Metamer mismatching produced by camera sensors

Metamer-mismatch volume in the CIE 1931 colorimetric space (Fig. 7.56) and its projection on the chromaticity plane (Fig. 7.57) were computed for the camera sensors depicted in Fig. 7.55 for

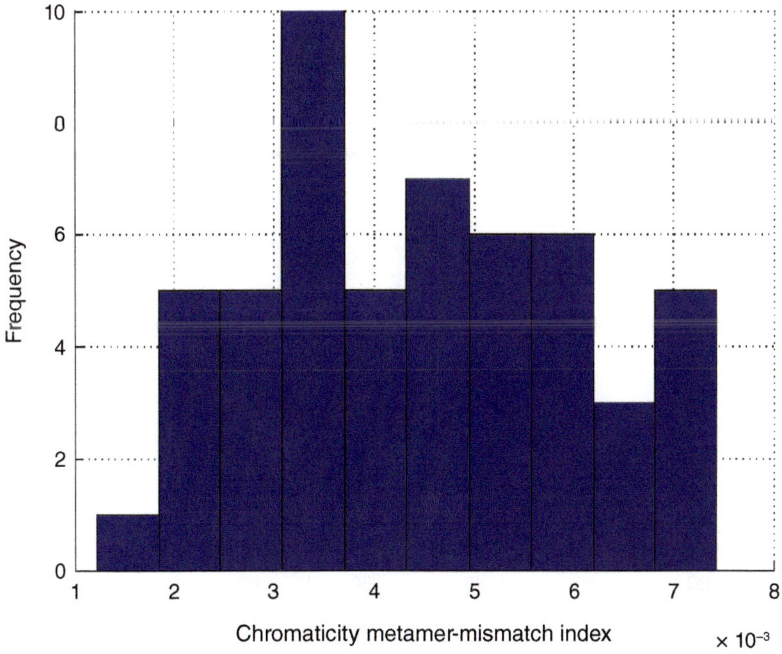

Figure 7.52 Chromaticity metamer-mismatch indices produced by the neutral density filter ND10 for 52 Stiles & Burch's observers.

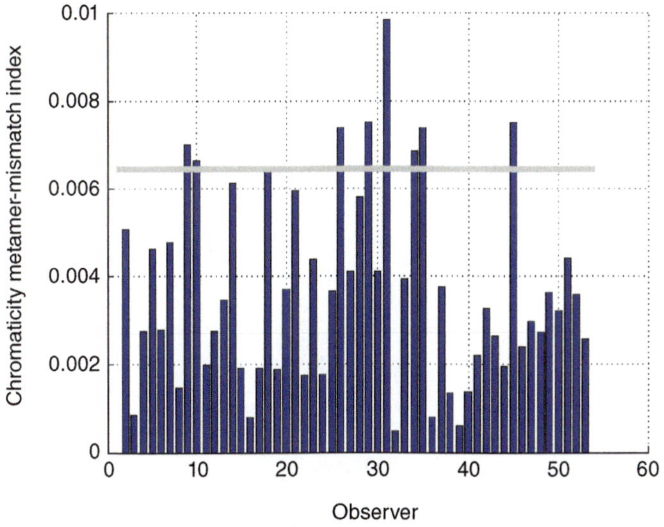

Figure 7.53 Chromaticity metamer-mismatch indices induced by the difference between Observer #1 and the other 52 Stiles & Burch's observers, as compared to that produced by the neutral density filter ND10 for Observer #1 (horizontal grey bar).

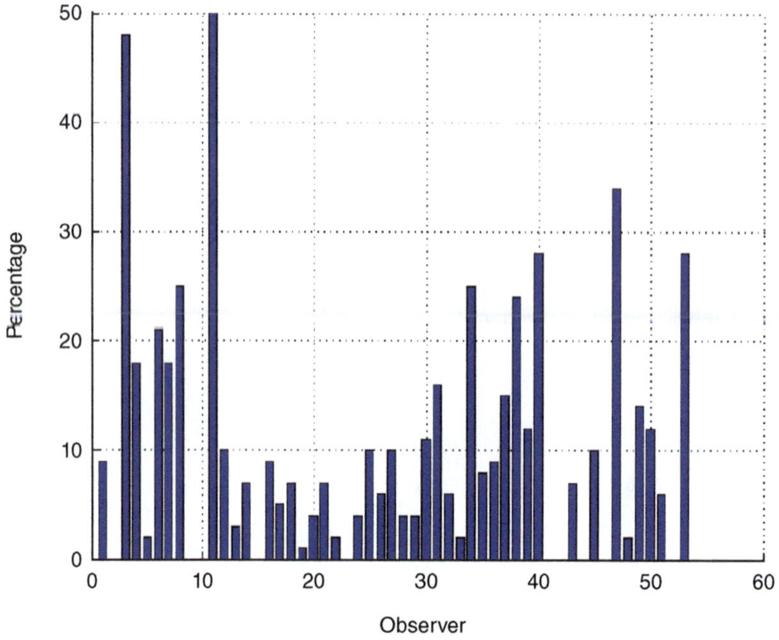

Figure 7.54 Along the vertical axis is the proportion of observers lying above the grey bar in a plot like Fig. 7.53, for each of 53 Stiles & Burch's observers (plotted along the horizontal axis).

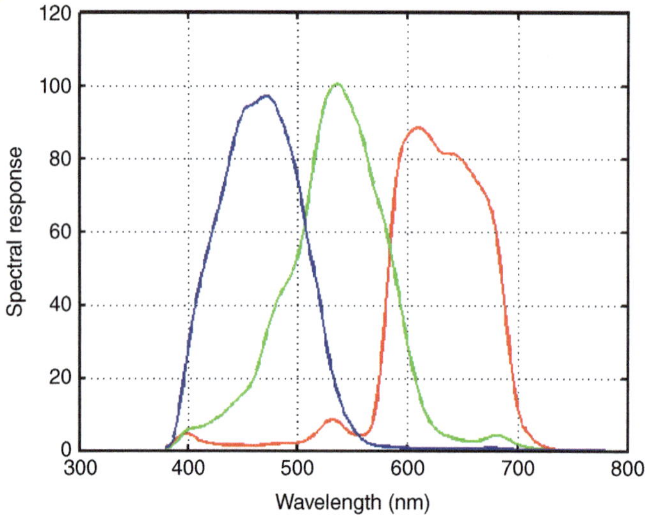

Figure 7.55 Spectral responses of the camera sensors.

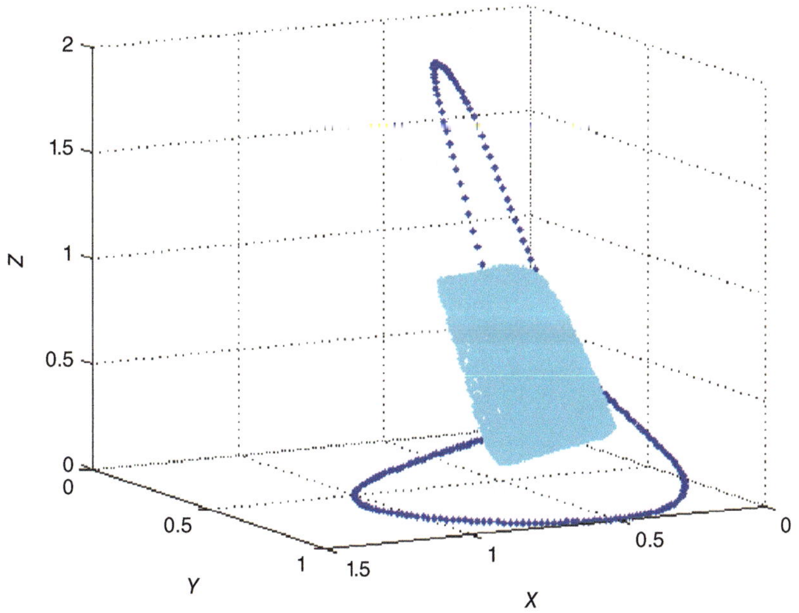

Figure 7.56 Metamer-mismatch volume in the CIE 1931 colorimetric space induced by the camera sensors shown in Fig. 7.55.

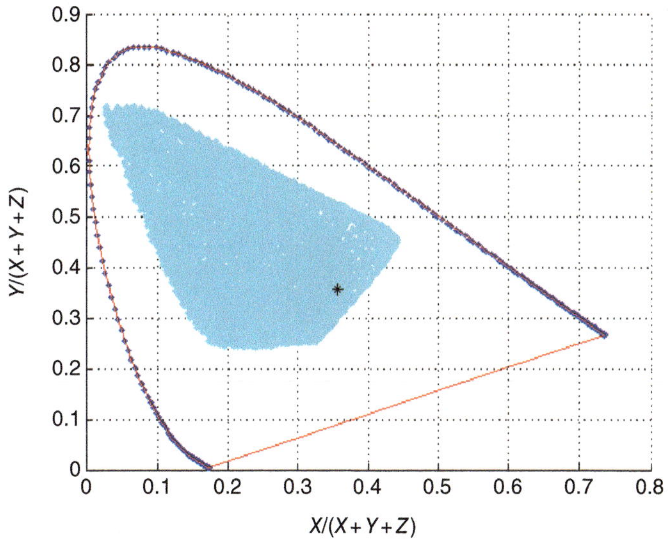

Figure 7.57 Chromaticity metamer-mismatch area for the camera sensors presented in Fig. 7.55. The asterisk marks the light for which this area was computed.

the equal energy light. Notably these are enormously large. The volumetric metamer-mismatch index is 0.0925. In the chromaticity plane, the neutral light (marked with the asterisk in Fig. 7.57) disperses due to metamer mismatching over one-third of the chromaticity gamut, the chromaticity metamer-mismatch index being 0.3333. The luminance metamer-mismatch index equals 0.7637.

8

Light-Colour Perception

The theory described in the previous chapters allows one to predict which lights (presented in the colorimetric paradigm) will be perceived as the same colour, but it does not tell us what exactly they will look like. Attempts to describe how lights look, lead us from colorimetry to the other area of colour research commonly referred to as 'colour appearance'. Finding this term rather misleading, we would better use the good old term 'colour perception'. Really, colorimetry is, in fact, a theory of light considered as a colour stimulus.[1] It does not predict how light stimuli are perceived. Perceiving implies experiencing, in a private way, visual images possessing certain attributes. Some of these are usually considered as pertinent to colour: *hue*, *saturation*, and *brightness* (Maxwell, 1860; Helmholtz, 1867; Evans, 1974; Kaiser & Boynton, 1996; Fairchild, 2005). At the first glance, there should not have been a problem to develop a theory or a model predicting for a particular light its hue, saturation, and brightness, further advancing the success of colorimetry. For instance, it seems an attractive idea to do that in the form of colour codes predicting these colour attributes. Indeed, several models have been put forward for predicting colour attributes. However, in order for such models to be successful, a more or less rigorous definition of these attributes is required. Without this definition, all attempts to create colour codes that serve as predictors of these attributes are doomed to failure.

Moreover, as will be shown below, these attributes make a specific structure. In other words, to answer the question: 'what makes a colour?' it is not enough to point out the colour attributes. One needs also to reveal the structure they manifest. This structure is perceptual in its nature, it cannot be derived from the colour signal, not to mention the physics of light.[2] In order to describe this perceptual structure of light colour, a new theoretical framework will be developed in this chapter. Notably, this theoretical framework for light colour perception will be developed independently of colorimetry, in parallel with the latter. However, before we proceed to presenting the new theory of light-colour perception, it is worth expounding how these three perceptual colour attributes have hitherto been treated in colour science.

As pointed out in the Outline (Chapter 1), although hue is, undoubtedly, the most important colour attribute, there is no formal definition of it. For example, Wyszecki & Stiles define hue as 'the attribute of a colour perception denoted by blue, green, yellow, red, purple, and so on' (Wyszecki & Stiles, 1982, p. 487). They simply reduce the definition of hue to verbal colour categories.

1 Recall that colour in colorimetry reduces to what is referred to in this book as 'colour signal'. 'Psychophysical colour is specified by the tristimulus values of the radiant power (colour stimulus) entering the eye' (Wyszecki & Stiles, 1982, p. 723).
2 Notably, the linear structure that colorimetry deals with is inherited from the linear structure of the light stimulus space.

Foundations of Colour Science: From Colorimetry to Perception, First Edition. Alexander D. Logvinenko and Vladimir L. Levin.
© 2023 John Wiley & Sons Ltd. Published 2023 by John Wiley & Sons Ltd.

Even less satisfactory is the 'definition' which practically identifies hue with the wavelength of a monochromatic light. The rationale for such a 'physicalist' definition is that the colour appearance of monochromatic lights of equal energy differs mainly in hue. However, in the visible part of the spectrum one can observe only a part of the whole range of hues. There are also *non-spectral hues* which can be observed by mixing the short and long wavelength monochromatic lights. So, the entire range of hues experienced by normal trichromatic observers[3] is much broader than that available in the rainbow (Fig. 8.1).

An observer is able to recognise lights of the same hue even though they differ in the other attributes. For example, the series of colour patches in Fig. 8.2 is usually perceived by normal trichromatic observers as having the same hue (apart from the leftmost one perceived as hueless, that is, as having no hue at all). The perceptual attribute along which the colours in Fig. 8.2 mainly vary is usually referred to as *purity* (or *saturation*[4]). This is the perceptual dimension which can be understood as the pronouncedness of the chromatic quale (i.e., hue), its subjective strength. Colours of low purity are usually called *pale*.

Yet another attribute with respect to which light colours can differ, is brightness. It is usually thought of as the subjective intensity of the light. Having displayed the colour patches depicted in Fig. 8.2 on the screen of a computer display, try to play around with the knob controlling brightness. You will see all the patches changing their colour appearance that can be described with such words as getting 'dimmer' or 'brighter'. Note that while changing their brightness these colours will practically retain their purity, the leftmost remaining hueless and the rightmost being most saturated.

Colour matching, generally, implies equality for each perceptual attribute. However, as stated above, lights can match each other in terms of one attribute, say, hue, differing in the others. For example, the two halves of a bipartite field (Fig. I.1) may appear the same in hue in spite of the presence of a distinct vertical border. In such a case, the two lights are said to make a *hue match*. Curiously enough, observers seem able to make hue matches despite the lack of a strict definition of hue; perhaps, because, being the most salient perceptual attribute, hue is usually taken as a synonym of colour outside of the colour science community. Another reason might be that lights

Figure 8.1 Spectral and non-spectral hues.

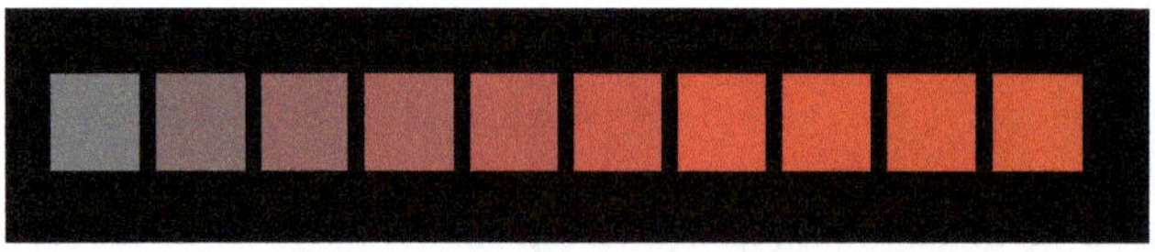

Figure 8.2 Colour patches varying mainly in saturation.

3 That is, their hue palette. This notion will be more formally defined in Section 8.6.
4 It should be borne in mind that the terms 'saturation' and 'purity' have, actually, somewhat different meanings in the colour literature (Wyszecki & Stiles, 1982). The difference between purity and saturation will be clarified in Section 8.6.3.

of different hues differ from each other *in kind*, whereas lights of the same hue, but of different brightness and/or purity, differ *in degree*.

It has been known for a long time that normal trichromatic observers can match in hue any light (except those which are perceived hueless) to either a monochromatic light or a mixture of the short and long wavelength monochromatic lights (Grassmann, 1853; Helmholtz, 1867; Wyszecki & Stiles, 1982). 'The colour impression produced by a certain quantity of mixed light *x* can always be reproduced by mixing a certain amount of white light *a* with a certain amount of some saturated colour *b* (spectrum colour or purple) of definite shade' (Helmholtz, 1867/1970, p. 94). It follows that, firstly, unless it is hueless, any light matches in hue one of the lights represented by the boundary points in the chromaticity diagram (Fig. 8.3). The hueless lights are commonly referred to as achromatic, the remainder of lights chromatic. Match in both hue and purity is said to be a match in *chromaticity*. In other words, loosely speaking, chromaticity is hue and purity. Secondly, the boundary of the chromaticity diagram represents the whole variety of hues.

As a matter of fact, amongst lights of equal energy there is only one light which is perceived by normal trichromatic observers as hueless. Such achromatic light is located approximately in the centre of the chromaticity diagram (Fig. 8.3). Moving from the chromaticity diagram boundary radially towards the centre brings about a change predominantly in purity,[5] though some hue change also might take place. Specifically, the loci of 'constant hue' in the Commission Internationale de l'Eclairage (CIE) chromaticity diagram have been examined in several experiments (S. A. Burns et al., 1984; Hung & Berns, 1995; Valberg et al., 1991; Pridmore, 2007). They were found to make a curved radial pattern with the neutral (hueless) point in the centre (Fig. 8.4).

Varying their intensity, lights can be made dimmer or brighter. Equating two lights in brightness makes a so-called *brightness match*. It must be said that it is an easy task to equate in brightness lights with the same relative radiant spectra (e.g., monochromatic lights of the same wavelength). However, brightness match becomes rather difficult when lights have different relative radiant

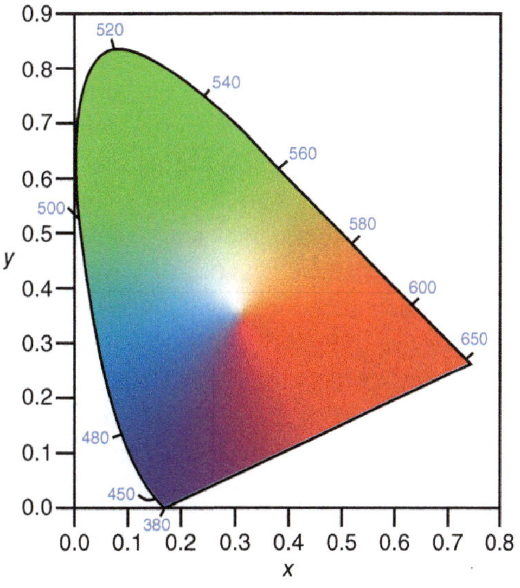

Figure 8.3 The CIE 1931 chromaticity diagram, as in Fig. 4.12, with the only difference that the points outside the colour gamut of the monitor (i.e., the triangle in Fig. 4.12) are filled with the colour of the corresponding points in the colour-gamut border. Source: BenRG, Wikipedia.

5 More specifically, approaching the centre of the chromaticity diagram is accompanied by decrease in purity. It is worth mentioning that low saturation in the central area of the chromaticity diagram makes locating the achromatic point a difficult experimental task (MacAdam, 1955).

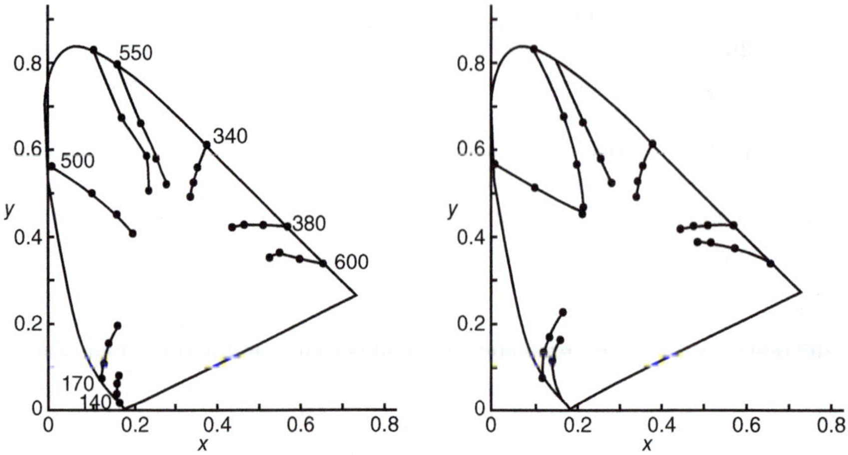

Figure 8.4 Constant hue loci in the chromaticity diagram for two observers. Source: Burns et al. (1984).

spectra, e.g., monochromatic lights of different wavelengths (referred to as *heterochromatic brightness match*). For example, it is not easy to decide whether, say, blue and yellow lights are equal in brightness.

The difficulty with direct subjective judgement of brightness could be got round via resorting to indirect criteria. Specifically, when two juxtaposed patches of lights are separated by a distinct border between them (as in Fig. I.1) there is an indirect criterion – subjective distinctness of the border – which is in a good correlation with our intuition concerning brightness difference. Specifically, it has been found that the relative intensity of two juxtaposed light patches can be adjusted so that the border between them is perceived to be minimally distinct (Boynton & Kaiser, 1968). As a result, the minimal distinct border criterion has been suggested as an operationalisation of brightness match (Kaiser, 1971). Thus, while the criterion of the colour match is the disappearance of the border between the two halves of the bipartite field (Fig. I.1), that of the brightness match is to minimise the distinctness of the border.

The description of light colour in terms of the perceptual attributes (i.e., hue, purity, and brightness) raises a question concerning such a geometrical representation of light colours where hue, purity, and brightness get a clear geometrical interpretation. In particular, it has been noticed a long time ago that the hues lend themselves for a circular representation (for a historical review see Kuehni & Schwarz, 2008). 'The various tints (i.e., hues – A. L. and V. L.) form a continuous series of such a kind, that when we start from one colour of this series and proceed forward, we finally arrive at the original colour' (Grassmann, 1853, p. 255). The hue circular representation leads naturally to an idea of a colour circle, credit for which is usually given to Newton (Newton, 1730; Mollon, 2003). 'The saturated colours constitute a closed series, and hence they must lie on a closed curve, which Newton chose to be a circle with white placed in the centre. The transition from white to any saturated colour at a point on the circumference of the circle lie along the radius drawn to this point, so that the paler shades of this hue are nearer the centre, and the more saturated ones nearer the circumference' (Helmholtz, 1867/1970, p. 95). Drawing rays from a point (representing the full dark) through the circle brings about the colour cone which can be found in many early classical papers on colour. 'When the different degrees of luminance are also to be taken into account, the third dimension of space has to be utilized, as was done by Lambert. The darkest colours, for which the number of discriminable shades continually gets less and less, may be made to culminate in a point corresponding to black. In this way, we get a *colour pyramid* or a colour cone' (Helmholtz, 1867/1970, p. 95). However, without a proper quantitative representation of the perceptual attributes (first of all, hue) the colour cone (and analogous geometrical representations such as the cylindrical one in Section 8.2) remains a metaphor, an artistic exercise because it is not specified, firstly, exactly what colour is located where in the cone; and secondly, which light maps to which point in the colour cone.

In this chapter, we define and quantitatively describe hue, purity, and brightness. Specifically, the same way as colour was defined in Chapter 2 (namely, as a class of the equivalence relation operationalised as colour matching), in Section 8.1 brightness will be defined as a class of the equivalence relation operationalised as brightness matching. Likewise, in Section 8.2 hue will be defined as a class of the equivalence relation operationalised as hue matching. Although purity matching is not defined, a formal definition of purity will also be given in Section 8.2.

As pointed out above, hue difference is a qualitative one, whereas brightness and/or purity difference a quantitative one. For that reason an observer can readily judge which of two lights is brighter, or which of them is more saturated; whereas, such ordinal judgements of hue do not make any sense (though, see Section 8.5.5). It follows that lights can be ordered in brightness in a similar way as in intensity. If this is the case, then one can measure brightness the same way as light intensity. Indeed, there are so-called *brightness scales* that are recommended to measure brightness of light (Wyszecki & Stiles, 1982). These brightness scales have a form of Grassmann codes. However, the necessary and sufficient conditions for the existence of a Grassmann code for brightness and/or purity has never been elucidated. One has to note that, generally, the existence of a Grassmann code for some perceptual attribute is in itself a problem. For instance, a code (not to mention a Grassmann code) for hue is impossible from principle (because the set of hues has the structure of a circle[6]). In Section 8.1, we will look into this problem as far as brightness is concerned.

8.1 Achromatic scales and achromatic codes

The departing point in this section is an empirical relation which can be tentatively called brightness ordering. It reveals itself as a simple fact that humans can judge, firstly, whether two lights match each other in brightness; and secondly, when the lights do not match in brightness, which of them is brighter, and which is dimmer. This empirical relation (i.e., brightness ordering) will be conceptualised as a binary relation which is reflexive (each light is equi-bright to itself) and transitive (if light a is brighter than light b, and b is brighter than light c, then a is brighter than c). We shall also assume that the brightness match meets the axioms of equivalence relation; and it is included in the brightness ordering.

Definition 8.1 *A* preorder $\succcurlyeq$ *on a set A is a* reflexive *($a \succcurlyeq a$ for all a in A) and* transitive *($a \succcurlyeq b$, $b \succcurlyeq c \Rightarrow a \succcurlyeq c$ for all a, b, c in A) binary relation. Given a preorder $\succcurlyeq$ on a set A, its symmetric $\sim$ and asymmetric $\succ$ parts are defined as follows: $a \sim b$ means $a \succcurlyeq b$ and $b \succcurlyeq a$, while $a \succ b$ means that $a \succcurlyeq b$ holds but $b \succcurlyeq a$ fails. Clearly $\sim$ is an equivalence relation, while $\succ$ is transitive and antireflexive. A binary relation (in particular, a preorder) $\succcurlyeq$ is called* complete *(or* total*) if any two elements of A are* comparable*: for every $a, b \in A$, one has $a \succcurlyeq b$ or $b \succcurlyeq a$. Sometimes a complete preorder is referred to as a* weak order. *An* order *is a preorder $\succcurlyeq$ satisfying the* antisymmetry axiom*: $a \sim b \Leftrightarrow a = b$. In another terminology, the complete order is called the* linear order.

6 Curiously enough, while recognising the circular structure of hues, Krantz used the term 'hue code' (Suppes et al., 1989).

Definition 8.2 *Given a complete preorder $\succcurlyeq_B$ on the light stimulus cone A, we will call it* brightness *(or achromatic)* ordering, *and* $\sim_B$ brightness matching relation *if it is compatible with the corresponding empirical relation, and for every non-zero light stimulus $a \in A$, (i) $a \succ_B 0$ and (ii) for any positive numbers t_1 and t_2, $t_1 a \sim_B t_2 a$ implies $t_1 = t_2$. In such a case, the pair $\langle A, \succcurlyeq_B \rangle$ will be called an* achromatically ordered structure, *and the elements of the quotient set $A/{\sim_B}$ (that is, the equivalence classes of $\sim_B$) the* equi-bright subsets.

8.1.1 Ordinal brightness scales

In this section, we shall establish numerical representations for the brightness ordering.

Definition 8.3 *Let $\langle A, \succcurlyeq_B \rangle$ be an achromatically ordered structure. A function $f_B : A \to \mathbf{R}_+$ satisfying $f_B(0) = 0$ and*

$$a \succcurlyeq_B b \Leftrightarrow f_B(a) \geq f_B(b) \tag{8.1}$$

will be referred to as the ordinal brightness (achromatic) scale. *The value of f_B at $a \in A$ will be called the* brightness value of the light *a*.

Remark 8.1 *If $f : A \to \mathbf{R}$ is an arbitrary function satisfying (8.1) then its shift $f_B(a) := f(a) - f(0)$ is an ordinal brightness scale.*

The ordinal brightness scale is a particular case of a more general notion of 'ordinal scale' which is of use in measurement theory[7] (Krantz et al., 1971, Chapter 4).

Definition 8.4 *Given a binary relation (ordering) $\succcurlyeq$ on some set A, a function $u : A \to \mathbf{R}$ is called its* ordinal scale *if*

$$a \succ b \Rightarrow u(a) > u(b), \tag{8.2}$$

$$a \sim b \Rightarrow u(a) = u(b). \tag{8.3}$$

If $\succcurlyeq$ is a *complete* binary relation, then the pair of conditions (8.2), (8.3) is obviously equivalent to the single condition

$$a \succcurlyeq b \Leftrightarrow u(a) \geq u(b). \tag{8.4}$$

In this case, (8.2) and (8.3) are rewritten as $a \succ b \Leftrightarrow u(a) > u(b)$ and $a \sim b \Leftrightarrow u(a) = u(b)$ respectively. (Moreover, from (8.4) it follows that a complete binary relation admitting a ordinal scale automatically proves to be a (complete) preorder.)

Below we present some classical results concerning the existence and uniqueness of ordinal scales (for a review see e.g., Krantz et al., 1971).

Definition 8.5 *Given a preorder $\succcurlyeq$ on a set A, a subset $A' \subseteq A$ is said to be* order dense *in A if for any a and b in A such that $a \succ b$, there is $c \in A'$ such that $a \succcurlyeq c \succcurlyeq b$.*

Given $a, b \in A$ such that $a \succ b$ and the set $\{c : a \succ c \succ b\}$ is empty, we say the pair $\{a, b\}$ defines a *jump* with the *endpoints* a and b. A simple fact follows directly from Definition 8.5, as follows.

7 In different branches of applied mathematics, different terminology is used for ordinal scale. For instance, in mathematical economics the term 'utility function' is widely accepted.

Lemma 8.1 *If $\{a, b\}$ is a jump and A' is order dense in A, then there exists $c \in A'$ such that $a \sim c$ or $b \sim c$.*

Two following Propositions 8.1 and 8.2 go back to Debreu (1954, 1964).

Proposition 8.1 *Given a complete preorder $\succcurlyeq$ on a set A, an ordinal scale for $\succcurlyeq$ exists if and only if $\succcurlyeq$ satisfies the Debreu order separability condition: there is a finite or countable order dense subset in A.*

Here, necessity follows from (8.4) and the separability of any subset in $\mathbf{R}$. Let us outline the proof of sufficiency. Clearly if $\{a, b\}$ is a jump and $a' \sim a, b' \sim b$, then $\{a', b'\}$ is a jump, as well. We say two jumps $\{a, b\}$ and $\{a', b'\}$ are *essentially different* if $a \nsim a'$ or $b \nsim b'$. If the Debreu order separability condition holds true, and A' is an at most countable order dense subset in A, then it follows from Lemma 8.1 that any maximal family of essentially different jumps is at most countable, too. We take such a family and, by adding to A' the corresponding endpoints, obtain a new at most countable set $A'' = \{c^1, c^2, \ldots\}$. Now, it is easily seen that

$$u(a) := \sum_{k\in N(a)} 2^{-k},$$

where $N(a) := \{k : a \succcurlyeq c^k\}$, is an ordinal scale for $\succcurlyeq$.

Corollary 8.1 *Given an achromatically ordered structure $\langle A, \succcurlyeq_B\rangle$, an ordinal brightness scale f_B exists if and only if $\succcurlyeq_B$ satisfies the Debreu order separability condition.*

Remark 8.2 *An achromatic ordering $\succcurlyeq_B$ on A induces an ordering $\succeq_B$ on the quotient set $A/{\sim_B}$ where, by definition, $\mathbf{a} \succeq_B \mathbf{b}$ for $\mathbf{a}, \mathbf{b} \in A/{\sim_B}$ if there are $a \in \mathbf{a}$ and $b \in \mathbf{b}$ such that $a \succcurlyeq_B b$. It follows from this definition that: (i) $\succeq_B$ is an order, that is, an antisymmetric preorder (see Definition 8.1); (ii) for any $a \in \mathbf{a}, b \in \mathbf{b}$ one has $\mathbf{a} \succeq_B \mathbf{b} \Leftrightarrow a \succcurlyeq b$; and (iii) two Debreu order separability conditions, for $\succcurlyeq_B$ on A and for $\succeq_B$ on $A/{\sim_B}$, are equivalent. The equivalence of these two Debreu conditions holds for every complete preorder on an arbitrary set A, and in Proposition 8.1 the Debreu condition for A can be replaced with a similar condition for $A/{\sim}$ (cf. Theorem 3.2 in Fishburn, 1970).*

Remark 8.3 *The function u in (8.4) (as well as the function f_B in (8.1)) is not uniquely determined. More specifically, if g is any strictly increasing function on the range of u, then $g \circ u$ defines another ordinal scale on A. (Similarly, if g is strictly increasing on the range of f_B in (8.1) and $g(0) = 0$, then $g \circ f_B$ defines another ordinal brightness scale.) Furthermore, let u and u' be two ordinal scales for a complete preorder $\succcurlyeq$ on the light stimulus cone A, and R and R' their respective ranges in $\mathbf{R}$. Then there exists a strictly increasing function g from R onto R', such that for all $a \in A$, $g(u(a)) = u'(a)$. The function g is defined as follows. For every $\alpha \in R$ we take an arbitrary $a \in u^{-1}(\alpha)$ and define $g(\alpha) := u'(a)$. Since u and u' are ordinal scales, the above definition does not depend on concrete choice of $a \in u^{-1}(\alpha)$, that is, g is well-defined. Clearly $g \circ u = u'$, and since u and u' satisfy (8.4), g is strictly increasing. Similarly, given two ordinal brightness scales, f_B and f_B' with ranges R and R', there exists a strictly increasing function g from R onto R' such that $g(0) = 0$ and $g\left(f_B(a)\right) = f_B'(a)$ for all $a \in A$ is well-defined.*

Definition 8.6 *An achromatic ordering $\succcurlyeq_B$ on A endowed with some topology is called* continuous *if for every $b \in A$ the sets $\{a \in A : a \succcurlyeq_B b\}$ and $\{a \in A : b \succcurlyeq_B a\}$ are closed in the topology of A. (An equivalent formulation: $\succcurlyeq_B$ is continuous if for every $b \in A$ the sets $\{a \in A : a \succ_B b\}$ and $\{a \in A : b \succ_B a\}$ are open.)*

This is a particular case of the known definition of continuity for an arbitrary preorder on a topological space (see e.g., Bridges & Mehta, 1995, p. 19): a preorder $\succcurlyeq$ on a topological space A is called *continuous* if for every $b \in A$ the sets $\{a : a \succcurlyeq b\}$ and $\{a : b \succcurlyeq a\}$ are closed. For a *complete* preorder (and, in particular, for achromatic ordering) the continuity of $\succcurlyeq$ is equivalent to the closedness in $A \times A$ of its graph $gr(\succcurlyeq) = \{(a, b) \in A \times A : a \succcurlyeq b\}$ (Levin, 1983; Bridges & Mehta, 1995, Proposition 1.6.2. at p. 19).

Remark 8.4 *In Levin (1983) the equivalence of the continuity of a complete preorder $\succcurlyeq$ and the closedness of its graph was formulated without proof. The corresponding proof is as follows. It is obvious that closedness implies continuity. Suppose now that $\succcurlyeq$ is continuous and show that its graph is then closed. Let $a_\nu \succcurlyeq b_\nu$ and the net (a_ν, b_ν) converge to $(a, b) \in A \times A$. We have to show that $a \succcurlyeq b$. It is easily seen by passing, if needed, to a subnet that it will be no loss of generality if we assume that the net a_ν is either increasing or decreasing with respect to $\succcurlyeq$. Let us fix ν and consider $\nu' > \nu$. If a_ν is increasing, then $a_{\nu'} \succcurlyeq a_\nu \succcurlyeq b_\nu$, and as $\succcurlyeq$ is continuous, we obtain $a = \lim_{\nu'} a_{\nu'} \succcurlyeq b_\nu$ hence $a \succcurlyeq \lim_\nu b_\nu = b$. If now a_ν is decreasing, then $a_\nu \succcurlyeq a_{\nu'} \succcurlyeq b_{\nu'}$ hence $a_\nu \succcurlyeq \lim_{\nu'} b_{\nu'} = b$ and $a = \lim_\nu a_\nu \succcurlyeq b$. The proof of this equivalence in Bridges & Mehta (1995) is different.*

Proposition 8.2 *Let A be a separable metrisable topological space with a complete preorder $\succcurlyeq$ on it. Then there exists a continuous ordinal scale for $\succcurlyeq$ if and only if the preorder is continuous.*

Corollary 8.2 *Let $A = \mathbf{L}^1_+$ be the light stimulus cone from Example 2.1(b), and $\succcurlyeq_B$ an achromatic ordering on it. Then there exists a continuous ordinal brightness scale if and only if $\succcurlyeq_B$ is continuous.*

Remark 8.5 *Proposition 8.2 cannot be applied when the light stimulus cone A is $\mathbf{C}^*_+$ with the weak* topology on it because although this space A is separable, it is not metrisable. Of course, the existence of a continuous ordinal scale implies the continuity of the preorder $\succcurlyeq_B$ but the opposite assertion generally needs not be true. Nevertheless, Corollary 8.2 can be partially generalized on the case where $A = \mathbf{C}^*_+$; see Proposition 8.3.*

As colour match is supposed to be a match with respect to all the colour attributes, including brightness, it seems reasonable to assume that for the human colour vision the metamerism $\sim$ is included in the equi-brightness $\sim_B$. (To what degree this assumption is supported by experimental data is discussed in the following Section 8.1.2.)

Definition 8.7 *Suppose $\langle A, \sim\rangle$ is a n-dimensional Grassmann structure, and $\langle A, \succcurlyeq_B\rangle$ is an achromatically ordered structure. If the metamerism $\sim$ is included in the equi-brightness $\sim_B$, i.e., $a \sim_B b$ whenever $a \sim b$, we say that $\langle A, \sim, \succcurlyeq_B\rangle$ is an* achromatically ordered Grassmann structure.

Remark 8.6 *If the metamerism $\sim$ is included in the equi-brightness $\sim_B$, and $\sim_B$ obeys Grassmann's laws (Definition 2.2), then $\langle A, \sim_B\rangle$ is a Grassmann substructure of $\langle A, \sim\rangle$ (Definition 2.3).*

When the metamerism $\sim$ is included in the equi-brightness $\sim_B$, the achromatic ordering $\succcurlyeq_B$ on A induces an ordering $\succeq_B$ on the colour-signal cone $\Phi(A)$ where, by definition, $z^1 \succeq_B z^2$ for $z^1, z^2 \in \Phi(A)$ if there exist $a, b \in A$ such that $\Phi(a) = z^1$, $\Phi(b) = z^2$, and $a \succcurlyeq_B b$. It follows from this definition that

$$\Phi(a) \succeq_B \Phi(b) \Leftrightarrow a \succcurlyeq_B b. \tag{8.5}$$

Indeed, here $\Leftarrow$ is a direct consequence of the definition. On the other hand, if $\Phi(a) \succeq_B \Phi(b)$ then there are a', b' such that $\Phi(a') = \Phi(a)$, $\Phi(b') = \Phi(b)$, and $a' \succcurlyeq_B b'$. Then $a \sim a'$, $b \sim b'$, therefore $a \sim_B a' \succcurlyeq_B b' \sim_B b$, and $\Rightarrow$ is thus established. Clearly, $\succeq_B$ is a complete preorder on $\Phi(A)$, and if it is continuous on $\Phi(A)$ (in such a case, the achromatic ordering $\succcurlyeq_B$ is continuous on A) then there exists a continuous ordinal scale c_B for it satisfying $c_B(0) = 0$. In such a case, $f_B := c_B \circ \Phi$ is a (weak*) continuous ordinal brightness scale on $A = \mathbf{C}_+^*$, and the ordinal representation (8.1) takes thus the form

$$a \succcurlyeq_B b \Leftrightarrow c_B(\Phi(a)) \geq c_B(\Phi(b)). \tag{8.6}$$

Note that weak* continuity of f_B obviously implies continuity of the brightness ordering $\succcurlyeq_B$.

Furthermore, if f_B is an arbitrary ordinal brightness scale for the achromatic ordering $\succcurlyeq_B$ on the light stimulus cone A, then the equality holds $f_B = c_B \circ \Phi$ where function c_B is defined as follows: for every $z \in \Phi(A)$ we take an arbitrary $a \in \Phi^{-1}(z)$ and let $c_B(z) := f_B(a)$. This definition is correct because if b is another point in $\Phi^{-1}(z)$ then, by (8.5), $a \sim_B b$ hence $f_B(a) = f_B(b)$. The ordinal brightness scale f_B is thus a colour code on A (Remark 2.9), which we will refer to as an *ordinal brightness code*.

Moreover, it follows from (8.5) and (8.6) that $\succeq_B$ is a complete preorder and c_B is an ordinal scale for it.

The above arguments are summarized in the following proposition.

Proposition 8.3 *Let A be the cone $\mathbf{C}_+^*$ endowed with the weak* topology, $\succcurlyeq_B$ an achromatic ordering on it such that the metamerism $\sim$ is included in $\sim_B$, and $\succeq_B$ the induced ordering on a colour-signal cone $\Phi(A)$. If the preorder $\succeq_B$ is continuous on $\Phi(A)$ then there exists a continuous ordinal scale for $\succcurlyeq_B$, and if there exists a continuous ordinal scale for $\succcurlyeq_B$ then the achromatic ordering $\succcurlyeq_B$ is continuous on A.*

If the light stimulus cone A is $\mathbf{L}_+^1$, then the ordinal brightness code f_B is continuous if and only if the achromatic ordering $\succcurlyeq_B$ is continuous on A (Corollary 8.2).

It must be said that human observers can not only evaluate which of the two lights is brighter, but also estimate in which of the two pairs of lights the brightness dissimilarity is more pronounced.[8] This can be formalised in terms of a binary relation $\succcurlyeq_*$ on the set of pairs of light stimuli, $A \times A$. We say the dissimilarity in brightness for a pair (a, b) is more pronounced than for (a', b') if $(a, b) \succcurlyeq_* (a', b')$.

Suppose $\succcurlyeq_*$ is a complete preorder on $A \times A$. Of great interest is the question of the existence of an ordinal scale for it, having the form of $u(a, b) = f(a) - f(b)$, in which case one has

$$(a, b) \succcurlyeq_* (a', b') \Leftrightarrow f(a) - f(b) \geq f(a') - f(b'). \tag{8.7}$$

(If, in addition, $(a, a) \sim_* (b, b)$ and $(a, b) \succcurlyeq_* (a, a) \Leftrightarrow a \succcurlyeq_B b$ for all $a, b \in A$ then f proves to be an ordinal scale for $\succcurlyeq_B$.)

Remark 8.7 *Given an abstract set A, various axiomatic characteristics of complete preorders on $A \times A$ that admit such ordinal scales were obtained by many authors; relevant results and references can be found in Fishburn (1970, Chapter 6) and Krantz et al. (1971, Chapter 4). Obvious necessary conditions are the implications $(a, b) \succcurlyeq_* (a', b') \Rightarrow (b', a') \succcurlyeq_* (b, a)$ and*

8 By using the words 'brightness dissimilarity' we in no way prejudge the existence in humans of the ability to quantify the dissimilarity in brightness. We simply indicate the ability to order pairs of lights in this regard.

$(a, b) \succcurlyeq_* (a', b') \Rightarrow (a, a') \succcurlyeq_* (b, b')$. *These conditions (axioms), supplemented by some others of algebraic or topological character, become not only necessary but also sufficient for representation (8.7) to hold. Moreover, such a function f is determined uniquely up to a positive multiplier and a shift: if f' is another function satisfying (8.7), then $f'(a) = \alpha f(a) + \beta$ for all $a \in A$, where $\alpha, \beta \in \mathbf{R}$, $\alpha > 0$.*

Remark 8.8 *According to the terminology adopted in measurement theory, scales for which the admissible transformations are exhausted by linear ones, are called* interval scales *(Krantz et al., 1971, Chapter 4). So, f in (8.7) is, actually, an interval scale.*

Unfortunately, the axioms that ensure the existence of an interval brightness scale have never been tested experimentally. Thus, this question remains open. Nevertheless, the possibility of measuring brightness in terms of an interval scale is usually taken for granted; possibly due to an underestimation of the complexity of the problem. For the sake of generality, when dealing with issues related to brightness, we will consider both types of brightness scales: ordinal and interval.

8.1.2 Grassmann brightness code. Luminance

In the case of the trichromatic human colour vision and $A = \mathbf{C}_+^*(\Lambda)$, the brightness code can be represented as

$$f_B(a) = c_B(\Phi(a)) = c_B(\varphi_S(a), \varphi_M(a), \varphi_L(a)),$$

where φ_S, φ_M, and φ_L stand for the fundamental colour mechanisms (the cone fundamentals) with the peak sensitivity in the short-, middle-, and long-wave regions of the visible spectrum wavelength interval respectively, and $\varphi_S(a) = \langle \varphi_S, a \rangle$, $\varphi_M(a) = \langle \varphi_M, a \rangle$, $\varphi_L(a) = \langle \varphi_L, a \rangle$. A natural question arises: is the brightness code a Grassmann code? If it is, then brightness matching should satisfy Grassmann's laws. More specifically, in this case $\langle A, \sim_B \rangle$ should be a one-dimensional Grassmann structure, that is, the additivity and proportionality laws (Definition 2.2) should hold true for brightness matching. Interestingly, taking for granted the additivity of brightness matching some authors include the additivity law for brightness matching as one of the laws of colour matching. Experimental support for the brightness additivity has been claimed by Abney in honour of whom the additivity of brightness matching is called *Abney's law.*

It must be said, however, that some deviations from Abney's law have been observed when a direct (subjective) criterion is used (see the Helmholtz–Kohlrausch effect below). Yet, despite the failure for direct brightness judgements, Abney's law has been confirmed by using the brightness matching method based on such indirect criteria as minimally distinct border and minimal flicker (heterochromatic flicker photometry) (Wyszecki & Stiles, 1982, pp. 413–420). Both the minimally distinct border method and the heterochromatic flicker photometry yield comparable results (Wagner & Boynton, 1972; Wyszecki & Stiles, 1982, pp. 394–410).

Abney's law (along with the proportionality law) entails the existence of a Grassmann code for brightness. When $\langle A, \sim_B \rangle$ is a one-dimensional Grassmann substructure of $\langle A, \sim \rangle$, the linear part of the Grassmann brightness code must be a linear combination of the cone fundamentals. It follows that there are three real numbers k_S, k_M, and k_L such that

$$f_B(a) = c_B(\langle k_S\varphi_S + k_M\varphi_M + k_L\varphi_L, a \rangle) \tag{8.8}$$

whenever $a \in A$.

The function on Λ

$$k_S\varphi_S(\lambda) + k_M\varphi_M(\lambda) + k_L\varphi_L(\lambda) = \langle k_S\varphi_S + k_M\varphi_M + k_L\varphi_L, \delta_\lambda \rangle$$

is called the *spectral luminous efficiency function* when it is normalised to have a unit as the maximum value (Wyszecki & Stiles, 1982, p. 256). The spectral luminous efficiency function can be determined in experiments on brightness matching. Only those methods which bring about results in accord with Abney's law (e.g., the minimally distinct border and heterochromatic flicker photometry methods) are usually used for determining the spectral luminous efficiency function. The spectral luminous efficiency functions obtained with these two methods were found to be in close agreement with each other (Wagner & Boynton, 1972; Lennie et al., 1993).

It should be borne in mind that similar to colour matching, brightness matching is subject to variability depending on the conditions of viewing, and the spatio-temporal characteristics (e.g., size and temporal modulation) of the test light patches (Wyszecki & Stiles, 1982; Smith & Pokorny, 2003). Observers, of course, are also an important cause of variations. Moreover, the intensity level of the light stimuli is an important factor determining the shape of the spectral luminous efficiency function as well. As a matter of fact, because of the duplicity of the human vision (that is, the presence of the two different types of photoreceptors – rods and cones – in the eye) the shape of the spectral luminous efficiency function obtained in a photopic range is essentially different from that obtained in a scotopic range (Wyszecki & Stiles, 1982). Specifically, a luminous efficiency function shifts its maximum toward the short wavelength end of the visible spectrum when the light intensity is reduced from a photopic to scotopic range (so-called *Purkinje shift*). This shift is thought to be caused by a transition from the mechanisms mediated by the cone receptors to those mediated by the rod receptors. The CIE has established two standard spectral luminous efficiency functions – for photopic and scotopic vision (Fig. 8.5). They specify the so-called standard photopic and scotopic CIE observers.

The CIE photopic spectral luminous efficiency function (referred to as $V(\lambda)$) is based on the averaged results of many experiments conducted by using the indirect criteria. It is identical to the CIE colour matching function $\bar{y}(\lambda)$ (Fig. 3.2). Note that it is rather a remarkable achievement to have two sets of experimental data (the colour and brightness matching data) obtained independently by using such different methods to be in agreement with each other, and in accord with Abney's law.

Although the more recent brightness matching results are somewhat different from those on which the CIE $V(\lambda)$ function is based, until recently, the CIE has been reluctant to upgrade $V(\lambda)$

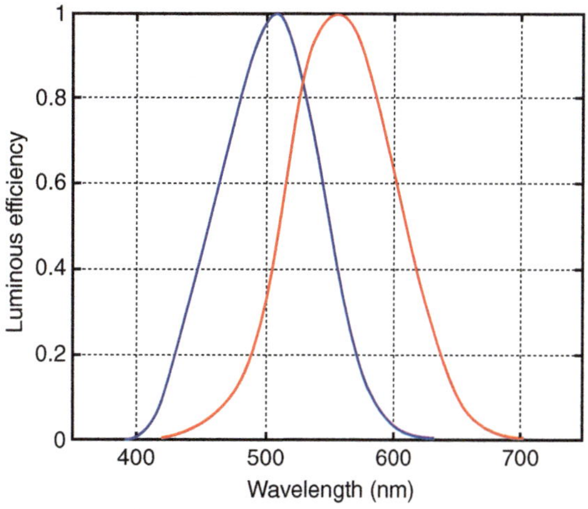

Figure 8.5 Spectral luminous efficiency functions, $V(\lambda)$ and $V'(\lambda)$, established by the CIE for photopic (red) and scotopic (blue) vision, respectively. The $V(\lambda)$ function takes the maximum value equal to one at $\lambda = 555$ nm, $V'(\lambda)$ at $\lambda = 507$ nm.

since many photometric units and standards are based on this function. It is worth noting, however, that, as mentioned in Section 3.3, Judd and then Vos modified the colour matching function so as to reconcile them with the updated measurements of the spectral luminous efficiency function (Fig. 3.12). Smith and Pokorny have derived a set of cone fundamentals by applying a particular linear transformation to the modified colour matching functions (Fig. 3.10). As a matter of fact, the CIE colour matching function $\bar{y}(\lambda)$ (Y in Fig. 3.12) modified by Judd and Vos (which is identical to the upgraded $V(\lambda)$ function) is equal to the sum of the M-, and L- Smith & Pokorny's cone fundamentals. In other words, k_S in (8.8) equals zero, and $k_M = k_L = 1$, that is, $V(\lambda) = \varphi_M(\lambda) + \varphi_L(\lambda)$. More recent independent measurements of luminous efficiency function showed that the mean $V(\lambda)$ function was consistent with the M- and L- Stockman & Sharpe's cone fundamentals (Fig. 3.13), being a linear combination of them (Sharpe et al., 2005).

Definition 8.8 *Let $G\left\langle \mathbf{C}_{+}^{*}, \sim, \mu_1, \mu_2, \mu_3 \right\rangle$ be a 3-pigment Grassmann–Govardovskii structure ($\lambda'_{\min} < \mu_1 < \mu_2 < \mu_3 < \lambda'_{\max}$) (Definition 3.4). A linear Grassmann code (written as ν) defined by the integral*

$$\nu(a) = \int_{\Lambda} \left(f_2(\lambda; \mu_2) + f_3(\lambda; \mu_3)\right) a(d\lambda), \quad a \in \mathbf{C}_{+}^{*}, \tag{8.9}$$

where $f_i(\lambda; \mu_i)$ is Govardovskii's cone fundamental (with $\lambda_i^{\max} = \mu_i$, $i = 2, 3$), and $\Lambda = \left[\lambda_{\min}, \lambda_{\max}\right]$, will be called a luminance code. *We will refer to $\nu(a)$ as the* luminance value *of the light a.*

Remark 8.9 *It must be said that the CIE defines the luminance in the form of an integral of $V(\lambda)$ multiplied by the spectral distribution function of a particular radiometric quantity – radiance (Wyszecki & Stiles, 1982, p. 259). Luminance can be understood as a photometric counterpart of radiance. Intuitively, luminance characterises visual efficiency of the light for the CIE standard observer. Therefore, luminance is not defined for individual observers. Defining a luminance code in terms of the M-, and L-cone fundamentals (8.9) makes it possible to have different luminance values for different observers.*

While to the first approximation luminance serves as a good estimate of brightness, there are essential differences between luminance values and subjective brightness judgements. Note that the luminance code (8.9) is a linear functional. Therefore, doubling light intensity has to result in doubling luminance. However, brightness is known to increase slower than a linear function of light intensity (Wyszecki & Stiles, 1982; Whittle, 1994). This discrepancy is usually taken into account by composing the luminance code and a monotonic negatively accelerating (transducer) function into a brightness code. It is usually either the logarithm or the power function with the exponent $0.3 \div 0.5$ that are taken as the transducer function. Note that such a composition remains to be a Grassmann code.

Fechnerian scaling

Gustav Theodor Fechner (1801–1887), the founder of psychophysics, proposed an original way to scale subjective magnitudes such as brightness, loudness and other sensation magnitudes (Fechner, 1860). He believed that humans were not able to directly estimate the magnitude of a sensation. Therefore, he developed an indirect ('objective', in his opinion) technique to measure it. This technique rests upon measuring the so-called *threshold*.

The notion of a threshold emerges from the fact that our sensations do not change with each and every change in physical stimulation. It is customary to distinguish between thresholds of two types: *absolute* and *difference thresholds*. The notion of absolute threshold is mainly related to our ability to perceive stimulus intensity. For example, the low absolute threshold is defined as the value that divides the stimulus intensity continuum into two subsets: those which we can perceive and those which are of so low intensity that we cannot perceive them. (The upper absolute threshold is defined as maximal stimulus intensity which can be adequately perceived without damaging the sense organ and feeling pain.) The smallest change in any physical stimulation that we can observe with our unaided senses is called the difference threshold for this sort of stimulation. In particular, the smallest change in light intensity, say, luminance that we can perceive with our eyes is called the luminance difference threshold. In other words, given two light intensities, $L_1 \leq L_2$, the difference $\Delta L = L_2 - L_1$ is the luminance difference threshold (with respect to stimulus L_1) if for any $\Delta L' < \Delta L$ the light stimuli L_1 and $L_1 + \Delta L'$ are visually indistinguishable.

Fechner also believed that there existed a psychophysical scale, i.e., a function relating subjective magnitude to its stimulus counterpart (e.g., brightness to light intensity; loudness to sound intensity; apparent distance to physical distance). Denote $B = f(L)$ the brightness/luminance psychophysical function. To measure brightness the same way as we measure luminance or any other physical magnitude, we need a yard-stick in the brightness continuum. As Fechner did not believe that we could trust observers even with judgements concerning equality of intervals in the brightness (and/or any other subjective) continuum, he had to circumvent the issue of yard-stick taking the following, as many believed, controversial assumption. Given $\Delta L = L_2 - L_1$ the luminance difference threshold, let us call the corresponding difference in the brightness continuum, $\Delta B = f(L_2) - f(L_1)$, the *just noticeable difference* (written as *jnd*). Fechner's assumption is that all the just noticeable differences are equal throughout the luminance continuum. If this is the case then the just noticeable difference can be taken as the yard-stick in the brightness continuum. Specifically, in order to measure the brightness of a light (L), he suggested to count the number of the just noticeable differences (Fig. 8.6) from the nil light ($L = 0$) to the light L.

The difference threshold is known to depend on the stimulus magnitude with respect to which it is determined (Brown & Mueller, 1965). Specifically, the light intensity difference threshold, ΔL, is proportional to the intensity, L, of the stimulus with respect to which it is determined:

$$\frac{\Delta L}{L} = const. \tag{8.10}$$

Equation (8.10) has been called by Fechner *Weber's Law* to honour Ernst Heinrich Weber (1795–1878), a German physician who described that relationship for weight discrimination.

Fechner assumed that, as the just noticeable difference is constant (that is, $\Delta B = const$) (8.10) entailed the following equation

$$\frac{\Delta L}{L} = k\Delta B, \tag{8.11}$$

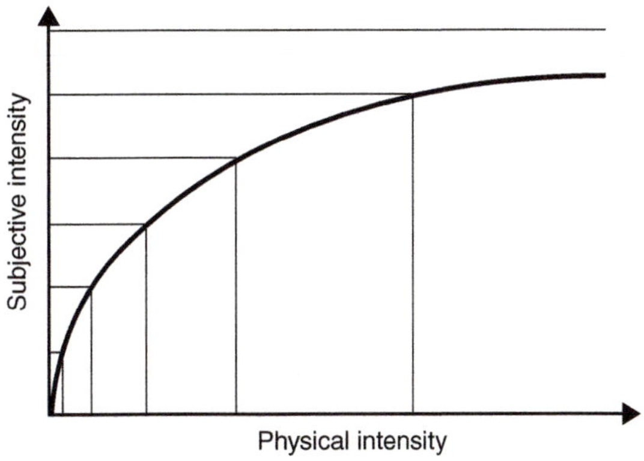

Figure 8.6 Counting just noticable differences.

where k is a constant of proportionality. Moreover, he also believed (another controversial assumption) that (8.11) remains to hold even after rewriting it in the differential form:

$$\frac{dL}{L} = kdB. \tag{8.12}$$

Solving the differential equation with respect to the unknown function $B = f(L)$ one has $kB = \ln(L) + C$. In order to determine the unknown constant of integration C, Fechner suggested that the sensation magnitude of the low absolute threshold (L_0) was zero, that is, $\ln(L_0) = 0$. (In other words, he assumed that in Fig. 8.6 the graph intersects the horizontal axis at L_0 rather than zero.) In this case one gets

$$B = \kappa \ln\left(\frac{L}{L_0}\right), \tag{8.13}$$

where κ is the constant of proportionality (equal to $1/k$).

Equation (8.13) is known as the *Fechner* (or *Weber–Fechner*) *Law* provided that L is taken as the stimulus magnitude, L_0 as the low absolute threshold, and B as the sensation magnitude (Krantz, 1971; Falmagne, 1985).

Fechner's Law (and, more generally, Fechner's threshold-based approach to measuring subjective magnitudes) was criticised by S. Stevens who noted that its prediction is not in line with direct subjective estimates (Stevens, 1961). He argued that the power function described better the relationship between the subjective and objective (physical) magnitudes of stimulation, not only in vision, but also in other sensory modalities (Stevens, 1957).

It should be said that although the Grassmann brightness code is in agreement with subjective estimates of the brightness of neutral lights, it cannot account for the brightness of chromatic lights. Specifically, when $V(\lambda)$ is evaluated by direct heterochromatic brightness matching, the brightness of equiluminant lights of complex radiant spectra can be different. For example, a monochromatic light was found to look generally brighter than a neutral light (e.g., daylight) of the same luminance. The effect is known as the *Helmholtz–Kohlrausch effect* (Wyszecki & Stiles, 1982, p. 410). It undermines brightness additivity for direct brightness judgements.

This effect is usually quantified in terms of so-called B/L ratio where B and L are the luminance of the brightness-matched test and reference (usually neutral) lights (Wyszecki & Stiles, 1982, p. 410). The mean B/L ratio obtained for 20 observers who equated in brightness nearly 100 test lights of different chromaticity to a neutral reference (the CIE 1931 chromaticity coordinates: $x = 0.342$; $y = 0.365$) of luminance 20 cd/m^{-2} varies between 1 and 1.76 (Wyszecki & Stiles, 1982, p. 412). Although significant, the effect is often ignored by those who want to have a Grassmann code for brightness that is a great deal more convenient for applications.

A few equations have been put forward to take into account the Helmholtz–Kohlrausch effect (Fairchild, 2005, p. 120). One example is the following equation for equi-bright lights in terms of their luminances – L_1 and L_2 – and chromaticities – (x_1, y_1) and (x_2, y_2):

$$\log(L_1) + F_1 = \log(L_2) + F_2, \tag{8.14}$$

where

$$F_i = 0.256 - 0.184y_i - 2.527x_iy_i + 4.656x_i^3y_i + 4.657x_iy_i^4, \quad i = 1, 2.$$

Lastly, it has long been known that the brightness of a patch of light can be affected by its immediate surroundings (Fig. 8.7). Such a phenomenon is commonly referred to as *simultaneous brightness contrast* (or *simultaneous lightness contrast* as far as object-colours are concerned).

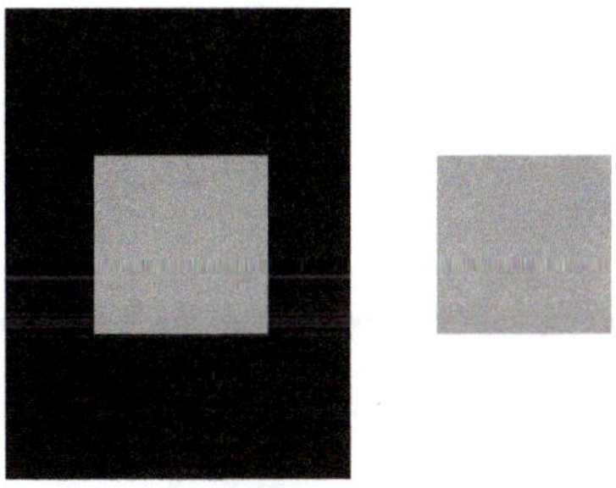

Figure 8.7 Simultaneous brightness contrast.

Simultaneous brightness/lightness contrast

As a rule, the simultaneous-contrast effect is rather small when observed on simple displays comprising a spatially uniform test against a spatially uniform background as in Fig. 8.7 (see for a review Gilchrist, 2006). However, when viewed against a structured background, the test may undergo more dramatic changes in brightness (respectively, lightness). For instance, while the luminance of the light emerging from all the horizontal rows of diamonds in Fig. 8.8 is equal, hardly anyone would insist that they are equi-bright. Observers usually judge the diamonds in the first, third, and the fifth rows (from top to bottom) as much brighter than in the second, fourth, and sixth rows (Logvinenko, 1999). The magnitude of the effect in Fig. 8.8 is so much larger than that of the classical effect (Fig. 8.7) that the question involuntarily comes to mind of whether these are in fact one and the same phenomenon (Logvinenko & Kane, 2004).

A distinctive feature of Fig. 8.8 is that it is a picture rendering quite a certain thing, namely a three-dimensional wall of blocks illuminated in such a way that the illumination intensity sinusoidally changes along the vertical dimension. The fact that it is all about the pictorial content, namely, what kind of illumination is depicted in it, follows from the following demonstration (Logvinenko, 2003a). Specifically, after turning upside down in Fig. 8.8 every other cube in each row, all the diamonds look almost the same (Fig. 8.9). Hence, although Fig. 8.9 consists of the same cubes as Fig. 8.8, the simultaneous-contrast effect for Fig. 8.9 practically disappears. Note that the pictorial illumination in Fig. 8.9 looks even, if anything, there is no impression of a sine wave of illumination. Interestingly, when the wall depicted in Fig. 8.8 was reproduced as a genuine three-dimensional wall of cardboard blocks, the effect was significantly reduced (Logvinenko & Kane, 2003). Furthermore, an analogous three-dimensional implementation of the celebrated picture of Adelson (1993), which Fig. 8.8 is actually the modification of, resulted in no simultaneous-contrast

Figure 8.8 A version of simultaneous brightness contrast. Source: Logvinenko (1999)/Sage Publications.

Figure 8.9 Rearranged version of Fig. 8.8. Source: Logvinenko (2003a)/Sage Publications.

effect at all (Logvinenko, Kane, & Ross, 2002). Generally speaking, this result is not surprising at all, since we know from everyday experience that the colour of things does not depend on what other objects are lying next to. (This fact is sometimes referred to as background-independent colour constancy (Foster et al., 1997).) It is the pronounced effect arising from the pictorial representation of these walls of blocks that requires explanation. However, it is difficult to hope for success unless we realise that an explanation is required for the phenomenon of pictorial, and not real, perception.

The point is not only, and not so much, that in one case the object of perception is real things, and in the other – their drawn pictures. There is evidence that the perceptual images of real things have different structure as compared to the perceptual images of pictures. Specifically, as shown by multidimensional scaling, perceptual images of real achromatic objects are two-dimensional (Logvinenko & Maloney, 2006) or even three-dimensional (Logvinenko, 2015a), whereas perceptual images of achromatic pictures are one-dimensional (Logvinenko et al., 2008). This means that, despite the distinct impression of uneven illumination in Fig. 8.8, it does not affect the observers' judgements, at least to the same extent as the lighting-brightness of real objects lit by real lights. Perhaps, this is because it is a 'pictorial illumination'. If this is the case, we need a theory of picture perception to explain why we see Figs 8.8 and 8.9 in such different ways.

8.2 Hue, purity, and brightness fibre bundles. Cylindrical and psychophysical colour coordinates

In this section, we consider various geometrical representations (that are often loosely called 'colour spaces') for light colour based on hue, purity, and brightness. For instance, in computer graphics, computer vision, and colour imaging cylindrical representations called the *HSB* (Hue–Saturation–Brightness), or, essentially the same thing, HSV (Hue–Saturation–Value) colour spaces are quite popular (Reinhard et al., 2008). It represents the whole colour palette as a cylinder in which brightness varies in the vertical dimension, each equi-bright set being represented as a disc where hue varies circularly and saturation radially (Fig. 8.10). Undoubtedly, such a cylindrical representation would be quite convenient provided that, firstly, it is based on the perceptual attributes (i.e., hue, saturation, and brightness), and secondly, it provides a rule how to compute hue, saturation, and brightness for each light. Unfortunately, the HSB values result from some transformations of the red, green, and blue primaries (e.g., the colour guns) of a particular device (e.g., a colour monitor). Therefore, depending on device, numerical HSB values will represent

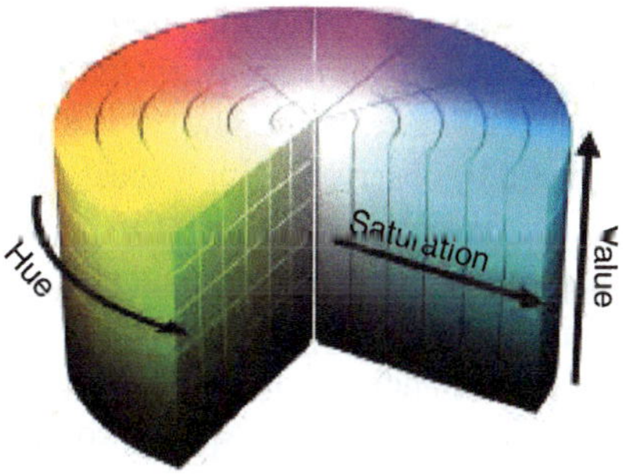

Figure 8.10 HSV cylindrical representation of colour.
Source: Wikimedia Commons.

different colours. Not to mention that there is no reason to expect human observers to see the colours lying along a radius as having the same hue. It would be more natural to plot along radii the lights which the human observer sees as equi-bright and having the same hue. Brightness value is provided by the brightness code (scale). However, as pointed out above, a code for hue does not exist. Nevertheless, below we propose a circular parametrisation of hue that allows for the cylindrical colour representation.

Let $\langle A, \sim, \succcurlyeq_B \rangle$ be a three-dimensional achromatically ordered weak* continuous Grassmann structure on the light stimulus cone $A = \mathbf{C}^*_+(\Lambda)$, satisfying the colour-signal-cone-boundary hypothesis (Definition 4.2) and having the strict convexity property (Definition 4.3), and $f_B : A \to \mathbf{R}_+$ a continuous brightness scale. Given a brightness value β, the inverse image $f_B^{-1}(\beta)$ is the equi-bright subset of light stimuli. Similarly, given the colour stimulus map π and the colour-signal map Φ, $\pi\left(f_B^{-1}(\beta)\right)$ and $\Phi\left(f_B^{-1}(\beta)\right)$ will be referred to as the equi-bright subset of the colour stimuli and the colour signals respectively. The colour stimulus cone $\pi(A)$ and the colour-signal cone $\Phi(A)$ are partitioned into equi-bright subsets. When assuming f_B to be a Grassmann code, each equi-bright subset $\pi(f_B^{-1}(\beta))$ in $\Phi(A)$ (except for $f_B^{-1}(0) = 0$) is a closed convex set $D'(H_\beta) = H_\beta \cap \Phi(\mathbf{C}^*_+)$ lying in some plane $H_\beta \in \mathcal{H}(\Phi)$ and limited by the closed curve $g_{H_\beta}(\Phi(\Delta'(\Lambda')))$ (see Chapter 4 for details).

In what follows, we restrict ourselves to considering a certain (arbitrary) equi-bright subset. As mentioned in the beginning of this chapter, a remarkable feature of the human colour vision is that, given an equi-bright subset of the colour stimuli $\pi\left(f_B^{-1}(\beta)\right)$ for some $\beta > 0$, for each but one colour stimulus, $x \in \pi\left(f_B^{-1}(\beta)\right)$, observers can point out a stimulus, say, x', in the boundary of $\pi\left(f_B^{-1}(\beta)\right)$ (for instance, the equi-bright stimulus in $\pi(con\Delta'(\Lambda'))$) which differs from x only in purity. To be more specific, given a light a presented in one half of the bipartite field (Fig. I.1), one can always find such a monochromatic light (or a light in the complementary light stimulus segment $I(\delta_{\lambda'_{min}}, \delta_{\lambda'_{max}})$) a' that the perceptual difference between $x = \pi(a)$ and $x' = \pi(a')$ is of the quantitative rather qualitative nature. For any other x'' in the boundary of $\pi\left(f_B^{-1}(\beta)\right)$ the perceptual difference between x and x'' will be of the qualitative nature. It is the term *hue* that is used to refer to such a qualitative difference. Respectively, the term purity refers to the quantitative difference between x and x' when they look of the same hue (that is, when the qualitative difference between x and x' has gone). For example, the difference between some shades of red and pink is mainly in purity, whereas, the difference between red and orange always includes the difference in hue. As stated above, when there is no qualitative difference between x and x' we say that x and x' make a *hue match*. The only colour stimulus in $\pi\left(f_B^{-1}(\beta)\right)$, for which there is no hue match with any stimulus in the boundary of $\pi\left(f_B^{-1}(\beta)\right)$, will be called *hueless* or *neutral*.

Hue matching implies that there is a map of the relative interior of an equi-bright subset of colour signals or colour stimuli (except one point) onto its boundary. All the points of the equi-bright subset that map to the same boundary point are experienced to have the same hue. For this reason, they are said to form a *constant hue set*. This map is not defined for one point which remains without hue – the neutral point. The color stimuli (colour signals) within a constant hue set can be naturally ordered by observers with respect to purity. We will refer to it as *purity ordering* (written as $\succcurlyeq_p$). It should be kept in mind that this ordering is defined for each constant hue set independently. The notions of constant hue set and purity make sense for the elements of any equi-bright subset of light stimuli as well. Specifically, the hue and purity of a light a are determined by the hue and purity of the corresponding colour stimulus $x = \pi(a)$ (or colour signal $\Phi(a)$).

The purity ordering $\succcurlyeq_p$ is a complete order (a linear order in another terminology). Indeed, reflexivity and transitivity are obvious, that is, purity ordering is a preorder. The completeness is obvious too: each two light stimuli a and b (or colour stimuli $x = \pi(a)$ and $y = \pi(b)$ or colour signals $\Phi(a)$ and $\Phi(b)$) in a constant hue set can be compared with respect to purity. Finally, $\succcurlyeq_p$ satisfies the antisymmetry axiom: if $a \succcurlyeq_p b$ and $b \succcurlyeq_p a$, then $a = b$. Otherwise one will have two different light stimuli which match each other in hue, brightness, and purity, that is possible only when the metamerism differs from colour matching. Although this is, in principle, possible (see Section 8.6.2 in this chapter and Chapter 10), in this section we assume that the metamerism does coincide with colour matching. Thus, purity ordering is actually an order (see Definition 8.1), therefore it may be referred to as a *purity order*. It is a matter of immediate observation that the spectral colours have maximum purity (in the equi-bright subset). Therefore, the monochromatic light (or the light from the complementary light stimulus segment $I(\delta_{\lambda'_{\min}}, \delta_{\lambda'_{\max}})$) in a constant hue set is the maximal element (with respect to $\succcurlyeq_p$). As the neutral light is hueless, one can formally assume that its purity is zero. Therefore, completing a constant hue set with the neutral light, we conclude that each constant hue set has the maximal and minimal elements.

The fact that purity ordering is a complete order is supported by the experimentally established fact that in the equi-bright subset in $\Phi(A)$ a constant hue set is a curve connecting the corresponding boundary point and the neutral point (Fig. 8.4). We will, therefore, refer to it as the *constant hue curve*. As can be seen in Fig. 8.4, while the colour stimuli of the same dominant wavelength have very close hue, the constant hue curves are not the radii of constant dominant wavelength. This is usually referred to as *the Abney effect* (Abney, 1910; Wyszecki & Stiles, 1982, p. 420). Moreover, the experimental data on loci of 'constant hue' show that it is safe to assume that the constant hue curves are continuous. Although the area around the neutral point in Fig. 8.4 is left without measurements, we will assume that the neutral point is the limiting point for every constant hue curve.

Each constant hue curve will be supposed to be homeomorphic to a real semi-interval (0, 1]. Thus, all the spectral hues will be assumed to have the same (unit) purity. This make an essential difference between purity, as defined here, and saturation. Saturation is, generally, understood as the degree to which a given patch of light differs from the equi-bright neutral light (Wyszecki & Stiles, 1982, p. 487). It is usually quantified in the number of just noticeable steps (see Section "Fechnerian scaling" at pp. 198–200) between the neutral point and the colour stimulus in question.[9] Such number is fewer for spectral yellow than, say, blue, that is in line with our immediate experience. Therefore, spectral yellow is less saturated than spectral blue, whereas according to our definition they are equal in purity. In colorimetry, purity is measured as the relative distance from the point representing the given light to the neutral point in the chromaticity diagram (Wyszecki & Stiles, 1982, p. 422).

9 We believe that counting the number of just noticeable steps should be accomplished along the constant hue curve. However, this condition has never been met in experiment to our knowledge.

At this point we drop the equi-bright restriction. It is well known that monochromatic lights of different brightness can readily be assessed whether they have the same or different hue. In other words, as pointed out above, hue matching can be done despite the difference in brightness. Moreover, it has been established that the wavelengths of monochromatic lights of different intensities which match each other in hue, are not always equal. This is known as the effect of luminance on hue (referred to as *Bezold–Brücke effect*) (Wyszecki & Stiles, 1982, p. 422). There are only three wavelengths at which no Bezold–Brücke effect has been observed (Purdy, 1931). These three spectral hues are sometimes called *permanent hues* (for review see Pridmore, 2007). Phenomenologically they are perceived as unique[10] blue, green and yellow (or very close to them) (Boynton & Gordon, 1965).

Hue matching of monochromatic lights of different brightness results in the constant hue lines in the spectral colour-signal conical surface $con\Phi\big(\Delta(\Lambda')\big) = \{p\Phi(\delta_\lambda) : p > 0, \lambda \in \Lambda'\}$. Because of the Bezold–Brücke effect, the constant hue lines are not, in general, straight rays. Experimental data collected so far on Bezold–Brücke effect show that the constant hue lines are continuous (Purdy, 1931). So is the loci of all the neutral points, which we will refer to as the *neutral line*. More formally, we assume that there exists a smooth map $n : \mathbf{R}_+ \to \Phi(A)$ (respectively, $\Phi_0^{-1} \circ n : \mathbf{R}_+ \to \pi(A)$, where $\Phi = \Phi_0 \circ \pi$; see the beginning of Chapter 5) the range of which comprises the hueless loci in $\Phi(A)$ (respectively, $\pi(A)$). Note that the domain of the map n can readily be thought to amount to the range of f_B. In this case, for each brightness value β the map n (respectively, $\Phi_0^{-1} \circ n$) determines the neutral point in the β-equi-brightness subset of $\Phi(A)$ (respectively, $\pi(A)$).

In what follows, we assume the existence of a light σ_0 such that $\Phi(\sigma_0)$ is an interior point of $\Phi(A)$ and the Grassmann structure $\langle A, \sim \rangle$ is standard with respect to σ_0 (see Definition 5.1). We will consider the colour stimulus manifold $\mathcal{A} = \mathcal{A}_{\sigma_0}$ and two subsets in it, $\mathcal{A}_{\sigma_0} \backslash \{q\sigma_0 : q \geq 0\}$ and $\mathcal{A}_{\sigma_0} \backslash \Phi_{|\mathcal{A}}^{-1}(n(\mathbf{R}_+))$. Clearly, both subsets are homeomorphic, hence each of them is homeomorphic to the colour stimulus cone without the ray through $\pi(\sigma_0)$, i.e., $\pi(A) \backslash \{q\pi(\sigma_0) : q \geq 0\}$. Below we describe all the homeomorphisms between both the sets; see Proposition 8.4.

Note the representation

$$\mathcal{A}_{\sigma_0} \backslash \{q\sigma_0 : q \geq 0\} = \{\sigma = p\sigma' + q\sigma_0 : p > 0, q \geq 0, \sigma' \in \Delta'(\Lambda')\},$$

that is, every $\sigma \in \mathcal{A}_{\sigma_0} \backslash \{q\sigma_0 : q \geq 0\}$ is uniquely represented in form $\sigma = p\sigma' + q\sigma_0$ where $\sigma' \in \Delta'(\Lambda'), p > 0, q \geq 0$. As the closed curve $\Delta'(\Lambda') = \Delta(\Lambda') \cup I(\delta_{\lambda'_{\min}}, \delta_{\lambda'_{\max}})$ is homeomorphic to the unit circle S^1, the set $\mathcal{A}_{\sigma_0} \backslash \{q\sigma_0 : q \geq 0\}$ endowed with the weak* topology is homeomorphic to a cylindrical body without a central axis in $\mathbf{R}^3$, $\mathcal{Z}_0 := \{(x, y, z) \in \mathbf{R}^3 : 0 < x^2 + y^2 \leq 1, z > 0\}$. (A concrete example of such a homeomorphism is given below.) Clearly $\mathcal{Z}_0$ is neither closed nor open in $\mathbf{R}^3$ as well as $\mathcal{A}_{\sigma_0} \backslash \{q\sigma_0 : q \geq 0\}$ is neither closed nor open in the light stimulus cone $A = \mathbf{C}_+^*(\Lambda)$ endowed with the weak* topology.

Let S^1 be identified with $[0, 2\pi] \bmod(2\pi)$ and $\Delta'(\Lambda') \ni \sigma' \mapsto \theta \in [0, 2\pi)$ be a homeomorphism of $\Delta'(\Lambda')$ onto S^1. There are many such homeomorphisms, so for definiteness we fix one of them, say

$$\theta(\sigma') = \begin{cases} \dfrac{\lambda - \lambda'_{\min}}{\lambda'_{\max} - \lambda'_{\min}}\pi & \text{if } \sigma' = \delta_\lambda, \lambda'_{\min} \leq \lambda \leq \lambda'_{\max}, \\ (1+\alpha)\pi & \text{if } \sigma' = \alpha\delta_{\lambda'_{\min}} + (1-\alpha)\delta_{\lambda'_{\max}}, \ 0 \leq \alpha \leq 1, \end{cases} \tag{8.15}$$

10 Instead of the term 'unique hue', which is widely used in the colour vision literature, we will henceforth use the term 'unitary hue', which is formally defined in Section 8.5.1 (Definition 8.9).

and notice that if $\lim_\nu \theta(\sigma'_\nu) = 2\pi$ then the net (σ'_ν) converges weakly* to $\delta_{\lambda'_{\min}}$. The inverse mapping $S^1 \to \Delta'(\Lambda')$ is given then by

$$\sigma'(\theta) = \begin{cases} \delta_{\lambda'_{\min} + \frac{\theta}{\pi}\left(\lambda'_{\max} - \lambda'_{\min}\right)} & \text{if } 0 \le \theta \le \pi, \\ \left(\frac{\theta}{\pi} - 1\right)\delta_{\lambda'_{\min}} + \left(2 - \frac{\theta}{\pi}\right)\delta_{\lambda'_{\max}} & \text{if } \pi \le \theta < 2\pi. \end{cases} \tag{8.16}$$

Every point σ in $\mathcal{A}_{\sigma_0}\backslash\{q\sigma_0 : q \ge 0\}$ is uniquely determined now by three numbers $p > 0$, $q \ge 0$, and $\theta \in [0, 2\pi)$, referred to as its *cylindrical coordinates*, and we obtain a homeomorphism g of $\mathcal{A}_{\sigma_0}\backslash\{q\sigma_0 : q \ge 0\}$ onto $\mathbb{Z}_0$ given by three functions connecting cylindrical coordinates (p, q, θ) on $\mathcal{A}_{\sigma_0}\backslash\{q\sigma_0 : q \ge 0\}$ with Cartesian coordinates (x, y, z) on $\mathbb{Z}_0$:

$$x = \frac{p}{p+q}\cos\theta, \quad y = \frac{p}{p+q}\sin\theta, \quad z = p + q. \tag{8.17}$$

Let us describe the inverse homeomorphism g^{-1} of $\mathbb{Z}_0$ onto $\mathcal{A}_{\sigma_0}\backslash\{q\sigma_0 : q \ge 0\}$, $g^{-1}(x, y, z) = (p(x, y, z), q(x, y, z), \theta(x, y, z))$. It follows from (8.17) that $\tan\theta = y/x$. There are two values θ_1 and θ_2 in $[0, 2\pi)$ satisfying this equality, and the correct choice of $\theta = \theta(x, y, z)$ is uniquely determined by the signs of x and y as follows. If $x \ge 0$, $y > 0$ then $\theta(x, y, z) \in (0, \pi/2]$; if $x < 0$, $y > 0$ then $\theta(x, y, z) \in (\pi/2, \pi)$; if $x \le 0$, $y < 0$ then $\theta(x, y, z) \in (\pi, 3\pi/2]$; if $x > 0$, $y < 0$ then $\theta(x, y, z) \in (3\pi/2, 2\pi)$. Finally, if $x > 0$, $y = 0$ then $\theta(x, y, z) = 0$, and if $x < 0$, $y = 0$ then $\theta(x, y, z) = \pi$. Taking into account (8.17), the value $p = p(x, y, z)$ is determined now from the equation $xz = p\cos\theta(x, y, z)$ if $\cos\theta(x, y, z) \neq 0$ or from the equation $yz = p\sin\theta(x, y, z)$ otherwise, and the value $q = q(x, y, z)$ is determined as $z - p(x, y, z)$. Of course, there exist other (different from g) homeomorphisms of $\mathcal{A}_{\sigma_0}\backslash\{q\sigma_0 : q \ge 0\}$ onto $\mathbb{Z}_0$.

Consider now the set $\mathcal{A}_{\sigma_0}\backslash\Phi^{-1}_{|\mathcal{A}}(n(\mathbf{R}_+))$. Being homeomorphic to $\mathcal{A}_{\sigma_0}\backslash\{q\sigma_0 : q \ge 0\}$, it is homeomorphic to $\mathbb{Z}_0$. There are many homeomorphisms between these sets. Now we will define one of them, $h : \mathcal{A}_{\sigma_0}\backslash\Phi^{-1}_{|\mathcal{A}}(n(\mathbf{R}_+)) \to \mathbb{Z}_0$.

From a formal point of view the possibility of hue matching means that there exists a continuous surjection $f_H : \mathcal{A}_{\sigma_0}\backslash\Phi^{-1}_{|\mathcal{A}}(n(\mathbf{R}_+)) \to \Delta'(\Lambda')$ which takes an element σ' in $\Delta'(\Lambda')$ to all the colour stimuli (i.e., elements of the colour stimulus manifold) that match σ' in hue. For example, $f_H(\sigma) = \delta_{\lambda_0}$ means that the monochromatic light δ_{λ_0} is a hue match[11] for the light σ.

Note that for any brightness value β, and any θ the subset $f_B^{-1}(\beta) \cap f_H^{-1}(\sigma'(\theta))$ comprises all the elements in the equi-bright subset $f_B^{-1}(\beta) \cap \mathcal{A}_{\sigma_0}\backslash\Phi^{-1}_{|\mathcal{A}}(n(\mathbf{R}_+))$ which match each other in hue. As mentioned above, we assume that for any θ the neutral point $\Phi^{-1}_{|\mathcal{A}}(n(\beta))$ is the limiting point of $f_B^{-1}(\beta) \cap f_H^{-1}(\sigma'(\theta))$. Moreover, we assume that $s_{\beta,\theta} = f_B^{-1}(\beta) \cap f_H^{-1}(\sigma'(\theta))$ is a continuous curve. This assumption may be justified if to pass to $\Phi(\mathcal{A})$ and notice that $\Phi(s_{\beta,\theta})$ is the intersection of two closed surfaces in $\mathbf{R}^3$. Let $\gamma_{\beta,\theta} : s_{\beta,\theta} \to (0, 1]$ be a continuous scale (ordinal scale) for the purity order on the curve $s_{\beta,\theta}$. Moreover, we will assume that the scales $\gamma_{\beta,\theta}$ corresponding to various pairs (β, θ) can be conformed in such a way that the function $(\beta, \gamma, \theta) \mapsto \gamma_{\beta,\theta}(\gamma)$ is continuous. Hence, we have defined a continuous map $f_P : \mathcal{A}_{\sigma_0}\backslash\Phi^{-1}_{|\mathcal{A}}(n(\mathbf{R}_+)) \to (0, 1]$ the values of which will be referred to as *purity*. The homeomorphism $h : \mathcal{A}_{\sigma_0}\backslash\Phi^{-1}_{|\mathcal{A}}(n(\mathbf{R}_+)) \to \mathbb{Z}_0$ is determined now by $h(\sigma) = (x', y', z')$,

$$x' = \gamma\cos\theta', \quad y' = \gamma\sin\theta', \quad z' = \beta,$$

where $\beta = f_B(\sigma)$ is the brightness, $\gamma = f_P(\sigma)$ is the purity, and $\theta' = \theta(f_H(\sigma))$ (here, the function $\Delta'(\Lambda') \ni \sigma' \mapsto \theta(\sigma') \in [0, 2\pi)$ is given by (8.15)) is the hue of the light $\sigma \in \mathcal{A}_{\sigma_0}\backslash\Phi^{-1}_{|\mathcal{A}}(n(\mathbf{R}_+))$.

11 As hue matching is less difficult for equi-bright stimuli, it would be more practical if it is conducted first for the equi-bright monochromatic stimuli, and then the equi-bright monochromatic hue match should be equated in hue with a monochromatic light of unit intensity, i.e., δ_{λ_0}.

The homeomorphism h of $\mathcal{A}_{\sigma_0}\backslash\Phi^{-1}_{|\mathcal{A}}(n(\mathbf{R}_+))$ onto $\mathbb{Z}_0$ determines thus a parametrisation of the topological space $\mathcal{A}_{\sigma_0}\backslash\Phi^{-1}_{|\mathcal{A}}(n(\mathbf{R}_+))$ in terms of brightness β, hue θ', and purity γ, where central axis corresponds to brightness β, angle around the axis corresponds to hue θ', and distance from the axis corresponds to purity γ. (In other words, every $\sigma \in \mathcal{A}_{\sigma_0}\backslash\Phi^{-1}_{|\mathcal{A}}(n\left(\mathbf{R}_+\right))$ is uniquely determined by three numbers: brightness β, purity γ, and hue θ'.) As this representation of the space $\mathcal{A}_{\sigma_0}\backslash\Phi^{-1}_{|\mathcal{A}}(n(\mathbf{R}_+))$ provides essential information about the colour appearance of the colour stimuli, we will call the variables β, θ', and γ the *psychophysical colour coordinates*. It must be said that although we have some empirical information about the functions f_B, $f'_H = \theta \circ f_H$, f_P, they are still to be determined in experiment.

Note that a subset of constant purity $f_P^{-1}(\gamma)$ is homeomorphic to a two-dimensional surface in $\mathbf{R}^3$ which is the product of the circle S^1 and $\mathbf{R}_+\backslash\{0\}$ for each purity value γ.

Note also that all the equi-bright subsets (for $\beta > 0$) are homeomorphic to each other. Each equi-bright subset in $\mathcal{A}_{\sigma_0}\backslash\Phi^{-1}_{|\mathcal{A}}(n(\mathbf{R}_+))$ can be considered then as a representation of the same topological space, namely the unit disc without the centre in $\mathbf{R}^2$, i.e., $\mathbb{S} := \{(x,y) \in \mathbf{R}^2 : 0 < x^2 + y^2 \leq 1\}$, $x = \gamma\cos\theta'$, $y = \gamma\sin\theta'$. We will call it the *chromaticity space*. Hue θ', and purity γ will be referred to as *chromaticity (coordinates)*.[12]

The continuous surjection $f_H : \mathcal{A}_{\sigma_0}\backslash\Phi^{-1}_{|\mathcal{A}}(n\left(\mathbf{R}_+\right) \to \Delta'(\Lambda')$ induces, in fact, a *fibre bundle*,[13] with the base space being $\Delta'(\Lambda')$, the total space being $\mathcal{A}_{\sigma_0}\backslash\Phi^{-1}_{|\mathcal{A}}(n\left(\mathbf{R}_+\right))$, and the fibre being the product of the open ray $\mathbf{R}_+\backslash\{0\}$ and the semi-interval $(0, 1]$. We will call the quadruplet

$$\left\langle \mathcal{A}_{\sigma_0}\backslash\Phi_{|\mathcal{A}}(n(\mathbf{R}_+)),\ \Delta'(\Lambda'),\ f_H,\ (\mathbf{R}_+\backslash\{0\})\times(0,1]\right\rangle \tag{8.18}$$

a *hue fibre bundle*.

Likewise, the function $f_P : \mathcal{A}_{\sigma_0}\backslash\Phi^{-1}_{|\mathcal{A}}(n(\mathbf{R}_+)) \to (0, 1]$ allows defining a *purity fibre bundle* with the product of S^1 and $\mathbf{R}_+\backslash\{0\}$ for the fibre. Finally, the brightness scale $f_B : \mathcal{A}_{\sigma_0}\backslash\Phi^{-1}_{|\mathcal{A}}(n(\mathbf{R}_+)) \to \mathbf{R}_+\backslash\{0\}$ determines a *brightness fibre bundle* over $\mathbf{R}_+\backslash\{0\}$ with a unit disc without the centre in $\mathbf{R}^2$ as a fibre. (Also, it determines a fibre bundle over $\mathbf{R}_+$ with a unit disc in $\mathbf{R}^2$ as a fibre.)

Thus, instead of the colour stimulus manifold, we deal with a more complicated mathematical object as a model of the human colour palette: three fibre bundles that induce the three colour attributes – hue, purity, and brightness – which are intuitively clear, but never have been formally defined so far.

The following representation result is a direct consequence of the above definitions.

Proposition 8.4 *Suppose f is a homeomorphism of $\mathcal{A}_{\sigma_0}\backslash\{q\sigma_0 : q \geq 0\}$ onto $\mathcal{A}_{\sigma_0}\backslash\Phi^{-1}_{|\mathcal{A}}(n(\mathbf{R}_+))$, then $\eta := h \circ f \circ g^{-1}$ is a homeomorphism of $\mathbb{Z}_0$ onto itself, and $f = h^{-1} \circ \eta \circ g$. Moreover, each homeomorphism η of $\mathbb{Z}_0$ onto itself determines a homeomorphism f of $\mathcal{A}_{\sigma_0}\backslash\{q\sigma_0 : q \geq 0\}$ onto $\mathcal{A}_{\sigma_0}\backslash\Phi^{-1}_{|\mathcal{A}}(n(\mathbf{R}_+))$ by the formula $f = h^{-1} \circ \eta \circ g$.*

In the above considerations, the neutral line comprising hueless points was supposed to be arbitrary continuous curve. Now we will assume that it is a ray passing through the light signal $\Phi(\sigma_0)$:

$$n(\mathbf{R}_+) = \{q\Phi(\sigma_0) : q \geq 0\}. \tag{8.19}$$

---

12 These should not be confused with the CIE chromaticity coordinates.

13 A quadruplet $\langle S, B, f, F\rangle$, where S, B, and F are topological spaces and $f : S \to B$ is a continuous surjection satisfying a so-called *local triviality condition* (which, in particular, means that each inverse image (fibre) $f^{-1}(b)$ is homeomorphic to F) is called a *fibre bundle*. The space B is called the *base space* of the bundle, S the *total space*, and F the *fibre*. The map f is called the *projection map* (or *bundle projection*).

This supposition looks rather plausible and acceptable. It means, in particular, that if the daylight σ_0 is neutral then changing the intensity cannot make it coloured. When condition (8.19) holds, both coordinate systems, cylindrical and psychophysical colour, are defined on one and the same set $\mathcal{A}_{\sigma_0} \backslash \{q\sigma_0 : q \geq 0\}$, hence functions f_B, f_P, and $f'_H = \theta \circ f_H$ are defined on that set, and Proposition 8.4 is applicable to the identical homeomorphism $f = id_{\mathcal{A}_{\sigma_0} \backslash \{q\sigma_0 : q \geq 0\}}$. Then there is a homeomorphism $\eta : \mathbb{Z}_0 \to \mathbb{Z}_0$ such that for every $\sigma = p\sigma' + q\sigma_0 \in \mathcal{A}_{\sigma_0} \backslash \{q\sigma_0 : q \geq 0\}$ one has $h(\sigma) = \eta(g(\sigma))$ or, in (Cartesian) coordinate form,

$$(\gamma \cos\theta', \gamma \sin\theta', \beta) = \eta\left(\frac{p}{p+q}\cos\theta, \frac{p}{p+q}\sin\theta, p+q\right),$$

where

$$\beta = f_B(\sigma) = f_B(p,q,\theta),\ \gamma = f_P(\sigma) = f_P(p,q,\theta),\ \theta' = f'_H(\sigma) = f'_H(p,q,\theta). \tag{8.20}$$

Here, Eq. (8.20) determine a continuous change of coordinates (cylindrical coordinates by psychophysical colour coordinates) on $\mathcal{A}_{\sigma_0} \backslash \{q\sigma_0 : q \geq 0\}$. Note that the Bezold–Brücke effect entails that for nearly all θ the inequality holds $f'_H(p,0,\theta) \neq \theta$ whenever $p > 0$. The monochromatic lights of three wavelengths, δ_{λ_i} $(i = 1,2,3)$ such that for $\theta_i = \theta(\delta_{\lambda_i})$ one has $f'_H(p,0,\theta_i) = \theta_i$ whenever $p > 0$ determine the permanent hues. Of course, exact representation of f_B, f_P, and f'_H as functions of (p,q,θ) is equivalent to finding the homeomorphism $\eta : \mathbb{Z}_0 \to \mathbb{Z}_0$ that provides (by Proposition 8.4) the identical homeomorphism on $\mathcal{A}_{\sigma_0} \backslash \{q\sigma_0 : q \geq 0\}$.

8.3 Colour transformation caused by media and metamer mismatching, as expressed in the psychophysical colour coordinates

In contrast to the colorimetric coordinates, the cylindrical coordinates (q,p,θ) and the psychophysical colour coordinates (β,γ,θ') (8.20) are suitable to represent the colour stimulus of the same light in different Grassmann structures. Specifically, assume $\langle A, \sim\rangle$ and $\langle A, \approx\rangle$ to be two Grassmann structures having the same colour stimulus manifold $\mathcal{A}$ (i.e., they satisfy the hypotheses of Theorem 5.1 with the same σ_0 and the same Λ'). Let $\kappa_\sim(a)$ and $\kappa_\approx(a)$ represent light a in $\mathcal{A}$ with respect to metamerisms $\sim$ and $\approx$ respectively. When both $\kappa_\sim(a)$ and $\kappa_\approx(a)$ belong to $\mathcal{A}_{\sigma_0} \backslash n(\mathbf{R}_+)$, there exist their cylindrical coordinates $(q_\sim(a), p_\sim(a), \theta_\sim(a))$ and $(q_\approx(a), p_\approx(a), \theta_\approx(a))$ respectively. The colour stimulus shift, that light a undergoes under transition from $\sim$ to $\approx$ (as defined by Eq. (5.13)), can be characterised by shifts in each of the three cylindrical coordinates: $|q_\approx(a) - q_\sim(a)|$, $|p_\approx(a) - p_\sim(a)|$, and $|\theta_\approx(a) - \theta_\sim(a)|$.

Although the θ-, q-, and p-shifts give some insight into how the colour of a changes because of the transition from $\sim$ to $\approx$, it is better to express the difference between $\kappa_\sim(a)$ and $\kappa_\approx(a)$ in terms of the psychophysical colour coordinates β, θ' and γ (8.20) for this purpose. Indeed, even if, for example, the shift occurs only along the dimension θ it does not mean that only the hue will change. In contrast, the coordinate shifts $|\theta'_\approx(a) - \theta'_\sim(a)|$, $|\beta_\approx(a) - \beta_\sim(a)|$, and $|\gamma_\approx(a) - \gamma_\sim(a)|$ will represent the shifts in hue, brightness and purity which the colour of a undergoes under the transition from $\sim$ to $\approx$. In this case, if non-zero shift takes place only along θ', the colours of $\kappa_\sim(a)$ and $\kappa_\approx(a)$ differ only in hue. When expressed in terms of the psychophysical colour coordinates (8.20), the shift (from $\kappa_\sim(a)$ to $\kappa_\approx(a)$) will be called the *psychophysical colour shift*.[14] When, for instance, $\langle A, \sim\rangle$

14 Although the θ'-component of the psychophysical colour shift indicates a change in hue, it is not quite informative regarding the qualitative aspect of this change. In order to overcome this shortcoming, one can use the perceptual colour coordinates introduced in Section 8.6.3.1. In this case we will use terms *colour transformation (caused by the transition from $\sim$ to $\approx$)* and *colour shift*.

and $\langle A, \approx\rangle$ represent the same observer in two different media, the psychophysical colour shift shows how the colour of a changes with the medium (cf. Section 5.2).

The variability of each of the θ-, q-, and p-shifts (induced by the transition from $\sim$ to $\approx$), depending on which metamer of a is taken, can be quantified by introducing the notions of maximal and relative maximal shifts similar to the way it was done in Section 5.2.1. Specifically, define the *maximal θ-shift* (induced by the transition from $\sim$ to $\approx$) as

$$\Delta\theta(\sim, \approx; a) := \max_{b\sim a} \left|\theta_{\approx}(b) - \theta_{\sim}(a)\right|, \tag{8.21}$$

and the *relative maximal θ-shift* as

$$\delta\theta(\sim, \approx; a) := \frac{\max_{b\sim a} \left|\theta_{\approx}(b) - \theta_{\sim}(a)\right|}{\theta_{\sim}(a)}. \tag{8.22}$$

Also, the *maximal* and *relative maximal q-* and *p-shifts* ($\Delta q(\sim, \approx; a)$, $\Delta p(\sim, \approx; a)$, $\delta q(\sim, \approx; a)$, and $\delta p(\sim, \approx; a)$) can defined as in Eqs. (5.16)–(5.19) respectively.

Likewise, define the *maximal θ'-shift* (induced by the transition from $\sim$ to $\approx$) as

$$\Delta\theta'(\sim, \approx; a) := \max_{b\sim a} \left|\theta'_{\approx}(b) - \theta'_{\sim}(a)\right|, \tag{8.23}$$

the *relative maximal θ'-shift* as

$$\delta\theta'(\sim, \approx; a) := \frac{\max_{b\sim a} \left|\theta'_{\approx}(b) - \theta'_{\sim}(a)\right|}{\theta'_{\sim}(a)}, \tag{8.24}$$

the *maximal β-shift* as

$$\Delta\beta(\sim, \approx; a) := \max_{b\sim a} \left|\beta_{\approx}(b) - \beta_{\sim}(a)\right|, \tag{8.25}$$

the *relative maximal β-shift* as

$$\delta\beta(\sim, \approx; a) := \frac{\max_{b\sim a} \left|\beta_{\approx}(b) - \beta_{\sim}(a)\right|}{\beta_{\sim}(a)}, \tag{8.26}$$

the *maximal γ-shift* as

$$\Delta\gamma(\sim, \approx; a) := \max_{b\sim a} \left|\gamma_{\approx}(b) - \gamma_{\sim}(a)\right|, \tag{8.27}$$

and the *relative maximal γ-shift* as

$$\delta\gamma(\sim, \approx; a) := \frac{\max_{b\sim a} \left|\gamma_{\approx}(b) - \gamma_{\sim}(a)\right|}{\gamma_{\sim}(a)}. \tag{8.28}$$

Note that these indices can be used to quantify metamer mismatching induced by replacing metamerism $\sim$ with $\approx$. The advantage over the metamer-mismatch indices put forward in Section 7.2 is rather obvious. The indices (8.23)–(8.28) show the maximal dispersion of hue, purity, and brightness caused by metamer mismatching. True, as pointed out above, the θ'-shift is a rather abstract measure for the hue shift. However, the use of perceptual rather than psychophysical colour coordinates will help get rid of this drawback.

The psychophysical colour coordinates can also be used to quantify light robustness in new ways (see Section 5.2.2). Specifically, the following indices of light robustness to medium transmittance change are defined as follows: the *hue robustness* (to medium transmittance change) *index*

$$r\theta'(\sim, \approx; a) := 1 - \frac{\left|\theta'_{\approx}(a) - \theta'_{\sim}(a)\right|}{\max_{b\sim a} \left|\theta'_{\approx}(b) - \theta'_{\sim}(b)\right|}, \tag{8.29}$$

the *brightness robustness* (to medium transmittance change) *index*

$$r\beta(\sim, \approx; a) := 1 - \frac{|\beta_{\approx}(a) - \beta_{\sim}(a)|}{\max\limits_{b\sim a} |\beta_{\approx}(b) - \beta_{\sim}(b)|}, \tag{8.30}$$

and the *purity robustness* (to medium transmittance change) *index*

$$r\gamma(\sim, \approx; a) := 1 - \frac{|\gamma_{\approx}(a) - \gamma_{\sim}(a)|}{\max\limits_{b\sim a} |\gamma_{\approx}(b) - \gamma_{\sim}(b)|}. \tag{8.31}$$

In Chapter 9, light robustness to medium transmittance change will be also expressed in terms of the perceptual light-colour coordinates (8.46)–(8.47) (see Eqs. (9.25)–(9.27)).

8.4 Light-colour perception in dichromats

The theory we are developing can just as well be used to study both trichromatic and dichromatic (or tetrachromatic) vision. True, for obvious reasons, the trichromatic case has, generally, been looked into more systematically and in more detail. Not surprisingly, the following question has been debated for a very long time. Can one say anything about the colour vision of dichromats based on knowledge of the colour vision of trichromats? (Wittgenstein, 1977; Byrne & Hilbert, 2010; Broackes, 2010b) In this section, we are going to give an affirmative answer if dichromat is understood as a reduced form of trichromat (as in Section 3.3.1).

More specifically, consider an observer, whose colour vision can be modelled by a 3-*pigment Grassmann-Govardovskii structure*, formed by three Govardovskii's cone fundamentals, $f_1(\lambda; \mu_1)$, $f_2(\lambda; \mu_2)$ and $f_3(\lambda; \mu_3)$ (Definition 3.4), and another observer, modelled by a 2-*pigment Grassmann-Govardovskii structure* formed by any pair of these three fundamentals: $f_i(\lambda; \mu_i)$ and $f_j(\lambda; \mu_j)$ where $i, j = 1, 2, 3; i \neq j$. In other words, the second observer misses cone photoreceptors of one particular type, being identical to the first observer in all other respects. The latter condition will be fulfilled if one assumes that the second observer is, actually, the first one, the outputs of the 'missing' cones of whom are blocked, so that they do not enter the brain further.

Such a dichromat will accept all the colour matches made by the Trichromat,[15] but the converse is not true. Specifically, there are lights, well distinguished by the trichromat, that the dichromat will confuse.[16] Indeed, given the metameric matching hypothesis, two lights match each other in colour when the cone responses of each type to these lights are equal. This condition is logically stronger for the trichromat than the dichromat.

A line parallel to the 'missing' coordinate axis in the trichromatic colour-signal space, and passing through the trichromatic colour-signal cone (Fig. 4.4) interior, generally, intersects this cone through an interval. The points of this interval represent the colours that the dichromat (lacking the cones corresponding to the coordinate axis in question) will confuse. This interval will be referred to as a *dichromatic confusion interval*. The dichromat will perceive all the lights represented in

15 The present consideration is confined to an ideal case when the trichromats and dichromats are assumed to have the common cones with the same peak sensitivity. Also the spectral transmittance of the ocular media and the macular pigment is supposed to be the same for trichromats and dichromats. Otherwise, even trichromatic observers will not always accept matches of each other.

16 A well known example is some shades of red and green which protanopes and deuteranopes cannot distinguish (Birch, 2001).

the trichromatic colour-signal cone by a dichromatic confusion interval as having the same colour. The big problem still remaining unsolved is what this colour is (Holmgren, 1881; Vienot et al., 1995; Broackes, 2010a; D. I. A. MacLeod, 2010).

The problem can be resolved if we accept the following two assumptions,[17] which look rather plausible if one remembers that the dichromat is the reduced form of the trichromat. (For brevity, these assumptions will be referred to as the *reductionist hypothesis*.) Firstly, we assume that the lights for which the dichromatic confusion interval is reduced to a point are perceived identically by both the trichromat and dichromat. Secondly, we assume that the phenomenological structure of the light-colour palette of the dichromat is based on the same structural principle as that of the trichromat. More specifically, as shown above, the light-colour palette of trichromatic colour vision is entirely determined by the trichromatic colour-signal cone boundary lights (i.e., the spectral light stimulus conical closure). Recall that for any light that maps to the trichromatic colour-signal cone interior, there is a trichromatic-colour-signal-cone-boundary light which has the same hue and brightness and differs only in purity (see Section 8.2). Moreover, the trichromatic hue palette (for light-colours) is, specified by a closed loop in the trichromatic colour-signal cone boundary – the spectral colour-signal closure. Any light mapping to the trichromatic colour-signal cone boundary is known to be perceived as having the same hue as one of the lights belonging to the spectral colour-signal closure (i.e., either the spectral curve or the purple interval) differing only in brightness.

A dichromatic colour-signal cone is, evidently, two-dimensional. It can be obtained as a projection of the trichromatic colour-signal cone parallel to the 'missing' coordinate axis on the plane made by the other two coordinate axes. Likewise, the dichromatic spectral colour-signal curve is the corresponding projection of the trichromatic spectral colour-signal curve (Fig. 8.11). Alternatively, using only two coordinate axes, one can independently plot a dichromatic spectral colour-signal curve (Fig. 8.12).

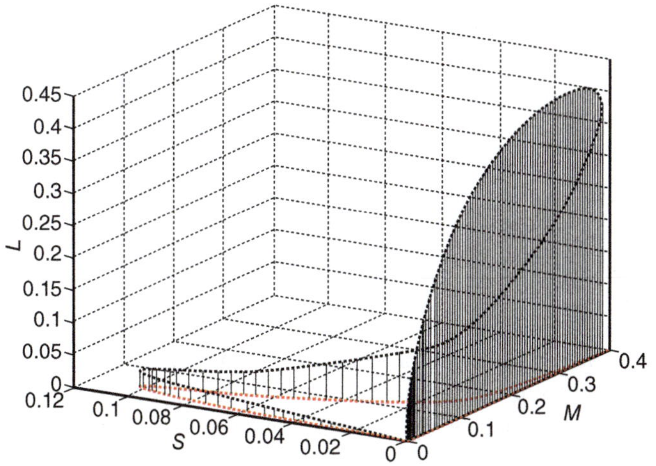

Figure 8.11 Trichromatic spectral colour signal curve in the cone excitation space. The coordinate axes are the responses of the *S*-, *M*-, and *L*-cones to equal energy monochromatic lights. Source: Logvinenko (2014)/John Wiley and Sons.

17 Similar assumptions will be adopted in Section 17.4 as well.

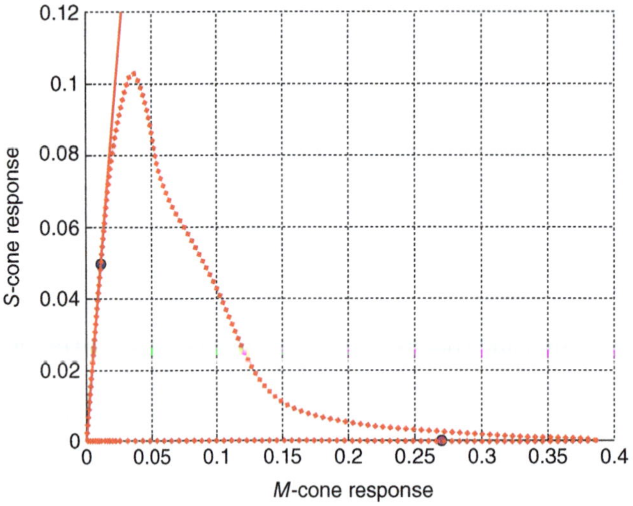

Figure 8.12 Protanopic spectral colour signal curve in the dichromatic cone excitation space. The protanopic colour cone is confined between the horizontal coordinate axis and the ray through the purple circle (see text). Source: Logvinenko (2014)/John Wiley and Sons.

If the second assumption is fulfilled, then the light-colour palette of dichromats should consist of only two hues, namely, hues of the monochromatic lights that map onto the dichromatic colour-signal cone boundary rays. (We will refer to these as the *dichromatic hues.*) All the other lights will be perceived by dichromats as having one of these two hues (or hueless) with various brightness and purity.

Interestingly, there is no need to postulate the existence of a hueless colour in the dichromatic light-colour palette – it follows from the principle of continuity. Indeed, the lights of equal intensity can be represented by a cross-section through the dichromatic colour-signal cone. Such a cross-section is an interval the points of which represent colours differing mainly in chromaticity (their brightness being approximately equal). It is a two-dimensional analogue of the chromaticity diagram. The endpoints of this chromaticity interval have different hues. Moving from one end will be accompanied by a gradual decrease in purity. There must be a point somewhere in the middle of the chromaticity interval which is perceived as having no hue. Otherwise one will experience a jump from one hue to the other. The discontinuity of this sort has not been reported either in trichromatic or in dichromatic colour vision, which confirms the principle of continuity.

Thus, the problem reduces to ascertaining the dichromatic hues. In other words, knowing how dichromats perceive lights represented by two boundary rays of a dichromatic colour-signal cone will allow us deducing the appearance of all other lights for dichromats. In this regard, it is worthy noting that the dichromatic confusion intervals for the lights that are mapped onto the boundary rays of the dichromatic colour-signal cone degenerate to a point (that is, for these light there is, actually, no confusion interval in the trichromatic colour-signal cone). Indeed, for each coordinate axis (let us consider for certainty the vertical axis) there are two (vertical) planes which are tangential to the trichromatic colour-signal cone. For each of the two tangent vertical planes there is the ray of tangency, that is, the boundary ray of the trichromatic colour-signal cone where the tangent plane and the trichromatic colour-signal cone meet. The vertical projection of these two rays of tangency is the boundary rays of the dichromatic colour-signal cone. Hence, the colours corresponding to the rays of tangency in the trichromatic colour-signal cone are not confused by the dichromats with any other colour (in the trichromatic light-colour palette). It follows from our first assumption that dichromats see the lights represented by the rays of tangency in the same way as trichromats.

The dichromatic hues (generating the dichromatic light-colour palette) can be derived from the trichromatic spectral colour-signal curve. Indeed, the ray of tangency touches the trichromatic spectral colour-signal curve at a point which is actually the point of tangency for the corresponding tangent plane and the trichromatic spectral colour-signal curve. Therefore, identifying the hue in question reduces to establishing the wavelength at which the trichromatic spectral colour-signal curve meets a plane parallel to the 'missing' axis and tangent to this curve.

In order to evaluate these wavelengths, Govardovskii's cone fundamentals (Definition 3.4) were used with the peak absorbance wavelength $\lambda_i^{\max}$ set at 430, 530, and 560 nm for the S-, M-, and L-cones respectively, the peak photopigment absorbance D being 0.3 (Logvinenko, 2014).

The plane tangent to the colour-signal cone at the point $\Phi(\delta_\lambda) = (f_1(\lambda), f_2(\lambda), f_3(\lambda))$ is defined by the following equation:

$$\begin{vmatrix} z_1 & z_2 & z_3 \\ f_1(\lambda) & f_2(\lambda) & f_3(\lambda) \\ f_1'(\lambda) & f_2'(\lambda) & f_3'(\lambda) \end{vmatrix} = 0, \tag{8.32}$$

where z_1, z_2, z_3 are the coordinates in the colour-signal space; and $f_i'(\lambda)$ is the derivative of f_i at λ. The plane (8.32) is parallel to the ith coordinate axis if

$$\begin{vmatrix} f_p(\lambda) & f_q(\lambda) \\ f_p'(\lambda) & f_q'(\lambda) \end{vmatrix} = 0, \tag{8.33}$$

where $p, q \neq i$.

Figure 8.13 shows the graph of the logarithm of the absolute value of the determinant (8.33) for $p = 2$ and $q = 3$ (i.e., for the M-, and L-cone fundamentals). It clearly indicates that Eq. (8.32) with respect to λ has exactly three roots. Numerical computation brings about the following values: 389, 466, and 700 nm. Therefore, the planes tangent to the colour-signal cone at the three points in the trichromatic spectral colour-signal curve corresponding to these wavelengths will be parallel to the coordinate axis corresponding to the S-cone fundamental. The first point (i.e., $\Phi(\delta_{\lambda=389})$) falls

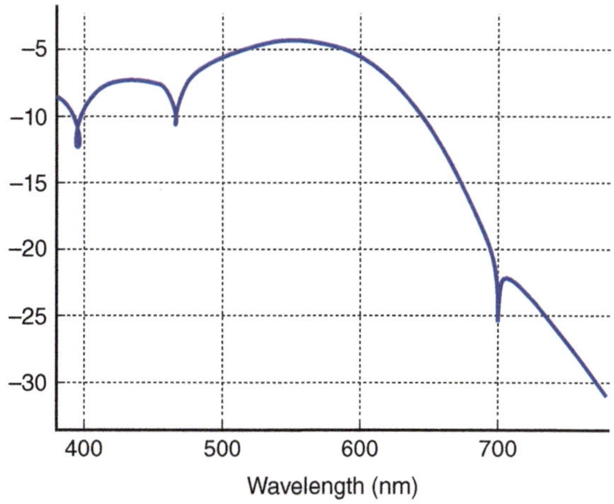

Figure 8.13 Logarithm of the absolute value of the determinant (8.33) for $p = 2$ and $q = 3$. Source: Logvinenko (2014)/John Wiley and Sons.

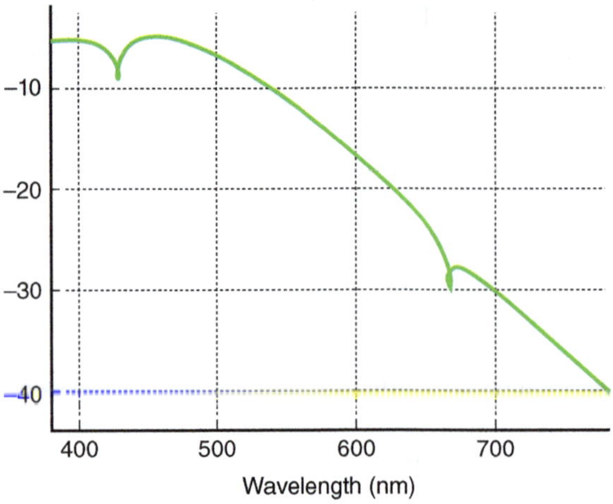

Figure 8.14 Logarithm of the absolute value of the determinant (8.33) for $p = 1$ and $q = 3$. Source: Logvinenko (2014)/John Wiley and Sons.

inside the colour-signal cone so it can be excluded. The other two wavelengths (466 and 700 nm) determine the tangency points in question for tritanopes.

As Fig. 8.14 shows, Eq. (8.32) for deuteranopes (i.e., for $p = 1$ and $q = 3$) has exactly two roots – at 433 and 669 nm. Likewise, Eq. (8.32) for protanopes (i.e., for $p = 1$ and $q = 2$) has the only two roots at 421 and 638 nm. In Fig. 8.12 the purple and red circles indicate the short and long wavelength tangency points for protanopes.

It is important to note that these estimates do not depend on the pre-receptoral filters such as the lens and macular pigment. Admittedly, they do depend on peak photopigment absorbance D, and peak absorbance wavelength λ_{max}. However, the variation of λ_{max} within the range of 10 nm has been found to result in the tangency point shift not exceeding ±1 nm (Logvinenko, 2014). A similar robustness has been found with respect to D as well.

It follows that the monochromatic light of 421 nm looks for protanopes the same way as for trichromats. So does the monochromatic light of 433 nm for deuteranopes; and 466 nm for tritanopes. Trichromats perceive these as various shades of purple-blue. Curiously enough, the last value is close to the wavelength which is usually reported by trichromats as 'monochromatic unique blue' (approximately 478 nm) (Wyszecki & Stiles, 1982, p. 423). Therefore, the reductionist hypothesis prediction is that tritanopes, who are sometimes, erroneously, referred to as individuals with 'no blue sensations', have almost unique blue hue as one of the two hues constituting their light-colour palette. The long wavelength part predictions – the monochromatic lights of wavelength of 638, 669, and 700 nm – are experienced by trichromats as orange or orange-red hues.

In the trichromatic colour-signal cone, the points representing the two dichromatic hues (more exactly, representing the two corresponding hue-fibres), together with the neutral line, form a continuous petal that partitions the cone into two halves. This petal can be thought of as a representation of the dichromatic light-colour palette in the trichromatic colour-signal cone. Each dichromatic confusion interval intersects this petal at some point. It follows that each dichromatic confusion interval contains a point which represents a colour from the dichromatic light-colour palette. Nevertheless, it is not the colour that dichromats will see. True, the hue will be the same, but the brightness and purity might be different. There is, generally speaking, no reason to believe that brightness (as well as purity) perception by dichromats would be a mirror replica of that by trichromats.

At first glance, the dichromatic hue palette can be established by resorting to so-called 'unilateral dichromats' – individuals whose one eye is trichromatic while the other eye is dichromatic.

Unilateral dichromats provide a unique opportunity to make a judgement of dichromatic colour sensations from a point of view of trichromatic observer. Furthermore, all the other predictions made above can be checked against the evidence gathered from unilateral dichromats. They have always interested colour scientists (Judd, 1948; Sloan & Wollach, 1948; Scheibner & Boynton, 1968; Alpern et al., 1983). However, the information from unilateral dichromats studies is, surprisingly, rather ambiguous and inconclusive (Broackes, 2010a).

On the one hand, most authors investigated into unilateral dichromatism concluded in favour of the two-hue palette of dichromatic vision, with purple-blue and orange-red being the two hues of the dichromatic light-colour palette (Judd, 1948; Alpern et al., 1983). On the other hand, there is ample evidence that dichromats (unilateral as well as bilateral) perceive more than two hues. Specifically, apart from the purple-blue and orange-red predicted by the reductionist hypothesis, protanopes and deuteranopes are systematically reported to see shades of red and, sometimes, green, though green being less pronounced (for a review see Jameson & Hurvich, 1978; Broackes, 2010a).

As to the light-colour palette of tritanope, every trichromatic observer can see, literally with their own eyes, what it is like. As a matter of fact, the central region (approximately 20 angular minutes) in the fovea of the trichromatic eye is known to be void of the S-cones (Curcio et al., 1991). As a result, a normal trichromat becomes a tritanope when the field of view is reduced to this region. This phenomenon is known as *tritanopia of the central foveola* (Williams et al., 1981; Mollon, 1982). Interestingly, a tritanopic colour transformation was reported to occur for small stimuli even when the stimulus is projected outside the central foveola – so-called *small-field tritanopia* (Hartridge, 1945; Mollon, 1982). Thus, observing a coloured object at a distance with one eye one can compare its tritanopic percept to the trichromatic percept of a copy of the same object presented at a shorter distance to the other eye. Although rather dramatic transformations of the colour of objects have been observed in such experiments, the hue palette has never been reported to reduce to only two hues (e.g., Willmer, 1944).

Furthermore, a tritanopic-like change of colour appearance can also be observed for a repetitive pattern such as a grid when the spatial frequency of the grid is high enough (referred to as *high-spatial-frequency tritanopia*) (Logvinenko, 2001). Once again, the illusory colours observed under high-spatial-frequency tritanopia consist of more than two hues (Logvinenko & Hutchinson, 2005, 2006, 2007).

At first glance, seeing by dichromats more than two hues undermines the reductionist hypothesis. Yet, this hypothesis can be readily reconciled with this fact. Firstly, it should be noted that the most striking evidence of this sort comes from the studies in which real objects (e.g., Munsell papers) rather than lights were used as stimuli (e.g., Montag & Boynton, 1987; Montag, 1994). For example, Montag & Boynton (1987, p. 2157) wrote: '... dichromats can distinctly categorize regions of colour space in all three dimensions using 11 basic colour names, and ... they do so in a manner that tends to resemble that used by normal subjects.' Secondly, the reductionist hypothesis predicts the two-hue dichromat colour palette only for light colours. As we will show in Part II, for object colours, the reductionist hypothesis predicts the dichromat colour palette with more than two hues.

8.5 Chromatic structures

In this section, we will consider the structure formed by all the hues, in more detail. Although the notion of the hue fibre bundle (8.18) reflects such an important aspect of the structure as its circularity (as well as the proximity between the hues), it does not capture some other aspects that seem very important.

Recall that textbooks treat hue as the quality of visual sensations (Section 8.2), which resembles the definition of pitch as a quality of auditory sensation. We will use this often-drawn parallel between vision and hearing, which is only true to some extent, to highlight some important distinctions between the two senses.

Pitch is defined as a subjective property of a sinusoidal sound wave closely related to its frequency. All the sine waves differ in pitch, the higher the frequency, the higher the pitch. The pitch of the complex (periodic) sound is also related to its repetition rate. The pitch of a complex (periodic) sound is usually measured by adjusting a sinusoidal sound wave so that it has the same pitch as the complex sound. With minor exceptions, there always exists a sinusoidal sound wave matching in pitch a complex (periodic) sound.

Likewise, hue is a subjective property of a sinusoidal electromagnetic wave (i.e., monochromatic light) closely related to its frequency (thus, wavelength). Also, the hue of a complex light is measured by adjusting a monochromatic light so that it has the same hue as this complex light. However, such a match not always exists. As pointed out above, there is no monochromatic light matching lights mapping into the purple interval in the chromaticity diagram. In other words, the set of hues is larger than the set of wavelengths in the visible part of the spectrum. So, the notion of hue has been generalised so as to encompass the *non-spectral hues*.[18] Yet, this does not resolve completely the issue since there are hueless lights (e.g., daylight). For the sake of generality, we will add another, albeit improper, hue: *achromatic*. It is improper because it is the subjective quality of a hueless light.

An important property of auditory perception is stated by Ohm's Acoustical Law[19] according to which humans can hear out the sinusoidal component of a complex (periodic) sound. In particular, being presented with two sinusoidal sound waves (with frequencies enough far apart), an observer will hear two different sounds (of two different pitches) rather than a sound of a single pitch. At this point textbooks commonly contrast hearing to vision since a mixture of two (and more) monochromatic lights is perceived being of a single hue.'... the sensation of colour appears to our consciousness as one and indivisible. We cannot directly recognize the elementary sensations of which it is composed, as we can distinguish the component notes of a musical chord' (Maxwell, 1872/1970, p. 78).

It must be said, however, that the first part of this quotation of Maxwell is not quite correct. In the reality some hues, such as pink, orange, violet are experienced as compound, that is, composed of other hues. For example, orange is experienced as a blend of red and yellow. On the contrary, one cannot see any shade of red or orange in yellow. A common way of expressing this is by saying that yellow is *unique* (or pure) hue (i.e., it cannot be phenomenologically reduced to other hues). As we are able to see out red and yellow hues in the orange hue, Ohm's law holds, in a sense, true in colour vision too. It is true in the sense that we can see out in a hue the component hues of which this hue is composed.[20] However, the number of component hues in vision is much smaller than the number of all hues. While the latter is infinite, the former is rather limited.

---

18 Sometimes the term 'non-spectral hues' is used to refer to some colours of object such as brown, olive and the like. In Part I we mean by non-spectral hues only the hues of light colours, that is, the hues of the lights mapping onto the purple interval.

19 This is the same German physicist Georg Ohm, but the law is, of course, different.

20 Admittedly, Ohm's law does not hold true in colour vision in the sense that the eye performs a sort of spectral analysis of light as the ear does of sound.

It should be noted that there has been no consensus on the number and nomenclature of unique hues. For instance, in the beginning of 18th century Thomas Young put forward the idea that the eye is equipped with the nervous fibres of three different types the stimulation of which produce three fundamental colour sensations (red, green, and violet), any other colour being a mixture of these three.[21] Although Young did not claim explicitly the fundamental colours to be unique, this is the way the textbook treat Young's fundamental colours (e.g., Kaiser & Boynton, 1996). In other words, an implication of Young's trichromatic theory is the existence of exactly three component hues. However, immediate introspective analysis shows that the number of component hues has to be more than three. For example, long before Young, Leondardo da Vinci stated that there were at least four unique hues – red, green, blue, and yellow[22] (Valberg, 2001; Kuehni & Schwarz, 2008). Ewald Hering was the first who explicitly stated that there were four component hues which could be observed 'in pure form' as 'unique hues'[23] from which all other colour sensations were composed (Hering, 1920/1964). Specifically, the other (compound) hues consist of various combinations of the component (unique) hues in different proportions.

Some scientists and artists believe that the component hues are not exhausted by the four. In particular, Munsell included in the list of component hues apart from red, green, blue, and yellow also purple. Curiously enough, accepting Young's idea in general, Helmholtz did not share the notion of fundamental colour. 'The notion of three fundamental colours as having any objective significance has no meaning' (Helmholtz, 1867/1970, p. 96). Rejecting the idea of fundamental (i.e., component) colour, Helmholtz believed that every hue is irreducible, that is, unique.

Though, introspective analysis can be rather misleading. A classical example is a dispute on the nature of green: is it unique, or not? For example, contrary to Hering, Goethe Goethe (1810/1970), Brentano (Nida-Rumelin & Suarez, 2009, p. 352), Holt (Boring, 1946, pp. 175–176) and some other authors (for a review see Nida-Rumelin & Suarez, 2009) believed that green was a phenomenological mixture of yellow and blue. The following passage from Boring (1946, p. 176) illustrate the controversy. '… there is always to be remembered that famous session of the Society of Experimental Psychologists in which Titchener, after hot debate with Holt, exclaimed: 'You can see that green is neither yellowish nor bluish!' And Holt replied: 'On the contrary, it is obvious that a green is that yellow-blue which is just exactly as blue as it is yellow.' That impasse was an ominous portent of the fate of introspection. When two distinguished experts could disagree vis-a-vis about so basic a matter as the nature of hue, some other method of approach was needed.'

An alternative method of approach to the nature of hue has been put forward recently (Logvinenko, 2012). This method (partial hue-matching) is aimed at assessing the nomenclature of hues into which each hue can be decomposed (referred to as *component hues*) without presupposing their number. The method allows for the derivation of component hues from observers' judgements of whether two colours have any hue in common; and without naming this hue. Moreover, the method allows for an evaluation of each colour with respect to which component hues it contains. Notably, with this method some colours have been found to comprise exactly one component hue (Logvinenko, 2012). By all standards such colours and/or hues can be thought of as unique. Hence, the partial hue-matching method allows for the derivation of unique hues without needing to explain to naive observers what the unique hues are.

---

21 Later on many scientists, amongst them Maxwell and Helmholtz, picked up this idea which has been known since then as the *trichromatic hypothesis* (or the *trichromatic theory of colour vision*).

22 Actually, Leondardo da Vinci claimed that in addition to the four chromatic unique hues, there were two more achromatic unique hues: white and black. However, as mentioned before we believe that black and white are perceptual attributes of object colour. So, we discuss these in Part II.

23 He called them 'primary colours' (*Urfarben*).

8.5.1 Partial hue-matching

When two lights differ in chromaticity, they may either share some common hues (as yellow and orange) or they may have nothing in common at all (as red and green). We shall say that, being well above colour discrimination threshold, two lights *partially match in hue* each other if they have common hues.

Definition 8.9 *Let* $\mathbb{S}$ *be a set of colour stimuli. We consider an abstract set* $\mathcal{H} = \{h_1, \dots, h_m\}$ *the elements of which will be referred to as* component hues. *Suppose there is a function* $\vartheta : \mathbb{S} \to 2^{\mathcal{H}} \backslash \{\emptyset\}$ *which assigns a non-empty subset of* $\mathcal{H}$ *to each colour stimulus* $x \in \mathbb{S}$. *We will call* ϑ a component hue function, *and* $\vartheta(x) = \left(h_{i_1}, \dots, h_{i_p}\right)$ the component-hue value *of* $x (1 \leq i_1 < \cdots < i_p \leq m)$. *We say that a colour stimulus x possesses a* unitary *hue, if* $\vartheta(x) = (h_i)$, *a* binary *hue if* $\vartheta(x) = \left(h_{i_1}, h_{i_2}\right)$, *a p*-nary *hue (for* $p > 2$*) if* $\vartheta(x) = \left(h_{i_1}, \dots, h_{i_p}\right)$. *The triplet* $\langle \mathbb{S}, \mathcal{H}, \vartheta \rangle$ *will be referred to as the* component hue structure. *If there is* $x \in \mathbb{S}$ *with p*-nary *hue, and* $\vartheta^{-1}\left(h_{i_1}, \dots, h_{i_q}\right) = \emptyset$ *for every* $q > p$, *we say that p is* the degree of chromaticity *of* $\langle \mathbb{S}, \mathcal{H}, \vartheta \rangle$. *A component hue structure* $\langle \mathbb{S}, \mathcal{H}, \vartheta \rangle$ *will be called* simple *if for each component hue* h_i $\vartheta^{-1}(h_i) \neq \emptyset$. *A component hue structure* $\langle \mathbb{S}, \mathcal{H}, \vartheta \rangle$ *will be called* complete *if the function* ϑ *is onto* $2^{\mathcal{H}} \backslash \{\emptyset\}$.

Remark 8.10 *This definition may be extended to the case where* $\mathcal{H}$ *contains yet one element* h_0 *such that* $\vartheta^{-1}(h_0) \neq \emptyset$ *and* $\vartheta^{-1}(h_0, h_{i_1}, \dots, h_{i_q}) = \emptyset$ *whenever* $q \geq 1$, $h_{i_1}, \dots, h_{i_q} \subset \mathcal{H} \backslash h_0$. *In this case, the set* $\mathbb{S}$ *can be the whole colour stimulus cone. The element* h_0 *will be referred to as the* improper *(hueless) component hue.*

Following the traditional terminology one can say that a component hue is unique if it is the unitary hue for some colour stimulus. The improper component hue is unique by definition. A component hue structure is simple if every component hue is unique. In a complete component hue structure, every combination of the component hues is assigned to some colour stimulus. There is every indication that a complete component hue structure cannot be a model of the human colour vision because such pairs of hues as red and green, and yellow and blue can hardly be experienced at the same place and in the same time (Hering, 1920/1964; Hurvich & Jameson, 1957).

Definition 8.10 *Given a component hue structure* $\langle \mathbb{S}, \mathcal{H}, \vartheta \rangle$, *let us define a binary relation* τ *on* $\mathbb{S}$ (partial hue-matching)

$$x \tau y \Leftrightarrow \vartheta(x) \cap \vartheta(y) \neq \emptyset, \quad \forall x, y \in \mathbb{S}. \tag{8.34}$$

The pair $\langle \mathbb{S}, \tau \rangle$ *will be called* a partial hue-matching structure associated with $\langle \mathbb{S}, \mathcal{H}, \vartheta \rangle$.

In other words, colour stimuli x and y partially match each other in hue if and only if they have the component-hue values with at least one common component hue. Partial hue-matching is a reflexive and symmetric binary relation, thus a tolerance (Zeeman, 1962). Note that in experiment one can observe only the tolerance τ, the component hue function ϑ remaining inaccessible. In this section, we investigate what component hue structures can be uniquely recovered from the tolerance τ.

In an experiment on partial hue-matching, observers are supposed to decide whether two colour stimuli partially match each other in hue, that is, if these stimuli have a shade of some component hue in common. Although this task is considered simple and easy by all observers, there is no guarantee that the observers' judgements are always based on sharing common component hues. Moreover, the results obtained in an experiment described below (Section 8.5.2) as well as in other studies (Logvinenko, 2012; Logvinenko & Beattie, 2011) indicate otherwise, that is, that some observers rest their judgements on different grounds. Therefore, it becomes important to ascertain, first, if the empirical tolerance registered in an experiment on partial hue-matching is really based on partial hue-matching; and then, if it is established that this is the case, to retrieve the component hues of the underlying component hue structure.

A departure point will be a tolerance τ defined on $\mathbb{S}$ which will be referred to as partial colour matching (see Definition 8.11). Every partial hue-matching is a partial colour matching, but the converse is certainly not true. The major objective is to find out the properties of partial colour matching which guarantee that it is induced by some component hue structure. To achieve this goal, the concepts of the theory of tolerance spaces (Zeeman, 1962; Shreider, 1970; Schreider, 1975) have proven useful.

Given a tolerance τ, define an equivalence relation, $\sim_\tau$, on $\mathbb{S}$ as

$$\forall\, x, y, z \in \mathbb{S} \quad x \sim_\tau y \;\Leftrightarrow (x\,\tau\, z \Leftrightarrow y\,\tau\, z). \tag{8.35}$$

It proves that $x \sim_\tau y \Leftrightarrow \tau(x) = \tau(y)$, where $\tau(x) = \{y \in \mathbb{S} : x\tau y\}$. The equivalence $\sim_\tau$ determines a partition of $\mathbb{S}$ into equivalence classes.

When τ is a partial hue-matching, $\tau(x)$ will be referred to as *the component hue match for x*. Therefore, for a partial hue-matching τ, $\sim_\tau$ takes place between two colour stimuli only if they have the same component hue match. It follows from (8.34) that stimuli have the same component hue match when they are assigned exactly the same component hues. For this reason, when τ is a partial hue-matching, the equivalence $\sim_\tau$ will be referred to as *component hue equivalence*. It determines a partition of $\mathbb{S}$ into equivalence classes (referred to as *component hue equivalence classes*). These prove to be non-empty pre-images of the component hue function ϑ.

It is clear that if a partial colour matching is induced by a component hue structure then the number of component hue equivalence classes is finite. Therefore, one can exclude from consideration those tolerances τ which result in an infinite number of $\sim_\tau$-equivalence classes.

Definition 8.11 *Given a colour stimulus set $\mathbb{S}$, we say that a reflexive and symmetric binary relation τ on $\mathbb{S}$ is* partial colour matching *and a pair $\langle \mathbb{S}, \tau \rangle$ is a* partial colour matching structure *if the quotient set $\mathbb{S}/\sim_\tau$ is finite, where $\sim_\tau$ is defined in (8.35).*

As τ is a tolerance and $\langle \mathbb{S}, \tau \rangle$ is a tolerance space, we can define tolerance classes and some other standard (for tolerance spaces) notions (Schreider, 1975).

Definition 8.12 *Given a partial colour matching structure $\langle \mathbb{S}, \tau \rangle$, a set $P \subseteq \mathbb{S}$ is called a τ-*preclass *in $\langle \mathbb{S}, \tau \rangle$ if each pair of colour stimuli x and y in P partially match each other, i.e., $x\tau y$.*

Given a tolerance τ, for any x and y to be in relation τ there must be a τ-preclass containing x and y. As such a preclass one can take $P = \{x, y\}$. Therefore, to be in the same τ-preclass is the necessary and sufficient condition to partially match each other.

Definition 8.13 *Let $\langle \mathbb{S}, \tau \rangle$ be a partial colour matching structure, a τ-preclass P is called a τ-class in $\langle \mathbb{S}, \tau \rangle$ if there is no other τ-preclass Q in $\langle \mathbb{S}, \tau \rangle$ such that $P \subset Q$.*

So, τ-classes are maximal τ-preclasses in $\langle \mathbb{S}, \tau \rangle$. It means that they cannot be extended without ceasing to be τ-preclasses. That is, given a τ-class K, for each $x \notin K$ there is $y \in K$ that does not partially match x (i.e., $y\tau x$ fails). Every τ-preclass is contained in at least one τ-class (this may be easily shown with the help of transfinite induction or the Zorn Lemma); therefore every element in $\mathbb{S}$ belongs to some τ-class. It follows that for any $x, y \in \mathbb{S}$ the existence of a τ-class containing both x and y is necessary and sufficient for x and y to partially match each other. This arguing is summarized in the following proposition (Schreider, 1975).

Proposition 8.5 *Let $\mathbb{S}$ be a colour stimulus set, and $\langle \mathbb{S}, \tau \rangle$ a partial colour matching structure on it. Two colour stimuli $x, y \in \mathbb{S}$ are partially match each other, i.e., $x\tau y$, if and only if there exists a τ-class containing both x and y.*

It follows from Proposition 8.5 that given a partial colour matching structure $\langle \mathbb{S}, \tau \rangle$, the partial colour matching τ can be defined in terms of τ-classes as in Definition 8.9. Specifically, let $\mathcal{K}$ be a set of all τ-classes. Let us define a multifunction $\upsilon : \mathbb{S} \to 2^{\mathcal{K}}$ which will assign to each $x \in \mathbb{S}$ the subset of $\mathcal{K}$ comprising all the τ-classes containing x. Since any $x \in \mathbb{S}$ belongs to at least one τ-class, $\upsilon(x) \neq \emptyset$ for every x in $\mathbb{S}$. Proposition 8.5 implies that for any x and y in $\mathbb{S}$ $x\tau y \Leftrightarrow \upsilon(x) \cap \upsilon(y) \neq \emptyset$. We will refer to this fact saying that the set of τ-classes $\mathcal{K}$ *induces* the partial colour matching τ.

As shown elsewhere (Schreider, 1975), given a tolerance τ, there may exist proper subsets of $\mathcal{K}$ which also induce τ (e.g., see Example 8.1).

Definition 8.14 *Let $\langle \mathbb{S}, \tau \rangle$ be a partial colour matching structure. A set $\mathcal{K}$ of τ-classes will be called a* basis *if it meets the following two conditions. First, for any $x, y \in \mathbb{S}$ that partially match each other (i.e., $x\tau y$), there exists a τ-class $K \in \mathcal{K}$ containing x and y. Second, $\mathcal{K}$ is the minimal set with this property, that is, removing any K in $\mathcal{K}$ will result in that for some x and y with $x\tau y$ there will be no τ-class in $\mathcal{K}$ which both x and y belong to.*

The next proposition is an immediate consequence of the definition of basis.

Proposition 8.6 *Let $\langle \mathbb{S}, \tau \rangle$ be a partial colour matching structure, and $\mathcal{K}$ its basis. Define a function $\upsilon : \mathbb{S} \to 2^{\mathcal{K}} \backslash \{\emptyset\}$ assigning to each $x \in \mathbb{S}$ the subset of $\mathcal{K}$ comprising all the τ-classes $K \in \mathcal{K}$ containing x. Then*

$$\forall\, x, y \in \mathbb{S} : \quad x\tau y \Leftrightarrow \upsilon(x) \cap \upsilon(y) \neq \emptyset. \tag{8.36}$$

It turns out that there is only one basis in a partial hue-matching structure associated with a simple component hue structure (see Proposition 8.7).

Definition 8.15 *In a component hue structure $\langle \mathbb{S}, \mathcal{H}, \vartheta \rangle$, a non-empty subset of stimuli each of which is assigned a component-hue value containing a particular component hue will be called the* component hue class. *More specifically, given a component hue $h_i \in \mathcal{H}$, the stimulus subset H_i defined as*

$$H_i := \{x \in \mathbb{S} : \vartheta(x) \ni h_i\} \tag{8.37}$$

will be called the component hue class *provided that it is not empty.*

All colour stimuli in a component hue class H_i partially match each other in hue because they share the component hue h_i. Every component hue class is thus a τ-preclass. Moreover, as follows from Proposition 8.7, component hue classes are τ-classes and the set of component hue classes in a simple component hue structure proves to be a minimal set of τ-classes inducing the partial hue-matching τ.

Proposition 8.7 *Suppose that $\mathcal{H} = \{h_1, \dots, h_m\}$ or $\mathcal{H} = \{h_0, h_1, \dots, h_m\}$ where h_0 is the improper component hue, and let $\langle \mathbb{S}, \mathcal{H}, \vartheta \rangle$ be a simple component hue structure. Then every H_i is a τ-class, and m component hue classes $H_1, \dots, H_m$ or $m+1$ component hue classes $H_0, H_1, \dots, H_m$ make a unique basis in the partial hue-matching structure $\langle \mathbb{S}, \tau \rangle$ associated with $\langle \mathbb{S}, \mathcal{H}, \vartheta \rangle$.*

So, the set of component hue classes in a simple component hue structure is the only basis of the partial hue-matching structure associated with this component hue structure. The component hue classes completely characterise the simple component hue structure. All the other τ-classes in it (as in the example below) can be ignored.

Example 8.1 *Let $\langle \mathbb{S}, \mathcal{H} = \{h_1, h_2, h_3\}, \vartheta \rangle$ be a simple complete component hue structure. Then the colour stimulus set*

$$\{\vartheta^{-1}(\{h_1, h_2\}) \cup \vartheta^{-1}(\{h_1, h_3\}) \cup \vartheta^{-1}(\{h_2, h_3\}) \cup \vartheta^{-1}(\{h_1, h_2, h_3\})\}$$

is a τ-class but it is not a component hue class.

The following proposition shows how the component hue equivalence classes of a component hue structure can be expressed in terms of its component hue classes.

Recall that the component hue equivalence is defined by (8.35), that is, a non-empty subset $C \subseteq \mathbb{S}$ is a component hue equivalence class if and only if $\vartheta(x) = \vartheta(y) \Leftrightarrow y \in C$ whenever $x \in C$. Therefore there is a non-empty subset I in $\{1, 2, \dots, m\}$ or in $\{0, 1, 2, \dots, m\}$ such that $\vartheta(x) = \{h_i : i \in I\}$ for all $x \in C$, which means that $h_i \in \vartheta(x)$ if $i \in I$, and $h_i \notin \vartheta(x)$ if $i \notin I$. Now, taking into account (8.37), this may be rewritten as

$$C = \left(\bigcap_{i \in I} H_i\right) \cap \left(\bigcap_{i \notin I} (\mathbb{S} \setminus H_i)\right). \tag{8.38}$$

Note that if $\mathcal{H}$ contains the improper component hue h_0 and $0 \in I$ then set (8.38) is non-empty if and only if $I = \{0\}$. In this special case, one has $H_0 = \{x \in \mathbb{S} : \vartheta(x) = (h_0)\} \subset \mathbb{S} \setminus H_i$ for every $i \in \{1, 2, \dots, m\}$, and (8.38) is rewritten then as $C = H_0$.

Taking into account (8.38) the above reasoning is summarized as follows.

Proposition 8.8 *Let $\langle \mathbb{S}, \mathcal{H}, \vartheta \rangle$ be a component hue structure, and $\{H_1, \dots, H_m\}$ the set of its component hue classes. A non-empty subset $C \subseteq \mathbb{S}$ is a component hue equivalence class if and only if there are component hue classes $H_{i_1}, \dots, H_{i_p}$ such that C consists of all such x in $\mathbb{S}$ that x belongs to these and only to these component hue classes (written as $C(H_i : i \in I) = C\left(H_{i_1}, \dots, H_{i_p}\right)$). If $\mathcal{H}$ contains the improper component hue h_0 and the set of component hue classes of the structure $\langle \mathbb{S}, \mathcal{H}, \vartheta \rangle$ is $\{H_0, H_1, \dots, H_m\}$ then a non-empty subset $C \subseteq \mathbb{S}$ is a component hue equivalence class if and only if $C = C(H_i : i \in I)$ for some $I \subset \{1, 2, \dots, m\}$ or $C = C(H_0) = H_0$.*

It follows that $\mathbb{S}$ is partitioned into subsets $C\left(H_{i_1}, \dots, H_{i_p}\right)$ each of which is assigned the same set of component hues.

As the existence of a unique basis is an important property of the partial hue-matching structure associated with a component hue structure, we will distinguish partial colour matching structures with this property as a special type.

Definition 8.16 *A partial colour matching structure* $\langle \mathbb{S}, \tau \rangle$ *will be called* m-chromatic *if there is a unique finite basis* $\mathbf{H} = \{H_1, \dots, H_m\}$ *in it. The* τ*-classes* $H_1, \dots, H_m$ *will be called* chromaticity classes.

A simple component hue structure with m component hues uniquely determines a m-chromatic partial colour matching structure. As the component hue classes make a basis, they constitute the chromaticity classes in this partial colour matching structure. Vice versa, a m-chromatic partial colour matching structure determines a simple component hue structure with m component hues. Indeed, let $\langle \mathbb{S}, \tau \rangle$ be a partial colour matching structure, and $\mathbf{H} = \{H_1, \dots, H_m\}$ its unique basis. Define a function $\upsilon : \mathbb{S} \to 2^{\mathbf{H}}$ assigning to each $x \in \mathbb{S}$ all the chromaticity classes containing x. Formally, $\langle \mathbb{S}, \mathbf{H}, \upsilon \rangle$ is a component hue structure, the chromaticity classes $\{H_1, \dots, H_m\}$ playing a role of component hues. As for each $x, y \in \mathbb{S}$ the equivalence holds $x\tau y \Leftrightarrow \upsilon(x) \cap \upsilon(y) \neq \emptyset$, $\langle \mathbb{S}, \tau \rangle$ is a partial hue-matching structure associated with $\langle \mathbb{S}, \mathbf{H}, \upsilon \rangle$. Likewise, let us call $\langle \mathbb{S}, \mathbf{H}, \upsilon \rangle$ the component hue structure *associated* with the m-chromatic partial colour matching structure $\langle \mathbb{S}, \tau \rangle$.

Note that, in general, the component hue structure $\langle \mathbb{S}, \mathbf{H}, \upsilon \rangle$ associated with the m-chromatic partial colour matching structure $\langle \mathbb{S}, \tau \rangle$ might differ from the component hue structure $\langle \mathbb{S}, \mathcal{H}, \vartheta \rangle$ which induces $\langle \mathbb{S}, \tau \rangle$. However, they coincide when $\langle \mathbb{S}, \mathcal{H}, \vartheta \rangle$ is simple. Therefore, a simple component hue structure and a m-chromatic partial colour matching structure associated with it are just the two sides of the same coin. We shall use 'm-chromatic structure' as a common term for both. As the chromaticity classes play the key role, we shall use the notation $\langle \mathbb{S}, H_1, \dots, H_m \rangle$ for a m-chromatic structure with the chromaticity classes $H_1, \dots, H_m$. When the number of component hues is not of main importance, the 'chromatic structure' will be used as a generic term.

Proposition 8.8 justifies the following definition.

Definition 8.17 *Let* $\langle \mathbb{S}, H_1, \dots, H_m \rangle$ *be a m-chromatic structure. Consider two complementary subsets of chromaticity classes:* $\{H_i : i \in I\} = \{H_{i_1}, \dots, H_{i_p}\}$, *and* $\{H_i : i \notin I\} = \{H_{i_{p+1}}, \dots, H_{i_m}\}$. *A subset*

$$C\left(H_{i_1}, \dots, H_{i_p}\right) := \left(\bigcap_{k=1}^{p} H_{i_k}\right) \cap \left(\bigcap_{k=p+1}^{m} (\mathbb{S} \setminus H_{i_k})\right)$$

will be called a chromatic class *in* $\mathbb{S}$ *unless it is empty. A chromatic class* $C(H_i)$ *will be called* unitary *(of the* degree of chromaticity *1), a chromatic class* $C\left(H_{i_1}, H_{i_2}\right)$ binary *(of the degree of chromaticity 2), a chromatic class* $C\left(H_{i_1}, H_{i_2}, H_{i_3}\right)$ ternary *(of the degree of chromaticity 3) and so on. If for each* p' *not more than some* $p \le m$ *there is a chromatic class with the degree of chromaticity* p', *but there is no chromatic class with the degree of chromaticity more than p, then we say that the degree of chromaticity for* $\langle \mathbb{S}, H_1, \dots, H_m \rangle$ *is p.*

It follows from Proposition 8.8 that, given a m-chromatic structure $\langle \mathbb{S}, H_1, \dots, H_m \rangle$, its chromatic classes are the component hue equivalence classes of the component hue structure $\langle \mathbb{S}, (H_1, \dots, H_m), \upsilon \rangle$ associated with $\langle \mathbb{S}, H_1, \dots, H_m \rangle$.

Definition 8.18 *Let* $\langle \mathbb{S}, H_1, \dots, H_m \rangle$ *be a m-chromatic structure. A chromaticity class* H_i *will be called* singular *if* $H_i = C(H_i)$.

Remark 8.11 *A particular type of singular chromaticity class – a so-called* neutral *chromaticity class – is defined in Section 8.5.4 (Definition 8.21). There is every indication that in the human trichromatic colour vision there is only one neutral chromaticity class (Logvinenko & Beattie, 2011).*

Now assume that we observe a subject in experiment on partial hue-matching. Let us also assume that the observer responses are based on an unknown component hue structure $\langle \mathbb{S}, \mathcal{H}, \vartheta \rangle$, and we want to identify this structure. The experiment allows for evaluating the partial colour matching structure $\langle \mathbb{S}, \tau \rangle$ associated with $\langle \mathbb{S}, \mathcal{H}, \vartheta \rangle$. If there exists a unique basis $\mathbf{H} = \{H_1, \dots, H_m\}$ in $\langle \mathbb{S}, \tau \rangle$ its elements $H_1, \dots, H_m$ will be chromaticity classes, and the component hue structure $\langle \mathbb{S}, \mathbf{H}, \upsilon \rangle$ associated with $\langle \mathbb{S}, \tau \rangle$ can be determined. If for each chromaticity class H_i there is a unitary chromatic class $C(H_i) \neq \emptyset$ the underlying component hue structure $\langle \mathbb{S}, \mathcal{H}, \vartheta \rangle$ amounts to $\langle \mathbb{S}, \mathbf{H}, \upsilon \rangle$, the basis $\mathbf{H}$ comprising its component hue classes.

When at least for one chromaticity class, H_i, a subset $C(H_i)$ is empty (i.e., there is no unitary chromatic class corresponding to this chromaticity class), one has to conclude that the underlying component hue structure is not simple.[24] In this case, one cannot suggest that the recovered in experiment component hue structure $\langle \mathbb{S}, \mathbf{H}, \upsilon \rangle$ is that used by the observer. However, one can assert that $\mathbf{H}$ is the minimal set of component hues resulting in such a partial colour matching structure that was obtained in experiment.

8.5.2 Experiment on partial hue-matching

In this section, we illustrate the main notions introduced above using the results from the following experiment (Logvinenko, 2012). Three trichromatic observers[25] were presented with a series of twenty coloured papers[26] selected from the Munsell book of colour[27] (Munsell, 1915). The task was to pick out those whose hue had something in common with that of one (out of the twenty papers) selected as a test.[28] Each of the twenty papers was used as a test 5 times. An averaged, symmetrised, and rounded matrix of results for one observer is presented in Fig. 8.15. White entries mark those pairs which partially matched each other (that is, which the observer judged as having a common component hue).

Four chromaticity classes have been found (Fig. 8.16): $H_1 = \{10R5/16 \text{ to } 10GY6/12\}$; $H_2 = \{10G5/10 \text{ to } 10RP5/14\}$; $H_3 = \{10Y8.5/12 \text{ to } 5B5/10\}$; $H_4 = \{5PB5/12 \text{ to } 10YR7/14\}$. The stimulus set proves to be partitioned into eight chromatic classes – four unitary and four binary. Each of the four unitary chromatic classes happens to contain just one Munsell paper: $C(H_1) = \{5Y8.5/14\}$, $C(H_2) = \{10B5/12\}$, $C(H_3) = \{5G5/10\}$, $C(H_4) = \{5R4/14\}$. Notably, these were exactly the papers that the observer verbally categorised as unique hues (after the experiment). Specifically, $C(H_1)$ contains the paper whose hue was named as 'yellow', $C(H_2)$ 'blue', $C(H_3)$ 'green' and $C(H_4)$ 'red', in agreement with Munsell's notation. The remaining four chromatic

24 An alternative conclusion might also be that the observer responses did not rest upon any chromatic structure at all. This case will be discussed below.
25 In subsequent similar experiments, more subjects were employed (Logvinenko & Beattie, 2011; Logvinenko & Geithner, 2015). Presented here are three prototypical patterns of the observed results. Most of trichromatic observers yielded the results similar to Observer 1. Although rather rare, the data produced by Observers 2 and 3 are of great theoretical importance, showing possible limitations of the method.
26 Actually, lights, not papers, should have been used as stimuli to ascertain the component hues of light colours. Such an experiment with light patches presented on the colour monitor screen was also carried out, though never published. The results were similar to those presented here.
27 Munsell notations of the papers can be seen in Fig. 8.16.
28 Strictly speaking, this experimental technique should have been called 'partial colour matching' since it is the partial colour matching structure $\langle \mathbb{S}, \tau \rangle$ that was observed in the experiment, not the underlying $\langle \mathbb{S}, \mathcal{H}, \vartheta \rangle$. However, as observers were instructed in terms of 'hue in common', this method was dubbed 'partial hue-matching' when published (Logvinenko & Beattie, 2011; Logvinenko & Geithner, 2015). Interestingly, in preliminary experiments it was found that the results were the same regardless of whether the subjects were asked about the presence of a 'hue in common' or 'colour in common'.

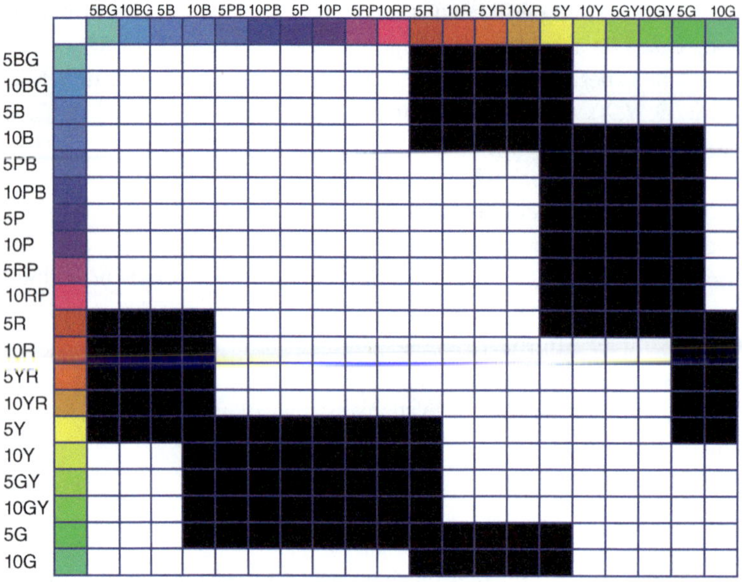

Figure 8.15 Averaged and rounded matrix of responses of Observer #1. White checks denote those pairs of Munsell papers which were judged as having common hue components. Source: Logvinenko (2012)/John Wiley and Sons.

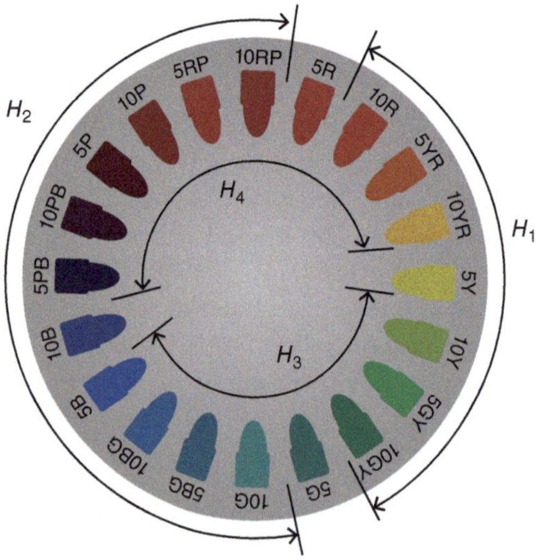

Figure 8.16 Chromaticity classes (marked with arcs) as derived from the response matrix in Fig. 8.15. Source: Logvinenko (2012)/John Wiley and Sons.

classes contain papers with binary hues: $C(H_1, H_3)$ – yellow-green, $C(H_1, H_4)$ – yellow-red, $C(H_2, H_3)$ – blue-green, and $C(H_2, H_4)$ – blue-red. Generally, this observer yielded results which were in accord with the classical view on human trichromatic colour vision (e.g., Hurvich & Jameson, 1957; De Valois & De Valois, 1993).

Note that the chromaticity classes H_1 and H_2 do not overlap, covering the opposite parts of the hue wheel in Fig. 8.16. Nor do the chromaticity classes H_3 and H_4. In Section 8.6.3, we introduce a terminology according to which the chromaticity classes H_1 and H_2 (as well as H_3 and H_4) are *opponent* to each other. This terminology is in line with the idea of colour opponency, enunciated by Hering, according to which blue and yellow, and red and green hues make perceptually opponent pairs. It should be borne in mind that Hering's colour opponency is based on introspective analysis whereas our definition of opponency is based on the behavioural criterion. Respectively, two component hues are called opponent if the corresponding chromaticity classes are opponent.

The second observer brought about different results (Fig. 8.17). As can be seen in Fig. 8.18 there are six chromaticity classes: $H_1 = \{10YR7/14 \text{ to } 10GY6/12\}$; $H_2 = \{10G5/10 \text{ to } 5PB5/12\}$; $H_3 = \{10Y8.5/12 \text{ to } 5B5/10\}$; $H_4 = \{5P4/12 \text{ to } 10R5/16\}$; $H_5 = \{5PB5/12 \text{ to } 10RP5/14\}$; $H_6 = \{10R5/16 \text{ to } 10YR7/14\}$. Six unitary chromatic classes have been found: $C(H_1) = \{5Y8.5/14\}$, $C(H_2) = \{10B5/12\}$, $C(H_3) = \{5G5/10\}$, $C(H_4) = \{5R4/14\}$, $C(H_5) = \{10PB4/12\}$, $C(H_6) = \{5YR6/14\}$.

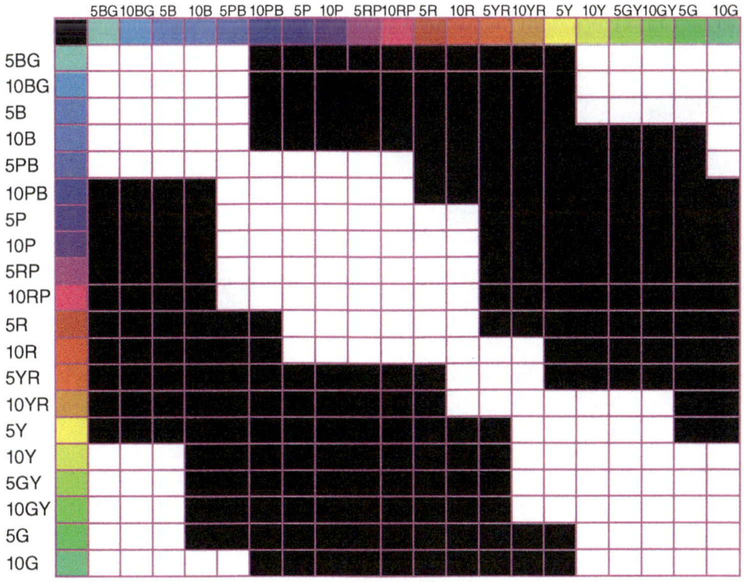

Figure 8.17 Averaged and rounded matrix of responses of Observer #2. Source: Logvinenko (2012)/John Wiley and Sons.

Figure 8.18 Chromaticity classes as derived from the response matrix in Fig. 8.17. Source: Logvinenko (2012)/John Wiley and Sons.

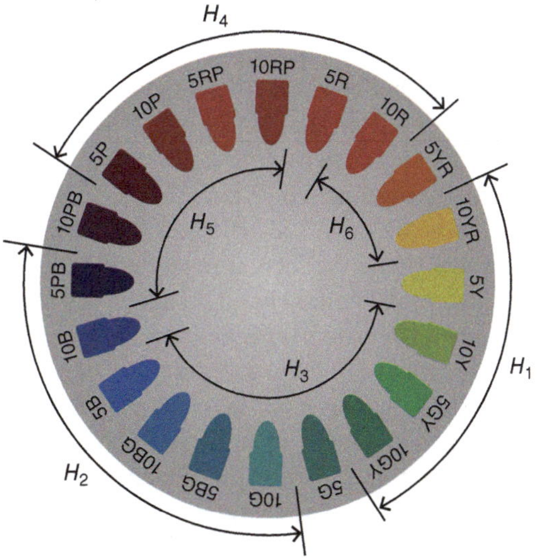

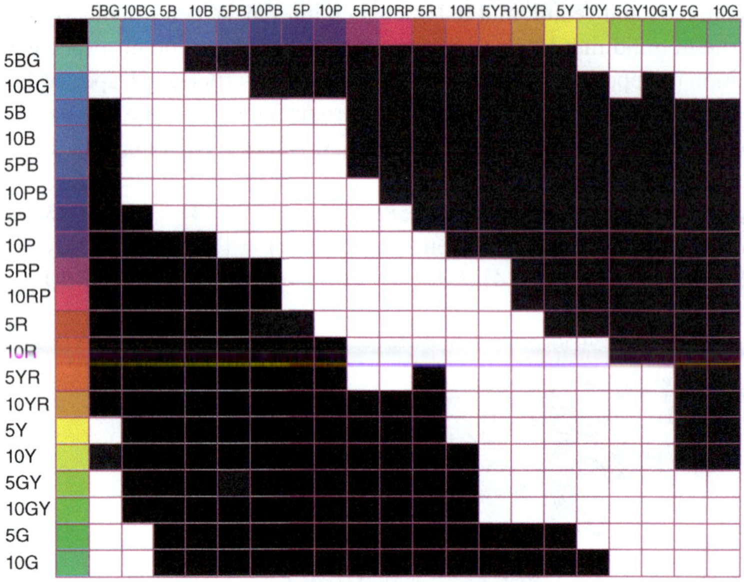

Figure 8.19 Averaged and rounded matrix of responses of Observer #3. Source: Logvinenko (2012)/John Wiley and Sons.

The first four chromaticity classes, H_1 to H_4, seem to be 'classical' yellow, blue, green, and red ones. The component hue common to the papers in the fifth chromaticity class, H_5, was verbally named by the observer afterwards as purple, and in the sixth, H_6, as orange.

The third observer (Fig. 8.19) produced a pattern of results which is rather different. Such a matrix of responses could be expected if the observer substituted the task of hue matching with that of similarity judgements. Indeed, let us assume that the observer yielded the yes-response when the similarity between a pair of Munsell papers exceeded some criterion value. In this case, the response matrix would be a narrow white band around the main diagonal, similar to what is observed in Fig. 8.19.

8.5.3 Colour categories

The partial hue-matching technique was devised on the assumption that an observer has the ability (i) to discern the component hues in a colour; and (ii) to decide whether there is a common component hue for a pair of colours. When this is the case the technique permits to derive the minimal set of component hues compatible with the results of the experiment on partial hue-matching (i.e., from the empirical partial colour matching relation (Definition 8.11) established in the experiment), and the set of unitary hues. If these two sets coincide (i.e., each component hue is at the same time a unitary hue), then it can be argued that the derived component hues are the very features that the observer used when making judgements.

If the number of the derived unitary hues is fewer than that of the derived component hues, there is a possibility that the observer might have used more component hues. However, these additional component hues did not affect the observer responses. Removing these does not change the response matrix, thus the partial colour matching relation. For example, using pink along with red and white as component hues results in the same partial colour matching relation as that induced by using only red and white. Therefore, the derivation of component hues from the response matrix accomplishes also removing redundant component hues (i.e., those which do not affect the observer response matrix).

Certainly, the assumption in question may fail, at least for some observers; that is, some observers may base their judgements not on the component hues. In other words, the partial colour matching relation registered in the experiment may happen to be induced by not the component hue structure, even if the latter exists, but by something else. For instance, the empirical partial colour matching relation may reflect the dissimilarity of colours. In particular, the observer can give the negative answer (in the experiment on partial hue-matching) whenever the dissimilarity in a pair of colours exceeds a certain criterion value. Note that Observer 3 appears to exhibit such a tendency to judge pairs with a large dissimilarity as 'not partially matched'. At any rate, Fig. 8.19 looks as if this observer was inclined to replace the task of hue judgement with the task of colour-similarity judgement.

It should be borne in mind that when the task of hue judgement is substituted with the task of colour-similarity judgement the empirical partial colour matching relation may have nothing to do with the component hues at all. More interesting is the case when observers did their best to follow the instruction to make a judgement on the basis of hue content but because of the lack of experience in introspective analysis they did not manage to base their judgements solely on component hues.

This might happen when observers are not experienced enough in evaluating the chromatic content of their colour sensations. Indeed, at first glance, it is hard to spot that, for instance, Munsell papers 5PB and 10YR have a common component hue (so small is the common shade of red in these Munsell papers). Still, after some training most trichromatic observers are rather confident in judging these Munsell papers as a partial match. Hence, we need an objective criterion to test whether observers base their judgements on the hue content.

Had one have such a criteria it could be applied not only to the results of Observer 3 but Observer 2 as well. In a sense, the Observer 2 results are even more controversial. While it is more or less clear that Observer 3 failed to follow to make a judgement of component hues, Observer 2 produced a matrix which can be interpreted as a graph of the partial colour matching relation τ with a unique basis consisting of six chromaticity classes. Moreover, for each chromaticity class there is a unitary chromatic class. Strictly speaking, one has to conclude that τ is induced by a chromatic structure with six component hues.

Yet, there is an alternative interpretation. One can suggest that Observer 2 rested his judgements not upon component hues but upon colour categories. In other words, Observer 2 could have answered 'yes' when the papers belonged to the same category. In this case, the results reveal six colour categories rather than six component hues.

More specifically, assume that there are m colour categories. Formally, it means that there is a cover $\mathbb{C}$ of $\mathbb{S}$, that is, a family of subsets $\{C_1, \dots, C_m\}$ of $\mathbb{S}$ whose union is the whole stimulus set $\mathbb{S}$. Each C_i is a collection of colour stimuli which belong to the ith colour category. Let us define a partial colour matching τ as follows. For every $x, y \in \mathbb{S}$ assume that $x\tau y$ if and only if there is such a colour category C_i that $x, y \in C_i$. The partial colour matching τ will be referred to as that *generated* by the colour categories $\mathbb{C}$.

It must be said that, generally, there is no simple relationship between the colour categories $\mathbb{C}$ and the τ-classes of the generated tolerance τ. For an arbitrary set of colour categories a colour category may happen even not to be a τ-class of the generated tolerance. Yet, it turns out that if the generated tolerance τ induces a m-chromatic structure, then each chromaticity class contains some colour category (Schreider, 1975, Theorem 3.7). However, the colour categories cannot be restored from the τ-classes of the generated tolerance.

Still, the set of chromaticity classes derived from the response matrix can be treated as a set of colour categories compatible with the response matrix (i.e., generating the matrix). If they make a basis then it is a minimal set of colour categories compatible with the response matrix. If this basis is unique then it is a unique minimal set of colour categories compatible with the response matrix. In this latter case one can say that irrespective of the colour categories that the observer really used in experiment (they can be rather redundant), the derived set of colour categories can be used to model the observer responses. An analogous situation takes place with applying the signal detection theory to psychophysics (D. M. Green & Swets, 1966). We do not believe that observers really evaluate the likelihood ratio. Yet, we assume that they behave as if they do so.

An important question then arises. How can one differentiate between the chromaticity classes representing component hues and those representing colour categories? For instance, Observer 2 provides a particular case when the set of colour categories and the basis of generated tolerance coincide. There seems to be no difference between the case when six colour categories give rise to a six-chromatic structure, and when six component (unique) hues give rise to the same component hue structure. A distinction between these two cases can only be made within a broader context.

In particular, such a distinction can be made based on whether an observer is able to linearly order the colours within chromaticity classes. If a chromaticity class comprises colours each of which contains a shade of one and the same hue, then it seems natural to expect that the observer will be able to arrange all these colours in accordance with the pronouncedness of this hue. If the chromaticity class is based on a different criterion, such an ability seems highly unlikely. For instance, if a chromaticity class is based on a colour category formed by a group of binary colours, such an ordering seems implausible since it is not clear what pronouncedness within this group might mean.[29] Thus, we believe that a prerequisite for a chromaticity class to represent a component hue is the existence of a chromatic ordering on this class.

Interestingly, it was discovered that observers are indeed capable of sorting Munsell papers within the chromaticity class by the strength of the component hue making up this class (Logvinenko & Beattie, 2006). We formalise this observers' ability by introducing the concept of *chromatic ordering* (see Definition 8.20). In fact, a chromatic ordering is implicitly assumed by the widely used hue scaling techniques which intend to describe colour in terms of the relative amount of component hues contained in this colour (Gordon & Abramov, 1988; Gordon et al., 1994; Abramov & Gordon, 1994).

It is safe to assume that the chromatic ordering is complete. Indeed, observers never refuse to compare any colours (within a chromaticity class). As to the definitive features of any preorder – transitivity, preliminary observations show that transitivity failures occur from trial to trial. Hence, a proper statistical test should be conducted to decide whether these failures are inconsistent with the transitivity hypothesis. For this purpose one can use a special test developed on purpose to test transitivity statistically (Logvinenko et al., 2004).

If the transitivity of chromatic ordering is confirmed, then the chromatic ordering is a complete preorder (a weak order in another terminology; see Definition 8.1). It follows that the features which observers used in partial hue-matching are one-dimensional variables (continua); most likely, component hues. If an observer failed to weakly order the elements of chromaticity classes it would testify for that the partial colour matching revealed in experiment is based on colour categories rather than component hues.

---

29 It is, of course, possible to order binary colours of the same hue (such an order underlies the notion of purity, as defined in Section 8.5.6), but not colours of different hue.

In particular, if Observer 2 is able to order the Munsell papers in the 'orange' and 'purple' chromaticity classes, then we have to conclude that there are six component hues in his hue palette. Otherwise, his results can be accounted for by simply using the 'orange' and 'purple' colour categories in the partial hue matching experiment.

8.5.4 Chromatically ordered structures

In view of the above, this section assumes that human observers are capable of lining up the colour stimuli in each chromaticity class according to the amount of the component hue constituting this chromaticity class. Such an ability will be conceptualised in the following definition as a preorder on the chromaticity class which will be referred to as the *chromatic ordering*. It will be convenient to give the corresponding definitions (see Definition 8.20) for light stimuli rather than colour stimuli. In Section 8.5.1 we examined m-chromatic structures on an arbitrary set of colour stimuli $\mathbb{S}$ (see Definition 8.16). In particular, one can take $\mathbb{S}$ to be the whole colour stimulus cone $\pi(A)$. Since the colour-signal cone $\Phi(A)$ may be considered as a concrete realisation of the colour stimulus cone $\pi(A)$, the corresponding notions make sense for $\mathbb{S}$ being a subset of colour-signal cone $\Phi(A)$, as well. Moreover, these definitions are naturally extended on the light stimulus cone $A = \mathbf{C}_+^*$ and its subsets $\mathbb{C}$ by passing from a light $a \in \mathbb{C}$ to its colour stimulus $\pi(a) = [a] \in \mathbb{S} = \pi(\mathbb{C})$ (or its colour signal $\Phi(a) \in \mathbb{S} = \Phi(\mathbb{C})$).

Definition 8.19 *Given a set of light stimuli, $\mathbb{C}$, we will say that a set $H \subset \mathbb{C}$ is a* chromaticity class *on it if $H = \pi^{-1}(H') \cap \mathbb{C}$ where $H' \subset \mathbb{S} = \pi(\mathbb{C})$ is a chromaticity class of colour stimuli on $\mathbb{S}$ in the meaning of Definition 8.16. We say $\langle \mathbb{C}, H_1, \dots, H_m \rangle$ is a* m-chromatic structure *on $\mathbb{C}$ if $H_i = \pi^{-1}(H_i') \cap \mathbb{C}$, $i = 1, 2, \dots, m$, and $\langle \mathbb{S}, H_1', \dots, H_m' \rangle$ is a m-chromatic structure on the colour stimulus set $\mathbb{S}$ in the meaning of Definition 8.16. Given a m-chromatic structure $\langle \mathbb{C}, H_1, \dots, H_m \rangle$, a chromaticity class H_i is said to be* singular *if the corresponding class H_i' on $\mathbb{S}$ is singular in the meaning of Definition 8.18 or, what is the same, if $H_i = C(H_i)$ where $C(H_i) := H_i \bigcap \left(\bigcap_{j \neq i} (\mathbb{C} \setminus H_j) \right)$.*

Definition 8.20 *Consider a m-chromatic structure, $\langle \mathbb{C}, H_1, \dots, H_m \rangle$, on a light stimulus set $\mathbb{C} \subseteq A = \mathbf{C}_+^*$. We will assume that there is a complete preorder (a weak order in another terminology) $\succsim_i$ on each H_i, which will be referred to as the* ith chromatic ordering, *the chromaticity classes being called* chromatically ordered. *The chromatic ordering $\succsim_i$ will be called* proper, *and the chromaticity class H_i* properly *chromatically ordered if, firstly, there is a maximal element in H_i, i.e., such a light $a^i_{\max}$ that $a^i_{\max} \succsim_i a$ for any $a \in H_i$, and secondly, there is no minimal element in H_i, i.e., such a light $a^i_{\min}$ that $a \succsim_i a^i_{\min}$ for any $a \in H_i$. It follows from these two conditions that the preorder $\succsim_i$ is non-trivial that is, there is at least one pair of lights $a, b \in H_i$ such that $a \succ_i b$. $\langle \mathbb{C}, H_1, \dots, H_m \rangle$ will be called* chromatically ordered structure *on $\mathbb{C}$ if each chromaticity class is chromatically ordered, and all, or all apart from one of H_i, are properly chromatically ordered.*

The rationale for assuming no minimal element in chromatically ordered classes is the fact that human observers are unable to single out a colour with the least amount of some proper component hue.

Definition 8.21 *Let $\langle A, H_0, H_1, \dots, H_m \rangle$ be a chromatically ordered structure on the light stimulus cone $A = \mathbf{C}_+^*$, where chromaticity classes $H_1, \dots, H_m$ are properly chromatically ordered, and a chromaticity class H_0 is singular (i.e., $H_0 = C(H_0)$). The class H_0 will be called* neutral *if, firstly, for* *every* $a, b \in H_0$ $a \sim_0 b$*, and secondly,* $C(H_0)$ *contains no light the colour signal of which belongs to the boundary of the colour-signal cone* $\Phi(A)$*. As* $H_0 = C(H_0)$*, the chromatic class* $C(H_0)$ *will also be called* neutral.

Given a chromatically ordered structure on $A = \mathbf{C}_+^*$, every proper chromatic ordering $\succsim_i$ can be extended on the whole light stimulus cone A letting by definition $a \succsim_i b$ for any light a in A and all the lights b outside H_i. It follows that $a \succ_i b$ whenever $a \in H_i, b \notin H_i$, and $a \sim_i b$ whenever $a \notin H_i$, $b \notin H_i$. Therefore, $A \backslash H_i$ is the set of lower bounds of H_i in the light stimulus cone A ordered by $\succsim_i$. From now on, all the chromatic orderings will be assumed to be defined on the whole light stimulus cone A unless it is said otherwise.

Note that any chromatically ordered structure $\langle A, H_0, H_1, \dots, H_m \rangle$ is fully specified by the families of its chromatic orderings. Indeed, for each $i \in \{1, 2, \dots, m\}$ the set $M(\succsim_i)$ of the lower bounds of the ordering $\succsim_i$ on A uniquely specifies the ith chromaticity class H_i in A, i.e., $H_i = A \backslash M(\succsim_i)$, while $H_0 = A \backslash \bigcup_{i=1}^m H_i = \bigcap_{i=1}^m M(\succsim_i)$. Therefore we shall use a notation $\langle A, \succsim_1, \dots, \succsim_m \rangle$ for a chromatically ordered structure $\langle A, H_0, H_1, \dots, H_m \rangle$.

Given a chromatically ordered structure $\langle A, \succsim_1, \dots, \succsim_m \rangle$, the intersection of all the equivalence relations $\sim_i$ on A will be called *chromatic equivalence* (denoted as $\approx$). When two light stimuli a and b are chromatically equivalent (i.e., $a \approx b$) it means that the observer cannot see any difference between them with respect to any of the m proper component hues.

Definition 8.22 *Let* $\langle A, \succsim_1, \dots, \succsim_m \rangle$ *be a chromatically ordered structure, and* $\approx$ *the chromatic equivalence. The quotient set* $A/\approx$ *with factor orderings* $\succsim_1/\approx, \dots, \succsim_m/\approx$ *on it*[30] *will be called the* chromaticity structure *(written as* $\langle A/\approx, \succsim_1/\approx, \dots, \succsim_m/\approx \rangle$*).*

8.5.5 Chromatic scales and chromatic codes

Let $\langle A, \succsim_1, \dots, \succsim_m \rangle$ be a chromatically ordered structure on the light stimulus cone $A = \mathbf{C}_+^*$. Each proper chromatic ordering can be represented with a proper scale the same way as it has been done for the achromatic ordering.

We will assume that there exists a continuous ordinal scale for each chromatic ordering (which will be referred to as an *ordinal chromatic scale*). That is, for each chromatic ordering $\succsim_i$ there exists a real-valued continuous function f_i on A such that for all $a, b \in A$,

$$a \succsim_i b \Leftrightarrow f_i(a) \geq f_i(b). \tag{8.39}$$

The existence of such a scale will follow from the continuity of the preorder $\succeq_i$ on $\Phi(A)$ induced by chromatic ordering $\succsim_i$ on A (cf. Proposition 8.3). An explicit definition of $\succeq_i$ is given below. Similar to an ordinal achromatic scale, an ordinal chromatic scale is unique up to a strictly increasing transformation (cf. Remark 8.3).

As already mentioned, human observers can not only compare two lights according to the strength of the common component hue in them, but also compare pairs of lights with respect to the difference in strength of the component hue between the lights in pairs. For instance, we can easily see that the difference in, say, yellowness between unique yellow and slightly greenish

30 A factor ordering $\succsim_i/\approx$ on $A/\approx$ is defined in a standard way. Two chromatically equivalent classes $\boldsymbol{a}, \boldsymbol{b} \in A/\approx$ are in the relation $\succsim_i/\approx$ if and only if there are such $a \in \boldsymbol{a}$ and $b \in \boldsymbol{b}$ that $a \succsim_i b$. Since $\approx$ is included in $\sim_i$, this definition is independent of the choice of a and b.

yellow is smaller than that between unique red and orange. In order to conceptualise such an ability, we will assume that for every component hue there exists a binary relation $\succcurlyeq_{*i} (i = 1, \ldots, m)$ on the set of pairs of light stimuli, $A \times A$. Introspectively, $(a, b) \succcurlyeq_{*i} (a', b')$ means that the difference in the strength of the ith component hue between the lights a and b is more pronounced than between a' and b'.

We will assume the existence of functions f_i on $A = C_+^*$ satisfying (8.39) and such that for all $a, b, a', b' \in A$ (cf. Eq. (8.7) in Section 8.1.1),

$$(a, b) \succcurlyeq_{*i} (a', b') \Leftrightarrow f_i(a) - f_i(b) \geq f_i(a') - f_i(b'). \tag{8.40}$$

Generally, under some natural conditions on $\succcurlyeq_{*i}$ (e.g., as those at the end of Section 8.1.1), such a function f_i is determined uniquely up to a positive multiplier and a shift: if f_i' is another function satisfying (8.40), then $f_i'(a) = \alpha f_i(a) + \beta$ for all $a \in A$, where $\alpha, \beta \in \mathbf{R}$, $\alpha > 0$; see Remark 8.7.

Remark 8.12 *Similar to the achromatic scale f in (8.7), f_i in (8.40) is an interval scale, according to the terminology in the measurement theory (Krantz et al., 1971, Chapter 4).*

Remark 8.13 *We will suppose that the functions f_i in (8.40) can be chosen so that their range is $[0, 1]$, and $f_i(a^i_{\max}) = 1$ (here $a^i_{\max}$ is any maximal element in H_i) and $f_i(x) = 0$ for $x \in M(\succsim_i)$ (recall that $M(\succsim_i)$ is the set of lower bounds of the order $\succsim_i$ on A, that is, $M(\succsim_i) = A \backslash H_i$). Admittedly, this is a rather strong supposition.*[31]

Definition 8.23 *Let $\langle A, \succsim_1, \ldots, \succsim_m \rangle$ be a chromatically ordered structure. The function f_i that is weak* continuous on $A = \mathbf{C}_+^*$ and satisfies (8.40) and the supposition of Remark 8.13 will be referred to as the* ith chromatic scale, *its value on $a \in A$ being called the* ith chromatic value of a. *The m-tuple $(f_1(a), \ldots, f_m(a))$ will be called the* chromaticness *of light a.*

Note that if a chromatic scale exists it is unique. Therefore, the chromaticness of a light is uniquely determined, if at all.

Until now, the chromatically ordered structures on the light stimulus set, A, have been considered regardless of whether the Grassmann structure is also defined on A or not. Although sometimes we implicitly assumed that, along with the chromatically ordered structure, the Grassmann structure exists on A, this was not crucial for our consideration. However, when a chromatically ordered structure is defined on a light stimulus set on which a Grassmann structure is also defined, a natural question arises: how are these two structures related? In particularly, how are the metamerism and the chromatic orderings related? In detail these questions will be discussed in Section 8.6.4. Here we only note the interest factor of a case where the two structures are related as per the following definition.

Definition 8.24 *Suppose $\langle A, \sim \rangle$ is a n-dimensional Grassmann structure, and $\langle A, \succsim_1, \ldots, \succsim_m \rangle$ is a chromatically ordered structure. If the metamerism $\sim$ is included in the symmetric part of each chromatic ordering $\succsim_i$, i.e., $a \sim_i b$ whenever $a \sim b$, we say that $\langle A, \sim, \succsim_1, \ldots, \succsim_m \rangle$ is a* chromatically ordered Grassmann structure.

In chromatically ordered Grassmann structure $\langle A, \sim, \succsim_1, \ldots, \succsim_m \rangle$ metamerism $\sim$ is included in the chromatic equivalence $\approx$ (i.e., the intersection of the symmetric parts of all $\succsim_i$).

---

31 It fails, for example, when $A = \boldsymbol{R}$ and $\succcurlyeq_*$ is given by $(a, b) \succcurlyeq_* (a', b') \Leftrightarrow b - a \geq b' - a'$.

If in chromatically ordered Grassmann structure $\langle A, \sim, \succsim_1, \dots, \succsim_m \rangle$ the symmetric part of a chromatic ordering $\succsim_i$, $\sim_i$, meets Grassmann's laws (Definition 2.2),then $\langle A, \sim_i \rangle$ is a Grassmann substructure of $\langle A, \sim \rangle$ (Definition 2.3).

Given a chromatically ordered Grassmann structure $\langle A, \sim, \succsim_1, \dots, \succsim_m \rangle$, each chromatic ordering $\succsim_i$ on A (in similar fashion as the achromatic ordering) induces an ordering (written as $\succeq_i$) on the colour-signal cone $\Phi(A)$ if we assume by definition that for any $z^1, z^2 \in \Phi(A)$ $z^1 \succeq_i z^2$ if there exist $a, b \in A$ such that $\Phi(a) = z^1$, $\Phi(b) = z^2$, and $a \succcurlyeq_i b$. It follows from this definition that each $\succeq_i$ is a complete preorder on $\Phi(A)$, and

$$\Phi(a) \succeq_i \Phi(b) \Leftrightarrow a \succcurlyeq_i b. \tag{8.41}$$

Suppose, in addition, that $\succeq_i$ is continuous. In this case, by Proposition 8.2, there exists a continuous real-valued function c_i on $\Phi(A)$ such that $f_i = c_i \circ \Phi$ (hence, f_i is weak* continuous on A), and ordinal representation (8.39) takes the form of

$$a \succcurlyeq_i b \Leftrightarrow c_i\,(\Phi(a)) \geq c_i\,(\Phi(b)). \tag{8.42}$$

Continuity of $\succeq_i$ implies weak* continuity of f_i (Proposition 8.3). Note that c_i is a continuous ordinal scale with respect to $\succeq_i$. Therefore the ordinal chromatic scale f_i is a (weak*) continuous colour code on A (Remark 2.9) which will be referred to as an *ordinal chromatic colour code*. It can be decomposed into the colour-signal map Φ and the colour-signal code c_i on $\Phi(A)$. As c_i is an ordinal scale with respect to $\succeq_i$, it will be referred to as an *ordinal chromatic colour-signal code* (Remark 2.9).

Analogously, a binary relation $\succcurlyeq_{*i}$ on $A \times A$ induces an ordering (written as $\succeq_{*i}$) on $\Phi(A) \times \Phi(A)$ such that

$$(\Phi(a), \Phi(b)) \succeq_{*i} (\Phi(a'), \Phi(b')) \Leftrightarrow (a, b) \succcurlyeq_{*i} (a', b'). \tag{8.43}$$

Furthermore, under appropriate conditions on $\succeq_{*i}$, there exist a real-valued function c_i on $\Phi(A)$ such that for all $a, b, a', b' \in A$,

$$(a, b) \succcurlyeq_{*i} (a', b') \Leftrightarrow c_i(\Phi(a)) - c_i(\Phi(b)) \geq c_i(\Phi(a')) - c_i\,(\Phi(b')). \tag{8.44}$$

The function c_i is determined uniquely up to a positive multiplier and a shift (Remark 8.7).

It follows that for any colour-signal map Φ one can exist at most one function c_i satisfying (8.44) and such that its range is $[0, 1]$, and $c_i\,(\Phi(a^i_{\max})) = 1$, $c_i\,(\Phi(a)) = 0$ for $a \in M\,(\succsim_i)$.

8.5.6 Hue, purity, and saturation in chromatic structures

There is every reason to expect that two light stimuli will match in hue when their chromaticnesses (Definition 8.23) are proportional to each other. This justifies the following definition. Let $\langle A, \succsim_1, \dots, \succsim_m \rangle$ be a chromatically ordered structure, and $f_1, \dots, f_m$ the corresponding chromatic scales. Two lights a and b in A will be said to have the same hue if there is such $k > 0$ that

$$(f_1(a), \dots, f_m(a)) = (kf_1(b), \dots, kf_m\,(b)). \tag{8.45}$$

Although this expectation looks more than plausible, strictly speaking, a special experiment should be carried out to verify it. In case of a positive outcome, Eq. (8.45) can be taken as an operational definition of hue. Furthermore, this definition takes precedence over the previous ones, because the latter imply that observers know what hue is.

True, definition (8.45) rests upon the partial hue-matching procedure that assumes observers are capable of establishing whether two light colours have a common hue. However, firstly, this assumption seems to be weaker than that underlying the hue matching. Actually, observers have only to discern the presence of common qualia in both lights, which seems to be a much easier task. They might successfully carry out this task without even knowing what hue means. Secondly, as mentioned earlier, a control experiment showed that observers yield the same results when asked to recognise if the lights have a common colour (rather than a common hue). So, if the hue matching procedure does not corroborate Eq. (8.45), we would rather give preference to the definition of hue based on the proportionality of chromaticnesses (i.e., Eq. (8.45)).

Note that chromatic equivalence (i.e., $a \approx b$) can be thought of as a particular case of Eq. (8.45) when $k = 1$. Let us define an equivalence relation (written as $\doteqdot$) on the chromaticity structure $A/\approx$ as follows: $(f_1(a), \ldots, f_m(a)) \doteqdot (f_1(b), \ldots, f_m(b))$ if and only if there is such $k > 0$ that $(f_1(a), \ldots, f_m(a)) = (kf_1(b), \ldots, kf_m(b))$. The quotient set $(A/\approx)/\doteqdot$ will be referred to as the *hue palette*, and the elements in $(A/\approx)/\doteqdot$ will be called *hues*. The class of $\doteqdot$-equivalence containing the neutral element (with zero chromaticness) of $A/\approx$ comprises this element only. This class will be called the *improper hue*, the rest of hues being referred to as *proper hues*. Respectively, the hue palette without the improper hue will be sometimes referred to as the *proper hue palette*.

Each proper hue (i.e., a class of $\doteqdot$-equivalence with non-zero chromaticness) can be naturally ordered if we define a binary relation $\gtrsim$ on it as follows. For any two $\doteqdot$-equivalent chromaticnesses $(\xi_1, \ldots, \xi_m)$ and $(\xi'_1, \ldots, \xi'_m)$, that is, such that $(\xi_1, \ldots, \xi_m) = (k\xi'_1, \ldots, k\xi'_m)$ we will assume $(\xi_1, \ldots, \xi_m) \gtrsim (\xi'_1, \ldots, \xi'_m)$ if $k \geq 1$; and $(\xi'_1, \ldots, \xi'_m) \gtrsim (\xi_1, \ldots, \xi_m)$ if $0 < k \leq 1$. Clearly, $\gtrsim$ is a complete order.

In the theory of chromaticity structures, the order $\gtrsim$ plays a similar role as the purity order (Section 8.2) in the theory of Grassmann structures. We will keep using the term 'purity' for the order $\gtrsim$ as well. However, there are important differences between the two to keep in mind. Firstly, the order $\gtrsim$ is defined on an arbitrary chromaticity structure, and it is applicable only for this structure. Secondly, this definition is not independent; it is derived from the chromatic orderings (Definition 8.20). Using the terminology of the measurement theory, one can say that chromatic value is a fundamental attribute (because 'its measurement does not depend on the measurement of anything else' (Krantz et al., 1971, p. 502) whereas purity, as defined above, a derived attribute.

Since the concept of colour has not been defined yet for chromaticity structures (this will be done in the following section), the purity determined via the order $\gtrsim$ applies to lights (not to colours). In other words, so defined purity is, strictly speaking, an attribute of light rather than colour. So is the hue defined by Eq. (8.45). However, just as Eq. (8.45) is suggested as an alternative definition of hue, the order $\gtrsim$ can also be thought of as an alternative basis for purity. Evidently, this is more of a declaration of intent than a strict definition. Until the light colour is properly defined for chromaticity structures (in the following section), terminological rigorism can be observed as follows. For any light in a chromaticity structure, its hue and purity is determined as above. Whatever the colour of this light, we will consider, by definition, that the hue and purity so defined are attributes of this colour. Thereby, we get an opportunity to apply the terminology, defined for chromaticity structures, to light colours. For instance, we will call a colour unitary or binary if its hue is unitary or binary respectively. Also, using this terminological convention, we introduce the following useful term. If there is the maximal element in a class of $\doteqdot$-equivalence, we will say that all lights comprising this element have a *full colour*.[32]

It should be pointed out that the purity of one hue is not related to the purity of another hue. This is a consequence of the independence of chromatic orderings. Really, Definition 8.20 assumes

32 Strictly speaking, this maximal element should be called '*full light-colour*'. However, since Part 1 is devoted to only light colours, we omit the word 'light' in this term.

by default that these orderings are not related to each other in any way. Accordingly, the chromatic scale values are not commensurate with each other. Because of that, all chromatic scales can be normalised (albeit rather arbitrarily) so as to have [0, 1] as their range. Hence, if, for instance, the chromaticness of both unitary blue and unitary yellow is maximal, this does not mean that these hues have equal purity. The equality of purity (in terms of the order $\succsim$) for different hues is simply not defined. At the same time, one cannot but recall the fact already mentioned earlier that human observers incline to judge the unitary blue as more saturated as the unitary yellow. Evidently, a special inquiry into this ability is required. In particular, one should find out whether it can be formalised. For example, one can check whether the axioms necessary and sufficient to conduct a conjoint measurement (Krantz et al., 1971, Ch. 6) of purity for unitary blue and yellow are satisfied. In case of a positive answer, one can introduce the notion of saturation, which will play the role of purity common to all hues (see Section 8.6.3).

Remark 8.14 *As a matter of fact, for human trichromatic vision there is the maximal element in each proper class of* $\mathring{=}$*-equivalence, that is, the full colour. Therefore, there is a one-to-one correspondence between the set of proper hues and the set of full colours. Furthermore, it seems plausible and consistent with the material in Section 8.2 to assume that, in a chromatically ordered Grassmann structure, for any class of* $\mathring{=}$*-equivalence (except for the improper hue) the maximal element in this class is represented by a light belonging to the spectral light stimulus conical closure.*

At last, with a view to simplifying the notation as well as for a reason that will become clear in the next section, the quotient set $A/\approx$ (i.e., the domain of the chromaticity structure according to Definition 8.22) will be also referred to as the *chromaticity palette*. Recall that each element of the chromaticity palette is uniquely represented by the chromaticness of its representatives. Note that the chromaticness of any neutral light (i.e., element of the chromaticity class H_0) is $0 \in \mathbf{R}^m$. Thus, the chromaticity palette contains just one element the chromaticness of which is zero ($0 \in \mathbf{R}^m$). This element will be referred to as the *neutral element*. The chromaticity palette without the neutral element (i.e., $(A \backslash H_0)/\approx$) will be referred to as the *proper chromaticity palette*.

8.6 Light-colour manifold

In this section, we will look into the chromaticity structures appropriate for describing perceiving light colour by human observers. These chromaticity structures have some remarkable features that allow to represent the light colours as the three-dimensional manifold with boundary (referred to as the light-colour manifold) which is, geometrically, of the same type as the colour stimulus manifold.

Recall that experiments on partial hue-matching (Section 8.5.2) reveal, usually, four properly chromatically ordered chromaticity classes: $H_1, \dots, H_4$; and one neutral chromaticity class H_0 such that $H_0 = \cap_{i=1}^{4} M(\succsim_i)$ (as above, $M(\succsim_i) = A \backslash H_i$). This gives grounds for conceptualising the colour perception of such the observers (as far as light colour is concerned) as a chromatically ordered structure: $\langle A, \succsim_1, \dots, \succsim_m \rangle$ where $m = 4$, and $\succsim_i$ is the chromatic ordering on the chromaticity class H_i extended onto A. We will assume that each chromatic ordering $\succsim_i$ can be represented with the chromatic scale f_i on A ($i = 1, \dots, 4$) (Definition 8.23).

Furthermore, as follows from the experiments on partial hue-matching, there are five unitary chromatic classes: $C(H_0), C(H_1), \ldots, C(H_4)$; and only four binary chromatic classes: $C(H_1, H_3)$, $C(H_1, H_4)$, $C(H_2, H_3)$, and $C(H_2, H_4)$. Therefore, as anticipated by Ewald Hering, apart from neutral colour, there are only four unitary and four binary colours,[33] there being neither ternary nor quaternary light colour in the human light-colour palette.[34]

Each of four proper unitary chromatic classes, $C(H_i)$ $(i = 1, \ldots, 4)$, consists of lights a with chromaticnesses $(f_1(a), f_2(a), f_4(a))$ where only $f_i(a)$ is non-zero. As any two such lights represent one and the same element of the hue palette (i.e., the same hue), the whole chromatic class $C(H_i)$ is identified with the sole class of $\mathring{=}$-equivalence. This class comprises chromaticnesses $(\xi_1, \xi_2, \xi_3, \xi_4)$ where $\xi_i = k\xi_i^{\max}$, $\xi_i^{\max}$ is some fixed positive number, $0 < k \leq 1$, and three other ξ are zero. Likewise, each of four binary chromatic classes $C(H_i, H_j)$ consists of lights a with chromaticnesses $(f_1(a), f_2(a), f_3(a), f_4(a))$ where only $f_i(a)$ and $f_j(a)$ are non-zero. Every binary chromatic class decomposes into classes of $\mathring{=}$-equivalence, and each such class comprises chromaticnesses $(\xi_1, \xi_2, \xi_3, \xi_4)$ where $(\xi_i, \xi_j) = (k\xi_i^{\max}, k\xi_j^{\max})$, $\xi_i^{\max}, \xi_j^{\max}$ are some fixed positive numbers, $0 < k \leq 1$, and two other ξ are zero. Thus, each class of $\mathring{=}$-equivalence corresponding to a proper hue can be considered as a topological space (with the order topology[35] induced by $\gtrsim$) homeomorphic to $(0, 1]$.

8.6.1 Hue cyclic order

The set of full colours (thus, the set of proper hues) can be cyclically ordered. Indeed, note first that a cyclic order naturally arises on the set of binary chromatic classes: $BHC = \{C(H_1, H_3), C(H_1, H_4), C(H_2, H_3), C(H_2, H_4)\}$. Recall that in order to define a cyclic order on a finite set N, one has to define on N which element immediately 'before' (i.e., a 'predecessor') and which element immediately 'following' (i.e., a 'successor') any given element in N. Let us assume that both the predecessor and successor for a given binary chromatic class must share a common component hue with it. Then one can define exactly two cyclic orders on the set of binary chromatic classes BHC. Indeed, let us start from the chromatic class $C(H_1, H_3)$. There are only two other binary chromatic classes sharing a component hue with it: $C(H_1, H_4)$ and $C(H_2, H_3)$. Defining $C(H_2, H_3)$ as the successor for $C(H_1, H_3)$ brings about one cyclic order on BHC. Alternatively, defining $C(H_1, H_4)$ as the successor for $C(H_1, H_3)$ brings about another cyclic order on BHC.

Cyclic ordering allows for arranging the set of binary chromatic classes in a circle, that is, arranging it as on a 4-hour clock face. Putting the unitary chromatic classes, $C(H_1), \ldots, C(H_4)$, each in between the binary chromatic classes containing the corresponding component hue (i.e., the unitary chromatic class $C(H_1)$ between $C(H_1, H_3)$ and $C(H_1, H_4)$; $C(H_2)$ between $C(H_2, H_3)$ and $C(H_2, H_4)$, and so on), we get all the eight chromatic classes arranged on a 8-hour clock face, that is, cyclically ordered.[36] (To be more exact, we have got two cyclic orders: clockwise and counterclockwise.)

---

33 It would be more accurate to say that there are only four types of binary colours, since the colours of one binary type (say, orange) can differ from each other in chromaticness.

34 Naturally, all this should be confirmed in further experiments that will need to be carried out.

35 Given an order $\gtrsim$ on a set C, the order topology on C induced by it is generated by order intervals $I(a, b) = \{c \in C : a > c > b\}, a, b \in C, a > b$.

36 Evidently, this cyclic order does not depend on the way we number the component hues.

Likewise, one can cyclically order the chromatic classes produced by Observer 2 (Fig. 8.18). All the twelve chromatic classes produced by this observer can be arranged on a 12-hour clock face. Note that the cyclical order is possible due to the fact that for Observer 2 each chromaticity class intersects with exactly two other chromaticity classes, making exactly two binary chromatic classes between which the unitary chromatic class, corresponding to this chromaticity class, is located on the 12-hour clock face.

Generally, consider a m-chromatic structure $\langle A, H_1, \dots, H_m \rangle$, which is chromatically ordered, the degree of chromaticity of which is 2, and the chromatic classes of which have the same property as those of Observer 2 (which will be referred to as the *two-concomitant-hue condition*). The two-concomitant-hue condition was observed for all the observers who took part in experiments on partial hue-matching conducted so far. When it takes place, the binary classes in $\langle A, H_1, \dots, H_m \rangle$ can be listed as $C(H_1, H_2), C(H_2, H_3), \dots, C(H_i, H_{i+1}), \dots, C(H_{m-1}, H_m), C(H_m, H_1)$. Clearly, these m unitary and m binary chromatic classes can be arranged on a $2m$-hour clock face so that the unitary chromatic classes are interleaved with the binary chromatic classes (i.e., so that between each pair of adjacent binary chromatic classes, e.g., $C(H_{i-1}, H_i)$ and $C(H_i, H_{i+1})$, there is exactly one unitary chromatic class, namely, $C(H_i)$).

There was found another condition (referred to below as the *concordance condition*) that, along with the two-concomitant-hue condition, permits generalising the cyclic order from the set of chromatic classes onto the whole hue palette.

Note that the full colours in each of binary chromatic classes can be ordered with respect to either of its two chromatic values. Specifically, given $a, b \in C(H_i, H_j)$, one can define either $a \geq_i b$ if and only if $f_i(a) \geq f_i(b)$, or $a \geq_j b$ if and only if $f_j(a) \geq f_j(b)$. A remarkable feature of human colour vision is that the following relation holds[37]: $f_i(a) \geq f_i(b)$ if and only if $f_j(a) \leq f_j(b)$. This means that if, for instance, the lights a and b belong to the yellow-green chromatic class, and the shade of yellow in a is not less pronounced than in b, then the shade of green in b is not less pronounced than in a. Therefore, either relation, $\geq_i$ and $\geq_j$, determines effectively the same order on $C(H_i, H_j)$.

Furthermore, since introspection indicates that the unitary colours are experienced as having the strongest shade of the corresponding component hue, we will assume that the chromatic value of the full unitary colour amounts to the maximum value of the corresponding chromatic scale.[38] It follows that there is just one full colour in each unitary chromatic class. The orders $\geq_i$ allow for extending the cyclic order defined on the set of the $2m$ chromatic classes onto the entire set of full colours.

Generally, a cyclic order on a (not necessarily finite) set, A, is determined by defining a ternary relation $[\cdot, \cdot, \cdot]$ on A satisfying the following axioms.

1. *Cyclicity*. If $[a, b, c]$ then $[b, c, a]$.
2. *Asymmetry*. If $[a, b, c]$ then not $[c, b, a]$.
3. *Transitivity*. If $[a, b, c]$ and $[a, c, d]$ then $[a, b, d]$.
4. *Totality*. If a, b, and c are distinct, then either $[a, b, c]$ or $[c, b, a]$.

The ternary relation $[a, b, c]$ can be interpreted as that b follows a and precedes c. The cyclic order defined above on the set of chromatic classes arranged on a clock face (i.e., the

37 Although this property seems self-evident, it, strictly speaking, should have been formulated as an axiom to be tested experimentally in the future.

38 This assumption also requires a thorough experimental verification. Note that the results of the hue cancellation experiment (Hurvich & Jameson, 1955) do not corroborate this assumption. Nor do the results of the hue scaling experiment (Abramov & Gordon, 1994) (see Figs 8.22 and 8.23). It must be pointed out, however, that this disaccord may be construed as evidence against these methods themselves (see our criticism in Section 8.6.3).

chromatic-class cyclic order) entails such a ternary relation. For example, chromatic classes $C(H_1, H_2)$, $C(H_3, H_4)$, and $C(H_1)$ occur after each other as one goes (clockwise) around the clock face, thus $[C(H_1, H_2), C(H_3, H_4), C(H_1)]$.

Now let us extend this ternary relation on the set of chromatic classes onto the whole set of full colours as follows (written as $\lfloor\cdot,\cdot,\cdot\rfloor$). Let a, b and c be full different colours belonging to chromatic classes C_i, C_j and C_k respectively. If i,j and k are distinct, then $\lfloor a, b, c\rfloor \Leftrightarrow [C_i, C_j, C_k]$. If $i = j$ but $i \neq k$, then C_i must be a binary chromatic class, say, $C(H_p, H_q)$. If $C_k \neq C(H_q)$ we have either $[C(H_q), C(H_p, H_q), C_k]$ or $[C_k, C(H_p, H_q), C(H_q)]$. If $[C(H_q), C(H_p, H_q), C_k]$, then $\lfloor a, b, c\rfloor \Leftrightarrow f_q(a) > f_q(b)$; if $[C_k, C(H_p, H_q), C(H_q)]$, then $\lfloor c, a, b\rfloor \Leftrightarrow f_p(a) > f_p(b)$. Assume now that $C_k = C(H_q)$. Recall that either $[C(H_p), C(H_p, H_q), C(H_q)]$ or $[C(H_q), C(H_p, H_q), C(H_p)]$. When $[C(H_p), C(H_p, H_q), C(H_q)]$, then $\lfloor a, b, c\rfloor \Leftrightarrow f_p(a) > f_p(b)$. If $[C(H_q), C(H_p, H_q), C(H_p)]$, then $\lfloor c, a, b\rfloor \Leftrightarrow f_q(a) > f_q(b)$. Finally, assume that $i = j = k$, that is, all the colour are in the same binary chromatic class, say, $a, b, c \in C(H_p, H_q)$. Then, when $[C(H_p), C(H_p, H_q), C(H_q)]$, then $\lfloor a, b, c\rfloor \Leftrightarrow f_p(a) > f_p(b) > f_p(c)$ and when $[C(H_q), C(H_p, H_q), C(H_p)]$, then $\lfloor a, b, c\rfloor \Leftrightarrow f_q(a) > f_q(b) > f_q(c)$.

Having fixed a particular full colour a, the remainder of the set of full colours becomes linearly ordered (written as $<_a$) if one assumes $b <_a c$ if and only if $[a, b, c]$. The linear order $<_a$ induces a corresponding interval topology. The set of subsets open in every linear order $<_a$ determines a topology on the set of full colours. This is the topology of the unit circle, S^1. Therefore, the set of full colours (thus, the set of proper hues) can be geometrically represented as a circle where the kth unitary colour ($k = 1, \ldots, n$) occupies the position corresponding to the value of the polar angle $\alpha = \frac{2k\pi}{n}$. For instance, in the case $n = 4$ the unitary red is at $\alpha = 0$, the unitary yellow at $\alpha = \pi/2$, the unitary green at $\alpha = \pi$, and the unitary blue $\alpha = 3\pi/2$. The binary colours lie in the sectors ended with the corresponding unitary colours. The chromatic values in the binary colours vary in each sector accordingly.

A chromatically ordered structure, $\langle A, H_1, \ldots, H_m\rangle$, with the degree of chromaticity of 2, the chromatic classes of which satisfy the two-concomitant-hue condition and the concordance condition, and for which the chromatic value of each full unitary colour is maximal, will be called a *circular chromatic structure*. As each proper hue in the circular chromatic structure (i.e., an element in $((A \backslash H_0)/\approx)/\doteq)$ can be considered as a topological space homeomorphic to $(0, 1]$, its proper chromaticity palette, $(A \backslash H_0)/\approx$, can be considered as a fibre bundle over the unit circle, S^1, each fibre being homeomorphic to $(0, 1]$. We will refer to this as the *hue fibre bundle.*[39] Furthermore, the chromaticity palette, $A/\approx$, is a topological space homeomorphic to a 2D unit disk, the neutral element mapping to the disk centre.

Thus, we see that the circular representation of hues, used by many authors, including Newton, can be arranged for any circular chromatic structure, irrespective of the number (m) of the component hues constituting it.

39 It is necessary to distinguish this concept from a notion with the same name that was introduced earlier (see Eq. (8.18)). These two notions have been defined independently of each other within the different conceptual frameworks (the Grassmann structure versus the chromatically ordered structures). Furthermore, the proper chromaticity palette, considered as a fibre bundle over the circle S^1, provides more information concerning hue than the hue fibre bundle defined in Section 8.2. Indeed, in the latter case, all one can say of the lights in some hue fibre is that they are of the same hue; whereas given a proper-chromaticity-palette fibre (i.e., a non-zero class of $\doteq$-equivalence in the chromaticity palette $A/\approx$), you can also say in what proportion the component hues are combined for the lights of this fibre. Still, one cannot exclude that these two concepts in fact define the same thing in the case of trichromatic human colour vision. Whether it is so or not, one can find out only in the experiment.

8.6.2 Light-colour manifold

Light stimuli falling into the same class of the chromatic equivalence $\approx$ are not experienced indistinguishable. They differ in what can be best described as subjective light intensity, which is usually called brightness. Although this term is widely used in the visual literature (including the previous chapters of this book), it has never been properly defined. In the framework of our approach, based on chromatically ordered structures and partial hue-matching, this can be done. For this it is sufficient to present to the observer the light stimuli forming the neutral chromaticity class H_0. Experiments show that, firstly, observers are able to order the light stimuli in H_0; and, secondly, it is the only order in H_0 that observers are capable of. Let us call the perceptual attribute with respect to which the light stimuli in H_0 can be ordered, *brightness*. So, any sample of light stimuli from H_0 that are perceived as differing from each other, will do manifest the perceptual attribute called brightness here. It is important to stress that, defining brightness this way, we first let observers experience this perceptual attribute (by presenting a sample all the items of which vary in this attribute only), and only afterwards we assign a term (brightness) for this attribute.[40]

After brightness is operationally defined in H_0 as described above, one can suggest the observer to operate with this perceptual attribute in the other chromaticity classes. Particularly, each element of the chromaticity palette, $A/\approx$, (that is, each class of the chromatic equivalence in A) can be suggested to be ordered with respect to brightness. Such an ordering can be independently made in every element of the chromaticity palette. Preliminary experiments show that it is not a problem when restricting to a single element of the chromaticity palette (or even to the same hue fibre). However, such brightness ordering is a much more difficult task within a sample encompassing different hue fibres. At present it is not quite clear if it is possible to compare the brightness of, say, unique yellow and unique blue.[41] Be that as it may, we will assume only that there exists a local brightness ordering in every element of the chromaticity palette, $A/\approx$, separately. We will call this family of orderings on the classes of the chromatic equivalence in A the *local achromatic orderings*. Once again, the local achromatic orderings do not allow for comparing the brightness of lights in different classes of the chromatic equivalence. In other words, we assume here only the ability to compare (with respect to brightness) lights of identical chromaticness. (If it proves that the heterochromatic brightness matching and ordering is possible, this can be conceptualised by postulating the existence of the global achromatic ordering, $\succcurlyeq_B$, on the whole light stimulus cone A as it is done in Definition 8.2. The local achromatic orderings, then, can be considered as the restriction of the achromatic ordering $\succcurlyeq_B$ onto the corresponding class of the chromatic equivalence in A.)

The same way as the chromaticity class H_0 manifests the neutral colour, brightness and achromatic (brightness) ordering (in fact, it defines these notions), any chromatic class $C(H_i)$ manifests the corresponding component hue, the perceptual attribute of its chromatic strength[42] (which is quantified via the corresponding chromatic scale), and the chromatic ordering $\succsim_i$. Indeed, being perceived as containing only one (component) hue, the light stimuli in $C(H_i)$ vary in brightness and yet one more attribute, which is defined as the chromatic strength of the corresponding component hue. It is an easy task for an observer to sort out the light stimuli in $C(H_i)$ according to this

40 Note that usually the situation is exactly the opposite. The experimenter suggests to perform some operation with (e.g., to match or to discriminate) the attribute that is associated with the meaning of the word 'brightness'. Curiously enough, but this situation is strikingly reminiscent of the experiment of 'free association' in psychoanalysis.

41 The known problem of heterochromatic matching of brightness is only part of this more general issue (i.e., the possibility of heterochromatic ordering of brightness).

42 Once again, the strength (pronouncedness or magnitude) of the component hue should not be confused with purity as it is defined in Section 8.2, not to mention in colorimetry.

attribute. This ordering can be taken as a case study to order (with respect to the chromatic strength of the ith component hue) the entire chromaticity class H_i. Note that such chromatic ordering in H_i (as well as the achromatic ordering in H_0) can be done without resorting to verbal categories (actually, without using words at all). This corroborates our belief that the approach developed here allows studying perceptual entities and their attributes irrespective of verbal associations.

The symmetric part of a local achromatic ordering will be referred to as the *local achromatic matching relation*. When the local achromatic ordering is assumed to coincide with the restriction of the achromatic ordering $\succcurlyeq_B$ onto the corresponding class of the chromatic equivalence, the local achromatic matching relation coincides with the restriction of the brightness matching relation $\sim_B$ onto the corresponding class of the chromatic equivalence.

The family of all the local achromatic matching relations (which includes the chromatic equivalence $\approx$) determines the finer equivalence relation on A, which will be called the *light-colour equivalence* and denoted by $\approxeq$. The quotient set $A/\approxeq$ is the set of all light colours. If the chromaticity palette, $A/\approx$, is a circular chromatic structure, the light colour set $A/\approxeq$ is homeomorphic to a closed convex cone (with nonempty interior) in $\mathbf{R}^3$ with the full dark mapping to the cone vertex. Particularly, $A/\approxeq$ is homeomorphic to the half-space $\mathbf{H}^3 = \{(x, y, z) \in \mathbf{R}^3 : z \geq 0\}$. Thus, for a circular chromatic structure the corresponding light colour set $A/\approxeq$ will be referred to as the *light-colour manifold* (with boundary).[43]

When the full dark is not included in the light stimulus set A, the light colour set $A/\approxeq$ is homeomorphic to a circular cylinder in $\mathbf{R}^3$. Therefore, in this case the light-colour manifold is topologically equivalent to a cylinder through the disc representing the circular chromatic structure. In the view of the impossibility of experimentally distinguishing between these two topological models in the vicinity of the full dark, both remain an option.[44]

8.6.3 Circular Hering structures, their representation and experimental identification

As was repeatedly noted above, in experiments on partial hue-matching most human observers yield the results compatible with a circular chromatic structure with exactly four chromaticity classes (thus, four component hues). Furthermore, for each chromaticity class H, there exists exactly one chromaticity class H', with which its intersection is empty ($H \cap H' = \emptyset$). We will call H and H' (and the corresponding component hues) *opponent* to each other. Therefore, the chromaticity classes break up into two pairs of mutually opponent chromaticity classes.[45] Such circular chromatic structures will be called *circular Hering structures*. We will consider a circular Hering structure $\langle A, \gtrsim_0, \gtrsim_Y, \gtrsim_B, \gtrsim_R, \gtrsim_G \rangle$ with two pairs of opponent chromaticity classes, H_Y, H_B, and H_R, H_G; H_0 being a neutral chromaticity class.[46] (Despite that the experiments were

43 Although there is every reason to refer to this cone as the colour cone, or the light-colour cone, we prefer the term 'light-colour manifold'.

44 Obviously, the circular cone (without a vertex) and the circular cylinder in $\mathbf{R}^3$ differ in metric. If the perceptual difference between colours is informally considered as a kind of 'distance' between them, then the cone model becomes more preferable. Indeed, the perceptual difference between the two lights decreases with decreasing their intensity. However, conceptualising the perceptual difference in terms of some metric is an independent work that goes beyond the scope of this book.

45 Note that the presence of opponent pairs is a concomitant circumstance resulting from the equality of the number of chromaticity classes to four. For a circular chromatic structure with a larger number of chromaticity classes, the very concept of opponency loses its meaning.

46 The notations Y, B, R, and G are used for the mnemonic purpose only. The chromaticity classes could have readily been just numbered as above.

conducted with a finite sample of lights, for the sake of generality we assume that $A = \mathbf{C}^*_+$.) They are arranged so as to make up five unitary chromatic classes, $C(H_0), C(H_Y), C(H_B), C(H_R)$ and $C(H_G)$, and four binary chromatic classes, $C(H_Y, H_R)$, $C(H_Y, H_G)$, $C(H_B, H_R)$, and $C(H_B, H_G)$. As the ability to make a heterochromatic brightness match is usually taken for granted by the colour scientists, we will assume that $\succsim_0$ is the global brightness ordering defined on the whole A. The data obtained by far testify that in each unitary chromatic class there exists the only maximal element (full colour). As argued above, there is every reason to believe that the chromatic and brightness orders can be represented by the chromatic, f_Y, f_B, f_R, f_G, and brightness, f_{Br}, scales. Their empirical identification amounts to identifying the light-colour manifold.

8.6.3.1 Opponent chromatic scales. Light-colour map. Perceptual light-colour coordinates. Light-colour gamut

The existence of the two opponent pairs of chromaticity classes allows a rather demonstrative representation of the light-colour manifold in the three-dimensional half-space.

Consider the circular Hering structure $\langle A, \succsim_0, \succsim_Y, \succsim_B, \succsim_R, \succsim_G \rangle$ described above, and two pairs of the chromatic scales on A representing the opponent pairs of chromaticity classes (thus, component hues) – yellow and blue: f_Y and f_B, and red and green: f_R and f_G. Let us define two new functions, which will be referred to as the *opponent chromatic scales*, f_{YB} and f_{RG}, as follows:

$$\forall a \in A \qquad f_{YB}(a) = \begin{cases} f_Y(a) & \text{if } f_Y(a) \neq 0 \text{ (thus, } f_B(a) = 0), \\ -f_B(a) & \text{if } f_B(a) \neq 0 \text{ (thus, } f_Y(a) = 0), \\ 0 & \text{if } f_Y(a) = 0 \text{ and } f_B(a) = 0. \end{cases} \tag{8.46}$$

$$\forall a \in A \qquad f_{RG}(a) = \begin{cases} f_R(a) & \text{if } f_R(a) \neq 0 \text{ (thus, } f_G(a) = 0), \\ -f_G(a) & \text{if } f_G(a) \neq 0 \text{ (thus, } f_R(a) = 0), \\ 0 & \text{if } f_Y(a) = 0 \text{ and } f_B(a) = 0. \end{cases} \tag{8.47}$$

Let $f_{Br} : A \to \mathbf{R}_+$ be a brightness scale. Now, consider a map $\Psi : A \to \mathbf{R}^3$, sending $a \in A$ to $(f_{YB}(a), f_{RG}(a), f_{Br}(a))$. This map will be referred to as the *light-colour map*. We will refer to the $\mathbf{R}^3$ equipped with the coordinate axes corresponding to the functions (8.46)–(8.47) and the brightness scale, as the *perceptual light-colour space*, $\Psi(A)$ going to be called the *light-colour gamut*. Clearly, $\Psi(A)$ lies in the positive half-space $\mathbf{H}^3 = \{(x, y, z) \in \mathbf{R}^3 : z \geq 0\}$. Triplet $(f_{YB}(a), f_{RG}(a), f_{Br}(a))$ will be referred to as the *perceptual light-colour coordinates* of light a. The coordinate plane $z = 0$ will be called the *chromaticness plane*. The projection of $\Psi(A)$ onto this plane will be referred to as the *chromaticness gamut*.

Since the range of a chromatic scale is $[0, 1]$ (Definition 8.23), the first two of the three perceptual light-colour coordinates are uniquely determined, whereas the last one is determined up to a positive scaling factor. Hence, although the chromaticness gamut is unique, the light-colour gamut $\Psi(A)$ is unique up to an arbitrary contraction along the brightness axis. By agreeing to assign a unit brightness value to a particular light, one can eliminate the uncertainty associated with the brightness scale. Given such the convention, the light-colour gamut (as well as the light-colour map and the perceptual light-colour coordinates) will be called *anchored*. The anchored light-colour map, $\Psi : A \to \mathbf{R}^3$, uniquely represents the circular Hering structure on A, the light-colour manifold induced by it, and the light-colour gamut, $\Psi(A)$.

The chromaticnesses constituting the same hue (and differing only in purity) fall on a ray from the origin in the chromaticness plane.[47] Indeed, each of the coordinate axes is the loci of the chromaticness of two opponent unitary light colours. The chromaticness of a binary light colour, $(\xi_1, \xi_2, \xi_3, \xi_4)$, where ξ_i stands for the chromatic value of the ith chromatic scale, has exactly two non-zero components. Assume these are ξ_1 and ξ_2. Then the chromaticness of the binary colours having the same hue have the following form: $(k\xi_1^{\max}, k\xi_2^{\max}, 0, 0)$, where $(\xi_1^{\max}, \xi_2^{\max}, 0, 0)$ is the chromaticness of the full colour of that hue, and $0 < k \leq 1$ stands for the purity. Purity k reaches its maximum at the full colour of that hue. Therefore, the full colours make the chromaticness gamut boundary, which will be, for brevity, referred to as the *full-colour contour*.

When we assume that observers are capable to compare light colours of different hues, just as they compare colours of the same hue (with respect to their purity), then this ability can be formalised by introducing the concept of *saturation order*. Specifically, within a hue fibre (i.e., an element of $((A \backslash H_0)/\approx)/\doteq$), the saturation order coincides with the purity order. However, unlike the purity order, the saturation order is defined on the entire proper chromaticity palette $(A \backslash H_0)/\approx$. It seems natural to expect that the saturation order can be represented by a saturation scale of the same type as the purity scale, that is, an interval scale.

There being the saturation scale on the chromaticity palette makes the same normalisation of chromatic scales unwarranted. Indeed, consider, for instance, the yellow and blue chromatic scales. Assigning equal purity to the unitary yellow and blue full colours has been justified by that their purities are impossible to compare. However, when observers believe that unitary blue full colour is saturated, say, twice as much as unitary yellow full colour, it seems reasonable to normalise the chromatic scales so that the maximal chromatic blue and yellow values are in a ratio of two to one. Hence, one chromatic scale, namely, that corresponds to the most saturated full unitary colour, can be normalised to have a range of $[0, 1]$. The range of the rest will look like $[0, k]$, where $k < 1$. As a result of such normalising, the full-colour contour will get compressed.

It should be noted that the full-colour contour derived from the chromaticness of full colours may well be different from that obtained by scaling the saturation. Whether these two different subjective estimation procedures lead to consistent data needs to be established in a separate study. If both procedures result in the same full-colour contour, then the saturation can be represented by the distance from the origin in the chromaticness plane. In this case, the light-colour gamut will acquire the features of the geometrical representation of light colour that the creators of the popular HSB and HSV colour spaces aspired to (see Fig. 8.10).

The notion of opponency introduced for component hues can be generalised on all hues as follows. Two hues will be called *opponent* if they are represented in the chromaticness plane by two complementary (i.e., forming a straight line) rays from the origin. Two light colours having opponent hues will be also called *opponent*. Note that opponent hues, as well as opponent colours, may have different purities. A pair of opponent colours of equal purity will be referred to as *complementary*.

For most brightness values the maximal purity for the hue (determined by pair $\xi_1^{\max}$ and $\xi_2^{\max}$) is considerably less than 1. This is definitely the case for low brightness values. Interestingly, immediate introspection testifies that the purity of very bright lights is less pronounced than for the lights of moderate intensity. Preliminary observations allow to suggest that for a given hue, its maximal purity, as a function of brightness, first, increases, reaches the maximum ($= 1$) and decreases when

47 Note that the mixtures of a monochromatic light δ_{λ_0} with the neutral light (mapping to the origin) will, generally, not lie on the radius to a point $\Psi\left(\delta_{\lambda_0}\right)$ for the same reason that the constant hue loci in the chromaticity diagram in Fig. 8.4 are bent (i.e., because of the Abney effect).

approaching the dazzling level. Hence, the shape of the light-colour gamut $\Psi(A)$ can be roughly outlined as a body like a double cone tapering down and up.

8.6.3.2 Hering–Grassmann structures. Opponent chromatic scales and codes

When a circular Hering structure, $\langle A, \gtrsim_0, \gtrsim_Y, \gtrsim_B, \gtrsim_R, \gtrsim_G \rangle$, is defined on the light stimulus set A along with a Grassmann structure $\langle A, \sim \rangle$, we will assume that $\langle A, \gtrsim_0, \gtrsim_Y, \gtrsim_B, \gtrsim_R, \gtrsim_G \rangle$ is a chromatically ordered Grassmann structure (Definition 8.24). Furthermore, we will assume that metamerism $\sim$ is included not only in chromatic equivalence (as follows from Definition 8.24) but also in light-colour equivalence. Such circular Hering structures will be called *Hering–Grassmann structures*.[48] They will be denoted by $\langle A, \sim, \approxeq, \gtrsim_0, \gtrsim_Y, \gtrsim_B, \gtrsim_R, \gtrsim_G \rangle$ where $\approxeq$ stands for the light-colour equivalence. (Sometimes for brevity, this notation will be abbreviated to $\langle A, \sim, \approxeq \rangle$ if there is no danger of confusion.)

In the case of a Hering–Grassmann structure, the functions ((8.46)–(8.47)) are, in fact, colour codes. So is the brightness scale f_{Br}. In order to remind that, f_{YB} and f_{RG} will be referred to as the *opponent chromatic codes*, and f_{Br} the *brightness code*.

The light-colour map, $\Psi : A \to \mathbf{R}^3$, naturally induces a map $\Psi_{\approxeq} : A/\approxeq \to \mathbf{R}^3$, assigning to a class of the light-colour equivalence $\approxeq$ the same value as to its members. When $\langle A, \sim \rangle$ is three-dimensional, the light-colour equivalence $\approxeq$ coincides with the metamerism $\sim$, and the light-colour manifold $A/\approxeq$ is homeomorphic to the colour stimulus manifold $\mathcal{A}$ (see Section 8.6.4 for the proof). In this case, map $\Psi_{|\mathcal{A}} : \mathcal{A} \to \mathbf{R}^3$ fully specifies the light-colour map, Ψ.

The continuity of the chromatic and brightness scales implies that the boundary of $\mathcal{A}$ (i.e., the spectral light stimulus conical closure (5.5)) maps onto the boundary of the light-colour gamut.[49] The map $\Psi_{|\mathcal{A}}$ can be represented by three functions, $\tilde{f}_{YB}$, $\tilde{f}_{RG}$, and $\tilde{f}_{Br}$, of three variables. As the latter, one can choose any of the coordinate variables we introduced before in the colour stimulus manifold. For example, taking the p, q, and μ coordinates (Section 5.1) we get: $\tilde{f}_{YB}(p, q, \mu)$, $\tilde{f}_{RG}(p, q, \mu)$, and $\tilde{f}_{Br}(p, q, \mu)$. Alternatively, one can take the cylindrical coordinates, p, q, and θ, of a light in $\mathcal{A}$ (Section 8.2). When these new coordinate functions (e.g., $\tilde{f}_{YB}(p, q, \mu)$, $\tilde{f}_{RG}(p, q, \mu)$, and $\tilde{f}_{Br}(p, q, \mu)$) are known, we will say that the colour stimulus manifold $\mathcal{A}$ is *colour graduated*.

Given the colour-graduated colour stimulus manifold, one can get the colour-signal cone to be colour-graduated too. As f_{YB}, f_{RG}, and f_{Br} are colour codes, each of them can be decomposed into the colour-signal map Φ (with respect to metamerism $\sim$) and a colour-signal code on $\Phi(A)$. Let us denote these colour-signal codes with the same subscript as the corresponding opponent chromatic (and/or brightness) code: c_{YB}, c_{RG}, and c_{Br}. So, we have: $f_{YB}(a) = c_{YB}(\Phi(a)) = c_{YB}(z_1, z_2, z_3)$, $f_{RG}(a) = c_{RG}(\Phi(a)) = c_{RG}(z_1, z_2, z_3)$, and $f_{Br}(a) = c_{Br}(\Phi(a)) = c_{Br}(z_1, z_2, z_3)$; where z_1, z_2, and z_3 stand for the colour-signal coordinates. Colour graduation of the colour-signal cone reduces to identifying colour-signal codes c_{YB}, c_{RG} and c_{Br}. Note that these colour-signal codes define a natural homeomorphism between the colour-signal cone $\Phi(A)$ and the light-colour gamut $\Psi(A)$. This homeomorphism $\Phi(A) \to \Psi(A)$ will be referred to as the *colour-graduation homeomorphism*.

Thus, identification of the light-colour manifold amounts to the colour graduation of either the colour stimulus manifold or the colour-signal cone. Though when the dimensionality of the underlying Grassmann structure on A is more than 3 (thus the colour stimulus manifold is not defined),

48 The relationship between the metamerism and light-colour equivalence in a Hering–Grassmann structure will be discussed in more detail later in Section 8.6.4.

49 This is in line with the aforementioned fact that the full colours belong to the spectral light stimulus conical closure (Remark 8.14). In other words, each chromatic scale achieves its maximum at some point in the spectral light stimulus conical closure.

the identification of the light-colour manifold can be conducted only in the form of the colour graduation of the colour-signal cone. In this case, the colour-signal codes c_{YB}, c_{RG}, and c_{Br} will be functions of more than three variables.

The light-colour gamut can be sliced into equi-bright level sets. Each of these will be referred to as the *iso-bright chromaticness gamut* (corresponding to a particular brightness). The identification of a number of the iso-bright chromaticness gamuts will, practically, suffice for identifying the light-colour gamut. The boundary of a iso-bright chromaticness gamut is made by the chromaticness of the (equi-bright) monochromatic lights completed by the chromaticness of the mixtures of the monochromatic lights making the endpoints of the effective visible spectrum (of the same brightness). There is every reason to call these two complementary fragments of the iso-bright chromaticness gamut boundary the spectral curve and the purple interval, despite the fact that the latter is not necessarily straight. Figure 8.24 gives an idea of the shape of the spectral curve of one of the iso-bright chromaticness gamuts when using Cartesian coordinate axes to represent the opponent chromatic codes. To be more exact, it depicts a fragment made up by the (iso-bright) monochromatic lights. Sometimes, for brevity, we will call the iso-bright chromaticness gamut boundary the *iso-bright full-colour contour*.

The light-colour gamut can be illustrated with a family of the iso-bright full-colour contours. Since the light-colour gamut has a form of the double cone tapering down and up, this family is likely to resemble a nested set of contours. (At least this will be so provided that the difference in brightness between the contours is large enough.) Its exact configuration should be established in the experiment. It must be said, however, that there is an uncertainty with regard to the shape of $\Psi(A)$ at the origin (in the perceptual light-colour space), which cannot be resolved in experiment. For the light intensity in the scotopic range we lose the very subject matter – the colour perception.[50]. Although, when considering the colour-signal cone, we assume the cone fundamentals to be defined for any light, this is just a convenient abstraction that does not lead to any serious inconsistency. However, it would be hardly justified to make any assumption concerning the chromaticness of a light for which the colour vision does not operate. Here we have a situation where the colorimetric paradigm ceases to work, showing its principal limitation.

8.6.3.3 Grassmann-opponent-chromatic-code hypothesis

In colour science, there is a widespread belief that there are some mechanisms in the human visual system whose functioning can be described by means of the opponent chromatic codes. These mechanisms are believed to be responsible for producing the four component hues. Such a view goes back to the doctrine of Hering (1920/1964) who believed that one such mechanism could produce a pair of opponent component hues thus representing the two opponent chromaticity classes. Elaborated later by Hurvich & Jameson (1957), who assumed further these hypothetical mechanisms to be linear, this idea has dominated colour science for many decades.

According to Hurvich and Jameson there exist two (yellow-blue and red-green) independent mechanisms in the human visual system which can be conceptualised as Grassmann colour codes (see Proposition 2.4). More specifically, the yellow-blue mechanism, implementing the opponent chromatic code, f_{YB}, (8.46), is supposed to be formally described as a Grassmann colour-signal code $c_{YB}(S,M,L) = g_{YB}(s_{YB}S + m_{YB}M + l_{YB}L)$ where, as above (Chapter 4), S, M, and L stand for the cone fundamentals with the peak sensitivity in the short, middle and long wavelength regions of the visible spectrum respectively; s_{YB}, m_{YB}, and l_{YB} stand for real

50 A similar problem also arises when considering the upper end of the light-colour gamut. The existence of the upper absolute light threshold prevents from studying the colour perception for lights exceeding this threshold.

constants; g_{YB} is the real increasing function. It is assumed that the yellow chromatic value is assigned when $g_{YB}(s_{YB}S + m_{YB}M + l_{YB}L) > 0$, and the blue when $g_{YB}(s_{YB}S + m_{YB}M + l_{YB}L) < 0$. When $g_{YB}(s_{YB}S + m_{YB}M + l_{YB}L) = 0$ neither yellow nor blue chromatic value is assigned. Respectively, the positive part of the red-green Grassmann colour-signal code, $c_{RG}(S, M, L) = g_{RG}(s_{RG}S + m_{RG}M + l_{RG}L)$, encodes the red chromatic value, and the negative part of the red-green Grassmann colour-signal code encodes the green chromatic value, $g_{RG}(s_{RG}S + m_{RG}M + l_{RG}L) = 0$ encoding neither red, nor green. Being the opponent chromatic codes according to our terminology, such hypothetical mechanisms are usually referred to in the colour vision literature as the yellow-blue and red-green *opponent chromatic channels* (Hurvich & Jameson, 1957).

It is worth pointing out that Hurvich and Jameson's hypothesis can be broken down to the following two separate assumptions. First, the opponent chromatic codes, f_{YB} and f_{RG} ((8.46)–(8.47)), are assumed to be Grassmann opponent chromatic codes; and second, some neural mechanisms are assumed to exist in the human visual system, which implement these Grassmann opponent chromatic codes. As argued below, there is strong evidence against the first assumption (it will be referred to as the *Grassmann-opponent-chromatic-code hypothesis*). If this hypothesis is incorrect then the search for the neural mechanisms implementing the Grassmann opponent chromatic codes (8.46)–(8.47) would be fruitless.[51]

8.6.3.4 Empirical status of the Grassmann-opponent-chromatic-code hypothesis

The Grassmann-opponent-chromatic-code hypothesis predicts that level sets of the colour-signal codes c_{YB} and c_{RG} in the colour-signal space are planar. Generally, a level set of the colour-signal code, c_{YB} (and/or c_{RG}), is a surface where the colour-signal code takes on a constant chromatic value. We will call it the *isochromatic surface*. The isochromatic surface of Grassmann opponent colour-signal codes are planes. Such a prediction is easy to test.

Figure 8.20 presents the experimental curves in the equiluminant plane which contain equal amount one of four component hues (referred to as the *isochromes*) (Logvinenko & Lu, 2008). The

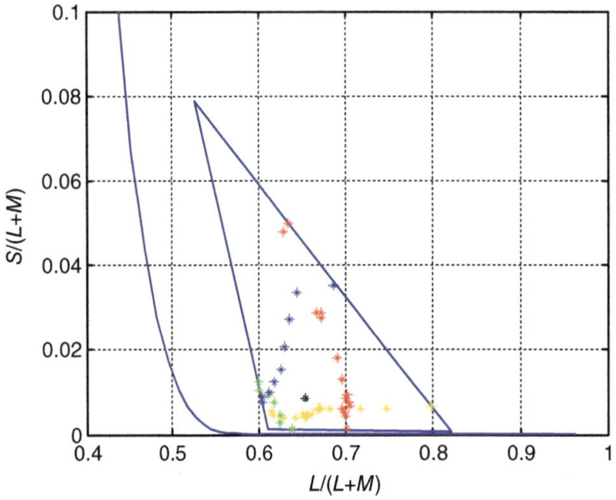

Figure 8.20 Iso-chromes presented in the MacLeod–Boynton chromaticity diagram. The colour of the symbols corresponds to the component hue (R, G, B, and Y) for which the iso-chromes were measured. The black asterisk in the centre stands for the neutral point.

51 Some physiological mechanisms relevant to human colour vision are discussed in more detail in Chapter 13.

isochromes can be interpreted as the intersection of the isochromatic surfaces with the equiluminant plane. If c_{YB} and c_{RG} are the Grassmann colour-signal codes, then the isochromes should be linear and parallel in pairs. However, as one can see, all the isochromes were found to be nonlinear. Moreover, parallelism was found between neither R- and G-isochromes, nor Y- and B-isochromes. This seriously undermines the Grassmann-opponent-chromatic-code hypothesis.

Of particular interest are zero-level isochromatic surfaces: $g_{YB}(s_{YB}S + m_{YB}M + l_{YB}L) = 0$ and $g_{RG}(s_{RG}S + m_{RG}M + l_{RG}L) = 0$. The former is the loci of the unitary (unique) red and unitary (unique) green light colours, the latter the unitary (unique) yellow and blue light colours. Indeed, for instance, the first equation determines the plane in the colour-signal space the points of which have neither yellow nor blue chromatic values. That is, they can be only assigned either red or green if any. Hence, this equation defines the constant hue surfaces for the unitary (unique) red and unitary (unique) green hues. The Grassmann-opponent-chromatic-code hypothesis predicts the planarity of these constant hue surfaces. It follows that in the chromaticity diagram the unique red and green hue should lie on colinear radii. However, as seen in Fig. 8.21, this is not the case. Specifically, the unique red and green are far from colinearity. This testifies against the linearity of the hypothetical YB-mechanism.

Strong evidence for the non-linearity of the YB-mechanism was also found in some earlier studies (Cicerone et al., 1975; Larimer et al., 1975b; Ejima & Takahashi, 1985). Specifically, Krantz (1975b) and Suppes et al. (1989) has developed a formal theory of Grassmann opponent chromatic codes based on the so-called *hue-cancellation technique* (Hurvich & Jameson, 1957). He and his colleagues tested this theory in experiment and found evidence for that the red-green opponent channel can be described as a Grassmann colour code (Larimer et al., 1975a). That was later confirmed by others Werner & Wooten (1979). However, strong evidence was found against that the yellow-blue channel was a Grassmann colour code (Larimer et al., 1975b).

In our opinion the hue cancellation technique is based on a quite arbitrary assumption that the ith chromatic value of a light can be measured in terms of the physical amount of the monochromatic light that (i) appears of the unitary hue opponent to the ith component hue, and (ii) being

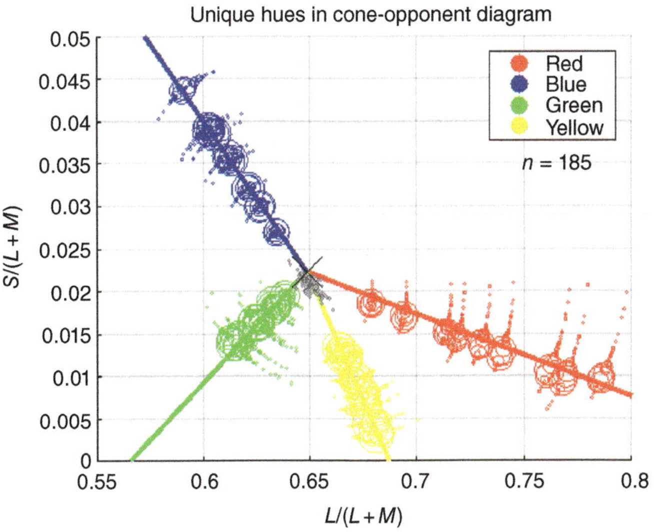

Figure 8.21 Unique hue settings in the MacLeod–Boynton chromaticity diagram made by 185 observers. Source: Wuerger & Xiao (2014).

mixed with the light in question, cancels the ith component hue (i.e., makes the ith chromatic value of the mixture equal zero). In particular, for the Y-chromatic value of a light a to be measured, one needs to take a monochromatic light that looks unitary (unique) blue, and to find what amount of this light is needed for its mixture with light a to contain no Y-component hue. However, there is no evidence substantiating this assumption. Any test of it requires the direct empirical measurement of the chromatic values of lights. However, on this score, there are very scanty data for today.

The chromatic scales have not been properly identified as yet.[52] The only relatively pertinent data come from the experiments conducted by the method of hue scaling mentioned above (Gordon & Abramov, 1988; Gordon et al., 1994; Abramov & Gordon, 1994). In one of such experiments, the observers were presented with monochromatic lights, and asked to 'state the percentages of their hue sensations using any combination of the four unique terms for a total of 100%. They also describe apparent saturation as the percentage of their entire sensation (chromatic and achromatic) that was chromatic. ... To combine hue and saturation, the hue values can be rescaled by their associated saturation values so that the sum of the hue values for each stimulus equals the saturation' (Abramov & Gordon, 1994, p. 460). Some of their results are depicted in Figs 8.22 and 8.23.

Unfortunately, the instruction to produce such percentages that make up in total 100%, seems not quite clear. Really, consider a light whose colour is binary, say, orange. If the observer's response is 70% yellow and 30% red, then one can ask 70% of what? It cannot be 70% of the maximum yellow

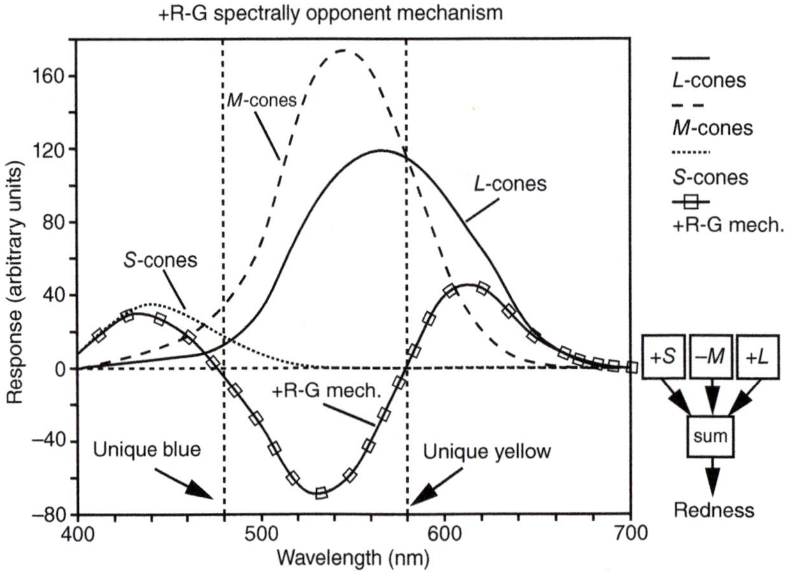

Figure 8.22 The square markers represent the rescaled estimate of the amount of red (above the horizontal axis) and green (below the horizontal axis) component hues in monochromatic lights. The direct estimates were rescaled by their associated saturation values so that the sum of the hue-component values for each light equals the saturation. The line through the markers is a linear combination of the responses of the S-, M-, and L-cones. Source: Based on Abramov and Gordon (1994).

52 In order to do this, one has to start from ascertaining whether the chromatic orders comply with the axioms, necessary for the existence of corresponding (order and/or interval) scales (Krantz et al., 1971). Then one can turn to constructing an empirical procedure for identifying chromatic scales.

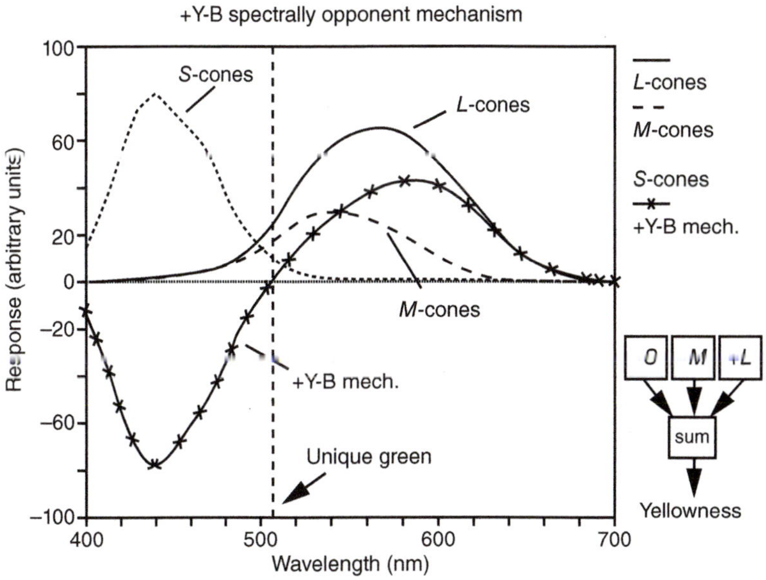

Figure 8.23 The markers represent the rescaled estimate of the amount of yellow (above the horizontal axis) and blue (below the horizontal axis) component hues in monochromatic lights. The direct estimates were rescaled by their associated saturation values so that the sum of the hue-component values for each light equals the saturation. The line through the markers is a linear combination of the responses of the *S*-, *M*-, and *L*-cones. Source: Based on Abramov and Gordon (1994).

chromatic value (i.e., the yellow chromatic value of the full colour of this hue), since in this case this would imposes a undesirable constraint on the observer's estimation of the red chromatic value (it must be 30% whereas it can be any percentage provided that it is with respect to the maximum red chromatic value). If 70/30 is an estimate of the ratio of the yellow and red chromatic values for this particular light, then one has to admit that the yellow and red chromatic values are estimated up to an unknown multiplicative constant. Specifically, the information of the purity of the light colour in question gets lost under this method. Even if we accept that observers are capable of estimating the ratio of the chromatic values for different component hues, how could the ambiguity of the estimated ratio be possibly resolved? In other words, how could the purity be possibly restored? The way Abramov & Gordon did it (see the quote above), i.e., to use the saturation for this purpose, seems rather controversial. So, strictly speaking, the curves in Figs 8.22 and 8.23 can hardly be interpreted as the responses of the hypothetical YB- (Fig. 8.22) and RG-mechanisms (Fig. 8.23). Still, in the absence of better these curves can give rough idea of what the restriction of the opponent chromatic codes f_{YB} and f_{RG} to the monochromatic lights might look like.

Notably, Abramov & Gordon (1994) considered the fact that the curves in Figs 8.22 and 8.23 could be fitted by a linear combination of the cone fundamentals as evidence supporting Hurvich & Jameson's model of the hypothetical YB- and RG-mechanisms (in our terminology, in favour of the Grassmann-opponent-chromatic-code hypothesis). At the first glance, the graphs in these figures are rather impressive. Really, the data obtained by the hue scaling technique prove to be in line with the data obtained by the rather different psychophysical (colour matching) technique. Such a remarkable convergence of the data from so different experiments, of course, deserves attention. Yet, it does not prove anything as far as the Grassmann-chromatic-opponent-code hypothesis is concerned.

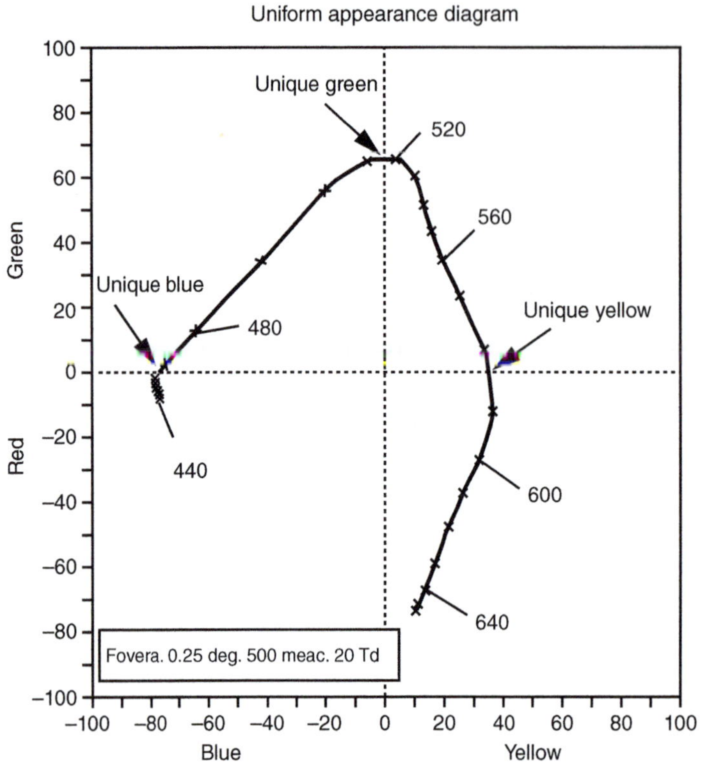

Figure 8.24 Hue-component values (from Figs 8.22 and 8.23) have been replotted on a two-dimensional diagram with orthogonal axes Y-B and R-G. The location of each monochromatic light defines its hue, and the distance from the origin (city-block metric) represents saturation. Source: Based on Abramov and Gordon (1994).

In the end note that if the curves in Figs 8.22 and 8.23 are interpreted as a rough approximation to the restriction of the opponent chromatic codes f_{YB} and f_{RG} to the monochromatic lights, then the following graph (Fig. 8.24) would be an outline of the fragment of the boundary of the equibright cross-section of the light-colour gamut, that is, one of the chromaticness gamuts (see Section 8.6.3).

8.6.4 Light-colour manifold vs. colour stimulus manifold

Given a light stimulus cone A, the colour stimulus manifold and the light-colour manifold can be defined independently of each other. Indeed, a light-colour equivalence, on which the light-colour manifold is based, can be defined on A irrespectively to whether a metamerism is defined on A or not.[53] These two concepts (i.e., the light-colour equivalence and the metamerism)

53 Moreover, the whole approach presented in this chapter is entirely independent of (i.e., not based on) the metamerism. This approach can be traced down to Goethe and Hering who tried to develop a psychological theory of colour in contrast to the physicalist approaches of Newton and Helmholtz. One has to admit that the tradition in colour science descended from Helmholtz and Newton has developed into modern colorimetry whereas the psychological approach by Goethe and Hering proves to remain a great deal less scientific. Hopefully, we have demonstrated in this chapter that the potential of the alternative (phenomenological) approach has not been exhausted as yet.

are not necessarily related to one another, neither theoretically nor operationally. However, when both the light-colour equivalence and the metamerism are defined on A, there might be a connection between these notions depending on how they are defined, theoretically and/or operationally.

The operational basis of the light-colour equivalence is the component-hue scaling (i.e., the hue-scaling based on the chromatic ordering in the chromaticity class). If being equi-bright two light patches contain the same amount of each component hue they are light-colour equivalent. If the metameric matching hypothesis (Chapter 3) is accepted, that is, if colour matching is assumed to be an empirical equivalence relation, and it is conceptualised as the metamerism, then it seems sensible to hypothesise that the metamerism implies the light-colour equivalence. Really, when an empirical criterion of a colour match is no border between the two halves in the bipartite field (as in Fig. I.1), one can hardly expect that, seeing no border, an observer will see a difference in redness or blueishness (or in any other component hue) between the halves. Whether the converse hypothesis holds true (i.e., the light-colour equivalence implies the metamerism) is to be established in experiment. It is quite possible, for instance, that the two halves of the bipartite field, appearing to contain equal amount of each component hue and to be equi-bright when separated apart, will not merge into a homogeneous patch without any border after joining them (i.e., after removing the gap between them). They will merge only if there is no other perceptual attribute except for component hues and brightness that affect colour matching. If this is the case then light-colour equivalent lights will match each other. In this case, one can consider colour matching as an empirical operationalisation of the light-colour equivalence.[54] Since in the preceding chapters (in line with the classical colour literature), colour matching has been also taken as an empirical operationalisation of the metamerism, the light-colour equivalence and the metamerism will be just one and the same thing.

The notion of chromatically ordered Grassmann structure (Definition 8.24) has been put forward keeping in mind simultaneous conceptualising in terms of both light-colour equivalence and metamerism. However, for that goal to be achieved, the definition has got to be strengthened[55] by requiring that the metamerism is included in the light-colour equivalence (i.e., the metamerism implies the light-colour equivalence). As shown in Section 8.5.5, if the chromatic and achromatic (brightness) orderings underlying the chromatically ordered Grassmann structure are continuous, the chromatic and achromatic (brightness) scales are colour codes. Since the continuity in question seems to be a safe assumption, we will restrict our further consideration to only continuous chromatic and achromatic orderings.

Consider a Hering–Grassmann structure $\langle A, \sim, \approxeq \rangle$ in which metamerism $\sim$ is defined as the fundamental metamerism generated by $n \geq 3$ fundamental mechanisms. When the number of fundamental colour mechanisms $n = 3$, then the light-colour equivalence $\approxeq$ coincides with the fundamental metamerism. In this case, the light-colour manifold $A/\approxeq$ is homeomorphic to the colour stimulus manifold $\mathcal{A}$. Moreover, there is a natural homeomorphism of $A/\approxeq$ onto $\mathcal{A}$. (However, this homomorphism does not necessarily put in correspondence the hue fibres in

54 Sometimes the simultaneous equality in hue, purity and brightness has been explicitly chosen as a criterion of colour match instead of the no-border criterion. However, the light-colour equivalence cannot be considered as a conceptualisation of colour matching. It is a theoretical notion arising within the theoretical framework of the chromatically ordered structures. Its empirical basis is the component-hue scaling rather than colour matching. Thus, under certain conditions, colour matching can be thought of as an operationalisation of the light-colour equivalence, but the latter cannot serve conceptualising the former. The light-colour equivalence has been defined in a different conceptual framework.

55 Recall that in a chromatically ordered Grassmann structure $\langle A, \sim, \succsim_1, \dots, \succsim_m \rangle$ metamerism $\sim$ was assumed to be included in the intersection of only the symmetrical parts of all $\succsim_i$ but not the achromatic order.

$A/\approxeq$ and $\mathcal{A}$ since they are based on different definitions of hue. As mentioned above, this issue requires separate experimental testing.) Indeed, given the colour-signal map Φ let us denote $\overline{\Phi}$ the map $\Phi_{|\mathcal{A}}$. Let us also denote $\Phi_{\approxeq}$ the map $A/\approxeq \to \mathbf{R}^3$ which is induced on $A/\approxeq$ by the colour-signal map Φ. (That is, the map $\Phi_{\approxeq}$ assigns to a class of the light-colour equivalence the same value as the map Φ assigns to each member of this class.) Then the map $\overline{\Phi}^{-1} \circ \Phi_{\approxeq}$ will be this natural homeomorphism between $A/\approxeq$ and $\mathcal{A}$. Hence, the colour stimulus manifold $\mathcal{A}$ and the light-colour manifold $A/\approxeq$ are the same in the case of trichromatic colour vision, that is, vision based on the three fundamental colour mechanisms.

If the number of fundamental colour mechanisms $n > 3$, the fundamental metamerism, $\sim$, does not coincide with the light-colour equivalence $\approxeq$, being only included in it (i.e., $a \sim b$ implies $a \approxeq b$). Although the notion of colour stimulus manifold has been defined in Section 5.1 only for three-dimensional Grassmann structures, the light stimulus subset $\mathcal{A}$ given by (5.4) remains being a three-dimensional manifold with boundary, homeomorphic to the half-space $\mathbf{H}^3$, for $n > 3$ as well.[56] Its image in the colour-signal space, $\Phi(\mathcal{A})$, being a n-dimensional subset in $\Phi(A)$, is also a three-dimensional manifold with boundary homeomorphic to the half-space $\mathbf{H}^3$. Since the light-colour manifold $A/\approxeq$ remains to be a three-dimensional manifold with boundary irrespective of the dimensionality of the Grassmann structure $\langle A, \sim\rangle$, $A/\approxeq$ is homeomorphic to $\Phi(\mathcal{A})$. Therefore, the latter (i.e., $\Phi(\mathcal{A})$) can be thought of as a homeomorphic representation of the light-colour manifold $A/\approxeq$ in the colour-signal cone $\Phi(A)$.

The assumption of the presence of more than three fundamental colour mechanisms is justified under at least two particular conditions. Firstly, it is pertinent when it comes to the human twilight vision. Indeed, in the mesopic range of light intensity both rods and cones are engaged. There is abundant evidence for the rod-cone interactions (Buck, 2004). Therefore, because of the rod involvement, the mesopic colour vision is, strictly speaking, tetrachromatic rather than trichromatic.[57] However, as we know from everyday life, we do not gain any new colours when entering into the mesopic range of light intensity. Although the colour appearance at dusk is a peculiar one (Pokorny et al., 2006), there seems to be no additional hues, not to mention a new colour dimension. The light-colour manifold seems to remain three-dimensional for the mesopic vision. Hence, Hering–Grassmann structure $\left\langle A, \sim, \approxeq, \gtrsim_0, \gtrsim_Y, \gtrsim_B, \gtrsim_R, \gtrsim_G \right\rangle$ seems to be a plausible model for the human twilight vision provided that $\langle A, \sim\rangle$ is the tetrachromatic Grassmann structure induced by the cones and rods, and the corresponding light-colour manifold is three-dimensional as described above.

Secondly, women whose sons exhibit deuteranomalous or protanomalous trichromatic vision also have the opportunity for tetrachromacy (Jordan et al., 2010). These women might have the fourth cone with the peak spectral sensitivity between those of the standard M- and L-cones. Although there are claims that such women show better chromatic discrimination in comparison to trichromats (K. Jameson et al., 2001), there is no convincing evidence in favour of existence of additional hues in their colour palette (Jordan et al., 2010; D. I. A. MacLeod, 2010). This is hardly surprising if the only difference between these women and normal trichromats is caused by an additional cone type (due to genetic mutations) without changing the further neural circuitry in the retina and brain. The geometry of the light-colour manifold, in general, and its dimensionality, in particular, is determined by the chromatic colour codes (which are believed to be implemented by some post-receptoral neural mechanisms) rather than the fundamental colour mechanisms (based on the photoreceptors) underlying the Grassmann structure. Moreover, it is the way the

56 All the technical assumptions of Theorem 5.1 concerning the fundamental mechanisms are assumed to be met.
57 Here tetrachromacy means that the dimensionality of the Grassmann structure induced by the fundamental colour mechanisms equals four.

chromatic colour codes are interrelated to each other rather than their number that determines the dimensionality of the light-colour manifold.[58]

The difference between the dimensionality of the light-colour manifold and that of the fundamental colour-signal space (in the case of the possible involvement of the fourth type of photoreceptor) leads to the following problem. Is colour match based on the equality of the colour signals, or the equality of the chromatic-colour-code outputs? In other words, can colour match be an operationalisation of the notion of fundamental metamerism, or that of light-colour equivalence?

The data available in the colour vision literature pertinent to this issue are inconclusive and contradictory. On the one hand, the rod participation in colour matching is a well established fact. For instance, colour matching in mesopic vision seems to be tetrachromatic (Bongard et al., 1958; Trezona, 1973, 1976). Specifically, for a complete colour match for all the monochromatic lights one needs four rather than three monochromatic primaries in mesopic vision (for a review see Brindley, 1960; Wyszecki & Stiles, 1982). This fact is in line with the hypothesis that colour match is based on the fundamental metamerism rather than the light-colour equivalence. On the other, the colour matches made by the female carriers of anomalous trichromacy mentioned above can be, by and large, described as trichromatic.[59] Certainly, this testifies in favour of the light-colour equivalence as the basis of colour matching. One might argue that if the colour match is underlain by the chromatic colour codes, which, as shown above, are essentially non-linear, then how can colour matching comply with Grassmann's laws? However, as mentioned before, the non-linearity of colour codes cannot violate either metamerism or Grassmann's laws, provided that the colour codes form a bijective transformation. Still, there are other questions that are much more difficult to answer.

For instance, for any point $z^0 = \left(z_1^0, z_2^0, z_3^0, z_4^0\right)$ in the three-dimensional manifold $\Phi(\mathcal{A})$ representing the light-colour manifold in the four-dimensional colour-signal space, there is the whole set of points in the colour-signal cone $\Phi(A)$ light-colour equivalent to z^0. This set consists of all the points $z = \left(z_1, z_2, z_3, z_4\right)$ satisfying the following equations:

$$\begin{aligned} c_{YB}\left(z_1, z_2, z_3, z_4\right) &= c_{YB}\left(z_1^0, z_2^0, z_3^0, z_4^0\right), \\ c_{RG}\left(z_1, z_2, z_3, z_4\right) &= c_{RG}\left(z_1^0, z_2^0, z_3^0, z_4^0\right), \\ c_{Br}\left(z_1, z_2, z_3, z_4\right) &= c_{Br}\left(z_1^0, z_2^0, z_3^0, z_4^0\right), \end{aligned} \tag{8.48}$$

where c_{YB}, c_{RG}, and c_{Br} stand for the colour-signal codes representing the opponent chromatic, f_{YB}, f_{RG}, and brightness, f_{Br}, codes (see Section 8.6.3). If c_{YB}, c_{RG}, and c_{Br} were linear (as many believe) then the solution for (8.48) would be the interval of a straight line contained in $\Phi(A)$. For a broad range of non-linear colour-signal codes (and for most of points z^0) Eq. (8.48) determine a curvilinear interval in $\Phi(A)$.

58 Recall that in a circular chromatic Grassmann structure, whose chromatic palette is two-dimensional, the number of chromatic codes can be arbitrary.

59 'Some 12% of women are carriers of the mild, X-linked forms of color vision deficiencies called 'anomalous trichromacy.' Owing to random X chromosome inactivation, their retinae must contain four classes of cone rather than the normal three; and it has previously been speculated that these female carriers might be tetrachromatic, capable of discriminating spectral stimuli that are indistinguishable to the normal trichromat. However, the existing evidence is sparse and inconclusive. ... Our results suggest that most carriers of color anomaly do not exhibit four-dimensional color vision, and so we believe that anomalous trichromacy is unlikely to be maintained by an advantage to the carriers in discriminating colors.' (Jordan et al., 2010)

In the colour vision literature, this controversy arising from the possible involvement of the fourth type of cones is usually discussed as the contradiction between the dimensionality of 'the sensory colour space' and 'the perceptual (or phenomenological) colour space' (for a review see MacLeod, 2010). We find such terminology rather misleading. First of all, it should be noted that there has been given no definition of the perceptual colour space in the works in question. What is usually meant by this term is nothing more than the three attributes in terms of which the colour appearance of light is commonly described (i.e., hue, purity, and brightness). Furthermore, these attributes are quite often erroneously understood as the dimensions of the hypothetical perceptual colour space. In fact, the intention behind such a terminology is to describe the set of light colours in the geometrical terms. That is, the term 'colour space' is not critically (and sometimes, metaphorically) used to refer to both the space in which the colours are represented as points, and the whole set of the points representing all the colours. In this chapter, we use the notion of light-colour manifold to refer to the latter, distinguishing between the light colour manifold and the ambient space into which the light colour manifold can be embedded. As shown above, the dimensionality of the ambient space can be larger than that of the light-colour manifold, that leads to the controversy in question.

If the colour match is an operationalisation of the fundamental metamerism, that is, if the inequality of the colour signals leads to colour mismatching, then how would an observer perceive the lights corresponding to this interval? On the one hand, they all are light-colour equivalent to each other, that is, they have the same chromaticness and brightness. On the other, they must appear differently because they mismatch each other.

In this context, of interest might be an observation made when studying detection of chromatic sinusoidal gratings (Logvinenko, 1996b). When the contrast of a chromatic unipolar grating is at threshold, one can observe a pattern as different from the background, however, one cannot recognise the colour of this pattern. The pattern is perceived correctly (i.e., the frequency of the grating is readily recognised) but its colour cannot be discerned (Logvinenko, 1996b). Hence, detecting the difference between two light stimuli is not the same as seeing a colour difference between the lights. In other words, colour detection is not always accompanied by colour recognition. Such the dissociation between detection and recognition has been revealed by using achromatic stimuli too.

Note, however, that colour detection without colour recognition is a paradox (a conceptual paradox) only within the colorimetric paradigm. It is implicitly assumed by this paradigm that seeing the border in the bipartite field implies seeing two different colours. However, colour perception in general, and colour recognition in particular, might belong to a higher (thus, more complicated) level of vision than simple detection. Therefore, one needs a more sophisticated paradigm, that distinguishes between detection (mismatching) and colour recognition, to incorporate this conceptual paradox. We will return to this and related issues in more detail in Chapters 11 and 12.

9

Typology of Light-Colour Perception. Inter-Individual Differences

The problem of the inter-individual differences in colour vision is without a doubt one of the most difficult. In this area, to put the question correctly is perhaps even more important than to answer it. To begin with, one should decide when the colour vision of two observers can be considered same and when different. The decision depends on in what form their colour vision is conceptualised. One can use the Grassmann structures, the chromatically ordered structures, and both.

If we confine ourselves to only Grassmann structures, which is characteristic of colorimetry, then it goes without saying, that the colour vision of two observers have to be recognised as identical only when described by the same Grassmann structure (or to be more exact, by the isomorphic Grassmann structures). When the Grassmann structures, in terms of which the colour vision of the observers is conceptualised, are different, the difference between their colour vision can be characterised, first of all, by the Grassmann-structure dimensionality. A difference of this kind is implied when, for example, dichromatic vision is contrasted with trichromatic vision. When the Grassmann structures are of equal dimension, the difference in colour vision can formally be characterised by the magnitude of the metamer mismatching.

One has to admit that, within this approach, it is almost impossible to find, even among trichromats, people with the same colour vision. Really, as shown above, the unavoidable difference in the photopigment peak sensitivity, as well as that in the lens and macular pigment spectral transmittance, will result in the metamer mismatching, thus, the difference in colour vision. However, as shown above (Chapter 8), trichromats with different colour vision (even potential tetrachromats) might well have identical colour perception. For this reason, and because the inter-individual differences between human observers in terms of metamer mismatching have been already discussed in Chapter 7 (Section 7.3.6), here we will focus on the differences between human observers in colour perception.[1]

From a formal point of view, those observers whose colour perception can be conceptualised in terms of one and the same chromatic structure, can be said to have identical colour perception. More specifically, let $\mathcal{A} = \langle A, \succsim_1, \dots, \succsim_m \rangle$ and $\mathcal{A}' = \langle A, \succsim'_1, \dots, \succsim'_m \rangle$ be the chromatically ordered structures (on the light stimulus set) representing two observers. We will say that they have the same colour perception if the identity map from A to A is an isomorphism between $\mathcal{A}$ and $\mathcal{A}'$ as the relational structures.

The observers with identical colour perception, therefore, have the same chromaticity classes (i.e., these coincide as the subsets of the light stimulus set). The corresponding

1 Especially since there are literature reviews on the inter-individual differences in colour vision from the traditional standpoint (e.g., Webster, 2015).

Foundations of Colour Science: From Colorimetry to Perception, First Edition. Alexander D. Logvinenko and Vladimir L. Levin.
© 2023 John Wiley & Sons Ltd. Published 2023 by John Wiley & Sons Ltd.

chromaticity classes of the two observers are identically chromatically ordered, i.e., the observers have the same number of chromatic orderings of the same kind. In particular, for Hering observers with identical colour perception (i.e., for isomorphic circular Hering structures $\mathcal{A}$ and $\mathcal{A}'$), their light-colour equivalencies are the same, as are their light-colour manifolds. The corresponding chromatic scales for these observers should be of the same type. If the chromatic scales are the interval ones (Remark 8.12), the chromaticness of every light should be the same for both observers. Furthermore, after proper anchoring the light-colour gamuts (i.e., assigning some light the same brightness value for both observers), the same lights will get not only the same chromaticness but also brightness value for such observers.

Strictly speaking, the colour perception of all the other observers should be considered as different. However, having taken this point of view, i.e., identifying colour perception with the underlying chromatically ordered structure, we have to admit that after putting on sun-protecting glasses our colour perception will change. Indeed, because of Bezold–Brücke shift (see Section 8.2), the glasses will to some extent change the chromaticness of almost every light. However, our intuition, supported by the language tradition, tells us that admitting this would be excessively scholastic pedantry. Therefore, our definition of 'sameness' for colour perception should be amended so as to encompass the colour perception of the same person under different viewing conditions. At the same time, it should not be unnecessarily expanded either. For example, deciding that all observers with Hering's circular structure have 'the same' colour perception, we thereby deprive ourselves of the subject matter. Indeed, there will be no individual differences between the vast majority of trichromatic observers, since their perception is the same in accordance with this definition.

It seems to us that a reasonable compromise might be to restrict our consideration to chromatically ordered structures of the same kind (not necessarily isomorphic), and analyse the differences in colour perception between such observers. In other words, we limit ourselves to studying the differences within a certain group of observers. Of course, such a group must be sufficiently representative. A natural choice are, in our opinion, all Hering observers (i.e., observers, whose colour perception can be conceptualised as a circular Hering structure). Indeed, on the one hand, a circular Hering structure can be viewed as a prototype of the trichromatic colour perception; on the other, as shown below, there are considerable differences between such observers. Specifically, first, the light-colour palettes[2] might be different; and second, the same lights might have different chromaticness and/or brightness. Still, it makes sense to consider the colour perception of such people, qualitatively, of the same kind. Furthermore, it is only within a population with colour perception of the same kind that is worth talking about inter-individual differences (i.e., variations from one individual to another within the same prototype).[3]

Let us look into the inter-individual differences between Hering observers at length. Amongst such observers there might be some whose light-colour gamuts coincide. Note that such a statement implies that we know which chromaticity class of one observer corresponds to which chromaticity class of another observer. However, what does it mean that a chromaticity class established for one observer is the same as some chromaticity class established for another observer?[4] The answer is self-evident when the chromaticity classes of these observers fully coincide. In this case, irrespective of how the lights are subjectively experienced by each observer, one can assume that they

2 As pointed out in the Outline (Chapter 1), the term 'light-colour palette' is loosely used to refer to all the light colour. Its conceptual counterpart for Hering structures is the 'light-colour gamut'.

3 Indeed, the differences between people can be so substantial that their colour perceptions are simply impossible to compare.

4 More generally, what does it mean to have the same colour palette for two human observers?

represent one and the same hue.[5] However, the complete coincidence of the chromaticity classes of human observers is quite a rare event. When the chromaticity classes do not coincide but overlap (considerable overlapping is what comes from most of the experiments on partial hue-matching), there is usually no problem with establishing a correspondence between the chromaticity classes of different observers. In fact, one needs to find a counterpart for just one chromaticity class. The rest will follow from the circular order.

Numerous experiments on unique hues showed that unique red is always found to be non-spectral. Therefore, one might expect that exactly one of the four full colours for Hering observers would be produced by non-spectral light, while the other three full colours by monochromatic lights. This peculiarity gives an opportunity to identify the only classes whose full colours are non-spectral. It is worth mentioning again that how the observers experience the lights in these classes, has no relevance in the present context.

Also, as has been confirmed in numerous experiments, the position of unitary (unique) yellow in the visible spectrum is subject to quite small variations between normal trichromatic observers. Therefore, for both observers, one can expect a chromatic class, $C(H_i)$, the maximum element of which is located in the narrow interval of the spectral curve (corresponding to approximately 570–580 nm). The colour of the lights in $C(H_i)$ is usually named as 'yellow', and this chromatic class is hard to confuse with any other. So, identifying the chromaticity classes with the maximum element positioned at approximately 570–580 nm, one can put in correspondence all the four chromaticity classes of two different observers. For the sake of convenience, we will keep referring to these chromaticity classes as Y, R, B, and G, meaning that, first, the Y-chromaticity class has its maximal element around 570–580 nm; and second, they are circularly ordered as a 4-hour clock face (e.g., R precedes Y and follows B; G precedes B and follows Y).

Consider two observers represented by circular Hering structures on the light stimulus cone A: $\langle A, H_0, H_Y, H_R, H_B, H_G \rangle$ and $\langle A, H'_0, H'_Y, H'_R, H'_B, H'_G \rangle$, the corresponding chromaticity classes being denoted with the same subscript. Let $\approxeq$ and $\approxeq'$ be their light-colour equivalencies. As stated above, both the light-colour manifolds, $A/\approxeq$ and $A/\approxeq'$, are homeomorphic to the half-space $\mathbf{H}^3 = \{(x, y, z) \in \mathbf{R}^3 : z \geq 0\}$, thus, to each other. There are an infinite number of homeomorphisms $A/\approxeq$ onto $A/\approxeq'$, and there is no reason to single out any of them. Still, if the light-colour gamuts, $\Psi(A)$ and $\Psi'(A)$, coincide, then one can construct, in a certain way natural, a homeomorphisms, establishing the correspondence between the classes of light-colour equivalencies $\approxeq$ and $\approxeq'$ by the equality of the perceptual light-colour coordinates assigned to the lights in these classes.

When the light-colour gamuts coincide for two observers this means that they share the same light-colour palette. It does not follow, however, that such observers will perceive all the lights the same way. There will be differences between these observers in perceiving the colour of particular lights. We will say that the colour differences of this sort are of *type I*. One could mark this case by saying that these observers, having the same light-colour palette, differ in perceiving particular lights. We will use the term 'inter-individual colour differences' only considering observers with the same light-colour gamut. When the light-colour gamuts for some observers differ, some colours, available in the light-colour palette (the light-colour gamut) of one observer, cannot be found in the light-colour palette (the light-colour gamut) of the other observer.[6] Variations between observers

5 The fact that a person, having the same chromaticity class as my 'red' one, but experiencing it the same way as I experience my 'green' chromaticity class, must not refuse him to have the same colour palette as mine, providing that this is the only difference between us as far as light-colour perception is concerned. In other words, such problems as 'inverted spectrum' (Locke, 1975) are left beyond the scope of this book.

6 The difference in the light-colour gamuts of the standard four-chromatic colour observers can be manifested in the fact that while one observer can see a light with some particular chromaticness the other observer is not capable to have such chromaticness at all. For instance, no human observer can experience a binary colour with chromaticness $(1, 1, 0, 0)$.

with different light-colour gamut will be referred to as perceptual differences,[7] or the colour differences of *type II*.

What we know about the colour vision of anomalous trichromats leads to a conjecture that the difference between their colour perception and that of normal trichromats is of type II. Indeed, on the one hand, the way they use colour names gives a hint that their colour perception can be represented by a circular Hering structure. Certainly, this assumption has yet to be verified using the partial hue-matching method. On the other hand, it would be unrealistic to expect anomalous trichromats to have the same light-colour gamut as normal trichromats.

Inter-individual colour differences between observers with the same light-colour gamut can be distinguished depending on their light-colour equivalencies and light-colour maps. When the light-colour maps are the same (thus, so are their light-colour equivalencies), we are dealing with the case of the absolute identity of the light-colour perception of the observers from which we begun this section.

When the light-colour maps differ, different light colours can be assigned to the same light by different observers (despite that the light-colour gamuts are identical). This case (i.e., when the light-colour gamuts are identical but the light-colour maps are different) is worth dividing into two subcases depending on whether the light-colour equivalencies are same or different. It should be borne in mind that the observers with the same light-colour equivalencies do not necessarily have the same light-colour maps, $\Psi : A \to \mathbf{R}^3$, induced by their four-chromatic opponent structures.[8] Really, although the light-colour equivalencies $\approxeq$ and $\approxeq'$ might be equal, the light-colour manifolds, $A/\approxeq$ and $A/\approxeq'$, can map to the perceptual light-colour space differently, that is, $\Psi_{\approxeq} : A/\approxeq \to \mathbf{R}^3$ might differ from $\Psi_{\approxeq'} : A/\approxeq' \to \mathbf{R}^3$. As the light-colour gamut is assumed to be common for both the observers, i.e., $\Psi_{\approxeq}(A/\approxeq) = \Psi_{\approxeq'}\left(A/\approxeq'\right)$, the following transformation of the light-colour gamut, T, is well-defined:

$$T(z) = \Psi_{\approxeq'}\left(\Psi_{\approxeq}^{-1}(z)\right), \tag{9.1}$$

where $z \in \Psi_{\approxeq}(A/\approxeq)$ is an arbitrary element of the light-colour gamut. Let us call T the *across-individual light-colour-gamut transformation*. It fully describes the differences between the observers with the same light-colour equivalencies (they will be referred to as the inter-individual colour differences of *type Ia*). Indeed, given the colour of a light in one observer, z, in order to predict the colour of this light in the other observer, z', suffices just to know the transformation T and apply it to z.

When the light-colour equivalencies differ (even if the light-colour gamuts are identical), the inter-individual colour differences will be said to be of *type Ib*. In this case, the knowledge of the colour of a light in the first observer is not enough to predict how the light will appear for the second observer because the perceptual light-colour coordinates of the second observer can differ for the light-colour equivalent (for the first observer) lights. Hence, the transformation T cannot be defined as above. Admittedly, it can be generalised as $T(z) = \Psi'\left(\Psi^{-1}(z)\right)$, where Ψ and Ψ' are the light-colour maps of the two observers. However, in this case $T(z)$ will be a subset in the Ψ'-perceptual light-colour space, T becoming a multi-valued map. In other words, when

7 It should be stressed that one can talk of perceptual differences in colour only as long as the observers share the same perceptual light-colour space (i.e., they exemplify the standard four-chromatic colour observer). When for some observers one needs different chromatic structures to represent them, we will say that their colour perception is of a different kind.

8 To verify this, it suffices to consider any bijective transformation of the light-colour gamuts of the first observer, as the light-colour gamuts of the second observer.

the light-colour equivalencies of the two observers differ the phenomenon analogous to metamer mismatching takes place. Let us call it the *light-colour dispersion*. Specifically, light-colour dispersion is said to take place when lights, colour-equivalent for one observer, become being not colour-equivalent for the other observer. Since light-colour dispersion might be a source of rather significant inter-individual colour differences (of type Ib), let us dwell on it in more detail.

Although the light-colour dispersion could be explored independently, from this moment till the end of this chapter, we will assume that along with the colour equivalencies $\approxeq$ and $\approxeq'$ the two metamerisms are also defined on A, and that these coincide with each other for both the observers. Specifically, consider two observers represented by Hering–Grassmann structures $\langle A, \sim, \approxeq \rangle$ and $\langle A, \sim', \approxeq' \rangle$ such that metamerism $\sim$ equals $\approxeq$, and metamerism $\sim'$ equals $\approxeq'$. In this case, light-colour dispersion and metamer mismatching are just two sides of the same coin. The metamer set for a light $a \in A$, defined in (7.1), can be rewritten in terms of the colour equivalencies $\approxeq$:

$$Met(\approxeq; a) = \{b \in A : b \approxeq a\}. \tag{9.2}$$

The Ψ'-image of $Met(\approxeq; a)$ in the Ψ'-perceptual light-colour space will be referred to as the *light-colour dispersion region*:

$$M(\Psi', \Psi; a) := \Psi'(Met(\approxeq; a)). \tag{9.3}$$

A quick comparison with Eq. (7.2) shows that the light-colour dispersion region is nothing more than the metamer-mismatch region represented in the Ψ'-perceptual light-colour space. So is $T(z) = \Psi'\left(\Psi^{-1}(z)\right)$. Really, given a colour-signal map Φ' representing Grassmann structure $\langle A, \sim' \rangle$, denote $C' : \Phi'(A) \to \Psi'(A)$ the *colour-graduation homeomorphism* for the second observer (Section 8.6.3). Then for any $z \in \Psi(A)$ we get $T(z) = \Psi'\left(\Psi^{-1}(z)\right) = C'\left(\Phi'(Met(\sim; a))\right)$ where a is such that $\Psi(a) = z$.

By the way, when the light-colour gamuts coincide, i.e., $\Psi(A) = \Psi'(A)$, there exists a natural homeomorphism between the (colour-graduated) colour-signal cones of the two observers even if the metamerisms $\sim$ and $\sim'$ are different: $C^{-1} \circ C'$. It puts in correspondence those points in the colour-signal cones $\Phi(A)$ and $\Phi'(A)$ that have the same perceptual light-colour coordinates. Let us call it the *colour-signal-conversion homeomorphism.*

More specifically, we define the colour-signal-conversion homeomorphism, $T_{\Phi'\Phi} : \Phi'(A) \to \Phi(A)$, as

$$T_{\Phi'\Phi} = C^{-1} \circ C', \tag{9.4}$$

where $C : \Phi(A) \to \Psi(A)$ and $C' : \Phi'(A) \to \Psi'(A)$ are the colour-graduation homeomorphisms for the colour-signal cones $\Phi(A)$ and $\Phi'(A)$; provided that the light-colour gamuts $\Psi(A)$ and $\Psi'(A)$ coincide. For any colour signal z' in $\Phi'(A)$, the colour signal $z = T_{\Phi'\Phi}\left(z'\right)$ in $\Phi(A)$ will have the same colour.

Moreover, by means of the equality of the perceptual light-colour coordinates one can also establish the natural homeomorphism between the (colour-graduated) colour stimulus manifolds of the two observers provided that their light-colour gamuts coincide. Specifically, define the *colour-stimulus-conversion homeomorphism*, $T_{\mathcal{A}'\mathcal{A}} : \mathcal{A}' \to \mathcal{A}$, as

$$T_{\mathcal{A}'\mathcal{A}} = C^{-1} \circ C', \tag{9.5}$$

where $C : \mathcal{A} \to \Psi(A)$ and $C' : \mathcal{A}' \to \Psi'(A)$ are the colour-graduation homeomorphisms for the colour stimulus manifolds $\mathcal{A}$ and $\mathcal{A}'$ (on condition that the light-colour gamuts $\Psi(A)$ and $\Psi'(A)$ coincide). As in the previous case (for two colour-signal cones), the colour-stimulus-conversion homeomorphism permits to find which colour stimuli in $\mathcal{A}$ and $\mathcal{A}'$ appear identical colour. Namely, for any colour stimulus a' in $\mathcal{A}'$, the colour stimulus $a = T_{\mathcal{A}'\mathcal{A}}(a')$ in $\mathcal{A}$ will have the same colour.

Let us now consider some quantitative measures by which one can characterise the difference between the two observers in the colour perception of a certain light. Denote the opponent chromatic codes and the brightness code for these observers as f_{YB}, f_{RG}, f_{Br} and f'_{YB}, f'_{RG}, f'_{Br} respectively (see (8.46)–(8.47)). The difference in perceiving the colour of a light $a \in A$ between these observers can be characterised by the perceptual-light-colour-coordinate difference (will be referred to as the *light-colour shift*):

$$\Psi'(a) - \Psi(a) = \left(f'_{YB}(a), f'_{RG}(a), f'_{Br}(a)\right) - \left(f_{YB}(a), f_{RG}(a), f_{Br}(a)\right). \tag{9.6}$$

The difference $f'_{YB}(a) - f_{YB}(a)$ (respectively, $f'_{RG}(a) - f_{RG}(a)$) represents the shift in colour along the yellow-blue (respectively, red-green) axis (in the perceptual light-colour space) induced by the light-colour shift between the observers. The positive difference indicates either a yellow (respectively, red) gain or a blue (respectively, green) loss. The difference $f'_{Br}(a) - f_{Br}(a)$ indicates the shift along the brightness axis.

If Ψ' describes the same observer Ψ, only in coloured glasses, Eq. (9.6) show the difference in colour perception induced by the coloured glasses. Generally, if the difference between Ψ and Ψ' is the difference in medium (e.g., in the pre-receptoral filter), then (9.6) will be referred to as the *light-colour shift* induced by the medium change.[9]

It must be said that, unlike the psychophysical colour shift (Section 8.3) which can be derived theoretically from the fundamental colour mechanisms, for the light-colour shift (9.6) to be evaluated one needs to know the chromatic and brightness scales, which cannot be derived from the fundamental colour mechanisms and have to be identified only in a separate experiment. To be more exact, one needs either the colour-graduated colour stimulus manifolds or the colour-graduated colour-signal cones.

As emphasised above, the light-colour shift (9.6) essentially depends on light a. Even if $a \approxeq b$ the colour shifts for a and b are most likely to be different (unless $\approxeq$ equals $\approxeq'$) because of light-colour dispersion. More specifically, the light-colour shift (induced by the transition from one observer to the other) is determined by the two factors: (i) the difference between the light-colour maps of these observers; and (ii) the light-colour dispersion. In a sense, these are independent of each other. When the observers have the same light-colour equivalence, there will be no light-colour dispersion; therefore, the light-colour shift will be solely determined by the difference in the light-colour maps. If, in addition, the light-colour gamuts are also identical, then the light-colour shift can be predicted based on the across-individual light-colour-gamut transformation (Eq. (9.1)). In the case of different light-colour equivalences, there occurs a light-colour dispersion, contributing to the light-colour shift.[10] Although the difference between the light-colour maps affects its quantitative expression, the nature of the light-colour dispersion is independent of the difference in the light-colour maps.

---

9 It should be said that the situation is more complicated. As a matter of fact, when seeing a light passed through two different mediae (e.g., coloured glasses), we are aware of the colour of this light and the colour of the mediae. Here we mean only the former. Generally, the colorimetric paradigm is oversimplified since it assumes the complete transparency of the media (as well as the absence of reflecting objects). As a result, it does not take into account this duality of colour experience in the presence of real (i.e., only partly transparent) media. We will deal with this issue in more detail in Part II.

10 Still, a difference in perceiving the colour of a monochromatic light in Hering–Grassmann structures can arise only due to the difference of the light-colour maps, since there can be no metamer mismatching (and hence, no colour dispersion) for monochromatic lights in such structures.

Thus, for an arbitrary element of the light-colour gamut, z, the light-colour shift induced by the difference between the observers (9.6) turns out to be fuzzy because of the light-colour dispersion. The volume of the light-colour-dispersion region $\Psi'\left(\Psi^{-1}(z)\right)$ can characterise the degree of this fuzziness. However, the volume of $\Psi'\left(\Psi^{-1}(z)\right)$ as a measure of light-colour dispersion can be misleading when the colour-graduated metamer-mismatch-region is rather flat (see Remark 7.5). For example, when the light-colour dispersion is very small along the brightness axis, the volume of $\Psi'\left(\Psi^{-1}(z)\right)$ can be close to zero despite the large light-colour dispersion along the *YB*- and/or *RG*-axis. It seems more reasonable, therefore, to quantify the light-colour dispersion more specifically, doing this separately along each coordinate axis in the perceptual light-colour space as follows.[11]

Define the *YB-range of the light-colour dispersion* (induced by the transition from $\approxeq$ to $\approxeq'$) as

$$\Delta YB\left(\approxeq, \approxeq'; a\right) := \max_{b\approxeq a} f'_{YB}(b) - \min_{b\approxeq a} f'_{YB}(b)\,; \tag{9.7}$$

the *YB-midpoint of the light-colour dispersion* (induced by the transition from $\approxeq$ to $\approxeq'$) as

$$M_\Delta YB\left(\approxeq, \approxeq'; a\right) := \frac{1}{2}\left(\max_{b\approxeq a} f'_{YB}(b) + \min_{b\approxeq a} f'_{YB}(b)\right); \tag{9.8}$$

and the *relative YB-range of the light-colour dispersion* (induced by the transition from $\approxeq$ to $\approxeq'$) as

$$\delta YB\left(\approxeq, \approxeq'; a\right) := \frac{\max_{b\approxeq a} f'_{YB}(b) - \min_{b\approxeq a} f'_{YB}(b)}{\max_{b\approxeq a} f'_{YB}(b) + \min_{b\approxeq a} f'_{YB}(b)}. \tag{9.9}$$

Likewise, define the *RG-range of the light-colour dispersion* (induced by the transition from $\approxeq$ to $\approxeq'$) as

$$\Delta RG\left(\approxeq, \approxeq'; a\right) := \max_{b\approxeq a} f'_{RG}(b) - \min_{b\approxeq a} f'_{RG}(b); \tag{9.10}$$

the *RG-midpoint of the light-colour dispersion* (induced by the transition from $\approxeq$ to $\approxeq'$) as

$$M_\Delta RG\left(\approxeq, \approxeq'; a\right) := \frac{1}{2}\left(\max_{b\approxeq a} f'_{RG}(b) + \min_{b\approxeq a} f'_{RG}(b)\right); \tag{9.11}$$

and the *relative RG-range of the light-colour dispersion* (induced by the transition from $\approxeq$ to $\approxeq'$) as

$$\delta RG\left(\approxeq, \approxeq'; a\right) := \frac{\max_{b\approxeq a} f'_{RG}(b) - \min_{b\approxeq a} f'_{RG}(b)}{\max_{b\approxeq a} f'_{RG}(b) + \min_{b\approxeq a} f'_{RG}(b)}. \tag{9.12}$$

Also, define the *brightness-range of the light-colour dispersion* (induced by the transition from $\approxeq$ to $\approxeq'$) as

$$\Delta Br\left(\approxeq, \approxeq'; a\right) := \max_{b\approxeq a} f'_{Br}(b) - \min_{b\approxeq a} f'_{Br}(b); \tag{9.13}$$

the *brightness midpoint of the light-colour dispersion* (induced by the transition from $\approxeq$ to $\approxeq'$) as

$$M_\Delta Br\left(\approxeq, \approxeq'; a\right) := \frac{1}{2}\left(\max_{b\approxeq a} f'_{Br}(b) + \min_{b\approxeq a} f'_{Br}(b)\right); \tag{9.14}$$

11 In what follows we will use some of the results obtained in Chapter 7 concerning the geometric properties of the metamer mismatch volumes, taking into account the continuity of the opponent chromatic scales and the brightness codes.

and the *relative brightness range of the light-colour dispersion* (induced by the transition from $\approxeq$ to $\approxeq'$) as

$$\delta Br(\approxeq, \approxeq'; a) := \frac{\max\limits_{b \approxeq a} f'_{Br}(b) - \min\limits_{b \approxeq a} f'_{Br}(b)}{\max\limits_{b \approxeq a} f'_{Br}(b) + \min\limits_{b \approxeq a} f'_{Br}(b)}. \tag{9.15}$$

Indices (9.7), (9.10), and (9.13) show how much the colour shift (induced by the transition from $\approxeq$ to $\approxeq'$) varies along the corresponding coordinate axes in the perceptual light-colour space depending on which light, light-colour equivalent to the given light a, is taken. Equations (9.9), (9.12), and (9.15) show the same but only in relative units.

It is worth noting that these indices can be also used to quantify the metamer mismatching provided that the colour equivalencies coincide with the metamerisms.[12] As pointed out earlier, the use of the perceptual rather than psychophysical light-colour coordinates seems more natural and straightforward for this purpose. Moreover, two points separated by the same interval in the colour-signal space render different perceptual colour differences depending on the location of this interval. In other words, the Euclidean metric in the colour-signal space does not reflect the subjective 'proximity' in the perceptual space. Thus, depending on its position in the colour-signal space, the same metamer-mismatch volume can induce rather different a light-colour dispersion as measured in terms of the indices (9.7)–(9.15).

As Eq. (9.6) represents the shift in colour for a particular light, a, it will be useful to generalise this formula so as to represent the shift for the particular colour (i.e., the particular point in the perceptual light-colour space) independently on the light mapping to this point. When a runs all the light metameric to each other, the subtrahend in Eq. (9.6) remains constant whereas the minuend varies (i.e., becomes fuzzy). Hence, the difference represented by Eq. (9.6) becomes fuzzy too. It can be characterised by the following indices. Let us define the *average*[13] *YB-shift* (induced by the transition from $\approxeq$ to $\approxeq'$) as

$$M_{sh}YB(\approxeq, \approxeq'; a) := M_{\Delta}YB(\approxeq, \approxeq'; a) - f_{YB}(a), \tag{9.16}$$

the *average RG-shift* (induced by the transition from $\approxeq$ to $\approxeq'$) as

$$M_{sh}RG(\approxeq, \approxeq'; a) := M_{\Delta}RG(\approxeq, \approxeq'; a) - f_{RG}(a), \tag{9.17}$$

and *the average brightness-shift* (induced by the transition from $\approxeq$ to $\approxeq'$) as

$$M_{sh}Br(\approxeq, \approxeq'; a) := M_{\Delta}Br(\approxeq, \approxeq'; a) - f_{Br}(a). \tag{9.18}$$

The range of the light-colour shift along the axes in the perceptual light-colour space can be specified by the following indices. Define the *maximal YB-shift* (induced by the transition from $\approxeq$ to $\approxeq'$) as

$$MAX_{sh}YB(\approxeq, \approxeq'; a) := \max_{b \approxeq a} f'_{YB}(b) - f_{YB}(a), \tag{9.19}$$

the *maximal RG-shift* (induced by the transition from $\approxeq$ to $\approxeq'$) as

$$MAX_{sh}RG(\approxeq, \approxeq'; a) := \max_{b \approxeq a} f'_{RG}(b) - f_{RG}(a), \tag{9.20}$$

and the *maximal brightness-shift* (induced by the transition from $\approxeq$ to $\approxeq'$) as

$$MAX_{sh}Br(\approxeq, \approxeq'; a) := \max_{b \approxeq a} f'_{Br}(b) - f_{Br}(a). \tag{9.21}$$

---

12 One could say that, as far as Hering–Grassmann structures are concerned, light-colour dispertion can be, generally speaking, thought of as colour-graduated metamers mismatching.

13 It should be noted that the term 'average' in the present context has no statistical connotation.

Similarly is defined the *minimal YB-shift* (induced by the transition from $\approxeq$ to $\approxeq'$):

$$MIN_{sh}YB\left(\approxeq, \approxeq'; a\right) := \min_{b \approxeq a} f'_{YB}(b) - f_{YB}(a), \tag{9.22}$$

the *minimal RG-shift* (induced by the transition from $\approxeq$ to $\approxeq'$):

$$MIN_{sh}RG\left(\approxeq, \approxeq'; a\right) := \min_{b \approxeq a} f'_{RG}(b) - f_{RG}(a), \tag{9.23}$$

and the *minimal brightness-shift* (induced by the transition from $\approxeq$ to $\approxeq'$):

$$MIN_{sh}Br\left(\approxeq, \approxeq'; a\right) := \min_{b \approxeq a} f'_{Br}(b) - f_{Br}(a). \tag{9.24}$$

(Notice that the shifts are called maximal (respectively, minimal) not because of the maximality (respectively, minimality) of their magnitudes but because they bring the maximal value to the corresponding component of $\Psi'(Met(\approxeq; a))$. In particular, it is because

$$\max_{b \approxeq a} f'_{YB}(b) = MAX_{sh}YB\left(\approxeq, \approxeq'; a\right) + f_{YB}(a)$$

that the shift (9.19) is called maximal.)

When the transition from $\approxeq$ to $\approxeq'$ is produced by a change in the medium transmittance, it is worthwhile to modify the robustness indices from Section 5.2.2 by using the perceptual light-colour coordinates ((8.46)–(8.47)). Specifically, we propose the following indices of light-colour robustness to medium transmittance change: the *YB robustness index*

$$rYB\left(\approxeq, \approxeq'; a\right) := 1 - \frac{f'_{YB}(b) - f_{YB}(a)}{\max_{b \approxeq a} \left|f'_{YB}(b) - f_{YB}(a)\right|}, \tag{9.25}$$

the *RG-robustness index*

$$rRG\left(\approxeq, \approxeq'; a\right) := 1 - \frac{\left|f'_{RG}(b) - f_{RG}(a)\right|}{\max_{b \approxeq a} \left|f'_{RG}(b) - f_{RG}(a)\right|}, \tag{9.26}$$

and the *brightness-robustness index*

$$rBr\left(\approxeq, \approxeq'; a\right) := 1 - \frac{\left|f'_{Br}(b) - f_{Br}(a)\right|}{\max_{b \approxeq a} \left|f'_{Br}(b) - f_{Br}(a)\right|}. \tag{9.27}$$

10

Colour Matching Structures and Matching Metamerism

In the previous chapters, colour matching was understood as an empirical binary relation satisfying a set of empirical laws, the most important of which were formulated as Grassmann's laws. Here we look into to what extent these empirical laws are consistent with the available colour matching data. It is not difficult to show examples contradicting Grassmann's laws.

For starters, let's take a close look at, for example, the transitivity of colour matching. In a real experiment, two monochromatic lights, δ_{λ_1} and δ_{λ_2}, may well be judged as matching each other in colour even when $\lambda_1 \neq \lambda_2$ provided that the difference between wavelengths λ_1 and λ_2 is small enough. Indeed, it is well established that for two monochromatic lights, δ_{λ_1} and δ_{λ_2}, to mismatch (i.e., to be discriminated) the wavelength difference, $|\lambda_1 - \lambda_2|$, should exceed some critical (threshold) value. In other words, there is a limit for wavelength discrimination (i.e., mismatching) for human observers – the wavelength difference less than 1 nm is unlikely to be resolved (Smith & Pokorny, 2003, p. 104).

An important consequence to this is, firstly, intransitivity of colour matching. For instance, in a series of monochromatic lights, $\delta_{\lambda_0}, \delta_{\lambda_0+\Delta\lambda}, \delta_{\lambda_0+2\Delta\lambda}, \ldots, \delta_{\lambda_0+n\Delta\lambda}$, each adjacent pair $\delta_{\lambda_0+i\Delta\lambda}$ and $\delta_{\lambda_0+(i+1)\Delta\lambda}$ may match each other provided $\Delta\lambda$ is small enough. However, the pair δ_{λ_0} and $\delta_{\lambda_0+n\Delta\lambda}$ will obviously be well discriminable for sufficiently large n.

Secondly, it follows that the colorimetric model of colour matching (see Section 3.3) is, strictly speaking, incorrect. Indeed, this model predicts that two lights will mismatch (i.e., be discriminated) if their colour signals are different, no matter how small the difference is. Particularly, two monochromatic lights, δ_{λ_1} and δ_{λ_2}, should always be discriminated unless $\lambda_1 = \lambda_2$. However, it is not the case for human colour matching.

Also there is evidence for invalidity of the Law of additivity (Definition 2.2) for colour matching. Let us consider two lights, a and b, such that $a \geq tb$ when $t > 0$ is small enough; then subtracting from a a small amount of light b gives a light: $a - tb \in A$. It is clear that for sufficiently small t, a will be indistinguishable from $a - tb$, that is, a will match $a - tb$. Then the Law of additivity entails that a should also match $a + tb$. However, that does not hold true for every pair of lights a and b. For example, when a is a neutral light, and $b = a$, then, there are $t_+, t_- > 0$ (called increment and decrement thresholds respectively) such that for every $t < t_+$, a will match $a + tb$, and for every $t < t_-$, a will match $a - tb$. However, as well known, there is so-called increment/decrement asymmetry of luminance threshold. Specifically, increment threshold t_+ can be more than twice as much as decrement threshold t_- (Boynton et al., 1964; Short, 1966; Patel & Jones, 1968; Cohn & Lasley, 1975; Kelly & Savoie, 1978; Kranda & King-Smith, 1979). A similar increment/decrement threshold asymmetry was also observed along the S-cone axis in the SML cone excitation space (Cole et al., 1993; Logvinenko, Tyurin, & Sawey, 2002). Evidently, such an increment/decrement threshold asymmetry is not compatible with the Grassmann additivity law.

Foundations of Colour Science: From Colorimetry to Perception, First Edition. Alexander D. Logvinenko and Vladimir L. Levin.
© 2023 John Wiley & Sons Ltd. Published 2023 by John Wiley & Sons Ltd.

The failure of colour matching to satisfy the laws of transitivity and additivity shows that the colorimetric model of colour matching is oversimplified. The major shortcoming of the colorimetric model of colour matching is that it is deterministic, that is, it does not take into account the stochastic nature of human responses in the colour matching experiment. In particular, this model implies that when the same lights are presented on two halves of a bipartite field, the fundamental colour mechanisms' responses to these lights will be equal, which, in turn, will lead to the equality of colour appearance of both lights. If this were the case then observers would always give a response 'same' to equal lights. However, in reality, equal lights not always cause equal responses. Quite often, observers judge the same lights as 'different'. Moreover, such erroneous judgements occur not only for identical lights. As a matter of fact, any lights which are close enough to each other may be acknowledged as matching (in colour) each other on one occasion and different on the other. Because of this uncertainty, in colour matching experiment one can establish only the probability of matching rather than matching as such. To put it the other way, human responses in the colour matching experiment are essentially random rather than deterministic events.[1]

There are physical, physiological, and psychological reasons for the stochastical nature of the colour matching responses. Firstly, because of the photon noise, the stimulus intensity on the two halves of the bipartite field can be strictly equal only on very rare occasions. (Though, the effect of photon noise becomes rather small in the light intensity range where the cone photoreceptors operate.) Secondly, photoreceptors, as well as other visual cells, are subject to internal (physiological) noise. Because of this physiological noise, the fundamental colour mechanisms' responses are practically never equal even when the number of photons emitted from the two halves is equal. Thirdly, as well established in numerous psychophysical studies (Blackwell, 1963; Graham, 1989), observers sometimes guess,[2] that is, they produce responses based on non-sensory information.

Thus, colour matching should be conceptualised in terms of random rather than deterministic variables. In fact, such a conceptualisation has been adopted in studying colour discrimination (Wyszecki & Stiles, 1982; Eskew et al., 1999; Smith & Pokorny, 2003; Stockman & Brainard, 2010). Recall that colour discrimination and colour matching are the two sides of the same coin. When two lights match each other there is no discrimination, and vice versa, when they do not match they are believed to be discriminated. Hence, any model of colour discrimination is a model of colour matching at the same time. (Colour discrimination models will be discussed at length in Chapter 12.) The theoretical framework for the psychophysical models of detection and discrimination (Luce & Green, 1974; Graham, 1989) is the signal detection theory (D. M. Green & Swets, 1966) which is a particular case of statistical decision theory. Below we summarise (very briefly and in the context of colour matching) the basic notions underlying the current psychophysical models of colour discrimination.

The core idea is that in the situation of discrimination between two lights observers behave as if they have to make a statistical decision on when to say 'same' and when 'different'. Indeed, let us accept, for the sake of certainty, that the effect of noise can be taken into account by assuming that each colour mechanism f_i (as in Example 2.2) is followed by a random variable with zero mean[3] (see Example 10.1). It follows that any pair of colour signals can be brought about (true, with different likelihood) by many pairs of lights (including identical lights). Therefore, given two sets of the colour mechanisms' outputs (e.g., from two halves of the bipartite filed), an observer

1 In fact, such a stochastic uncertainty is inherent in not only colour matching, but all other human judgements.
2 Observers are particularly prone to guessing when they are presented pairs of light stimuli for a short period of time with an instruction to decide whether they match.
3 In many models of visual detection and discrimination, these random variables are assumed to be Gaussian, and independent of each other (Graham, 1989).

has, in fact, to make a decision: whether these have been produced by the same lights or different. Any decision will lead to two types of errors: 'false alarm' ('different' when the lights are, in reality, same), and 'miss' ('same' when the lights are, in reality, different). There is no decision rule which eliminates the errors of both types. Still, as established in the theory of statistical decision, it is possible to minimise the effect of errors provided a function specifying the cost of the error of each type is defined.

A decision rule, generally, instructs what response ('same' or 'different') should be assign to any pair of the colour signals. Formally, a decision rule in colour matching experiment implies a partition of the Cartesian square of the colour-signal cone $\Phi(A) \times \Phi(A)$ into two regions: one resulting in the response 'same', and the other 'different'. We will call the former the *matching criterion region* (written as $\Theta \subseteq \Phi(A) \times \Phi(A)$). The decision rule based on the equality of colour signals amounts to assuming that the matching criterion region is the diagonal of $\Phi(A) \times \Phi(A)$ (written as $\Theta = \Delta_{\Phi(A)}$). Numerous experiments on colour discrimination convince that observers do not base their response 'same' solely on the equality of colour signals.[4] It is safe to assume, however, that $\Delta_{\Phi(A)} \subset \Theta$. There is also every indication that Θ is symmetric (that is, $\Theta = \Theta^{-1}$). Indeed, it seems reasonable to believe that if a pair of colour signals (x, y) entails the response 'same' so does the pair (y, x).

It is clear that the likelihood of the response 'same' is different for different pairs of colour signals (x, y). Specifically, it is natural to expect that the more separated x and y in the colour-signal space, the more probable the response 'different'. This can be formalised as follows. Let $\langle A, \sim \rangle$ be a n-dimensional Grassmann structure, and Φ a colour-signal map. We will assume that for any pair of colour signals x and y there exists a probability measure μ_{xy} on $\mathbf{R}^n \times \mathbf{R}^n$, and a measurable subset $\Theta \subseteq \Phi(A) \times \Phi(A)$ (matching criterion region) such that the integral

$$P_S(x, y) := \int_\Theta \mu_{xy}(dz), \ z \in \mathbf{R}^{2n} \tag{10.1}$$

can be interpreted as the probability of the response 'same' on condition that any lights $a \in \Phi^{-1}(x)$ and $b \in \Phi^{-1}(x)$ are presented.[5] Respectively, $P_D(x, y) =: 1 - P_S(x, y)$ is the probability of the response 'different' on the same condition. In other words, the conditional probabilities, $P(S|\, a, b)$ and $P(D|\, a, b)$, of the responses 'same' and 'different' (on condition that lights a and b are presented) are assumed to be equal to $P_S(x, y)$, and $P_D(x, y)$ respectively, provided $\Phi(a) = x$, and $\Phi(b) = y$. Thus, the probability of the match between lights a and b is $P(S|\, a, b)$ times the probability of presentation[6] of the pair of lights a and b.

Example 10.1 *Let $\langle A, \sim \rangle$ be a n-dimensional weak* continuous Grassmann structure, and $\varphi_1(\lambda), \dots, \varphi_n(\lambda)$ its fundamental colour mechanisms. We will assume that a noise is added to the response of each mechanism. Specifically, the noisy response of the ith mechanism φ_i to a light σ is assumed to be represented as a random variable $\xi_i = \langle \varphi_i, \sigma \rangle + u_i$, where u_i is a random variable*

---

4 If it were the case the probability of the 'same' response would be zero. However, in reality the latter is always positive.

5 A pair of light stimuli a and b such that $\Phi(a) = x$ and $\Phi(b) = y$, is assumed to be capable (with different likelihood, though) to invoke any pair of colour signals. It is taken for granted that this likelihood does not change when a (or/and b) is replaced with a' (respectively, b') such that $\Phi(a') = \Phi(a)$ and $\Phi(b') = \Phi(b)$. This explains why measure μ_{xy} (which reflects this likelihood) has subscripts xy rather than ab. Still, the integral on the right-hand side of Eq. (10.1) actually represents the likelihood of the response 'same' to light stimuli a and b. Note that the matching criterion region is assumed to be independent of light stimuli.

6 In psychophysical experiments on discrimination, the stimulus presentation schedule is usually supposed to be random. It means that only the probability of presentation of each stimulus on each presentation is determined but not the very sequence of presentation (e.g., Bush et al., 1963).

distributed as $N(0, \zeta^2)$, *i.e., a normally distributed random variable with zero mean and variance* ζ^2. *We will also assume that the matching criterion region* Θ *is defined as follows*

$$(x_1, x_2) \in \Theta \Leftrightarrow \|x_1 - x_2\|_2 \leq \varepsilon, \tag{10.2}$$

where ε *is a fixed positive real number,* $x_1, x_2 \in \mathbf{R}^n$, *and* $\|\cdot\|_2$ *is the Euclidian norm.*

When x_1 *and* x_2 *are the noisy colour signals in response to not metameric lights* σ_1 *and* σ_2 *respectively, that is,* $x_i (i = 1, 2)$ *is a normal random vector with mean* $(\langle \varphi_1, \sigma_i \rangle, \ldots, \langle \varphi_n, \sigma_i \rangle)$ *and the diagonal covariance matrix with* ζ^2 *on the diagonal,* $(2\zeta^2)^{-1} \|x_1 - x_2\|_2^2$ *has the noncentral chi-square distribution with n degrees of freedom, and noncentrality parameter* v

$$v = (2\zeta^2)^{-1} \|(\langle \varphi_1, \sigma_1 - \sigma_2 \rangle, \ldots, \langle \varphi_n, \sigma_1 - \sigma_2 \rangle)\|_2^2. \tag{10.3}$$

Thus, if $F_{\chi^2}(\cdot | n, v)$ *denotes the cumulative distribution function of the noncentral chi-square distribution with n degrees of freedom and noncentrality parameter* v, *the probability of the response 'same', on condition that the not metameric lights* σ_1 *and* σ_2 *are presented, is expressed as*

$$P(\|x_1 - x_2\|_2 \leq \varepsilon) = F_{\chi^2}\left(\varepsilon^2 (2\zeta^2)^{-1} \middle| n, v\right).$$

If lights are metameric (i.e., $\langle \varphi_i, \sigma_1 - \sigma_2 \rangle = 0$ *for every i),* $(2\zeta^2)^{-1} \|x_1 - x_2\|_2^2$ *has the chi-square distribution with n degrees of freedom. For metameric lights the probability of the response 'same' equals* $F_{\chi^2}\left(\varepsilon^2 (2\zeta^2)^{-1} \middle| n\right)$, *where* $F_{\chi^2}(\cdot | n)$ *stands for the cumulative distribution function of the (central,* $v = 0$*) chi-square distribution with n degrees of freedom. Bear in mind that* $F_{\chi^2}(x | n) > F_{\chi^2}(x | n, v)$ *for any* $x > 0$ *and* $v > 0$.

Note that in Example 10.1 $P_S(x, y) = P_S(y, x)$ for any x and y. It follows that $P(S | a, b) = P(S | b, a)$. Notice also that $P(S | a, a)$ is the same for any a. Moreover, for any metameric lights a and b, $P(S | a, b) = P(S | a, a)$. If a is not metameric to b then $P(S | a, a) > P(S | a, b)$. In other words, in Example 10.1 $P(S | a, b) = P(S | a, a)$ takes place if and only if $a \sim b$.

All this inspires the following extension of the notion of colour matching into the stochastic domain. We will assume that for any pair of lights a and b two non-negative real numbers $P(S | a, b)$ and $P(D | a, b)$ (the probabilities of the responses 'same' and 'different' on condition that lights a and b are presented respectively) are defined such that

(i) $P(S | a, b) + P(D | a, b) = 1$;
(ii) $P(S | a, b) = P(S | b, a)$;
(iii) $P(S | a, a) = P(S | b, b) > 0$;
(iv) $P(S | a, b) \leq P(S | a, a)$.

Let us define a binary relation

$$a \sim b \Leftrightarrow P(S | a, b) = P(S | a, a), \quad \forall a, b \in A. \tag{10.4}$$

Binary relation $\sim$ is obviously an equivalence. In the view of Example 10.1, it seems plausible to consider (10.4) as an operational definition of colour match when the colour matching judgements are inherently uncertain. In other words, two lights are assumed to match if they cause the response 'same' with the same probability as a pair of physically identical lights.

Such a definition is essentially based on properties (i)–(iv). However, there is evidence that at least one of them, namely, the equality in (iii), can fail for human observers. One reason for fluctuation of the probability of the response 'same' to a pair of physically identical lights is guessing. The tendency to guess may change during experiment.

Admittedly, in the psychophysical methodological repertoire there is a remedy for guessing: the so-called *forced choice technique* (D. M. Green & Swets, 1966). In a forced choice experiment on colour discrimination, an observer is informed that one light stimulus, say a, will be presented on the one half of a bipartite field, and another stimulus, b, on the other. The observer's task is to locate a and b. In the case of uncertainty, the observer is forced to guess. When both halves are physically identical, the observer's responses (i.e., 'a is on the right hand side' and 'a is on the left hand side') are distributed in equal proportions no matter what the guessing likelihood is. Thus, the probability of correct stimulus locating will be equal to 0.5. If we interpret $P(S|\, a, a)$ as the probability of correctly locating a stimulus a when both stimuli are equal (i.e., $a = b$), then $P(S|\, a, a)$ will always be 0.5 irrespective of the tendency to guess.

Another reason is a change of the matching criterion region during experiment. In other words, fluctuation of the probability of the response 'same' to a pair of physically identical lights may be due to alterations of the matching criterion region. Admittedly, observers in psychophysical experiments on detection and discrimination are usually encouraged to maintain their criterion (i.e., the matching criterion region) constant during experiment. After some period of training observers are believed to be able to do this. Yet, there is no guarantee that the matching criterion region remains unchanged between experimental sessions which may take place in different days or weeks.[7]

Although one cannot expect that the probability of the response 'same' to a pair of physically identical lights remains constant for all lights and throughout the experiment, it is safe to assume that it is within certain fixed limits. In this case, the definition (10.4) can be generalised as follows. Given some $\varepsilon \geq 0$, let us define a relation (written as ς_ε) on the light stimulus cone A as

$$a\varsigma_\varepsilon b \Leftrightarrow P(S|\, a, b) \geq \varepsilon, \quad \forall a, b \in A. \tag{10.5}$$

In other words, we assume that two lights match each other if the proportion of times where the observer does not see the difference between them is high enough. Note that the definition (10.4) implies that this proportion is the maximum possible (due to properties (iii) and (iv)).

We will refer to (10.5) as a *stochastic model of colour matching*, and to ς_ε as a *stochastic colour matching relation*. This definition is in line with the long psychophysical tradition according to which two stimuli are assumed to be discriminated if the probability $P(D|\, a, b)$ exceeds some fixed level (Blackwell, 1972; Luce & Green, 1974; Graham, 1989).

The stochastic colour matching relation ς_ε in (10.5) is reflexive and symmetric but not transitive. Therefore, an essential difference between the definitions (10.4) and (10.5) is that the colour matching relation as defined by (10.4) is an equivalence whereas that defined by (10.5) is a tolerance.

When colour matching is an intransitive tolerance Grassmann's laws cannot hold true for it.[8] Still, as we show below, if a tolerance meets some other axioms then there is an equivalence associated with this tolerance which satisfies Grassmann's laws, and thus, this equivalence can be, generally, interpreted as metamerism. In this chapter, we develop a theory of Grassmann structures departing from colour matching defined as an intransitive tolerance on the light cone A.

It must be said that although the rationale for generalising the notion of colour matching relation from equivalence to tolerance is stochastic uncertainty of human responses in the colour matching experiment, our consideration in this chapter will be purely deterministic. Specifically, we will look into the following problem. Given a Grassmann structure $\langle A, \sim \rangle$, we will assume that the metamerism $\sim$ is not accessible. Instead, we will assume that there is an accessible tolerance

---

7 It is believed that the forced choice technique makes it possible to circumvent the issue of fluctuations in the subject's criterion in detection experiments as well (more detail see in Green & Swets, 1966).

8 It will be shown below that if a tolerance satisfies the Grassmann laws then it is transitive, that is, an equivalence.

ς containing $\sim$. We will investigate conditions under which one can recover $\sim$ from ς. The fact that both $\sim$ and ς can be defined in probabilistic terms (as in (10.4) and (10.5)) is of secondary importance.

10.1 Colour matching structures

In this section, we study colour matching structures as well as colour matching relations (Definition 10.1) and indiscriminate relations (Definition 10.3) on light stimulus cones A and on colour stimulus cones $A(\sim) := \pi(A)$. Since most results hold true when A is an arbitrary set and since these and similar results may be of use in different contexts, we formulate them in a maximally general way, by treating A as an *abstract light stimulus set* and its elements as *abstract light stimuli* or *abstract lights*. When A is a proper stimulus cone, the word *abstract* will be omitted.

Definition 10.1 *Recall (see, e.g., Sections 5.2 and 8.5) that a reflexive and symmetric binary relation on an arbitrary set A is called* a tolerance. *Given a tolerance ς on A, we define, for each $a \in A$, $\varsigma(a) := \{b \in A : a\varsigma b\}$ and consider on A the binary relation* (matching equivalence) $\sim_\varsigma$,

$$a \sim_\varsigma b \text{ if and only if, for each } c \in A,\ a\varsigma c \Leftrightarrow b\varsigma c. \tag{10.6}$$

Proposition 10.1 *Given a tolerance ς on A, the binary relation $\sim_\varsigma$ as given by (10.6) is an equivalence on A such that if $a\varsigma b$ then $a'\varsigma b$ whenever $a' \sim_\varsigma a$. Moreover, $\sim_\varsigma$ is the largest equivalence relation with that property: if $\sim$ is an equivalence on A and $a\varsigma b$ implies $a'\varsigma b$ whenever $a' \sim a$, then, for every $a, b \in A$, $a \sim b \Rightarrow a \sim_\varsigma b$.*

Remark 10.1 *Note that (10.6) may be rewritten as $a \sim_\varsigma b \Leftrightarrow \varsigma(a) = \varsigma(b)$.*

Definition 10.2 *Given a tolerance ς on A, we take the matching equivalence $\sim_\varsigma$ associated with it and consider the set* (abstract colour stimulus set) $A(\sim_\varsigma)$ *of the corresponding equivalence classes, $A(\sim_\varsigma) := \{[a] : a \in A\}$, where $[a]$ stands for the equivalence class of the element $a \in A$,*

$$[a] := \{b \in A : a \sim_\varsigma b\} = \{b \in A : \varsigma(a) = \varsigma(b)\}.$$

We denote by π the canonical map (abstract colour stimulus map) *of A onto $A(\sim_\varsigma)$, $\pi(a) = [a]$, $a \in A$.*

Definition 10.3 *Given a tolerance ς and the matching equivalence $\sim_\varsigma$ on A, we define a binary relation* (indiscriminate relation) v_ς *on $A(\sim_\varsigma)$ as follows. Given two abstract colour stimuli, $x, y \in A(\sim_\varsigma)$, we write $xv_\varsigma y$ if and only if there are abstract lights $a \in \pi^{-1}(x)$ and $b \in \pi^{-1}(y)$ such that $a\varsigma b$.*

Proposition 10.2 *The relation v_ς is a tolerance on $A(\sim_\varsigma)$, and for any $x, y \in A(\sim_\varsigma)$ the equivalence holds:*

$$xv_\varsigma y \Leftrightarrow a\varsigma b \text{ for all } a \in \pi^{-1}(x), b \in \pi^{-1}(y). \tag{10.7}$$

Remark 10.2 *Note that (10.7) may be rewritten as follows:*

$$a\varsigma b \Leftrightarrow \pi(a)v_\varsigma\pi(b) \text{ whenever } a, b \in A, \tag{10.8}$$

or which is the same,

$$\varsigma(a) = \pi^{-1}(v_\varsigma(\pi(a))), \quad a \in A.$$

Definition 10.4 *We say a tolerance* ν *on a set* M *is* simple *if* $\nu(x) = \nu(y) \Leftrightarrow x = y$ *for all* $x, y \in M$, *where* $\nu(x) = \{z \in M : x\nu z\}$. *In other words,* ν *is simple if and only if, for any* $x, y \in M$, $x \neq y$, *there exists* $z \in M$ *such that* $x\nu z$ *holds but* $y\nu z$ *fails or* $y\nu z$ *holds but* $x\nu z$ *fails.*

Corollary 10.1 *The tolerance* ν_ς *from Definition 10.3 is simple.*

Proposition 10.3 *Suppose* $\sim$ *is an equivalence on* A, *and* ν *is a tolerance on the set* $A(\sim)$ *of equivalence classes for* $\sim$, *and define on* A *a binary relation (tolerance)* ς,

$$a\varsigma b \Leftrightarrow \pi(a)\nu\pi(b), \tag{10.9}$$

where π *is the canonical map of* A *onto* $A(\sim)$. *Then* $\sim_\varsigma$ *is larger than* $\sim$ *(i.e.,* $a \sim b \Rightarrow a \sim_\varsigma b$, $a, b \in A$*), and* $\sim_\varsigma$ *coincides with* $\sim$ *(hence* ν_ς *coincides with* ν*; see (10.8)) if and only if* ν *is simple.*

Remark 10.3 *Given a map* Φ *of* A *onto some set* M, *the equivalence* $\sim_\Phi$ *is determined on* A *as follows:*

$$a \sim_\Phi b \Leftrightarrow \Phi(a) = \Phi(b), \quad a, b \in A.$$

Every $x \in M$ *determines an equivalence class* $[a] = \Phi^{-1}(x) \in A(\sim_\Phi)$, *and every equivalence class* $[a] \in A(\sim_\Phi)$ *is obtained in such a way; hence* $\pi(a) = \pi(b) \Leftrightarrow \Phi(a) = \Phi(b)$, $a, b \in A$. *Therefore,* $A(\sim_\Phi)$ *and* $\pi : A \to A(\sim_\Phi)$ *can be identified with* M *and* $\Phi : A \to M$. *Note that if* M *is an externally determined (concrete) set, then* M *and* Φ *may be considered as concrete representations of* $A(\sim_\Phi)$ *and* π. *If, in particular,* $M \subset \mathbf{R}^n$ *then* $\Phi = (\varphi_1, \dots, \varphi_n)$, *where* $\varphi_i : A \to \mathbf{R}$, *and* $a \sim_\Phi b \Leftrightarrow \varphi_i(a) = \varphi_i(b)$, $i = 1, \dots, n$.

The next result is a direct consequence of Proposition 10.3.

Corollary 10.2 *Given a map* Φ *of* A *onto* M *and a tolerance* ν *on* M, *we define a tolerance* ς,

$$a \,\varsigma\, b \Leftrightarrow \Phi(a)\nu\Phi(b), \quad a, b \in A, \tag{10.10}$$

and consider on A *the matching equivalence* $\sim_\varsigma$. *Then* $\sim_\varsigma$ *is larger than* $\sim_\Phi$, *and* $\sim_\varsigma$ *coincides with* $\sim_\Phi$ *if and only if* ν *is simple.*

Suppose now that two sets, E, $E \supset A$, and M', are given together with a map Φ of E onto M'. We will distinguish the equivalence $\sim_\Phi$ on E,

$$e_1 \sim_\Phi e_2 \Leftrightarrow \Phi(e_1) = \Phi(e_2), \quad e_1, e_2 \in E,$$

and its restriction to A denoted as $\approx_\Phi$.

Let ν' be a tolerance on M' and ν its restriction to the set $M = \Phi(A)$. Two tolerances, ς' on E and ς on A, are then determined: $e_1\varsigma' e_2 \Leftrightarrow \Phi(e_1)\nu'\Phi(e_2)$, $e_1, e_2 \in E$, and $a\varsigma b \Leftrightarrow \Phi(a)\nu\Phi(b)$, $a, b \in A$. They generate the corresponding matching equivalencies, $\sim' = \sim_{\varsigma'}$ on E ($e_1 \sim' e_2 \Leftrightarrow \varsigma'(e_1) = \varsigma'(e_2)$, $e_1, e_2 \in E$) and $\sim_\varsigma$ on A ($a \sim_\varsigma b \Leftrightarrow \varsigma(a) = \varsigma(b)$, $a, b \in A$). Clearly, ς is the restriction of ς' to A, that is, for every $a \in A$, $\varsigma(a) = \varsigma'(a) \cap A$. It follows that $a \sim' b$ implies $a \sim_\varsigma b$ for any $a, b \in A$. However, it can occur that, for some $a, b \in A$, $\varsigma(a) = \varsigma(b)$ but $\varsigma'(a) \neq \varsigma'(b)$. In such a case, $\sim_\varsigma$ is not the restriction of $\sim'$ to A: $a \sim_\varsigma b$ holds but $a \sim' b$ fails.

The next result follows immediately from Proposition 10.3.

Proposition 10.4 *Suppose* ν' *is simple. Then* $\approx_\Phi$ *coincides with the restriction of* $\sim'$ *to* A. *Furthermore, the following statements are equivalent:*

(i) $\sim_\varsigma$ *is the restriction of* $\sim'$ *to* A: $a \sim_\varsigma b \Leftrightarrow a \sim' b$ *whenever* $a, b \in A$;
(ii) $\sim_\varsigma$ *coincides with* $\approx_\Phi$;
(iii) ν *is simple.*

Remark 10.4 *Sometimes, it is easy to see that* ν' *is simple but it is difficult to check whether* ν *is simple or not. For example, suppose that E is a vector space, A is a convex cone in it,* $M' = \mathbf{R}^n$*, and* $\Phi = (\varphi_1, \ldots, \varphi_n)$*, where* $\varphi_1, \ldots, \varphi_n$ *are linear independent linear functionals on E. To show that a given tolerance* ν' *on* $M' = \mathbf{R}^n$ *is simple, one has to find for any* $x, y \in \mathbf{R}^n, x \neq y$*, a vector* $z \in \mathbf{R}^n$ *such that* $x\nu'z$ *holds but* $y\nu'z$ *fails or* $y\nu'z$ *holds but* $x\nu'z$ *fails. Often such a check can be easily performed for a particular* ν'*; however, for the corresponding* ν *this might become much more difficult. This is because in the latter case one has to take x and y* $(x \neq y)$ *from the cone* $M = \Phi(A)$ *and to find the corresponding z also in M; but it may be a rather difficult task to find out whether such a vector* $z \in M$ *exists or not. We shall see that in some cases* ν' *is simple but* ν *is not.*

Let us consider two examples. In both examples, A is a convex cone in a vector space E, $E = A - A$, $M' = \mathbf{R}^n$, $M = \Phi(A)$, and $\Phi = (\varphi_1, \ldots, \varphi_n)$, where $\varphi_1, \ldots, \varphi_n$ are linear independent linear functionals on E.

Example 10.2 *Given a number* $\varepsilon > 0$ *and a function* $d : \mathbf{R}^n \times \mathbf{R}^n \to \mathbf{R}_+$ *such that* $d(x,y) = d(y,x)$ *for all* $x, y \in \mathbf{R}^n$ *and* $d(x,x) = 0$ *for all* $x \in \mathbf{R}^n$*, we define a binary relation* ν' *by the convention that* $x\nu'y$ *if and only if* $d(x,y) \leq \varepsilon$*. Clearly,* ν' *is a tolerance: it is reflexive since* $d(x,x) = 0$ *and symmetric since* $d(x,y) = d(y,x)$*.*

Example 10.3 *Given an integer m,* $m < n$*, and a number* $\varepsilon \geq 0$*, we define a tolerance* ν' *by the convention that* $x\nu'y$ *for* $x = (x_1, \ldots, x_n), y = (y_1, \ldots, y_n)$ *if and only if there are m indices* $i_1, \ldots, i_m$ *such that* $|x_{i_k} - y_{i_k}| \leq \varepsilon, k = 1, \ldots, m$*.*

We now examine these examples in detail.

Proposition 10.5 *The tolerance* ν' *from Example 10.2 is simple if and only if for any* $x, y \in \mathbf{R}^n$*,* $x \neq y$*, there is a vector* $z \in \mathbf{R}^n$ *such that*

$$\min(d(x,z), d(y,z)) \leq \varepsilon, \quad \max(d(x,z), d(y,z)) > \varepsilon. \tag{10.11}$$

Similarly, ν *is simple if and only if for any* $x, y \in M, x \neq y$*, there is a vector* $z \in M$ *such that (10.11) is satisfied.*

The next result is a direct consequence of Proposition 10.5.

Corollary 10.3 *Suppose d is some distance function (metric) on* $\mathbf{R}^n$*. Then the tolerance* ν' *(respectively,* ν*) from Example 10.2 is simple if and only if for any* $x, y \in \mathbf{R}^n$ *(respectively,* $x, y \in M$*),* $x \neq y$*, the corresponding* ε*-balls with centres at x and y cannot coincide:*

$$x \neq y \Leftrightarrow B_\varepsilon(x) \neq B_\varepsilon(y), \quad x, y \in \mathbf{R}^n \tag{10.12}$$

and

$$x \neq y \Leftrightarrow B_\varepsilon(x) \cap M \neq B_\varepsilon(y) \cap M, \quad x, y \in M, \tag{10.13}$$

where $B_\varepsilon(x) = \{z \in \mathbf{R}^n : d(x,z) \leq \varepsilon\}, x \in \mathbf{R}^n$*.*

Remark 10.5 *Note that condition (10.12) is satisfied when* $d(x,y) = ||x - y||$ *where* $||\cdot||$ *is an arbitrary norm on* $\mathbf{R}^n$*; however, (10.13) can fail. A simplest counterexample is as follows:* $n = 2$*,* $||z|| = |z_1| + |z_2| \, \forall z = (z_1, z_2)$*,* $\varepsilon = 5$*,* $M = \{z = (z_1, z_2) \in \mathbf{R}^2_+ : z_1/2 \leq z_2 \leq 2z_1\}$*,* $x = (2,1), y = (1,2)$*. A direct calculation yields* $B_5(x) \cap M = B_5(y) \cap M = \{z = (z_1, z_2) \in M : z_1 + z_2 \leq 8\}$*, that is, (10.13) fails. Note that condition (10.13) holds (for any convex cone M in* $\mathbf{R}^n$*) when* $||\cdot||$ *is a rotund* *norm*[9] *such as p-norm* $\|z\| = \|(z_1, \dots, z_n)\| = \left(\sum_1^n |z_i|^p\right)^{1/p}$, $1 < p < \infty$. *At the same time, there exist distances for which both (10.12) and (10.13) fail. Indeed, given* $\varepsilon_0 > 0$ *and a norm on* $\mathbf{R}^n$, $\|\cdot\|$, *the function* $d(x, y) := \frac{\varepsilon_0\|x-y\|}{1+\|x-y\|}$ *is a (continuous) distance on* $\mathbf{R}^n$, *and for any* $\varepsilon \geq \varepsilon_0$ *one has* $B_\varepsilon(x) = \mathbf{R}^n$ *whenever* $x \in \mathbf{R}^n$. *For such distances, the corresponding* ν' *and* ν *are not simple.*

Corollary 10.4 *Suppose* $d(x,y) = g(x - y)$, *where the function* $g : \mathbf{R}^n \to \mathbf{R}_+$ *satisfies the following conditions:* $g(0) = 0$, *for every* $x \in \mathbf{R}^n$ *the equality holds* $g(-x) = g(x)$ *and the function on* $\mathbf{R}_+$, $t \mapsto g(tx)$, *is non-decreasing and* $\lim_{t\to\infty} g(tx) = \infty$. *Then the tolerance* ν' *from Example 10.2 is simple.*

Remark 10.6 *Note that every convex function* $g : \mathbf{R}^n \to \mathbf{R}_+$ *such that* $g(0) = 0$, $g^{-1}(0)$ *is bounded, and* $g(-x) = g(x)$ $\forall x$, *satisfies the hypotheses of Corollary 10.4.*

Proposition 10.6 *The tolerance* ν' *from Example 10.3 is simple.*

We now turn to the case where A is the *light stimulus cone* and E is the corresponding *light stimulus space*, that is, E is a vector space, and A is a convex cone in it such that $E = A - A$ and A does not contain a line (see Chapter 2).

Definition 10.5 *Given a light stimulus cone A and a tolerance* ς *on it, we say that a pair* $\langle A, \varsigma\rangle$ *is a* colour matching structure *(defined on A), if two axioms are satisfied as follows:*

1. *For any a, b, and c in A,* $\varsigma(a) = \varsigma(b)$ *if and only if* $\varsigma(a + c) = \varsigma(b + c)$.
2. *For any a, b in A and any positive real number t,* $\varsigma(a) = \varsigma(b)$ *implies* $\varsigma(ta) = \varsigma(tb)$.

In such a case, we call ς *a* colour matching relation *and say that a is a* colour match *for b (or that a and b* match each other*) if* $a\varsigma b$. *We will call* $\varsigma(a)$ *the* colour matching set *for a.*

Definition 10.6 *When* $\langle A, \varsigma\rangle$ *is a colour matching structure, the corresponding matching equivalence* $\sim_\varsigma$ *on A (see Definition 10.2) will be called* matching metamerism. *(In such a case, the abstract colour stimulus set* $A(\sim_\varsigma)$ *(see Definition 10.2) becomes the* colour stimulus cone *as it was defined in Chapter 2; see Definition 2.5.)*

As follows from (10.6), two lights are metameric if and only if they can substitute each other without breaking the colour matching relation.

The next statement is an immediate consequence of Definitions 10.5 and 10.6.

Proposition 10.7 *Suppose that* $\langle A, \varsigma\rangle$ *is a colour matching structure, and* $\sim_\varsigma$ *is the matching metamerism determined by it. Then* $\langle A, \sim_\varsigma\rangle$ *is a Grassmann structure on A.*

Definition 10.7 *Let* $\langle A, \varsigma\rangle$ *be a colour matching structure, and* $\sim_\varsigma$ *the matching metamerism determined by it. We will say that the Grassmann structure* $\langle A, \sim_\varsigma\rangle$ *is* induced *by the colour matching structure* $\langle A, \varsigma\rangle$ *(or by the colour matching relation* ς*), and the colour matching structure* $\langle A, \varsigma\rangle$ *is* compatible *with the Grassmann structure* $\langle A, \sim_\varsigma\rangle$.

Definition 10.8 *A colour matching structure* $\langle A, \varsigma\rangle$ *is called n-dimensional if the induced Grassmann structure* $\langle A, \sim_\varsigma\rangle$ *is n-dimensional.*

---

9 A norm $\|\cdot\|$ is called *rotund* if there are no line segments in the unit sphere or, equivalently, if the strict inequality holds $\|u_1 + u_2\| < \|u_1\| + \|u_2\|$ whenever $u_1 \neq u_2$, $\|u_1\| = \|u_2\| = 1$. Sometimes, rotund norms are referred to as *strictly convex*, but we abstain from using this terminology because *no norm* is a strictly convex *function*.

Given a colour matching structure $\langle A, \varsigma \rangle$, the indiscriminate relation ν_ς is defined on the colour stimulus cone $A(\sim_\varsigma)$ (see Definition 10.3). When the structure $\langle A, \varsigma \rangle$ is n-dimensional, we can take a particular *colour-signal map* $\Phi : A \to \mathbf{R}^n$ (see Remark 2.5), identify the colour stimulus cone $A(\sim_\varsigma)$ with the *colour-signal cone* $\Phi(A)$, which is its concrete representation in $\mathbf{R}^n$ (cf. Chapter 4), and consider ν_ς as the indiscriminate relation on $\Phi(A)$:

$$x\nu_\varsigma y \Leftrightarrow a\varsigma b \text{ for some (for all) } a \in \Phi^{-1}(x), b \in \Phi^{-1}(y)$$

(cf. Proposition 10.2 and Remark 10.2). Note that in both cases (for colour stimuli and for colour signals) $x \neq y, x\nu_\varsigma y$ means that two colour stimuli (colour signals) may be different but an observer cannot discriminate between them.

Definition 10.9 *Suppose $\langle A, \varsigma \rangle$ is a colour matching structure, and $\langle A, \sim_\varsigma \rangle$ is a Grassmann structure induced by it. The set $\nu_{\sim_\varsigma}(x) := \left\{ y \in A/\sim_\varsigma \, : x\nu_{\sim_\varsigma} y \right\}$ of all colour stimuli indiscriminate from a particular colour stimulus $x \in A/\sim_\varsigma$ will be called an* indiscriminable set *of x. The same name and notation will be used for its concrete representation in a particular colour-signal cone $\Phi(A)$,*

$$\nu_{\sim_\varsigma}(x) := \left\{ y \in \Phi(A) : x\nu_{\sim_\varsigma} y \right\}, \; x \in \Phi(A).$$

Hence, a matching relation ς on the light stimulus cone A is associated with two binary relations: the matching metamerisms $\sim_\varsigma$ on A and the indiscriminate relation $\nu_{\sim_\varsigma}$ on the colour stimulus cone $A/\sim_\varsigma$. Two lights a and b match each other if either they are metameric (i.e., $a \sim_\varsigma b$), or, being not metameric, bring about indiscriminable colour signals (i.e., $\pi(a) \, \nu_{\sim_\varsigma} \, \pi(b)$).

It is worth noting the important distinction between these two relations (i.e., $\sim_\varsigma$ and $\nu_{\sim_\varsigma}$). The matching metamerism is a fundamental property of the visual system eventually determined by the spectral sensitivity of the photoreceptors (or the sensors if we deal with artificial devices such as digital cameras and colour monitors). The indiscriminate relation is the conceptualisation of a task-specific performance of a certain kind (colour discrimination or colour matching), which is subject to improvement due to training. It is also known that colour discrimination (as well as other types of discrimination, e.g., pattern discrimination) depends on what subjective criteria are adopted by the observer, which makes colour discrimination rather flexible and subject to significant inter-individual differences. It follows that observers with the same matching metamerism may yield different results under colour discrimination (thus, matching) task.

Of great theoretical importance is, then, to be able to study matching metamerism and indiscriminate relation separately, independent of each other. It is not an easy task since in experiment one can directly observe only matching responses in which there is no indication whether they were caused by matching metamerism or by indiscriminability of colour signals. Nevertheless, in Chapter 11 we shall show that it is possible to investigate matching metamerisms when the indiscriminate relation is not known.

The next result is a particular case of Proposition 10.3.

Proposition 10.8 *Suppose that $\langle A, \sim \rangle$ is a Grassmann structure, and ν is a tolerance on the colour stimulus cone $A/\sim$. Let us consider a tolerance ς on A as given by (10.9). The following statements are then equivalent:*

(i) $\langle A, \varsigma \rangle$ is a colour matching structure compatible with the Grassmann structure $\langle A, \sim \rangle$;
(ii) ν is simple.

If these equivalent statements hold true, then $\sim_\varsigma = \sim$ and $\nu_\varsigma = \nu$.

Corollary 10.5 *Suppose that $\langle A, \sim\rangle$ is a n-dimensional Grassmann structure, $\Phi : A \to \mathbf{R}^n$ is a colour-signal map, and ν is a tolerance on the colour-signal cone $\Phi(A)$. The statements (i) and (ii) of Proposition 10.8 are then equivalent, and if they hold true, then $\sim_\varsigma = \sim$ and $\nu_\varsigma = \nu$.*

The next two results are particular cases of Proposition 10.4.

Proposition 10.9 *Suppose $\langle A, \sim\rangle$ is a Grassmann structure, ν' is a tolerance on the colour stimulus space $E/E(\sim)$, and ν is its restriction to the colour stimulus cone $A(\sim)$. Let ς' be a tolerance on E as given by*

$$e_1\varsigma' e_2 \Leftrightarrow \pi(e_1)\nu'\pi(e_2), \quad e_1, e_2 \in E,$$

and ς a tolerance on A as given by

$$a\varsigma b \Leftrightarrow \pi(a)\nu\pi(b), \quad a, b \in A.$$

The tolerance ς' induces the matching equivalence on E, $\sim' = \sim_{\varsigma'}$,

$$e_1\sim' e_2 \Leftrightarrow \varsigma'(e_1) = \varsigma'(e_2), \quad e_1, e_2 \in E, \tag{10.14}$$

and the tolerance ς induces the matching equivalence on A, $\sim_\varsigma$,

$$a \sim_\varsigma b \Leftrightarrow \varsigma(a) = \varsigma(b), \quad a, b \in A. \tag{10.15}$$

Suppose ν' is simple, then the original metamerism $\sim$ on A coincides with the restriction of $\sim'$ to A, and the following statements are equivalent:

(i) $\sim_\varsigma$ coincides with the restriction of $\sim'$ to A;
(ii) $\langle A, \varsigma\rangle$ is a colour matching structure compatible with the Grassmann structure $\langle A, \sim\rangle$;
(iii) ν is simple.

This is an immediate consequence of Proposition 10.4 for $M' = E/E(\sim)$ and $\Phi = \pi$ if one takes into account that $\approx_\Phi = \sim$. Likewise, the next result is a particular case of Proposition 10.4 for $M' = \mathbf{R}^n$.

Proposition 10.10 *Suppose $\langle A, \sim\rangle$ is a n-dimensional Grassmann structure, $\Phi : A \to \mathbf{R}^n$ is a colour-signal map (hence Φ is uniquely extended to a linear map of $E = A - A$ onto $\mathbf{R}^n$), ν' is a tolerance on $\mathbf{R}^n$, and ν is its restriction to the colour-signal cone $\Phi(A)$. Let ς' be a tolerance on E as given by $e_1\varsigma' e_2 \Leftrightarrow \Phi(e_1)\nu'\Phi(e_2)$, $e_1, e_2 \in E$, and ς a tolerance on A as given by $a\varsigma b \Leftrightarrow \Phi(a)\nu\Phi(b)$, $a, b \in A$. The tolerance ς' induces the matching equivalence on E, $\sim' = \sim_{\varsigma'}$ as given by (10.14) and the tolerance ς induces the matching equivalence on A, $\sim_\varsigma$, as given by (10.15). Suppose ν' is simple, then the original metamerism $\sim$ on A coincides with the restriction of $\sim'$ to A, and statements (i), (ii), (iii) of Proposition 10.9 are equivalent.*

Remark 10.7 *Note that statement (i) of Proposition 10.9 means the equivalence $\nu'(\pi(a)) = \nu'(\pi(b)) \Leftrightarrow \nu'(\pi(a)) \cap A(\sim) = \nu'(\pi(b)) \cap A(\sim)$ whenever $a, b \in A$, and statement (i) of Proposition 10.10 means the equivalence $\nu'(\Phi(a)) = \nu'(\Phi(b)) \Leftrightarrow \nu'(\Phi(a)) \cap \Phi(A) = \nu'(\Phi(b)) \cap \Phi(A)$ whenever $a, b \in A$. When ν' is simple but ν is not simple, these equivalencies fail.*

Taking into account conditions for indiscriminate relations ν and ν' to be simple (see Proposition 10.5, Corollaries 10.3 and 10.4, and Proposition 10.6), various particular results on connections between Grassmann and colour matching structures may be derived from Propositions 10.8, 10.9, and 10.10. We omit their formulations.

Remark 10.8 *When ν' is simple but ν is not,*[10] *the Grassmann structure $\langle A, \sim\rangle$ is not induced by the colour matching relation ς, $\varsigma(a) = \{b \in A : \Phi(a)\nu\Phi(b)\}$. Moreover, as follows from the next example, in such a case $\langle A, \varsigma\rangle$ need not be a colour matching structure at all.*

Example 10.4 *Suppose $E = \mathbf{R}^m$, $A = \{a = (z_1, \dots, z_m) \in \mathbf{R}^m_+ : z_1/2 \le z_2 \le 2z_1\}$, where $m \ge 2$, $n = 2$, $\Phi = (\varphi_1, \varphi_2) : E \to \mathbf{R}^2$, where φ_1 and φ_2 are linear functionals on E defined by $\varphi_1(e) := z_1$, $\varphi_2(e) := z_2$ for every $e = (z_1, \dots, z_m)$, and the equivalence $\sim\, = \,\sim_\Phi$ is defined on A by*

$$a \sim b \Leftrightarrow \Phi(a) = \Phi(b), \quad a, b \in A.$$

Then $\langle A, \sim\rangle$ is a two-dimensional Grassmann structure, $\Phi(E) = \mathbf{R}^2$, and $\Phi(A) = \{z = (z_1, z_2) \in \mathbf{R}^2_+ : z_1/2 \le z_2 \le 2z_1\}$ is the corresponding colour-signal cone. Consider in $\mathbf{R}^2$ the norm $\|z\| = |z_1| + |z_2|$ and define ν' and ν by

$$\nu'(z) := \{z' \in \mathbf{R}^2 : \|z - z'\| \le \varepsilon\}, \quad z \in \mathbf{R}^2,$$

$$\nu(z) := \nu'(z) \cap \mathbf{R}^2_+ = \{z' \in \mathbf{R}^2_+ : \|z - z'\| \le \varepsilon\}, \quad z \in \mathbf{R}^2_+,$$

where $\varepsilon > 0$. The tolerance ν' is simple (this follows from Corollary 10.3), however ν is not simple. Indeed, for $x = (2\varepsilon/5, \varepsilon/5)$, $y = (\varepsilon/5, 2\varepsilon/5)$, one has $\nu(x) = \nu(y) = \{z = (z_1, z_2) \in \mathbf{R}^2_+ : z_1 + z_2 \le 8\varepsilon/5\}$ (cf. Remark 10.5); therefore, ν is not simple, and $\varsigma(a) = \varsigma(b)$ whenever $a \in \Phi^{-1}(x)$, $b \in \Phi^{-1}(y)$. At the same time, a direct calculation shows that the vector $z = (11\varepsilon/15, 22\varepsilon/15)$ belongs to $\nu(2x)$ and does not belong to $\nu(2y)$, hence $\nu(2x) \ne \nu(2y)$. Therefore, $\varsigma(2a) \ne \varsigma(2b)$ whenever $a \in \Phi^{-1}(x)$, $b \in \Phi^{-1}(y)$, that is, ς does not satisfy Axiom 2 from the definition of a colour matching structure (see Definition 10.5). (One can check that Axiom 1 is not satisfied as well.)

Of course, this example is rather artificial; but constructing a more realistic example with the light stimulus cone A from Example 2.1 is not easy because of difficulty in explicit describing the colour-signal cone $\Phi(A)$ for a given colour-signal map Φ; cf. Remark 10.4.

Nevertheless, under certain conditions on the function $d(x, y)$, by means of which the tolerance ν is defined, the latter will be simple (see Theorems 10.1 and 10.2).

Theorem 10.1 *Suppose $d(x, y) = g(x - y)$, where $g : \mathbf{R}^n \to \mathbf{R}_+$ is a convex function satisfying the following conditions:*

(a) $g(0) = 0$,
(b) for every $x \in \mathbf{R}^n$ the equality holds $g(-x) = g(x)$, and
(c) for any linearly independent vectors $u_1, u_2 \in \mathbf{R}^n$, g is strictly convex on the segment $[u_1, u_2] := \{(1 - t)u_1 + tu_2 : 0 \le t \le 1\}$, that is, $g((1 - t)u_1 + tu_2) < (1 - t)g(u_1) + tg(u_2)$.

Then the tolerance ν, $\nu(x) = \{y \in M = \Phi(A) : g(x - y) \le \varepsilon\}$, $x \in M$, is simple.

---

10 For the sake of definiteness, we suppose that ν' and ν are considered on a colour-signal space and a colour-signal cone. The case where they are defined on the colour-stimulus space and the colour-stimulus cone is similar.

Remark 10.9 *Note that every strictly convex function $g : \mathbf{R}^n \to \mathbf{R}$ satisfies hypothesis (c), and every rotund norm in $\mathbf{R}^n$ satisfies all the hypotheses of the theorem.*

In the next example, we consider a function $d : \mathbf{R}^n \times \mathbf{R}^n \to \mathbf{R}_+$ satisfying on $\Phi(A)$ the following monotonicity condition:

$$d(x+y+z, x+y+z') \le d(x+z, x+z'), \quad x, y, z, z' \in \Phi(A). \tag{10.16}$$

(In other words, if colour signals z and z' are indistinguishable against a background x then they will be indistinguishable against another background $x+y$ too.)

Example 10.5 *Suppose the colour-signal cone $\Phi(A)$ has nonempty interior and is contained in the positive orthant $\mathbf{R}^n_+$, and let*

$$d(x,y) = \frac{\|x-y\|}{\|x\|_1 + \|y\|_1 + 1}, \quad x, y \in \mathbf{R}^n, \tag{10.17}$$

where $\|\cdot\|$ means the Euclidean norm, $\|z\| = \left(\sum_1^n z_i^2\right)^{1/2}$, while the 1-norm $\|\cdot\|_1$ is given by $\|z\|_1 := |z_1| + \cdots + |z_n|$, $z = (z_1, \ldots, z_n) \in \mathbf{R}^n$.

Lemma 10.1 *The function d as given by (10.17) satisfies monotonicity condition (10.16).*

Given a number $\varepsilon > 0$, the function d generates a tolerance ν on $\Phi(A)$,

$$x\nu y \Leftrightarrow d(x,y) \le \varepsilon, \quad x, y \in \Phi(A). \tag{10.18}$$

Lemma 10.2 *For every $x \in \Phi(A)$, the set $\nu(x)$ is convex and contains the set $B_\varepsilon(x) \cap \Phi(A) = \{y \in \Phi(A) : \|x-y\| \le \varepsilon\}$.*

Remark 10.10 *Note that for the corresponding tolerance ν', determined by the function $d(x,y)$ on the whole of $\mathbf{R}^n$, the set $\nu'(x)$ is not convex.*

Theorem 10.2 *Suppose $\varepsilon < \frac{1}{\sqrt{n}}$, then the tolerance ν generated by equation (10.18) and function (10.17) is simple.*

Remark 10.11 *Since $\|\cdot\| \le \|\cdot\|_1$, one has*

$$d(x,y) \le \frac{\|x\|_1 + \|y\|_1}{\|x\|_1 + \|y\|_1 + 1} < 1$$

whenever $x, y \in \mathbf{R}^n$; therefore, if $\varepsilon \ge 1$ then $x\nu y$ for all x, $y \in \Phi(A)$.

Remark 10.12 *Given $\alpha = (\alpha_1, \ldots, \alpha_n) \in \text{int}\, \mathbf{R}^n_+$, one can consider on $\mathbf{R}^n$ a norm*

$$\|x\|_{1,\alpha} := \alpha_1|x_1| + \cdots + \alpha_n|x_n|, \; x = (x_1, \ldots, x_n) \in \mathbf{R}^n.$$

Theorem 10.2 remains true (with practically the same proof) when function (10.17) is replaced by

$$d(x,y) = \frac{\|x-y\|}{\|x\|_{1,\alpha} + \|y\|_{1,\alpha} + 1}$$

and $\varepsilon < 1/\|\alpha\| = \left(\alpha_1^2 + \cdots + \alpha_n^2\right)^{-1/2}$. Also, Lemmas 10.1 and 10.2 hold true for these d and ε.

In connection with Example 10.5 and Theorem 10.2, the following definition may be useful.

Definition 10.10 *Say a colour matching relation (tolerance) ς on A meets the* monotonicity condition *if, for any $a, b, c, c' \in A$, $(a + c)\varsigma(a + c')$ implies $(a + b + c)\varsigma(a + b + c')$.*

The monotonicity condition means that if light stimuli c, c' match each other against background a, then they will also match each other against background $a + b$. For instance, if in Fig. I.1 the yellow-greenish patch (c) is not discriminated from the yellow patch (c') against the neutral background (a) then casting an additional light (b) on the whole display will not make the border between the two halves of the bipartite field reveal itself. Still, two stimuli that are discriminable from each other against a can become indistinguishable against $a + b$.

The next result is an immediate consequence of Proposition 10.2.

Proposition 10.11 *If a colour matching relation ς on A meets the monotonicity condition, then the indiscriminate relation ν_ς on $A(\sim_\varsigma)$ meets it as well: for any $x, y, z, z' \in A(\sim_\varsigma)$, $(x + z)\nu_\varsigma(x + z')$ implies $(x + y + z)\nu_\varsigma(x + y + z')$.*

In conclusion, we note that although colour matching has never been explicitly conceptualised as intransitive relation; implicitly, in the context of colour discrimination, such attempts have previously been made. In particular, an idea to introduce a metric on the colour stimulus cone so as to predict the human ability to distinguish between colour stimuli (thus, colours), goes back the middle of the 19th century. Note that the Euclidean metric is not suitable for this purpose. Indeed, the Euclidean metric d interacts with the linear structure on the linear space L where it is defined by virtue of being translation invariant:

$$\forall x, y, z \in L \qquad d(x + z, y + z) = d(x, y). \tag{10.19}$$

Hence, if colour discrimination is based on a translation invariant metric then the discriminability of a pair of colour stimuli would depend only on the distance between them. It follows that the discriminability of two lights would be the same irrespective of a background against which they are presented. However, this is obviously not the case. For instance, being well discriminated from the sky in nights, stars are not distinguishable in daylight. Such a dependence of light intensity discriminability on the background's light intensity is a direct consequence of Weber's law (Eq. (8.10)).

A great impact on the direction of the search for a suitable metric had the idea of Helmholtz, who conjectured that it might be a Riemannian metric that colour discrimination is based on.[11] The particular form of function d, as in Theorem 10.1, proposed by Helmholtz (and his followers inspired by this idea) is discussed in Section 12.3.

10.2 Matching metamerism

The theory of Grassmann structures developed in Chapter 2 is fully applicable to the induced Grassmann structure $\langle A, \sim_\varsigma \rangle$. Here we summarise the main results of Chapter 2 at a slightly different angle, introducing a new terminology that will be used in subsequent chapters.

11 Furthermore, he believed that large colour differences were based on the same metric as small colour differences (underlying colour discrimination). It should be noted, however, that three different problems are unduly confounded here: (i) the evaluation of colour mechanisms representing the matching metamerism; (ii) mathematical modelling of indiscriminate relations; and (iii) mathematical modelling of large subjective differences between colours. Only the first two of these will be the focus in this book. These two issues are actually related to the fundamental problem of identification of a colour matching structure which will be dealt with in the next two chapters.

The matching metamerism $\sim_\varsigma$ determines a partition of A into non-overlapped classes (colour stimuli) which constitute the colour stimulus cone $A/\sim_\varsigma$. As in Chapter 2, the equivalence relation $\sim_\varsigma$ can be extended to the ambient space $E = A - A$. The equivalence classes for $\sim_\varsigma$ are the affine subspaces in E parallel to $E(\sim_\varsigma)$. Recall that given fixed $a \in E$ the affine subspaces in E parallel to $E(\sim_\varsigma)$ through a is the subset $a + E(\sim_\varsigma) := \{a + b : b \in E(\sim_\varsigma)\}$. The set of all affine subspaces parallel to $E(\sim_\varsigma)$ becomes a linear (quotient) space $E/E(\sim_\varsigma)$. The quotient space $E/E(\sim_\varsigma)$ is the colour stimulus space, and the canonical map π of E onto $E/E(\sim_\varsigma)$ defined by $\pi(a) := a + E/E(\sim_\varsigma)$ is the *colour stimulus map*. The linear subspace $E(\sim_\varsigma)$ is thus the *null-space (kernel)* of the colour stimulus map, and $\pi(A)$ is the *colour stimulus cone* $A/\sim_\varsigma$.

There exist subspaces M such that $E = M \oplus E(\sim_\varsigma)$. Any such subspace M is called a *complementary* subspace of $E(\sim_\varsigma)$ in E. Given a fixed M, for every $a \in E$ there is a unique representation $a = b(a) + c(a), b(a) \in M, c(a) \in E(\sim_\varsigma)$. Since $b(a) \sim_\varsigma a$ we will call $b(a)$ a *quasi-metamer* of a. While in colour literature $b(a) \in M$ is sometimes referred to as the *fundamental metamer* of a (Cohen & Kappauf, 1982; Wyszecki & Stiles, 1982), according to our terminology, $b(a)$ will be a metamer of a only when $b(a)$ is in the light stimulus cone A. Also, contrary to the fundamental metamer as defined by Cohen and colleagues,[12] $b(a)$ is not uniquely determined, being depended on the choice of the complementary subspace M. Let us define a linear map $P_{\sim_\varsigma} : E \to M$ by $P_{\sim_\varsigma}(a) =: b(a)$. The map $P_{\sim_\varsigma}$ is the *projection* of E on M along $E(\sim_\varsigma)$.

Any complementary subspace M of $E(\sim_\varsigma)$ is isomorphic to the colour stimulus space $E/E(\sim_\varsigma)$ (indeed, a linear map $M \ni a \to \pi(a) \in E/E(\sim_\varsigma)$ is clearly one-to-one). Therefore, all the complementary subspaces M have the same dimension: $\dim E/E(\sim_\varsigma)$. This dimension, is by definition the *codimension* of $E(\sim_\varsigma)$ in E. The induced Grassmann structure $\langle A, \sim_\varsigma \rangle$ is n-dimensional if and only if, first, the codimension of $E(\sim_\varsigma)$ in E equals n; and second, there is a complementary subspace (of $E(\sim_\varsigma)$) M such that $M = A' - A'$ and $A' \subset A$. Any n linearly independent vectors $a_1, \ldots, a_n$ in M will be primaries, that is, any light a will be metameric to some (uniquely determined) linear combination of these vectors $a \sim_\varsigma \xi_1 a_1 + \cdots + \xi_n a_n$. In the view of the isomorphism $M \ni a \to \pi(a) \in E/E(\sim_\varsigma)$, the coefficients $\xi_1, \ldots, \xi_n$ can be considered as the coordinates of the colour stimulus $\pi(a)$ with respect to the basis $\pi(a_1), \ldots, \pi(a_n)$ in the colour stimulus space $E/E(\sim_\varsigma)$. Following the tradition established in colour vision literature, we will call them *n-colour stimulus coordinates*.[13]

Alternatively, n-colour stimulus coordinates can be operationalised in terms of the notion of algebraic dual space. The set of all linear functionals on E (written as E') is called the *algebraic dual space* of E. Alongside with a conventional notation $\varphi(a)$, $\varphi \in E', a \in E$ we will also write $\langle a, a' \rangle$ $a' \in E', a \in E$ for the value the linear functional $a' \in E'$ takes on $a \in E$. Note that n-colour stimulus coordinates can be considered as elements of the algebraic dual space of the colour stimulus space $(E/E(\sim_\varsigma))'$. As known, for any $a' \in (E/E(\sim_\varsigma))'$ $\langle \pi(a), a' \rangle = \langle a, \pi'(a') \rangle$, where $\pi' : (E/E(\sim_\varsigma))' \to E'$ is the *conjugate colour stimulus map*. Hence, for each linear functional a' on the colour stimulus space there exists a linear functional $\varphi = \pi'(a')$ on the light stimulus space E which takes on a the same value as a' on $\pi(a)$, that is, $\varphi(a) = \langle \pi(a), a' \rangle$. The linear functional $\varphi(a)$ is obviously a colour mechanism (Definition 2.9). It turns out that the converse is also true, namely, any linear colour mechanism belongs to the range of π'.

12 They considered E to be endowed with the Euclidean structure. The subspace M was chosen so as to be orthogonal to $E(\sim_\sigma)$ (Cohen & Kappauf, 1982).

13 In colorimetry, the colour signal produced by a particular triplet of colour matching functions is referred to as the *tristimulus values* of a colour stimulus (Wyszecki & Stiles, 1982, p. 723).

Definition 10.11 *Let X be a linear subspace of the linear space E. The subset of E' consisting of such linear functionals φ on E that $\varphi(a) = 0$ for every $a \in X$ is called the* (algebraic) annihilator *of X (written as X^0).*

The annihilator X^0 is obviously a linear subspace in E'. The following proposition is a well-known fact in linear functional analysis (Kantorovich & Akilov, 1982).

Proposition 10.12 *The range of the conjugate colour stimulus map is exactly the annihilator of the null-space of the colour stimulus map, that is, the range of π' is $E(\sim_\varsigma)^0$.*

It follows that a linear functional on the light stimulus space E is a colour mechanism if and only if it belongs to the annihilator of $E(\sim_\varsigma)$. Therefore, the annihilator $E(\sim_\varsigma)^0$ is a subspace in E' which consists of all linear colour mechanisms. This justifies the following definition.

Definition 10.12 *The annihilator of the null-space of the colour stimulus map $E(\sim_\varsigma)^0$ is called* the linear colour mechanism space.

For any n linearly independent colour mechanisms there exist primaries relative to which these mechanisms can be considered as the coordinate functionals. Specifically, for any $\varphi_1, \dots, \varphi_n$ in $E(\sim_\varsigma)^0$ there exist primaries $a_1, \dots, a_n$ such that $\varphi_i(a_j) = \delta_{ij}$. Therefore, the n-stimulus coordinates $\xi_1, \dots, \xi_n$ of any light a relative to these primaries are $\xi_i = \varphi_i(a)$. In such a case, we say that the colour mechanisms $\varphi_1, \dots, \varphi_n$ and the primaries $a_1, \dots, a_n$ are mutually *dual*.

Any n linearly independent colour mechanisms $\varphi_1, \dots, \varphi_n$ in $E(\sim_\varsigma)^0$ define a colour-signal map $\Phi : A \ni a \to (\varphi_1(a), \dots, \varphi_n(a)) \in \mathbf{R}^n$. In Chapter 11, we will see how colour-signal maps can be derived from colour discrimination data.

11

Identification of Grassmann Structures Induced by Colour Matching Structures

Identification of any Grassmann structure amounts to the identification of its linear colour mechanism space (Definition 10.12). As shown in Chapter 3, when Grassmann structure is based on metameric matching (i.e., when colour matching obeys Grassmann's laws) this reduces to the evaluation of colour matching functions with respect to some set of primaries (Section 3.1). However, when colour matching does not obey Grassmann's laws (e.g., when colour matching relation is not transitive), one cannot use colour matching functions to identify the Grassmann structure induced by the intransitive colour matching relation. Still, a Grassmann structure induced by the (nontransitive) colour matching relation ς which is a tolerance (that is, based on the matching metamerism $\sim_\varsigma$ (Definition 10.7)) can be, in principle, identified. Specifically, we show in Section 11.3 that the linear colour mechanism space for the induced Grassmann structure $\langle A, \sim_\varsigma \rangle$ can be derived from the colour matching set of any arbitrary light provided that the matching metamerism $\sim_\varsigma$ satisfies some (rather technical) conditions (Section 11.2). Moreover, we explore under what conditions and with what accuracy linear colour mechanisms representing the matching metamerism $\sim_\varsigma$ can be derived from the geometrical shape of a colour matching set. Thus, experimental identification of colour matching sets becomes a central concern. This task is quite complicated due to a number of circumstances, the most obvious of which is the infinite-dimensionality of colour matching sets. In this regard, special attention will be paid to those additional conditions that must be imposed on colour matching relations so that colour matching sets can be identified in the experiment.

As noted in the previous chapter, data available in literature, possibly related to colour matching sets, were obtained in colour discrimination experiments rather than colour matching. Although we treat colour matching and colour discrimination as mutually complementary actions, we adopt in this and Chapter 12 the conceptual framework based on discrimination. In particular, below colour matching relations and colour matching sets are expressed by using the response 'different' (rather than 'same' as in the previous chapter). However, it is not so important since our main consideration will remain deterministic. True, in the beginning (Section 11.1) we introduce stochastic terminology to illustrate some theoretical notions at the level of experimental implementation. Yet, the methods of identification of colour mechanisms underlying the Grassmann structure induced by some colour matching relation, which we develop in this chapter, are applicable irrespective of whether this colour matching relation is stochastic or deterministic (as described in Chapter 10).

Foundations of Colour Science: From Colorimetry to Perception, First Edition. Alexander D. Logvinenko and Vladimir L. Levin.
© 2023 John Wiley & Sons Ltd. Published 2023 by John Wiley & Sons Ltd.

11.1 Colour matching set, threshold set, and sensitivity function

Here we introduce terminology regarding colour discrimination, which is broadly consistent with what is used in psychophysics. Given the stochastic colour matching relation (10.5), the colour matching set for light a (according to Definitions 10.5) is

$$\varsigma(a) = \{b \in A : P(S|a, b) \geq \varepsilon\}.$$

As the probability $P(S|\, a, b)$ of matching lights a and b is complementary to the probability $P(D|\, a, b)$ of discrimination between these lights, the colour matching set can be rewritten as

$$\varsigma(a) = \{b \in A : P(D|\, a, b) \leq \delta\}, \tag{11.1}$$

where $\delta = 1 - \varepsilon$. The positive number δ will be referred to as the *discriminability level.*

One of the two lights to be discriminated is usually chosen as a *background* (or *reference*) and kept constant during the experiment. It will be denoted by a^0. The other will be represented as $a^0 + a$, with a playing the role of a *target.*[1] In such an experimental condition (when the background is fixed), we deal with a particular case of colour discrimination – *colour detection.* To be more exact, to detect the target, a, on the background a^0, means, by definition, to discriminate between light stimuli a^0 and $a^0 + a$. Using this terminology, the colour matching set for a^0 is fully specified by the set of targets a which are detected on the background a^0 with the probability not more than the discriminability level, i.e.,

$$\varsigma\left(a^0\right) = \left\{a^0 + a \in A : P\left(D|\, a^0, a^0 + a\right) \leq \delta\right\}. \tag{11.2}$$

There are a few common experimental displays to study colour detection. Since detection is a particular case of discrimination, one can use the bipartite display as in Fig. I.1. The background light stimulus a^0 is presented in one-half of the bipartite field, and the stimulus $a^0 + a$ in the other half. Alternatively, a large homogeneous field can be used as the background light stimulus, and a small target is superimposed on it (as in Fig. 11.1). An advantage of such a stimulus display is that the constant state of chromatic adaptation (see Section 13.1) can be maintained because it is mainly determined by the spatially large background which does not change during the experiment. The observer is supposed to decide whether the target is present or not. In this case, D in (11.2) should be understood as a *detection response*, that is, in favour of the presense of the target. In the present chapter, we will restrict ourselves mainly to the bipartite-field based paradigm (Fig. I.1), whereas the stimulus display as in Fig. 11.1 will be focus in Chapter 12 where colour detection will be considered in more detail.

In experiments on colour detection, the two-alternative forced choice technique is often used. The observers would familiarise themselves with the background light stimulus in advance. Then, when being presented with, say, the bipartite field (as in Fig. I.1), they are asked in which half the target is presented. For example, assume that a^0 has a neutral colour, and it is presented on the left hand side. Then the right hand side may happen to look tinged with some hue, say, greenish, that guides the observer that the target is on the right. However, it may happen that the whole filed look

1 Note that target a need not be a positive Borel measure provided that the sum $a^0 + a$ remains positive. In other words, the family of all possible targets for a given background light stimulus is larger than the light stimulus cone A. Moreover, for the sake of generality we will put into consideration also those targets a in the ambient linear space E for which $\left(a^0 + a\right) \notin A$ (i.e., $a^0 + a$ are imaginary light stimuli). Such targets will be referred to as *imaginary targets.* (Note that for the target to be imaginary, it is not enough for it to be an imaginary light stimulus. This requires the background plus the target to be an imaginary light stimulus.)

Figure 11.1 Colour detection display. A small yellow patch of light (the target light stimulus, a) is to be detected against the large yellow-greenish patch (the background light stimulus, a^0), both being presented on the neutral remote background.

homogeneously neutral. In this case, the observer is forced to guess where the target is. When both halves are physically identical (i.e., $a = 0$), the observer's responses (i.e., 'on the right' and 'on the left') are distributed in equal proportions.[2] In the case of the display in Fig. I.1, it is presented to the observer twice, sequentially in two time intervals. Only in one of these intervals there may be the target. The observers is suggested to indicate in which time interval the target was. When target is absent in both intervals the probability of both responses are equal to 0.5.

When the target a belongs to the null-space of the colour map $E(\sim)$, the background light stimulus a^0 will be metameric to $a^0 + a$. In this case, the probability of both responses will be equal to 0.5 because the colour signals from both halves of the bipartite field will be equally distributed. Thus, there will be no sensory basis for distinguishing between the halves. When the light stimuli a^0 and $a^0 + a$ are not metameric (i.e., $a \notin E(\sim)$), the probability of the correct response will be more than 0.5.

Thus, the probability of the correct response can be taken as a dependent variable common to both types of experiment. Then, the colour matching set for a^0 takes the following form:

$$\varsigma(a^0) = \{a^0 + a \in A : P(C| a^0, a^0 + a) \leq \delta\}, \tag{11.3}$$

where C stands for the correct response. Clearly, the discriminability level δ in (11.3) should exceed 0.5.

We say that the target a is at *threshold* if the probability of the response (either D as in (11.2) or C as in (11.3)) equals the discriminability level δ. If the probability of the response is lower than the discriminability level (i.e., when a^0 and $a^0 + a$ match each other), the target is said to be *subthreshold*. Otherwise (i.e., when a^0 and $a^0 + a$ are discriminated) a target is called *superthreshold*. All the subthreshold and threshold targets for a given background constitute the subset of E which we shall call the *threshold set*. It is clear that the threshold set contains the intersection of $A - a^0$ with the null-space of the colour map $E(\sim)$.[3] The set of threshold stimuli constitutes a boundary of the threshold set and will be referred to as the *threshold (or equi-discrimination) surface*.

There is every indication that any proper (not imaginary) target can be made subthreshold when decreasing its intensity. Specifically, for any target a there is a positive number k (intensity) such that the target ka is subthreshold. For every superthreshold target a there exists the lowest intensity,

---

2 In reality, some observers may demonstrate bias in favour of one of two responses. We will leave this case without consideration, referring the reader to the relevant literature on psychophysics, where this problem is discussed.
3 When imaginary targets are included in consideration, the whole $E(\sim)$ consists of all (both, proper and imaginary) threshold targets, that is $E(\sim)$ is contained in the *extended* threshold set (see Definition 11.3).

written as $k_{th}(a)$, at which the probability of detection of $k_{th}(a)\,a$ is still not less than the discriminability level. In psychophysics, the number $k_{th}(a)$ is called the *threshold* for target a. When the intensity k is less than the threshold $k_{th}(a)$ the target ka becomes subthreshold.

Given an arbitrary (not necessarily proper) target a, we define its *threshold* to be the infimum of all numbers $k > 0$ for which the probability of detection of ka is not less than the discriminability level:

$$k_{th}(a) := \inf\{k > 0 : P(D|a^0, a^0 + ka) \geq \delta\}.$$

If the set of such numbers k is empty, that is, $P(D|a^0, a^0 + ka) < \delta$ whatever $k > 0$, then, by definition, $k_{th}(a) = \infty$. In particular, any target $a \in E(\sim)$ has the infinite threshold.

Not every subthreshold target can be made superthreshold by increasing its intensity. Firstly, for $a \in E(\sim)$, ka also belongs to $E(\sim)$. Secondly, for some negative (or signed) measures a and some $k > 0$ the sum $a^0 + ka$ may become not positive as well. (We will call such k *inadmissible*.) When for a subthreshold target a there exists the highest intensity, written as k', at which the probability of detection of $k'a$ is still not more than the discriminability level, k'/k equals to the threshold for any target ka ($k > k'$).

There are established psychophysical methods to statistically estimate threshold (Guilford, 1956; Treutwein, 1995; Gescheider, 1997). It is well established that the probability of detection of the target ka monotonically increases with k for any $a \notin E(\sim)$ unless k is inadmissible. A monotonic function $P\left(D|\, a^0, a^0 + ka\right)$ of k can be evaluated in experiment for any target $a \notin E(\sim)$. In psychophysics, it is called *psychometric function*.[4] A threshold for a target $a \notin E(\sim)$ is just such k that $P\left(D|\, a^0, a^0 + ka\right)$ equals the discriminability level.

A reciprocal of the detection threshold for a stimulus is commonly used in psychophysics as a measure of *sensitivity* to this stimulus (Falmagne, 1985).

Definition 11.1 *We define the* sensitivity function *on the light stimulus space E to be the function* $a \mapsto k_{th}^{-1}(a)$.

Note that if the *extended threshold set* (i.e., the set of all subthreshold and threshold stimuli in E, including imaginary ones) is a body then the sensitivity function obviously coincides with its gauge (Minkovski function); see Appendix A.2 for definitions. For the definition of sensitivity function in general (not necessarily stochastic) situation, see Definition 11.5.

11.2 Regular and strongly regular tolerance extensions

Let $\langle A, \varsigma\rangle$ be a colour matching structure, and $\left\langle A, \sim_{\varsigma}\right\rangle$ an induced Grassmann structure.

Definition 11.2 *Given a colour matching structure $\langle A, \varsigma\rangle$ and a background light stimulus $a^0 \in A$, the translated colour matching set $B_{a^0} := \varsigma(a^0) - a^0$ will be referred to as the* threshold set *of a^0.*

Definition 11.3 *Given a colour matching structure $\langle A, \varsigma\rangle$, we say a tolerance ς' on E is an* extension *of ς if, for all $a, b \in A$, $a\varsigma' b$ is equivalent to $a\varsigma b$. Given a background light stimulus $a^0 \in A$ and a tolerance extension ς' of ς, we define the* extended colour matching set *for a^0,*

$$\varsigma'\left(a^0\right) := \{e \in E : a^0\varsigma' e\},$$

4 The notion of psychometric function will further be generalised (see Eq. (12.1)).

and the extended threshold set *of* a^0,

$$B'_{a^0} := \varsigma'(a^0) - a^0.$$

When ς' is an extension of ς, one has

$$\varsigma(a^0) = \{a \in A : a^0 \varsigma a\} = \varsigma'(a^0) \cap A \tag{11.4}$$

(and $\varsigma(a) = \varsigma'(a) \cap A$ whenever $a \in A$); therefore

$$B_{a^0} = B'_{a^0} \cap (A - a^0). \tag{11.5}$$

Definition 11.4 *We say a tolerance extension ς' of a given colour matching relation ς is* regular *(with respect to $a^0 \in A$), if $E(\sim_\varsigma) \subset B'_{a^0}$. If, in addition, $\alpha e \in B'_{a^0}$ for all $\alpha > 0$ imply $e \in E(\sim_\varsigma)$, we will say that ς' is* strongly regular.

Proposition 11.1 *Suppose ς' is a regular tolerance extension of ς. If $a \in \varsigma(a^0)$, φ is a linear functional on E, and $\varphi(e) \leq \varphi(a)$ for all $e \in \varsigma'(a^0)$, then $\varphi(e) = 0$ for all $e \in E(\sim_\varsigma)$, where $\sim_\varsigma$ is the matching metamerism induced by ς.*

Proposition 11.2 *Let $\langle A, \varsigma \rangle$ be a colour matching structure such that the induced Grassmann structure $\langle A, \sim_\varsigma \rangle$ is continuous (that is, $E = A - A$ is a Hausdorff locally convex space and $E(\sim_\varsigma)$ is closed in E; see Definition 2.10). Given a background light stimulus $a^0 \in A$ and a tolerance ς' on E, which is a regular extension of ς, we suppose that the extended colour matching set $\varsigma'(a^0)$ is closed convex and its interior in E (denoted as $int(\varsigma'(a^0))$) is nonempty. Then for every boundary point e of $\varsigma'(a^0)$ there is a functional $\varphi \in E^*$ such that $\langle e', \varphi \rangle < \langle e, \varphi \rangle$ whenever $e' \in int(\varsigma'(a^0))$.*

Let $E(\sim_\varsigma)^0$ denote the linear colour mechanism space that is, the annihilator of $E(\sim_\varsigma)$ in E^*,

$$E(\sim_\varsigma)^0 := \{\varphi \in E^* : \langle e, \varphi \rangle = 0 \quad \forall e \in E(\sim_\varsigma)\}.$$

The next result is an immediate consequence of Propositions 11.1 and 11.2.

Corollary 11.1 *Suppose that the hypotheses of Proposition 11.2 are satisfied; then for every $e \in bdry(\varsigma'(a^0))$ there is a non-zero functional $\varphi \in E(\sim_\varsigma)^0$ supporting $\varsigma'(a^0)$ at e: $\langle e', \varphi \rangle \leq \langle e, \varphi \rangle$ whenever $e' \in \varsigma'(a^0)$.*

Let $\langle A, \varsigma \rangle$ be a colour matching structure such that the induced Grassmann structure $\langle A, \sim_\varsigma \rangle$ is continuous, and let $\Phi : E \to \mathbf{R}^n$ be a colour-signal map[5]. Suppose that the colour matching relation ς is given by

$$a \varsigma b \Leftrightarrow d(\Phi(a), \Phi(b)) \leq \varepsilon \quad (a, b \in A), \tag{11.6}$$

where $\varepsilon > 0$ and a function $d : \mathbf{R}^n \times \mathbf{R}^n \to \mathbf{R}_+$ satisfies two conditions as follows: $d(x, y) = d(y, x)$ for all $x, y \in \mathbf{R}^n$, and $d(x, x) = 0$ for all $x \in \mathbf{R}^n$. As follows from Corollary 10.1, the indiscriminate relation ν_ς on $\Phi(A)$ is then simple. Note that $\nu_\varsigma = \nu$, where

$$x \nu y \Leftrightarrow d(x, y) \leq \varepsilon \quad (x, y \in \Phi(A)). \tag{11.7}$$

5 In such a case, Φ is necessarily continuous; see Proposition 2.6.

Furthermore, if $E = A - A$ is a Hausdorff locally convex space, the cone A is closed in E, Φ is a continuous linear operator that maps E onto $\mathbf{R}^n$, and the tolerance relation ν on $\Phi(A)$ given by (11.7) is simple, then tolerance relation (11.6) determines on A a colour matching structure such that the induced Grassmann structure $\langle A, \sim_{\varsigma}\rangle$ is continuous (relative to the original topology in E), Φ is a colour-signal map for it, and $\nu_{\varsigma} = \nu$. This follows from Corollary 10.2. Also it follows from Corollary 10.2 that $E(\sim_{\varsigma}) = \Phi^{-1}(0)$ if and only if ν is simple. Moreover, in this case the following proposition holds true.

Proposition 11.3 *Let $a^0 \in A$ and $x^0 = \Phi(a^0)$. Suppose that: (i) the set $\{x \in \Phi(A) : d(x^0, x) \leq \varepsilon\}$ is convex and closed in $\mathbf{R}^n$; (ii) the function $\mathbf{R}^n \ni x \mapsto d(x^0, x)$ is continuous; (iii) the equality holds*

$$\{x \in \mathbf{R}^n : d(x^0, x) \leq \varepsilon\} = \{x \in \Phi(A) : d(x^0, x) \leq \varepsilon\}. \tag{11.8}$$

Consider the relation ς on A given by (11.6) and the relation ν on $\Phi(A)$ given by (11.7), and suppose that ν is simple. Then the relation ς' on E,

$$e_1 \varsigma' e_2 \Leftrightarrow d(\Phi(e_1), \Phi(e_2)) \leq \varepsilon \quad (e_1, e_2 \in E),$$

is a regular extension of ς, the set $\varsigma'(a^0)$ is closed and convex, its interior in E is nonempty, and the equality holds

$$\varsigma'(a^0) = \varsigma(a^0) + E(\sim_{\varsigma}). \tag{11.9}$$

Remark 11.1 *Note that (11.9) can be rewritten as*

$$B'_{a^0} = B_{a^0} + E(\sim_{\varsigma}). \tag{11.10}$$

Corollary 11.2 *Suppose (11.9) holds, ς' is a regular extension of ς, the set $\varsigma'(a^0)$ is closed and convex, and its interior in E is nonempty; then, given an element $e \in bdry(\varsigma'(a^0))$ and a linear functional φ supporting $\varsigma'(a^0)$ at e, there exists an element $a \in A \cap bdry(\varsigma'(a^0))$ such that φ supports $\varsigma'(a^0)$ at a.*

Theorem 11.1 *Suppose $\langle A, \sim\rangle$ is a n-dimensional continuous Grassmann structure, $\Phi : E \to \mathbf{R}^n$ is a colour-signal map for it, $a^0 \in A$, and $g : \mathbf{R}^n \to \mathbf{R}_+$ is a convex function satisfying the following conditions: (i) $g(x) = 0$ if and only if $x = 0$; (ii) $g(x) = g(-x)$ for all $x \in \mathbf{R}^n$; (iii) for any linearly independent vectors $x, y \in \mathbf{R}^n$, g is strictly convex on the segment $\{(1-t)x + ty : 0 \leq t \leq 1\}$. Consider a colour matching structure $\langle A, \varsigma\rangle$, where the tolerance ς is given by*

$$a \varsigma b \Leftrightarrow g(\Phi(a) - \Phi(b)) \leq \varepsilon \quad (a, b \in A).$$

The following assertions hold then true: (i) For every $\varepsilon > 0$, the Grassmann structure $\langle A, \sim\rangle$ is induced by the colour matching structure $\langle A, \varsigma\rangle$, and the tolerance ς' on E,

$$e_1 \varsigma' e_2 \Leftrightarrow g(\Phi(e_1 - e_2)) \leq \varepsilon \quad (e_1, e_2 \in E),$$

is a strongly regular extension of ς. (ii) If $x^0 = \Phi(a^0) \in int(\Phi(A))$, then there is a number $\varepsilon(x^0) > 0$ such that, for every $0 < \varepsilon \leq \varepsilon(x^0)$, the extended colour matching set $\varsigma'(a^0)$ is closed, convex and has the nonempty interior. In such a case, for every $e \in bdry(\varsigma'(a^0))$ there is a non-zero functional $\varphi \in E(\sim_{\varsigma})^0$ supporting $\varsigma'(a^0)$ at e: $\langle e', \varphi\rangle \leq \langle e, \varphi\rangle$ whenever $e' \in \varsigma'(a^0)$.

Theorem 11.2 *Suppose that $\langle A, \sim\rangle$ is a n-dimensional continuous Grassmann structure, $\Phi : E \to \mathbf{R}^n$ is a colour-signal map for it, $a^0 \in A$, and $d : \mathbf{R}^n \times \mathbf{R}^n \to \mathbf{R}_+$ is given by*

$$d(x,y) = \frac{\|x-y\|}{\|x\|_{1,\alpha} + \|y\|_{1,\alpha} + 1} \quad (x, y \in \mathbf{R}^n),$$

where $\|x\|_{1,\alpha} := \alpha_1|x_1| + \cdots + \alpha_n|x_n|$, $x = (x_1, \ldots, x_n) \in \mathbf{R}^n$ (see Remark 10.12). Consider a colour matching structure $\langle A, \varsigma\rangle$, where the tolerance ς is given by (11.6). The following assertions hold then true: (i) If $\varepsilon > 0$ is small enough ($\varepsilon < 1/\|\alpha\| = \left(\alpha_1^2 + \cdots + \alpha_n^2\right)^{-1/2}$; see Remark 10.12), then the Grassmann structure $\langle A, \sim\rangle$ is induced by the colour matching structure $\langle A, \varsigma\rangle$, and the tolerance ς' on E,

$$e_1\varsigma' e_2 \Leftrightarrow \frac{\|\Phi(e_1 - e_2)\|}{\|\Phi(e_1)\|_{1,\alpha} + \|\Phi(e_1)\|_{1,\alpha} + 1} \leq \varepsilon \quad (e_1, e_2 \in E),$$

is a strongly regular extension of ς. (ii) If $x^0 = \Phi(a^0) \in int(\Phi(A))$, then there is a number $0 < \varepsilon(x^0) < 1/\|\alpha\|$ such that, for every $0 < \varepsilon \leq \varepsilon(x^0)$, the extended colour matching set $\varsigma'(a^0)$ is closed, convex and has the nonempty interior. In such a case, for every $e \in bdry(\varsigma'(a^0))$ there is a non-zero functional $\varphi \in E(\sim_\varsigma)^0$ supporting $\varsigma'(a^0)$ at e: $\langle e', \varphi\rangle \leq \langle e, \varphi\rangle$ whenever $e' \in \varsigma'(a^0)$.

Remark 11.2 *If $x^0 = \Phi(a^0) \notin int(\Phi(A))$, then the set $\varsigma'(a^0)$ fails to be convex; cf. Remark 10.10.*

We now give general definitions of sensitivity function, most sensitive colour mechanism and colour detector that will be systematically used further.

Definition 11.5 *Let $\langle A, \varsigma\rangle$ be a colour matching structure, ς' a strongly regular extension of the tolerance ς. Given a background light stimulus $a^0 \in A$, we consider the extended threshold set B'_{a^0} and suppose that it is a body (i.e., absorbing closed[6] convex set) in $E = \mathbf{C}^*$. The* sensitivity function *is defined then to be its gauge (Minkovski function $\mu(\cdot; B'_{a^0})$; see Appendix A.2). Given a light stimulus a in the boundary of the extended threshold set ($a \in bdry(B'_{a^0})$), a colour mechanism $\varphi \in \mathbf{C}$ supporting B'_{a^0} at it ($\varphi \in N(a|B'_{a^0})$) and satisfying $\varphi(a) = \mu(a; B'_{a^0})(= 1)$ is said to be* most sensitive *(for a). Clearly, such a mechanism lies in the threshold body polar $(B'_{a^0})^0$ (cf. (A.18)). If for a given stimulus $a \in bdry(B'_{a^0})$ there is exactly one such a mechanism then it called a* colour detector.

11.3 Identification of Grassmann structures induced by colour matching tolerance relations

In this section, we describe three methods of evaluating the linear colour mechanism space. All the three require the shape of the colour matching set to be known for some light a. Specifically, the presumption is that the detection threshold (with respect to a) for any arbitrary target can be measured (in experiment). No explicit assumption concerning the indiscriminate relation (Definition 10.3), apart from those imposed by condition (I) below, is made for the first two methods (Sections 11.3.1 and 11.3.2). The third method implies the particular indiscriminate relation determined by a particular way of combining the outputs of some (though unknown) colour mechanisms. We will keep referring to them as the fundamental colour mechanisms despite that the latter have been defined only for the colorimetric colour matching model. Since these colour mechanisms are not assumed to be known in advance in Sections 11.3.2 and 11.3.3, the formal

6 As usually, the space $\mathbf{C}^*$ is considered with the weak* topology on it.

generalisation of the notion of fundamental colour mechanisms is deterred to Section 12.1.2. True, in Section 11.3.1 the fundamental colour mechanisms are indeed assumed to belong to some finite family of linear functionals. However, as explained in the Outline (Chapter 1), this is made to give consideration to an important particular case arising from the obvious fact that the colour vision begins from the excitation of the finite number of the photoreceptors the spectral sensitivity profile of which is genetically predetermined. In other words, the method presented in Section 11.3.1 is applicable when the linear colour mechanism space is determined by an unknown number of the cone fundamentals with known spectral sensitivities.

11.3.1 Identification of the linear colour mechanism space as a subspace in the linear span of a given set of linearly independent functionals

In what follows, A is the cone $\mathbf{C}_+^* = \mathbf{C}_+^*(\Lambda)$ and $E = A - A$ is the space $\mathbf{C}^* = \mathbf{C}^*(\Lambda)$ endowed with the weak* topology. The method is as follows. We take a large positive integer N and fix N linearly independent measures $\sigma_1, \dots, \sigma_N \in A = \mathbf{C}_+^*$ and N linearly independent functions $\varphi_1, \dots, \varphi_N \in E^* = \mathbf{C}$. We will approximate $E = \mathbf{C}^*$ and $E^* = \mathbf{C}$ by their finite-dimensional subspaces $E_N = Lin(\sigma_1, \dots, \sigma_N)$ and $E_N^* = Lin(\varphi_1, \dots, \varphi_N)$, and suppose that the restriction to $E_N^* \times E_N$ of the canonical bilinear form on $\mathbf{C} \times \mathbf{C}^*$,

$$\langle \varphi, \sigma \rangle = \int_\Lambda \varphi(\lambda)\ \sigma(d\lambda),$$

is *non-degenerate* in both arguments, that is: (i) for every non-zero $e \in E_N$ there is an element $\varphi \in E_N^*$ such that $\langle \varphi, e \rangle = \int_\Lambda \varphi(\lambda) e(d\lambda) \neq 0$, and (ii) for every non-zero $\varphi \in E_N^*$ there is an element $e \in E_N$ such that $\langle \varphi, e \rangle \neq 0$. Note that both these conditions, (i) and (ii), are satisfied when $\sigma_i = \delta_{\lambda_i}$, $i = 1, \dots, N$, all λ_i are different (i.e., $\lambda_i \neq \lambda_j$ for $i \neq j$), and $\varphi_1, \dots, \varphi_N \in \mathbf{C}$ are arbitrary functions that are linearly independent as functions on the set $\{\lambda_1, \dots, \lambda_N\}$.[7]

Lemma 11.1 *Given $E_N = Lin(\sigma_1, \dots, \sigma_N) \subset \mathbf{C}^*$, $E_N^* = Lin(\varphi_1, \dots, \varphi_N) \subset \mathbf{C}$, the following assertions are equivalent: (i) the non-degeneracy condition is satisfied: the restriction to $E_N^* \times E_N$ of the canonical bilinear form $\langle \varphi, \sigma \rangle$ is non-degenerate in both arguments; (ii) $\sigma_1, \dots, \sigma_N$ are linearly independent as functionals on E_N^* (that is, $\langle \varphi, \sum_1^N \alpha_i \sigma_i \rangle = 0$ for all $\varphi \in E_N^* \Rightarrow \alpha_i = 0,\ i = 1, \dots, N$) and $\varphi_1, \dots, \varphi_N$ are linearly independent as functionals on E_N (that is, $\langle \sum_1^N \alpha_i \varphi_i, e \rangle = 0$ for all $e \in E_N \Rightarrow \alpha_i = 0,\ i = 1, \dots, N$); (iii) there exist elements $e_1, \dots, e_N \in E_N$, $\varphi_1', \dots, \varphi_N' \in E_N^*$ satisfying the biorthogonality condition:*

$$\langle \varphi_i', e_j \rangle = \begin{cases} 1 & for\ i = j \\ 0 & for\ i \neq j \end{cases}$$

(iv) the $N \times N$-matrix $(\langle \varphi_i, \sigma_j \rangle)_{ij}$ is non-singular: $det(\langle \varphi_i, \sigma_j \rangle)_{ij} \neq 0$.

Remark 11.3 *Assertion (ii) means $E_N \cap (E_N^*)^0 = \{0\}$, $E_N^* \cap (E_N)^0 = \{0\}$, where $(E_N^*)^0$ is the annihilator of E_N^* in $\mathbf{C}^*$ and $(E_N)^0$ is the annihilator of E_N in $\mathbf{C}$.*

---

7 The latter means $det(\varphi_i(\lambda_j)) \neq 0$. In particular, one can take $\varphi_i(\lambda) = p(\lambda) e^{-(\lambda - \lambda_i)^2}$, $i = 1, \dots, N$, where p is a positive continuous function on Λ (cf. Example 4.1). Moreover, as follows from Lemma 2.1, for *any* linearly independent functions $\varphi_1, \dots, \varphi_N \in \mathbf{C}$ one can find N different points $\lambda_1, \dots, \lambda_N \in \Lambda$ such that $det(\varphi_i(\lambda_j)) \neq 0$ hence conditions (i) and (ii) are satisfied for $\sigma_i = \delta_{\lambda_i}$.

Remark 11.4 *In (iii) one can fix $e_j = \sigma_j$ and find $\varphi_i' \in E_N^*$ or fix $\varphi_i' = \varphi_i$ and find $e_j \in E_N$.*

Remark 11.5 *Note that for $E_N = Lin(\delta_{\lambda_1}, \dots, \delta_{\lambda_N})$, $\varphi_1, \dots, \varphi_N$ are linearly independent as functionals on E_N if and only if they are linearly independent as functions on $\{\lambda_1, \dots, \lambda_N\}$.*

Given a background light stimulus $a^0 \in A = \mathbf{C}_+^*$, it follows from Propositions 11.1 and 11.2 that, under the hypotheses of Proposition 11.2, every functional $\varphi \in \mathbf{C}$ supporting $\varsigma'(a^0)$ at some point $a \in bdry(\varsigma'(a^0))$ belongs to the linear colour mechanism space $E(\sim_\varsigma)^0$. If we know that the linear colour mechanism space $E(\sim_\varsigma)^0$ is n-dimensional and if we can find n points $a_1, \dots, a_n \in bdry(\varsigma'(a^0))$ and n linearly independent functionals $\phi_1, \dots, \phi_n \in \mathbf{C}$ supporting $\varsigma'(a^0)$ at these points, then $E(\sim_\varsigma)^0$ turns out to coincide with the subspace $Lin(\phi_1, \dots, \phi_n)$ spanned over $\phi_1, \dots, \phi_n$. If the dimension of $E(\sim_\varsigma)^0$ is unknown, one can take a large number $N_1 > N$ of points $a_1, \dots, a_{N_1} \in bdry(\varsigma'(a^0))$ and find $N_2 \geq N_1$ functionals ϕ_j supporting $\varsigma'(a^0)$ at a_i, $i = 1, \dots, N_1$. If the space $Lin(\phi_1, \dots, \phi_{N_2})$ spanned over these functionals is n-dimensional and remains n-dimensional when N_1 and N_2 increase and new points a_i and supporting functionals ϕ_j are added to $\{a_1, \dots, a_{N_1}\}$ and $\{\phi_1, \dots, \phi_{N_2}\}$, then it is natural to assume that $E(\sim_\varsigma)^0$ coincides with $Lin(\phi_1, \dots, \phi_{N_2})$, hence is n-dimensional. Thus, in order to determine the linear colour mechanism space $E(\sim_\varsigma)^0$, one has to collect many boundary points of the extended colour matching set $\varsigma'(a^0)$, to find supporting functionals at these points, and to take the subspace spanned over these functionals. The practical implementation of this plan faces some difficulties caused by the fact that the extended colour matching set is inaccessible to an observer. The observer is able to judge whether or not a given light stimulus a is a colour match for a^0, that is, s/he knows the colour matching set $\varsigma(a^0)$. As for the extended colour matching set $\varsigma'(a^0)$, only some indirect information can be available.

Given a light stimulus $a \in \varsigma(a^0)$, we will consider the target stimulus $e = a - a^0$. Note that if $\varsigma(a^0)$ is assumed to be convex, then $a^0 + te \in \varsigma(a^0)$ for all $0 < t < 1$. Now, given an arbitrary $e \in E$, $e \neq 0$, we define

$$t_e = \max\{t \geq 0 : a^0 + te \in \varsigma(a^0)\} \tag{11.11}$$

(maximum is attained since $\sup\{t \geq 0 : a^0 + te \in \varsigma(a^0)\}$ is finite[8] and $\varsigma(a^0)$ is closed) and consider two sets, Q and Q', as follows:

$$Q := \{e \in E : t_e > 0\}, \tag{11.12}$$

$$Q' := \{e \in Q : a^0 + te \in A \text{ for some } t > t_e\}. \tag{11.13}$$

Proposition 11.4 *Suppose that ς' is a strongly regular extension of ς, the extended colour matching set $\varsigma'(a^0)$ is closed and convex, its interior (relative to the weak* topology) in $E = \mathbf{C}^*$ is nonempty, and equality (11.9) is satisfied; then:*

1. *The extended colour matching set $\varsigma'(a^0)$ is represented as*

$$\bigcap_{a,\varphi} \{e \in E : \langle \varphi, e \rangle \leq \langle \varphi, a \rangle\},$$

where the intersection is taken over $a \in bdry(\varsigma'(a^0))$ and $\varphi \in \mathbf{C}$ supporting $\varsigma'(a^0)$ at a.

---

8 Indeed, if $e \notin A$, then $a^0 + te \notin A$ for large $t > 0$, hence $a^0 + te \notin \varsigma(a^0)$. If $e \in A$ and $a^0 + te \in \varsigma(a^0)$ whenever $t > 0$, then $te \in B_{a^0} \subset B'_{a^0}$ for all $t > 0$, and as ς' is assumed to be a strongly regular extension of ς, this implies $e \in E(\sim_\varsigma)$ but the latter is impossible because $e \in A$.

2. *If $a \in A$ is a boundary point of $\varsigma'(a^0)$, then $e = a - a^0 \in Q$ and $t_e = 1$.*
3. *If $e \in Q'$, then $a = a^0 + t_e e \in A \cap bdry(\varsigma'(a^0))$.*

It is easy to see that $A \subset Q'$. Given a non-zero $e \notin A$, there is a number $t_{1e} \geq 0$ such that $a^0 + te \in A$ for $t \leq t_{1e}$ and $a^0 + te \notin A$ for $t > t_{1e}$.[9] Clearly, $t_e \leq t_{1e}$. Furthermore, it is easily seen that, for $e \notin A$, the inequality $t_{1e} > 0$ holds true if and only if: (i) e is absolutely continuous with respect to a^0 and (ii) its density is an a^0-measurable a^0-almost everywhere bounded function of λ: $\frac{de}{da^0} \in \mathbf{L}^\infty(\Lambda, a^0)$. In this case, $a^0 + t_{1e}e = (1 + t_{1e}\frac{de}{da^0}(\lambda))a^0$; therefore, $\sup\{\alpha : \frac{de}{da^0}(\lambda) \geq \alpha \;\; a^0\text{-a.e.}\} < 0$ and $t_{1e} = -\left(\sup\{\alpha : \frac{de}{da^0}(\lambda) \geq \alpha \;\; a^0\text{-a.e.}\}\right)^{-1}$.

The next result follows directly from the definitions of t_e and t_{1e}.

Lemma 11.2 *$A \subset Q'$, and for $e \notin A$, $e \in Q'$ if and only if $t_e < t_{1e}$.*

Remark 11.6 *Thus, from Proposition 11.4 and Lemma 11.2 it follows that $a^0 + t_e e$ is a boundary point of $\varsigma'(a^0)$ when $e \in A$ or $e \notin A$ but $t_e < t_{1e}$. If $e \notin A$ and $t_e = t_{1e}$, then $a^0 + t_e e \in A$ and $a^0 + te \notin A$ whenever $t > t_e$. In this case, we cannot say whether $a^0 + t_e e$ is a boundary or an interior point of $\varsigma'(a^0)$. Note in this connection that if the extended colour matching relation ς' is given by*

$$e_1 \varsigma' e_2 \Leftrightarrow d(\Phi(e_1), \Phi(e_2)) \leq \varepsilon,$$

where $d : \mathbf{R}^n \times \mathbf{R}^n \to \mathbf{R}_+$ is a function with appropriate properties (see Theorem 11.1 or Theorem 11.2), and ε is small enough, then $\varsigma'(a^0)$ and t_e depend on ε (they decrease as ε decreases), but t_{1e} is determined by A and e only and does not depend on ε. (Note that the corresponding matching metamerism $\sim_\varsigma$ is completely determined by Φ and does not depend on ε.) Therefore, it can occur that $t_e(\varepsilon) = t_{1e}$ for a given ε but $t_e(\varepsilon') < t_{1e}$ when $\varepsilon' < \varepsilon$.

We shall need some a priori assumptions concerning the extended colour matching relation ς', the background light stimulus a^0 and the colour-signal map Φ determined by (unknown) fundamentals f_i. We will assume that:

(I) ς' is a strongly regular extension of ς, the extended colour matching set $\varsigma'(a^0)$ is closed and convex, its interior (relative to the weak* topology) in $E = \mathbf{C}^*$ is nonempty, and equality (11.9) is satisfied.

(II) Although neither the number n of fundamentals f_i nor their exact shape are known, it is assumed that they are among N functionals $\varphi_1, \dots, \varphi_N$ and belong to $E_N^* = Lin(\varphi_1, \dots, \varphi_N)$.

(III) The indiscriminate relation $\nu = \nu_\varsigma$ on the colour-signal cone $\Phi(A)$ is simple, $x^0 = \Phi(a^0)$ is an interior point of $\Phi(A)$, and $\nu'(x^0) = \nu(x^0)$.

It follows from (II) that if n is the maximum amount of linearly independent functionals $\phi_i \in E_N^*$ that are supporting to $\varsigma'(a^0)$, then $E(\sim_\varsigma)^0 = Lin(\phi_1, \dots, \phi_n) = Lin(\varphi_{i_1}, \dots, \varphi_{i_n})$ for some $1 \leq i_1 < i_2 < \dots < i_n \leq N$. In such a case, $\varphi_{i_1}, \dots, \varphi_{i_n}$ prove to be the fundamentals.

We now describe an approach to finding the functionals $\phi_1, \dots, \phi_n$ and $\varphi_{i_1}, \dots, \varphi_{i_n}$. Note that every non-zero $e \in E_N$ is uniquely represented as $e = t(\alpha_1\sigma_1 + \dots + \alpha_N\sigma_N)$, where $t > 0$, $|\alpha_1| + \dots + |\alpha_N| = 1$, and similarly, every non-zero $\varphi \in E_N^*$ is uniquely represented as

9 Note that if the signed measure $e \notin A$ is not absolutely continuous with respect to a^0, then $t_{1e} = 0$. In particular, it is the case where a^0 is the Lebesgue measure and $e = -\delta_\lambda$ is any Dirac measure with a minus sign.

$\varphi = t(\alpha_1\varphi_1 + \cdots + \alpha_N\varphi_N)$, where $t > 0$, $|\alpha_1| + \cdots + |\alpha_N| = 1$. We will consider E_N as a target space with respect to the background a^0. Let us fix a positive integer m and define

$$S_m := \left\{ s = (\alpha_1, \dots, \alpha_N) : \alpha_i = \epsilon_i m_i/m, \epsilon_i = \pm 1, m_i \in \mathbf{N}_0, \sum_1^N m_i = m \right\},$$

where $\mathbf{N}_0$ is the set of non-negative integers. Every $s = (\alpha_1, \dots, \alpha_N) \in S_m$ determines an element $e_s = \alpha_1\sigma_1 + \cdots + \alpha_N\sigma_N \in E_N$, and we define a *target grid* in E_N, $TG_m := \{e = e_s : s \in S_m\}$. Also we associate every $s = (\alpha_1, \dots, \alpha_N) \in S_m$ with a functional $\varphi_s = \alpha_1\varphi_1 + \cdots + \alpha_N\varphi_N$ and consider a *functional grid* in E_N^*, $FG_m := \{\varphi = \varphi_s : s \in S_m\}$. Note that $s = (\alpha_1, \dots, \alpha_N) \in S_m$ implies $|\alpha_1| + \cdots + |\alpha_N| = 1$, therefore $e_s \neq 0$ and $\varphi_s \neq 0$ whenever $s \in S_m$. The target grid TG_m approximates the set $\{e \in E_N : e = \sum_1^N \alpha_i\sigma_i, \sum_1^N |\alpha_i| = 1\}$, the functional grid FG_m approximates the set $\{\varphi \in E_N^* : \varphi = \sum_1^N \alpha_i\varphi_i, \sum_1^N |\alpha_i| = 1\}$, and the larger m, the more accurate both approximations. According to Proposition 11.4, every $e \in TG_m \cap Q'$ determines an element $a = a^0 + t_e e$ in $\varsigma(a^0)$ which is a boundary point of $\varsigma'(a^0)$. Let $\{a_1, \dots, a_{N_1}\}$ be the set of all such elements. Their total number N_1 equals the number of elements in $TG_m \cap Q'$ (denoted as $\#(TG_m \cap Q')$), which is much larger than N. (Indeed, since $A \subset Q'$, $e_s \in TG_m \cap Q'$ whenever $s = (\alpha_1, \dots, \alpha_N) \in S_m \cap \mathbf{R}_+^N$.) Let $\{\phi_1, \dots, \phi_{N_2'}\}$ be the set of all functionals φ_s in FG_m. Their general number is $N_2' = \#(FG_m) = \#(S_m) = \#(TG_m) \geq N_2 \geq N_1$, where N_2 is the general number of those φ_s from FG_m that are supporting functionals of $\varsigma'(a^0)$. Let us define

$$F(a_i) := \{\phi_j : \langle \phi_j, a_k \rangle \leq \langle \phi_j, a_i \rangle \quad \forall k \neq i\}, \quad i = 1, \dots, N_1.$$

Since a_i is a boundary point of $\varsigma'(a^0)$, there is a functional $\varphi \in E(\sim_\varsigma)^0$ supporting $\varsigma'(a^0)$ at a_i. Then $\varphi \in Lin(f_1, \dots, f_n)$ where $f_1, \dots, f_n$ are the fundamentals, and as $\{f_1, \dots, f_n\} \subset \{\varphi_1, \dots, \varphi_N\}$ (see (II)), it follows that $\varphi = \alpha_1\varphi_1 + \cdots + \alpha_N\varphi_N$, where at most n coefficients α_i are non-zero; however, we don't know what the number of fundamentals (i.e., n) is. We assume without loss of generality that $|\alpha_1| + \cdots + |\alpha_N| = 1$, and as $\langle \varphi, a_k \rangle \leq \langle \varphi, a_i \rangle$ for all $k \neq i$, it can be accepted that φ belongs to $F(a_i)$ (strictly speaking, φ is approximated with a good precision by one of functionals from $F(a_i)$). Thus, for each i, at least one of functionals ϕ_j from $F(a_i)$ is supporting for $\varsigma'(a^0)$ hence belongs to the linear colour mechanism space $E(\sim_\varsigma)^0$. In general, we cannot reveal which functionals from $F(a_i)$ are supporting because the validity of inequalities $\langle \phi_j, a_k \rangle \leq \langle \phi_j, a_i \rangle$ for $k \neq i$ does not imply that $\langle \phi_j, e \rangle \leq \langle \phi_j, a_i \rangle$ *for all* $e \in \varsigma'(a^0)$. Nevertheless, if $F(a_i)$ consists of the *unique* functional ϕ_j, then this functional will necessarily be supporting.[10] Note that, up to a given precision, the requirement of each $\phi_j \in F(a_i)$ to belong to E_N amounts to it being a supporting functional of $B'_{a^0} \cap E_N$ at the point $e = a_i - a^0 \in bdry(B'_{a^0} \cap E_N)$. Since the space E_N is finite-dimensional, it follows from Theorems 18.8 and 25.3 in Rockafellar (1970) that the set of *tangency points*[11] of $B'_{a^0} \cap E_N$ is dense in $bdry(B'_{a^0} \cap E_N)$ and massive in various meanings (it is a dense G_δ set of full N-dimensional Lebesgue measure). This makes plausible the hypothesis that many of $e_i = a_i - a^0$ are tangency points of $B'_{a^0} \cap E_N$, therefore, each of the corresponding $F(a_i)$ comprises the unique functional $\phi_{j_i} = \varphi_s$ supporting $\varsigma'(a^0)$ at a_i. Let $S_m^0 \subset S_m$ be the set of all such s. We know that each $\phi_{j_i} = \varphi_s$, $s \in S_m^0$, is a linear combination of fundamentals, and as we assume (II), the general number of fundamentals n

10 In this case, the hyperplane $\{e \in E : \langle \phi_j, e \rangle = \langle \phi_j, a_i \rangle\}$ is tangential to $\varsigma'(a^0)$ at a_i.

11 Given a closed convex set B with the nonempty interior in a (Hausdorff) locally convex space L, its boundary point e is called a tangency point if there is a unique functional $\varphi \in L^*$ supporting B at e.

equals the rank of the system $\{\varphi_s : s \in S_m^0\}$ and there are indices $i_1, \dots, i_n$ such that: (i) $\alpha_i = 0$ whenever $s = (\alpha_1, \dots, \alpha_N) \in S_m^0$, $i \notin \{i_1, \dots, i_n\}$,[12] and (ii) each $F(a_i)$ contains at least one functional $\phi_j = \alpha_1\varphi_1 + \cdots + \alpha_N\varphi_N$ with $\alpha_i = 0$ for $i \notin \{i_1, \dots, i_n\}$. It follows that $\varphi_{i_1}, \dots, \varphi_{i_n}$ are the fundamentals.

The above reasonings lead to the following procedure of determining the fundamentals, which is entirely based on observer's judgement on discrimination of test light stimuli $a^0 + te$ against the background stimulus a^0.

1. Fix $e \in TG_m$ and consider various vectors $a^0 + te \in a^0 + E_N$ such that $t > 0$ and $a^0 + te$ is a proper light stimulus: $a^0 + te \in A$. When $a^0 + te \notin A$ whenever $t > 0$ (this means $t_e = 0$), we pass to the next $e \in TG_m$. By comparing $a^0 + te \in A$ with the background light stimulus a^0, we determine the number $t_e > 0$ such that: (i) $a^0 + t_e e$ is a proper light stimulus indistinguishable from a^0, (ii) if $t > t_e$ and $a^0 + te$ is a proper light stimulus, then it is discriminated from a^0. If $a^0 + te$ is a proper light stimulus for some $t > t_e$ (this means $t_{1e} > t_e$ that is, $e \in Q'$), then $a^0 + t_e e$ is a boundary point of $\varsigma'(a^0)$. In such a case, we mark $a^0 + t_e e$ as one of a_i and pass to the next $e \in TG_m$. If the stimulus $a^0 + te$ is improper whenever $t > t_e$ (this means $t_{1e} = t_e$ that is, $e \notin Q'$), we skip over this target e and move on to the next $e \in TG_m$. Looking over all $e \in TG_m$ gives us the points $a_1, \dots, a_{N_1} \in bdry(\varsigma'(a^0))$.
2. Fix $s = (\alpha_1, \dots, \alpha_N) \in S_m$, take the functional

$$\phi_j = \varphi_s = \alpha_1\varphi_1 + \cdots + \alpha_N\varphi_N \in FG_m,$$

   and calculate N_1 numbers: $\langle \phi_j, a_i \rangle$, $i = 1, \dots, N_1$. Looking over all $\phi_j \in FG_m$ gives us the $(N_1 \times N_2')$-matrix $(\langle \phi_j, a_i \rangle)_{ij}$.
3. For each row i of that matrix we find the set $J(i)$ of all columns j satisfying $\langle \phi_j, a_i \rangle \geq \langle \phi_k, a_i \rangle$ whenever $k = 1, \dots, N_2'$, and determine the set $F(a_i) = \{\phi_j : j \in J(i)\}$. Consider the set I of those i, for which $F(a_i)$ comprises the unique functional ϕ_{j_i}, the set J of all such functionals, and the set S_m^0 of $s = (\alpha_1, \dots, \alpha_N) \in S_m$ corresponding to functionals $\phi_j, j \in J$,

$$\begin{aligned} S_m^0 &= \{s = (\alpha_1, \dots, \alpha_N) \in S_m : \varphi_s = \phi_j,\ j \in J\} \\ &= \{s = (\alpha_1, \dots, \alpha_N) \in S_m : \phi_j = \alpha_1\varphi_1 + \cdots + \alpha_N\varphi_N,\ j \in J\}. \end{aligned}$$

4. Take the submatrix $(\langle \phi_j, a_i \rangle)_{i \in I, j \in J}$, determine its rank n, and verify that there is a subset $\{i_1, \dots, i_n\} \subset \{1, \dots, N\}$ comprising n different indices and such that:
   (i) if $s = (\alpha_1, \dots, \alpha_N) \in S_m^0$ then $\alpha_i = 0$ for all $i \notin \{i_1, \dots, i_n\}$;
   (ii) for every $i = 1, \dots, N_1$ there exists $s = (\alpha_1, \dots, \alpha_N) \in S_m$ such that $\alpha_i = 0$ for all $i \notin \{i_1, \dots, i_n\}$.[13]

If both conditions, (i) and (ii), are satisfied, one can accept that $\varphi_{i_1}, \dots, \varphi_{i_n}$ are fundamentals. Note that negligible violations of (i) and (ii) caused by random errors and noise can occur, and to eliminate them one can raise the approximation precision by increasing m. However, there being substantial and systematic violations of (i) and (ii) would indicate that our basic hypothesis (II), which states that the set of fundamentals is contained in $\{\varphi_1, \dots, \varphi_N\}$, is not correct. Thus the above procedure may be considered as a test for that hypothesis.

---

12 Since the number of linearly independent functionals from $\{\varphi_s : s \in S_m^0\}$ is n, it follows from (i) that for each i_k there is $s = (\alpha_1, \dots, \alpha_N) \in S_m^0$ with $\alpha_{i_k} \neq 0$.
13 This means that the functional $\varphi_s \in F(a_i)$ supports $\varsigma'(a^0)$ at a_i.

11.3.2 Deriving the linear colour mechanism space from the colour matching set (the method of tangential hyperplane)

In this section, we present an alternative method of deriving the colour mechanism space $E(\sim_\varsigma)^0$ from colour matching sets. Given N linearly independent measures $\sigma_1, \dots, \sigma_N \in A = \mathbf{C}_+^*$ and N linearly independent functions $\varphi_1, \dots, \varphi_N \in E^* = \mathbf{C}$, we consider N-dimensional subspaces $E_N = Lin(\sigma_1, \dots, \sigma_N)$ in $\mathbf{C}^*$ and $E_N^* = Lin(\varphi_1, \dots, \varphi_N)$ in $\mathbf{C}$ and assume that the colour mechanism space $E(\sim_\varsigma)^0$ is contained in E_N^*.[14] Moreover, we will suppose that the restriction to $E_N^* \times E_N$ of the canonical bilinear form on $\mathbf{C} \times \mathbf{C}^*$,

$$\langle \varphi, \sigma \rangle = \int_\Lambda \varphi(\lambda)\, \sigma(d\lambda),$$

is non-degenerate in both arguments[15] and the biorthogonality condition

$$\langle \varphi_i, \sigma_j \rangle = \delta_{ij}, \quad i, j \in \{1, \dots, N\} \tag{11.14}$$

holds true (see Lemma 11.1). Given a background light stimulus $a^0 \in A$, we consider the extended colour matching set $\varsigma'(a^0)$ and the extended threshold set $B'_{a^0} = \varsigma'(a^0) - a^0$, and suppose that hypotheses (I) and (III) in Section 11.3.1 are satisfied.[16]

Let us briefly recall some facts from convex analysis that will be needed hereinafter (for more details see Appendix A.2 and references there). Given a (Hausdorff) locally convex space E (in what follows, E is $\mathbf{C}^*$ endowed with the weak* topology), and a closed absorbing convex set (body) B in it, the *Minkowski function (gauge)* of B is defined as the following sublinear function on E:

$$\mu(e; B) := \inf \left\{ \alpha > 0 : \frac{e}{\alpha} \in B \right\}.$$

A functional $\varphi \in E^*$ is supporting to B at a point $e_0 \in bdry(B)$ if and only if there is a number $\alpha > 0$ such that $\alpha\varphi \in \partial\mu(e_0; B)$, where $\partial\mu(e_0; B)$ stands for the subdifferential of $\mu(\cdot; B)$ at e_0,

$$\begin{aligned}
\partial\mu(e_0; B) &= \{\varphi' \in E^* : \langle \varphi', e - e_0 \rangle \le \mu(e; B) - \mu(e_0; B) \quad \forall e \in E\} \\
&= \{\varphi' \in E^* : \langle \varphi', e \rangle \le \mu(e; B)\ \forall e \in E,\ \langle \varphi', e_0 \rangle = \mu(e_0; B)\} \\
&= \{\varphi' \in \partial\mu(0; B) : \langle \varphi', e_0 \rangle = \mu(e_0; B)\},
\end{aligned}$$

and e_0 is a tangency point of B if and only if the Minkowski function of B is (Gâteaux) differentiable at e_0, in which case its subdifferential $\partial\mu(e_0; B)$ becomes the singleton coinciding with the gradient $\nabla\mu(e_0; B)$.

Proposition 11.5 *Given a functional $\varphi \in \mathbf{C}$ and an (extended) target stimulus $e_0 \in E_N \cap bdry(B'_{a^0})$, the following statements are equivalent: (a) e_0 is a tangency point of B'_{a^0} and $\{e \in E : \langle \varphi, e \rangle = \langle \varphi, e_0 \rangle\}$ is the corresponding tangential hyperplane of B'_{a^0} at e_0; (b) $\varphi \in E_N^*$, e_0 is a tangency point of the convex set $B'_{a^0} \cap E_N$ in E_N, and the hyperplane in E_N, $\{e \in E_N : \langle \varphi, e \rangle = \langle \varphi, e_0 \rangle\}$ is tangential to $B'_{a^0} \cap E_N$ at e_0.*

We now describe briefly an approach to get the linear colour mechanism space $E(\sim_\varsigma)^0$. Let $e \in E_N$ be a proper target stimulus[17] lying in the boundary of the extended threshold set B'_{a^0}, hence

---

14 This is a weakening of supposition (II) from Section 11.3.1.

15 That is, for every non-zero $e \in E_N$ there is an element $\varphi \in E_N^*$ such that $\langle \varphi, e \rangle = \int_\Lambda \varphi(\lambda) e(d\lambda) \neq 0$, and for every non-zero $\varphi \in E_N^*$ there is an element $e \in E_N$ such that $\langle \varphi, e \rangle \neq 0$. The N-dimensional spaces E_N and E_N^* are then mutually dual with respect to this bilinear form, that is, every linear functional on E_N is determined by a unique $\varphi \in E_N^*$ and every linear functional on E_N^* is determined by a unique $e \in E_N$.

16 In Section 11.3.1, we delt with the extended colour matching set $\varsigma'(a^0)$, but now it will be more convenient to use the extended threshold set B'_{a^0}.

17 Recall that a target stimulus e is proper if $a^0 + e \in A$.

$a^0 + e \in A \cap bdry(\varsigma'(a^0))$. This relation holds true when $e \in Q$, $t_e = 1$, and either $e \in A$ or $e \in Q' \backslash A$ (for the definitions and properties of Q, Q', and t_e, see (11.11), (11.12), (11.13), Proposition 11.4, and Lemma 11.2). Note that checking the last condition is possible for an observer.

We take a large number $N_1 > N$ of such targets, $e_i \in bdry(B'_{a^0}) \cap E_N, i = 1, \dots, N_1$, and find around each e_i a cloud of proper target stimuli $e_{ij} \in bdry(B'_{a^0}) \cap E_N$, $j = 1, \dots, N_2$ where $N_2 > N$ is large enough. All these target stimuli lie in the threshold surface $bdry(B'_{a^0}) \cap E_N$ in E_N and are uniquely represented as linear combinations of $\sigma_1, \dots, \sigma_N$,

$$e_i = x_{i1}\sigma_1 + \cdots + x_{iN}\sigma_N, \quad i = 1, \dots, N_1, \tag{11.15}$$

$$e_{ij} = x_{ij1}\sigma_1 + \cdots + x_{ijN}\sigma_N, \quad i = 1, \dots, N_1, \; j = 1, \dots, N_2. \tag{11.16}$$

Recall that e_i is a tangency point of $B'_{a^0} \cap E_N$ in E_N if and only if the Minkowski function of $B'_{a^0} \cap E_N$,

$$\mu(e; B'_{a^0} \cap E_N) := \inf \left\{ \alpha > 0 : \frac{e}{\alpha} \in B'_{a^0} \cap E_N \right\}, \quad e \in E_N,$$

is differentiable at e_i, in what case

$$\{ e \in E_N : \langle \nabla\mu(e_i; B'_{a^0} \cap E_N), e \rangle = \langle \nabla\mu(e_i; B'_{a^0} \cap E_N), e_i \rangle \}$$

is the tangential hyperplane of the threshold surface $bdry(B'_{a^0}) \cap E_N$ at e_i. Furthermore, since the finite-dimensional spaces E_N and E_N^* are in duality with respect to the bilinear form $E_N^* \times E_N \ni (\varphi, e) \mapsto \langle \varphi, e \rangle$, there is the unique vector $y^* = (y_1^*, \dots, y_N^*) \in \mathbf{R}^N$ such that $\nabla\mu(e_i; B'_{a^0} \cap E_N)$ is the restriction to E_N of the functional $y_1^*\varphi_1 + \cdots + y_N^*\varphi_N$, and it follows from Proposition 11.5 that e_i is a tangency point of B'_{a^0} in E and

$$\nabla\mu(e_i; B'_{a^0}) = y_1^*\varphi_1 + \cdots + y_N^*\varphi_N.$$

Identifying elements $e = \sum_1^N x_i\sigma_i \in E_N$ and $\varphi = \sum_1^N y_i\varphi_i \in E_N^*$ with the corresponding N-dimensional vectors $x = (x_1, \dots, x_N) \in \mathbf{R}^N$ and $y = (y_1, \dots, y_N) \in \mathbf{R}^N$ respectively, and taking into account the biorthogonality condition (11.14), we get

$$\langle \varphi, e \rangle = \langle y, x \rangle = \sum_{i=1}^{N} x_i y_i.$$

Since, in view of hypotheses (I) and (III), $B'_{a^0} \cap E_N$ is a closed convex subset in the finite-dimensional space E_N and the origin 0 is its interior point, it follows from convex analysis (e.g., see Theorems 18.8 and 25.3 in Rockafellar (1970)) that the set of tangency points of $B'_{a^0} \cap E_N$ is dense in the threshold surface $bdry(B'_{a^0} \cap E_N) = bdry(B'_{a^0}) \cap E_N$ and massive in various meanings. Moreover, if a representation holds $B'_{a^0} \cap E_N = \left\{ e = \sum_1^N x_i\sigma_i : g(x) \le 1 \right\}$ where g is a convex function on $\mathbf{R}^N$ differentiable on $\mathbf{R}^N \backslash \{0\}$,[18] then the *whole* threshold surface in E_N is the set of tangency points of $B'_{a^0} \cap E_N$. The hypothesis that every e_i is a tangency point of $B'_{a^0} \cap E_N$ is thus fairly plausible. We believe that, knowing the cloud of proper target stimuli $e_{ij} \in bdry(B'_{a^0}) \cap E_N$, $j = 1, \dots, N_2$, one can test this hypothesis for a given e_i (in experiment) and find the corresponding tangential hyperplane. In other words, given the N-dimensional vectors $x^i = (x_{i1}, \dots, x_{iN})$ and $x^{ij} = (x_{ij1}, \dots, x_{ijN})$, $j = 1, \dots, N_2$ (see (11.15) and (11.16)), and based solely on this information, one has to decide whether there is the

18 Particularly, it is a case when the Minkowski function $\mu(\cdot; B'_{a^0} \cap E_N)$ is differentiable on the threshold surface $bdry(B'_{a^0}) \cap E_N$.

unique (hence, tangential) hyperplane in E_N supporting $B'_{a^0} \cap E_N$ at e_i.[19] This is a separate problem and various methods of solving it can be proposed.

If a given e_i is known to be a tangency point of the threshold surface in E_N, then the corresponding tangential hyperplane can be determined as in the preceding section, or by using the least squares technique. In the latter case, the vector $y^i = (y_{i1}, \dots, y_{iN})$, $\|y^i\| = 1$, which determines the corresponding tangential hyperplane[20]

$$S^{y^i} := \left\{ e = x_1\sigma_1 + \cdots + x_N\sigma_N \in E_N : \sum_{j=1}^{N} x_j y_{ij} = \sum_{j=1}^{N} x_{ij} y_{ij} \right\},$$

can be found from the condition that the function of $y = (y_1, \dots, y_N) \in \mathbf{R}^N$, $\|y\| = 1$,

$$\sum_{j=1}^{N_2} dist^2(x^{ij}, S^y) = \sum_{j=1}^{N_2} \min\{\|x^{ij} - z^j\|^2 : z^j \in S^y\}$$

attains its minimum at $y = y^i$. The condition on $z^j = (z_{j1}, \dots, z_{jN})$, $dist(x^{ij}, S^y) = \|x^{ij} - z^j\|$, means that $x^{ij} - z^j$ is collinear to y,

$$x^{ij} - z^j = \alpha_j y, \tag{11.17}$$

and as $z^j \in S^y$, one has

$$\sum_{k=1}^{N} z_{jk} y_k = \sum_{k=1}^{N} x_{jk} y_k. \tag{11.18}$$

We have thus $N + 1$ linear equations (11.17) and (11.18) to determine $N + 1$ variables $z_{j1}, \dots, z_{jN}, \alpha_j$ as functions of y, and minimizing in y, $\|y\| = 1$ the sum

$$\sum_{j=1}^{N_2} dist^2(x^{ij}, S^y) = \sum_{j=1}^{N_2} \alpha_j^2(y)\|y\|^2$$

gives us the desired vector y^i determining the tangential hyperplane S^{y^i}. One can assume without loss of generality that $y^i = t_i \nabla\mu(e_i; B'_{a^0} \cap E_N)$ where $t_i > 0$. If our model is correct (that is, the basic suppositions (I) and (III) are fulfilled) and all e_i are tangency points,[21] then the inequalities hold true

$$\langle y^i, x^{ij} \rangle \le \langle y^i, x^i \rangle, \quad j = 1, \dots, N_2, \tag{11.19}$$

and the functionals on E, $\nabla\mu(e_i; B'_{a^0}) = \sum_{j=1}^{N} t_i y_{ij} \varphi_j$, $i = 1, \dots, N_1$, belong to the colour mechanism space $E(\sim_\varsigma)^0$. We get N_1 vectors y^i, $i = 1, \dots, N_1$, and N_1 is much larger than N. If the rank of that system of N-dimensional vectors is n and $\{y^{i_1}, \dots, y^{i_n}\}$ is a subsystem consisting of linearly independent vectors, then we can assume that

$$E(\sim_\varsigma)^0 = Lin(\nabla\mu(e_{i_1}; B'_{a^0}), \dots, \nabla\mu(e_{i_n}; B'_{a^0})).$$

19 Any hyperplane in E_N supporting $B'_{a^0} \cap E_N$ at e_i may be represented as $\{e = \sum_1^N x_k\sigma_k \in E_N : \langle x, y \rangle = \langle x^i, y \rangle\}$ where y is normalized by $\|y\| = 1$ and the inequalities hold true $\langle y, x^{ij} \rangle \le \langle y, x^i \rangle$, $j = 1, \dots, N_2$. There are infinitely many vectors $y \in \mathbf{R}^N$, $\|y\| = 1$, satisfying these inequalities. If any two of them, y^1 and y^2, are close enough ($\|y^1 - y^2\| \le \varepsilon$ where ε is a chosen closeness criterion), then one can assume that all these y approximate one and the same vector y^* that determines the tangential hyperplane of the threshold surface $bdry(B'_{a^0}) \cap E_N$ at e_i. If there are y^1 and y^2 with $\|y^1 - y^2\| > \varepsilon$, then one decides that e_i is not a tangency point.

20 Hence, the restriction to E_N of the functional $y_{i1}\varphi_1 + \cdots + y_{iN}\varphi_N$ is collinear to the gradient $\nabla\mu(e_i; B'_{a^0} \cap E_N)$.

21 When some e_i is not a tangency point, we will eliminate it.

If, for some i, inequalities (11.19) fail to be true, then the set B'_{a^0} is not convex, that is, condition (I) is not satisfied.

Thus, the procedure for determining the colour mechanism space is as follows.

1. Fix N and consider N-dimensional subspaces, $E_N^* = Lin(\varphi_1, \dots, \varphi_N)$ in $\mathbf{C}$ and $E_N = Lin(\sigma_1, \dots, \sigma_N)$ in $\mathbf{C}^*$, such that $\sigma_i \in \mathbf{C}_+^*$, $i = 1, \dots, N$, and the biorthogonality condition is satisfied. Also, we assume that N is large enough and that the colour mechanism space is contained in $Lin(\varphi_1, \dots, \varphi_N)$.
2. Take a very large number of proper target stimuli e_i lying in the threshold surface $bdry(B'_{a^0}) \cap E_N$.
3. Around each e_i find a cloud of proper target stimuli e_{ij} lying in the threshold surface $bdry(B'_{a^0}) \cap E_N$. Using these e_{ij}, test the hypothesis that e_i is a tangency point of $B'_{a^0} \cap E_N$ in E_N, remove the points e_i that fail to satisfy this hypothesis, and make sure that the number N_1 of remaining e_i is yet large enough (much more than N). Thus, we can find $N_1 > N$ of proper target stimuli e_i lying in the threshold surface $bdry(B'_{a^0}) \cap E_N$ and being its tangency points.
4. For every such e_i take the corresponding functional $\sum_{j=1}^N y_{ij}\varphi_j = t_i \nabla\mu(e_i; B'_{a^0})$, $t_i > 0$, determining the tangential hyperplane in E_N and test inequalities (11.19). If these inequalities are satisfied, then any maximal linearly independent subsystem of the vectors

$$y^{i_1} = (y_{i_1 1}, \dots, y_{i_1 N}), \dots, y^{i_n} = (y_{i_n 1}, \dots, y_{i_n N})$$

determines the functionals $\sum_{j=1}^N y_{i_k j}\varphi_j = t_{i_k} \nabla\mu(e_{i_k}; B'_{a^0})$, $k = 1, \dots, n$, and the colour mechanism space is given by

$$E(\sim_\varsigma)^0 = Lin(\nabla\mu(e_{i_1}; B'_{a^0}), \dots, \nabla\mu(e_{i_n}; B'_{a^0})).$$

If inequalities (11.19) are substantially violated for at least one i,[22] then B'_{a^0} is not convex, and the basic model (in particular, assumptions (I) and (III)) must be rejected.

11.3.3 Deriving the fundamental colour mechanisms from the colour matching set that they generate (the method of quadratic approximation)

In Section 11.3.2, based on some general information on the extended colour matching relation ς' and the background light stimulus a^0, we have derived the colour mechanism space, making no assumption concerning the very existence of the fundamental colour mechanisms and their relation to ς'. In the present section, we shall assume that the boundary of the extended threshold set of a^0 can be represented as a level set of the function d (the 'detection function', according to the terminology introduced in Chapter 12 (see Eq. (12.2))):

$$bdry\left(B'_{a^0}\right) = \{e \in E : d(e) = 1\}, \tag{11.20}$$

which has the following form:

$$d(e) = G\left(\sum_{i=1}^n F_i(\langle f_i, e\rangle)\right), \tag{11.21}$$

where $f_1, \dots, f_n \in \mathbf{C}$ are the colour mechanisms determining the colour matching relation ς, and hence the colour matching metamerism $\sim_\varsigma$, $F_i : \mathbf{R} \to \mathbf{R}$, $i = 1, \dots, n$, are some functions, and $G : \mathbf{R} \to \mathbf{R}$ is a strictly monotonic function. (For the reason which will be explained in Section 12.1.2,

22 Minor violations caused by random errors and noise can be ignored.

$f_1, \dots, f_n$ will be further treated to as the fundamental colour mechanisms.) Although we suppose that such a representation holds true, we don't know the exact form of functions F_i and G. Also, the functionals $f_1, \dots, f_n$ and even their general number n are unknown to us. Nevertheless, we shall assume that each f_i can be (at least approximately) represented as a linear combination of a set of known linear (coordinate) functionals $\varphi_1, \dots, \varphi_N$,

$$f_i = \sum_{j=1}^{N} u_{ij}\varphi_j. \tag{11.22}$$

The functionals $\varphi_1, \dots, \varphi_N$ are supposed to be linearly independent, and their number N is assumed to be much larger than the (unknown) number of the fundamental colour mechanisms, n. The problem of identification of the fundamental colour mechanisms is, thus, reduced to identification of the $n \times N$ matrix $U = \left(u_{ij}\right)$.

It must be said that, strictly speaking, Eq. (11.22) cannot be uniquely resolved with respect to matrix U because representation (11.21) is not unique. Indeed, for any numbers $d_i \neq 0, i = 1, \dots, n$, one has

$$F_i(\langle f_i, e\rangle) = F_{d_i i}(\langle f_{d_i i}, e\rangle), \quad i = 1, \dots, n,$$

where $F_{d_i i}(\alpha) := F_i(\alpha/d_i)\ \forall \alpha \in \mathbf{R}$, $f_{d_i i} := d_i f_i$; therefore

$$d(e) = G\left(\sum_{i=1}^{n} F_{d_i i}(\langle f_{d_i i}, e\rangle)\right),$$

and one cannot distinguish between pairs: $(f_1, \dots, f_n)$, U and $(f_{d_1 1}, \dots, f_{d_n n})$, $DU = (d_i u_{ij})$, where $D = D\{d_1, \dots, d_n\}$ is the diagonal $n \times n$ matrix with the elements d_i on the diagonal. Nevertheless, we shall see that, in many cases, it is the only lack of uniqueness; that is, the matrix U can be determined uniquely up to multiplying from the left by a non-singular diagonal $n \times n$ matrix D.[23]

We now describe a method of finding such matrices U, and hence the functionals f_i. As in Section 11.3.2, we approximate $E^* = \mathbf{C}$ by its N-dimensional subspace $E_N^* = Lin(\varphi_1, \dots, \varphi_N) \subset \mathbf{C}$ spanned over $\varphi_1, \dots, \varphi_N$, and find N linearly independent signed measures $e_1, \dots, e_N \in E = \mathbf{C}^*$ such that the biorthogonality condition $\langle \varphi_j, e_k\rangle = \delta_{jk}$ is satisfied (see Lemma 11.1 and the footnote to its proof). We will consider $E_N = Lin(e_1, \dots, e_N)$ as a target space with respect to the background a^0. Every $e \in E_N$ is uniquely represented as a linear combination of e_j, $e = x_1e_1 + \dots + x_Ne_N$, where $x_j = \langle \varphi_j, e\rangle$, $j = 1, \dots, N$. Now, by identifying e with the N-dimensional vector $x = (x_1, \dots, x_N)$, we derive from (11.20), (11.21) and (11.22) that

$$bdry(B'_{a^0}) \cap E_N = \{x = (x_1, \dots, x_N) : F(x_1, \dots, x_N) = \varepsilon\}, \tag{11.23}$$

where

$$F(x_1, \dots, x_N) = \sum_{i=1}^{n} F_i\left(\sum_{j=1}^{N} u_{ij}x_j\right) \tag{11.24}$$

and $\varepsilon = G^{-1}(1)$. Henceforth we will restrict our consideration to the subspace E_N identified with $\mathbf{R}^N$.

---

23 Such indeterminacy reflects the arbitrariness of the choice of a scaling factor when measuring the outputs of the colour mechanisms. In the other words, the results of identification do not depend on the units of measurements in which we measure these outputs.

We will assume that all F_i, $i = 1, \dots, n$, are convex, assumptions (I) and (III) from Section 11.3.1 are satisfied, and the global shape of the extended threshold surface (11.23) is smooth enough to justify the following assumption: all the pooling functions F_i are twice differentiable.

It follows, then, that there exist the first and second derivatives of the function F, i.e., the gradient $F' = \left(\frac{\partial F}{\partial x_1}, \dots, \frac{\partial F}{\partial x_N}\right)$ considered as a row vector and the Hessian matrix $F'' = \left(\frac{\partial^2 F}{\partial x_j \partial x_k}\right)$. Given a point $x = (x_1, \dots, x_N) \in bdry(B'_{a^0}) \cap E_N$, we derive from (11.24) that

$$\frac{\partial F(x)}{\partial x_j} = \sum_{i=1}^{n} F_i'\left(\sum_{l=1}^{N} u_{il}x_l\right) u_{ij}, \tag{11.25}$$

$$\frac{\partial^2 F(x)}{\partial x_j \partial x_k} = \sum_{i=1}^{n} F_i''\left(\sum_{l=1}^{N} u_{il}x_l\right) u_{ij}u_{ik}. \tag{11.26}$$

In matrix form, Eq. (11.26) is rewritten as

$$F''(x) = U^{\top} D(x) U, \tag{11.27}$$

where the symbol $\top$ stands for transposition (hence $U^{\top}$ is the $N \times n$ matrix, $U^{\top} = (v_{ji})$, $v_{ji} = u_{ij}$), $D(x)$ is the diagonal $n \times n$ matrix

$$D\left\{ F_1''\left(\sum_{1}^{N} u_{1l}x_l\right), \dots, F_n''\left(\sum_{1}^{N} u_{nl}x_l\right)\right\}.$$

We will suppose that the rank of the $N \times N$ matrix $F''(x)$ equals n. This implies that the matrix $D(x)$ is non-singular, and since all the functions F_i are convex, the numbers $F_i''\left(\sum_{l=1}^{N} u_{il}x_l\right)$, $i = 1, \dots, n$, must be strictly positive.

Recall some facts from Section 11.3.1. Given a point $e = x_1e_1 + \cdots + x_Ne_N$ such that $t_e = \max\{t \geq 0 : te \in B_{a^0}\} = \max\{t \geq 0 : a^0 + te \in \varsigma(a^0)\} = 1$ and either $e \in A$ or $e \notin A$ and $t_{1e} = \max\{t : a^0 + te \in A\} > 1 = t_e$, it follows from Proposition 11.4 and Lemma 11.2, along with definitions of Q and Q' (see (11.12) and (11.13)), that e is a proper target light stimulus (i.e., $a^0 + e \in A$) and $e \in bdry(B'_{a^0})$.[24] Take such a point e and consider the corresponding vector $x = (x_1, \dots, x_N)$. Then, for any y in a small neighbourhood of x, the value $F(y)$ of the function F in (11.24) can be approximated by the Taylor polynomial of the second degree:

$$F(y) = F(x) + F'(x)(y - x) + \frac{1}{2}(y - x)^{\top} F''(x)(y - x) + o(\|y - x\|^2), \tag{11.28}$$

where $y - x$ is considered as a column vector and $(y - x)^{\top}$ stands for the corresponding (transposed) row vector. Then, near x, the extended threshold surface $bdry(B'_{a^0}) \cap E_N = \{y : F(y) = \varepsilon\}$ is approximated by the quadratic surface $S^x = \left\{y : F(x) + F'(x)(y - x) + \frac{1}{2}(y - x)^{\top} F''(x)(y - x) = \varepsilon\right\}$. Now, since $F(x) = \varepsilon$ and $y^{\top} F''(x)x = x^{\top} F''(x)y$, one has

$$S^x = \left\{y : h(x)y + \frac{1}{2}y^{\top} H(x)y = h_0(x)\right\},$$

where $H(x) = F''(x)$, $h(x) = (h_1(x), \dots, h_N(x)) = F'(x) - x^{\top} F''(x)$, $h_0(x) = F'(x)x - \frac{1}{2}x^{\top} F''(x)x$. Taking into account the symmetry of the matrix $H(x)$, the quadratic surface S^x is determined by $N + \frac{N(N+1)}{2} + 1 = \frac{N(N+3)}{2} + 1$ parameters $h_j(x)$ $(j = 1, \dots, N)$, $h_{jk}(x)$ $(j \leq k, k = 1, \dots, N)$, $h_0(x)$.

24 The above conditions on e mean that: (i) $a^0 + te$ is a proper light stimulus for all $0 \leq t \leq t_{1e}$, where $t_{1e} > 1$, and (ii) it is indistinguishable from a^0 if and only if $0 \leq t \leq 1$. These conditions are thus observable hence verifiable. Note in this connection that for an *arbitrary* point $e \in E_N$ the observer cannot say whether it belongs to the boundary of the extended threshold set or not.

Of course, multiplying all the parameters by a constant $c > 0$ gives the same surface S^x. (Here, the sign of c is determined by the fact that the new matrix $cH(x)$ must remain positive semi-definite.) We believe that if one takes a great number $N_1 > \frac{N(N+3)}{2} + 1$ of observable points[25] $y^1, \dots, y^{N_1}$ belonging to the extended threshold surface $bdry(B'_{a^0}) \cap E_N = \{y : F(y) = \varepsilon\}$ and lying near x, then S^x may be determined as (or approximated by) a quadratic surface that minimally deviates from the points $y^1, \dots, y^{N_1}$ in an appropriate sense, say in terms of the least squares. In particular, we can thus assume that the Hessian $H(x) = F''(x)$ can be, in principle, experimentally evaluated up to a positive multiple c.

Remark 11.7 *In order to determine S^x (hence $H(x)$) by the method of least squares, one has to take N_1 points $z^1, \dots, z^{N_1} \in S^x$ such that $\|y^l - z^l\| = dist(y^l, S^x) = \min\{\|y^l - z\| : z \in S^x\}$, $l = 1, \dots, N_1$, and to find the values of parameters h_j, h_{jk}, h_0 (determining the desired quadratic surface S^x), on which the sum $\sum_1^{N_1} \|y^l - z^l\|^2$ as a function of these parameters attained its minimum. The condition $\|y^l - z^l\| = dist(y^l, S^x)$ means that the vector $y^l - z^l$ is collinear to the gradient of the function $y \mapsto hy + \frac{1}{2}y^\top Hy$ at the point $y = z^l$: there is a non-zero number α_l such that*

$$y^l - z^l = \alpha_l(h^\top + Hz^l). \tag{11.29}$$

Now, (11.29) together with the condition that z^l lies on the quadratic surface S^x gives a system of $N + 1$ equations with $N + 1$ variables $z^l_1, \dots, z^l_N, \alpha_l$,

$$y^l_j - z^l_j = \alpha_l\left(h_j + \sum_{k=1}^{N} h_{jk} z^l_k\right), \quad j = 1, \dots, N,$$

$$\sum_{j=1}^{N} h_j z^l_j + \frac{1}{2}\sum_{j=1}^{N}\sum_{k=1}^{N} h_{jk} z^l_j z^l_k = h_0.$$

Solving this system gives us $z^l_1, \dots, z^l_N$ as functions of h_j, h_{jk}, h_0, and minimizing $\sum_1^{N_1} \|y^l - z^l\|^2$ enables to find the values $h_j(x)$, $h_{jk}(x)$, $h_0(x)$ corresponding to S^x. (If some collection h_j, h_{jk}, h_0 is a minimizer for $\sum_1^{N_1} \|y^l - z^l\|^2$, then ch_j, ch_{jk}, ch_0 is a minimizer as well whenever $c > 0$, and any of two these collections may be taken as $h_j(x)$, $h_{jk}(x)$, $h_0(x)$.)

Taking into account (11.27), we get

$$U^\top D(x) U = H(x),$$

where the matrices $D(x)$ and U are unknown, and the matrix $H(x)$ is known up to a multiple $c > 0$.[26] We will show that, in many cases, two such matrix equations

$$U^\top D(x^1) U = H(x^1), \quad U^\top D(x^2) U = H(x^2) \tag{11.30}$$

with the matrices $H(x^1)$ and $H(x^2)$ estimated for two different points x^1 and x^2 enable either to find a matrix U and to obtain some information about matrices $D(x^1)$ and $D(x^2)$ satisfying (11.30) or to find

25 Recall that a point $y = (y_1, \dots, y_N) \in bdry(B'_{a^0})$ is observable if, for some $t > 1$, $a^0 + t(y_1 e_1 + \cdots + y_N e_N)$ is a proper light (i.e., belongs to A).

26 If one takes $cH(x)$ instead of $H(x)$, then $D(x)$ is replaced with the diagonal matrix $cD(x)$ while U does not change.

out that our basic model ((11.20) and (11.21)) does not hold true. To this end, we will need some auxiliary information concerning the so-called *pseudo-inverse* or generalised (Moore–Penrose) inverse matrices. For reader's convenience, we briefly recall here some basic facts about these matrices.[27]

On pseudo-inverse matrices

For any $n_1 \times n_2$matrix H there exists the unique $n_2 \times n_1$matrix $H^{\#}$such that $HH^{\#}H = H$ and $H^{\#} = VH^{\top} = H^{\top}W$, where V is a $n_2 \times n_2$ matrix, and $W$is a $n_1 \times n_1$matrix.[28] The matrix $H^{\#}$ is called the pseudo-inverse matrix of H. Let us list (without proofs) some properties of the pseudo-inverse matrix $H^{\#}$.

(a) If $n_1 = n_2$ and U is non-singular, then $H^{\#} = H^{-1}$;
(b) if r is the rank of H and $H = U_1 U_2$ where U_1 is a $n_1 \times r$ matrix and U_2 is a $r \times n_2$ matrix, then

$$H^{\#} = U_2^{\#} U_1^{\#} = U_2^{\top}(U_2 U_2^{\top})^{-1}(U_1^{\top} U_1)^{-1} U_1^{\top}; \tag{11.31}$$

(c) the following equalities hold true:

$$(H^{\top})^{\#} = (H^{\#})^{\top}, \quad (H^{\#})^{\#} = H,$$
$$(HH^{\#})^{\top} = (HH^{\#})^2 = HH^{\#}, \quad (H^{\#}H)^{\top} = (H^{\#}H)^2 = H^{\#}H.$$

It follows from (11.31) that if the rank of the $n_1 \times n_2$ matrix H equals n_1, then

$$H^{\#} = H^{\top}(HH^{\top})^{-1}, \tag{11.32}$$

and if the rank of H equals n_2, then

$$H^{\#} = (H^{\top}H)^{-1}H^{\top}. \tag{11.33}$$

We now return to matrix equations (11.30) and consider the matrix

$$H(x^1)H(x^2)^{\#} = U^{\top}D(x^1)U(U^{\top}D(x^2)U)^{\#}.$$

Taking into account that the rank of $H(x^2)$ is n and applying (11.31) with $H = H(x^2)$, $U_1 = U^{\top}$, $U_2 = D(x^2)U$, one gets

$$H(x^2)^{\#} = (U^{\top}D(x^2)U)^{\#} = U^{\top}(UU^{\top})^{-1}D(x^2)^{-1}(UU^{\top})^{-1}U;$$

therefore,

$$\begin{aligned} H(x^1)H(x^2)^{\#} &= U^{\top}D(x^1)UU^{\top}(UU^{\top})^{-1}D(x^2)^{-1}(UU^{\top})^{-1}U \\ &= U^{\top}D(x^1)D(x^2)^{-1}(UU^{\top})^{-1}U. \end{aligned} \tag{11.34}$$

Now, multiplying both sides of (11.34) by $U^{\top}$ from the right, one gets

$$H(x^1)H(x^2)^{\#}U^{\top} = U^{\top}D(x^1/x^2), \tag{11.35}$$

where $D(x^1/x^2) := D(x^1)D(x^2)^{-1}$ is the $n \times n$ diagonal matrix with numbers

$$\frac{F_1''\left(\sum_{l=1}^{N} u_{1l}x_l^1\right)}{F_1''\left(\sum_{l=1}^{N} u_{1l}x_l^2\right)}, \ldots, \frac{F_n''\left(\sum_{l=1}^{N} u_{nl}x_l^2\right)}{F_n''\left(\sum_{l=1}^{N} u_{nl}x_l^2\right)}.$$

---

27 For more details (including proofs), see e.g., Section 1.5 in Gantmacher (2004) or Campbell & Meyer (1979). In addition, in Gantmacher (2004) one can find some equivalent definitions of pseudo-inverse matrices, as well as methods for their numerical computation.
28 All matrices under consideration are supposed to be real. In case of complex matrices, the transposition has to be replaced with conjugation.

on the diagonal. All these numbers must be strictly positive provided that our suppositions about the functions F_i, $i = 1, \dots, n$, are satisfied. (Furthermore, the equality holds

$$H(x^1)H(x^2)^\# = U^\top D(x^1/x^2)(U^\top)^\#, \tag{11.36}$$

following from (11.34) if one takes into account that $(UU^\top)^{-1}U = (U^\top)^\#$, which, in turn, follows from identity (11.33) for $H = U^\top$ and the fact that the rank of $U^\top$ is n.)

Equation (11.35) means that the numbers on the diagonal of $D(x^1/x^2)$ are the eigenvalues of the matrix $H(x^1)H(x^2)^\#$ while the columns of $U^\top$ (thus, the rows of U) are its eigenvectors.

Now, taking into account the above reasonings, the method of testing our basic model ((11.20) and (11.21)) and determining the matrix U can be described as follows.

1. Fix N, consider N-dimensional subspaces, $E_N^* = Lin(\varphi_1, \dots, \varphi_N)$ in $\mathbf{C}$ and $E_N = Lin(e_1, \dots, e_N)$ in $\mathbf{C}^*$, where the biorthogonality condition $\langle \varphi_j, e_k \rangle = \delta_{jk}$ is satisfied, and take two different observable points, x^1 and x^2, in the threshold surface (11.23).
2. Around each of them, fix a cloud of observable points lying in the threshold surface (11.23), and determine, by using the method of least squares (or some other regular approximation method), the corresponding $N \times N$ matrices $H(x^1)$ and $H(x^2)$.
3. Calculate the $N \times N$ matrix $H(x^1)H(x^2)^\#$, determine its rank n, and find its characteristic values (eigenvalues), α_j, $j = 1, \dots, N$, which are the solutions (roots) of the characteristic equation

$$\det\left(H(x^1)H(x^2)^\# - \alpha I\right) = 0,$$

   where I stands for the unit $N \times N$ matrix. (We believe that N is large enough, and the matrix $H(x^1)H(x^2)^\#$ is singular.) Since the rank of the matrix $H(x^1)H(x^2)^\#$ is $n < N$, $\alpha = 0$ is its eigenvalue of multiplicity $N - n$.
4. Let $\alpha_{j_1}, \dots, \alpha_{j_n}$ be non-zero eigenvalues of $H(x^1)H(x^2)^\#$. If each of them is positive and has multiplicity one (hence all α_j are different), then the corresponding eigenvector is determined uniquely up to multiplying by a scalar, and (11.35) is satisfied with the $n \times n$ diagonal matrix $D(x^1/x^2) = D\{\alpha_{j_1}, \dots, \alpha_{j_n}\}$ and the $n \times N$ matrix U composed of these eigenvectors as its rows. Multiplying the rows of U by non-zero numbers d_i, $i = 1, \dots, n$, gives us a new matrix $U_1 = DU$, where $D = D\{d_1, \dots, d_n\}$. Taking into account that U satisfies (11.35), one has

$$\begin{aligned} H(x^1)H(x^2)^\# U_1^\top &= H(x^1)H(x^2)^\# U^\top D = U^\top D(x^1/x^2)D \\ &= U^\top DD(x^1/x^2) = U_1^\top D(x^1/x^2), \end{aligned}$$

   that is, U_1 satisfies (11.35) as well, hence it may be accepted instead of U. (Also, (11.36) implies the equality $U_1^\top D(x^1/x^2)(U_1^\top)^\# = U^\top DD(x^1/x^2)(U^\top D)^\# = U^\top DD(x^1/x^2)D^{-1}(U^\top)^\# = U^\top D(x^1/x^2)(U^\top)^\# = H(x^1)H(x^2)^\#$. Note that all d_i must be non-zero, because the rank of U_1 equals n.) Thus, if all non-zero eigenvalues of $H(x^1)H(x^2)^\#$ are positive and different, then the matrix U is determined uniquely up to multiplying from the left by a non-singular diagonal matrix.

   If now the multiplicity of some eigenvalue is $q > 1$ (hence q of α_{j_i} coincide), then the eigenvectors corresponding to it make up a q-dimensional vector subspace, any q linearly independent vectors from this subspace may be taken as the corresponding q rows of matrix U, and any matrix U obtained in such a way satisfies the requirements of the basic model. In order to eliminate this indeterminacy and obtain more information about desired matrix U, one can try to repeat the above procedure with other pairs of observable points x^1, x^2 in the threshold surface (11.23).[29]

5. Though the matrix $H(x^1)H(x^2)^\#$ is real, it can occur that some of its eigenvalues are complex. In such a case, our basic model ((11.20) and (11.21)) fails to be true. Furthermore, if all non-zero eigenvalues of $H(x^1)H(x^2)^\#$ are real but some of them are negative, then at least one of the diagonal matrices $D(x^1)$ and $D(x^2)$ has negative elements. In that case, the corresponding quadratic surface S^x approximating near x the threshold surface (11.23) is of hyperbolic type; therefore, the convexity assumption (I) cannot be true, and the model ((11.20) and (11.21)) must be rejected.

29 This recommendation is being made under assumption that the indeterminacy in question arises due to measurement errors and inaccurate computation. Of course, when the indeterminacy is inherent to the model it cannot be overcome. For instance, if each function F_i in Eq. (11.21) is a quadratic (as in Eq. (12.5)) then all eigenvalues of $H(x^1)H(x^2)^\#$ equal 1. In such a case, the method of quadratic approximation is obviously inapplicable.

12

Identification of Indiscriminate Relations. Colour Detection and Discrimination

A colour matching structure is determined by the colour matching relation which can be represented as a composition of the matching metamerism and the indiscriminate relation. The methods of tangential hyperplane and quadratic approximation allow one to identify the matching metamerism independently from the indiscriminate relation. In this chapter, we will deal with the indiscriminate relation, assuming that the colour stimulus map has already been identified. The indiscriminate relation is completely specified if the indiscriminable set (Definition 10.9) is specified for each colour stimulus.

Identification of the indiscriminate relation can be performed in the form of identification of the discrimination function (see Eq. (12.1)). The conditions for the existence of such a function were outlined in Chapters 10 and 11. There are two approaches to identifying the discrimination function which are complementary to each other. First, one can concentrate on experimental studying the indiscriminable set (using threshold techniques) with an intention to construct an analytical discrimination function that would fit the experimental data. The experimentally discovered geometric properties of the indiscriminable set will impose certain constraints on the class suitable models. For example, as an indiscriminable set is a projection of the corresponding (extended) threshold set parallel to the null-space of the colour stimulus map, all the geometrical properties of the extended threshold set established in the previous chapter (e.g., absorptivity, convexity) hold true for indiscriminable set too. As we shall see below, such a property of an indiscriminable set as its convexity considerably restricts the class of discrimination functions that can predict the indiscriminable set. Second, one can start from a theoretical model of colour discrimination, and then to test its prediction in experiment.

It must be said that detection and discrimination are among the major subjects in psychophysics (e.g., see Luce & Green, 1974; Falmagne, 1985). In psychophysics, discrimination models are generally designed to predict the probability of discrimination. Therefore, from a psychophysical point of view any model of colour discrimination is an analytical representation of the probability P_{ab} that the subject judges the light stimuli a and b to be different. The most general form of the representation of P_{ab} is:

$$P_{ab} = F(\mathbf{d}(a, b)), \quad a, b \in A, \tag{12.1}$$

where A is the light stimulus cone; $\mathbf{d} : A \times A \to \mathbf{R}$ will be referred to as a *discrimination function*; and $F : \mathbf{R} \to \mathbf{R}$ is called a *psychometric function*. The discrimination function $\mathbf{d}$ takes into account the different discriminability of different light stimuli. Psychometric function F is assumed to be an increasing function with the real interval $[0, 1]$ as its range.

Foundations of Colour Science: From Colorimetry to Perception, First Edition. Alexander D. Logvinenko and Vladimir L. Levin.
© 2023 John Wiley & Sons Ltd. Published 2023 by John Wiley & Sons Ltd.

When one of the two light stimuli to be discriminated is fixed as the background, a^0 (Fig. 11.1), the discrimination function becomes a one-variable function which will be referred to as the *detection function* (written as $\mathbf{d}_{a^0}$):

$$\mathbf{d}_{a^0}(a) := \mathbf{d}\left(a^0, a^0 + a\right), \quad a \in A - a^0 \subset E, \tag{12.2}$$

where $E = A - A$ is the ambient linear space; a^0 and a are the background and target light stimuli respectively. (Moreover, if the discrimination function $\mathbf{d}$ is determined on the whole of $E \times E$ then the detection function (12.2) makes sense whenever $a \in E$.)

If the discriminability level is fixed at certain value, say ϵ, then an (extended) threshold set can be represented as a sublevel set of the detection function[1]:

$$B = L_\epsilon\left[F(\mathbf{d}_{a^0}(\cdot)\right] = \left\{a \in E : F(\mathbf{d}_{a^0}(a)) \leq \epsilon\right\} - L_\varepsilon\left[\mathbf{d}_{a^0}\right], \tag{12.3}$$

where $\varepsilon = \sup\{\alpha \geq 0 : F(\alpha) \leq \epsilon\}$. If F is a strictly monotone continuous function then $\varepsilon = F^{-1}(\epsilon)$. The value ε will be called the *detection criterion*. As one can see, detection process is entirely predetermined by the detection function, the psychometric function being just a scale securing a proper range for the model output.

Studying colour discrimination by humans can be, generally, reduced to studying how colour detection depends on the background light stimulus. The target and background stimuli involved in a colour discrimination experiment are supposed to be spatially homogeneous light spots of the same size and shape so as to make the discrimination base solely on differences in the spectral composition of the radiant power. The rest of the visual field, usually, filled with spatially homogeneous light, play a role of the *remote background*. Since the remote background is usually considerably larger than the target (see Fig. 11.1) it mainly determines the adaptation state of the visual system.

Although the studies where the adaptation state was controlled by means of the background stimulus, and the adaptation effect was registered by measuring the detection performance, are an important source of experimental data on human colour discrimination, they (usually referred to as masking and/or adaptation studies) will be left out of consideration in this chapter. We will restrict ourselves to only the role of the immediate background (which will be called just the background). We will consider, first, colour detection models, and then we will turn to colour discrimination models.

Detection models predicting a convex extended threshold set will be called *convex*. If, in addition, the extended threshold set is a body,[2], then its gauge (the Minkowski function; see Appendix A.2) can be taken as a suitable detection function. In such a case, we will refer to the extended threshold set as the *threshold body*. The corresponding detection model will be referred to as *sublinear*. Since sublevel sets of a convex function are convex, the convexity (or merely, quasi-convexity) of a detection function is a sufficient condition for the threshold set convexity. Likewise, lower semicontinuity and sublinearity (see Appendix A.2) of a detection function is a sufficient condition for sublinearity of the corresponding detection model.

When an extended threshold set can be represented as a sublevel set of some (detection) function, we will say that it is *compatible* with this (detection) function. Given an arbitrary *body* B in E, there exist obviously many functions on E compatible with B. (Indeed, if B is compatible with a function $f : E \to \mathbf{R}_+$, and a function $F : \mathbf{R}_+ \to \mathbf{R}_+$ is strictly monotone and satisfies $F(0) = 0$, then B is compatible with the function $F \circ f$, as well. In particular, if a function $\|\Phi(\cdot)\|$ where $\|\cdot\|$ is a

---

1 Recall that for $\alpha \in \mathbf{R}$ the α-sublevel set, $L_\alpha\left[f\right]$, of function $f : E \to \mathbf{R}$ is defined as a subset in E $\{a \in E : f(a) \leq \alpha\}$.
2 That is, a closed, convex, and absorbing set (see Appendix A.2).

norm on $\mathbf{R}^n$ can be taken as a detection function for B, then the function $||\Phi(\cdot)||^2$ can also be taken as such.) This fact justifies the following definition.

Definition 12.1 *Any two detection functions, and respectively the detection models they determine, which induce one and the same extended threshold set (threshold body), will be called* psychophysically equivalent.[3]

Unfortunately, this definition of psychophysical equivalence depends on the detection criterion. Indeed, for two detection functions to be psychophysically equivalent it is sufficient for the ε-sublevel sets of these functions to be same, regardless of how the other sublevel sets of these relate. So, it may occur that another criterion value ε' would imply the (extended) threshold sets which are different.

One can circumvent such an ambiguity by considering only those detection functions, for each of which all its sublevel sets are in proportion to each other. A detection model with the detection function which meets this condition will be called *criterion independent* model. For such models the definition of psychophysical equivalence may be used without danger of contradiction.

It should be noted that the criterion independence is a strong, though quite natural, requirement. It is interesting that in principle it could be subjected to experimental verification. Indeed, in psychophysics, a technique is widely used that implies rating the observer's confidence in detection (D. M. Green & Swets, 1966). Using this technique one can experimentally estimate threshold sets corresponding to different rates of confidence. If they are similar (i.e., proportional) to each other, this will support the use of criterion independent models.

A broad class of criterion independent models is given by detection functions of the form $F \circ f$, where $f : E \to \mathbf{R}_+$ is a gauge, while $F : \mathbf{R}_+ \to \mathbf{R}_+$ is strictly monotone and satisfies $F(0) = 0$. In particular, a sublinear model is criterion independent, and its gauge (the sensitivity function) can be taken as the corresponding detection function. Interestingly, most of the detection models that have been proposed so far to account for not only chromatic, but also spatial and temporal discrimination, fall into the class of sublinear ones. They will be discussed in Section 12.1.

12.1 Colour detection models

12.1.1 Single-channel detection models

Given a colour-signal map Φ, detection function[4] $\mathbf{d}(a)$ can be represented as $\mathbf{d}(a) = t(\Phi(a))$, where t is a real function of n real variables. In Chapter 11, we assumed that for two lights to be matched the colour signals produced by these lights should be close enough in terms of some proximity measure. Function t is such a proximity measure. Given n linear colour mechanisms $\varphi_1, \dots, \varphi_n$ representing the colour-signal map Φ, the detection function $\mathbf{d}$ takes the form:

$$\mathbf{d}(a) = t\big(\xi_1, \dots, \xi_n\big), \tag{12.4}$$

where $\xi_i = \varphi_i(a)$ $(i = 1, \dots, n)$. We shall call t in (12.4) *n-colour stimulus coordinate representation* of the detection function.

3 It is important to note that the psychophysical equivalence of the detection models does not imply that the models are indistinguishable relative to all possible experiments. They are only so with respect to psychophysical experiments determining whether the stimulus is under threshold or not. Certainly, one might invent, for example, a physiological experiment to distinguish between two psychophysically equivalent models of detection. But there is no point in distinguishing psychophysically equivalent models under a psychophysical level of consideration.

4 When the background light stimulus is not important, thus not specified, we will omit the subscript.

The n-colour stimulus coordinate representation of the detection function (12.4) can be relatively simple in terms of some special linear mechanisms. In particular, in the visual literature there are many detection models where t in (12.4) is a norm.

Example 12.1 *Let $\varphi_1, \dots, \varphi_n$ be linear colour mechanisms underlying the matching metamerism. A model will be referred to as a* p-norm colour detection model *if its detection function is as follows*

$$\mathbf{d}(a) = t(\xi_1, \dots, \xi_n) = \left(\sum_{i=1}^{n} |\xi_i|^p\right)^{\frac{1}{p}} = \|x\|_p, \quad x = (\xi_1, \dots, \xi_n), \quad (1 \le p < \infty), \tag{12.5}$$

where $\xi_i = \langle \varphi_i, a \rangle$ $(i = 1, \dots, n)$.

A particular case of (12.5) when $p = 1$,

$$\mathbf{d}(a) = t(\xi_1, \dots, \xi_n) = \sum_{i=1}^{n} |\xi_i| = \|x\|_1, \tag{12.6}$$

is called detector-summator model.

For $p = \infty$, the detection function is defined by the Chebyshev norm $t(x) = \|x\|_\infty = i \max_i |\xi_i|$:

$$\mathbf{d}(a) = \max_i \{ |\langle \varphi_i, a \rangle| \}, \tag{12.7}$$

and the corresponding model is called a special version of the peak-detector model.

A general form of the peak-detector model is given by the detection function

$$\mathbf{d}(a) = \max_i \{ \langle \varphi_i, a \rangle \}. \tag{12.8}$$

When all φ_i are non-negative, the sign of module in (12.7) can be omitted, and both forms of the peak-detector model coincide.

The p-norm model implies that all the colour mechanisms $\varphi_1, \dots, \varphi_n$ contribute into making a decision whether the stimulus $a^0 + a$ is different from the background. On the contrary, the peak-detector model (12.8) assumes that the detection of a is based on the response of a single colour mechanism, the one whose response to a is largest.

The rational for using a p-norm is that it naturally arises when so-called probability summation[5] between stochastically independent mechanisms is taken into account. Indeed, let us assume (i) a number of the fundamental colour mechanisms whose responses are stochastically independent variables; and (ii) two lights a^0 and $a^0 + a$ are discriminated when, at least for one fundamental colour mechanism, the probability of its response to the target a exceeds some critical value. Following the terminology widely used in psychophysical literature (e.g., Graham, 1989), we say that ith fundamental colour mechanism, φ_i, detects target a if the probability that its response (written as $P_i(a)$) to a exceeds the critical value. In other words, lights a^0 and $a^0 + a$ match each other when no mechanism detects a. Following Quick (1974), let us assume that $P_i(a)$ is related to the mechanism's output $\varphi_i(a)$ as follows:

$$P_i(a) = 1 - \exp\left(-|\varphi_i(a)|^p\right), \tag{12.9}$$

where a positive number p is a parameter.

5 A notion of probability summation in the visual system is discussed, e.g., by Graham (1989).

Because of stochastic independence of the fundamental colour mechanisms, the probability $P(a)$ of detection of target a equals to $1 - \prod_i (1 - P_i(a))$. In the view of (12.9)

$$P(a) = 1 - \exp\left(-\sum_i |\varphi_i(a)|^p\right), \tag{12.10}$$

or

$$P(a) = 1 - \exp\left(-(t(\Phi(a)))^p\right), \tag{12.11}$$

where $t(\xi_1, \ldots, \xi_n) = \left(\sum_{i=1}^{n} |\xi_i|^p\right)^{\frac{1}{p}}$.

A number of researchers claimed that the p-norm model fitted their colour discrimination data. The only problem which was debated was what exponent p in (12.5) fitted the data best. While some researchers suggested the threshold set to have an ellipsoidal shape that corresponds to the exponent $p = 2$ (Noorlander & Koenderink, 1983; Poirson et al., 1990; Poirson & Wandell, 1996; Knoblauch & Maloney, 1996), others reported that their data are better fitted with a parallelepiped with rounded corners as predicted by model (12.5) with $p > 2$ (Cole et al., 1990, 1993, 1994; Sankeralli & Mullen, 1996).

There is, however, strong experimental evidence against models predicting a sensitivity function being a norm. The fact is that such models imply that the threshold set is *balanced.*

Definition 12.2 *A subset B in a linear space is* balanced *if $\alpha B \subseteq B$ whenever $|\alpha| \le 1$, where $\alpha B := \{\alpha x : x \in B\}$.*

As follows from the definition, the balanced threshold set is symmetric relative to the background light stimulus. It means that multiplying a threshold target stimulus by a coefficient, absolute value of which is not greater than 1, leaves the target stimulus under threshold. It seems to hold true for positive coefficients, that is, the threshold target stimulus remains under threshold when its intensity is reduced to some fraction of its former value. However, this may not be the case when multiplying by a negative coefficient. As pointed out at the beginning of Chapter 10, the human visual system is more sensitive to luminance decrements than to luminance increments. This kind of asymmetry could not be observed if the threshold bodies were balanced. A detailed study of this issue showed that some colour detection contours (for a neutral light stimulus as a background) were not balanced: they looked like ellipsoids oblate at the poles (Logvinenko, Tyurin, & Sawey, 2002). Therefore, the threshold set (at least for some backgrounds) cannot be represented as a sub-level set of any norm.

As following example shows, p-norm model can be easily generalised to take into account the threshold set unbalancedness.

Example 12.2 *Let $\varphi_1, \ldots, \varphi_n$ be linear colour mechanisms underlying the matching metamerism. A colour detection model with the following detection function*

$$\mathbf{d}(a) = t(x) = g\left(\sum_{i=1}^{n} f_i(\xi_i)\right), \quad x = (\xi_1, \ldots, \xi_n), \tag{12.12}$$

where $\xi_i = \langle \varphi_i, a \rangle$ $(i = 1, \ldots, n)$, $g : \mathbf{R} \to \mathbf{R}$ is a strictly monotonic function, can be considered as a natural generalisation of the p-norm model (12.5).

A proper choice of non-linear function f_i following each linear colour mechanism φ_i in (12.12) allows taking into account, first, a possible non-linearity of the transduction process (e.g., from photoreceptors to the other visual cells), and second, the increment-decrement asymmetry (i.e., the threshold set asymmetry with respect to the coordinate axes). Also, the pooling function g (which equals $(\cdot)^{1/p}$ in (12.5)) can have a more general form.

12.1.2 Fundamental colour mechanisms revisited

The definition of fundamental colour mechanisms given in Section 3.3 rests upon a particular (colorimetric) model of colour matching, which can be considered as a degenerated case of the p-norm colour detection model when the detection criterion ε equals zero. In this case, the indiscriminable set degenerates to a point, colour matching becomes transitive, and the fundamental colour mechanisms are impossible to derive from colour matching as well as colour discrimination data. Such indeterminacy arises from the fact that in this case any linearly independent combination of the fundamental colour mechanisms can also serve as fundamental colour mechanisms. As a matter of fact, given the colorimetric model of colour matching, the only criterion for distinguishing the fundamental colour mechanisms among all the other colour mechanisms is a physiological one. At the psychophysical level, the fundamental colour mechanisms do not manifest themselves.

If the detection criterion ε is not zero, the shape of the indiscriminable set might happen to contain certain indications of what the fundamental colour mechanisms are. For instance, in the case of the peak-detector model with $\varepsilon > 0$ (Example 12.1), the indiscriminable set has a shape of parallelepiped the sides of which are determined by the colour mechanisms $\varphi_1, \dots, \varphi_n$ appearing in Eq. (12.8). In other words, these colour mechanisms can be derived from the shape of the indiscriminable parallelepiped. It seems natural to call $\varphi_1, \dots, \varphi_n$ the fundamental colour mechanisms (as far as the peak-detector model with $\varepsilon > 0$ is concerned), generalising the previous definition valid only in the context of the colorimetric model of colour matching (i.e., for the peak-detector model with $\varepsilon = 0$).

For the p-norm colour detection model with $2 < p < \infty$ the indiscriminable set has a shape of parallelepiped with rounded vertices. Still, the colour mechanisms $\varphi_1, \dots, \varphi_n$ (in terms of which the model is defined by Eq. (12.5)) remain derivable from the indiscriminable-set shape. So do the colour mechanisms underlying the more general colour detection model (12.12). Really, as shown in Section 11.3.3, the indiscriminable set induced by this model is such that the tangential quadratic surface estimated for different points x have the same principal axes. In other words, for any pair of points x^1 and x^2, the matrix $A = H(x^1)\, H(x^2)^{\#}$ can be diagonalised by the same linear transformation; that is, it has the same eigenvectors. Moreover, the colour mechanisms $\varphi_1, \dots, \varphi_n$ in (12.12) can be uniquely derived from these eigenvectors (provided that all the eigenvectors are different[6]). As the derivation of colour mechanisms $\varphi_1, \dots, \varphi_n$ do not depend on points x^1 and x^2, these are a characteristic of the indiscriminable set as such. Thus, these colour mechanisms $\varphi_1, \dots, \varphi_n$ can be taken as the fundamental colour mechanisms, that is, those underlying colour detection, thus, colour matching.[7]

To summarise, generalising the colorimetric model of colour detection to the model defined by Eq. (12.12), the fundamental colour mechanisms are defined as n continuous functions $\varphi_1(\lambda), \dots, \varphi_n(\lambda)$ such that colour match of two lights, a and b, means that the value of the function (12.12) does not exceed $\varepsilon > 0$.

---

6 As mentioned in Section 11.3.3, this is not the case for the p-norm colour detection model with $p = 2$. Therefore, for this model the fundamental colour mechanisms cannot be derived from a single indiscriminable set.

7 Admittedly, colour matching when one colour stimulus to be matched is fixed.

Note that the condition that the eigenvectors of matrix $H(x^1)\, H(x^2)^{\#}$ do not depend on points x^1 and x^2, can be tested experimentally. Passing the test will mean that the indiscriminable-set shape is consistent with the model (12.12). This will, in turn, be a justification of the notion of fundamental colour mechanisms obtained at the psychophysical level.

Were it found that the matrix A is significantly different for different regions of the threshold surface, that would testify against this model. Hence, a more sophisticated model of colour matching will be needed (see Section 12.1.3).

12.1.3 Multi-channel detection models

One may happen that, being inconsistent with all the colour detection data, model (12.12) can account for the colour discrimination data collected locally, that is, in the vicinity of some light stimulus. In this case, one can use a model composed of a collection of the individual colour detection models as in (12.12) acting in parallel. Such a model will be referred to as *multi-channel*. To be more exact, what we call a *multi-channel model* here is a set of detection models with individual detection functions (i.e., channels) combined together via some pooling function.[8] Multi-channel models of detection have been widely used to account for spatial and temporal pattern detection (R. De Valois & De Valois, 1988; Graham, 1989). The method of quadratic approximation can be used to identify a multi-channel model if one assumes that, first, each channel detects a light stimulus independently of the output of the other channels; and second, an observer detects a light stimulus if at least one channel detects it. In this case, the channels interact only due to probability summation between them. The effect of probability summation between channels can be taken into account by assuming that the channels' detection functions are summed up as in the p-norm colour detection model:

$$\mathbf{d}(a) = \left(\sum_{i=1}^{n} |\mathbf{d}_i(a)|^p \right)^{1/p}, \tag{12.13}$$

where $\mathbf{d}_i$ is the detection function of the ith channel. The target light stimulus, a, is assumed to be detected if the resultant detection function $\mathbf{d}$ in (12.13) exceeds the discriminability level.

When $p = \infty$, (12.13) is replaced by:

$$\mathbf{d}(a) = \max_{i=1,\ldots,n} \{\mathbf{d}_1(a), \ldots, \mathbf{d}_n(a)\}. \tag{12.14}$$

It will be shown in Section 12.2.2 that the threshold set as determined by (12.14) can be represented as the intersection of the threshold sets defined by each individual channel constituting a multi-channel model. So the threshold surface will be composed of fragments of individual threshold surfaces. Therefore, the eigenvectors of the matrix $H(x^1)\, H(x^2)^{\#}$ will remain the same unless one crosses a border between adjacent fragments. Thus, a distinctive feature of a model based on independent multiple channels with the maximum-of-output decision rule (12.14) is that it predicts invariance of the eigenvectors over some areas of the threshold surface, and sharp difference in the eigenvectors between these areas.

When $p \neq \infty$ there will be no sharp borders between the individual areas of the threshold surface. The exponent p is usually interpreted as an index of steepness of the psychometric function which is as a rule not less than 4–5 (for more details see Graham, 1989, pp. 172–177). For such an exponent

8 The increment-decrement asymmetry can be accounted for if one assumes that luminance increments and decrements are processed by different colour channels. This idea has been implemented for temporal waveforms detection in vision (Krauskopf, 1980; Watson, 1986).

the effect of rounding of the border between the individual areas will be quite local. In other words, this effect will be negligibly small for the most of an individual area apart from a relatively narrow region around the border. At any rate, one might expect that there will be regions in the threshold surface where the eigenvectors of the matrix $H(x^1)\,H(x^2)^{\#}$ will be practically constant. Therefore, it will be possible to evaluate the matrix A for each individual part of the threshold surface.

For further consideration, it is important to know how to determine whether a particular multi-channel model is sublinear or not. Although it is not a trivial problem, even if individual channels themselves represent sublinear models, in one particular case a solution can be easily achieved. Namely, as the family of sublinear functionals is closed under operations of summation and taking the supremum, it follows that any combination of sublinear models via these operations belongs to the class of sublinear models as well.

12.2 Peak-detector model equivalent to a sublinear colour detection model

As mentioned above, there are many analytical representations for a given particular threshold set (body). And it is not easy to decide whether two particular representations, are psychophysically equivalent (that is, determine the same body) or not, and if not, how essential the difference between them. In this regard, it will be helpful if some canonical representation is chosen. The sensitivity function can certainly play the role of such a representation in terms of which various detection models can be compared because it is uniquely determined by the extended threshold set (threshold body). However, the sensitivity function can be explicitly evaluated only in very few simple cases.

Yet, the sensitivity function can be quasi-linearised and then evaluated in the form of support function (see Eq. (A.10)), i.e., as envelope of the family of the linear colour mechanisms. Such a representation is none other than some infinite version of what was above called the peak-detector model. A peak-detector model (12.8) is rather simple model of colour detection. It is completely, though not uniquely, characterized by the set of linear colour mechanisms $\varphi_1, \ldots, \varphi_n$ which in the present context we will refer to as *colour preprocessor*. As easily seen, the same sensitivity function can be induced by the peak-detector models with different colour preprocessors. Of special interest here is the peak-detector model based on the (infinite) set of colour mechanisms making up the polar (see Eq. (A.6)) of the extended threshold set which we will call a *polar peak-detector model* (see Eq. (12.17)). It is maximal one, that is, it contains all the linear colour mechanisms included in any equivalent peak-detector model. For every sublinear detection model there exists exactly one polar peak-detector model psychophysically equivalent to it.

The threshold body polar cannot always be analytically expressed. In this case, a peak-detector model based on a smaller set of colour mechanisms (or colour detectors if they exist) and equivalent to the original sublinear (or convex) model can be evaluated. Indeed, as follows from the bipolar theorem (see Appendix A.2) along with the duality formula (A.11), the following equality holds

$$\mu(\cdot; B'_{a^0}) = s(\cdot; M), \tag{12.15}$$

where M is a subset of the threshold body polar $(B'_{a^0})^0$ such that its convex hull $co\,M$ is weakly* dense in $(B'_{a^0})^0$. Any such M can be considered as a colour preprocessor of the corresponding peak-detector model equivalent to the original sublinear detection model.

Below we evaluate the polar peak-detector model and the peak-detector model based on some set of most sensitive colour mechanisms (or colour detectors if they exist), which are equivalent to the sublinear models of detection considered above. This evaluation actually comes down to evaluating the most sensitive colour mechanism for a given particular light stimulus.

12.2.1 Sublinear colour detection models

The sensitivity function for a sublinear model is expressed in terms of the gauge specifying the model:

$$\mu(\cdot;B) = \mu\left(\cdot;L_\varepsilon[t\circ\Phi]\right) = \mu\left(\cdot;L_1[\varepsilon^{-1}t\circ\Phi]\right) = \mu(\cdot;\varepsilon L_1[t\circ\Phi]) = \varepsilon^{-1}t(\Phi(a)). \tag{12.16}$$

Here, $B = L_\varepsilon[\mathbf{d}]$ is the extended threshold set (threshold body), $\mu(\cdot;B)$ is its gauge, $\mathbf{d} = t\circ\Phi$ is the detection function, $\Phi : E \to \mathbf{R}^n$ is a colour-signal map, t is a gauge on $\mathbf{R}^n$, ε is a detection criterion, and $L_\varepsilon[\mathbf{d}]$ is the ε-sublevel set of the detection function. The polar representation of sensitivity function (12.16) following from the bipolar theorem (see formula (A.7) in Appendix A.2) is given by the duality between Minkowski and support functions (A.11):

$$\mu(\cdot;B) = s(\cdot;B^0). \tag{12.17}$$

We must thus evaluate the polar of B (the polar colour preprocessor).

Let us consider the set

$$B_1 := \Phi(B) = \{x \in \mathbf{R}^n : t(x) \le \varepsilon\} = L_\varepsilon[t] = \varepsilon L_1[t]$$

in $\mathbf{R}^n$. Since obviously $B = B + \Phi^{-1}(0)$, one has $B = \Phi^{-1}(B_1)$, and formula (A.12) from Appendix A.2 can be used. Taking into account that $(\alpha C)^0 = \alpha^{-1}C^0$ whenever $\alpha > 0$, $C \subset \mathbf{R}^n$, we get

$$B^0 = \Phi^*(B_1^0) = \Phi^*(\varepsilon L_1[t])^0 = \varepsilon^{-1}\Phi^*(L_1[t])^0 = \Phi^*(L_{\varepsilon^{-1}}[t])^0, \tag{12.18}$$

and the polar representation of the sensitivity function is thus as follows:

$$\mu(\cdot;B) = s(\cdot;B^0) = \varepsilon^{-1}\sup\{\langle\cdot,\Phi^*x^*\rangle : x^* \in (L_1[t])^0\}. \tag{12.19}$$

Recall that since the range of the conjugate operator $\Phi^* : (\mathbf{R}^n)^* \to E^*$ is the subspace $Lin(\varphi_1,\dots,\varphi_n)$ in E^*, for every $x^* = (\eta_1,\dots,\eta_n) \in \left(\mathbf{R}^n\right)^*$ Φ^*x^* may be represented as a linear combination of the functionals $\varphi_1,\dots,\varphi_n$, i.e.,

$$\Phi^*x^* = \sum_{i=1}^{n}\eta_i\varphi_i, \tag{12.20}$$

and since the functionals $\varphi_1,\dots,\varphi_n$ are assumed to be linearly independent the representation (12.20) is unique.

If now the gauge t is some norm on $\mathbf{R}^n$, $t(\cdot) = \|\cdot\|$, then $L_1[t]$ is the corresponding unite ball in $\mathbf{R}^n$, and its polar $(L_1[t])^0$ is the unite ball (with respect to the dual norm $\|\cdot\|_*$) in $(\mathbf{R}^n)^*$. In such a case, the polar preprocessor is

$$B^0 = \varepsilon^{-1}\Phi^*(L_1[t])^0 = \{\Phi^*x^* : \|x^*\|_* \le \varepsilon^{-1}\} \tag{12.21}$$

$$= \left\{\sum_{i=1}^{n}\eta_i\varphi_i : \|(\eta_1,\dots,\eta_n)\|_* \le \varepsilon^{-1}\right\}. \tag{12.22}$$

Consider, in particular, the p-norm colour detection model (12.5). The dual of p-norm for $1 < p < \infty$ is known to be the q-norm $(p+q=1)$ (e.g., Kantorovich & Akilov, 1982, p. 191). In this case, (12.21) takes the form:

$$B^0 = \left(L_\varepsilon\left[\|\Phi(\cdot)\|_p\right]\right)^0 = \Phi^*L_{\varepsilon^{-1}}\left[\|\cdot\|_q\right] \tag{12.23}$$

$$= \left\{\sum_{i=1}^{n}\eta_i\varphi_i : \left(\sum_{i=1}^{n}|\eta_i|^q\right)^{\frac{1}{q}} \le \varepsilon^{-1}\right\}, \quad p+q=1. \tag{12.24}$$

A smaller preprocessor consisting of some most sensitive colour mechanisms will be considered in Section 12.2.3.

12.2.2 Multi-channel sublinear models

Consider first a finite set of sublinear models connected in parallel by means of a logical function 'or'. Let B_i and $\mu_i(\cdot; B_i)$ be the extended threshold set (threshold body) and the sensitivity function of ith model respectively. Then the resulting threshold body is the intersection of individual bodies, $B = \bigcap_i B_i$, and sensitivity function of the resultant model, $\mu(\cdot; B)$, is expressed as supremum of the individual sensitivity functions:

$$\mu(\cdot; B) = \sup_i \mu(\cdot; B_i). \tag{12.25}$$

In view of (A.11), the support functions are related in the same way:

$$s(\cdot; B^0) = \sup_i s(\cdot; B_i^0). \tag{12.26}$$

Furthermore, it follows from (A.14) that

$$\left(\bigcap_i B_i\right)^0 = cl\ co \bigcup_i B_i^0; \tag{12.27}$$

therefore $B^0 = cl\ co \bigcup_i B_i^0$, that is, the polar preprocessor of a multi-channel detection model with the maximum-output decision rule is the closed convex hull of the union of the individual polar preprocessors.

Suppose now that $B_i = L_\varepsilon[t_i \circ \Phi]$ where t_i is a gauge on $\mathbf{R}^n$ specifying the ith model, and consider in $\mathbf{R}^n$ the sets $B_{1i} = \Phi(B_i) = L_\varepsilon[t_i] = \varepsilon L_1[t_i]$ and $B_1 := \bigcap_i B_{1i}$. Since for all i, $B_i = B_i + \Phi^{-1}(0)$, it follows that $B_i = \Phi^{-1}(B_{1i})$ and $B = B + \Phi^{-1}(0)$, hence $B = \Phi^{-1}(B_1)$. It follows from general polar formula (A.14) that

$$\left(\bigcap_i B_{1i}\right)^0 = cl\ co \bigcup_i B_{1i}^0. \tag{12.28}$$

Being the finite union of compact sets, $\bigcup_i B_{1i}^0$ is compact, and as its convex hull is taken in a finite dimensional space, it is compact (hence closed) as well. And formula (12.28) can be rewritten now as

$$B_1^0 = \left(\bigcap_i B_{1i}\right)^0 = co \bigcup_i B_{1i}^0,$$

where co stands for the convex hull. Now, applying (A.12) yields

$$B^0 = \Phi^*(B_1^0) = \Phi^*\left(co \bigcup_i B_{1i}^0\right) = co \bigcup_i \Phi^*(B_{1i}^0) = co \bigcup_i B_i^0, \tag{12.29}$$

that is, the polar preprocessor of a multi-channel detection model with individual threshold bodies of the form $B_i = L_\varepsilon[t_i \circ \Phi]$ and the maximum-output decision rule is the convex hull of the union of the individual polar preprocessors.

When the individual channels detection functions are added up before comparing with the discriminability level, so are the sensitivity functions and thus individual support functions. And since the sum of the support functions of the individual threshold set's polars is the support function of the sum of the individual polars (e.g., Rockafellar, 1970, p. 146) we have:

$$\sum_{i=1}^{n} \mu(\cdot; B_i) = \sum_{i=1}^{n} s(\cdot; B_i^0) = s\left(\cdot; \sum_{i=1}^{n} B_i^0\right). \tag{12.30}$$

This means that summation of the detection functions results in summation of the individual polar preprocessors.

Lastly, we show that the operation on sensitivity functions which induces union of threshold sets is infimal convolution.

Definition 12.3 *Given convex functions $f_1, \ldots, f_n$ on E, their* infimal convolution $f_1 \oplus \cdots \oplus f_n$ *is defined by*

$$(f_1 \oplus \cdots \oplus f_n)(a) := \inf \left\{ \sum_{i=1}^{n} f_i(a_i) \; : \; a_i \in E, \sum_{i=1}^{n} a_i = a \right\}. \tag{12.31}$$

Indeed, it is known that infimal convolution corresponds to intersection of polars (e.g., Rockafellar, 1970, p. 146; Holmes, 1975, p. 113):

$$s\left(\cdot; B_1^0\right) \oplus \cdots \oplus s\left(\cdot; B_n^0\right) = s\left(\cdot; \bigcap_i B_i^0\right),$$

and applying polar formula (A.13) yields

$$\mu\left(\cdot; B_1\right) \oplus \cdots \oplus \mu\left(\cdot; B_n\right) = \mu\left(\cdot; \bigcup_i B_i\right). \tag{12.32}$$

12.2.3 The most sensitive colour mechanisms

Consider now a general case when a convex model is psychophysically equivalent to a peak-detector model based on a colour preprocessor M comprising some most sensitive colour mechanisms. In what follows, we suppose that the tolerance extension ς' is strongly regular (Definition 11.4) hence the threshold body B'_{a^0} has the following property:

$$\alpha a \in B'_{a^0} \; \forall \alpha > 0 \Leftrightarrow a \in E(\sim).$$

This property holds when the detection function is of the form $\mathbf{d}(\cdot) = t(\Phi(\cdot))$ where t is a gauge on $\mathbf{R}^n$ such that $t(x) = 0 \Leftrightarrow x = 0$.

Proposition 12.1 *Consider the peak-detector model determined by the colour preprocessor*

$$M = \{\varphi_e \; : \; e \in C\}, \tag{12.33}$$

where C is an arbitrary weakly dense subset in the threshold body boundary bdry(B'_{a^0}), and for every $e \in C$, φ_e is some most sensitive colour mechanism for e. Then this peak-detector model is psychophysically equivalent to the original detection model.*

The corresponding peak-detector model is needed to be evaluated for various concrete colour preprocessors (12.33). This task is reduced to evaluating the most sensitive colour mechanisms for an arbitrary threshold stimulus. As follows from Definition 11.5, if a functional (colour mechanism)

φ is supporting the threshold body B'_{a^0} at a threshold stimulus $e_0 \in E$ then the colour mechanism $\frac{1}{\varphi(e_0)}\varphi$ is most sensitive for e_0 hence lies in the threshold body polar $(B'_{a^0})^0$. Let $ms\left(e_0|(B'_{a^0})^0\right)$ be the set of all colour mechanisms that are most sensitive for e_0.

As follows from the general formula (A.18) in Appendix A.2, the colour mechanisms in the threshold body polar most sensitive to a given light stimulus e_0 coincide with the subdifferential of the sensitivity function at e_0 (cf. also Proposition 11.5):

$$ms\left(e_0|(B'_{a^0})^0\right) = \partial\mu(e_0; B'_{a^0}). \tag{12.34}$$

Given a detection function $\mathbf{d}$ on E and a detection criterion ε so that the threshold body is given by $B'_{a^0} = L_\varepsilon[\mathbf{d}] = \{e \in E : \mathbf{d}(e) \leq \varepsilon\}$, we consider the function

$$\mathbf{d}^\varepsilon(e) := \begin{cases} \mathbf{d}(e) & \text{if } \mathbf{d}(e) \leq \varepsilon, \\ +\infty & \text{otherwise.} \end{cases} \tag{12.35}$$

Clearly, for $e_0 \in bdry(B'_{a^0})$ one has

$$\partial\mathbf{d}^\varepsilon(e_0) = \{\varphi \in \mathbf{C} : \varphi(e) - \varphi(e_0) \leq \mathbf{d}(e) - \mathbf{d}(e_0)\ \forall e \in B'_{a^0}\}.$$

The foregoing proposition shows that when the sensitivity function is not given explicitly many most sensitive colour mechanisms can be evaluated by subdifferentiating function (12.35).

Proposition 12.2 *Let $\mathbf{d}$ be a convex detection function continuous at a (extended) threshold point e_0, and ε a detection criterion, determining the threshold body $B'_{a^0} = L_\varepsilon[\mathbf{d}]$. Then*

$$\left\{\frac{\varphi}{\varphi(e_0)} : \varphi \in \partial\mathbf{d}^\varepsilon(e_0)\right\} \subseteq ms\left(e_0|(B'_{a^0})^0\right). \tag{12.36}$$

Remark 12.1 *As obviously $\partial\mathbf{d}(e_0) \subset \partial\mathbf{d}^\varepsilon(e_0)$, any subgradient φ of the detection function at e_0 determines a most sensitive colour mechanism for e_0, $\frac{\varphi}{\varphi(e_0)}$.*

Corollary 12.1 *If $\mathbf{d}$ is differentiable at e_0, then*

$$\frac{\nabla\mathbf{d}(e_0)}{\langle e_0, \nabla\mathbf{d}(e_0)\rangle} \in ms\left(e_0|(B'_{a^0})^0\right), \tag{12.37}$$

(Here, gradient $\nabla\mathbf{d}$ means the Gâteaux differential.)

Remark 12.2 *The mechanisms $\{\frac{\varphi}{\varphi(e_0)} : \varphi \in \partial\mathbf{d}^\varepsilon(e_0)\}$ apparently don't exhaust the set $ms\left(e_0|(B'_{a^0})^0\right)$. Indeed, if $\phi \in ms\left(e_0|(B'_{a^0})^0\right)$ then, by definition, ϕ belongs to the threshold body polar and equals 1 at e_0:*

$$\phi(e) \leq \phi(e_0) = 1, \quad \forall e \in B'_{a^0}. \tag{12.38}$$

Then $\phi \in N(e_0|B'_{a^0})\backslash\{0\}$, and as $B'_{a^0} = L_\varepsilon[\mathbf{d}]$, we have: $\phi(e) - \phi(e_0) \leq 0$ whenever $\mathbf{d}(e) - \mathbf{d}(e_0) < 0$. Let us consider the interior of the threshold body, $int\ (B'_{a^0}) = \{e : \mathbf{d}(e) < \varepsilon = \mathbf{d}(e_0)\}$, and define

$$\alpha := \inf\left(\frac{\phi(e) - \phi(e_0)}{\mathbf{d}(e) - \mathbf{d}(e_0)} : e \in int\ (B'_{a^0})\right). \tag{12.39}$$

Since for every $e \in int\ (B'_{a^0})$ both $\phi(e) - \phi(e_0)$ and $\mathbf{d}(e) - \mathbf{d}(e_0)$ are negative, we derive from (12.39) that $\alpha \geq 0$ and

$$\phi(e) - \phi(e_0) \leq \alpha(\mathbf{d}(e) - \mathbf{d}(e_0)) \tag{12.40}$$

whenever $e \in int\,(B'_{a^0})$. *Moreover, as the function* $\mathbf{d}$ *is continuous, inequality (12.40) holds for all* e *from the threshold body. Now, if* $\alpha > 0$ *then* $\varphi := \alpha^{-1}\phi \in \partial\mathbf{d}^{\varepsilon}(e_0)$ *and* $\phi = \alpha\varphi$, *and applying (12.38) yields* $\alpha = \frac{1}{\varphi(e_0)}$. *Thus if* $\alpha > 0$, *the mechanism* ϕ *is of the form (12.36). However, if* $\alpha = 0$ *we cannot assert that* ϕ *is of the form (12.36).*

Now, we see from Propositions 12.1 and 12.2 that a rather general colour preprocessor for a given convex detection model is as follows: we fix a dense subset C in the threshold body boundary and take for every $e \in C$ an (arbitrary) subgradient $\varphi_e \in \partial\mathbf{d}^{\varepsilon}(e)$; then the set $M = \{\frac{\varphi_e}{\varphi_e(e)} : e \in C\}$ is a colour preprocessor. Moreover, if the function $\mathbf{d}^{\varepsilon}$ is differentiable on C, then every φ_e is the gradient (Gâteaux differential) $\nabla\mathbf{d}^{\varepsilon}(e)$, $e \in C$, and the colour preprocessor M consists of colour detectors for $e \in C$, $M = \{\frac{\nabla\mathbf{d}^{\varepsilon}(e)}{\langle e, \nabla\mathbf{d}^{\varepsilon}(e)\rangle} : e \in C\}$

In the following example, a class of peak-detection models admitting a finite colour preprocessor is described.

Example 12.3 *Suppose the threshold body* B'_{a^0} *is of the form*

$$B_m(\psi_1, \ldots, \psi_m) = \{a : \psi_i(a) \leq 1,\ i = 1, \ldots, m\}, \tag{12.41}$$

where $\psi_1, \ldots, \psi_m$ *are linear colour mechanisms* ($\psi_i \in Lin(\varphi_1, \ldots, \varphi_n) \subset \mathbf{C}(\Lambda)$, $i = 1, \ldots, m$). *Then* $M = \{\psi_1, \ldots, \psi_m\}$ *is a colour preprocessor for the corresponding peak-detection model (12.8). Indeed, obviously* $(B_m(\psi_1, \ldots, \psi_m))^0 = conv\,(0, \psi_1, \ldots, \psi_m) = conv\,(M \cup \{0\})$; *therefore* $M^0 = B_m(\psi_1, \ldots, \psi_m)$. *Furthermore, the boundary of this threshold body consists of* m *pieces* $C_i := \{a : \psi_i(a) = 1, \psi_j(a) \leq 1 (j \neq i)\}$, $i = 1, \ldots, m$, *and Proposition 12.1 can be applied with* $C = C_1 \cup \cdots \cup C_m$ *and* $\varphi_e = \psi_i$ *for any* $i \in I(e) := \{i : e \in C_i\}$. *Moreover, if we take a smaller* C, $C = C'_1 \cup \cdots \cup C'_m$, *where* $C'_i := \{a : \psi_i(a) = 1, \psi_j(a) < 1 (j \neq i)\}$, $i = 1, \ldots, m$, *then this smaller* C *is dense in the threshold body boundary and every* ψ_i *becomes a detector for all* $e \in C'_i$.

Notice that the threshold body

$$B := \{a : -\beta_i \leq \psi_i(a) \leq \alpha_i,\ i = 1, \ldots, m\},$$

where all α_i *and* β_i *are positive, is of type (12.41):*

$$B = B_{2m}(\alpha_1^{-1}\psi_1, \ldots, \alpha_m^{-1}\psi_m; \beta_1^{-1}(-\psi_1), \ldots, \beta_m^{-1}(-\psi_m)).$$

When we deal with a multi-channel model the detection function is a composition of individual detection functions and the pooling function. It should be noted that convexity of channels' detection functions and pooling function does by no means guarantee convexity of the whole model. Hence, it is important to know how to determine whether a given particular multi-channel model is convex or not.

Consider a set of detection models with detection functions $\mathbf{d}_1, \ldots, \mathbf{d}_n$ combined into a multi-channel model via the pooling function u. The resultant detection function

$$\mathbf{d}(\cdot) = u\big(\mathbf{d}_1(\cdot), \ldots, \mathbf{d}_n(\cdot)\big) \tag{12.42}$$

can be represented as a composition $\mathbf{d} = u \circ \Psi$, where Ψ is a nonlinear map

$$\Psi(a) = \big(\mathbf{d}_1(a), \ldots, \mathbf{d}_n(a)\big). \tag{12.43}$$

It is obvious that the composition of a convex function and an arbitrary non-linear mapping is, generally, not convex. Yet, this is the case for one particular class of convex functions provided that the map Ψ is convex.

The latter notion is a natural generalization of convex function.

Let B be a convex subset of a (Hausdorff) locally convex space E (in particular, one can take E to be $\mathbf{C}^*$ with the weak* topology) and let E_1 be a (Hausdorff) locally convex space with an order $\geq_1$ on it defined by a closed convex cone (*positive cone*) $E_{1+} = \{x \in E_1 \,:\, x \geq_1 0\}$. We will also use the notation $x \leq_1 y$ for $y \geq_1 x$.

Definition 12.4 *We call the map* $\Psi : B \to E_1$ convex *if*

$$\Psi(\alpha a + (1-\alpha)b) \leq_1 \alpha\Psi(a) + (1-\alpha)\Psi(b), \quad \forall a, b \in B, \quad 0 \leq \alpha \leq 1. \tag{12.44}$$

A subdifferential may be also defined for a convex mapping.

Definition 12.5 *A subdifferential of a convex mapping* $\Psi : B \to E_1$ *at* $a \in B$ *is defined as*

$$\partial\Psi(a) := \left\{T \in \mathcal{L}\left(E, E_1\right) \,:\, Te - Ta \leq {}_1\Psi(e) - \Psi(a),\ \forall e \in B\right\}, \tag{12.45}$$

where $\mathcal{L}(E, E_1)$ *stands for the set of all continuous linear mappings from E into* E_1. *When* Ψ *is continuous at a, every linear mapping T satisfying (12.45) is continuous (that is, the subdifferential* $\partial\Psi(a)$ *coincides with the* algebraic subdifferential*).*

The following theorem says when the image of a convex function under non-linear mapping is convex, and how to calculate its subdifferential. Before to state the theorem, we will need some notions concerning vector lattices.[9]

Thus, suppose now that the space E_1 with the order $\geq_1$ defined on it by the positive closed cone E_{1+} is a locally convex *vector lattice* that is, for any pair of elements x, y in E_1 there exists their supremum, i.e., the sole element $x \vee y = \sup_{E_1}(x, y) \in E_1$ such that $x \vee y \geq_1 x$, $x \vee y \geq_1 y$ and if $z \geq_1 x$, $z \geq_1 y$ then $z \geq_1 x \vee y$. The space of measures $\mathbf{C}^* = \mathbf{C}^*(\Lambda)$ with the weak* topology is an example of a locally convex lattice, and the corresponding positive cone is the cone of positive measures $\mathbf{C}^*_+$. The non-negative orthant of $\mathbf{R}^n$

$$\left\{x = \left(\xi_1, \dots, \xi_n\right) \,:\, \xi_1 \geq 0, \dots, \xi_n \geq 0\right\} \tag{12.46}$$

and the cone of non-negative functions in a linear space of functions (or classes of Lebesgue equivalent functions) defined on a set Λ are another example of positive cones E_{1+}. We will say that these induce the *natural pointwise vector order*.

A subset Q in a lattice E_1 is called *order bounded* if there are elements $z, z' \in E_1$ such that $z \geq_1 x \geq_1 z'$ for all $x \in Q$. A vector lattice E_1 is called *order complete* if there exists supremum for every order bounded set in it. If E_1 is order complete then for every order bounded set Q in it there exists infimum, i.e., the (sole) element $\wedge(Q) = \inf_{E_1}(Q)$ such that $Q \geq_1 \wedge(Q)$ (that is, $x \geq_1 \wedge(Q)$ for all $x \in Q$) and $Q \geq_1 z \Rightarrow \wedge(Q) \geq_1 z$. (Such element is unique because $\geq_1$ is an order.) Moreover, the equality holds: $\inf_{E_1}(Q) = -\sup_{E_1}(-Q)$. The vector lattices $\mathbf{R}^n$, $\mathbf{L}^1$, $\mathbf{L}^2$, $\mathbf{L}^\infty$ with natural vector orders, and $\mathbf{C}^*$ are order complete. The vector lattice $\mathbf{C} = \mathbf{C}(\Lambda)$ (with the natural order) is not order complete.

Definition 12.6 *An order complete locally convex lattice* E_1 *satisfies* condition (A) *if for all* $x_0 \in E_{1+}$ *the set* $\{x \in E_1 \,:\, |x| \leq_1 x_0\}$ *is weak compact. (Here,* $|x| := \sup_{E_1}(x, -x)$.*) For an order complete Banach lattice condition (A) is equivalent to the following one:* $\|x_n\| \to 0$ *whenever* $x_1 \geq_1 x_2 \geq_1 \cdots \geq_1 x_n \geq_1 \cdots$ *and* $\inf_{E_1}\{x_1, x_2, \dots, x_n, \dots\} = 0$.

9 For more details see Levin (1985).

The locally convex lattice $\mathbf{C}^*$ with the weak* topology, the Banach lattice $\mathbf{L}^1$, and any reflexive Banach lattice satisfy condition (A). The Banach lattice $\mathbf{L}^\infty$ does not satisfy this condition: the unite ball in $\mathbf{L}^\infty$, $\{x : |x| \leq_1 1$, is weak* compact but not weak compact. (Note also that if $x_n = \chi_{[0,1/n]}$ is the indicator (see Eq. (14.3)) of $[0,1/n]$ then obviously $x_1 \geq_1 x_2 \geq_1 \cdots \geq_1 x_n \geq_1 \cdots$ and $\inf_{\mathbf{L}^\infty}\{x_1, x_2, \ldots, x_n, \ldots\} = 0$ but $\|x_n\| = 1$.)

Theorem 12.1 *(Levin, 1985, p. 101) Suppose B is an open convex set in a (Hausdorff) locally convex space E, E_1 is an order complete locally convex vector lattice satisfying condition (A), $\Psi : B \to E_1$ is a convex map, $t : E_1 \to \mathbf{R}^1 \cup \{+\infty\}$ is a convex function with the effective domain $dom(t) := \{x \in E_1 : t(x) < +\infty\}$. Suppose that int $dom(t) \supset \Psi(B)$ and t is non-decreasing on int $dom(t)$ (in the sense that for any $x, y \in$ int $dom(t)$, $y - x \in E_{1+} \Rightarrow t(x) \leq t(y)$), and $\mathbf{d}(\cdot) = t(\Psi(\cdot))$. Then the function $\mathbf{d}$ is convex, and if furthermore Ψ is continuous at a point $a \in B$ and t is continuous at $\Psi(a)$, then the subdifferential of $\mathbf{d}$ at a is non-empty and equals*

$$\partial\mathbf{d}(a) = \{\varphi \circ T : \varphi \in \partial t(\Psi(a)), T \in \partial\Psi(a)\}. \tag{12.47}$$

When Ψ is differentiable at a we have

$$\partial\mathbf{d}(a) = \Psi'^* \partial t(\Psi(a)), \tag{12.48}$$

where Ψ'^* is conjugate to the Fréchet derivative of Ψ at a. Furthermore, if t is differentiable we have

$$\partial\mathbf{d}(a) = \Psi'^* \nabla t(\Psi(a)). \tag{12.49}$$

Equations (12.47)–(12.49) along with (12.36) and (12.37) enable the evaluation of the most sensitive colour mechanisms (and colour detectors) to be made for quite a broad class of detection models.

12.3 Colour discrimination models

A colour detection model (specifically, its detection function) aims at accounting for a particular indiscriminable set defined for a particular background colour. If the latter changes, the shape of the indiscriminable set will, generally speaking, change too. A colour discrimination model (specifically, its discrimination function) is supposed to take into account the way the indiscriminable set varies with the background colour. Although systematic research on this subject has not yet been carried out, some data are still available. Historically, these have been first obtained in studying colour discrimination of monochromatic lights (loosely dubbed in the literature as 'wavelength discrimination').

It was in order to take into account the wavelength discrimination curves available by that time that Helmholtz proposed the first colour discrimination model. Specifically, he suggested the following discrimination function:

$$\mathbf{d}\left(a^0, a^0 + a\right) = \left(\sum_{i=1}^{n}\left(\frac{\varphi_i(a)}{\varphi_i(a^0)}\right)^2\right)^{\frac{1}{2}}, \tag{12.50}$$

where a^0 and a are the background and target light stimuli respectively, and $\varphi_1, \ldots, \varphi_n$ are the cone fundamentals.

When the background a^0 is fixed, the discrimination function (12.50) becomes a detection function as in (12.5) with $p = 2$, and $\xi_i = k_i\varphi_i(a)$, where $k_i^{-1} = \varphi_i\left(a^0\right)$. Hence Helmholtzian model

(12.50) predicts the indiscriminable sets being ellipsoids the axes of which are proportional to the coordinates of the background colour signal. In other words, the effect of the background on the shape of the indiscriminable set is taken into account by normalising the cone fundamental mechanism response to the target a with that to the background a^0. The physiological substratum of this normalisation can be the process of light adaptation in the photoreceptors. Indeed, the photoreceptors change their gain in inverse proportion to their response to the stimulus light (Section 13.1). Hence, $\frac{\varphi_i(a)}{\varphi_i(a^0)}$ in Eq. (12.50) can be considered as an effective response of the ith cone fundamental mechanism to the target a against the background a^0. Alternatively, one can use the term 'cone contrast' for $\frac{\varphi_i(a)}{\varphi_i(a^0)}$. In this, widely spread, terminology, Helmholtzian model predicts that all the indiscriminable sets had the same shape in the cone contrast space – the Euclidean ball.

As the cone spectral sensitivities had not been know by the time, Helmholtz assumed that $\varphi_1, \ldots, \varphi_n$ are some linear transformation of the colour matching functions, and tried to find the coefficients of this transformation which bring about the best fit to the wavelength discrimination data. Unfortunately, the resulting 'cone fundamentals' had a rather implausible form. Not surprisingly, a few attempts have been made since to improve the discrimination function (12.50).

In particular, Stiles modified Helmholtzian discrimination function as follows

$$\mathbf{d}\left(a^0, a^0 + a\right) = \left(\sum_{i=1}^{n} \left(k_i f\left(\varphi_i(a^0)\right) \varphi_i(a)\right)^2\right)^{\frac{1}{2}}, \tag{12.51}$$

where $f(\xi) = \frac{9}{1+9\xi}$, and the constants k_i are adjusted so as to get the best fit to the experimental data on colour discrimination available at the time.

Schrödinger put forward more radical modification of Helmholtzian discrimination function:

$$\mathbf{d}\left(a^0, a^0 + a\right) = \left(\frac{1}{\sum_{i=1}^{n} \varphi_i(a^0)} \sum_{i=1}^{n} \frac{k_i}{\varphi_i(a^0)} \left(\varphi_i(a)\right)^2\right)^{\frac{1}{2}}, \tag{12.52}$$

where k_i are empirical constants.

As many other modifications of Helmholtzian model of this sort (Wyszecki & Stiles, 1982, pp. 654–689), Eqs. (12.51) and (12.52) predict the equi-discrimination surfaces of the elliptic shape in the colour-signal space, the principal axes of these equi-discrimination ellipsoids being parallel to the coordinate axes determined by the cone fundamentals. This prediction is in the obvious discord with the more recent data on colour discrimination. Actually, it could hardly be otherwise since all these models amounts locally (i.e., for a fixed background a^0) to a p-norm colour detection model (12.5) with $p = 2$. However, as mentioned above, this model is compromised by the facts that, firstly, the equi-discrimination surfaces have more complex shape than the ellipsoid (Cole et al., 1993, 1994). Secondly, the indiscriminable sets are found to be unballanced (i.e., asymmetric) (Logvinenko, Tyurin, & Sawey, 2002), which is incompatible with not only the p-normcolour detection model, but any detection model based on the norm and/or metric. Hence, a model that would account for all the data colour discrimination currently available, is yet to be established.

13

In Search of Colour Mechanisms in the Eye and the Brain

As has been repeatedly noted earlier, the light processing in the human visual system begins from light absorption in the cone photopigments. The resultant photopigment quantum catch determines the excitation of the cone photoreceptors. The absorption of light in a photoreceptor is generally agreed to be a linear process (in the broad range of light intensities). So the cone excitation induced by light σ can be modelled as a linear functional $\langle \varphi, \sigma \rangle$, where φ is the photopigment absorptance spectrum (Section 3.3). The cone excitation induces a Grassmann structure (provided the metamerism is defined as the equality in excitation for all the cone photoreceptors). Hence, in the trichromatic case, the triplet of the cone excitations produced by a light specifies it as a colour stimulus (i.e., it indicates to which metamer set this light belongs). Or, to put it another way, the cone excitation triplet implements the colour signal (see Remark 2.5). It seems natural to expect the visual system to register this colour signal and carefully pass on it to the brain. However, strange as it may seem, this does not happen. The cone response (i.e., the electrical signal which the cone photoreceptor sends to the next neural layers in the retina (Fig. 13.3)) does not appear to encode the cone excitation (i.e., the quantum catch).

If the colour signals, registered in the form of cone excitation triplets, are not transferred further, does this mean that the colour signals (thus, the colour stimuli) are not encoded in the visual system in any form at all? And what happens next with the cone excitation triplets in the visual system? What sort of processing they undergo in the retina and the brain? What is the meaning of this processing in regard to colour perception? These questions are so important that they should be given detailed consideration, which is what this chapter is about.

13.1 Do the cone photoreceptor responses encode the colour stimulus?

There are three major factors determining the relationship between the cone excitations and cone responses: (i) non-linear transduction in a photoreceptor; (ii) light adaptation in a photoreceptor; and (iii) lateral interactions between photoreceptors.

13.1.1 Local non-linearity of the photoreceptor response

The photoreceptor excitation is only the first stage of the process of converting the input light into the output photoreceptor response. The photoreceptor response is understood here as the magnitude of hyperpolarisation of its outer membrane as measured in volts. The absorption of light in the

Foundations of Colour Science: From Colorimetry to Perception, First Edition. Alexander D. Logvinenko and Vladimir L. Levin.
© 2023 John Wiley & Sons Ltd. Published 2023 by John Wiley & Sons Ltd.

photopigment is followed by the process of transduction of the photoreceptor excitation into the membrane hyperpolarisation. This second stage is non-linear and require a non-linear function for its description. As the output of a photoreceptor (as well as any real neuron) cannot indefinitely increase, one might expect some compressive non-linearity to follow the first stage.

Specifically, the response, f, of a photoreceptor to light σ can be modelled as a composition of a linear functional with some spectral sensitivity function $\varphi(\lambda)$, and a non-linear function $g : \mathbf{R}_+ \to \mathbf{R}_+$:

$$\mathsf{f}_\varphi(\sigma) = g\left(\langle \varphi, \sigma \rangle\right). \tag{13.1}$$

The form of the subsequent non-linearity in photoreceptors was revealed out in the experiments where the intensity of input light was varied, whereas its spectral composition remained same. It has been found that for a fixed light, σ, the photoreceptor response to $t\sigma$ ($t > 0$) as a function of t, when normalised to the maximum response value (usually referred to as the *response-intensity relationship*), is fitted well by the so-called *Naka–Rushton equation*:

$$g(t) = \frac{t^q}{t^q + a^q}, \tag{13.2}$$

where a and q are positive constants. The best fit to the experimental response–intensity curve is reported to have been achieved with $q = 0.7 \div 1.0$ (Fig. 13.1).

Applying the local non-linearity at the second stage (g in Eq. (13.2)) itself retains the cone photoreceptor as a colour mechanism with the same linear part (Definition 2.9 and Proposition 2.4) – the photoreceptor excitation $\langle \varphi, \sigma \rangle$. Thus, in principle, the cone responses f_φ induce the same Grassmann structure as that induced by φ at the first stage (i.e., by the cone excitations).

It should be pointed out that function g results in quite a strong compression of the effective photoreceptor response range, which can be seen better when the graphs in Fig. 13.1 are replotted in the semi-logarithmic coordinates (Fig. 13.2). Unavoidable contamination of the photoreceptor responses with noise makes the very small responses in the left nearly horizontal branch of the graph (as well as the very large ones in the right nearly horizontal branch of the graph), practically indistinguishable from each other, thus reducing the effective response range to a narrower band (approximately 4 log units) as compared to that within which the eye can in principle

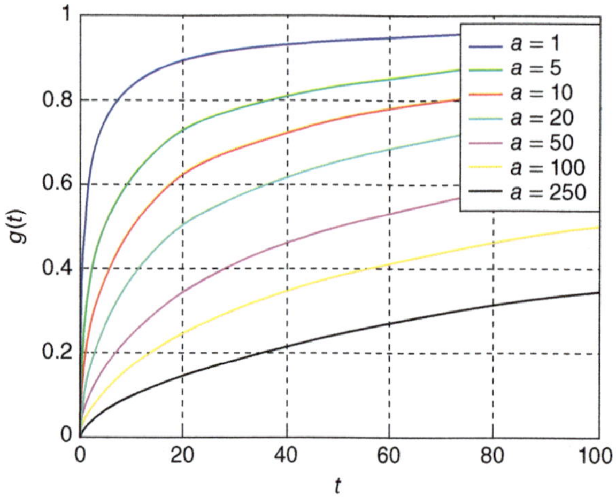

Figure 13.1 Naka–Rushton curves (Eq. (13.2)) for different values of a ($q = 0.7$).

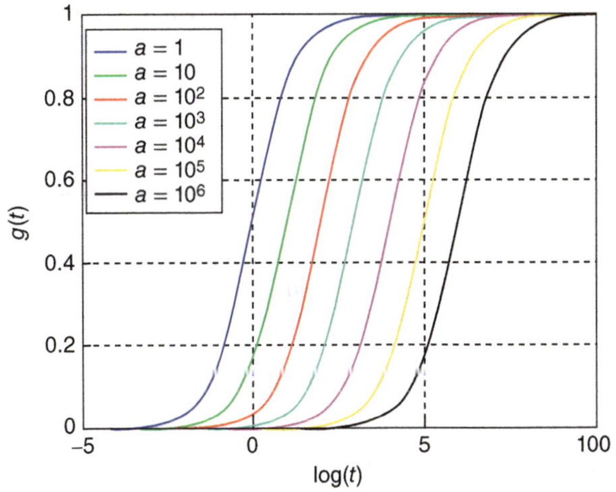

Figure 13.2 Naka–Rushton curves, as in Fig. 13.1, with a logarithmic scale on the *x*-axis, and a linear scale on the *y*-axis.

operate (12 ÷ 14 log units). Still, the existence of adaptive gain control in photoreceptors, discussed in Section 13.1.2, makes it possible to expand their effective response range.

13.1.2 Light adaptation in photoreceptors

Apart from the response compression (Eq. (13.2)), there is another source of non-linearity in photoreceptors, namely, *gain control* (or *light adaptation*) (Shapley & Enroth-Cugell, 1984). Gain control means, in particular, that parameter a in (13.2) increases with stimulus light intensity (Valeton & Van Norren, 1983). More specifically, the adapting photoreceptor response to light σ can be described as:

$$\mathrm{f}_{\varphi}(\sigma) = \frac{k\langle\varphi,\sigma\rangle^{q}}{\langle\varphi,\sigma\rangle^{q}+\alpha^{q}}, \quad \alpha = f(\langle\varphi,\sigma\rangle), \tag{13.3}$$

where α is referred to as *adaptation state*, and f is some monotonic real-valued function, k being a scaling factor.

This explains why moving from a room to a darkened cellar or under direct sun light might result in temporary 'blindness'. It takes some time to restore a clear vision. During this period of time, a new adaptation state is being established. Adapting to a brighter light results in a shift of the photoreceptor response-intensity curve to the right along the light intensity axis (plotted in logarithmic units) establishing a new range of light intensities which can be effectively discriminated (Fig. 13.2). As a result, the light intensity range which can be processed by the human eye considerably exceeds the effective response range of a photoreceptor without gain control.

Adaptive change in gain breaks the one-to-one relationship between the photoreceptor response, $\mathrm{f}_{\varphi}(\sigma)$, and the photoreceptor excitation, $\langle\varphi,\sigma\rangle$. Same $\langle\varphi,\sigma\rangle$ might result in different $\mathrm{f}_{\varphi}(\sigma)$ depending on the adaptation state. Therefore, the photoreceptors with such gain control cannot, strictly speaking, encode the colour signal for the light stimulus (i.e., $\langle\varphi,\sigma\rangle$). True, the colour signal[1] could, in principle, be restored provided that not only the photoreceptor response, i.e., $\mathrm{f}_{\varphi}(\sigma)$ in Eq. (13.3),

1 Unless otherwise indicated, in this chapter, this term will mean the colour signal determined by the cone excitations.

but also the adaptation state parameter, α, is available. However, as we argue below, there is no evidence that this actually happens.

It must be said that, under free viewing (i.e., without eye movement restrictions) such a display as in Fig. I.1, the large constant background secures an approximately equal adaptation state for all the photoreceptors (including those stimulated by the target semicircles), thus, practically ruling out light adaptation as a factor. For light adaptation to come into play, the background light is usually varied. In this case, the colour of a light seems to be determined by two factors: (i) the colour signal of this light and (ii) the colour signal of the background light which basically determines the adaptation state. There have been attempts to incorporate light adaptation into colorimetry (Wyszecki & Stiles, 1982; Fairchild, 2005) taking into account not only the colour signal evoked by the test light but also that evoked by the background. However, they are usually restricted to very simple displays (as in Fig. I.1) where one can point out the light determining the adaptation state. So, they cannot apply to spatially more articulated displays, not to mention real scenes and images where it is not easy, if at all possible, to control the adaptation state of all the photoreceptors involved.

13.1.3 Spatial interaction between the cone photoreceptors

There are two types of intercellular communication in the retina. The major one is mediated by well-known chemical synaptic transmission. Also there exists electrical synaptic transmission between adjacent retinal cells named as *gap junctions* (Bloomfield & Völgyi, 2014). Due to these, a current applied externally to a cone in the turtle retina was demonstrated to induce a response in neighbouring cones (Baylor et al., 1971). Electrical coupling of this kind was shown to take place between photoreceptors in many species including monkey (Tsukamoto et al., 1992).

A benefit of electrical coupling of adjacent cone photoreceptors is believed to be significant reducing the level of noise to which individual cones are subject to (because of random photon absorptions and some other sources of fluctuation in the process of phototransduction) (DeVries et al., 2002). An unavoidable side effect of the cone electrical coupling is some ‘neural blurring’ (see below more about this notion) which is added up to the optical blur of the retinal image.

13.1.4 Why the colour stimulus cannot be derived from the cone photoreceptor responses

As we can see, different responses of a single photoreceptor can be evoked by the same number of quanta (caught by its photopigment) depending on its adaptation state (determined by the prehistory of stimulation) even when the photoreceptor is isolated (thus not affected by other photoreceptors). Hence, instead of a functional relationship between the quantum catch, considered as an input, and the photoreceptor response, considered as an output, we have a more complicated situation when the output is determined by both the input and the adaptation state, the latter being affected by the input as well.[2] Therefore, even in this simplest (laboratory) situation, one cannot uniquely restore the quantum catch from the cone response unless the adaptation state is known.

As to natural perception of real scenes, the relationship between the quantum catch and the photoreceptor response becomes even more complicated. Even if we abstract from the fact that the

2 In fact, there is every indication that the situation is even more complicated. Namely, the adaptation state is determined by not directly the input but by the output. Such input–output relationships mediated by state variables are the subject matter of the system theory. Attempts to study photoreceptors from the system theory point of view were undertaken back in the 1970s (Marmarelis & Marmarelis, 1978).

real scenes consist of objects illuminated by multiple light sources,[3] there are two features of such perception with important implications. Namely, the light coming to the eye varies over space; and the eyes freely move. It follows that the light stimulating a particular photoreceptor not only differs from photoreceptor to photoreceptor but, because of the eye movements, varies in time[4]. This makes the adaptation state of different photoreceptors different and varying in time. Furthermore, the lateral effects from neighbouring photoreceptors and the effect from the horizontal cells, which are subject to light adaptation on their own, makes the photoreceptor adaptation state very difficult, if not impossible to derive from the stimulating light. Hence, it could be possible to derive the colour stimulus from the cone photoreceptor response only if along with this response the signal encoding the photoreceptor adaptation state is transmitted to the brain. However, in this case, the amount of information to be sent through the optic nerve would double. There is no sign of any signals encoding the photoreceptor adaptation state in the optic nerve. What is transmitted through the optic nerve are the neural signals produced by the retinal ganglion cells. In Section 13.2, we discuss a possible role of the ganglion cells in encoding the colour stimulus.[5]

13.2 Do cone-opponent neural cells encode the opponent chromatic codes?

Information from the cone photoreceptors travel further to the brain through three anatomically separate pathways named as magnocellular (M-), parvocellular (P-), and koniocellular (K-) (Gegenfurtner, 2003; Kaplan, 2014). On its way to visual cortex (V1) the photoreceptor output is processed in succession by the following inter-relay neural cells: (i) bipolar and (ii) ganglion cells in the retina; and then (iii) the lateral geniculate nucleus (LGN) cells in thalamus. The magnocellular pathway comprises (i) diffuse bipolar cells, (ii) M-ganglion cells, (iii) M-geniculate cells and ends in (iv) the magnocellular layers of the cortical area V1. The parvocellular pathway comprises (i) midget bipolar cells, (ii) P-ganglion cells, (iii) P-geniculate cells and ends in (iv) the parvocellular layers of the cortical area V1. The diffuse and midget bipolar cells receive input from M- and L-cones. The koniocellular pathway is made by (i) the diffuse bipolar cells and so-called 'blue-cone' bipolar cells receiving input from S-cones, (ii) small bistratified ganglion cells, (iii) LGN cells (different from the M- and P-geniculate cells) and ends in (iv) some layers of area V1 different from the M- and P-layers.

Unlike photoreceptors, the neural cells in the visual pathways process colour signals not locally but over some area. Specifically, a cell is driven by a number of adjacent photoreceptors. The area in the visual field corresponding to the retinal area covered by these photoreceptors is called the *receptive field* of the cell. The ganglion and subsequent cells respond with spikes of electrical activity maintaining some level of such activity even in the absence of stimulation. Due to such maintained level of activity, they respond to stimulation not only with increasing the frequency of firing (i.e., excitation) but also with its decreasing (i.e., inhibition). The firing rate is usually taken as a measure of the cells' response. The excitatory response is taken with positive sign, and the inhibitory one with negative sign.

---

3 This will be taken into consideration in Part II.

4 Actually, one has to consider as an input the quantum catch as a function of time and space (i.e., the retinal location). Respectively, the output should be also thought of as the cone response considered as a function of time and space. As far as the spatial variable is concerned, this will be accomplished in Section 13.3.2.

5 Notably, the photoreceptors do not seem to form any signal to transmit further, since they do not have axons and they do not send any spikes to any neurons. The array of photoreceptors appears to lend itself as a kind of syncytium to be sampled by the bipolar and then ganglion cells, which do send trains of spikes via the optic nerve.

The signals from M- and L-cones make the same effect on a M-ganglion cell. Either in the receptive field centre they evoke excitation, and in the receptive field periphery inhibition, or in the receptive field centre inhibition, and in the receptive field periphery excitation. More specifically, the M-ganglion cells have concentric receptive fields: either an excitatory central circular region surrounded by an inhibitory annulus-like area, or an inhibitory central circular region surrounded by an excitatory annulus-like area. Such centre/surround antagonism within the receptive field of M-cells is often called *spatial opponency*.

The P-ganglion cells exhibit, apart from spatial also cone opponency. They also have concentric receptive fields which receive input from M- and L-cones. However, the effect of M-cones is opposite to that of the L-cones. If the M-cones result in excitatory cell's response, the L-cones evoke inhibitory response, and vice versa. Because of this, these cells are called *cone-opponent* (Fig. 13.3).

The K- (small bistratified) ganglion cells get the inputs from the 'blue-cone' bipolar and diffuse bipolar cells with opposite signs, thus, also exhibiting, generally, cone opponency.

The nomenclature of the LGN cells and their receptive fields are basically the same as those of the ganglion cells.

There is a widespread belief that the M-, P-, and K- (ganglion and/or LGN) cells implement colour mechanisms that can be thought as a linear transformation of the cone fundamentals considered as the colour mechanisms implemented by the cone photoreceptors. Specifically, the M-cells are usually modelled as the sum of the M-, and L-cone fundamentals: $(M + L)$. They are believed to be involved in luminance processing. The P-cells are modelled as the difference of the M-, and L-cone fundamentals: $(M - L)$ or $(L - M)$; the K-cells being modelled as $S - (M + L)$.[6] While the corresponding directions in the fundamental colour-signal space differ from the unique hue loci, the P- and K-cells are believed to be a neural implementation of the opponent chromatic codes supporting Hering's idea. However, if the absolute colour signal gets lost in the photoreceptors, one cannot expect it to reappear in the subsequent retinal layers, and/or in the other visual areas of the brain. As the absolute colour signal is necessary to compute the opponent chromatic codes, it is, generally, unrealistic to expect the existence of neurons implementing the opponent chromatic codes. Apart from this general argument, there are some other important reasons why

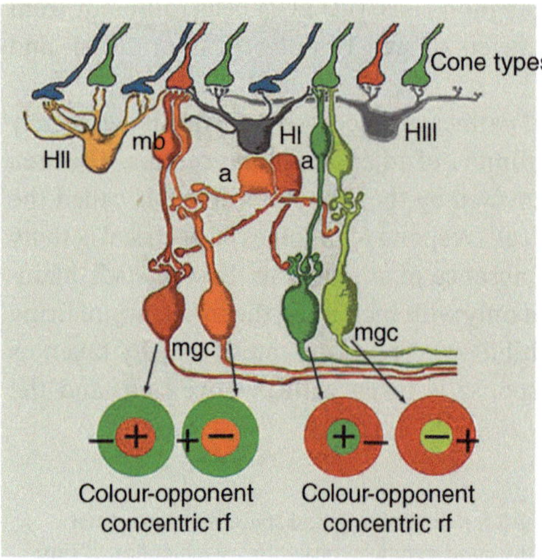

Figure 13.3 Summary diagram of the organisation of midget pathways of the primate retina. Source: Webvision by Kolb, Fernandez, and Nelson.

6 Actually, the weights with which the cones of different types contribute to the ganglion cells of a particular type vary from cell to cell. So, these notations should be taken as symbolic designations indicating the cell type, rather than exact formulae.

the cone-opponent ganglion (and/or LGN) cells cannot play the role of neural substratum for a colour code (in particular, a colour mechanism) not to mention an opponent chromatic code.

First of all, as a matter of fact, the cone-opponent cells process not only in the chromatic but also in spatial domain. That is, they are spatio-chromatic processors. These cells are thought to integrate the output signals from the cone photoreceptors across their receptive fields with weights depending on the cones' location and type. As it is only linear cells that can be considered for the role of colour mechanism,[7] in what follows we consider only cells for which a linear model can be adopted. Such cells can be fully characterised by their spatio-chromatic sensitivity function (Shapley & Hawken, 2011):

$$r(\lambda, x, y) = a_S S(\lambda)\, r_S(x,y) + a_M M(\lambda)\, r_M(x,y) + a_L L(\lambda)\, r_L(x,y), \tag{13.4}$$

where $S(\lambda)$, $M(\lambda)$, and $L(\lambda)$ are the spectral sensitivity of the S-, M-, and L-cones, respectively; $r_S(x,y)$, $r_M(x,y)$, and $r_L(x, y)$ are the weights for location (x, y) in the receptive field for each cone type; a_S, a_M, and a_L are some parameters. The response of a (linear) cell with such sensitivity function to a stimulus $p(\lambda, x, y)$ can be evaluated by integrating $p(\lambda, x, y)\, r(\lambda, x, y)$ over the receptive field of the cell.

Note that it is only if the spatial integration within the receptive field is the same for each cone type, i.e.,

$$r_S(x, y) = r_M(x, y) = r_L(x, y), \tag{13.5}$$

one can consider a cell (13.4) as a chromatic processor. Indeed, when Eq. (13.5) holds true $r(\lambda, x, y)$ breaks down into separate chromatic and spatial sensitivity functions. In this case, the former specifies the chromatic properties of the cell as such, irrespective to spatial location. Correspondingly, the latter specifies the spatial characteristics of the cell irrespective of wavelength. Thus, the equality of the weight functions in Eq. (13.4), securing the spatio-chromatic separability of $r(\lambda, x, y)$, is a necessary condition for a cell to be a neural substratum of a colour code (e.g., a colour mechanism). However, such spatio-chromatic separability is very rare for ganglion and LGN cone-opponent cells. The input signals from the cone photoreceptors are mostly processed by these cells across their receptive fields in a different way depending on the cone type.

Consider, for example, one particular type of the cone-opponent cells in the retinal ganglion layer (and in the LGN) which are often pointed out as a possible neural substratum of an opponent chromatic code. Their 'cone-opponency' is usually discerned in that their response is proportional to the difference of the L- and M-cone inputs. Since such the cells subtract the M-component (of the colour signal produced by the light stimulating the receptive field of the cell) from the L-component, they are referred to as *$L - M$ cells* (Valberg, 2005). For these cells, $a_S = 0$, $a_L > 0$, and $a_M < 0$ in Eq. (13.4).

As a matter of fact, almost all of these cells do not obey Eq. (13.5) necessary for them to implement a colour code. Although both $r_M(x, y)$ and $r_L(x, y)$ can be approximated by circular Gaussian functions of the same type:

$$r_M(x, y) = \exp\left(-\left(x^2 + y^2\right)/2\pi\sigma_M^2\right) \tag{13.6}$$

and

$$r_L(x, y) = \exp\left(-\left(x^2 + y^2\right)/2\pi\sigma_L^2\right), \tag{13.7}$$

7 Though, as follows from Proposition 2.4, a local non-linearity following the linear part can also be admitted.

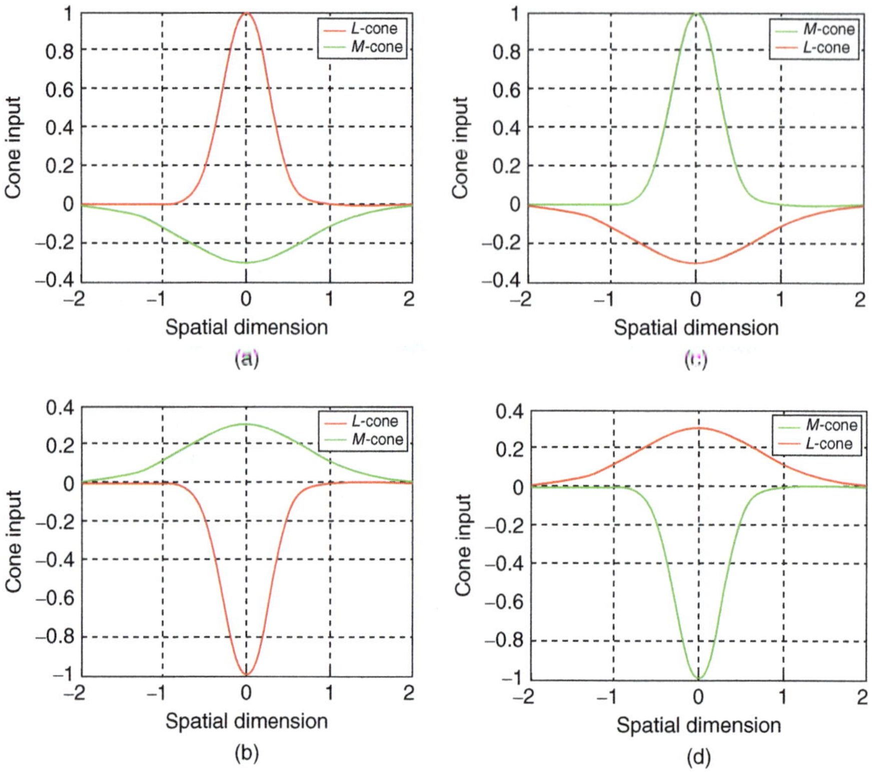

Figure 13.4 Spatial weighting functions, $r_M(x, y)$ (green line) and $r_L(x, y)$ (red line) (modelled as in Eqs. (13.6) and (13.7)) for different cell types. Distance from the centre of the Gaussian, $\sqrt{x^2 + y^2}$, is plotted on the horizontal axis. (See further explanation in text.)

parameters σ_M and σ_L are essentially different[8] (Fig. 13.4a). Hence, there is no spatio-chromatic separability for their spatio-chromatic sensitivity function, $r(\lambda, x, y)$. As rightly noted by Marr (1982), for this reason alone these cells cannot serve as the substratum of a colour code in general, and an opponent chromatic code in particular.

Another property of the $L - M$ cell in Fig. 13.4a preventing it to be taken as a substratum for an opponent chromatic code is that it produces only an excitatory response to the L-cone input, and only an inhibitory response to the M-cone input.[9] However, any cell implementing an opponent chromatic code should be able to produce a response of each of the two (excitatory and inhibitory) polarities to each of the two (L- and M-cone) inputs. More specifically, given that (S_A, M_A, L_A) is the cone response triplet to the adapting stimulus, a $L - M$ cell is expected to respond with excitation to a L-incremental light (i.e., whose cone coordinates are $(S_A, M_A, L_A + \Delta L)$), and inhibition to a L-decremental light (i.e., whose cone coordinates are $(S_A, M_A, L_A - \Delta L)$) in order to be a candidate to a neural substratum of an opponent chromatic code.

8 True, cells with $\sigma_M = \sigma_L$ were found in LGN (Type II according to Wiesel & Hubel, 1966) but very seldom (Wiesel & Hubel, 1966).

9 It should be borne in mind that the inhibitory range of the cone-opponent cells is usually much narrower than the excitatory range, thus being far too small to represent the opposite colour. Thus, 'neutralisation' of one colour by its opponent cannot correspond to the balance between excitation and inhibition of the same opponent cell type.

Interestingly, there is another type of the $L - M$ cells which indeed get inhibited when a L-decremental light is placed onto the receptive field (Fig. 13.4b). They have a higher maintained firing rate, which allows them to have a larger inhibitory response range. Valberg (2005) referred to them as the $L - M$ cells of *D-type*, referring to the former (Fig. 13.4a) as *I-type*. The periphery of the I-type of $L - M$ cells respond with inhibition to a M-incremental light (i.e., with cone coordinates $(S_A, M_A + \Delta M, L_A)$), the periphery of the D-type of $L - M$ cells responding with excitation to a M-decremental light (i.e., with cone coordinates $(S_A, M_A - \Delta M, L_A)$). As one can see, the I-type of $L - M$ cells could, in principle, implement the positive part of a colour-opponent code, and the D-type the negative part of the same code provided that condition (13.5) holds. So, combining the I- and D-type of $L - M$ cells one could possibly get an implementation of an opponent chromatic code (provided that condition (13.5) is met). However, neither of these cells alone can be suitable for this purpose.

In the primate retina, there are also another two (D- and I-) types of cone-opponent cells differencing the M- and L-cone inputs (Fig. 13.4c,d). They subtract the L-component from the M-component (referred to as *$M - L$ cells* (Valberg, 2005)). Also there are cone-opponent cells of D- and I-type differencing on the one hand the S-, and on the other, some combination of the M- and L-cone inputs (e.g., $M + L$) (Valberg, 2005). Although many consider these cells as possible neural substratum of some hypothetical colour-opponent mechanisms, they cannot be such for the same reason as outlined above for the $L - M$ cells.

In brief, for a number of reasons (an extended list of which and detailed discussion can be found in Valberg (2005)), first of all, because of their spatio-chromatic inseparability, the cone-opponent cells in the retina and LGN cannot be considered as a neural implementation of colour mechanisms. What is then the role of these cone-opponent cells? First, an obvious advantage of processing the colour signals which takes place in the cone-opponent cells is decorrelation of the highly correlated components of the colour signal. By subtracting the response of one class of cone (e.g., M) from that from another (e.g., L), the information that is common to both (and which is therefore redundant) is effectively removed (Buchsbaum & Gottschalk, 1983). This is a well-known strategy (difference encoding) in storing and/or transmitting data.

Second, many believe that, performing differencing the cone outputs from different (though adjacent) locations in the retina, the cells of this kind estimate local contrast. For instance, the $M + L$ ganglion cells are believed to evaluate the local luminance contrast. As known, the brightness of an achromatic patch of light presented against the achromatic homogeneous background can remain the same in spite of a change of its luminance provided that the luminance contrast between the patch and the background is kept constant. A number of computational models based on such cells have been proposed to account for brightness perception of the targets on the various backgrounds of this kind (Gilchrist, 2006).

Likewise, lots of observations have been reported that indicate that the colour of a chromatic patch of light on the chromatic (homogeneous) background is determined by the cone contrast between the patch and the background rather than the colour signal induced by the patch. An experiment of this sort is described below (Section 13.3.1). It should be said that this experiment was performed in a different experimental paradigm as compared to the colorimetric one. The main difference is the variation of the background chromaticity. Although this is certainly a step forward towards natural vision, it is still far too impoverished stimulus conditions as compared to natural scenes where the light varies across the whole scene.

It should be also said that the notion of contrast implies two spatially homogeneous areas the contrast at the border between which is being estimated. It is not quite clear how to apply this notion to more complicated, spatially non-homogeneous stimuli, let alone natural scenes. One possibility is to consider local luminance contrast encoding as a step towards approximate estimation of spatial derivative of the retinal luminance distribution (Young, 1987). In this case, the $M + L$ ganglion cells can be thought of as the devices evaluating the spatial derivatives and sending them further to the brain (Young, 1987).

Such an approach seems attractive; however, it implies a transition to a different conceptual paradigm which will be outlined next (Section 13.3). Furthermore, as rightly argued by Marr (1982), the cone-opponent cells, differencing the M- and L- (or S- and $M + L$) cone outputs, could not be considered as mechanisms of approximate differentiating the spatial colour-signal distribution. Still, as we argue in Section 13.4, in the visual cortex there are neurons which are more suitable for this kind of interpretation.

13.3 Transition to a different paradigm

Until now, the presentation in this book has actually revolved around a certain experimental paradigm: the colorimetric one. The light patches were assumed to be spatially homogeneous and not changing in time. This allowed to abstract from the spatial and temporal dimensions when considering light stimuli. Furthermore, the light patches have been assumed to be presented against the same background as in Fig. I.1. For such conditions one can assume that, as mentioned above, being determined mainly by this background, the adaptation state is approximately the same for all the photoreceptors. In this case, there is approximately a one-to-one relationship between the cone excitations and the cone responses. This is why the theory works for these particular conditions. However, it fails for more complicated conditions, letting alone natural vision.

For example, the theory based on colour matching obtained when the light patches to be matched are presented against the same background as in Fig. I.1 fails to predict colour matches when the backgrounds are different. Colour matching in such condition is called *asymmetric*,[10] the case with equal backgrounds being referred to as *symmetric colour matching*.

13.3.1 From symmetric to asymmetric colour matching

Let us consider briefly one exemplary experiment on asymmetric colour matching (Chichilnisky & Wandell, 1995). Observers were dichoptically presented with (i) a light patch a_t against a background b (to one eye) (referred to as a test target) and (ii) another light patch a_m (match target) against another background b' (to the other eye). Due to the dichoptical presentation, each eye was stimulated independently. An adaptation state of each eye was independently determined by the corresponding background. The task was to establish a match between a_t and a_m. Therefore, it was a typical case of asymmetric colour matching (as the targets are presented on different backgrounds). It has been well established long before that for different backgrounds the asymmetric colour match is achieved when the colour signals in response to a_t and a_m are not equal. It is hardly a surprise because due to light adaptation, equal a_t and a_m result in unequal responses of the cone photoreceptors.

The colour appearance of a small target was found to depend on its incremental colour signal, that is, the difference between the target and background colour signals (Walraven, 1976).

10 The notion of asymmetric match will be also used in Part II though in the different context.

More specifically, suppose a test target a_t on a background b is matched by some match target a_m on a background b'. Let $(\varphi_1(a_t), \varphi_2(a_t), \varphi_3(a_t))$ and $(\varphi_1(a_m), \varphi_2(a_m), \varphi_3(a_m))$ be the colour signals of the target a_t and its match a_m. Let also $(\Delta\varphi_1(a_t), \Delta\varphi_2(a_t), \Delta\varphi_3(a_t))$ and $(\Delta\varphi_1(a_m), \Delta\varphi_2(a_m), \Delta\varphi_3(a_m))$ (where $\Delta\varphi_i(a_t) = \varphi_i(a_t) - \varphi_i(b)$ and $\Delta\varphi_i(a_m) = \varphi_i(a_m) - \varphi_i(b')$) be the incremental colour signals of the test and match targets, respectively. Suppose that the gain of the colour mechanisms φ_i in the eye exposed to background b is given by $g_i(b)$, and that of the colour mechanisms φ_i in the eye exposed to background b' is $g_i(b')$. Chichilniski & Wandell's working hypothesis was that the test and match targets appear the same when the scaled incremental receptor signals are equal:

$$g_i(b')\,\Delta\varphi_i(a_m) = g_i(b)\,\Delta\varphi_i(a_t) \quad (i = 1, 2, 3).$$

This hypothesis predicts that a plot of the incremental colour signals of the match target vs. the incremental colour signals of the test target forms a straight line through the origin:

$$\Delta\varphi_i(a_m) = \frac{g_i(b)}{g_i(b')}\Delta\varphi_i(a_t) \quad (i = 1, 2, 3). \tag{13.8}$$

The results of Chichilniski & Wandell's experiment were in an excellent agreement with this prediction. They varied the test target a_t and the match background b' and found a linear relationship between the incremental colour signals of the match and test targets. These results are in line with the results of similar experiments conducted by other researchers (e.g., Shepherd, 1997).

Note that the model of asymmetric colour matching (13.8) implies the principle of persistence of the colour match according to which a symmetric colour match is unaffected by light adaptation (Judd & Wyszecki, 1975, pp. 361–362). Indeed, it follows from (13.8) that two lights, making a colour match against one background, will remain to be such against any other background. This is exactly what is to be expected if the light adaptation in photoreceptors does not break the metamerism based on the equality of the cone photopigment excitations.

In the case of asymmetric colour match, Eq. (13.8) predict that a change in the background is most likely to result in a mismatch of the lights hitherto matching each other. So, lights will change their colour depending on the background, which is supported by the introspective reports of the participants in these experiments. Therefore, the results of the experiments on asymmetric colour matching show that the colour of a light is not based on the colour signals determined by the cone photopigment excitations. The most likely reason for this is that, as claimed above, the cone responses do not encode the colour signals determined by the cone photopigment excitations.

Many attempts have been made to extend the theory of symmetric colour matching to asymmetric colour matching taking into account the role of the differential colour signal.[11] However, they have hardly been successful by far (Suppes et al., 1989). One of the problems is the colour of the background.[12] It presents a problem for any account based on the differential colour signals

11 To mark its importance Paul Whittle coined the terms 'contrast-brightness' (Whittle, 1994) and 'contrast-colour' (Whittle, 2003).

12 It should be noted, however, that in Chichilniski & Wandell's experiment (as in all experiments with haploscopic presentation), the binocular fusion of the two backgrounds took place. It follows that, in a single trial, the apparent colours of the test and match backgrounds were equal. Therefore, perceptually, matching was not quite asymmetrical in their experiment. However, when the match background varied, the apparent colour of the binoculary fused background might have varied too. As known, the apparent colour of the background essentially affects the colour of the target (e.g., Gilchrist, 1977; Whittle, 2003; Logvinenko, 2005a, 2009c). The common apparent background colour allowed to split the two effects of the background colour. On the one hand, the background was used to control the adaptation state; on the other, the effect of an apparent difference between the background colours was eliminated.

because, while the differential colour signal determines the colour of the test, it is not clear what specifies the colour of the background. Quite popular is a belief that after a proper period of adaptation (20–30 minutes), a background begins to be perceived achromatic no matter what its spectral power distribution is. However, this assumption does not help when there are more than two backgrounds, letting alone that many find this assumption rather controversial.

We believe that the lack of a successful theory of asymmetric colour matching is not least due to the fact that the experimental paradigm implies spatial heterogeneity of the stimulation, while the conceptual framework does not explicitly include spatial domain. One has to realise that, admitting the spatial heterogeneity of the light stimulus, we had better this done, if possible, in a fairly general form. In other words, it is worth to generalise the notion of light (and colour) stimulus including in it the spatial (and generally speaking, temporal) dimension.[13]

13.3.2 From light stimulus to light stimulus array

Generally, it is a light with the radiant spectral density function varying across space and time that should be considered as visual stimulus.[14] For the sake of brevity, we will restrict further consideration to only the spatial domain, assuming that light stimuli do not vary in time. (In principle, the temporal variable can be taken into account in the same way as the spatial one.) More specifically, we will extend now our definition of a stimulus for vision as follows. Consider a bundle of light rays arriving at an observation point (potentially occupied by the observer's eye) from all directions in the 3D space (cf. the notion of an *ambient optic array* in Gibson (1979), pp. 50–51). As each direction can be represented by a point in the unit sphere around the point of observation, S^1, this bundle can be formally represented as a map $S^1 \to A$, where A is the light stimulus cone (Definition 2.1). Given a colour stimulus manifold $\mathcal{A}_{\sigma_0}$, and/or a colour-signal cone $\Phi(A)$, this map naturally gives rise also to the following maps: $S^1 \to \mathcal{A}_{\sigma_0}$ and $S^1 \to \Phi(A)$.

For the stationary eye, one can ignore the optical directions from which light cannot enter the eye chamber, and restrict consideration to only the part, V, of S^1 that corresponds to what is usually referred to as the *visual field*. To denote such a restriction of these maps to $V \subset S^1$, the term *light stimulus array* will be used. Specifically, given an open connected subset $V \subset S^1$, we will refer to a map $p : V \to A$ (respectively, $V \to \mathcal{A}_{\sigma_0}$) as a *light-* (respectively, *colour-*) *stimulus array*. Therefore, given a direction $\overline{\alpha} = (\alpha_1, \alpha_2)$, where α_1 and α_2 stand for the azimuth and zenith in the spherical coordinate system with the origin at the point of observation, $p(\alpha_1, \alpha_2)$ is a light coming to the eye from this direction.

Given a colour-signal map Φ, a map composition $\Phi \circ p$ (i.e., $V \to \Phi(A)$; it will be referred to as the *colour-signal array*) can be considered as an immediate visual response (at the level of the cone excitation) to light stimulus array p. For each direction $\overline{\alpha}$, it specifies the colour signal induced

13 Here we come, though from a different direction, to what we already know. Namely, that we are unable to perceive colour as such. As stated in the Outline (Chapter 1), what we see is the spatio-temporal distribution of colour. It is this distribution that makes up the *sensory fabric* of mind. (For more detail on the notion 'sensory fabric', see Section 21.2.) Of course, using the analytical introspection, the sensory fabric can be further decomposed into its constituent elements, just as molecules can be decomposed into atoms. Colour will be one of the elements constituting the sensory fabric; hue, saturation, and brightness being colour's constituent elements. The sensory fabric, the colour, and the hue have different modes of existence in mind, just as the modes of existence of molecules and atoms are different. We believe that a 'molecule' of perception is the sensory fabric.

14 As mentioned in the Outline (Chapter 1), a steady spatially homogeneous light is perceived as colourless. Therefore, being a necessary condition for colour perception, a radiant spectral distribution varying across space and time is a proper stimulus for colour vision.

by the light coming to the point of observation from this direction.[15] The corresponding colour stimulus array (i.e., $V \to \mathcal{A}_{\sigma_0}$) can be uniquely restored from the colour-signal array.

Given the cone fundamentals, $S(\lambda), M(\lambda)$, and $L(\lambda)$, the colour-signal array can be thought of as a triplet of functions:[16]

$$\left[f_S(\overline{\alpha}), f_M(\overline{\alpha}), f_L(\overline{\alpha})\right], \tag{13.9}$$

where $f_S(\overline{\alpha}) = \int_\Lambda S(\lambda)\, p(\overline{\alpha}, \lambda)\, d\lambda$, $f_M(\overline{\alpha}) = \int_\Lambda M(\lambda)\, p(\overline{\alpha}, \lambda)\, d\lambda$, and $f_L(\overline{\alpha}) = \int_\Lambda L(\lambda)\, p(\overline{\alpha}, \lambda)\, d\lambda$ provided that $p(\overline{\alpha}, \lambda)$ denotes the spectral power distribution function of the light entering the eye from the direction $\overline{\alpha} = \left(\alpha_1, \alpha_2\right)$.

Of particular interest are such light-stimulus arrays $p(V)$ that induce smooth colour-signal arrays $\Phi \circ p\ ;\ V \to \Phi(A)$, that is, those for which there exists a Jacobian matrix:

$$J_p(\overline{\alpha}) = \begin{Bmatrix} \frac{\partial f_S}{\partial \alpha_1}(\overline{\alpha}) & \frac{\partial f_S}{\partial \alpha_2}(\overline{\alpha}) \\ \frac{\partial f_M}{\partial \alpha_1}(\overline{\alpha}) & \frac{\partial f_M}{\partial \alpha_2}(\overline{\alpha}) \\ \frac{\partial f_L}{\partial \alpha_1}(\overline{\alpha}) & \frac{\partial f_L}{\partial \alpha_2}(\overline{\alpha}) \end{Bmatrix} \tag{13.10}$$

for each direction $\overline{\alpha} = (\alpha_1, \alpha_2)$. When $J_p(\overline{\alpha})$ has a full rank the range $\Phi \circ p(V)$ is a two-dimensional submanifold in $\Phi(A)$. It will be referred to as the *colour-signal surface*. The dimensionality of submanifold $\Phi \circ p(V)$ will be called the dimensionality of the colour-signal (respectively, colour stimulus) array $\Phi \circ p$. As rows in (13.10) are the gradients of the cone fundamentals, Eq. (13.10) can be rewritten using conventional notation, ∇, for gradient:

$$J_p(\overline{\alpha}) = \begin{Bmatrix} \nabla f_S(\overline{\alpha}) \\ \nabla f_M(\overline{\alpha}) \\ \nabla f_L(\overline{\alpha}) \end{Bmatrix}. \tag{13.11}$$

In the preceding chapters, we implicitly assumed that light stimuli were spatially homogeneous, that is, we assumed the light stimulus arrays the range of which was a single point. The Jacobian for such light stimulus arrays is the zero matrix. The dimensionality of such arrays is, therefore, zero.

An example of a one-dimensional light stimulus array is presented in Fig. 13.5 which depicts a so-called spatio-chromatic sinusoidal grating (a stimulus display widely used in psychophysical and neurophysiological experiments on spatio-chromatic pattern vision). In this display, the light distributions along the vertical spatial dimension (β) is constant. Therefore, this display can be

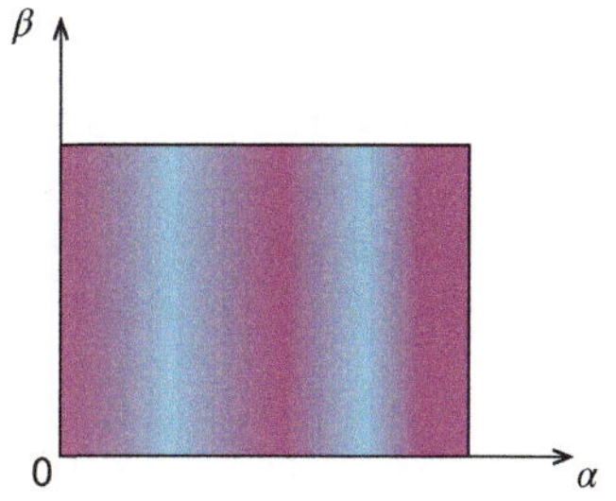

Figure 13.5 Spatio-chromatic sinusoidal pattern.

15 Note that what has been said remains valid even when the point of observation is not occupied by a real observer, and only an abstract colour-signal map is given.
16 For simplicity, we assume that here $A = \mathbf{L}^1_+$ (see Example 2.1b).

fully specified as a map $p : [\alpha_1', \alpha_1''] \to A$, where $[\alpha_1', \alpha_1'']$ is a interval on the spatial (horizontal) dimension along which the light distribution varies. This map defines a curve $p([\alpha_1', \alpha_1''])$ in A (respectively, in $\Phi(A)$). A black/white photograph (reflecting only neutral light) is another example of a one-dimensional light stimulus array despite that as a spatial pattern it is not one-dimensional the way a sine-wave grating is.[17] We will refer to a light stimulus array p such that its range $p(V)$ is included in the neutral line in $\mathcal{A}_{\sigma_0}$ (see Section 8.2) as *achromatic light stimulus array*.

Note that the columns in matrix $J_p(\overline{\alpha})$ in (13.10) represent two vectors tangent to the colour-signal surface at point $\overline{\alpha}$. They determine the tangent surface at $\overline{\alpha}$. The cross product of them is a vector that is perpendicular to the tangent plane, thus to the colour-signal surface at $\overline{\alpha}$. Therefore, the cross product of the columns in (13.10) represents a normal to the colour-signal surface $\Phi \circ p(V)$ at a point $\overline{\alpha}$.[18]

As there is an arbitrariness in choice of the colour-signal space to represent the colour stimulus space, a similar arbitrariness exists between the colour stimulus array (i.e., $V \to \mathcal{A}_{\sigma_0}$) and the colour-signal array (i.e., $V \to \Phi(A)$) representing it. Still, although choosing different linear colour mechanisms (instead of the cone fundamentals) will bring about a different triplet of functions in (13.9), and thus, a different Jacobian matrix in (13.10), the colour-signal surface will, geometrically, remain same since the new linear colour mechanisms must be linearly related to the cone fundamentals.

13.3.3 On the notion of 'neural image'

When the input to the visual system is considered as a spatial array of light, it seems natural to describe the output also as the spatial distribution of some output variable (or variables), that is, as a kind of 'image' (or a pile of 'images'). For example, a pattern of the cone photoreceptor responses can be thought of as the 'cone photoreceptor image'. Since the retina, the LGN and striate cortex (V1) are organised according to the principle of *retinotopy*,[19] one can also talk of the 'ganglion-cell image', the 'LGN image' and the 'cortical image'.

Likening the eye to an optical device (e.g., a camera) and considering the visual process as a cascade of the transformations of the stimulus pattern into a 'neural image' at each cascade was a key idea of the physiological optics, which had gained a second wind in the 1970–1980s (Robson, 1980). The notion of 'photoreceptor image' naturally arises at the very first stage of this cascade. As there are three different types of cone photoreceptors, one can actually consider three different 'cone photoreceptor images'.

While the conceptual framework for image processing proves to be useful, allowing us to understand the role of various neural cells in retina and cortex better than under the local approach based on analysis of individual units, it should not be taken literally. Indeed, first of all, the notion of image implies an observer to watch it.[20] The lack of such undermines the very notion (Gibson, 1979). Second, there are 11 types of bipolar cells, each type having contacts with the cones of all the three types. It looks as if the 3 'cone photoreceptor images' split into 11 'bipolar-cell images'. And if one considers that in total there are more than 60 distinct types of neural cells in the retina (Masland, 2012), then the shortcoming of understanding of vision as the consecutive transformation of

17 In particular, the sinusoidal grating in Fig. 13.5 is separable with respect to the spatial variables whereas for an arbitrary black/white photograph this is not the case.
18 Alternatively, a colour-signal surface normal at $\overline{\alpha}$ can be expressed as the gradient (at $\overline{\alpha}$) of an implicit function, $F(S, M, L) = 0$, determining the colour-signal surface.
19 This means, in particular, that the receptive fields of neighbouring cells in the retina (as well as in the LGN and V1) cover adjacent (and/or overlapping) points in the visual field.
20 That is, a 'little man' or *homunculus* inside the brain 'looking at' the image.

one neural image into another will become obvious. Finally, the farther one moves from the photoreceptors to the brain, the less pronounced the resemblance between the neural images becomes. Eventually, the 'quality of image' gets so bad that it hardly deserves the very term of 'image'. Figuratively speaking, the neural image eventually dissolves somehow over the brain.

Yet, after some refinement, the concept of 'neural image' can still quite successfully serve in the present context. Specifically, the following conditions need be met for the 'neural image' to be more than just a metaphor. First, there must be an ensemble of cells of the same type, the receptive fields of which densely cover the whole visual field. Second, these cells have to be arranged according to the retinotopic principle. Third, the cells should be linear (perhaps, being followed by a local non-linearity). Fourth, the spatio-chromatic separability has to apply for these cell. In this case, the chromatic processing carrying out by the cells can be taken out of the brackets, and the 'neural image' can be treated as the pattern formed by the colour-signal-code values determined by the chromatic processing conducted by these neurons.

More specifically, consider a linear neuron the receptive field of which is centred at a point, $\overline{\alpha}_0$. Assume its spatio-chromatic sensitivity function, $h(\overline{\alpha}_0;\overline{\alpha},\lambda)$, is factored into separate spatial, $h_\alpha(\overline{\alpha}_0;\overline{\alpha})$, and chromatic, $h_\lambda(\lambda)$, sensitivity functions. The linearity of the neuron means that its response, $r(\overline{\alpha}_0)$, to a light stimulus array, $p(\overline{\alpha},\lambda)$, (here p is a real function of three variables) is expressed as:

$$r(\overline{\alpha}_0) = \int_V\int_\Lambda h(\overline{\alpha}_0;\overline{\alpha},\lambda)\, p(\overline{\alpha},\lambda)\, d\overline{\alpha} d\lambda, \tag{13.12}$$

or, as $h(\overline{\alpha}_0;\overline{\alpha},\lambda) = h_\alpha(\overline{\alpha}_0;\overline{\alpha})\, h_\lambda(\lambda)$,

$$\begin{aligned} r(\overline{\alpha}_0) &= \int_V\int_\Lambda h_\alpha(\overline{\alpha}_0;\overline{\alpha})\, h_\lambda(\lambda)\, p(\overline{\alpha},\lambda)\, d\overline{\alpha} d\lambda \\ &= \int_V \left(\int_\Lambda h_\lambda(\lambda)\, p(\overline{\alpha},\lambda)\, d\lambda\right) h_\alpha(\overline{\alpha}_0;\overline{\alpha})\, d\overline{\alpha}. \end{aligned} \tag{13.13}$$

Denoting the internal integral $f(\overline{\alpha})$, i.e.,

$$f(\overline{\alpha}) = \int_\Lambda h_\lambda(\lambda)\, p(\overline{\alpha},\lambda)\, d\lambda, \tag{13.14}$$

one has

$$r(\overline{\alpha}_0) = \int_V f(\overline{\alpha})\, h_\alpha(\overline{\alpha}_0;\overline{\alpha})\, d\overline{\alpha}. \tag{13.15}$$

Assume now that there are many neurons with the same chromatic sensitivity function $h_\lambda(\lambda)$ and, perhaps, different spatial sensitivity functions $h_\alpha(\overline{\alpha}_0;\overline{\alpha})$. Moreover, assume that the neuron population is so dense that one can presume that for every $\overline{\alpha}_0$ there is a neuron centred at it.[21] In this case, $r(\overline{\alpha}_0)$ in (13.15) will be defined for entire visual field,[22] and it can be thought of as the neural image represented by this population. In other words, the neural image is a spatial distribution of the neuron responses determined by (13.15).

As all visual neurons are fed by the cone photoreceptors in the first instance, it seems natural to assume that the chromatic sensitivity function can be represented as a linear combination of the cone fundamentals:

$$h_\lambda(\lambda) = k_S S(\lambda) + k_M M(\lambda) + k_L L(\lambda). \tag{13.16}$$

21 Of course, the number of neural cells is finite. However, when their number is large enough, such an abstraction as 'centred at each point' seems quite feasible.

22 The boundary value problem that arises for points $\overline{\alpha}_0$ sufficiently close to the visual field boundary is left without consideration.

In other words, we assume that, as far as its chromatic property is concerned, each neuron in the population implements a colour-signal code. To stress this, $f(\overline{\alpha})$ in Eq. (13.14) will be referred to as the *colour-signal-code pattern*.

Thus, because of the spatio-chromatic separability of the neuron sensitivity functions, the neural image formation can be considered as a sequence of two independent operations. First, the light stimulus array, $p(\overline{\alpha}, \lambda)$, is being transformed into the colour-signal-code pattern, $f(\overline{\alpha})$, according to Eq. (13.14). Second, the colour-signal-code pattern undergoes a spatial transformation described by Eq. (13.15). In other words, the network of the neurons in question can be considered as a chromatic processor (specified by Eq. (13.14)) followed by a spatial processor (specified by Eq. (13.15)). The former converts the light stimulus array into the colour-signal-code pattern which the latter transforms into the neural image. The properties of the spatial processor are fully determined by the spatial sensitivity functions of each neuron, $h_\alpha(\overline{\alpha}_0; \overline{\alpha})$. When the spatial processor can be described as a one-to-one linear operator, the colour-signal-code pattern can be uniquely restored from the neural image.[23] In this case, we will say that the neural image encodes the colour-signal-code pattern.

We will use different notation for the neural image depending on the context. In order to stress that it is just the neural response (13.15) as a function of the spatial variable, Eq. (13.15) will be simply rewritten without the subscript 0:

$$r(\overline{\alpha}) = \int_V f(\overline{\beta})\, h_\alpha(\overline{\alpha}; \overline{\beta})\, d\overline{\beta}. \tag{13.17}$$

Alternatively, to emphasise that the neural image is the colour-signal-code pattern, $f(\overline{\alpha})$, spatially transformed, the following notation will be used: $r = f * h_\alpha$ (which is nothing more than a short version of Eq. (13.17)), h_α being termed as the kernel of the spatial transformation in question. Finally, when it is clear what spatial sensitivity function (h_α) is meant, and in order to emphasise that the neural image is just an encoded colour-signal-code pattern, we will use an even shorter notation: $r = \tilde{f}$.

Further calculations will be simplified if we expand the visual field V to a plane, which will henceforth be called the *visual plane*. Such an idealisation allows us to identify V with the arithmetic vector space $\mathbf{R}^2$, provided that a basis is chosen in the plane V. In this case, if the shape of the spatial sensitivity function $h_\alpha(\overline{\alpha}_0; \overline{\alpha})$ is the same for each point $\overline{\alpha}_0$, then the integral (13.17) can be replaced by a two-dimensional convolution of $f(\overline{\alpha})$ and $h_\alpha(\overline{\alpha})$, where the latter is $h_\alpha(0; \overline{\alpha})$:

$$r(\overline{\alpha}) = \left(f * h_\alpha\right)(\overline{\alpha}) = \int_{\mathbf{R}^2} f(\overline{\beta})\, h_\alpha(\overline{\alpha} - \overline{\beta})\, d\overline{\beta}. \tag{13.18}$$

Hopefully, the context will make it clear when the asterisk $*$ stands for convolution and when it is just a shorthand version of Eq. (13.17).

When the neuron population comprises the cone photoreceptors of one type, the colour-signal-code patterns will be referred to as the *cone fundamental patterns*. We will denote them as in Eq. (13.9) with the corresponding subscript letter (L, M, or L). For instance, when in Eq. (13.16) $k_S = k_M = 0$, thus $h_\lambda(\lambda) = k_L L(\lambda)$, we get the *L cone fundamental pattern* $f_L(\overline{\alpha})$. The corresponding neural images will be referred to as the *cone photoreceptor images*.

For an arbitrary (linear) neuron with chromatic sensitivity function $h_\lambda(\lambda)$ as in (13.16), the colour-signal-code pattern induced by it can be thought of as a linear combination of the three cone fundamental patterns:

$$f(\overline{\alpha}) = k_S f_S(\overline{\alpha}) + k_M f_M(\overline{\alpha}) + k_L f_L(\overline{\alpha}). \tag{13.19}$$

It should be borne in mind that the colour-signal-code pattern is an abstract notion arising as a result of extending the consideration into the spatial domain. In a sense, it is a colorimetric

23 Yet, one should not expect a sort of 'backward' transformation of the neural image somewhere in the brain. Just as we do not believe that a 'backward inversion' of the inverted retinal image is necessary for upright vision.

characteristic of the light stimulus array.[24] It does not imply a real observer in the point of observation. In contrast, the cone photoreceptor images do imply the observer. Because of this they differ from the cone fundamental patterns. In particular, due to the inevitable optical blur created by the lens and the other optical components of the eye, even the spatial pattern of cone photoreceptor excitation in the retina (referred to as the *cone excitation pattern*) differ from the corresponding cone fundamental pattern. To take into account the effect of optical blurring, one needs to use its point spread function as $h_\alpha(\overline{\alpha};\overline{\beta})$ in Eq. (13.17).[25]

As mentioned above, the gap junctions between the photoreceptors cause a kind of 'neural blurring' which makes the cone photoreceptor image significantly different not only from the corresponding cone fundamental pattern, but also from the corresponding cone excitation pattern. So, not surprisingly, cone photoreceptors were found to possess non-trivial receptive fields.[26] It follows that when evaluating the cone photoreceptor image, one has to use such a spatial sensitivity function $h_\alpha(\overline{\alpha}_0;\overline{\alpha})$ that it reflects the optical blur, the lateral interactions between the cone photoreceptors, and an the inhibitory effect from the horizontal cell.

Let us denote $h_{\alpha S}(\overline{\alpha};\overline{\beta})$, $h_{\alpha M}(\overline{\alpha};\overline{\beta})$, and $h_{\alpha L}(\overline{\alpha};\overline{\beta})$ the spatial sensitivity functions for the *S*-, *M*-, and *L*-cones (describing the receptive fields for the *S*-, *M*-, and *L*-cones respectively). Then, the *S*, *M*, and *L* cone photoreceptor images, $\tilde{f}_S(\overline{\alpha})$, $\tilde{f}_M(\overline{\alpha})$, and $\tilde{f}_L(\overline{\alpha})$, can be expressed as $\tilde{f}_S(\overline{\alpha}) = \int_V f_S(\overline{\beta})\, h_{\alpha S}(\overline{\alpha};\overline{\beta})\, d\overline{\beta}$, $\tilde{f}_M(\overline{\alpha}) = \int_V f_M(\overline{\beta})\, h_{\alpha M}(\overline{\alpha};\overline{\beta})\, d\overline{\beta}$, $\tilde{f}_L(\overline{\alpha}) = \int_V f_L(\overline{\beta})\, h_{\alpha L}(\overline{\alpha};\overline{\beta})\, d\overline{\beta}$, where $f_S(\overline{\beta})$, $f_M(\overline{\beta})$, and $f_L(\overline{\beta})$ are the *S*-, *M*-, and *L* cone fundamental patterns.[27]

So, each cone photoreceptor image is the result of the linear spatial transformation of the corresponding cone fundamental pattern. This transformation is often treated as a sort of spatial filtration, $h_{\alpha S}(\overline{\alpha};\overline{\beta})$, $h_{\alpha M}(\overline{\alpha};\overline{\beta})$, and $h_{\alpha L}(\overline{\alpha};\overline{\beta})$ being assumed to be spatially homogeneous (i.e., being the same for all $\overline{\alpha}$). Moreover, it is believed to be a one-to-one transformation. In this case, in principle, the cone fundamental pattern can be recovered from the corresponding cone photoreceptor image. Naturally, the techniques of the linear system analysis (e.g., the Fourier analysis) has been extensively used to specify the transformation determining the photoreceptor image. Obviously, the non-linearity of the transduction of the cone excitation into the cone response puts certain limitations to the linear analysis in this context. Still, the problem of this sort can be, in principle, resolved by resorting to appropriate techniques of non-linear analysis (e.g., Logvinenko, 1990).

Now the problem posed in the beginning of this chapter can be reformulated as follows. Are the cone fundamental patterns encoded and transmitted further to the brain? In particular, can one restore the cone fundamental patterns from the ganglion neuron images transmitted through the optic nerve to the brain? Generally, is the colour-signal array[28] encoded and transmitted to the brain? Is there any information of the colour stimulus array in the brain?

Surprisingly, even the cone photoreceptor images do not seem to be transmitted to the brain, not to mention the cone fundamental patterns. As pointed out above, it is only the M-ganglion cells that obey the spatio-chromatic separability condition. Therefore, only these cells could implement

---

24 Needless to say, the complete colorimetric characteristic of the light stimulus array is provided by the whole colour-signal array rather than a single colour-signal-code pattern.

25 As known, the chromatic aberrations in the eye cannot be described by the two separable (with respect to α and λ) functions. That is, the relation between the cone fundamental patterns and the cone excitation pattern even more complicated. However, taking this into consideration would lead us far beyond the scope of the book.

26 A photoreceptor response can be obtained by illuminating an area in the visual field that is not directly projected onto that photoreceptor.

27 To be exact, these are just the linear parts of the cone responses, that should be, generally, followed by the local non-linearity (as in Eq. (13.1)) subject to gain control (as in Eq. (13.3)). However, for the sake of simplicity, we restrict consideration below to the linear parts of the cone responses.

28 It is worth recalling that a colour-signal array cannot simply be reduced to three cone fundamental patterns. To recover the colour-signal array three cone fundamental patterns are required, recorded in the same coordinate system (in the visual plane). Therefore, if the cone fundamental patterns are stored separately in different regions of the brain, this is not enough to encode the colour-signal array.

a colour-signal code. It is certainly not enough to encode the colour-signal array since at least two more independent colour-signal-code patterns are needed. Nevertheless, it seems natural to expect that these cells implement the luminance code. The following question arises then: how are related the luminance pattern and the M-ganglion cell image? Putting aside the gain control in the cone photoreceptors, can one restore the luminance pattern from the M-ganglion cell image provided that the light adaptation is same for all the cone photoreceptors? As a matter of fact, the answer is negative.

The luminance pattern is a particular case of the colour-signal-code pattern (13.14) when in Eq. (13.16), $k_S = 0$ and $k_M = k_L$. Specifically, assuming $k_M = k_L = 1$, and denoting the luminance pattern as $f_{M+L}(\overline{\alpha})$, we get

$$f_{M+L}(\overline{\alpha}) = f_M(\overline{\alpha}) + f_L(\overline{\alpha}). \tag{13.20}$$

Therefore, neurons fed by only M- and L-cones in equal proportion will result in the neural image, $\tilde{f}_{M+L}(\overline{\alpha})$, which will, in principle, encode the luminance pattern.[29] As argued above, since $\tilde{f}_{M+L}(\overline{\alpha})$ is a 'blurred' version of $f_{M+L}(\overline{\alpha})$, the latter can be derived from the former by using the inverse transformation. Even if the M-ganglion cells added up more 'blurring', this would not prevent from retaining the luminance pattern from the pattern of their responses.

However, the spatial sensitivity function of the M-ganglion cells is essentially different from that of the cone photoreceptors. As pointed out above, it is such that the spatial processing at the level of the M-ganglion cells can be understood as some smoothing (blurring) of the luminance pattern followed by differentiation (Young, 1987). It follows that there is no one-to-one correspondence between the luminance pattern and the M-ganglion cell image. Therefore, the M-ganglion cell image does not, generally speaking, encode the luminance pattern since the latter cannot be uniquely recovered from the former. Still, the luminance pattern can be recovered from the M-ganglion cell image up to unknown additive constant.

Since, apart from the M-ganglion cells, there are no other neurons in the retina (and in the LGN) that exhibit the spatio-chromatically separability, it can be concluded that no other colour-signal-code pattern (let alone the cone fundamental pattern) are encoded there, even up to an unknown additive constant. Yet, neurons with the property of spatio-chromatically separability were found in striate cortex (V1). Furthermore, these neurons also seem to perform spatial differentiation. The next section (Section 13.4) will examine in detail the possible role of these neurons in encoding the cone fundamental patterns and colour-signal arrays.

13.4 Spatio-chromatic processing in the visual cortex

Two types of neurons in cortical area V1 are of main interest in the present context. Namely, these are *simple cells*, according to the terminology of Hubel & Wiesel (1962) and Callaway (2014); and *colour luminance double-opponent cells*, in terms of Shapley et al. (2014).

13.4.1 Estimating luminance pattern gradient using simple cortical cells

The linearity of simple cells (for a fixed average illumination) is well documented. Their chromatic sensitivity function $h_\lambda(\lambda)$ is described by Eq. (13.16) with $k_M = k_L$ and $k_S = 0$. Quite often the simple cells are referred to as 'luminance neurons'. Being insensitive to the spectral variation of the light

29 Of course, the effect of the difference in gain control is assumed to be insignificant here.

stimulus not affecting its luminance, the simple cells are usually not considered in the context of the colour encoding. Nevertheless, they implement a colour-signal code of a particular type and most likely play an important role in colour perception.

As the retinotopic mapping takes place in V1 for the simple cells, one can talk of the neural image made up by these cells. However, the relationship between this neural image and the luminance pattern is not quite straightforward. Strictly speaking, the simple cells do not encode the luminance pattern. Still, in a sense, the simple cells represent the luminance pattern in the cortex, but they do that not uniquely, in a rather sophisticated way.

Indeed, the spatial sensitivity function of a simple cell was found to vary first and foremost along one specific direction (let us call it the *principal direction* for such a cell), being practically constant along the direction orthogonal to the principal direction.[30] It is convenient to choose these two directions as the coordinate axes when describing the cell. Such a coordinate frame will be referred to as *associated with a simple cell*. In this coordinate system, the spatial sensitivity function of a simple cell, centred at some point $\overline{\alpha}^0$, can be factored into two functions of one variable:

$$h_\alpha(\overline{\alpha}^0,\overline{\alpha}) = h_1(\overline{\alpha}^0,\alpha_1)\, h_2(\overline{\alpha}^0,\alpha_2), \quad \overline{\alpha} = (\alpha_1,\alpha_2).$$

Let us agree that α_1 is the variable corresponding to the cell's principal direction. In this case, $h_2(\overline{\alpha}^0,\alpha_2)$ practically does not vary with α_2 within the cell's receptive field. Hence, to characterise the spatial sensitivity of a simple cell, only the function $h_1(\overline{\alpha}^0,\alpha_1)$ is sufficient. Sometimes, for brevity, it is $h_1(\overline{\alpha}^0,\alpha_1)$ that will be meant under the term of the spatial sensitivity function of the simple cell.

The spatial-sensitivity-function profile of the simple cells is such that they do not respond at all to the spatially homogeneous light stimulus arrays. This itself makes impossible for the simple-cell neural image to be a neural substratum of the luminance pattern.

In the 1970–1980s, the simple cells were suggested to implement decomposing the luminance pattern over a set of basic functions (e.g., Gabor functions). However, this idea has remained nothing more than an elegant hypothesis, having received neither enough experimental support nor proper theoretical elaboration.[31]

It seems a more insightful idea, repeatedly expressed in vision literature, that the simple cells can be considered as performing two consecutive operations on the luminance pattern: smoothing and spatial differentiation (Koenderink & Doorn, 1987; Georgeson et al., 2007). This idea was inspired by the fact that, firstly, the form of the spatial sensitivity function of the simple cell, $h_1(\overline{\alpha}^0,\alpha_1)$, is practically the same for each $\overline{\alpha}^0$, thus it can be expressed as a function of one variable $h(\alpha_1) = h_1(0,\alpha_1)$. Secondly, $h(\alpha_1)$ resembles the derivative of a Gaussian (Fig. 13.6). If $h(\alpha_1)$ is the derivative of some function of α_1, say, $s(\alpha_1)$, that is, $h_1(\alpha_1) = s'(\alpha_1)$, or, more generally,

$$h_\alpha(\overline{\alpha}) = \frac{\partial S(\alpha_1,\alpha_2)}{\partial \alpha_1},$$

30 The notion of orthogonality here implies that the visual plane V is regarded as the Euclidean plane. Since we do not know in what form the Euclidean properties are rendered in the visual cortex and whether they are represented there at all, we will assume V to be just a linear (or affine) plane, without endowing it with any norm or scalar product. In the rare cases where the latter will still be mentioned in relation to V, this will be done for illustrative purposes only.

31 Alternatively, the simple cells are thought of as detectors of some specific features such as lines, edges and the like. However, it should be borne in mind that, being linear, these cells cannot successfully function as feature detectors because they would confound pattern and its contrast. Any feature detector must be non-linear.

where $S(\alpha_1, \alpha_2) = s(\alpha_1)\, h_2(0, \alpha_2)$, then, in the view of Eq. (13.18),

$$r(\overline{\alpha}) = \left(f_{M+L} * h_\alpha\right)(\overline{\alpha}) = \left(f_{M+L} * \frac{\partial S}{\partial \alpha_1}\right)(\overline{\alpha}).$$

Due to the well-known property of convolution,

$$\left(f_{M+L} * \frac{\partial S}{\partial \alpha_1}\right)(\overline{\alpha}) = \frac{\partial}{\partial \alpha_1}\left(f_{M+L} * S\right)(\overline{\alpha}).$$

Hence, we have

$$r(\overline{\alpha}) = \frac{\partial}{\partial \alpha_1}\left(f_{M+L} * S\right)(\overline{\alpha}). \tag{13.21}$$

The operation of smoothing ($*$) can be considered as part of the differentiation technique implemented by the cell in question. As is well known, differentiation is not robust to noise, so in image processing it is usually preceded by smoothing to minimise the effect of noise. Thus, instead of interpreting the cell response $r(\overline{\alpha})$ in Eq. (13.21) as the partial derivative of the smoothed luminance pattern $\left(f_{M+L} * S\right)$, we will interpret it as an approximate estimate of the partial derivative of the luminance pattern itself (f_{M+L}). Therefore, the symbol of convolution in Eq. (13.21) can be safely omitted if we consider $r(\overline{\alpha})$ as the result of numerical differentiation of the luminance pattern:

$$r(\overline{\alpha}) = \frac{\partial f_{M+L}}{\partial \alpha_1}(\overline{\alpha}). \tag{13.22}$$

Note that, with such an interpretation, it does not really matter what kind of smoothing precedes differentiation unless it significantly affects the derivative estimation.[32] A number of authors are inclined towards Gaussian smoothing, believing that the derivative of a Gaussian is a satisfactory approximation of the spatial sensitivity function of simple cells (Fig. 13.6).

Note that the formula (13.22) assumes a coordinate system associated with the simple cell in question, which presents some inconvenience. This inconvenience can be avoided by remembering that a partial derivative is a particular case of a directional derivative. Specifically, let **v** be the basis

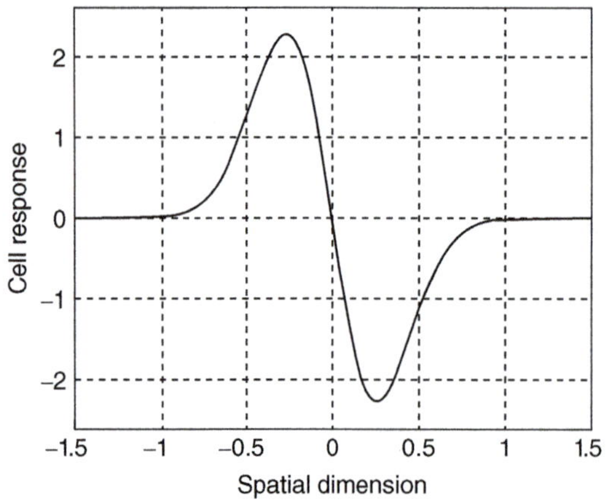

Figure 13.6 Derivative–Gaussian model of the spatial sensitivity function of simple cells.

32 Generally speaking, the kernel of smoothing can be inferred from the cell's spatial sensitivity function. Given α_1 being the variable corresponding to the cell's principal direction, the kernel of smoothing is an antiderivative (i.e., an indefinite integral) of $h_1(\overline{\alpha}^0, \alpha_1)$ times $h_2(\overline{\alpha}^0, \alpha_2)$.

vector determining the principal direction for a simple cell centred at $\overline{\alpha}^0$ corresponding to α_1. Then its response, $r(\overline{\alpha}^0)$, to luminance pattern $f_{M+L}(\overline{\alpha})$, as presented in (13.22), can be rewritten as:

$$r(\overline{\alpha}^0) = \frac{\partial f_{M+L}}{\partial \alpha_1}(\overline{\alpha}) \mid_{\overline{\alpha}=\overline{\alpha}^0} = \partial_{\mathbf{v}} f_{M+L}(\overline{\alpha}) \mid_{\overline{\alpha}=\overline{\alpha}^0}. \tag{13.23}$$

Here, we use a conventional notation $\partial_{\mathbf{v}} f(\alpha) \mid_{\alpha=\alpha^0}$ for the directional derivative of a multivariate function f along the vector $\mathbf{v}$ at a point α^0.

As the integrals (13.17) and (13.18) do not depend on the choice of basis vectors, $r(\overline{\alpha}^0)$ in Eq. (13.23) also does not depend on the choice of the basis. However, for an arbitrary differentiable function, its directional derivative along a vector $\mathbf{v}$ does depend on $\mathbf{v}$. For example, if its domain is assumed to be endowed with a norm, then the directional derivative along a vector $\mathbf{v}$ depends on its length, $\|\mathbf{v}\|$. Sometimes, the definition of directional derivative presupposes a preliminary normalisation of the vector along which it is defined. In other words, this definition reduces the evaluation of the directional derivative for a vector $\mathbf{v}$ to evaluating the directional derivative along the unit vector collinear to $\mathbf{v}$. In this case, the directional derivative depends only on the direction of the vector $\mathbf{v}$, but not on its length. That is, in this case, we get a truly directional derivative.

Thus, we come to the conclusion (which was also expressed by other authors; e.g., see Lindeberg (2013) that the response of a simple cell can be interpreted as an estimate of the truly directional derivative of the luminance pattern along the cell's principal direction. However, this interpretation of the performance of simple cells will remain little but a verbal exercise, fruitless reflection, unless we clarify the following questions. What good is such directional derivatives of the luminance pattern to the brain? How could they be possibly used in estimating the luminance pattern?

The fact is that the directional derivative of a function f at a point $\overline{\alpha}^0$ is closely related to the total derivative of f at $\overline{\alpha}^0$, which is the best linear approximation to f near $\overline{\alpha}^0$. Specifically, a linear functional $f'_{\overline{\alpha}^0} : \mathbf{R}^2 \to \mathbf{R}$ is the total derivative of f at $\overline{\alpha}^0$ if

$$f(\overline{\alpha}^0 + \overline{\Delta\alpha}) \approx f(\overline{\alpha}^0) + f'_{\overline{\alpha}^0}(\overline{\Delta\alpha}) \tag{13.24}$$

for small $\overline{\Delta\alpha}$. If the total derivative of the luminance pattern is known for every $\overline{\alpha}^0$, then the luminance pattern can be reconstructed up to an arbitrary additive constant.

The directional derivative of f along a vector $\mathbf{v}$ at the point $\overline{\alpha}^0$ is equal to the value that the total derivative of f takes on $\mathbf{v}$:

$$\partial_{\mathbf{v}} f(\overline{\alpha}) \mid_{\overline{\alpha}=\overline{\alpha}^0} = f'_{\overline{\alpha}^0}(\mathbf{v}). \tag{13.25}$$

Conversely, the total derivative can be specified in terms of the directional derivatives. Indeed, given a basis $\{\overline{\alpha}_1, \overline{\alpha}_2\}$ in V and an arbitrary vector $\mathbf{v} = v_1\overline{\alpha}_1 + v_2\overline{\alpha}_2$, the total derivative on $\mathbf{v}$,

$$f'_{\overline{\alpha}^0}(\mathbf{v}) = v_1 f'_{\overline{\alpha}^0}(\overline{\alpha}_1) + v_2 f'_{\overline{\alpha}^0}(\overline{\alpha}_2) = v_1 \partial_{\mathbf{v}} f(\overline{\alpha}_1) \mid_{\overline{\alpha}=\overline{\alpha}^0} + v_2 \partial_{\mathbf{v}} f(\overline{\alpha}_2) \mid_{\overline{\alpha}=\overline{\alpha}^0}. \tag{13.26}$$

Thus, two directional derivatives (along different vectors) of the luminance pattern suffice to compute the total derivative of the luminance pattern f. If the total derivative of the luminance pattern is known at every point $\overline{\alpha}$, $f_{M+L}(\overline{\alpha})$ can be recovered up to an unknown additive constant. Hence, a set of pairs of simple cells centred at each point, with their principal directions coinciding with the coordinate axes (in the visual plane), can encode the luminance pattern (up to an unknown additive constant).[33]

33 It must be kept in mind that for every direction two directional derivatives are required: one for each of the two unit vectors defining that direction. Consequently, along with the cells with the spatial sensitivity profile shown in Fig. 13.6, there should be available a set of their counterparts with an inverted polarity as in Fig. 13.7. Indeed, among the simple cells in V1, cells of both polarities are reported to exist.

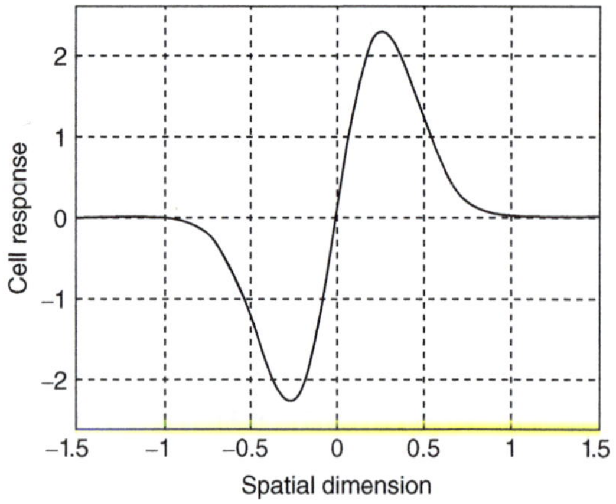

Figure 13.7 Derivative of a Gaussian with negative sign.

Although such encoding requires just two simple cells (with different principal directions) centred at each point $\overline{\alpha}$, the remarkable fact is that for each particular $\overline{\alpha}^0$ there are many simple cells with various principal directions. The receptive fields of all these cells are approximately the same, varying only in their orientation and the location of their centre. Therefore, for each $\overline{\alpha}^0$ the estimation of the directional derivative of the luminance pattern is performed for various directions. Since the principal directions of the simple cells vary widely, it looks as if the directional derivative is computed by the simple cells for almost every direction in the visual field. Since any pair of cells with different principal directions suffice to estimate the total derivative of the luminance pattern (f_{M+L}) at $\overline{\alpha}^0$, why such redundancy?

In search of an answer to this question, the following consideration turns out to be useful. Note that the described method for estimating the total derivative is a coordinate one, that is, it relies heavily on the coordinate representation of vectors in the chosen basis. With all the conventionality of the computational metaphor applied to the brain, which after David Marr has become a common way of reasoning in the neuroscience, it is somewhat difficult to imagine in what form the coordinate representation could be implemented in the brain. Furthermore, for such a method of estimating the total derivative, it is very important that the directional derivatives are calculated with respect to the same pair of vectors. In other words, pairs of simple cells with the same principal directions should be used at every point $\overline{\alpha}^0$. Any misalignment of the cell's principal directions will bring an error in the total derivative estimation. Thus, being correct from a theoretical point of view, such an approach to estimating the total derivative turns out to be difficult to implement in the brain and fraught with significant inaccuracy. Clearly, this is not the kind of computation that we think the visual system is capable of performing.

In this context, it is worth noting that calculating directional derivatives for all directions, as simple cells seem to do, is, in a sense, a coordinateless representation of the total derivative of the luminance pattern. Really, the response of a simple cell does not depend on the choice of the basis vector in terms of which its spatial sensitivity function is defined. (This choice will affect the unit of measurement along its principal direction, but the integral (13.17) is independent of the choice of the measurement unit.) The geometrical meaning of the total derivative of the luminance pattern $f_{M+L}(\overline{\alpha})$ at $\overline{\alpha}^0$ can be illustrated as follows. This total derivative determines the plane tangent to the graph of $f_{M+L}(\overline{\alpha})$ at point $\overline{\alpha}^0$. Any direction, say, along a vector $\mathbf{v}$, through $\overline{\alpha}^0$ determines a vertical plane through it, the cross section of the graph of $f_{M+L}(\overline{\alpha})$ made by which can be considered as a graph of a function of one variable. The derivative of this function at $\overline{\alpha}^0$, firstly, coincides with the directional derivative $\partial_{\mathbf{v}} f_{M+L}(\overline{\alpha})\,|_{\overline{\alpha}=\overline{\alpha}^0}$, and secondly, determines a line in the tangent plane at $\overline{\alpha}^0$. Hence, a set of directional derivatives determines a bunch of tangent lines through $\overline{\alpha}^0$ in the tangent plane. Clearly, this bunch of straight lines through $\overline{\alpha}^0$ determines the tangent plane, thus, the total derivative at $\overline{\alpha}^0$. Although two lines are enough to uniquely determine the tangent plane, the visual system choses much more redundant way to specify it. Evidently, this way is coordinate-free. Clearly, we cannot be sure whether the visual system really implements this way or not. It all depends on how the visual system further uses the information about the luminance pattern, potentially contained in the directional derivative set.

Another way in which the visual system can use the information obtained by the simple cells in order to estimate the total derivative of the luminance pattern (thus, the pattern itself) is as follows. The directional derivative, defined only for unit vectors in $\mathbf{R}^2$, has an important extreme property. Namely, the directional derivative at a point has a minimum and a maximum.[34] The absolute minimum of the directional derivative is obviously zero. Really, a tangent line at a point $\overline{\alpha}^0$ which is a level set for the linear functional determining the tangent plane (i.e., the total derivative f'_{M+L} at $\overline{\alpha}^0$) specifies the direction along which the directional derivative is zero, that is, the absolute minimum of the directional derivative. Therefore, in the vicinity of $\overline{\alpha}^0$ a local fragment of such a tangent line approximates a level set for the luminance pattern $f_{M+L}(\overline{\alpha})$, that is, the set of all points $\overline{\alpha}$ where $f_{M+L}(\overline{\alpha}) = f_{M+L}(\overline{\alpha}^0)$. Therefore, the knowledge of the direction with zero directional derivative for each $\overline{\alpha}^0$ implicitly specifies the isoluminance contours for $f_{M+L}(\overline{\alpha})$. Although the contour map for $f_{M+L}(\overline{\alpha})$ is important source of information about the luminance pattern, it does not suffice its unique reconstruction. The directions bringing maximum to directional derivatives turn out to be more informative.

The unit vector $\mathbf{v}'$, for which the directional derivative reaches its absolute maximum, can be determined from Eq. (13.26). Indeed, since $f'_{\overline{\alpha}^0}(\mathbf{v}) = \partial_{\mathbf{v}} f(\alpha)\,|_{\overline{\alpha}=\overline{\alpha}^0}$ and the directional derivatives along the basis vectors are just the partial derivatives, i.e., $\partial_{\mathbf{v}} f(\overline{\alpha}_i)\,|_{\overline{\alpha}=\overline{\alpha}^0} = \partial f/\partial\alpha_i(\overline{\alpha}^0)$ $(i = 1, 2)$, Eq. (13.26) can be rewritten as:

$$\partial_{\mathbf{v}} f(\overline{\alpha})\,|_{\alpha=\alpha^0} = v_1 \frac{\partial f(\overline{\alpha})}{\partial\alpha_1}\,|_{\overline{\alpha}=\overline{\alpha}^0} + v_2 \frac{\partial f(\overline{\alpha})}{\partial\alpha_2}\,|_{\overline{\alpha}=\overline{\alpha}^0}. \tag{13.27}$$

It follows from (13.27) that, for all unit vectors $\mathbf{v}$, $\partial_{\mathbf{v}} f(\overline{\alpha})\,|_{\overline{\alpha}=\overline{\alpha}^0}$ reaches its maximum for a unit vector $\mathbf{v}'$ whose components are proportional to the partial derivatives of f. A vector whose coordinates are equal to the partial derivatives of $f(\overline{\alpha})$ at the point $\overline{\alpha}^0$ is nothing else than the *gradient* of f at $\overline{\alpha}^0$, according to its standard definition. Therefore, $\mathbf{v}'$ is collinear with the gradient. This can be expressed by using the standard denotation for the gradient of f, $\nabla f(\overline{\alpha}^0)$, as $\mathbf{v}' = k\nabla f(\overline{\alpha}^0)$, where $k = \left(\partial_{\mathbf{v}'} f_{M+L}(\overline{\alpha})\,|_{\overline{\alpha}=\overline{\alpha}^0}\right)^{-1}$. To put this the other way round, the gradient $\nabla f(\overline{\alpha}^0)$ is a vector collinear with $\mathbf{v}'$, the length of which is equal to the value attained by the directional derivative along $\mathbf{v}'$:

$$\nabla f(\overline{\alpha}^0) = k'\mathbf{v}', \tag{13.28}$$

where $k' = \left(\partial_{\mathbf{v}'} f_{M+L}(\overline{\alpha})\,|_{\overline{\alpha}=\overline{\alpha}^0}\right)$.

34 The set of one-dimensional subspaces of a two-dimensional real vector space is a compact manifold. Consequently, the bundle of straight lines passing through a point on the plane is compact. The directional derivative (13.23) is continuous (see the integrals (13.17) and (13.18)). Therefore, it must reach the minimum and maximum in the set of all directions.

Equation (13.27) can be rewritten in a general form that show explicitly the relation between the gradient and the directional derivatives. Specifically, the directional derivative of the luminance pattern (f_{M+L}) at $\overline{\alpha}^0$ in the direction given by the vector **v**, ($\partial_{\mathbf{v}} f_{M+L}(\overline{\alpha}) \,|_{\overline{\alpha}=\overline{\alpha}^0}$), can be expressed through the gradient of the luminance pattern at $\overline{\alpha}^0$ ($\nabla f_{M+L}(\overline{\alpha}^0)$) as follows:

$$\partial_{\mathbf{v}} f_{M+L}(\overline{\alpha}) \,|_{\overline{\alpha}=\overline{\alpha}^0} = \left\langle \nabla f_{M+L}\left(\overline{\alpha}^0\right), \mathbf{v} \right\rangle, \tag{13.29}$$

where $\langle \cdot, \cdot \rangle$ stands here for the dot product of vectors.

Therefore, if we know the direction for which directional derivative of the luminance pattern f_{M+L} at $\overline{\alpha}^0$, as defined by Eq. (13.17), reaches maximum, then the gradient of the luminance pattern, $\nabla f_{M+L}\left(\overline{\alpha}^0\right)$, will be a vector with this direction whose length is equal to the value attained by the directional derivative along this direction.

It is important to note that although we come to the notion of gradient using a vector basis, the magnitude and direction of the gradient vector do not dependent on the choice of this basis. The direction of the gradient of the luminance pattern at a point $\overline{\alpha}^0$ is the direction in which the luminance increases most rapidly when moving away from $\overline{\alpha}^0$, and the magnitude of the luminance gradient is the rate at which the luminance increases in that direction. When f_{M+L} is assumed to be defined on the Euclidean plane, then the gradient $\nabla f_{M+L}\left(\overline{\alpha}^0\right)$ is orthogonal to the isoluminance contour at a point $\overline{\alpha}^0$. In other words, it is the *normal derivative*, that is, a directional derivative taken in the direction orthogonal to the isoluminance contour at $\overline{\alpha}^0$.

The total derivative and the gradient are closely related. Indeed, in the view of (13.25) and (13.29), we have

$$f'_{\overline{\alpha}^0}(\mathbf{v}) = \left\langle \nabla f\left(\overline{\alpha}^0\right), \mathbf{v} \right\rangle.$$

It is important to realise that the total derivative and the gradient belong to different linear spaces. The gradient vector belongs to the two-dimensional linear space, which can be considered the same as the domain of $f_{M+L}(\overline{\alpha})$, only shifted so that its origin is at $\overline{\alpha}^0$. This is why the vector $k'\overline{\mathbf{v}}'$ in Eq. (13.28) can be considered as a representative of $\nabla f_{M+L}\left(\overline{\alpha}^0\right)$. The total derivative belongs to the dual linear space (i.e., the space of linear functionals). Fixing a basis in the domain of $f_{M+L}(\overline{\alpha})$, we get a coordinate representation not only for the vectors in the domain of $f_{M+L}(\overline{\alpha})$ (which permits identifying it with the arithmetic space $\mathbf{R}^2$) but also for the dual space (which permits identifying it with the arithmetic space $\mathbf{R}^2$ too). The fact is that the coordinates of both the total derivative and the gradient a point $\overline{\alpha}^0$ are the same sets of numbers, namely, the partial derivatives of f_{M+L} at $\overline{\alpha}^0$. Because of this, firstly, Eq. (13.24) can be rewritten as

$$f\left(\overline{\alpha}^0 + \overline{\Delta\alpha}\right) \approx f\left(\overline{\alpha}^0\right) + \left\langle \nabla f\left(\overline{\alpha}^0\right), \overline{\Delta\alpha} \right\rangle, \tag{13.30}$$

secondly, the luminance gradients at all points $\overline{\alpha}^0$ form a vector field that encodes the luminance pattern up to an additive constant. Therefore, if the visual system is capable of evaluating this gradient vector field, it can restore the luminance pattern up to an additive constant.

For the visual system evaluates the directional derivatives of the luminance pattern at every point for every direction, it can, potentially, evaluate the gradient of the luminance pattern at each point. Specifically, the direction of the gradient at $\overline{\alpha}^0$ can be estimated from the principal direction of the simple cell with maximal absolute response (compared to all other simple cells centred at $\left(\overline{\alpha}^0\right)$).[35] The length of the gradient vector can be encoded by the magnitude of the cell response. We will call this the *directional gradient encoding*.

35 It seems that here we are dealing with what in vision literature is referred to as 'population coding' (Wu et al., 2002).

Two properties of *directional gradient encoding* are worth mentioning. First, this method (of the gradient evaluation) is 'coordinate-free', that is, it does not imply a particular coordinate system. As mentioned above, evaluating the gradient in terms of a fixed coordinate system would require strict alignment of all simple cells (participating in the evaluation) with each other, which is difficult to achieve in practice. Inevitable misalignments will reduce the accuracy of the gradient evaluation. Also, any damage and/or malfunction of a pair of such cells centred at some point $\overline{\alpha}^0$ will result in a loss of the gradient estimate at this point, with corresponding consequences for the luminance pattern encoding. In contrast, the directional gradient encoding, not relying on any particular coordinate system, appears to be robust to any accidental damage and/or loss of individual simple cells.

Second, the direction of the gradient is directly encoded by the principal direction of the simple cell with the maximum response at the point $\overline{\alpha}^0$. Its accuracy is limited by the number of simple cells centred at the point $\overline{\alpha}^0$. If the distribution of the principal directions of these cells is more or less uniform, then the more cells, the higher the accuracy. The variability of the cells responses does not affect the accuracy of the gradient direction encoding; it might affect only the accuracy of the gradient magnitude encoding. If the gradient were encoded via estimating directional derivatives with respect to two fixed directions (corresponding to two basis vectors), then an unavoidable contamination of the responses of the simple cells (with principal directions coinciding with these fixed directions) with noise would affect both the direction of the gradient and its magnitude.

It is not possible to say with certainty which of these two methods for evaluating the total derivatives of the luminance pattern is really used by the visual system, and it is not even possible to be completely sure whether they are used at all. It should be remembered that simple cells, like other types of cortical cells, can simultaneously take part in various systems of neural activity in the cerebral cortex. We can only say with certainty what role they cannot play. For example, simple cells are often considered orientation-, and/or edge detectors, that is, feature detectors. However, as noted earlier, due to their linear properties, these cells cannot perform this function. Still, there is every indication that one of their functions might be to compute the gradient of the luminance pattern, and hence its total derivative, in order to restore the luminance pattern. Evidently, it is just a hypothesis. We will call it the *directional gradient encoding hypothesis*. Naturally, it needs appropriate experimental verification. Future generations of researchers will, probably, clarify whether simple cells are involved in gradient encoding or not. In this context, it is sufficient to claim that these cells can, in principle, perform luminance gradient coding. At any rate, we have established that the luminance pattern does not disappear in the visual pathways to the brain, as it might seem. Information about it exists implicitly in V1 in the form of the gradient vector field. In other words, there is enough information in V1 to restore the luminance pattern (up to an additive constant). From this, we will proceed as from a plausible hypothesis in the subsequent presentation.

13.4.2 Directional gradient encoding with double-opponent cells

Another type of cortical (V1) cells with spatio-chromatic separability are those which Shapley and coworkers called *colour luminance double-opponent cells* (Johnson et al., 2001, 2008).

13.4.2.1 $M-L$-double-opponent cells

One class of such cells, abundantly represented in V1, will be called $M-L$-double-opponentcells. As with the simple cells described above, the responsitivity of the $M-L$-double-opponent cells varies mostly along one spatial direction (the principle direction) being practically constant along the orthogonal one. For this reason, they lend themselves to be linearly modelled similarly to the simple cells (Eqs.(13.12)–(13.15)) with the following major distinction. In contrast to the simple cells, they are driven by the L- and M-cones in the opponent way, that is, their chromatic sensitivity function has the form of

$$h_\lambda(\lambda) = k_M M(\lambda) - k_L L(\lambda), \tag{13.31}$$

where k_M and k_L are positive weights. Their spatial sensitivity function resembles the spatial sensitivity profile of the simple cells. While the exact form of this function is not critical, there is every indication that the derivative of the difference between the Gaussians might be a good model for them, at least to a first approximation (see the sketch in Fig. 13.8).

The colour-signal code implemented by the $(M - L)$-double-opponent cells will be referred to as the *$(M - L)$-colour-signal code*. It induces a corresponding colour-signal-code pattern:

$$f_{M-L}(\overline{\alpha}) = k_M f_M(\overline{\alpha}) - k_L f_L(\overline{\alpha}), \tag{13.32}$$

which shows how the $(M - L)$-colour-signal value varies across the visual field. It will be referred to as the *$(M - L)$ pattern.*

Because of the resemblance of the spatial sensitivity profiles of the simple cells and the $(M - L)$-double-opponent cells,[36] it is safe to assume that $(M - L)$-double-opponent cells evaluate the directional derivative of the $(M - L)$ pattern, $f_{M-L}(\overline{\alpha})$, along their principle directions. Furthermore, since, in common to the simple cells, a large number of $(M - L)$-double-opponent cells with various principle directions are centred at each point $\overline{\alpha}$, the directional gradient encoding hypothesiscan be extended to cells of this type. That is, it can be suggested that the $(M - L)$-double-opponent cells perform gradient evaluation for $f_{M-L}(\overline{\alpha})$ in the same manner as the simple cells perform gradient evaluation for $f_{M+L}(\overline{\alpha})$. The existence of the $(M - L)$-double-opponent cells with the reversed spatial sensitivity profile,[37] as in Fig. 13.9, (Johnson et al., 2004) corroborates such a vision of the possible role of the $(M - L)$-double-opponent cells in colour perception. Indeed, as indicated above, directional gradient encoding is only possible with both types of cells (shown in Figs 13.8 and 13.9).

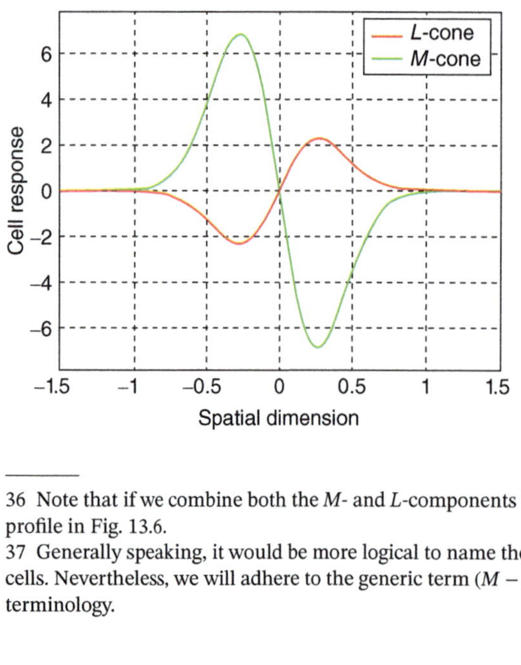

Figure 13.8 The spatial sensitivity function profiles of a $(M - L)$-double-opponent cell plotted separately for the M- (green line) and L- (red line) components along the principle direction. They have different amplitudes, which is determined by the ratio of k_M and k_L in (13.31). After equalising in amplitude, the red curve is a mirror image of the green one.

36 Note that if we combine both the M- and L-components in Fig. 13.8, we get a profile identical in shape to the profile in Fig. 13.6.

37 Generally speaking, it would be more logical to name these cells $(L - M)$- rather than $(M - L)$-double-opponent cells. Nevertheless, we will adhere to the generic term $(M - L)$-double-opponent cells so as not to complicate the terminology.

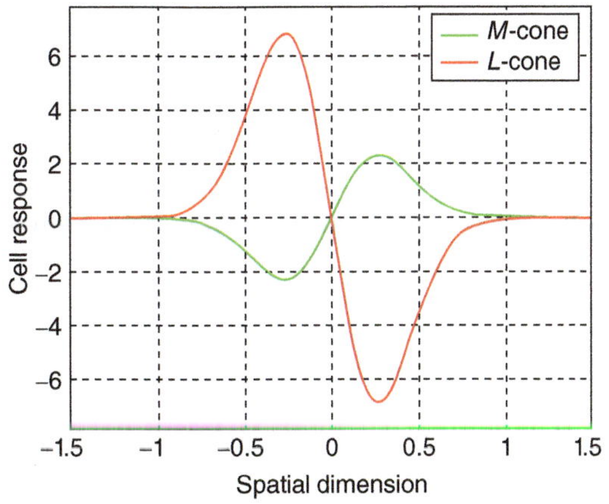

Figure 13.9 The spatial-sensitivity-function profiles of a $(L - M)$-double-opponent cell plotted separately for the M- (green line) and L- (red line) components along the principle direction, similarly to how it is done in Fig. 13.8.

There is, however, one fact concerning $(M - L)$-double-opponent cells that impeeds such an interpretation of their role. As a matter of fact, the experiments showed significant differences in the weights k_M and k_L (in Eq. (13.31)) for different cells (Johnson et al., 2004; Shapley, 2019). At the first glance, such a variability of the weights contradicts the directional gradient encoding hypothesis. Really, cells with different weights k_M and k_L will compute the directional derivative of different colour-signal-code patterns. If the visual system 'knew' the weights for each cell, then it would not be a problem since the gradient, $\nabla f_{M-L}(\overline{\alpha})$, obtained for one pair of weights can be converted to that for another pair. However, the suggestion of such 'knowledge' seems to be rather implausible.

Alternatively, the weights variation could be reconciled with the directional gradient encoding hypothesis by assuming that the visual system 'averages' the data from the entire ensemble of the $(M - L)$-double-opponent cells centred at a particular point $\overline{\alpha}$ for each direction. If the distributions of the weights k_M and k_L do not change significantly from point to point in the visual field, as well as from one direction to another, then one can assume that at each point $\overline{\alpha}$ the gradient of the following colour-signal-code (rather than $f_{M-L}(\overline{\alpha})$) is evaluated:

$$f_{\overline{M-L}}(\overline{\alpha}) = \overline{k}_M f_M(\overline{\alpha}) - \overline{k}_L f_L(\overline{\alpha}), \tag{13.33}$$

where $\overline{k}_M$ and $\overline{k}_L$ denote the weights k_M and k_L, 'averaged' for a given point $\overline{\alpha}$.

From what we now know about the weights distribution (Johnson et al., 2004; Shapley, 2019), we can conjecture that the averaged weight ratio, $\overline{k}_M/\overline{k}_L$, will, probably, not deviate significantly from 1. In this case, the colour-signal-code (13.33) will be close to (13.32). Though, the specific form of the code (13.33) is of no fundamental importance.

13.4.2.2 $S - (M + L)$-double-opponent cells

Also, there exist double-opponent cells of another type that get the input from all the three types of cones. Their chromatic sensitivity function can be expressed as a linear combination with positive weights of the following form:

$$h_\lambda(\lambda) = k_S S(\lambda) - \big(k_M M(\lambda) + k_L L(\lambda)\big). \tag{13.34}$$

These will be referred to as the $S - (M + L)$-double-opponent cells. Although this type of cells have been studied to a lesser extent than the previous two, it seems safe to assume that they perform the same computation. In other words, we believe that the $S - (M + L)$-double-opponent cells are quite suitable for computing the gradient of the following pattern, which is defined by analogy with Eq. (13.33):

$$f_{\overline{S-(M+L)}}(\overline{\alpha}) = \overline{k}_S f_S(\overline{\alpha}) + \left(\overline{k}_M f_M(\overline{\alpha}) + \overline{k}_L f_L(\overline{\alpha})\right). \tag{13.35}$$

Evidently, this belief is only justified if the variability of the weights in Eq. (13.35) is of the same kind as those in Eq. (13.32), that is, the weight distribution does not depend on the points $\overline{\alpha}$ and the cell principal directions.

13.4.3 Difference in spatial sensitivity of $(M + L)$-, $(M - L)$-, and $S - (M + L)$-cells, and its implication for colour perception

It is worth noting that little is known about the spatial sensitivity function of the $S-(M+L)$-double-opponent cells. As the S-cones are very sparse in the retina, it is naturally expect that blurring produced by these cells is stronger than that for the $(M + L)$- and $(M - L)$-double-opponent cells. In other words, the blur in the pattern $f_{\overline{S-(M+L)}}(\overline{\alpha})$ is higher than in the patterns $f_{\overline{M-L}}(\overline{\alpha})$ and $f_{M+L}(\overline{\alpha})$.

There are psychophysical studies corroborating this expectation. For instance, in some studies the 'yellow-blue' mechanism was claimed to have narrower spatial frequency characteristics than the 'red-green' and 'luminance' mechanisms (e.g., Green, 1968; Kelly, 1974; Cavonius & Estevez, 1975; Hess et al., 1989). Though, in some other the yellow-blue and red-green mechanisms were claimed to have nearly the same spatial resolution, somewhat lower than that of the luminance mechanism[38] (e.g., Sekiguchi et al., 1993; McKeefry et al., 2001).

The difference in the spatial sensitivity functions of the three colour-opponent mechanisms is commonly believed to be a cause of a change in colour induced by a change in size (and/or distance of observation) of some patterns. That the colour of a pattern might change when its size gets smaller has been known for a long time (Willmer, 1944). For example, Wandell and collaborators found that the coloured strips changed their colour appearance with spatial frequency (i.e., the inter-strip distance) (Poirson & Wandell, 1993; Bauml & Wandell, 1996). They argued that this effect resulted from the difference in the spatial frequency characteristics of the colour-opponent mechanisms.

More specifically, Poirson & Wandell (1993) found that, firstly, the colour matching (for the square-wave gratings employed in their experiment) is best accounted for by a model comprising three colour-opponent mechanisms, whose response is separable with respect to chromatic and spatial dimensions. It is noteworthy that this is the only evidence for spatial-chromatic separability, obtained by psychophysical methods, known to the authors so far. Secondly, the chromatic sensitivity functions of these mechanisms were claimed to be close to the yellow-blue and red-green opponent mechanisms derived from the hue cancellation experiment. Thirdly, the spatial sensitivity functions of the mechanisms prove to be different, the yellow-blue mechanism having the poorest spatial resolution.

The dependence of the colour of an object on its size can be observed in Fig. 13.10. The colour illusion in Fig. 13.10 can be accounted for as a result of poorer spatial resolution of the yellow-blue opponent mechanism (Logvinenko & Hutchinson, 2005, 2007).

38 It must be said that these claims should be treated with caution because their authors, though like many others, did not specify how they understood the mechanisms in questions. And most importantly, they did not provide sufficiently convincing arguments in favour of that the experimental techniques they used are indeed capable to reveal the spatial frequency characteristic of these mechanisms.

Figure 13.10 As viewing distance increases, the neutral horizontal diamonds become coloured. Neutral diamonds on the pink and yellow background become tinged blue and green respectively. Source: Logvinenko (2001)/Sage Publications.

13.4.4 Representation of the colour-signal surface in the form of its tangent bundle

According to the directional gradient encoding hypothesis, the $(M+L)$, the $(M-L)$, and the $S-(M+L)$-double-opponent cells evaluates, for each point $\overline{\alpha}$, the gradients of the three colour-signal codes f_{M+L}, $f_{\overline{M-L}}$ and $f_{\overline{S-(M+L)}}$, considered as the functions of the variable $\overline{\alpha}$ (see Eqs. (13.20), (13.33), and (13.35)), i.e., $\nabla f_{M+L}(\overline{\alpha})$, $\nabla f_{\overline{M-L}}(\overline{\alpha})$, and $\nabla f_{\overline{S-(M+L)}}(\overline{\alpha})$, respectively. Since this evaluation is a differential (thus, local) operation, knowing the gradient at only one point $\overline{\alpha}$ does not allow recovering the value of the corresponding colour-signal code at this point. Recall that the gradient in question is a vector in the plane to which $\overline{\alpha} = (\alpha_1, \alpha_2)$ belongs (namely, the visual plane V), not a number (which the corresponding colour-signal-code value is).[39] In a sense, the situation in the cortex (for the adherents of the idea of local colour encoding) is even more mysterious than in the retina. Definitely, for the reasons stated above (Section 13.1.4), the response of the photoreceptor is ambiguous, and this ambiguity prevents the recovery of the colour-signal code from the photoreceptor response. However, the gradient that the cortical cells could possibly evaluate is not a scalar value at all. Thinking in local terms, we are faced not only with ambiguity in the recovery (of the colour-signal code) but with the fundamental impossibility of its recovering.

A different picture opens up when we begin thinking in holistic terms, that is, when we consider the colour-signal-code patterns $f_{M+L}(\overline{\alpha})$, $f_{\overline{M-L}}(\overline{\alpha})$ and $f_{\overline{S-(M+L)}}(\overline{\alpha})$. In this case, as pointed out above, knowing the gradients $\nabla f_{M+L}(\overline{\alpha})$, $\nabla f_{\overline{M-L}}(\overline{\alpha})$, $\nabla f_{\overline{S-(M+L)}}(\overline{\alpha})$ for each $\overline{\alpha}$ allows us to recover each of these patterns up to an arbitrary additive constant (of course, its own for each pattern).

Note that from a formal point of view, a gradient considered as a function of $\overline{\alpha}$ is a vector field in the visual plane V. Importantly, the elements of this field (i.e., the gradients) are vectors from the two-dimensional vector space associated with this plane. As a matter of fact, the directional gradient encoding is, effectively, a spatial operation that involves detecting a certain direction in the visual plane V (namely, the one that maximises the response of the corresponding cells). In fact, colour coding appears to be implemented in the form of direction coding (i.e., spatial dimension coding). Colour is involved here only in that form of the spectral selectivity of the colour-signal code, the gradient of which is estimated. In other words, this operation is related to colour only insofar as it indirectly involves the chromatic sensitivity of this colour signal code. Moreover, the specific

39 It should be borne in mind that two partial derivatives, as real numbers, can play the role of coordinates not only of some vector (in the visual plane V), but also of a covector (i.e., of a linear functional). For this reason, it is always necessary to indicate the coordinates of which space (vector space or its dual space) are meant. In this context, it is the visual plane V.

form of the chromatic sensitivity function is not particularly important since it determines just one of the three coordinate axes in the colour signal space, with respect to which the colour-signal codes are expressed. The visual system can afford not to even know what exactly they are, provided they are fixed.

It must be emphasised that, strictly speaking, just the availability of the gradients $\nabla f_{M+L}(\overline{\alpha})$, $\nabla f_{\overline{M-L}}(\overline{\alpha})$, and $\nabla f_{\overline{S-(M+L)}}(\overline{\alpha})$ for each $\overline{\alpha}$ is not enough for recovering the colour-signal-code patterns $f_{M+L}(\overline{\alpha})$, $f_{\overline{M-L}}(\overline{\alpha})$ and $f_{\overline{S-(M+L)}}(\overline{\alpha})$. These gradients should be in the same spatial order as the corresponding elements in the colour-signal-code patterns. Recall that each type of the double-opponent cells belongs to a separate visual pathway (from the retina to V1): the M (magnocellular) pathway for the $(M + L)$ type, the P (parvocellular) pathway for the $(M - L)$ type, and K (koniocellular) pathway for the $S - (M + L)$ type (Kaplan, 2014; Shapley et al., 2014). Moreover, each type occupies its own layer (different from the other two) (Gegenfurtner, 2003; Shapley et al., 2014). Within each of this pathways, the cells are located according the principle of the retinotopic mapping, which implements the above requirement, that is, the estimated gradients $\nabla f_{M+L}(\overline{\alpha})$, $\nabla f_{\overline{M-L}}(\overline{\alpha})$, and $\nabla f_{\overline{S-(M+L)}}(\overline{\alpha})$ are in the same spatial order as the corresponding elements in the colour-signal-code patterns $f_{M+L}(\overline{\alpha})$, $f_{\overline{M-L}}(\overline{\alpha})$, and $f_{\overline{S-(M+L)}}(\overline{\alpha})$.

It is worth emphasising once again that in the context considered here, nothing takes place in each of the three pathways other than spatial differentiation. If there exsited only one such a pathway, there would be no colour rendering at all (even though the cones of all the three types feed to the cells forming this pathway). The presense of the three pathways with different chromatic sensitivity opens up an opportunity to render colour. Within each pathway, no cell encodes colour in any way. And it is unlikely that such cells exist anywhere else in the visual system. The very idea of a so-called 'colour centre' somewhere in the brain (Zeki, 1973; Tootell et al., 2004) is certainly misleading. It is not surprising that all attempts to find such a centre in the cerebral cortex have so far been unsuccessful. From the holistic standpoint, the multiple spatio-chromatic channels with differ chromatic sensitivity (as well as with spatio-chromatic separability) may well serve as the basis for colour vision.

So, we see that the processing of the light stimulus array is carried out by the three spatio-chromatic channels, as a result of which the three colour-signal-code patterns are simultaneously and independently created (with the aforementioned uncertainty up to an additive constant). Aside from the inevitable differences between these three patterns (considered, first of all, as spatial patterns) caused by the difference between the channels in chromatic sensitivity, each of them, by and large, renders quite successfully the spatial structure of the light stimulus array.[40] What is the need for such a triple representation of the light stimulus array? Is not this triplicating far too redundant? While the answers to these questions are known only to the Grand designer (i.e., Mother Nature), a few conjectures are noteworthy. Firstly, the spatial structure of the light stimulus array is so important from an ecological point of view that its reliable representation becomes literally a matter of life and death for animals. Its triple representation in the visual cortex contributes to increased reliability – three representations are obviously more reliable than one. Secondly, the spatial structure of the light stimulus array is ecologically important not itself but insofar as it reflects the spatial layout of the animal environment. In Gibsons' words, it carries information about the spatial layout of the environment. However, the ambient light (thus, the light stimulus array) is structured not only by the spatial layout of surfaces in the environment but also by the light, illuminating these surfaces, and the surfaces reflectivity. In other words,

40 As a rough analogy, you can consider three photographs of the same scene taken through different coloured filters.

the information of the spatial layout is confounded in the light stimulus array with information of the illumination and the surfaces reflectivity. To select out the information concerning the spatial layout (which is the top priority), the visual system has to disentangle these three different sources of information. As shown in Section 21.1, even with three colour-signal-code patterns, exact disentangling in question remains quite problematic (at least, from the physical standpoint).

Known from immediate introspection, the unitarity of our perceptual images implies that the three colour-signal-code patterns must somehow come together to result in a unitary percept. How, then, are the outputs of the three pathways combined together? Specifically, do they coalesce somewhere in the brain before becoming a perceptual image, or are the three colour-signal-code patterns simply experienced simultaneously and independently? In the case of affirmative answer to the last question, simultaneously and independently of each other, we should perceive three colour images. To be more exact, simultaneously and independently, we perceive three patterns with similar spatial structure, each of which is tinged with its own one-dimensional colour-opponent palette. (In accord with Hering's idea, these patterns can be thought of as 'achromatic image', 'yellow-blue image', and 'red-green image'.) By the way, experiencing them as of different colour seems to be of secondary importance, it simply indicates which pattern is produced by which channel.

There are several observations indicating that the independent phenomenological status of the three colour-signal-code patterns is highly unlikely. First of all, as already mentioned, most people incline to consider colour as the unitary phenomenological entity, hard to split into its components. It takes some experience and some training to learn how to introspectively discern hue components. Second, as pointed out by a number of researchers, although the spatial resolution of the colour-signal-code patterns are different (due to the difference in spatial sensitivity in the three pathways), we do not experience this fact directly. In particular, the spatial quality of the yellow-blue image is not perceived as worse than that of the achromatic image. The difference in spatial resolution of the pathways could reveal itself in a change in the colour of the whole pattern, as in Fig. 13.10, but not in the poor spatial distinctness of the yellow-blue image.[41] Finally, the fact of immediate introspection testifies that the three images are experienced as perceptually aligned and congruous with each other, which can hardly be achieved by itself.

All this leads to the conjecture that even if the three colour-signal-code patterns independently acquire phenomenological status on an individual basis, a special process must take place, a kind of perceptual fusion, so that the fused image is more than a mechanistic combination of them. In this regard, an analogy arises with the problem of fusing two monocular images into a binocular one. Due to the disparity caused by the interocular shift, the monocular images are slightly different, which can sometimes be experienced in the form of diplopia (namely, when the disparity exceeds the diplopia threshold). Nevertheless, we usually experience single binocular images, presumably resulting from the binocular fusion process. The latter seems to be more than just superimposing and aligning of monocular images. For example, when viewing a repetitive striped pattern, binocular fusion of 'wrong' stripes may occur, resulting in an 'illusory' depth shift of the stripes, known as the wallpaper phenomenon (Woodworth, 1938). Recording of binocular eye movements while observing the wallpaper phenomenon showed that the actual disparity between the 'fused' monocular images of stripes far exceeded the known thresholds for diplopia (Logvinenko & Belopolsky, 1994; Logvinenko et al., 2001). Therefore, the term 'binocular fusion' seems to us to be somewhat oversimplifying. Rather, the wallpaper phenomenon reveals the creative nature of the process of generating binocular images, usually hidden from our introspection. The pseudoscopic

41 An elegant demonstration can be found in Wandell (1995, Figure 9.23) where it is shown that spatial blurring of an achromatic component image leads to perceptual blur, while spatial blurring of red-green or yellow-blue component images has no apparent effect.

transformations of binocular images described in Chapter 21 clearly corroborate this idea. Moreover, it is from the pseudoscopic vision that follows that combining the three colour-signal-code patterns cannot reduce to just fusing the corresponding three images. It is more creative process than simple fusion. Figuratively speaking, one can say that the binocular image is being built, the monocular images being used as a raw material for this building process.

Howbeit, the achromatic component image, the yellow-blue component image and the red-green component image appear to be perfectly aligned and fitted to each other. Such an alignment would be incomprehensible unless we assume there being a common frame of reference for all three pathways. More specifically, the visual system should 'know' how to identify the triplet of the gradient estimations, one in each of the pathways, that are made for the same point $\overline{\alpha}$. If this is the case, then the colour-signal surface is represented in the visual cortex up to an arbitrary additive colour signal.

Indeed, the functions $f_{M+L}(\overline{\alpha})$, $f_{\overline{M-L}}(\overline{\alpha})$ and $f_{\overline{S-(M+L)}}(\overline{\alpha})$ define a map, $\overline{p} : V \rightarrow T \circ \Phi(A)$, where T is the linear operator relating the cone fundamentals, S, M, and L, and the colour-signal codes f_{M+L}, $f_{\overline{M-L}}$, and $f_{\overline{S-(M+L)}}$. The map range of $\overline{p}\,(V)$ determines a surface in the colour-signal cone (with respect to the colour-signal codes f_{M+L}, $f_{\overline{M-L}}$, and $f_{\overline{S-(M+L)}}$) which is nothing more than the colour-signal surface, only in the new coordinate system. (For simplicity, we will assume that the colour-signal surface is smooth.) According to the directional gradient encoding hypothesis, for each point $\overline{\alpha}$ the gradients of these functions, $\nabla f_{M+L}(\overline{\alpha})$, $\nabla f_{\overline{M-L}}(\overline{\alpha})$, and $\nabla f_{\overline{S-(M+L)}}(\overline{\alpha})$, are evaluated by the $(M+L)$, the $(M-L)$, and the $S-(M+L)$-double-opponent cells, respectively. In view of (13.11), one can say that these three types of cortical cells compute the Jacobian matrix, $J_{\overline{p}}(\overline{\alpha})$, of map $\overline{p}$ specifying the colour-signal surface. If the visual system encodes $J_{\overline{p}}(\overline{\alpha})$ at every point $\overline{\alpha}$, this amounts to encoding the colour signal surface up to an unknown additive vector (i.e., a colour-signal) constant. If we also assume that the visual system does not 'know' the 'average' weights, that is, it does not 'know' exactly with respect to which colour-signal map (i.e., which concrete representation of the colour stimulus cone), the Jacobian matrix, $J_{\overline{p}}(\overline{\alpha})$, is evaluated, then the Jacobian matrix will be evaluated up to an unknown linear transformation. Thus, in this case, the colour-signal surface will be retained up to an unknown affine transformation.[42]

As pointed out in Section 13.3.2, Jacobian matrix $J_{\overline{p}}(\overline{\alpha})$ determines the two-dimensional tangent space to the colour-signal surface at point $\overline{\alpha}$. Interestingly, due to the retinotopic mapping in the primary visual cortex, all these tangent planes can be thought to fit ('glue') together to form another manifold (of the double dimension). It is nothing more than the tangent bundle[43] of the colour-signal surface. Therefore, if the colour-signal surface is represented in the visual cortex at all, it is present there as its tangent bundle rather than in its genuine form.

Eventually, one cannot help but wonder why the visual system resorts to a method of encoding the colour-signal surface based on the gradient estimation, if it is burdened with such ambiguity? At least one reason lies on the surface: the photoreceptors are not capable of encoding the colour signals within the whole range of intensities. Yet, one can make up from them a device performing the spatial differentiation of the colour-signal array, i.e., computing the Jacobian matrix, $J_{\overline{p}}(\overline{\alpha})$ (13.10).

Another reason why the visual system can afford to encode the colour-signal surface with such an indeterminacy will become clear in Part II (see Section 20.3). The fact is that encoding the colour signal is just the beginning of the story, not its end. It is only for light colours (which can be observed

---

42 Note that even though the photoreceptor response did encode the cone excitation and send it further, the colour signal would be given to the brain up to a linear transformation anyway unless the brain 'knows' the cone spectral sensitivities.

43 Tangent bundle is a particular case of the fibre bundle. The tangent spaces are the fibres here.

only in the specific conditions of a laboratory) that one can expect that the colour signal, produced by the light, determines its colour. In the case of object colour, the relation between the colour signal of the light reflected by objects and the object colour becomes more complicated. The indeterminacy of the object colour (as far as the colour signal is concerned) qualitatively increases (as compared to light colours) (see Chapter 21). In view of this new kind of indeterminacy, it becomes almost unimportant whether the colour-signal surface itself or the tangent bundle thereof is represented in the brain.

Moreover, after Gibson's criticism of the very concept of a retinal image and the idea of transferring it to the brain for subsequent storage there, the lack of representation of the colour-signal surface in the visual cortex can hardly be very surprising. Following up this criticism, in Chapter 21, we argue that the optical metaphor does not seem to us to be appropriate at all, and that the linguistic one seems to be more fruitful for perception. The linguistic approach to perception discussed in this chapter does not require the representation of the colour-signal surface in the brain, although it does not exclude it. The evidence presented in this section in favour of representing the colour-signal surface in the form of its tangent bundle should be regarded as nothing more than a claim that this representation is, in principle, possible. We argue that the colour-signal surface is not hopelessly lost, that the visual cortex contains information that, in principle, allows it to be restored, in the worst case, up to an unknown affine transformation. Another question requiring more specific research is whether the brain actually makes use of this information.

As a concluding remark, we note that the described method of representing the colour-signal surface cannot be accurate due to the assumptions made in determining $\overline{p}$. Particularly, in reality, the distribution of the weights k_M and k_L is hardly exactly the same for all points and directions in the visual field, as assumed above (Section 13.4.2). Any variability in this distribution will inevitably lead to a difference between the colour-signal surface and its estimate derived from the Jacobian matrix $J_{\overline{p}}(\overline{\alpha})$. It is tempting to conclude that, as a result, the colour of the light stimulus array will be encoded inaccurately. However, such a conclusion will be correct only if we identify the colour of the light with the colour signal induced by this light (i.e., with the photopigment excitation triplet), which is typical for colorimetry and its applications. However, as light itself is colourless, one could just as well identify the colour with the colour signal derived from $J_{\overline{p}}(\overline{\alpha})$. Since there is no 'true' colour, the very term 'encoding the colour' is rather misleading. Colour is assigned to light; it is created, not decoded (or encoded). True, the term 'encoding' is quite justified when it comes to the colour signal and the colour-signal surface, but not the colour as such.

The variability of the distribution of the weights k_M and k_L may well manifest itself in a mismatch between the colour-signal surface and the resulting colour image. For example, the colour signal at some point $\overline{\alpha}$ may be such that the assigned colour is, say, unitary yellow, whereas the estimated colour signal at $\overline{\alpha}$ gives rise to a greenish yellow. It would be interesting to find out to what extent the variability of k_M and k_L can bring about this colour mismatch. Unfortunately, the available data on k_M and k_L are too scarce to assess their impact in question. However, the colour mismatch arising from the variability of k_M and k_L can be expected to be mainly limited to moderate changes in chromatic value without changing chromaticness type. In other words, if the assigned colour is, say, orange then the estimated colour will also be orange, only with different chromatic values of yellow and red in it. Anyway, the subjective colour difference is unlikely to be large. Even if a unitary colour turns to binary (as in the above example with unitary yellow and greenish yellow), the difference between the chromatic values is not expected to be large. Certainly, the ratio of yellow to green can sometimes have some ecological value. For example, the more yellow the pear is, the more likely it is ripe. However, in general, it is the spatial distribution of colour, i.e., colour texture that is important from an ecological point of view, rather than what colour a particular pixel of this texture has.

It follows from the above that the colour texture derived from the Jacobian matrix $J_{\overline{p}}(\overline{\alpha})$ may differ from the colour texture corresponding to the colour-signal surface. However, the ecological validity of texture arises from the fact that it renders the spatial layout (Gibson, 1979). If the estimated colour texture renders the spatial layout in accord with the luminance texture (i.e., the luminance pattern (see Eq. (13.20))), then the specific colours that make up that texture are of secondary ecological importance. Still, these are all just speculations, the decisive word in this matter should belong to experimentation and modelling.

Part II

Object Colour

In Part I, we assumed that light from a certain source goes directly into the eye, possibly passing through some kind of medium. However, in a real environment full of opaque objects, light comes to the eye mainly from these objects. So, in this Part, the following chain of events will be assumed to precede the visual process. A light source – the *illuminant* – (or many sources) emits light and passes it through the medium to an opaque object, which interacts with it. As a result of this interaction (between the light and the object surface), the physical nature of which is not completely clear yet, the light is re-emitted by the object. Using more conventional terminology, we say that the object *reflects* the incident light. The reflected light then enters the eye and is ultimately processed by distinct classes of photoreceptors (or, in general, artificial photosensors) with different spectral sensitivity, in the way discussed in Part I.

The way an object appears for us is determined by many factors, amongst which the main one is, undoubtedly, its illumination. The illumination of an object surface at a point $\mathbf{x}$ can be characterised by the light incident at this point from each direction. It can be specified by means of a directional radiometric quantity such as *spectral intensity*, i.e., radiant intensity (the radiant flux (or power) received at $\mathbf{x}$ per unit solid angle) taken as a function of wavelength. The total radiant flux received by the object surface at $\mathbf{x}$ per unit area considered as a function of wavelength (i.e., spectral irradiance of the surface at $\mathbf{x}$) can be computed by integrating over all the directions from which light comes to $\mathbf{x}$, taking into account the local orientations of the surface at $\mathbf{x}$ relative to the incident directions. Hence, the radiant power received by a surface at a given point depends not only on its illumination at this point but also on its local shape and orientation (in the vicinity of $\mathbf{x}$).

Still, the main factor that determines the appearance of an object is its reflectivity. Due to the complexity of the reflection process, its quantitative characterisation is usually carried out under some simplifying assumptions. Specifically, like many colour scientists, we will restrict ourselves to diffuse reflection only, assuming no specular reflection. Also, in what follows, we assume that an electromagnetic wave does not change its wavelength during re-emission (i.e., that true reflection takes place), and that there is no internal scattering (which allows us to consider the point of incidence and the point of reflection as the same point). An area of reflecting surface around the point in question will be assumed to be uniformly lit by the collimated beam of incident light.

Foundations of Colour Science: From Colorimetry to Perception, First Edition. Alexander D. Logvinenko and Vladimir L. Levin.
© 2023 John Wiley & Sons Ltd. Published 2023 by John Wiley & Sons Ltd.

Under the assumptions made, our intuitive understanding of reflectance as the fraction of the incident flux that is re-emitted can be justified by relating the radiance, reflected in some direction at the point **x**, to the flux, incident on **x**, through a proportionality factor. When expressing this in differential form, we arrive at the quotient of two differentials: reflected radiance and incident irradiance. As this quotient depends on both the incident and reflected directions, it is commonly referred to as the *bidirectional reflectance distribution function* (*BRDF*) (Nicodemus et al. 1977). It follows that the radiance $I_r(\omega_r)$ of the light reflected from the object-surface at **x** along the direction ω_r is related to the radiance $I_i(\omega_i)$ of the light received by the object-surface at **x** from the direction ω_i as:

$$dI_r(\omega_r, \mathbf{x}) = R(\omega_i, \omega_r, \mathbf{x}) \cos(\theta_i(\mathbf{x}))\, I_i(\omega_i, \mathbf{x})\, d\omega_i, \tag{II.1}$$

where $R(\omega_i, \omega_r, \mathbf{x})$ stands for BRDF (at **x**), and $\theta_i(\mathbf{x})$ is the angle between ω_i and the surface normal at **x**.

Note, firstly, that, being a function of the incident and reflecting directions, ω_i and ω_r, BRDF cannot be considered as an intrinsic property of the object material. The directions of incidence and reflection of light also matter. Secondly, as reflectivity depends on the spectral composition of incident light, the quotient R in Eq. (II.1) should be considered as a function of wavelength as well. In this regard, of special interest are objects for which the quotient R in Eq. (II.1) can be factored into chromatic and geometrical components, only one of which is a function of wavelength:

$$R(\lambda; \omega_i, \omega_r, \mathbf{x}) = R_m(\lambda; \mathbf{x})\, R_s(\omega_i, \omega_r, \mathbf{x}). \tag{II.2}$$

We will call $R_m(\lambda; \mathbf{x})$ the *material spectral reflectance function* (or, just *spectral reflectance function*) of the object at **x**. For such objects, with geometrico-chromatically separable reflectivity, the effect of the directions of incidence and reflection reduces to just scaling their the material spectral reflectance function. Admittedly, the reflectivity of real objects often does not satisfy the geometrico-chromatical separability (Eq. (II.2)) even in the first approximation. Nevertheless, we will consider only such objects as described above because otherwise the notion of spectral reflectance function simply becomes meaningless.[1]

When the spectral selectivity of the surface reflectivity is not of interest, the surface reflectance can be specified by a single index – *albedo* – which is the mean of the spectral reflectance function (i.e., its average value over the spectrum). The surface albedo shows how much of the light, reflected by the perfect reflector with the same geometric factor R_s, is reflected by the surface in question.

Consider now a (potential) point of view, O, the ambient optic array around which is formed by the light reflected by an object illuminated by some (other) light. We will assume that no two points on the object surface can send direct light towards that point of view in the same direction. Therefore, we can parameterise with the same angular variable, $\overline{\alpha} = (\alpha_1, \alpha_2)$ (see Section 13.3.2), both the ambient optic array and the object surface.

Combining Eqs. (II.1) and (II.2) we get

$$dI_r(\lambda; \omega_r, \overline{\alpha}) = R_m(\lambda; \overline{\alpha})\, R_s(\omega_i, \omega_r, \overline{\alpha}) \cos(\theta_i(\overline{\alpha}))\, I_i(\lambda; \omega_i, \overline{\alpha})\, d\omega_i. \tag{II.3}$$

As we can see, the four factors contribute to the spectral radiance of the light coming to O along a direction $\overline{\alpha}$: (i) the material spectral reflectance function $R_m(\lambda; \overline{\alpha})$ at $\overline{\alpha}$; (ii) the local slant of the surface at $\overline{\alpha}$ determining $\cos(\theta_i(\overline{\alpha}))$; (iii) the local illumination at $\overline{\alpha}$ as represented by $I_i(\lambda; \omega_i, \overline{\alpha})$; and (iv) the geometric factor $R_s(\omega_i, \omega_r, \overline{\alpha})$. The first two can be construed as intrinsic properties of the

1 Although very few authors patently explicate this property (e.g., Shafer, 1985; Maloney, 1999), the very use of the term 'spectral reflectance function' tacitly implies the geometrico-chromatically separable reflectivity (II.2).

object. They are determined by object material and shape. The third is determined by illumination conditions, the fourth is dependent on the viewing geometry.

It does not take a vision scientist to realise that we are always aware of object shape, with great accuracy, and regardless of object material, illumination and viewing conditions. True, it is not quite clear how exactly the shape of an object is represented in our perception. We will discuss this issue in Section 20.2.3.

As mentioned in the Outline (Chapter 1), the basic truth, which can be found in any textbook on perception, states that the colour of an object correlates with the object spectral reflectance function and remains approximately independent of object shape and illumination, as well as of viewing condition. In other words, it is taken for granted that object colour is an apparent form for the perceptual representation of the object spectral reflectance function, and it is relatively constant with regard to the object shape and the illumination and viewing conditions. However, here we run into the following problem. As pointed out above, the material spectral reflectance function does not exist for all objects but only for special ones (the BRDF of which meets Eq. (II.2)). What does the colour of such objects represent then? Unlike the dependence of BRDF on wavelength, the dependence of BRDF on the direction of incidence, ω_i, does not seem to reveal itself in any perceptual attribute of the colour of an object. Despite the importance of this question, its exceptional difficulty compels us to postpone any attempt to answer it until the last chapter (21).

Curiously, in literature on colour vision, this issue (about the directional nature of reflectivity) is largely ignored due to additional simplifying assumptions and consideration of even more idealised reflective objects such as Lambertian surfaces. The latter are ideal diffuse (matte) reflecting objects, for which factor $R_s(\omega_i, \omega_r, \mathbf{x})$ in Eq. (II.2) is constant. Although we do not share the view that 'many surfaces are approximately Lambertian over a range of viewing conditions' (Barrow & Tenenbaum, 1978, p. 8), we adopt the Lambertian model. After all, it is just one of a series of idealisations typical of colour science.

Another common truth, moving from one textbook to another, is that while perceiving the colour of an object we are also aware of the colour of illumination. At first glance, it might seem that the latter could be a perceptual counterpart of the factor $I_i(\lambda; \omega_i, \overline{\alpha})$ in (II.3). However, the spectral radiance depends on the direction of incidence, ω_i, whereas we are hardly capable of seeing this dependence. (True, sometimes we may be aware of the direction of the prevailing illumination, but this seems to be the result of cognitive inference, a kind of visual experience of a higher order.) This leads to a conjecture that maybe it is the spectral irradiance of the light incident at some point of the object surface that cumulatively determines the apparent lighting of the object at this point.

The rationale for using Lambertian objects is that they allow exploring the role of these two main factors – the spectral reflectance and the spectral irradiance of the incident light – independently, and in isolation from the confounding effect of the incident direction ω_i within the following idealised experimental paradigm. Consider a Lambertian surface illuminated by light sufficiently collimated to be considered to be falling from one direction.[2] The light-stimulus array at O (see Section 13.3.2) will be proportional to the following expression:

$$I(\lambda; \overline{\alpha})\, r(\lambda; \overline{\alpha}) \cos(i(\overline{\alpha})), \tag{II.4}$$

where $I(\lambda; \overline{\alpha})$ is the total radiant flux falling at $\overline{\alpha}$ per unit area (perpendicular to the illumination direction); $r(\lambda; \overline{\alpha})$ is the spectral reflectance function at $\overline{\alpha}$; and $\theta_i(\overline{\alpha})$ stands for the angle of incidence of the illumination with respect to the surface normal at $\overline{\alpha}$.

---

2 Such an illumination can be produced, for example, by a remote point source.

The spectral irradiance of the incident light equals $I(\lambda;\overline{\alpha})\cos\left(\theta_i(\overline{\alpha})\right)$. This product, considered as a function of $\overline{\alpha}$, will be called the *lighting pattern* to emphasise that it depends on not only the illumination, but also on the shape and orientation of the object. A particular case of the lighting pattern, when it is created by changing only the object shape and orientation, will be called the *shading pattern*. That is, when the illumination is spatially homogeneous (i.e., $I(\lambda;\overline{\alpha}) = I_0(\lambda)$), the product $I(\lambda;\overline{\alpha})\cos\left(\theta_i(\overline{\alpha})\right)$ $(= I_0(\lambda)\cos\left(\theta_i(\overline{\alpha})\right))$ will be referred to as the shading pattern. For the sake of generality, $r(\lambda;\overline{\alpha})$ and $I(\lambda;\overline{\alpha})$ will be referred to as the *reflectance pattern* and the *illumination pattern* respectively.

A fact of fundamental importance, presenting a great challenge for vision science, is that reflectance, lighting, and shading patterns are perceptually experienced in qualitatively different ways. This means that, firstly, each of reflectance, lighting, and shading patterns alone results in a qualitatively different percept. Specifically, the same light-stimulus array can be obtained using a homogeneous reflectance pattern accompanied by some lighting pattern, or using a non-homogeneous reflectance pattern accompanied by a homogeneous lighting pattern.[3] We will see these two scenes differently. For example, we see a sheet of paper with a painted colour design, and the same paper, with the same design made on its surface by a digital projector, in different ways. Also, our ability to see cast and attached shadows in different ways (Logvinenko, 2015a) indicates that the same lighting pattern yields different percepts depending on whether it arises from spatial variations of illumination or from changing the object shape and orientation. It is to refer to this perceptual difference that we introduced the notions of *material*, *lighting*, and *shading colours*, and *material*, *lighting*, and *shading images* (as the spatial distribution of the corresponding colours) (Section 20.2.2).

Secondly, despite the fact that the reflectance, lighting, and shading patterns are irreversibly entwined in the light-stimulus array, we can see simultaneously the material, lighting, and shading images, each of which, by and large, perceptually represents the reflectance, lighting, and shading patterns respectively.

Thirdly, we see the material, lighting, and shading images not only in such simplest case, as is made up on purpose, as in Eq. (II.4), where the reflectance, lighting, and shading patterns can be separated from each other (thus, if desired, they could be considered as stimulus correlates for their perceptual counterparts), but in any natural scene, no matter how complex it may be. This simple fact, familiar to all from introspection, is a great challenge for colour science.

When dealing with this challenge, the most common approach is (in full keeping with the spirit of psychophysics) to further simplify the experimental paradigm in an attempt to firstly explore the role of each of the three patterns individually, keeping in mind the goal of subsequently beginning to return to natural scenes, as far as possible within the framework of the chosen strategy. In particular, only flat surfaces, usually, are taken into consideration. In this case θ_i in Eq. (II.4) becomes the angle between the surface and the direction of illumination, the entire surface-geometry contribution reducing to one number, $\cos\left(\theta_i(\overline{\alpha})\right)$.

This idealisation is known as the 'flat patchy world hypothesis' (Hurlbert, 1998; Maloney, 1999, 2003), which implies the simplest experimental paradigm for studying the colour perception of objects. Specifically, it assumes a spatially homogeneous atmosphere (medium), and a single dotted distant light source uniformly illuminating a flat matte surface partitioned into 'patches,' the reflective properties of each of which are spatially homogeneous. In such a model, the surface-patch reflectivity turns out to depend on only just one variable, namely, the wavelength. Furthermore, as for the sake of simplicity, the medium (atmosphere) is, usually, supposed to be ideally transparent,

3 See the celebrated exemplification of this situation in Adelson & Pentland (1996).

the illumination proves to be determined only by a sole light source, thus becoming also a function of one variable (i.e., wavelength).

Although this model becomes far too simple (for example, it actually equates illumination and illuminant), we adopt it as a starting point for a number of reasons. (While scenes illuminated by more than one light source – multiple-illuminant scenes – will be given consideration in Chapter 19, until then only single-illuminant scenes will be our focus.) Firstly, it is something that, explicitly or implicitly, lies behind most publications on object colour. Secondly, despite its simplicity, this model leads to a number of interesting mathematical problems. Thirdly, their solution shows that even with such an oversimplified approach the colour of an object cannot be understood as its 'intrinsic' property.

More specifically, in this Part we will consider only such illuminants, which can be formally represented by absolutely continuous positive measures (Example 2.1 b).[4] Any such light, σ, can be identified by its spectral density p, which will be considered as an element of the positive cone $\mathbf{L}^1_+ = \mathbf{L}^1_+(\Lambda)$, so that $\sigma(d\lambda) = p(\lambda)d\lambda$. We shall also assume that the spectral reflectance function for each surface patch $x(\lambda)$ is a positive essentially bounded measurable function, that is, $x(\lambda) \in \mathbf{L}^\infty_+ = \mathbf{L}^\infty_+(\Lambda)$. Thus, denoting the incident light $p(\lambda)$, the light reflected by the object is an absolutely continuous positive measure with the spectral density $p(\lambda)x(\lambda)$. Hence, we come to the following representation of the interaction between the kth surface patch and the illuminant:

$$s_k(\lambda) = p(\lambda)\, x_k(\lambda), \tag{II.5}$$

where $p(\lambda)$ and $s_k(\lambda)$ denote the radiant spectral density functions of the incident and reflected lights respectively; and $x_k(\lambda)$ stands for the spectral reflectance function of the kth patch.

Since fluorescent surfaces are left outside the scope of this inquiry, we take for granted that a surface cannot reflect more light than it receives from the illuminant. In other words, reflecting surfaces can only subtract power at every wavelength of the visible spectrum wavelength interval from the incident light. Thus, the surface spectral reflectance functions $x_k(\lambda)$ in (II.5) are assumed to be larger than or equal to zero, and smaller than or equal to one at each wavelength: $0 \le x_k(\lambda) \le 1$.

4 It means, in particular, that monochromatic illuminants are excluded from consideration.

14

Object-Colour Solid

We will begin with a further simplification, assuming that there are only two surface patches[1] – a test $x_t(\lambda)$ and a background $x_b(\lambda)$. Moreover, we shall assume that the background is black (i.e., $x_b(\lambda) \equiv 0$) so that all the variations of the reflected light are due to the test surface reflectance $x_t(\lambda)$. As $x_t(\lambda)$ is going to be the only spectral reflectance to consider we shall omit the subscripts, $x(\lambda)$ standing for the (test) surface spectral reflectance and $s(\lambda)$ for the radiant spectral density function of the reflected light.

In this chapter, we are concerned with the following issue. What set of the colour signals is produced by all the surfaces provided that the illuminant is fixed? We shall refer to this set of the colour signals as the *object-colour solid* (see the formal Definition 14.1). The object-colour solid is a proper subset of the colour-signal cone. When the incident light is fixed, not every light can be produced by varying the surface which reflects it. As the surface spectral reflectance functions $x(\lambda)$ in (II.5) are assumed to be bounded: $0 \leq x(\lambda) \leq 1$, the radiant power spectral distributions of the reflected light $s(\lambda)$ in (II.5) will always be majorated by $p(\lambda)$. Therefore, in contrast with the colour-signal cone, the object-colour solid must be bounded. Nevertheless, the colour signals of all chromaticities are present in the object-colour solid (see Proposition 14.2).

As above, our consideration will be mainly devoted to linear colour mechanisms $\varphi_1, \dots, \varphi_n$ which will be identified with their spectral sensitivity functions considered as elements of $\mathbf{C}$ or a broader Banach space of measurable functions (say, $\mathbf{L}^\infty$ or $\mathbf{L}^1$). Since we assume throughout this chapter that the illuminant is fixed, this can be taken into account by formal redefining the spectral sensitivity functions by letting

$$\varphi_i'(\lambda) = \varphi_i(\lambda)\, p(\lambda). \tag{14.1}$$

However, as the choice of colour mechanisms is arbitrary, we shall further omit the prime, using the same notations for colour mechanisms as before.[2]

Also the same notation as before will be used for the colour signal, $z = \left(z_1, \dots, z_n\right)$, considered as the output of the (redefined) colour mechanisms φ_i in response to spectral reflectance $x(\lambda)$ (the latter will be referred to as the *object stimulus*):

$$z_i = \left\langle x, \varphi_i \right\rangle = \int_\Lambda \varphi_i(\lambda)\, x(\lambda)\, d\lambda. \tag{14.2}$$

1 The presence of two surfaces is essential. A single spatially (and temporarily) homogeneous surface is, in fact, impossible to perceive (see more on this matter, in connection with the notion of 'Ganzfeld', in Section 20.4).
2 Recall that $\varphi_i p \in \mathbf{L}^1$ whenever $\varphi_i \in \mathbf{L}^\infty$ and $p \in \mathbf{L}^1$ or $\varphi_i \in \mathbf{L}^1$ and $p \in \mathbf{L}^\infty$. Thus, the redefined colour mechanisms are elements of $\mathbf{L}^1$.

Foundations of Colour Science: From Colorimetry to Perception, First Edition. Alexander D. Logvinenko and Vladimir L. Levin.
© 2023 John Wiley & Sons Ltd. Published 2023 by John Wiley & Sons Ltd.

Since we assume that $x \in \mathbf{L}^\infty_+$ and $\varphi_i \in \mathbf{L}^1$, integral (14.2) is well-defined. It defines a linear functional on $\mathbf{L}^\infty$. As a rule, the functions (hence the linear functionals generated by them) $\varphi_1, \dots, \varphi_n$ are assumed to be linearly independent.

14.1 General properties of the object-colour solid

The set of linear functionals $\varphi_1, \dots, \varphi_n$ as defined by (14.2), determines a linear operator (the colour-signal map) $\Phi : \mathbf{L}^\infty \to \mathbf{R}^n, \Phi(x) = z = (z_1, \dots, z_n)$. It follows from its definition that the operator Φ is weak* continuous (that is, continuous with respect to the weak* topology in $\mathbf{L}^\infty = \mathbf{L}^{1*}$).

Definition 14.1 *Let $\mathcal{X}$ stand for the set of the object stimuli, that is, $\mathcal{X} = \{x \in \mathbf{L}^\infty : 0 \le x(\lambda) \le 1\}$. Its Φ-image in $\mathbf{R}^n$, $S = \Phi(\mathcal{X})$ will be called the* object-colour solid *(with regard to a given colour-signal map Φ).*

Definition 14.2 *We say the subset $[0, \Phi(\chi_\Lambda)] := \{t\Phi(\chi_\Lambda) : 0 \le t \le 1\}$ of the object-colour solid is the* achromatic segment.[3] *Its ends, $0 = \Phi(\chi_\emptyset)$ and $z_* := \Phi(\chi_\Lambda)$ are called the* black *and the* white poles *respectively. The point $\frac{1}{2}z_* = \Phi\left(\frac{1}{2}\chi_\Lambda\right)$ is called the* centre of the object-colour solid.

Since $x \in \mathcal{X}$ if and only if $\chi_\Lambda - x \in \mathcal{X}$, it follows that $z = \Phi(x) \in S$ if and only if $z_* - z \in S$. In other words, the object-colour solid is symmetric with regard to its centre: for every $z \in \mathbf{R}^n$ one has $z = \frac{1}{2}z_* + (z - \frac{1}{2}z_*) \in S \Leftrightarrow z_* - z = \frac{1}{2}z_* - (z - \frac{1}{2}z_*) \in S$.

Furthermore, being closed, convex, and norm-bounded subset of a dual Banach space $\mathbf{L}^\infty = \mathbf{L}^{1*}$, $\mathcal{X}$ is weak* compact. And as Φ is a weak* continuous linear operator, the set $S = \Phi(\mathcal{X})$ is a compact convex subset in $\mathbf{R}^n$. Let $Z = Lin(S)$ be the linear subspace in $\mathbf{R}^n$ spanned over S,

$$Z = \{\alpha_1\Phi(x_1) + \cdots + \alpha_n\Phi(x_n) : \alpha_i \in \mathbf{R}, x_i \in \mathcal{X}, i = 1, \dots, n\}.$$

It is clear that the centre of S (and other points of the achromatic segment with the exception of its poles) belongs to the relative interior of S that is, to its interior relative to Z.

The above reasonings are summarized in the following statement.

Proposition 14.1 *Given a colour-signal map $\Phi = (\varphi_1, \dots, \varphi_n)$ where all φ_i are arbitrary elements of $\mathbf{L}^1$, the object-colour solid S is a compact convex subset in $\mathbf{R}^n$. It is symmetric with regard to its centre, and any point of the achromatic segment differing from the poles is a relative interior point of S.*

A question arises, when the object-colour solid S is of full dimension in $\mathbf{R}^n$, that is, its interior in $\mathbf{R}^n$ is non-empty. The following result answers this question.

Theorem 14.1 *Suppose $\varphi_1, \dots, \varphi_n \in \mathbf{C}$; the following statements are then equivalent:*

(i) the functions $\varphi_1, \dots, \varphi_n$ are linearly independent;
(ii) the object-colour solid $S = \Phi(\mathcal{X})$ has a nonempty interior in $\mathbf{R}^n$.

If these equivalent statements hold true, then any point of the achromatic segment other than its poles is an interior point of S.

---

3 Here χ_Λ is the indicator of the wavelength interval Λ (see Eq. (14.3)).

In the next result, we establish a connection between the object-colour solid and the colour-signal cone which was defined and studied in Chapter 4.

Proposition 14.2 *Suppose $\varphi_1, \dots, \varphi_n \in \mathbf{C}$, then the colour-signal cone is exactly the closed conic hull of the object-colour solid:*

$$\Phi(\mathbf{C}_+^*) = cl\ con\ S,$$

where cl stands for the closure and

$$con\ S = \{\alpha z : \alpha \geq 0, z \in S\}.$$

By the Liapunov convexity theorem (e.g., see Holmes, 1975, p. 108), for every $x \in \mathcal{X}$ there is a measurable subset B in Λ such that $\Phi(x) = \Phi\left(\chi_B\right)$, where χ_B stands for the *indicator* (*characteristic function*) of B, that is

$$\chi_B(\lambda) = \begin{cases} 1, & \text{if} \quad \lambda \in B; \\ 0, & \text{if} \quad \lambda \notin B. \end{cases} \tag{14.3}$$

The indicators represent objects which reflect all the light (100% reflectivity) within some particular regions of the visible spectrum wavelength interval $\left[\lambda_{\min}, \lambda_{\max}\right]$ and none (0% reflectivity) outside these regions. While objects with such spectral reflectance functions do not exist in reality, they play an important role in the theoretical analysis. The indicators belong to the boundary of $\mathcal{X}$. Furthermore, they actually constitute the set of extreme points of $\mathcal{X}$ (written as $Ext(\mathcal{X})$).

Not every indicator is mapped by Φ into the boundary[4] of the object-colour solid (written as ∂S). Many of them will be mapped inside the object-colour solid. Since knowing the boundary of the object-colour solid allows us to restore the whole solid, the description of those indicators which are mapped into ∂S is of particular interest.

Definition 14.3 *Given a colour-signal map Φ, indicators mapped by Φ on the boundary of the object-colour solid will be called* optimal object stimuli *(with respect to the colour-signal map Φ). The set of optimal object stimuli with respect to the colour-signal map Φ will be denoted by $Opt\ (\mathcal{X} | \Phi\ )$, or $Opt(\mathcal{X})$ if it is clear which colour-signal map is meant.*

In the following section, we show how to evaluate the optimal object stimuli.

14.2 Optimal object stimuli

The following lemma provides a useful geometrical criterion for the fact that the indicator $x_0 \in \mathcal{X}$ is optimal object stimulus. Let $\ker(\Phi)$ be the kernel (null-space) of the colour-signal map Φ,

$$\ker(\Phi) := \{x \in \mathbf{L}^\infty : \Phi(x) = 0\}.$$

It is clear that

$$\ker(\Phi) = \bigcap_{i=1}^{n} \ker\left(\varphi_i\right),$$

where $\ker(\varphi_i) = \{x \in \mathbf{L}^\infty : \langle \varphi_i, x \rangle = 0\}$.

4 When $Z = Lin(S)$ is a proper subspace in $\mathbf{R}^n$, the interior and the boundary of S are considered relative to Z.

Lemma 14.1 *An indicator $x_0 \in \mathcal{X}$ is an optimal object stimulus if and only if the affine subspace passing through x_0 parallel to $\ker(\Phi)$ (written as $x_0 + \ker(\Phi)$) has no common points with the interior of $\mathcal{X}$:*

$$\Phi\left(x_0\right) \in \partial S = \partial\Phi(\mathcal{X}) \Leftrightarrow \left(x_0 + \ker(\Phi)\right) \cap int(\mathcal{X}) = \emptyset. \tag{14.4}$$

The following theorem provides a complete description of the set of optimal object stimuli.

Theorem 14.2 *An indicator $x_0 = \chi_B$ is an optimal object stimulus if and only if there exist numbers $\xi_1, \dots, \xi_n \in \mathbf{R}$ such that at least one of them is non-zero, and*

$$\sum_{i=1}^{n} \xi_i \varphi_i(\lambda) \geq 0 \quad a.e. \text{ on } B, \tag{14.5}$$

$$\sum_{i=1}^{n} \xi_i \varphi_i(\lambda) \leq 0 \quad a.e. \text{ on } \Lambda \backslash B. \tag{14.6}$$

Here and below, the abbreviation *a.e.* stands for 'almost everywhere relative to the Lebesgue measure on Λ'.

Corollary 14.1 *Let $\Phi = \{\varphi_1, \dots, \varphi_n\}$ be a colour map, $0 < k < n$, and $\Phi' = \{\varphi_{i_1}, \dots, \varphi_{i_k}\}$ is a subset of Φ. If an indicator χ_B is an optimal object stimulus with respect to the colour map Φ', i.e., $\chi_B \in Opt\left(\mathcal{X}\,|\Phi'\right)$, then $\chi_B \in Opt(\mathcal{X}\,|\Phi)$.*

It follows from Theorem 14.2 that any non-zero vector in $\mathbf{R}^n$ determines, generally, a set of optimal object stimuli. Indeed, consider a non-zero vector $\left(\xi_1, \dots, \xi_n\right) \in \mathbf{R}^n$ and let us designate $B_+\left(\xi_1, \dots, \xi_n\right) = \left\{\lambda : \sum_{i=1}^n \xi_i\varphi_i(\lambda) > 0\right\}$ and $B_0\left(\xi_1, \dots, \xi_n\right) = \left\{\lambda : \sum_{i=1}^n \xi_i\varphi_i(\lambda) = 0\right\}$. Then, the indicator of any set $B\left(\xi_1, \dots, \xi_n\right) = B_+\left(\xi_1, \dots, \xi_n\right) \cup B_0'\left(\xi_1, \dots, \xi_n\right)$, where $B_0'(\xi_1, \dots, \xi_n) \subseteq B_0\left(\xi_1, \dots, \xi_n\right)$, will be an optimal object stimulus. This shows how to practically generate optimal object stimuli. Given a non-zero vector $\left(\xi_1, \dots, \xi_n\right)$, we evaluate the sets $B_+\left(\xi_1, \dots, \xi_n\right)$, $B_0\left(\xi_1, \dots, \xi_n\right)$, and $B_-\left(\xi_1, \dots, \xi_n\right) = \{\lambda : \sum_{i=1}^n \xi_i\varphi_i(\lambda) < 0\}$. Then, any indicator which equals 1 on $B_+\left(\xi_1, \dots, \xi_n\right)$ and 0 on $B_-\left(\xi_1, \dots, \xi_n\right)$ will be an optimal object stimulus. It should be mentioned that for any $\theta > 0$ $B_+\left(\xi_1, \dots, \xi_n\right) = B_+\left(\theta\xi_1, \dots, \theta\xi_n\right)$ and $B_0\left(\xi_1, \dots, \xi_n\right) = B_0\left(\theta\xi_1, \dots, \theta\xi_n\right)$; hence, when generating optimal object stimuli, one can restrict oneself to only vectors $\left(\xi_1, \dots, \xi_n\right)$ Euclidean length of which $\|\left(\xi_1, \dots, \xi_n\right)\|$ equals 1.

The subset $B_0\left(\xi_1, \dots, \xi_n\right)$ has a simple geometrical meaning: it is the set of wavelengths λ for which the spectral colour-signal curve $\Phi\left(\delta_\lambda\right) = \left(\varphi_1(\lambda), \dots, \varphi_n(\lambda)\right)$ intersects the hyperplane $H(\xi_1, \dots, \xi_n; 0)$ defined by the following equation in the colour-signal space:

$$H(\xi_1, \dots, \xi_n; 0) = \left\{\left(\xi_1, \dots, \xi_n\right) : \xi_1 z_1 + \cdots + \xi_n z_n = 0\right\}. \tag{14.7}$$

Likewise, $B_+\left(\xi_1, \dots, \xi_n\right)$ and $B_-\left(\xi_1, \dots, \xi_n\right)$ are such the subsets in Λ that the spectral colour-signal curve $\Phi\left(\delta_\lambda\right) = \left(\varphi_1(\lambda), \dots, \varphi_n(\lambda)\right)$ remains off the same side of the hyperplane $H(\xi_1, \dots, \xi_n; 0)$.

Note that the subsets $\left\{\lambda : \sum_{i=1}^n \xi_i\varphi_i(\lambda) \geq 0\right\}$ and $\left\{\lambda : \sum_{i=1}^n \xi_i\varphi_i(\lambda) \leq 0\right\}$ remain the same if the functions φ_i are divided by a function everywhere positive on Λ. Therefore, if the radiant power of the incident light $p(\lambda)$ is everywhere positive on Λ (which is a safe assumption for most broad band natural illuminants), the optimal object stimuli are solely determined by the spectral sensitivity functions $\varphi_1(\lambda), \dots, \varphi_n(\lambda)$. In other words, the optimal object stimuli are independent of the illuminant provided that its radiant power function is everywhere positive. Furthermore, for the same reason the set of the optimal object stimuli does not depend on the macular pigment and lens (Figs 3.7 and 3.6), or any other pre-receptoral filter (including medium (atmosphere)), which affects the incident light equally for all photoreceptors.

It is clear that the set of optimal object stimuli is determined by the linear subspace $Lin(\varphi_1, \dots, \varphi_n)$ rather than the spectral sensitivity functions $\varphi_1, \dots, \varphi_n$ as such. Any n linear independent functions from $Lin(\varphi_1, \dots, \varphi_n)$ can be used to derive the set of optimal object stimuli. Therefore, the colour matching functions are as good as the cone fundamentals for evaluating the optimal object stimuli provided that the cone-fundamentals hypothesis (Section 3.3) holds true. If the cone-fundamentals hypothesis fails, and the method of tangential hyperplane should be used to evaluate the colour-signal map, then any n linear independent colour mechanisms obtained by this method can be used to calculate the optimal object stimuli.

If we presume that the linear colour mechanisms are followed by local non-linearities as suggested by (4.1), or, more generally, that colour signals are transformed by a continuous invertible map $G : \mathbf{R}^n_+ \to \mathbf{R}^n_+$, then the shape of the object-colour solid will be transformed by G. However, the set of the optimal object stimuli will remain the same since the object-colour solid boundary will be mapped by G onto the boundary of the new object-colour solid. Therefore, the optimal object stimuli, and only they, will be mapped onto the boundary of the new object-colour solid.

14.3 Elementary step functions as optimal object stimuli

When the spectral colour-signal curve $\Phi(\delta_\lambda)$ intersects each hyperplane $H(\xi_1, \dots, \xi_n; 0)$ in (14.7) at a finite number of points, the optimal object stimuli have very simple form provided that the illuminant has everywhere-positive continuous spectrum. Specifically, they have the form of step functions defined as follows.

Definition 14.4 *We say that the spectral reflectance function is an* elementary step function *of degree 0 if it takes the same value, either 0 or 1, on the whole segment* $[\lambda_{\min}, \lambda_{\max}]$. *Spectral reflectance functions*

$$x_1(\lambda; \lambda_1) = \begin{cases} 0, & \text{if} \quad \lambda < \lambda_1; \\ 1, & \text{if} \quad \lambda \geq \lambda_1 \end{cases} \tag{14.8}$$

and $1 - x_1(\lambda; \lambda_1)$ *will be called elementary step functions of degree 1. Spectral reflectance functions* $x_m(\lambda; \lambda_1, \dots, \lambda_m) = \sum_{i=1}^{m} (-1)^{i-1} x_1(\lambda; \lambda_i)$ *and* $1 - x_m(\lambda; \lambda_1, \dots, \lambda_m)$, *where* $\lambda_{\min} < \lambda_1 < \lambda_2 < \dots < \lambda_m < \lambda_{\max}$, *will be called the elementary step functions of degree m,* $\lambda_1, \dots, \lambda_m$ *being referred to as* transition points. *The elementary step functions* $x(\lambda) \equiv 1$, *and* $x_m(\lambda; \lambda_1, \dots, \lambda_m)$ $(m \geq 1)$ *will be referred to as those of* type I, *the rest (i.e.,* $x(\lambda) \equiv 0$, *and* $1 - x_m(\lambda; \lambda_1, \dots, \lambda_m)$ $(m \geq 1)$*) as* type II.

Example 14.1 shows that the number of colour mechanisms imposes, generally, no restriction on the degree of elementary step functions that can serve as optimal object stimuli.

Example 14.1 *Let us consider n colour mechanisms, the spectral sensitivity functions of which are defined as follows:* $\varphi_i(\lambda) = -a_i(\lambda - \lambda_i)^{2m} + b_i$, $a_i > 0, b_i > 0, i = 1, \dots, n$. *To secure the positivity of* φ_i *on* Λ, *we assume that* $b_i \geq a_i \max\left\{(\lambda_{\min} - \lambda_i)^{2m}, (\lambda_{\max} - \lambda_i)^{2m}\right\}$. *Since* $\sum_{i=1}^{n} \xi_i \varphi_i(\lambda)$ *is a polynomial of degree 2m, it has 2m zeros, which can be made real, different, and lying in* Λ *by properly adjusting the parameters* a_i, b_i, *and* λ_i. *Therefore, in principle, an elementary step function of degree 2m can be an optimal object stimulus for such colour mechanisms. As n and m are arbitrary positive integers, there is no connection between them.*

Schrödinger (1920) implicitly, and later Maximov (1984) explicitly assumed that the human spectral sensitivity functions $\varphi_1, \dots, \varphi_n$ satisfy a certain condition that limits the maximal degree of elementary step functions in $Opt(\mathcal{X} \mid \Phi)$. In what follows, $\varphi_1, \dots, \varphi_n$ are supposed to be continuous functions of λ.

Condition 14.3 *We say that colour mechanisms* $\varphi_1, \dots, \varphi_n$ *meet the* Schrödinger–Maximov constraint *if, for every set of different* $\lambda_1, \dots, \lambda_n \in \Lambda$,

$$\det\left(\varphi_i(\lambda_j)\right) \neq 0, \quad i, j = 1, \dots, n. \tag{14.9}$$

(Here $\det(\varphi_i(\lambda_j))$ *stands for the determinant of the matrix* $(\varphi_i(\lambda_j))$*).*

As follows from Lemma 2.1, the Schrödinger–Maximov constraint means that, when being considered as functions on *every* subset in Λ which contains at least n points, $\varphi_1, \dots, \varphi_n$ are linearly independent.

The Schrödinger–Maximov constraint guarantees that, for every non-zero vector $\left(\xi_1, \dots, \xi_n\right) \in \mathbf{R}^n$, the equation

$$\sum_{i=1}^{n} \xi_i \varphi_i(\lambda) = 0 \tag{14.10}$$

has at most $n-1$ solutions on the segment Λ. In other words, any hyperplane in $\mathbf{R}^n$ passing through the origin has at most $n-1$ common points with the spectral colour-signal curve. It follows that, for every non-zero vector $\left(\xi_1, \dots, \xi_n\right) \in \mathbf{R}^n$, the indicator $\chi_{B(\xi_1,\dots,\xi_n)}$, where $B\left(\xi_1, \dots, \xi_n\right) = \left\{\lambda : \sum_{i=1}^{n} \xi_i \varphi_i(\lambda) \geq 0\right\}$, is an elementary step function of degree $m \leq n-1$ with the transition points being solutions of (14.10). Any Chebyshev system of functions satisfies the Schrödinger–Maximov constraint. However, it is a rather strong constraint; so one can hardly expect that the set of biologically plausible spectral sensitivity functions will meet it.

14.4 Optimal object stimuli for trichromatic human observers

Since the pre-receptoral filters do not affect the set of optimal object stimuli, for trichromatic human observers, it is fully determined by the spectral sensitivity of their cone photopigments (Fig. 3.5). Thus, a good insight into the optimal object stimuli of the trichromatic human observers can be made by studying the optimal object stimuli for a 3-pigment Grassmann structure (Definition 3.2). These can be evaluated by substituting $\varphi_i(\lambda)$ in (14.10) with $g(\lambda; \lambda^{\max})$ from Eq. (3.17), and solving then the equation with respect to λ.

Specifically, the transition wavelengths of all the optimal stimuli for trichromatic colour vision are determined explicitly by the following equation:

$$\xi_1 g\left(\lambda; \lambda_S^{\max}\right) + \xi_2 g\left(\lambda; \lambda_M^{\max}\right) + \xi_3 g\left(\lambda; \lambda_L^{\max}\right) = 0, \tag{14.11}$$

where ξ_1, ξ_2, and ξ_3 are arbitrary real numbers; $g(\lambda; \lambda^{\max})$ is the spectral sensitivity of the cone photopigments (Eq. (3.17)); and $\lambda_S^{\max}$, $\lambda_M^{\max}$, and $\lambda_L^{\max}$ are the peak absorbance wavelengths for the S-, M-, and L-cones respectively. For each triplet ξ_1, ξ_2, and ξ_3, the zero-crossings in the graph of the function in Eq. (14.11) specify the transition wavelengths of some optimal stimulus.

The number of solutions for such equation amounts to the number of intersections which the hyperplane in the colour signal space (defined by the photopigment spectral sensitivities) determined by vector $\left(\xi_1, \xi_2, \xi_3\right)$ has with the spectral colour-signal curve (defined by the photopigment spectral sensitivities). This hyperplane (passing through the origin) determines a line in a chromaticity plane, the number of solutions being equal to the number of intersections which this line makes with the spectral locus.

Figure 4.6 depicts a chromaticity diagram for Govardovskii's cone fundamentals with the peak absorbance wavelength $\lambda^{\max}$ at 430, 530, and 560 nm, D being equal to 0.3. As one can see in Fig. 4.6, there exist many straight lines intersecting the spectral locus in more than two points. Specifically, a line passing through one of the two 'beaks' and a point belonging to the effective visible spectrum intersects the spectral locus in three points. A line passing through both the two 'beaks' (i.e., containing two points in both the tails outside the effective visible spectrum) intersects the spectral locus in four points.

14.5 Condition for all step functions of degree $m < n$ to be optimal object stimuli

Theorem 14.2 along with the Schrödinger–Maximov constraint guarantees that every optimal object stimulus is an elementary step function of degree $m < n$; however, one cannot be sure that every such a function is an optimal object stimulus. Indeed, for example, it can occur that, for a given non-zero vector $(\xi_1, \dots, \xi_n) \in \mathbf{R}^n$, $\lambda = \lambda_1$ is the unique solution to (14.10), and the curve $\{\sum_1^n \xi_i \varphi_i(\lambda) : \lambda_{\min} \leq \lambda \leq \lambda_{\max}\}$ is tangent to the abscissa axis at λ_1 but does not intersect it; hence, the set $B(\xi_1, \dots, \xi_n) = \{\lambda : \sum_1^n \xi_i \varphi_i(\lambda) \geq 0\}$ is the whole segment Λ. Moreover, if this holds for every non-zero $(\xi_1, \dots, \xi_n) \in \mathbf{R}^n$ such that λ_1 is the unique solution to (14.10), then the elementary step functions $x_1(\lambda; \lambda_1)$ and $1 - x_1(\lambda; \lambda_1)$ are not optimal object stimuli.

For *all* the elementary step functions of degree $m < n$ to be optimal object stimuli, the Schrödinger–Maximov constraint must be replaced with a stronger regularity assumption. Since trichromatic colour vision is of the main interest for us, the assumption will be formulated in detail for the case $n = 3$ only.

Definition 14.5 *Let $n = 3$. We say that the* regularity assumption *is satisfied if: (i) the Schrödinger–Maximov constraint holds true, (ii) the functions $\varphi_1(\lambda)$, $\varphi_2(\lambda)$, $\varphi_3(\lambda)$ are continuously differentiable, and (iii) for any two different λ_1, λ_2 in Λ, the three-dimensional vectors*

$$(\varphi_1(\lambda_1), \varphi_2(\lambda_1), \varphi_3(\lambda_1)),$$
$$(\varphi_1(\lambda_2), \varphi_2(\lambda_2), \varphi_3(\lambda_2)),$$
$$\left(\frac{d\varphi_1}{d\lambda}(\lambda_1), \frac{d\varphi_2}{d\lambda}(\lambda_1), \frac{d\varphi_3}{d\lambda}(\lambda_1)\right)$$

are linearly independent. (Note that (iii) means

$$\det \begin{pmatrix} \varphi_1(\lambda_1) & \varphi_1(\lambda_2) & \frac{d\varphi_1(\lambda_1)}{d\lambda} \\ \varphi_2(\lambda_1) & \varphi_2(\lambda_2) & \frac{d\varphi_2(\lambda_1)}{d\lambda} \\ \varphi_3(\lambda_1) & \varphi_3(\lambda_2) & \frac{d\varphi_3(\lambda_1)}{d\lambda} \end{pmatrix} \neq 0, \quad \det \begin{pmatrix} \varphi_1(\lambda_1) & \varphi_1(\lambda_2) & \frac{d\varphi_1(\lambda_2)}{d\lambda} \\ \varphi_2(\lambda_1) & \varphi_2(\lambda_2) & \frac{d\varphi_2(\lambda_2)}{d\lambda} \\ \varphi_3(\lambda_1) & \varphi_3(\lambda_2) & \frac{d\varphi_3(\lambda_2)}{d\lambda} \end{pmatrix} \neq 0$$

whenever $\lambda_1 \neq \lambda_2$.) The linear colour mechanisms meeting the regularity assumption (as well as the observer equipped with such mechanisms) will be called regular.

Given $\lambda_1, \lambda_2 \in \Lambda$, $\lambda_1 < \lambda_2$, the indicator of the set $[\lambda_1, \lambda_2]$, considered as an element of the function space $\mathbf{L}^\infty = \mathbf{L}^\infty(\Lambda)$, will be denoted by $x_2^{\lambda_1, \lambda_2}$. This is the elementary step function of degree 2,

$x_2(\lambda; \lambda_1, \lambda_2)$, if $\lambda_{min} < \lambda_1 < \lambda_2 < \lambda_{max}$, and this is the elementary step function of degree 1, $x_1(\lambda; \lambda_1)$ or $1 - x_1(\lambda; \lambda_2)$, if respectively $\lambda_2 = \lambda_{max}$ or $\lambda_1 = \lambda_{min}$.[5] Finally, this is the elementary step function of degree 0, $\chi_\Lambda = 1$ if $\lambda_1 = \lambda_{min}, \lambda_2 = \lambda_{max}$.

Proposition 14.3 *Suppose that $n = 3$ and the regularity assumption is satisfied. Then all elementary step functions of degree $m \leq 2$ are the optimal object stimuli, and the boundary of the object-colour solid, ∂S, consists of two mutually symmetric (relative to the centre $\frac{1}{2}z_*$) parts, $\partial S = Q \cup Q'$, where Q comprises all the colour signals $\Phi(x_2^{\lambda_1,\lambda_2})$ and Q' comprises all the colour signals $\Phi(1 - x_2^{\lambda_1,\lambda_2})$, $\lambda_1 < \lambda_2$.*

Note that $Q \cap Q'$ comprises the colour signals produced by all elementary step functions of degree 1.

5 Generally speaking, the function $1 - x_1(\lambda; \lambda_2)$ and the indicator of $[\lambda_{min}, \lambda_2]$ differ at λ_2 (see Eq. (14.8)). However, since elements of $\mathbf{L}^\infty$ are functions defined up to their values on Lebesgue negligible sets (see Appendix A.1), they both define the same optimal stimulus.

15

Trichromatic Regular Object-Colour Solid

Given a trichromatic colour-signal map Φ, $\Phi(x_2^{\lambda_1,\lambda_2}) \cup \Phi(1 - x_2^{\lambda_1,\lambda_2})$ generates a volume in $\mathbf{R}^3$ which we shall call the *trichromatic regular object-colour solid* for Φ. The set $\Phi(x_2^{\lambda_1,\lambda_2}) \cup \Phi(1 - x_2^{\lambda_1,\lambda_2})$ itself servers as the boundary of the regular object-colour solid. It is clear that the regular object-colour solid is included in the proper object-colour solid, that is, $\Phi(x_2^{\lambda_1,\lambda_2}) \cup \Phi(1 - x_2^{\lambda_1,\lambda_2}) \subseteq \Phi(Opt(\mathcal{X}|\Phi))$. Obviously, for regular observers (Definition 14.5) the regular object-colour solid coincides with the proper object-colour solid.

When the colour mechanisms $\varphi_1, \dots, \varphi_n$ are linearly independent the regular object-colour solid has a nonempty interior in $\mathbf{R}^n$. Moreover, for Govardovskii's cone fundamentals, the boundaries of the regular and proper object-colour solids are so close to each other that for practical purposes the regular object-colour solid can be considered as a good approximation to the proper object-colour solid (Logvinenko, 2009b).

Thus, the notion of regular object-colour solid will have twofold meaning depending on the context. It is either the object-colour solid for a regular observer, or the volume, delimited by $\Phi(x_2^{\lambda_1,\lambda_2}) \cup \Phi(1 - x_2^{\lambda_1,\lambda_2})$, inside the object-colour solid in the case when the colour-signal map Φ does not satisfies the regularity assumption (Definition 14.5). As the concept of regular object-colour solid will play an important role in our further analysis, we will consider in detail its geometrical properties.

15.1 Meridians of the trichromatic regular object-colour solid

Let, for every $\lambda_1 \in \Lambda$, $x_1^{\lambda_1}$ denote the element of the function space $\mathbf{L}^\infty = \mathbf{L}^\infty(\Lambda)$ which represents the elementary step function of degree 1, $x_1(\lambda; \lambda_1)$, if $\lambda_{\min} < \lambda_1 < \lambda_{\max}$, or the elementary step function of degree 0, $\chi_\Lambda = 1$ or $\chi_\emptyset = 0$, if, respectively, $\lambda_1 = \lambda_{\min}$ or $\lambda_1 = \lambda_{\max}$. The colour signals $\Phi(x_1^{\lambda_1})$ constitute a continuous curve $g : \Lambda \to \partial S$ on the regular object-colour solid boundary[1]

$$g(\lambda_1) := \Phi(x_1^{\lambda_1}) = \left(\int_{\lambda_1}^{\lambda_{\max}} \varphi_1(\lambda)\, d\lambda, \int_{\lambda_1}^{\lambda_{\max}} \varphi_2(\lambda)\, d\lambda, \int_{\lambda_1}^{\lambda_{\max}} \varphi_3(\lambda)\, d\lambda \right). \tag{15.1}$$

This curve will be referred to as the *main meridian* (Fig. 15.1). Its ends, $g(\lambda_{\min})$ and $g(\lambda_{\max})$, are respectively the white and the black poles of the regular object-colour solid. All the points of the main meridian except its ends lie in $Q \cap Q'$.

1 We will use the same notations for the regular object-colour solid and its boundary as for their ordinary counterparts, S and ∂S, when it is clear from the context what is meant.

Foundations of Colour Science: From Colorimetry to Perception, First Edition. Alexander D. Logvinenko and Vladimir L. Levin.
© 2023 John Wiley & Sons Ltd. Published 2023 by John Wiley & Sons Ltd.

Also, we shall consider a continuous curve $g' : \Lambda \to \partial S$ on the regular object-colour solid boundary, which is symmetric to g relative to the centre of S, $\frac{1}{2}z_* = \frac{1}{2}\Phi(1)$,

$$g'(\lambda_1) := \Phi(1 - x_1^{\lambda_1}) = z_* - g(\lambda_1).$$

This curve will be called the *opposite meridian*. Its ends, $g'(\lambda_{\min})$ and $g'(\lambda_{\max})$, are respectively the black and the white poles of the regular object-colour solid. Apart from its ends, all the points of the opposite meridian lie in $Q \cap Q'$. The white and the black poles are the only points that belong to both the meridians (Fig. 15.2).

The meridians g and g' generate two one-parameter families of (mutually symmetric) curves, $g_{\lambda_2} : [\lambda_{\min}, \lambda_2] \to \partial S$ and $g'_{\lambda_2} : [\lambda_{\min}, \lambda_2] \to \partial S$, where the parameter λ_2 ranges over Λ,

$$\begin{aligned} g_{\lambda_2}(\lambda_1) &:= g(\lambda_1) - g(\lambda_2) = \Phi\left(x_2^{\lambda_1,\lambda_2}\right) \\ &= \left(\int_{\lambda_1}^{\lambda_2} \varphi_1(\lambda)\, d\lambda, \int_{\lambda_1}^{\lambda_2} \varphi_2(\lambda)\, d\lambda, \int_{\lambda_1}^{\lambda_2} \varphi_3(\lambda)\, d\lambda\right), \end{aligned} \tag{15.2}$$

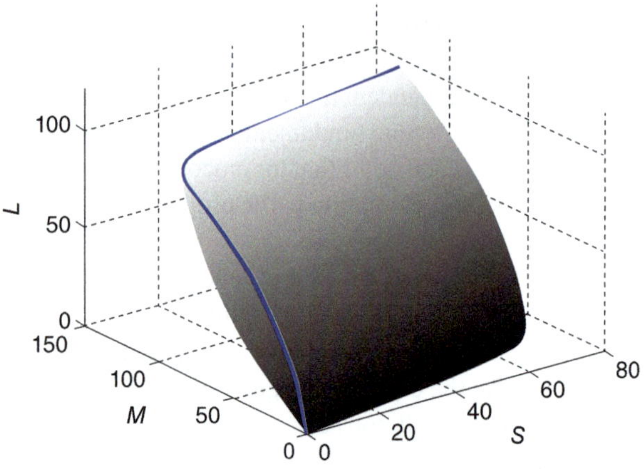

Figure 15.1 The main meridian of the regular object-colour solid.

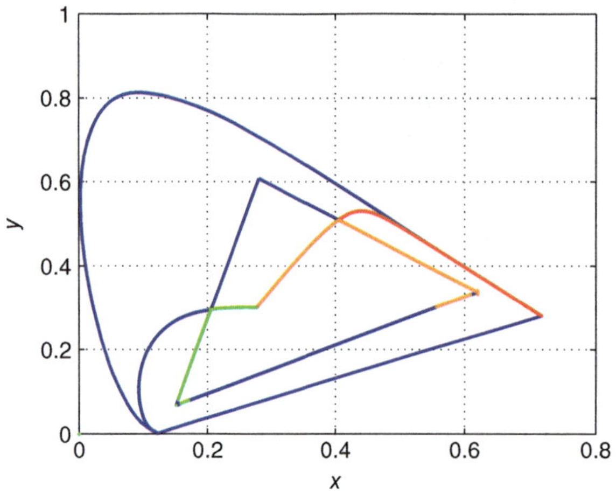

Figure 15.2 CIE 1931 chromaticity coordinates of the main (red line) and opposite (blue line) meridian colours. They were simulated using a colour monitor (see Figs 15.3 and 15.4), the colour-gamut boder of which is represented by a triangle. The part of the main (respectively, opposite) meridian that falls within the monitor colour-gamut is marked in orange (respectively, cyan).

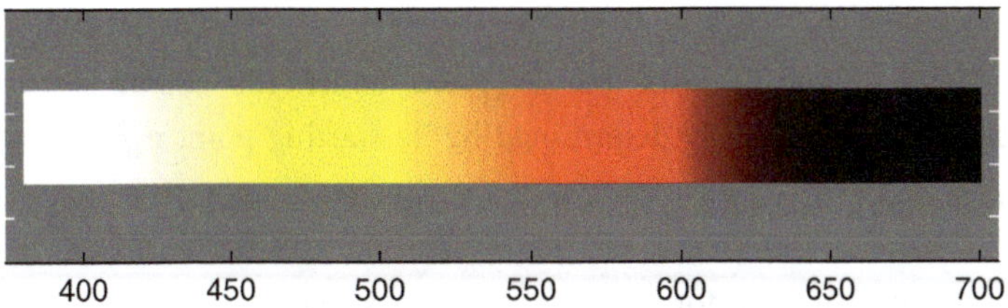

Figure 15.3 Simulation of the main-meridian colours on the monitor screen. Those colours which are out of the monitor colour-gamut were replaced with the closest to the achievable colours (the orange fragments of the monitor colour-gamut boder in Fig. 15.2). The horizontal axis is the wavelength in nm.

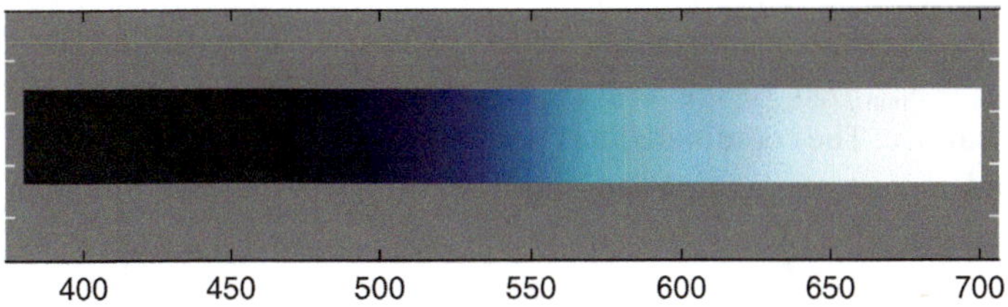

Figure 15.4 Simulation of the opposite-meridian colours on the monitor screen. Those colours which are out of the monitor colour-gamut were replaced with the closest to the achievable colours (the cyan fragment of the monitor colour-gamut boder in Fig. 15.2). The horizontal axis is the wavelength in nm.

$$\begin{aligned} g'_{\lambda_2}(\lambda_1) &: = g'(\lambda_1) + g(\lambda_2) = \Phi\left(1 - x_2^{\lambda_1,\lambda_2}\right) \\ &= \left(\int_{\lambda_{\min}}^{\lambda_1} \varphi_1(\lambda)\, d\lambda,\ \int_{\lambda_{\min}}^{\lambda_1} \varphi_2(\lambda)\, d\lambda,\ \int_{\lambda_{\min}}^{\lambda_1} \varphi_3(\lambda)\, d\lambda\right) \\ &+ \left(\int_{\lambda_2}^{\lambda_{\max}} \varphi_1(\lambda)\, d\lambda,\ \int_{\lambda_2}^{\lambda_{\max}} \varphi_2(\lambda)\, d\lambda,\ \int_{\lambda_2}^{\lambda_{\max}} \varphi_3(\lambda)\, d\lambda\right). \end{aligned} \tag{15.3}$$

It follows from (15.2) that the curve $\{g_{\lambda_2}(\lambda_1) : \lambda_{\min} \le \lambda_1 \le \lambda_2\}$ may be considered as a fragment of the main meridian (from $g(\lambda_{\min})$ to $g(\lambda_2)$), which is displaced by the vector $-g(\lambda_2)$. Similarly, it follows from (15.3) that the curve $\{g'_{\lambda_2}(\lambda_1) : \lambda_{\min} \le \lambda_1 \le \lambda_2\}$ may be considered as a fragment of the opposite meridian (from $g'(\lambda_{\min})$ to $g'(\lambda_2)$), which is displaced by the vector $g(\lambda_2)$.

When λ_1 varies from $\lambda_{\min}$ to λ_2, $g_{\lambda_2}(\lambda_1)$ goes from the colour signal $g'(\lambda_2)$ produced by the object stimulus $1 - x_1(\lambda; \lambda_2)$ to the black pole 0, and $g'_{\lambda_2}(\lambda_1)$ goes from the colour signal $g(\lambda_2)$ produced by the object stimulus $x_1(\lambda; \lambda_2)$ to the white pole z_*.

Thus, every curve g_{λ_2} is congruent to a fragment of the main meridian corresponding the values of λ_1 from $\lambda_{\min}$ to λ_2. The curve starts from the point $g'(\lambda_2)$ of the opposite meridian and ends at the black pole. Similarly, every curve g'_{λ_2} is congruent to a fragment of the opposite meridian corresponding the same values of λ_1. The curve starts from the point $g(\lambda_2)$ of the main meridian and ends at the white pole.

The colour signals of elementary step functions, $x_2^{\lambda_1,\lambda_2}$ and $1 - x_2^{\lambda_1,\lambda_2}$, $(\lambda_1 < \lambda_2)$, can be represented, respectively, as $g_{\lambda_2}(\lambda_1)$ and $g'_{\lambda_2}(\lambda_1)$. When λ_2 varies from $\lambda_{\min}$ to $\lambda_{\max}$ the curves $\{g_{\lambda_2}(\lambda_1) : \lambda_{\min} \le \lambda_1 \le \lambda_2\}$ fill in the part $Q \cup \{0\}$ of the boundary surface of the regular object-colour solid, while the curves $\{g'_{\lambda_2}(\lambda_1) : \lambda_{\min} \le \lambda_1 \le \lambda_2\}$ fill in the other (symmetric) part $Q' \cup \{z_*\}$ of the boundary surface.

For any elementary step function of the form $x_2^{\lambda_1,\lambda_2} := x_2(\lambda; \lambda_1, \lambda_2)$, $g_{\lambda_2}(\lambda_1)$ specifies two coordinates on Q, λ_1 and λ_2. Here, λ_2 specifies the curve g_{λ_2} on which the colour signal $\Phi\left(x_2^{\lambda_1,\lambda_2}\right) = g_{\lambda_2}(\lambda_1)$ lies (this curve, in turn, is fully determined by its starting point $g_{\lambda_2}(\lambda_{\min}) = g'(\lambda_2)$ which belongs to the opposite meridian), and λ_1 determines the position of $\Phi\left(x_2^{\lambda_1,\lambda_2}\right)$ at the curve.

Likewise, for any elementary step function of the form $1 - x_2^{\lambda_1,\lambda_2}$, $g'_{\lambda_2}(\lambda_1)$ specifies two coordinates on Q', λ_1 and λ_2. Here, λ_2 specifies the curve g'_{λ_2} on which the colour signal $\Phi\left(1 - x_2^{\lambda_1,\lambda_2}\right) = g'_{\lambda_2}(\lambda_1)$ lies (this curve, in turn, is fully determined by its starting point $g'_{\lambda_2}(\lambda_{\min}) = g(\lambda_2)$ which belongs to the main meridian), and λ_1 determines the position of $\Phi\left(1 - x_2^{\lambda_1,\lambda_2}\right)$ at the curve.

Figure 15.5 presents the regular object-colour solid for Govardovskii's cone fundamentals. The curves on its boundary surface are the lines of constant λ_1 and λ_2. The space between the coordinate lines corresponds to $(\lambda_{\max} - \lambda_{\min})/30$. As one can see the coordinate lines are not evenly distributed across the boundary surface. The coordinate lines become equally spaced after natural reparametrisation of the main and opposite meridians. The length s of the arc of the main meridian $g(\lambda)$ (15.1) between the north pole $\left(\Phi\left(x_1^{\lambda_{\min}}\right)\right)$ and the point $\Phi(x_1^{\lambda})$ $(\lambda_{\min} \leq \lambda \leq \lambda_{\max})$ is given by

$$s(\lambda) = \int_{\lambda_{\min}}^{\lambda} \sqrt{(\varphi_1(\mu))^2 + (\varphi_2(\mu))^2 + (\varphi_3(\mu))^2}\, d\mu, \tag{15.4}$$

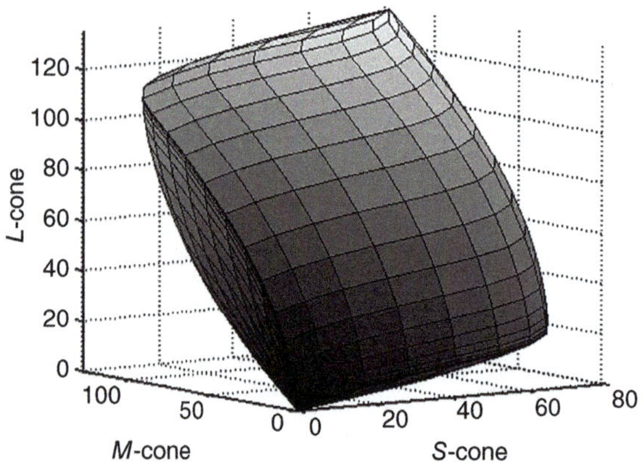

Figure 15.5 Regular object-colour solid as evaluated for Govardovskii's fundamentals with the peak sensitivities at $\lambda = 420$, 530 and 560 nm, for the Plankian radiator (6500 K).

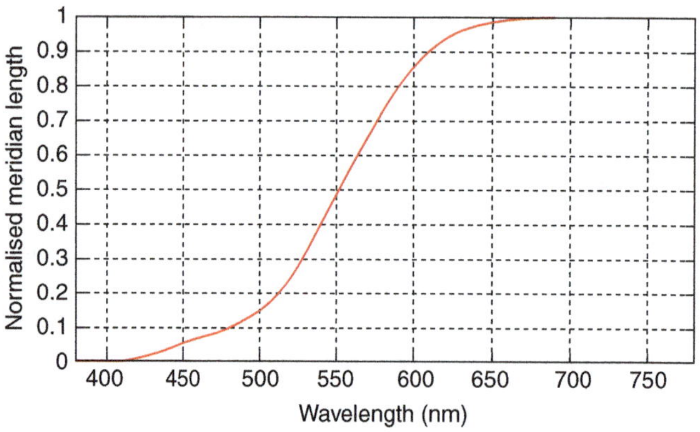

Figure 15.6 Normalised length of the main meridian for the regular object-colour solid in Fig. 15.5.

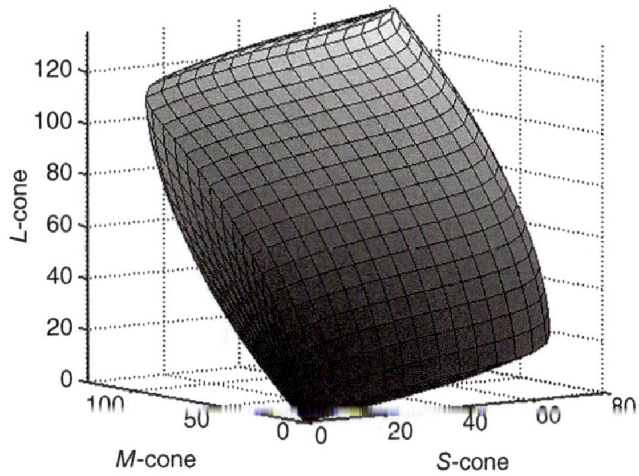

Figure 15.7 Regular object-colour solid, the same as in Fig. 15.5, with evenly spaced coordinate lines.

and presented in Fig. 15.6. As one can see, the meridian length does not practically differ from zero for $\lambda < 420$ nm (specifically, $s(420) = 0.01$), and it reaches its maximum after 650 nm ($s(650) = 0.99$).

In Fig. 15.7, the same colour solid as in Fig. 15.5 is presented with the coordinate lines equally spaced in terms of the meridian length parameter s. The space between the coordinate lines Fig. 15.7 corresponds to $s(\lambda_{\max})/30$.

15.2 Equator of the trichromatic object-colour solid and strictly optimal object stimuli

In this section, we introduce and study an important subset of the object-colour solid which will be referred to as its equator.[2] Let us consider in $\mathbf{R}^n$ various hyperplanes parallel to the achromatic segment $[0, \Phi(\chi_\Lambda)] = \{tz_* : 0 \le t \le 1\}$, where $z_* = (z_{*1}, \dots, z_{*n}) = \Phi(\chi_\Lambda)$ is the white pole of S (see Definition 14.2). Any such a hyperplane is of the form

$$H(\xi_1, \dots, \xi_n; c) := \left\{z = (z^1, \dots, z^n) \in \mathbf{R}^n \ : \ \xi_1 z_1 + \cdots + \xi_n z_n = c\right\},$$

where $\xi_i \in \mathbf{R}$, $i = 1, \dots, n$, at least one of $\xi_1, \dots, \xi_n$ is non-zero, and

$$\xi_1 z_{*1} + \cdots + \xi_n z_{*n} = 0. \tag{15.5}$$

Furthermore, for any $(\xi_1, \dots, \xi_n) \neq (0, \dots, 0)$ satisfying (15.5) there exists a unique $c = c(\xi_1, \dots, \xi_n)$ such that the hyperplane $H(\xi_1, \dots, \xi_n; c)$ is *supporting* to the object-colour solid, that is, the intersection of $H(\xi_1, \dots, \xi_n; c)$ with the solid is nonempty and c is the maximum number having that property. In such a case, that intersection is contained in the boundary of the solid, hence $H(\xi_1, \dots, \xi_n; c) \cap S = H(\xi_1, \dots, \xi_n; c) \cap \partial S$. Let $\mathcal{H}$ denote the set of all supporting hyperplanes parallel to the achromatic segment.

Definition 15.1 *We say the set*

$$Eqt(S) := \bigcup_{H(\xi_1,\dots,\xi_n;c)\in\mathcal{H}} H(\xi_1, \dots, \xi_n; c) \cap S$$

is the equator *of the object-colour solid* $S = \Phi(\mathcal{X})$.

2 Note that this notion is defined in the general case, that is, for the proper object-colour solid.

Remark 15.1 *Notice that, because of (15.5), for every c, $H(-\xi_1, \dots, -\xi_n; c)$ is symmetric to $H(\xi_1, \dots, \xi_n; c)$ with regard to the centre of the object-colour solid:*

$$z = \Phi(\frac{1}{2}\chi_\Lambda) + \left(z - \Phi(\frac{1}{2}\chi_\Lambda)\right) \in H(\xi_1, \dots, \xi_n; c) \Leftrightarrow \Phi(\chi_\Lambda) - z$$
$$= \Phi(\frac{1}{2}\chi_\Lambda) - \left(z - \Phi(\frac{1}{2}\chi_\Lambda)\right) \in H(-\xi_1, \dots, -\xi_n; c),$$

and as the object-colour solid is symmetric as well (see Proposition 14.2), it follows that $H(\xi_1, \dots, \xi_n; c) \in \mathcal{H}$ if and only if $H(-\xi_1, \dots, -\xi_n; c) \in \mathcal{H}$, and $c(-\xi_1, \dots, -\xi_n) = c(\xi_1, \dots, \xi_n)$.

Every hyperplane $H(\xi_1, \dots, \xi_n; c) \in \mathcal{H}$ determines a closed half-space

$$H(\xi_1, \dots, \xi_n; c)^{\leq} := \left\{z = (z_1, \dots, z_n) \in \mathbf{R}^n : \ \xi_1 z_1 + \dots + \xi_n z_n \leq c\right\},$$

and the intersection of all such half-spaces is a cylindrical closed convex body that contains S and is symmetric with regard to the centre of S. As follows from the definition of the equator, $Eqt(S)$ is exactly the intersection of two boundaries, of S and of that cylindrical body. (This is the same as the intersection of S with the boundary of the cylindrical body or as the intersection of the cylindrical body with the boundary of S.) The equator $Eqt(S)$ is thus a $(n-2)$-dimensional surface on the boundary of S. (For $n = 3$ it becomes a closed curve on ∂S in accordance with the traditional usage of the word 'equator'.) If the boundary of S does not contain flat pieces then the intersection of S with any hyperplane from $\mathcal{H}$ consists of a single point. If the boundary of S is smooth then for every point z of the equator there is a unique hyperplane supporting S at z (the tangent hyperplane) and that hyperplane belongs to $\mathcal{H}$.

Remark 15.2 *If the boundary of S comprises flat pieces parallel to the achromatic segment, then for some hyperplane $H(\xi_1, \dots, \xi_n; c) \in \mathcal{H}$ its intersection with S consists of more than one point.*

Remark 15.3 *For $n = 3$ an equivalent definition of the equator may be given as follows. Every two-dimensional plane E through the achromatic segment cuts out a plane body $E \cap S$ of the object-colour solid. We consider all points on the boundary of $E \cap S$ with the maximum Euclidean distance from the achromatic segment. These points make up a symmetric set $M(E) \subset E \cap S$ (it consists of two points if the boundary of $E \cap S$ does not contain flat pieces parallel to the achromatic segment), and the equator of S is exactly the union of all M(E), $Eqt(S) = \cup_E M(E)$.*

In what follows, we suppose for the sake of simplicity that the boundary of S does not contain flat pieces parallel to the achromatic segment. Then for every $(\xi_1, \dots, \xi_n) \neq (0, \dots, 0)$ satisfying (15.5) an element $z(\xi_1, \dots, \xi_n) \in Eqt(S)$ is uniquely defined by $z(\xi_1, \dots, \xi_n) := H(\xi_1, \dots, \xi_n; c(\xi_1, \dots, \xi_n)) \cap S$, and all elements of the equator are obtained in such a way.

Definition 15.2 *We say an indicator $x = \chi_B$ is a* strictly optimal *object stimulus if it is mapped by Φ into the equator of $S = \Phi(\mathcal{X})$.*

The following proposition describing strictly optimal object stimuli is established by the same argument as Theorem 14.2.

Proposition 15.1 *An indicator $x = \chi_B$ is a strictly optimal object stimulus if and only if there exist $(\xi_1, \dots, \xi_n) \neq (0, \dots, 0)$ satisfying (15.5) and such that*

$$\xi_1\varphi_1(\lambda) + \dots + \xi_n\varphi_n(\lambda) \geq 0 \text{ a.e. on } B, \tag{15.6}$$

$$\xi_1\varphi_1(\lambda) + \dots + \xi_n\varphi_n(\lambda) \leq 0 \text{ a.e. on } \Lambda \backslash B. \tag{15.7}$$

The next result is a direct consequence of Proposition 15.1.

Corollary 15.1 *The class of strictly optimal stimuli remains the same if the linear colour mechanisms* $\varphi_1(\lambda), \ldots, \varphi_n(\lambda)$ *undergo a non-singular linear transformation. In other words, the strictly optimal stimuli are determined by the linear colour mechanism space, and they do not depend on the choice of the colour mechanisms from Lin* $(\varphi_1, \ldots, \varphi_n)$.

Although the set of optimal stimuli $Opt(\mathcal{X}\,|\Phi)$ does not depend on illuminant $p(\lambda)$ provided the latter is strictly positive for all $\lambda \in \Lambda$, the set of strictly optimal stimuli, generally speaking, does, because so does condition (15.5). Really, it involves the coordinates of the north pole of the object colour solid: $z_* = (z_{*1}, \ldots, z_{*n}) = \Phi(x_\Lambda)$. As $\varphi_1, \ldots, \varphi_n$ include the illuminant (see Eq. (14.1)), changing illuminant will change z_*, thus the strictly optimal stimuli. However, rescaling the colour-signal coordinates so that the location of the north pole of the object-colour solid, z_*, remains constant, will remove the dependence of the set of strictly optimal stimuli on illuminant. Specifically, let us multiply each $\varphi_i(\lambda)$ by $k_i = \left(\int_\Lambda p(\lambda)\varphi_i(\lambda)d\lambda\right) / \left(\int_\Lambda p'(\lambda)\varphi_i(\lambda)d\lambda\right)$ when replacing illuminant $p(\lambda)$ with $p'(\lambda)$. Then in the colour-signal space determined by $k_1\varphi_1, \ldots, k_n\varphi_n$ the location of the north pole of the object-colour solid, z_*, will be the same. Hence, condition (15.5) will define the same subset in the set of optimal stimuli $Opt(\mathcal{X}\,|\Phi)$ irrespective of illuminant.

The photoreceptor adaptation as described by von Kries (Section 13.1) can secure the rescaling of this sort. Hence, the dependence of strictly optimal stimuli on the illuminant proclaimed in colour literature (see e.g., Koenderink, 1987; Koenderink & Doorn, 2003) takes place only without the adaptation in photoreceptors.

For the same reason the transmittance of the ocular media do not affect the strictly optimal stimuli provided that the photoreceptors are susceptible to adaptation. Therefore, one can derive the strictly optimal stimuli from the photopigment spectral sensitivity functions (Fig. 15.8). We found that not all the elementary step functions of degree 1, $x_1(\lambda; \lambda_1)$, belong to the class of strictly optimal stimuli, as it was claimed before (Koenderink & Doorn, 2003), but only those with transition point $469\,\text{nm} < \lambda_1 < 548\,\text{nm}$. We also found that those strictly optimal stimuli which were the elementary step functions of degree 2 differed from Ostwald's semichroms.[3] More specifically, we found that only the elementary step functions of degree 2 $x_2(\lambda; \lambda_1, \lambda_2)$ with $429\,\text{nm} < \lambda_1 < 470\,\text{nm}$, and $549\,\text{nm} < \lambda_2 < 699\,\text{nm}$ were the strictly optimal stimuli. Surprisingly, we found that an essential part of the equator is formed by the image of the elementary step functions of degree 3 (red dots in Fig. 15.8).

We conclude this section by presenting an approach to evaluate points of the equator. In what follows, $n = 3$ and all elementary step functions of degree $m \leq 2$ are supposed to be optimal object stimuli.[4] We will consider various two-dimensional planes

$$E = \{z = (z_1, z_2, z_3) : \xi_1 z_1 + \xi_2 z_2 + \xi_3 z_3 = 0\}$$

$((\xi_1, \xi_2, \xi_3) \neq 0)$ passing through the achromatic segment:

$$\xi_1 z_{*1} + \xi_2 z_{*2} + \xi_3 z_{*3} = 0.$$

---

3 Though, it is hardly surprising because Ostwald determined his semichromes as such $x_2(\lambda; \lambda_1, \lambda_2)$ that λ_1 and λ_2 complementary relative to the illuminant $p(\lambda)$. Since we already know that the strictly optimal stimuli do not depend on the illuminant (provided that the photoreceptors are subject to light adaptation), this conjecture of Ostwald's - which is still supported by some authors (e.g., Koenderink & Doorn, 2003) must have been wrong.
4 It suffices to satisfy the regularity condition; see Proposition 14.3. Though it is not a case of the human colour vision, this supposition may be of interest when dealing with artificial (technical) colour vision systems.

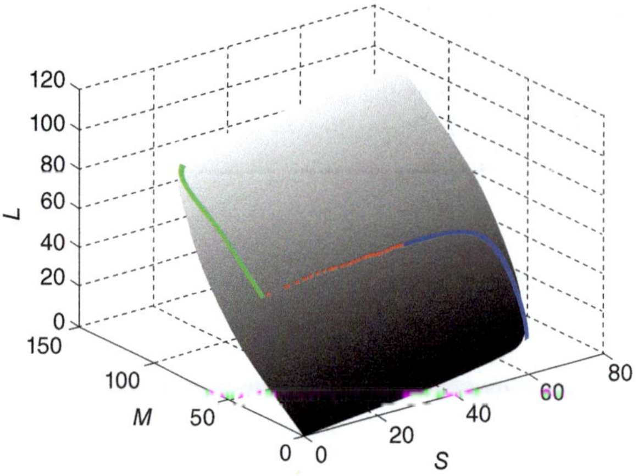

Figure 15.8 The regular object-colour solid from Fig. 15.1 with the equator plotted with coloured dots. The green, blue, and red dots represent fragments of the equator produced by elementary step functions of degree 1, 2, and 3 respectively.

Given such a plane, we take its intersection with S and look for the points z in $E \cap \partial S$ with the maximum Euclidean distance from the achromatic segment. Exactly they are all the points of the equator $Eqt(S)$ lying in E. Since $\partial S = Q \cup Q'$ where

$$Q = \{\Phi(x_2^{\lambda_1,\lambda_2}) : \lambda_{\min} \leq \lambda_1 < \lambda_2 \leq \lambda_{\max}\},$$

and

$$Q' = \{\Phi(1 - x_2^{\lambda_1,\lambda_2}) : \lambda_{\min} \leq \lambda_1 < \lambda_2 \leq \lambda_{\max}\}$$

are mutually symmetric relative to the centre of S (cf. Proposition 14.3), we shall restrict ourselves to points $z \in E \cap Q$ with the maximum Euclidean distance from the achromatic segment. Any such $z = (z_1, z_2, z_3)$ is of the form

$$z_i = \int_{\lambda_1}^{\lambda_2} \varphi_i(\lambda)\, d\lambda, \quad i = 1, 2, 3, \tag{15.8}$$

where $\lambda_{\min} \leq \lambda_1 < \lambda_2 \leq \lambda_{\max}$ and at most one of λ_1, λ_2 may be the end point of Λ. Also, since $z \in E$, one has $\xi_1 z_1 + \xi_2 z_2 + \xi_3 z_3 = 0$, i.e.,

$$\sum_{i=1}^{3} \xi_i \int_{\lambda_1}^{\lambda_2} \varphi_i(\lambda)\, d\lambda = 0. \tag{15.9}$$

Furthermore, the Euclidean distance between z and the achromatic segment is attained at a point αz_* of the achromatic segment, where $0 < \alpha < 1$ is determined from the condition that $z - \alpha z_*$ is perpendicular to z_*: $\alpha = (z, z_*)/(z_*, z_*)$, where $(\cdot\,,\cdot)$ stands for the scalar product in $\mathbf{R}^n$. The squared distance between z and the achromatic segment is then

$$(z, z) - \frac{1}{(z_*, z_*)}(z, z_*)^2,$$

which equals

$$g(\lambda_1, \lambda_2) := \sum_{i=1}^{3} \left(\int_{\lambda_1}^{\lambda_2} \varphi_i(\lambda)\, d\lambda \right)^2 - \frac{1}{(z_*, z_*)} \left(\sum_{i=1}^{3} z_{*i} \int_{\lambda_1}^{\lambda_2} \varphi_i(\lambda)\, d\lambda \right)^2, \tag{15.10}$$

where $z_{*i} = \int_\Lambda \varphi_i(\lambda)\, d\lambda$. Thus one has to maximise function (15.10) subject to constraints (15.9) and $\lambda_{\min} \le \lambda_1 < \lambda_2 \le \lambda_{\max}$.

First suppose that z is produced by an elementary step function of degree 2, i.e., $\lambda_{\min} < \lambda_1 < \lambda_2 < \lambda_{\max}$. Since z is a maximiser to the formulated problem, there exists a Lagrange multiplier μ such that

$$\frac{\partial L(\lambda_1, \lambda_2, \mu)}{\partial \lambda_1} = 0, \qquad \frac{\partial L(\lambda_1, \lambda_2, \mu)}{\partial \lambda_2} = 0, \qquad \frac{\partial L(\lambda_1, \lambda_2, \mu)}{\partial \mu} = 0, \tag{15.11}$$

where the Lagrangian L is given by

$$L(\lambda_1, \lambda_2, \mu) := g(\lambda_1, \lambda_2) + \mu \sum_{i=1}^{3} \xi_i s_i(\lambda_1, \lambda_2),$$

$$s_i(\lambda_1, \lambda_2) = \int_{\lambda_1}^{\lambda_2} \varphi_i(\lambda)\, d\lambda, \quad i = 1, 2, 3,$$

Taking into account (15.10), the first two equations in (15.11) are rewritten equivalently as

$$\sum_{i=1}^{3} s_i(\lambda_1, \lambda_2)\varphi_i(\lambda_1) - \frac{1}{(z_*, z_*)}\left(\sum_{i=1}^{3} z_{*i} s_i(\lambda_1, \lambda_2)\right)\left(\sum_{i=1}^{3} z_{*i}\varphi_i(\lambda_1)\right) + \mu \sum_{i=1}^{3} \xi_i \varphi_i(\lambda_1) = 0, \tag{15.12}$$

$$\sum_{i=1}^{3} s_i(\lambda_1, \lambda_2)\varphi_i(\lambda_2) - \frac{1}{(z_*, z_*)}\left(\sum_{i=1}^{3} z_{*i} s_i(\lambda_1, \lambda_2)\right)\left(\sum_{i=1}^{3} z_{*i}\varphi_i(\lambda_2)\right) + \mu \sum_{i=1}^{3} \xi_i \varphi_i(\lambda_2) = 0, \tag{15.13}$$

while the third equation is exactly (15.9). We have thus a system of three equations with three variables, λ_1, λ_2, and μ. Excluding μ gives us for finding λ_1 and λ_2 a system of two equations, and if $z = \Phi(x_2^{\lambda_1,\lambda_2}) \in Eqt(S) \cap E$ then λ_1 and λ_2 satisfy that system (Fig. 15.9).

If z is produced by an elementary step function of degree 1, then $\lambda_1 = \lambda_{\min}$ or $\lambda_2 = \lambda_{\max}$, and arguing as above, we get a system of two equations with two variables, μ and one of λ_i (λ_2 if $\lambda_1 = \lambda_{\min}$ or λ_1 if $\lambda_2 = \lambda_{\max}$). The equations are, respectively, (15.9) and (15.13) with $\lambda_1 = \lambda_{\min}$ or (15.9) and (15.12) with $\lambda_2 = \lambda_{\max}$. Excluding μ gives us one equation for finding one variable, λ_2 or λ_1 respectively. It follows that if $z = \Phi(x_1^{\lambda_2}) \in Eqt(S) \cap E$ (or $z = z_* - \Phi(x_1^{\lambda_1}) \in Eqt(S) \cap E$) then λ_2 (resp., λ_1) satisfies the corresponding equation.

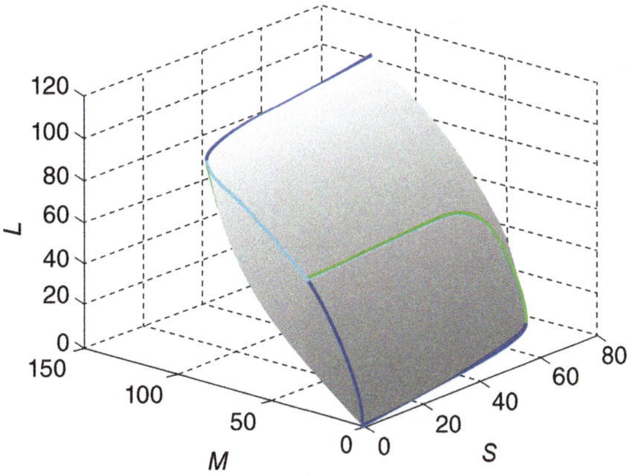

Figure 15.9 The equator (green line) and the meridian (blue line) of the regular object-colour solid. The cyan line shows where they overlap.

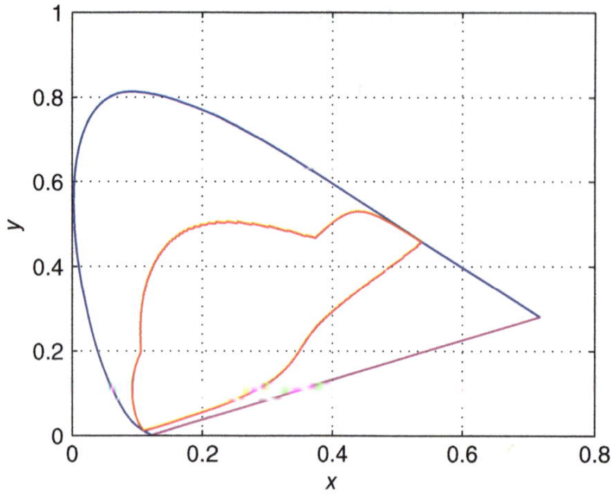

Figure 15.10 CIE 1931 chromaticity coordinates of the equator colours (red line).

When the functions $\varphi_1, \varphi_2, \varphi_3$ are good enough, the system of two equations arising from (15.9), (15.12), and (15.13) after excluding μ has a finite number of solutions (λ_1, λ_2). Likewise, in such a case the number of solutions for each of two equations with one variable (λ_2 or λ_1) is also finite. We obtain a finite set of points $z = \Phi(x_2^{\lambda_1,\lambda_2})$, $z = \Phi(x_1^{\lambda_2})$, and $z = z_* - \Phi(x_1^{\lambda_1})$ which provides the desired solutions. For each of them we compute its distance from the achromatic segment (the squared value of that distance is given by (15.10)). The set of all z with maximum distance is exactly $Eqt(S) \cap E \cap Q$, and adding to them the symmetric points $z_* - z$ gives us the whole of $Eqt(S) \cap E$ (see Figs. 15.9 and 15.10).

16

Object-Colour Stimulus Manifold

Although the colour of an object is something different from the colour of the light reflected by this object, it is generally assumed that when two objects reflect metameric lights they are perceived to be of the same object colour. This assumption certainly fails when the scene is illuminated by many light sources (see Chapter 19). However, for a single-illuminant scene it is quite acceptable (at least, in the first approximation). So, in this chapter we will assume that objects reflecting metameric lights match each other in (object) colour. Such objects themselves will be called metameric (see Definition 16.1). This allows defining an object-colour stimulus as a class of metameric objects by analogy with the notion of colour stimulus defined for lights (Definition 2.5). Also, the object-colour stimulus manifold can be defined applying the same line of reasoning as when defining the notion of the colour stimulus manifold for lights (in Chapter 5).

16.1 Object metamerism

Objects with different spectral reflectance functions may happen to reflect metameric lights. Such objects are called metameric.[1]

Definition 16.1 *Two objects with spectral reflectance functions $x(\lambda)$ and $x'(\lambda)$ are called* metameric *with respect to the illuminant $p(\lambda)$ if the colour signals $\left(\int_\Lambda x(\lambda)\, p(\lambda)\, \varphi_1(\lambda)\, d\lambda, \ldots, \int_\Lambda x(\lambda)\, p(\lambda)\, \varphi_n(\lambda)\, d\lambda\right)$ and $\left(\int_\Lambda x'(\lambda)\, p(\lambda)\, \varphi_1(\lambda)\, d\lambda, \ldots, \int_\Lambda x'(\lambda)\, p(\lambda)\, \varphi_n(\lambda)\, d\lambda\right)$ are equal.*

Metamerism as defined is an equivalence relation on the object stimulus set $\mathcal{X}$. Given a metamerism $\sim$, the quotient set of $\mathcal{X}$ by $\sim$ (i.e., the set of all the classes of metamerism in $\mathcal{X}$) will be denoted by $\mathcal{X}/\sim$ and called the object-colour stimulus set, *its elements being referred to as* object-colour stimuli.

Much effort has been made to find out how to generate spectral reflectance functions metameric to each other (Wyszecki & Stiles, 1982, pp. 185-194). Still, it remains as a theoretical problem to describe all the objects metameric to a given one. Addressing this problem we start from the following definition.

1 When using the term 'metamerism' in relation to objects, the same ambiguity takes place as in relation to lights, which was noted in the Outline (Chapter 1). Specifically, 'object metamerism' has at least two rather different meanings: (i) object metamerism as metamerism of reflected lights; and (ii) object metamerism as the phenomenon that different objects may reflect metameric lights under one illumination and not-metameric lights under the other.

Foundations of Colour Science: From Colorimetry to Perception, First Edition. Alexander D. Logvinenko and Vladimir L. Levin.
© 2023 John Wiley & Sons Ltd. Published 2023 by John Wiley & Sons Ltd.

Definition 16.2 *Let x_0 be a point in $\mathcal{X} = \{x \in \mathbf{L}^\infty : 0 \leq x(\lambda) \leq 1\}$ such that $\Phi(x_0) \in int(S) = int(\Phi(\mathcal{X}))$. We define the set of metamers of x_0 (or the* object metamer set*) as*

$$\mathcal{X}_0 := \{x \in \mathcal{X} : \Phi(x) = \Phi(x_0)\}. \tag{16.1}$$

Since Φ is a weak* continuous linear operator $\mathbf{L}^\infty \to \mathbf{R}^n$, and $\mathcal{X}$ is weak* compact and convex, $\mathcal{X}_0$ is weak* compact and convex as well. By analogy with Definition 6.1 the set $\mathcal{X}_0$ of objects metameric to x_0 with respect to the colour mechanisms $\varphi_1, \dots, \varphi_n$ will be also denoted by $Met(\varphi_1, \dots, \varphi_n; x_0)$.

The following theorem describes the extreme points of the object metamer set (16.1).

Theorem 16.1 *A point $x \in \mathcal{X}_0$ is an extreme point of $\mathcal{X}_0$ if and only if x is an indicator, that is, $x = \chi_B$ where B is a measurable subset of Λ.*

So, the problem of generating metamers for objects is reduced to generating metameric indicators. It follows from Theorem 16.1 that any convex combination of indicators metameric to a particular object will be metameric to it as well. Moreover, from Theorem 16.1 and the Krein–Milman theorem it follows that the object metamer set $\mathcal{X}_0$ is exactly the weak* closed convex hull of all the indicators metameric to x_0.

The following theorem shows when there is no metamerism on the boundary of the object-colour solid.

Theorem 16.2 *Suppose that, for every non-zero vector $(\xi_1, \dots, \xi_n) \in \mathbf{R}^n$,*

$$mes\left\{\lambda : \sum_{i=1}^{n} \xi_i \varphi_i(\lambda) = 0\right\} = 0, \tag{16.2}$$

where mes stands for the n-dimensional Lebesgue measure. Let $x_0 = \chi_B$ and $\Phi(x_0) \in \partial S$. Then $Met(\varphi_1, \dots, \varphi_n; x_0) = \{x_0\}$.

Condition (16.2) does certainly hold true for Govardovskii's cone fundamentals. However, this condition fails if some $\varphi_i(\lambda)$ in (16.2) takes zero within some interval in the visible spectrum. For instance, this is the case for the CIE (2006) 2° LMS cone fundamentals. Indeed, for $\lambda \geq 620$ nm, the S-cone fundamental in this set is not defined because the corresponding values are negligibly small. As such property seems unrealistic from a biological point of view, we will restrict ourselves to only those fundamental colour mechanisms for which condition (16.2) is satisfied.

For such colour mechanisms the restriction of the map Φ to the set $Opt(\mathcal{X}) = \Phi^{-1}(\partial S)$ of optimal object stimuli is one-to-one. Moreover, the following consequence of Theorem 16.2 holds true.

Corollary 16.1 *The set $Opt(\mathcal{X})$ is weak* compact in $\mathbf{L}^\infty$, and $\Phi : Opt(\mathcal{X}) \to \partial S$ is a homeomorphic map of $Opt(\mathcal{X})$ onto ∂S, that is, both Φ and $\Phi^{-1} : \partial S \to Opt(\mathcal{X})$ are continuous.*

Thus, when condition (16.2) holds true,[2] there is a natural homeomorphism between the object-colour solid boundaries induced by different illuminants. More specifically, let Φ_1 and Φ_2 be the colour-signal maps induced by the same set of colour mechanisms under two illuminants, S_1 and S_2 being the corresponding object-colour solids. In this case, Corollary 16.1 entails that a map $\Phi_1 \circ \Phi_2^{-1}$ is a homeomorphism between ∂S_1 and ∂S_2. It is natural in the sense that the boundary

2 In particular, in the case of Govardovskii's cone fundamentals (Definition 3.4).

points map to each other if and only if they are the colour signals produced by the same object stimulus. Unfortunately, one cannot extend this homeomorphism to the whole object-colour solids, S_1 and S_2. Indeed, a natural attempt to generalise this homeomorphism would be as in the following definition.

Definition 16.3 *Given colour mechanisms $\varphi_1, \dots, \varphi_n$, we consider two colour-signal maps Φ' and Φ'' determined by illuminations $p'(\lambda)$ and $p''(\lambda)$ respectively. Let $S' = \Phi'(\mathcal{X})$ and $S'' = \Phi''(\mathcal{X})$. Let us define a multifunction (set-valued map) $\rho : S' \to S''$,*

$$\rho(z') := \bigcup_{x \in \Phi'^{-1}(z')} \Phi''(x), \; z' \in S'. \tag{16.3}$$

Correspondence ρ in (16.3) is a many-to-many map because of object metamer mismatching (see Chapter 18). Since object metamerism, as defined in Definition 16.1, depends on illuminant, if the illuminant $p(\lambda)$ is replaced by the other $p'(\lambda)$ the hitherto metameric objects may cease to be metameric.[3] And vice versa, reflectance spectra which produce different colour signals under one illuminant may become metameric under the other. To put it the other way, after an illumination change some reflectance spectra will remain in the same metameric class, the other will fall into different classes. Therefore, as defined above, object metamerism as the equivalence relation is different for different illuminants. To be more exact, the metamerism $\sim$ with respect the illuminant $p(\lambda)$, and the metamerism $\sim'$ with respect the illuminant $p'(\lambda)$ are not isomorphic as equivalence relations on $\mathcal{X}$.

It follows that there is no meaningful objective way to put in a bijective correspondence the metameric classes produced under one illumination with those that arose under another. At first glance, it seems natural to establish such a correspondence by mapping a metameric class under one illumination to the metameric class under another illumination that has the same 'colour'. However, paradoxical as it may seem, it is not easy to answer the question of whether the differently illuminated objects have the same colour. We will return to this issue in Chapter 19. Not to mention the fact that the establishment of the correspondence between two object-colour solids on the grounds of the equality of object colours requires the existence of measurement techniques similar to those described in Chapter 8 for light colours. Such techniques will be considered in Chapter 17.

Note that since object metamerism is defined above in terms of (reflected) light metamerism, multifunction ρ in (16.3) can be considered as a particular case of multifunction γ in (5.2). So, just as multifunction γ was approximated by a one-to-one function by restricting it onto a particular subset of lights ($\mathcal{A}$ as defined by Eq. (5.4)), multifunction ρ can also be approximated by a one-to-one function by restricting it onto a subset of objects each of which represents just one class of metameric objects. Of course, such the subset of objects has to have some of those essential properties that the subset $\mathcal{A}$ is endowed with. All this leads us to a notion of object atlas.

Colour atlas

In colour literature, there is a closely related, but still a different concept – colour atlas. A colour atlas is commonly understood as a sample of coloured papers representing the object-colour palette. The members of a colour atlas are usually arranged in the 3D space in some particular order so as to manifest the perceptual attributes of object colour. Depending on the colour appearance model used as the basis for the colour atlas, the list of the attributes is different for different atlases (see Chapter 17). Among most popular colour atlases

3 Surely, the same may happen if the spectral sensitivity of the colour mechanisms changes.

one can mention the Munsell book of colour, the Natural Colour System (NCS), the Optical Society of America (OSA) colour atlas (see e.g., Kuehni, 2003, Ch. 7). While earlier attempts were made to conceptualise the term 'colour atlas' (e.g., Koenderink, 1987), our notion 'object atlas' differs significantly in that it is a sample of objects rather than object colours. According to our terminology an object atlas becomes an object-colour atlas (i.e., represents object colours) only after a procedure we name a colour-charting (see Section 17.1). To avoid confusion in terminology, we will hereinafter refer to the established collections of coloured papers produced on a commercial basis (such as the Munsell book of colour and the like) as *colour collections* or *colour charts*.

16.2 Object atlas

Let us make a collection of object stimuli picking up just one stimulus from a metamer set. All the stimuli in such a selected sample will produce different colour signals. Each object stimulus collection with such a feature will be referred to as an object atlas.

Definition 16.4 *A subset $\mathcal{A} \subset \mathcal{X}$ such that $\Phi(x) \neq \Phi(y)$ whenever $x, y \in \mathcal{A}, x \neq y$, is called an* object atlas. *When $\Phi(\mathcal{A}) = S$ the atlas $\mathcal{A}$ is called* full. *If $\mathcal{A}$ is finite (infinite) subset of $\mathcal{X}$ the atlas $\mathcal{A}$ is called* finite (infinite).

A full object atlas is a sample of the object stimuli such that all they have different colour signals, and for each colour signal there is exactly one member of the atlas. Any full atlas is obviously infinite.

Generally, a subset of the object stimuli which is an object atlas under one illumination may cease being such under the other illumination. Of special interest are selected samples of object stimuli each of which will be an object atlas under any illumination.

Definition 16.5 *A subset $\mathcal{A} \subset \mathcal{X}$ is called an* illuminant-invariant object atlas *if for any illumination p and for any $x, y \in \mathcal{A}, x \neq y$, one has $\Phi'(x) \neq \Phi'(y)$, where Φ' is a colour-signal map determined by $\varphi_1', \ldots, \varphi_n'$, where $\varphi_i'(\lambda) = \varphi_i(\lambda)\, p(\lambda)$.*

Example 16.1 *Let us pick all optimal object stimuli, and fix an arbitrary stimulus x_0 in the interior of $\mathcal{X}$. If we complete such a subset of the object stimuli with the stimuli $(1-\alpha)x_0 + \alpha x$, where x is any optimal object stimulus and $0 \leq \alpha \leq 1$, we will get a full illuminant-invariant object atlas. Indeed, it is an illuminant-invariant atlas since $(1-\alpha)x_0 + \alpha x \neq (1-\beta)x_0 + \beta y$ implies either $x \neq y$ or $\alpha \neq \beta$. If $x \neq y$ then $\Phi'(x) \neq \Phi'(y)$, and the colour signals $\Phi'((1-\alpha)x_0 + \alpha x)$ and $\Phi'((1-\beta)x_0 + \beta y)$ cannot coincide since they lie inside two segments, $[x_0, \Phi'(x)]$ and $[x_0, \Phi'(y)]$, that join x_0 with two different boundary points of S'. When $\alpha \neq \beta$ we get $\Phi'((1-\alpha)x_0 + \alpha x) \neq \Phi'((1-\beta)x_0 + \beta y)$ even if $x = y$. Further, any colour signal z is uniquely represented as $z = (1-\alpha)\,\Phi'(x_0) + \alpha z'$, where z' is a boundary point of S' and $0 \leq \alpha \leq 1$; therefore, $z = \Phi'((1-\alpha)x_0 + \alpha x')$, where x' is the only optimal object stimulus that is mapped by Φ' to z' (see Theorem 16.2). Thus we have proved that our atlas is illuminant-invariant and full under all illuminations.*

A similar reasoning yields yet one example of an illuminant-invariant atlas which is full under all illuminations.

Example 16.2 *If we pick all optimal object stimuli and add to them the object stimuli αx, where $0 < \alpha < 1$ and x is any optimal object stimulus, then we will get an object atlas which is illuminant-invariant and full under all illuminations.*

Let $S' = \Phi'(\mathcal{X})$ and $S'' = \Phi''(\mathcal{X})$ be two object-colour solids defined by the same colour mechanisms with respect to two different illuminants $p'(\lambda)$ and $p''(\lambda)$. Consider $\Phi'(\mathcal{A})$ and $\Phi''(\mathcal{A})$ where $\mathcal{A}$ is a (not necessarily full) object atlas. We have $\Phi'(\mathcal{A}) \subseteq S', \Phi''(\mathcal{A}) \subseteq S''$ (if $\mathcal{A}$ is full, these relations hold with the equality sign), and as $\mathcal{A}$ is an atlas, a one-to-one map $\Phi'|_{\mathcal{A}}^{-1} : \Phi'(\mathcal{A}) \to \mathcal{A}$ is well-defined. Then, given ρ as in (16.3), the map $\rho_{\mathcal{A}} = \Phi'' \text{o} \Phi'|_{\mathcal{A}}^{-1} : \Phi'(\mathcal{A}) \to \Phi''(\mathcal{A})$ is a selection of ρ: $\rho_{\mathcal{A}}(z') \in \rho(z')$ for all $z' \in \Phi'(\mathcal{A})$.

Proposition 16.1 *Suppose an atlas $\mathcal{A}$ is a weak* closed (hence weak* compact) subset in $\mathcal{X}$ (atlases from Examples 16.1 and 16.2 satisfy that condition). The map $\rho_{\mathcal{A}}$ is then a homeomorphism of $\Phi'(\mathcal{A})$ onto $\Phi''(\mathcal{A})$.*

Both atlases from Examples 16.1 and 16.2 will be referred to as *optimal* object atlases. Of particular interest are the following two optimal object atlases. One is given in Example 16.2. This atlas will be denoted by $\mathcal{A}_0$. Every element in $\mathcal{A}_0$ can be represented as αx_{opt} where x_{opt} is some optimal object stimulus, and $0 \le \alpha \le 1$. Another one is a particular case of Example 16.1 when $x_0 = 0.5\chi_\Lambda$. For reasons which will become clear later we will pay much attention to this optimal atlas.

Definition 16.6 *Let Φ be a colour-signal map and $Opt(\mathcal{X}\,|\Phi)$ an optimal object stimulus set induced by Φ. A set of object stimuli*

$$\left\{\left((1-c)x_{0.5} + cx_{opt}(\lambda)\right) \,:\, x_{opt} \in Opt\,(\mathcal{X}\,|\Phi), 0 \le c \le 1\right\}, \tag{16.4}$$

where $x_{0.5} = 0.5\chi_\Lambda$, will be referred to as an optimal object atlas *$\mathcal{A}_{0.5}$. When the dimensionality of the colour-signal space, n, matters, the notation $\mathcal{A}^n_{0.5}$ will be used.*[4]

An element of the optimal atlas, $\mathcal{A}$, that is metameric to the object stimulus x, will be called the *optimal metamer* of x (with respect to $\mathcal{A}$). (Admittedly, the optimal metamer depends on the illuminant.) For instance, for a fixed illuminant radiant spectral density function, for each reflectance spectrum $x(\lambda)$ there are a unique optimal reflectance spectrum x_{opt}, and a unique number c such that $(1-c)x_{0.5} + cx_{opt}(\lambda)$ is metameric to $x(\lambda)$. Element $(1-c)x_{0.5} + cx_{opt}(\lambda)$ in $\mathcal{A}_{0.5}$ is the *optimal metamer* of $x(\lambda)$ with respect to $\mathcal{A}_{0.5}$. The optimal reflectance spectrum $x_{opt}(\lambda)$ and the number c in (16.4) will be referred to as the *spectral waveform* and *chromatic contrast* of the optimal metamer (with respect to $\mathcal{A}_{0.5}$) respectively.[5] The justification for such a terminology becomes transparent if $(1-c)x_{0.5} + cx_{opt}(\lambda)$ is rewritten as $x_{0.5} + c\left(x_{opt}(\lambda) - x_{0.5}\right)$.

Definition 16.7 *Recall that given the object stimulus set $\mathcal{X}$, $x \in \mathcal{X}$ if and only if $\chi_\Lambda - x \in \mathcal{X}$. The object stimulus $\chi_\Lambda - x$ will be called* complementary *to x.*

As $x_{0.5} + c\left(x_{opt}(\lambda) - x_{0.5}\right) = x_{0.5} - c\left(\left(\chi_\Lambda - x_{opt}(\lambda)\right) - x_{0.5}\right)$ one can extend formally the range of chromatic contrast treating an optimal metamer with negative chromatic contrast and the spectral waveform $x_{opt}(\lambda)$ as that of the positive chromatic contrast of the same magnitude and the complementary spectral waveform.

4 The object atlas $\mathcal{A}_0$ can be, formally, represented as in (16.4), if instead of $x_{0.5}$ we take the perfect absorber. This explains the designation of this atlas (i.e., the subscript 0).

5 Hereafter, unless otherwise stated, we will keep in mind $\mathcal{A}_{0.5}$ when using the term 'optimal object atlas' and/or 'optimal metamer'.

When the optimal object stimuli set comprises only elementary step functions, using negative chromatic contrast, one can restrict oneself to only elementary step functions of the form $x_m(\lambda; \lambda_1, \ldots, \lambda_m)$ (Definition 14.4).

Remark 16.1 *For practical purpose one can pick only a finite number of object stimuli. It is clear that such a finite sample of the object atlas should be made uniformly in some sense to make the sample as representative as possible. Moreover, in the existing colour charts (such as the Munsell book and NCS collection) their elements are arranged in three-dimensional space in some particular order so as to manifest the perceptual attributes of object colours such as hue, purity and some others depending on the colour appearance model underlying them (see Chapter 17). In this regard, we note that an object atlas can be arranged in the same way as the corresponding colour signals if one locates each its member, $x \in \mathcal{A}$, in the 3D space at the same position as its colour signal, $\Phi(x)$. In this case, the spatial array of the atlas $\mathcal{A}$ will be similar to the spatial array of the corresponding colour signals $\Phi(\mathcal{A})$. Such an arrangement will be, in a sense, robust to a change in illuminant. Indeed, while a change in illuminant will inevitably change the object-colour solid, thus the spatial array of the atlas, it follows from Proposition 16.1 that the spatial configuration of the atlas members will actually remain the same. In particular, if the colour signal of some atlas member lies on the achromatic segment (or on the meridian) of a given object-colour solid, this property will be well preserved under a change in illuminant. Moreover, given a weak* continuous map $g : [0, 1] \to \mathcal{A}$ and three atlas members $x_1, x_2, x_3 \in \mathcal{A}$ such that $x_1 = g(0)$, $x_2 = g(1/2)$, and $x_3 = g(1)$, the configuration of the corresponding colour signals in the 3D space does not depend on the illuminant: $\Phi'(x_2)$ lies between $\Phi'(x_1)$ and $\Phi'(x_3)$ on the continuous curve $\Phi' o g : [0, 1] \to \mathbf{R}^3$ whenever $\Phi' = p\Phi$, $p(\lambda) > 0$.*

16.3 Object-colour stimulus manifold. Illuminant-induced non-linear object-colour stimulus map

Given the object stimulus set $\mathcal{X}$, consider linear colour mechanisms $\Phi = \{\varphi_1, \ldots, \varphi_n\}$ satisfying the condition in Theorem 16.2. Any illuminant $p(\lambda)$ induces an object metamerism as an equivalence relation $\sim_p$ determined by the set of the linear colour mechanisms $\Phi_p = \{p\varphi_1, \ldots, p\varphi_n\}$. Each object-colour stimulus set $\mathcal{X}/\sim_p$ is homeomorphic to the corresponding object-colour solid $\Phi_p(\mathcal{X})$. Consider the optimal object atlas $\mathcal{A}^n_{0.5}$. It is full and illuminant-invariant. As it satisfies the condition in Proposition 16.1, $\mathcal{A}^n_{0.5}$ is homeomorphic to $\Phi_p(\mathcal{X})$ for each illuminant p. Moreover, the map $\rho_{\mathcal{A}^n_{0.5}} = \Phi_{p''} o \Phi_{p'}|^{-1}_{\mathcal{A}^n_{0.5}}$ establishes a homeomorphisms between the object-colour solids $\Phi_{p'}(\mathcal{X})$ and $\Phi_{p''}(\mathcal{X})$ induced by any two illuminants p' and p''. It follows that for a given observer (i.e., a set of linear colour mechanisms) the object-colour stimulus sets induced by any two (thus all) illuminants are homeomorphic. Also, all they are homeomorphic to the object atlas $\mathcal{A}^n_{0.5}$ (with respect to the weak* topology). As the object-colour solid $\Phi_p(\mathcal{X})$ is homeomorphic to the unit ball in $\mathbf{R}^n$, each $\Phi_p(\mathcal{X})$ (and $\mathcal{A}^n_{0.5}$) can be considered as a topological manifold with boundary homeomorphic to the unit ball in $\mathbf{R}^n$. Therefore, $\mathcal{A}^n_{0.5}$ can be considered as a representation of all the object-colour stimulus sets $\mathcal{X}/\sim_p$ considered as topological manifolds. For this reason we shall sometimes refer to the atlas $\mathcal{A}^n_{0.5}$ as the *object-colour stimulus manifold*.

Let us consider a map $\pi_p : \mathcal{X} \to \mathcal{X}/\sim_p$ which sends each object stimulus to its class of $\sim_p$-metamerism. The restriction $\pi_p|_{\mathcal{A}_{0.5}}$ of π_p to the atlas $\mathcal{A}_{0.5}$ (i.e., the function from $\mathcal{A}_{0.5}$ to $\mathcal{X}/\sim_p$ such that $\pi_p|_{\mathcal{A}_{0.5}}(x) = [x]_p$ for all x in $\mathcal{A}_{0.5}$, where $[x]_p$ is the metameric class of x) is a bijection because the atlas $\mathcal{A}_{0.5}$ is full. Therefore, there exists an inverse function $\pi_p|^{-1}_{\mathcal{A}_{0.5}}$ for $\pi_p|_{\mathcal{A}_{0.5}}$.

The function composition $\kappa_p := \pi_p|^{-1}_{\mathcal{A}_{0.5}} \circ \pi_p$ mapping the object stimulus set $\mathcal{X}$ to the optimal object atlas $\mathcal{A}_{0.5}$ will be called the *non-linear object-colour stimulus map* induced by the illuminant p.

This map determines which colour will be assigned to a given reflectance spectrum under a given illumination by a particular observer. Once again, the map does not tell anything about colour appearance as such. Assigning colour is understood in the present context only as the prediction that the colour of a given surface will be the same as that of the optimal metamer into which the reflectance spectrum of the surface maps.

When illumination changes a given object stimulus may move from one class of metamerism to another. In other words, a given object stimulus can be metameric to different optimal metamers under different illuminants. Hence, the colour of an object can change when illumination changes, as manifested by the shift of its optimal metamer from one metameric class to the other. Such a shift of a proximal stimulus (for the object colour) induced by a change in illumination (referred to an *illuminant-induced object-colour stimulus shift*) can be characterised by using the non-linear object-colour stimulus map. Given two illuminants p' and p'', a pair $\kappa_{p'}(x)$ and $\kappa_{p''}(x)$ shows to what extent the shift in illuminant changes the object-colour stimulus.

16.4 Trichromatic object-colour stimulus manifold

This section will be devoted to trichromatic observers. While we will keep in mind mostly 3-pigment Grassmann–Govardovskii structures (Definition 3.4), the main results in the section will be true for arbitrary trichromatic observers with continuously differentiable fundamental colour mechanisms satisfying the conditions set forth in Section 16.4.2.

16.4.1 Trichromatic regular object-colour stimulus manifold and its spherical representation

Consider first colour mechanisms $\Phi = (\varphi_1(\lambda), \varphi_2(\lambda), \varphi_3(\lambda))$ satisfying the regularity assumption (Definition 14.5). Denote $X^I_2 =: \left\{x_2^{\lambda_1,\lambda_2} : \lambda_{\min} \le \lambda_1 < \lambda_2 \le \lambda_{\max}\right\}$ and $X^{II}_2 =: \left\{1 - x_2^{\lambda_1,\lambda_2} : \lambda_{\min} \le \lambda_1 < \lambda_2 \le \lambda_{\max}\right\}$, where $x_2^{\lambda_1,\lambda_2}$ stands for the indicator of the set $[\lambda_1, \lambda_2] \subseteq \Lambda$. As follows from Proposition 14.3, in this case $Opt(\mathcal{X}\,|\Phi) = X^I_2 \cup X^{II}_2$, that is, the optimal object stimulus set $Opt(\mathcal{X}\,|\Phi)$ consists of all the elementary step functions of degree $m \le 2$. As claimed in Corollary 16.1, $Opt(\mathcal{X}\,|\Phi)$ is homeomorphic to $\partial S = \Phi(Opt(\mathcal{X}|\Phi))$ where $S = \Phi(\mathcal{X})$. Moreover, $Opt(\mathcal{X}|\Phi)$ is homeomorphic to the unit sphere S^2 in $\mathbf{R}^3$. The optimal object atlas $\mathcal{A}^3_{0.5}$ (16.4) induced by $Opt(\mathcal{X}|\Phi)$, where $\Phi = (\varphi_1, \varphi_2, \varphi)$ meets the regularity assumption, will be referred to as the *rectangular object atlas* (written as $\mathcal{AR}^3_{0.5}$). It is homeomorphic to the unit ball B^3_1 in $\mathbf{R}^3$. Given an object stimulus, $x \in \mathcal{X}$, the element in $\mathcal{AR}^3_{0.5}$ which is metameric to x will be referred to as its *rectangular metamer*.

Let us introduce the spherical coordinates on $\mathcal{AR}^3_{0.5}$. Turn the visible spectrum wavelength interval Λ into the *visible spectrum circle* (written as C_Λ) making no difference between $\lambda_{\min}$ and $\lambda_{\max}$. Two points $\lambda_1 < \lambda_2$ determine two complementary arcs on this circle: $[\lambda_1, \lambda_2]$ and $\overline{[\lambda_1, \lambda_2]}$. The function $x_2^{\lambda_1,\lambda_2}$ takes 1 on $[\lambda_1, \lambda_2]$, and the function $1 - x_2^{\lambda_1,\lambda_2}$ takes 1 on $\overline{[\lambda_1, \lambda_2]}$. So, both the optimal object stimuli of type I and II on the circle C_Λ may be described in the same way.[6]

6 Specifically, if we assume that points λ_1 and λ_2 belong not only to the arc $[\lambda_1, \lambda_2]$ but also to the arc $\overline{[\lambda_1, \lambda_2]}$ (i.e., $[\lambda_1, \lambda_2] \cap \overline{[\lambda_1, \lambda_2]} = \{\lambda_1; \lambda_2\}$), then each of the optimal object stimuli (of both types) can be considered as the indicator of the corresponding arc. Since the optimal object stimuli are treated as elements of the function space $\mathbf{L}^\infty$, such an interpretation is quite acceptable.

For a pair of optimal object stimuli determined by wavelengths $\lambda_1 < \lambda_2$, define two numbers – $\Delta\lambda$ and $\overline{\lambda}$ – which we will call the *spectral bandwidth* and *central wavelength* of these stimuli respectively, as follows:

$$\Delta\lambda = \begin{cases} |\lambda_1 - \lambda_2|, & \text{for } x_2^{\lambda_1,\lambda_2}; \\ (\lambda_{\max} - \lambda_{\min}) - |\lambda_1 - \lambda_2|, & \text{for } 1 - x_2^{\lambda_1,\lambda_2}; \end{cases} \tag{16.5}$$

$$\overline{\lambda} = \begin{cases} 0.5(\lambda_1 + \lambda_2), & \text{for } x_2^{\lambda_1,\lambda_2}; \\ 0.5((\lambda_1 + \lambda_2) + (\lambda_{\max} - \lambda_{\min})), & \text{for } 1 - x_2^{\lambda_1,\lambda_2} \text{ if } (\lambda_1 + \lambda_2) < (\lambda_{\max} - \lambda_{\min}); \\ 0.5((\lambda_1 + \lambda_2) - (\lambda_{\max} - \lambda_{\min})), & \text{for } 1 - x_2^{\lambda_1,\lambda_2} \text{ if } (\lambda_1 + \lambda_2) > (\lambda_{\max} - \lambda_{\min}). \end{cases} \tag{16.6}$$

Note that $\Delta\lambda$ is the length of the arc where the corresponding stimuli ($x_2^{\lambda_1,\lambda_2}$ or $1 - x_2^{\lambda_1,\lambda_2}$) takes 1, and $\overline{\lambda}$ is the arc centre. Note also that for $x_2^{\lambda_{\min},\lambda_{\max}}$ (i.e., the spectral reflectance $x(\lambda) \equiv 1$) we have $\Delta\lambda = \lambda_{\max} - \lambda_{\min}$ but $\overline{\lambda}$ is not defined. Nor is it for $1 - x_2^{\lambda_{\min},\lambda_{\max}}$.

Conversely, given $0 < \Delta\lambda < (\lambda_{\max} - \lambda_{\min})$, and $\lambda_{\min} < \overline{\lambda} < \lambda_{\max}$, a pair $(\Delta\lambda, \overline{\lambda})$ uniquely defines one of the two optimal object stimuli – $x_2^{\lambda_1,\lambda_2}$ or $1 - x_2^{\lambda_1,\lambda_2}$. Their transition wavelengths λ_1, and λ_2 are evaluated as

$$\lambda_1 = \begin{cases} \overline{\lambda} - 0.5\Delta\lambda, & \text{if } (\overline{\lambda} - 0.5\Delta\lambda) > \lambda_{\min}; \\ \overline{\lambda} - 0.5\Delta\lambda + (\lambda_{\max} - \lambda_{\min}), & \text{if } (\overline{\lambda} - 0.5\Delta\lambda) < \lambda_{\min}; \end{cases} \tag{16.7}$$

$$\lambda_2 = \begin{cases} \overline{\lambda} + 0.5\Delta\lambda, & \text{if } (\overline{\lambda} + 0.5\Delta\lambda) < \lambda_{\max}; \\ \overline{\lambda} + 0.5\Delta\lambda - (\lambda_{\max} - \lambda_{\min}), & \text{if } (\overline{\lambda} + 0.5\Delta\lambda) > \lambda_{\max}. \end{cases} \tag{16.8}$$

Equations (16.7) and (16.8) establish a one-to-one correspondence between the optimal object stimuli $\left\{ Opt(\mathcal{X}|\Phi) \backslash \left\{ x_2^{\lambda_{\min},\lambda_{\max}}, 1 - x_2^{\lambda_{\min},\lambda_{\max}} \right\} \right\}$ and the pairs $(\Delta\lambda, \overline{\lambda})$ $(0 < \Delta\lambda < (\lambda_{\max} - \lambda_{\min}), \lambda_{\min} < \overline{\lambda} < \lambda_{\max})$. We will assume, by definition, that zero spectral bandwidth ($\Delta\lambda = 0$) defines the optimal object stimulus $x(\lambda) \equiv 0$. Likewise, the spectral bandwidth $\Delta\lambda = \lambda_{\max} - \lambda_{\min}$, defines the optimal object stimulus $x(\lambda) \equiv 1$.

Consider now $\mathbf{R}^3$ endowed with the spherical coordinate system (ρ, θ, β) related to the Cartesian coordinates (x, y, z) as

$$x = \rho \sin\beta \cos\theta; \quad y = \rho \sin\beta \sin\theta; \quad z = \rho\cos\theta.$$

Let us define a map $f : Opt(\mathcal{X}|\Phi) \to S^2$ sending an optimal object stimulus with the spectral bandwidth $\Delta\lambda (0 < \Delta\lambda < (\lambda_{\max} - \lambda_{\min}))$ and central wavelength $\overline{\lambda}$ to a point on the sphere S^2 with the azimuth angle

$$\beta = 2\pi \frac{\overline{\lambda} - \lambda_{\min}}{\lambda_{\max} - \lambda_{\min}}, \tag{16.9}$$

and the zenith angle

$$\theta = \pi \frac{\Delta\lambda}{\lambda_{\max} - \lambda_{\min}}. \tag{16.10}$$

The stimuli $x(\lambda) \equiv 0$ and $x(\lambda) \equiv 1$ are being sent to the south ($\theta = 0$) and north ($\theta = \pi$) poles respectively. The map f is one-to-one and weak* continuous, that is, homeomorphism between $Opt(\mathcal{X}|\Phi)$ and S^2.

Complete Eqs. (16.7) and (16.8) with the following

$$\rho = c, \tag{16.11}$$

where c is the chromatic contrast (16.4). Then sending a rectangular metamer $(1-c)x_{0.5}(\lambda) + cx_{opt}$ ($x_{opt} \in Opt(\mathcal{X} \mid \Phi)$) to a point in the unit ball B_1^3 in $\mathbf{R}^3$ determined by the point $f\left(x_{opt}\right)$ on its surface (S^2) and the radius equal to c, we define a bijection $\mathcal{AR}_{0.5}^3 \to B_1^3$ which is actually a homeomorphism.

This homeomorphism provides a spherical representation of the rectangular object atlas, $\mathcal{AR}_{0.5}^3$. Since the parameters c, $\Delta\lambda$, and $\overline{\lambda}$ of a rectangular metamer uniquely determine the position of the rectangular metamer in B_1^3, these can be thought of as coordinates in $\mathcal{A}_{0.5}^3$ considered as the object-colour stimulus manifold. We will refer to these as $c\Delta\lambda\overline{\lambda}$-coordinates. For any object stimulus x, the $c\Delta\lambda\overline{\lambda}$-coordinates of its rectangular metamer will be considered, by definition, as the $c\Delta\lambda\overline{\lambda}$-coordinates of x.

The usual differential structure on the sphere S^2 induces that on $Opt(\mathcal{X}|\Phi)$. Likewise a differential structure of a manifold with boundary on B_1^3 induces that on $\mathcal{AR}_{0.5}^3$. Equations (16.9)–(16.11) provide the spherical coordinates in $\mathcal{AR}_{0.5}^3$. Note that the azimuth angle ϕ is not defined for the object colour stimuli $\left\{t\chi_\Lambda : 0 \le t \le 1\right\}$. We will refer to the latter as the *achromatic segment of* $\mathcal{AR}_{0.5}^3$ (should not be confused with the achromatic segment of the object-colour solid $[0, \Phi(\chi_\Lambda)]$ in Definition 14.2).

The restriction $\Phi|_{\mathcal{AR}_{0.5}^3}$ of the colour-signal map Φ on $\mathcal{AR}_{0.5}^3$ embeds $\mathcal{AR}_{0.5}^3$ as a topological submanifold with boundary in $\mathbf{R}^3$. However, the object-colour solid $\Phi(\mathcal{X}) = \Phi\left(\mathcal{AR}_{0.5}^3\right)$ is not a smooth submanifold with boundary in $\mathbf{R}^3$ because the north and south poles are the critical values of $\Phi|_{\mathcal{AR}_{0.5}^3}$.

Given an illuminant $p(\lambda)$, the colour signal $z = (z_1, z_2, z_3)$ produced by a rectangular metamer, $x\left(c, \Delta\lambda, \overline{\lambda}\right)$, determined by the coordinate triplet $(c, \Delta\lambda, \overline{\lambda})$ (i.e., $z = \Phi_p|_{\mathcal{AR}_{0.5}^3}\left(x\left(c, \Delta\lambda, \overline{\lambda}\right)\right)$) is evaluated as

$$z_i = 0.5(1-c)\int_{\lambda_{\min}}^{\lambda_{\max}} p(\lambda)\,\varphi_i(\lambda)\,d\lambda + c\int_{\overline{\lambda}-0.5\Delta\lambda}^{\overline{\lambda}+0.5\Delta\lambda} p(\lambda)\,\varphi_i(\lambda)\,d\lambda \quad (i = 1, 2, 3). \tag{16.12}$$

Here we interpret the functions $\varphi_i(\lambda)$ as periodical on $\mathbf{R}$ with period $\lambda_{\max} - \lambda_{\min}$.

Differentiating $z_i (i = 1, 2, 3)$ with respect to c, $\Delta\lambda$, and $\overline{\lambda}$ we get

$$\begin{aligned}
\frac{\partial z_i}{\partial c}\left(c, \Delta\lambda, \overline{\lambda}\right) &= \int_{\overline{\lambda}-0.5\Delta\lambda}^{\overline{\lambda}+0.5\Delta\lambda} p(\lambda)\,\varphi_i(\lambda)\,d\lambda - 0.5\int_{\lambda_{\min}}^{\lambda_{\max}} p(\lambda)\,\varphi_i(\lambda)\,d\lambda, \\
\frac{\partial z_i}{\partial(\Delta\lambda)}\left(c, \Delta\lambda, \overline{\lambda}\right) &= 0.5c\left(p\left(\overline{\lambda}+0.5\Delta\lambda\right)\varphi_i\left(\overline{\lambda}+0.5\Delta\lambda\right)\right. \\
&\quad \left. + p\left(\overline{\lambda}-0.5\Delta\lambda\right)\varphi_i\left(\overline{\lambda}-0.5\Delta\lambda\right)\right), \\
\frac{\partial z_i}{\partial\overline{\lambda}}\left(c, \Delta\lambda, \overline{\lambda}\right) &= c\left(p\left(\overline{\lambda}+0.5\Delta\lambda\right)\varphi_i\left(\overline{\lambda}+0.5\Delta\lambda\right) - p\left(\overline{\lambda}-0.5\Delta\lambda\right)\varphi_i\left(\overline{\lambda}-0.5\Delta\lambda\right)\right).
\end{aligned} \tag{16.13}$$

One can see that the Jacobian of $\Phi_p|_{\mathcal{AR}_{0.5}^3}$ does not have full rank at the points of the achromatic segment of $\mathcal{AR}_{0.5}^3$. Also $\Phi_p\left(\mathcal{AR}_{0.5}^3\right)$ is not smooth along both the meridians. Being continuous on $\left[\lambda_{\max}, \lambda_{\min}\right]$, $\varphi_i(\lambda)$ considered as a periodical function on $\mathbf{R}$ has a jump discontinuity at $\lambda_{\min}$ and $\lambda_{\max}$ unless $\varphi_i\left(\lambda_{\min}\right) = \varphi_i\left(\lambda_{\max}\right)$. Therefore, the partial derivatives are not defined for the elementary step functions $x_1^{\lambda_1}$ for which $\overline{\lambda} + 0.5\Delta\lambda = \lambda_{\max}$. Nor are they defined for the elementary step functions $1 - x_1^{\lambda_1}$ for which $\overline{\lambda} - 0.5\Delta\lambda = \lambda_{\min}$.

Apart from the achromatic segment and meridians the Jacobian of $\Phi_p|_{\mathcal{A}\mathcal{R}^3_{0.5}}$ has full rank provided that the spectral line $\Phi_p(\delta_\lambda : \lambda \in \Lambda)$ does not have self-intersections. Therefore, one can use the inverse of the Jacobian of $\Phi_p|_{\mathcal{A}\mathcal{R}^3_{0.5}}$ to locally resolve Eq. (16.12) with respect to z. Specifically, let us fix some $x^0\left(c^0, \Delta\lambda^0, \overline{\lambda}^0\right)$, and let $z^0 = (z_1^0, z_2^0, z_3^0)$ be such colour signal that $z^0 = \Phi_p|_{\mathcal{A}\mathcal{R}^3_{0.5}}\left(x^0\left(c^0, \Delta\lambda^0, \overline{\lambda}^0\right)\right)$. Consider an incremental colour signal $\Delta z = (\Delta z_1, \Delta z_2, \Delta z_3)$. It follows from the inverse function theorem that in some neighbourhood of z^0 the following equation approximately holds:

$$\Phi_p^{-1}|_{\mathcal{A}\mathcal{R}^3_{0.5}}(z^0 + \Delta z) \approx x^0 + J^{-1}(\Delta z), \tag{16.14}$$

where J^{-1} is the inverse of the Jacobian of $\Phi_p|_{\mathcal{A}\mathcal{R}^3_{0.5}}$ at $x^0\left(c^0, \Delta\lambda^0, \overline{\lambda}^0\right)$, and $\Delta z = (\Delta z_1, \Delta z_2, \Delta z_3)$ is the incremental colour signal.

Equation (16.14) shows how rectangular metamer $x(c, \Delta\lambda, \overline{\lambda})$ in (16.12) will vary in response to a small variation of colour signal z in the left hand side of Eq. (16.12).

Since Govardovskii's cone fundamentals (Fig. 3.8) take at $\lambda_{\min}$ and $\lambda_{\max}$ values very close to zero, it seems reasonable to assume that each φ_i vanishes outside of the visible spectrum. Specifically, let us extend the domain of definition of each φ_i to $\mathbf{R}_+$ so that the extension $\tilde{\varphi}_i(\lambda)$ is everywhere continuous and positive, $\lim\limits_{\lambda \to +\infty} \tilde{\varphi}_i(\lambda) = \tilde{\varphi}_i(0) = 0$, and $\left(\int_0^{\lambda_{\min}} \sqrt{(\tilde{\varphi}_1(\lambda))^2 + (\tilde{\varphi}_2(\lambda))^2 + (\tilde{\varphi}_3(\lambda))^2}\, d\lambda + \int_{\lambda_{\max}}^{+\infty} \sqrt{(\tilde{\varphi}_1(\lambda))^2 + (\tilde{\varphi}_2(\lambda))^2 + (\tilde{\varphi}_3(\lambda))^2}\, d\lambda\right) \le \varepsilon$ $(\varepsilon > 0)$. Respectively, let us extend $s(\lambda)$ in (15.4) to $\tilde{s}(\lambda)$ $(0 \le \lambda \le +\infty)$ as

$$\tilde{s}(\lambda) = \int_0^\lambda \sqrt{(\tilde{\varphi}_1(\mu))^2 + (\tilde{\varphi}_2(\mu))^2 + (\tilde{\varphi}_3(\mu))^2}\, d\mu.$$

Letting $\hat{s}(\lambda) =: \tilde{s}(\lambda)/\tilde{s}(+\infty)$, define functions $\hat{\varphi}_i(\omega) =: \tilde{\varphi}_i \circ \hat{s}^{-1}(\omega)$ which map $[0, 1] \to \mathbf{R}$. The functions $\hat{\varphi}_i$ can be considered as periodical on $\mathbf{R}$ with period 1. They preserve their continuity on $\mathbf{R}$.

The spectral bandwidth and central wavelength of optimal object stimuli can be redefined in terms of the new parameter ω. Specifically, given optimal object stimuli $x_2^{\lambda_1,\lambda_2}$ and $1 - x_2^{\lambda_1,\lambda_2}$, define

$$\Delta\omega = \begin{cases} |\omega_1 - \omega_2|, & \text{for } x_2^{\lambda_1,\lambda_2}; \\ 1 - |\omega_1 - \omega_2|, & \text{for } 1 - x_2^{\lambda_1,\lambda_2}; \end{cases} \tag{16.15}$$

where $\omega_i = \hat{s}(\lambda_i)$ $(i = 1, 2)$. Note that $\Delta\omega \ne \hat{s}(\Delta\lambda)$. Similarly, define

$$\overline{\omega} = \begin{cases} 0.5(\omega_1 + \omega_2), & \text{for } x_2^{\lambda_1,\lambda_2}; \\ 0.5((\omega_1 + \omega_2) + 1), & \text{for } 1 - x_2^{\lambda_1,\lambda_2} \text{ if } (\omega_1 + \omega_2) < 1; \\ 0.5((\omega_1 + \omega_2) - 1), & \text{for } 1 - x_2^{\lambda_1,\lambda_2} \text{ if } (\omega_1 + \omega_2) > 1. \end{cases} \tag{16.16}$$

Let us define now a function $\hat{f} : Opt(\mathcal{X}\,|\Phi) \to S^2$ which maps an optimal object stimulus with the spectral bandwidth $\Delta\omega$ and central wavelength $\overline{\omega}$ to the point on the unit sphere with the azimuth angle $\hat{\beta} =: 2\pi\overline{\omega}$ and the zenith angle $\hat{\theta} := \pi\Delta\omega$. Note that now $\Phi \circ \hat{f}^{-1}$ is a homeomorphism between the unit sphere S^2 and the object-colour-solid boundary ∂S which is a local diffeomorphism everywhere except the poles. Letting $\rho = c$ as in (16.11) this homeomorphism is smoothly extended to the whole object-colour solid and the unit ball in $\mathbf{R}^3$. We will call it the *spherical representation* of the object-colour solid.

As for all the triplets of regular linear colour mechanisms $\Phi = (\varphi_1(\lambda), \varphi_2(\lambda), \varphi_3(\lambda))$ (i.e., those satisfying the regularity assumption in Definition 14.5) the set of optimal object stimuli $Opt(\mathcal{X}\,|\Phi)$

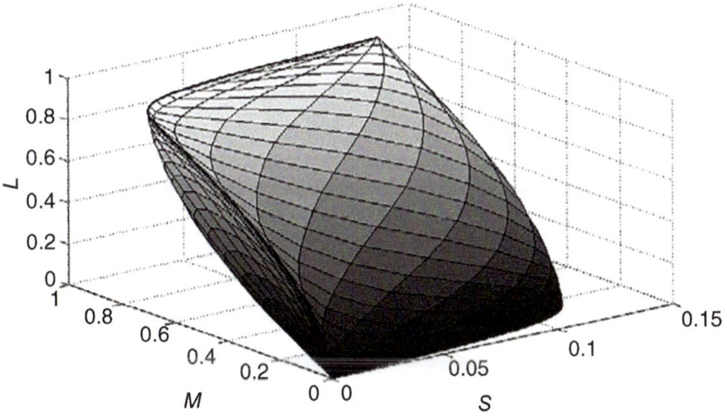

Figure 16.1 Regular object-colour solid with coordinate lines on its boundary surface representing loci of constant spectral bandwidth and central wavelength.

is the same, the regular optimal object atlas $\mathcal{AR}^3_{0.5}$ established above will serve all of them. Therefore, for all the regular linear colour mechanisms the object-colour stimulus manifold will be the same. It will referred to as the *trichromatic regular object-colour stimulus manifold*. All the difference between two trichromatic observers satisfying the regularity assumption can be described as that in their non-linear object-colour stimulus maps.

16.4.2 Spherical representation of the trichromatic object-colour stimulus manifold and the object-colour stimulus gamut

In the general case (i.e., when the regularity assumption does not hold true), the optimal object atlas $\mathcal{A}^3_{0.5}$ may be different for different colour mechanisms. In other words, $\mathcal{A}^3_{0.5}$ does not always coincide with $\mathcal{AR}^3_{0.5}$. Still, one can eliminate from the atlas the optimal metamers with more than two transitions if one extends the range of chromatic contrast magnitudes to values more than 1. More specifically, given linear colour mechanisms $\Phi = \left(\varphi_1(\lambda), \varphi_2(\lambda), \varphi_3(\lambda)\right)$, an illuminant $p(\lambda) > 0$, and an object stimulus $x(\lambda) \in \mathcal{X}$, consider simultaneous equations

$$\int_{\lambda_{\min}}^{\lambda_{\max}} \left((1-c)x_{0.5}(\lambda) + cx_2(\lambda;\lambda_1,\lambda_2)\right) p(\lambda)\,\varphi_i(\lambda)\,d\lambda = \int_{\lambda_{\min}}^{\lambda_{\max}} x(\lambda)\,p(\lambda)\,\varphi_i(\lambda)\,d\lambda, \tag{16.17}$$

where $x_2(\lambda;\lambda_1,\lambda_2)$ is the indicator of the set $[\lambda_1,\lambda_2]$ ($\lambda_{\min} \le \lambda_1 < \lambda_2 \le \lambda_{\max}$), $i = 1, 2, 3$; or

$$c\int_{\lambda_1}^{\lambda_2} p(\lambda)\,\varphi_i(\lambda)\,d\lambda = \int_{\lambda_{\min}}^{\lambda_{\max}} (x(\lambda) - 0.5(1-c))\,p(\lambda)\,\varphi_i(\lambda)\,d\lambda. \tag{16.18}$$

This system of equations can always be resolved with respect to c, λ_1, and λ_2.

Indeed, the regular object-colour solid $\Phi_p\left(\mathcal{AR}^3_{0.5}\right)$ is a closed neighbourhood of the centre in the colour solid $\Phi_p(\mathcal{X})$. Hence, the line through the centre and the colour signal $\Phi_p(x)$ intersects the boundary of $\Phi_p\left(\mathcal{AR}^3_{0.5}\right)$ at the two points the inverse Φ_p-images of which will contain at least one $x_2(\lambda;\lambda_1,\lambda_2)$ such that along with a proper c it brings a solution to the system (16.18).

The problem here is that the solution may be not unique. In other words, for some colour mechanisms these stimuli $x_2(\lambda;\lambda_1,\lambda_2)$ may be metameric to each other.

Definition 16.8 *We say that linear colour mechanisms* $\Phi = \left(\varphi_1(\lambda), \varphi_2(\lambda), \varphi_3(\lambda)\right)$ *meet the* weak regularity assumption *if there are no such* $x_2^{\lambda_1',\lambda_2'}$ *and* $x_2^{\lambda_1'',\lambda_2''}$ *in* X_2^I *that* $\Phi\left(x_2^{\lambda_1',\lambda_2'}\right) = \Phi\left(x_2^{\lambda_1'',\lambda_2''}\right)$.

Conjecture 16.3 *Govardovskii's templates (3.16) meet the weak regularity assumption.*

It may happen that for some illuminant $p(\lambda)$ $\Phi_p\left(x_2^{\lambda_1',\lambda_2'}\right) = \Phi_p\left(x_2^{\lambda_1'',\lambda_2''}\right)$ even if Φ meets the weak regularity assumption.

Definition 16.9 *Given colour mechanisms* $\Phi = (\varphi_1, \varphi_2, \varphi_3)$ *meeting the weak regularity assumption, an illuminant* $p(\lambda)$ *will be called* admissible *with respect to* Φ *if for any* $x_2^{\lambda_1',\lambda_2'}, x_2^{\lambda_1'',\lambda_2''} \in X_2^I$ *we have* $\Phi_p\left(x_2^{\lambda_1',\lambda_2'}\right) \neq \Phi_p\left(x_2^{\lambda_1'',\lambda_2''}\right)$.

Problem 16.1 *Describe the set of all admissible illuminants with respect to Govardovskii's templates (3.16).*

We will restrict our further consideration in this section to only the linear colour mechanisms satisfying the weak regularity assumption and the admissible illuminants with respect to these mechanisms. For these the system (16.18) has the unique solution with respect to c, λ_1, and λ_2. If the colour signal $\Phi_p(x)$ lies within the regular object-colour solid the solution will be such that $0 \leq |c| \leq 1$. For those colour signals which are between the boundary surfaces of the real and regular colour solids we will get $|c| > 1$.

The function $0.5 + c\left(x_2(\lambda; \lambda_1, \lambda_2) - 0.5\right)$ takes two values: $0.5(1 \pm c)$. When one of these exceeds 1 the other becomes negative. While such functions (i.e., for $|c| > 1$) cannot be treated as reflectance spectra, we will refer to them as *imaginary object stimuli*. Taking them into consideration allows one to make up an *extended rectangular atlas* (comprising all the ordinary rectangular metamers and some imaginary rectangular metamers).

Using the extended rectangular atlas one can uniformly represent the trichromatic object-colour stimulus manifolds, $\mathcal{A}_{0.5}^3$, in terms of one and the same spherical coordinate system based on the $c\Delta\lambda\bar{\lambda}$-coordinates. Indeed, for each element of any $\mathcal{A}_{0.5}^3$ one can evaluate its $c\Delta\lambda\bar{\lambda}$-coordinates solving Eq. (16.18). Then, using Eqs. (16.7)–(16.11), this element can be represented as a point in $\mathbf{R}^3$. In this spherical coordinate system, longitude will encode central wavelength, latitude spectral bandwidth, and the distance from the centre will encode chromatic contrast.

Generating its own extended rectangular atlas, each $\mathcal{A}_{0.5}^3$ will be represented in this space as a volume encompassing the unit ball B_1^3. This volume depends on the colour mechanisms and, generally speaking, on the illuminant. We will call it the *object-colour stimulus gamut* and denote Γ_{Φ_p}. Recall that although the extended rectangular atlas, generally, depends on colour mechanisms, this dependence reduces to variations in the chromatic contrast range. The object-colour stimulus gamut can be considered as a continuous transformation of the object colour solid $\Phi_p(\mathcal{X})$. In contrast with the object-colour solid which changes its shape as a result of the linear transformation of the colour mechanisms, the object-colour stimulus gamut is invariant with respect to any non-singular linear transformation of the colour mechanisms.

Consider a set of the linear colour mechanisms satisfying the weak regularity assumption and the admissible illuminants with respect to these mechanisms. All the regular linear colour mechanisms have the same the object-colour stimulus gamut (the unit ball B_1^3). Hence, the set of all the object-colour gamuts is partially ordered by inclusion. The unit ball B_1^3 is the least element for any set of the object-colour gamuts.

16.4.3 Object-colour stimulus shift induced by the illuminant change

Since the set of optimal object stimuli is independent of illumination, an optimal object atlas for a particular observer remains the same for all illuminations. However, the non-linear object-colour stimulus map does change with illumination. Because the non-linear object-colour stimulus map determines to which element of object atlas $\mathcal{A}_{0.5}$ an object stimulus will be metameric, the optimal metamer of one and the same object stimulus varies with the illuminant. So do the $c\Delta\lambda\overline{\lambda}$-coordinates of the object stimulus. The illuminant-induced object-colour stimulus shift, introduced in Section 16.3 so as to characterise the effect of the illuminant change on the object-colour stimulus, can be expressed in terms of $c\Delta\lambda\overline{\lambda}$-coordinates. In this section, we shall study the effect of illumination on $c\Delta\lambda\overline{\lambda}$-coordinates.

Let p' and p'' be two illuminants, and $\kappa_{p'}(x)$ and $\kappa_{p''}(x)$ the object-colour stimuli of the object stimulus x under these illuminants ($\kappa_{p'}$ and $\kappa_{p''}$ being the non-linear object-colour stimulus maps corresponding to these illuminants). When the linear colour mechanisms satisfy the weak regularity assumption and the illuminants are admissible with respect to these mechanisms, $\kappa_{p'}(x)$ and $\kappa_{p''}(x)$ can be characterised by $c\Delta\lambda\overline{\lambda}$-coordinates. Let assume that $\kappa_{p'}(x) = x\left(c', \Delta\lambda', \overline{\lambda}'; \lambda\right)$ and $\kappa_{p''}(x) = x\left(c'', \Delta\lambda'', \overline{\lambda}''; \lambda\right)$. Then the object-colour stimulus shift induced by a change from p' to p'' can be decomposed into (i) the *chromatic-contrast shift*: $c'' - c'$; (ii) the *spectral-bandwidth shift*: $\Delta\lambda'' - \Delta\lambda'$; and (iii) the *central-wavelength shift*: $\overline{\lambda}'' - \overline{\lambda}'$.

When the difference between p' and p'' is small, the components of the object-colour stimulus shift induced by a change from p' to p'' can be explicitly approximated. Specifically, consider linear colour mechanisms $\Phi = (\varphi_1(\lambda), \varphi_2(\lambda), \varphi_3(\lambda))$ satisfying the weak regularity assumption and an illuminant $p(\lambda)$ admissible with respect to Φ. Given an object stimulus $x(\lambda)$, let c, $\Delta\lambda$ and $\overline{\lambda}$ be the chromatic contrast, spectral bandwidth, and central wavelength of $x(\lambda)$ when illuminated by $p(\lambda)$ respectively. Let us define a function $F(u, v) := (F_1(u, v), F_2(u, v), F_3(u, v))$ where $u =: (c, \Delta\lambda, \overline{\lambda})$, $v = p(\lambda)$, and for each $i = 1, 2, 3$

$$F_i(u, v) =: c \int_{\overline{\lambda}-0.5\Delta\lambda}^{\overline{\lambda}+0.5\Delta\lambda} p(\lambda)\,\varphi_i(\lambda)\,d\lambda + \int_{\lambda_{\min}}^{\lambda_{\max}} (0.5(1-c) - x(\lambda))\,p(\lambda)\,\varphi_i(\lambda)\,d\lambda. \tag{16.19}$$

Consider particular $x(\lambda)$ and $p^0(\lambda)$, denoting chromatic contrast, spectral bandwidth, and central wavelength of $x(\lambda)$ under the illuminant $p^0(\lambda)$ as c^0, $\Delta\lambda^0$ and $\overline{\lambda}^0$ respectively. Note that $F\left(u^0, v^0\right) = 0$, where $u^0 =: \left(c^0, \Delta\lambda^0, \overline{\lambda}^0\right)$, $v^0 =: p^0(\lambda)$.

If the illuminant $p^0(\lambda)$ is a continuous function, then $F(u, v)$ is Fréchet differentiable. Indeed, $F(u, v)$ is linear with respect to v. When v is fixed the partial derivative of F with respect to u at $\left(u^0, v^0\right)$ comprises the partial derivatives of the component functions $F_i (i = 1, 2, 3)$:

$$\begin{aligned}
\frac{\partial F_i}{\partial c}\left(u^0, v^0\right) &= \int_{\overline{\lambda}^0-0.5\Delta\lambda^0}^{\overline{\lambda}^0+0.5\Delta\lambda^0} p^0(\lambda)\,\varphi_i(\lambda)\,d\lambda - 0.5 \int_{\lambda_{\min}}^{\lambda_{\max}} p^0(\lambda)\,\varphi_i(\lambda)\,d\lambda, \\
\frac{\partial F_i}{\partial(\Delta\lambda)}\left(u^0, v^0\right) &= c\Big(p^0\left(\overline{\lambda}^0 + 0.5\Delta\lambda^0\right)\varphi_i\left(\overline{\lambda}^0 + 0.5\Delta\lambda^0\right) \\
&\quad + p^0\left(\overline{\lambda}^0 - 0.5\Delta\lambda^0\right)\varphi_i\left(\overline{\lambda}^0 - 0.5\Delta\lambda^0\right)\Big), \\
\frac{\partial F_i}{\partial \overline{\lambda}}\left(u^0, v^0\right) &= c\Big(p^0\left(\overline{\lambda}^0 + 0.5\Delta\lambda^0\right)\varphi_i\left(\overline{\lambda}^0 + 0.5\Delta\lambda^0\right) \\
&\quad - p^0\left(\overline{\lambda}^0 - 0.5\Delta\lambda^0\right)\varphi_i\left(\overline{\lambda}^0 - 0.5\Delta\lambda^0\right)\Big).
\end{aligned} \tag{16.20}$$

We will assume that $\Delta\lambda^0 \neq 0$. For the sake of brevity we will also assume[7] that $\overline{\lambda}^0 + 0.5\Delta\lambda^0 \neq \lambda_{\max}$, and $\overline{\lambda}^0 - 0.5\Delta\lambda^0 \neq \lambda_{\min}$. In this case, the matrix $\frac{\partial F}{\partial u}\left(u^0, v^0\right)$ is invertible unless the line

7 Though this assumption is not necessary as one can extend the visible interval to the circle, and change the variable λ to ω as it was done above. Yet, it would make the formulae more complicated.

$\Phi_p(\delta_\lambda : \lambda \in \Lambda)$ self-intersects. Therefore, the function F implicitly defines locally a function $u = g(v)$ such that $F(g(v), v) = 0$. Differentiating the equation $F(u, v) = 0$ at (u^0, v^0) results in

$$du = -\left(\frac{\partial F}{\partial u}(u^0, v^0)\right)^{-1} \circ \frac{\partial F}{\partial v}(u^0, v^0) \circ dv. \tag{16.21}$$

Equation (16.21) can be used to predict how much the coordinates $\left(c^0, \Delta\lambda^0, \overline{\lambda}^0\right)$ of $x(\lambda)$ will alter as a result of a small change of the illuminant $p^0(\lambda)$. For example, consider $p(\lambda) = p^0(\lambda) + \Delta p(\lambda)$, where $\|\Delta p(\lambda)\|_1 < \varepsilon$, and B_ε is the ball where the explicit function $u = g(v)$ is defined. The partial derivative $\frac{\partial F}{\partial v}(u^0, v^0)$ is a linear map $DF : \mathbf{L}^1 \to \mathbf{R}^3$ defined by the three linear functionals DF_i:

$$\begin{aligned} DF_i(q(\lambda)) = c &\int_{\overline{\lambda}^0-0.5\Delta\lambda^0}^{\overline{\lambda}^0+0.5\Delta\lambda^0} q(\lambda)\,\varphi_i(\lambda)\,d\lambda \\ &+ \int_{\lambda_{\min}}^{\lambda_{\max}} (0.5(1-c) - x(\lambda))\,q(\lambda)\,\varphi_i(\lambda)\,d\lambda \quad (i = 1, 2, 3). \end{aligned}$$

Applying the 3×3 matrix $-\left(\frac{\partial F}{\partial u}(u^0, v^0)\right)^{-1}$ to the vector $DF(\Delta p(\lambda)) = \big(DF_1(\Delta p(\lambda)), DF_2(\Delta p(\lambda)), DF_3(\Delta p(\lambda))\big)$ one gets the alteration of the coordinates $\left(c^0, \Delta\lambda^0, \overline{\lambda}^0\right)$ induced by $\Delta p(\lambda)$.

Note that the linear map DF determines an equivalence relation which can be interpreted as a metamerism on the set of illuminants (i.e., $\mathbf{L}^1$). Of course it is essentially dependent on the stimulus object $x(\lambda)$. Any two metameric alterations $\Delta p'(\lambda)$ and $\Delta p''(\lambda)$ of the illuminant $p^0(\lambda)$, that is, such that $DF(\Delta p'(\lambda)) = DF(\Delta p''(\lambda))$, will produce equal alteration of the coordinates $\left(c^0, \Delta\lambda^0, \overline{\lambda}^0\right)$. In other words, they will have the same effect on the colour of the stimulus object $x(\lambda)$.

Of practical interest is the effect of some known parametric families of illuminants (e.g., black-body radiators and/or natural daylight) on object colour. Consider, for instance, such colour mechanisms for which the black-body radiator (5.54) is admissible. When its colour temperature varies within an interval $[T_1, T_2]$ equations

$$\begin{aligned} F_i(c, \Delta\lambda, \overline{\lambda}, T) =: c &\int_{\overline{\lambda}-0.5\Delta\lambda}^{\overline{\lambda}+0.5\Delta\lambda} p(\lambda; T)\,\varphi_i(\lambda)\,d\lambda \\ &+ \int_{\lambda_{\min}}^{\lambda_{\max}} (0.5(1-c) - x(\lambda))\,p(\lambda; T)\,\varphi_i(\lambda)\,d\lambda \quad (i = 1, 2, 3) \end{aligned} \tag{16.22}$$

implicitly defines a curve $\mathbf{c}_x(T) =: \left(c(T), \Delta\lambda(T), \overline{\lambda}(T)\right)$ in the trichromatic object-colour stimulus manifold for each object stimulus $x(\lambda)$. The curve $\mathbf{c}_x(T)$ describes the illuminant-induced object-colour stimulus shift which $x(\lambda)$ undergoes due to the colour temperature change.

As known, as few as only three basis functions suffice to approximate by their linear combination the radiant spectral density function of daylight under various conditions with quite reasonable accuracy (Judd et al., 1964). A volume in the three-dimensional space of the parameters specifies some variations of daylight. An equation similar to (16.22) will determine a corresponding volume in the trichromatic object-colour stimulus manifold specifying the illuminant-induced object-colour stimulus shift (for a particular object stimulus) induced by these daylight variations.

17

Object-Colour Perception in a Single-Illuminant Scene

Since the notion of colour perception is applicable exclusively to human subjects, the material in this chapter concerns only human observers – mainly trichromatic observers (with the exception of Section 17.4). We will look into the internal structure of the object-colour palette, in much the same way as it was done for the light-colour palette in Chapter 8. Generally speaking, the colorimetric paradigm is so impoverished that one can hardly speak of perception in this experimental condition. Bringing back real objects lit by a real light source is an important step forward toward natural (genuine) perception – the ultimate destination of any perceptual theory. Yet, a single-illuminant scene is also an abstraction, as unrealistic as the colorimetric experimental paradigm. Even in the presence of a single light source the illumination of different objects in the scene (even various parts of one and the same object) will usually differ, for example, due to mutual reflections. However, to begin our consideration with this abstraction is a good departing point, given the exclusive complexity of the multi-illuminant case. The conceptual framework that will be developed in this chapter will be then used as an ancillary stage in Chapter 19.

It will not be superfluous to remind that, as emphasised in the Outline (Chapter 1), although beyond all doubt we perceive things in colour because they reflect light, it would be wrong to jump from this indisputable truth to the conclusion that the colour of a reflecting object is nothing more than the colour of the light that it reflects.[1] In order to make sure that it is not so, it is sufficient to resort to an obscuring screen. If you look at a dandelion through a wee hole in a sheet of paper, you will see that the flower (or more precisely, part of it accessible to observation through the hole) is not as yellow (or rather it is not yellow at all) as it appears without the screening paper. The relevance of this observation to the present context is that what is seen through the hole is no object colour. It is usually referred to as the aperture colour (Kaiser & Boynton, 1996). We believe that while, strictly speaking, being neither light colour nor object colour, it is closer to light colour rather than to object colour. In colour literature, one can find lots of other demonstrations testifying that the same colour signal can evoke different colours depending on whether it is produced by light as such or by the light reflected from an object. It is important to emphasise that the difference in question is not just the difference between two colours within the same light-colour palette. It is the difference between two colours belonging to two different colour palettes. In fact, the palette of object colours is rather different from that of light colours.

1 'Light emitted from a source of radiant power, such as the sun or a tungsten filament lamp, strikes a given object and thus, makes the object visible to the observer. The color stimulus **Q** emerging from the object and entering the observer's eye produces a sensation of color that observer identifies as the color of the object.' (Wyszecki & Stiles, 1982, p. 143).

Foundations of Colour Science: From Colorimetry to Perception, First Edition. Alexander D. Logvinenko and Vladimir L. Levin.
© 2023 John Wiley & Sons Ltd. Published 2023 by John Wiley & Sons Ltd.

If the colour of a reflecting object were the colour of the reflected light, one could expect the object-colour palette to be poorer than the light-colour palette (under a single illuminant) since the object-colour solid is a proper subset of the colour-signal cone. To the contrary, the former is richer than the latter. Indeed, as well-known, things might appear to be brown, khaki, olive and many other colours which are not available in the light-colour palette.

The reason for this is the fact that the object-colour palette comprises six rather than four component hues. Indeed, Da Vinci, Hering and many other scientists and artists included in the list of elemental (i.e., component) colours along with yellow, blue, red, and green also white and black. This conjecture has been verified by using the partial hue-matching technique (Section 8.5.1). Specifically, the partial hue-matching experiment with a sample of Munsell papers as in Fig. 8.16 completed with black, white, and blackish and whitish Munsell papers yielded the same four (yellow, blue, red, and green) chromaticity classes plus two more, one containing black and blackish papers, the other white and whitish papers (Logvinenko & Beattie, 2011). These two chromaticity classes lend themselves to be named the black and white chromaticity classes. The existence of such chromaticity classes makes black and white included in the list component hues despite the fact that this is somewhat contrary to the usual use of this colour term (i.e., hue).

When the sample is completed with a grey Munsell paper, a new chromaticity class comes out which contains only this grey paper. In other words, grey was found to be included in no other chromaticity class (Logvinenko & Beattie, 2011). Interestingly, this special status of grey was observed even when the grey paper was accompanied by Munsell papers with lowest Munsell Chroma looking very bleak (Logvinenko & Beattie, 2011). Such the papers map close to the centre of the object-colour solid (where the grey paper maps to). Nevertheless, neither of these papers fell into the same chromaticity class as the grey paper. All this leads to a conclusion that grey plays in the object-colour palette the same special role as the neutral light colour in the light-colour palette. Grey is a hueless object-colour; the only hueless object-colour.

It is important to note that such the property of the grey chromaticity class contradicts a widespread view that grey is a mixture (a phenomenological composition) of black and white (e.g., Gilchrist, 2006). If it were the case the grey chromaticity class would be included in the intersection of the black and white chromaticity classes. However, the black and white chromaticity classes do not intersect. Moreover, the grey chromaticity class does not intersect with either of these.

As in the case of light colours, observers prove capable of lining up object-colours in any chromaticity class with respect to the strength of the component hue that underlies this class (Logvinenko & Beattie, 2006). This suggests that there is an order on each chromaticity class (which will be referred to as the *chromatic ordering* for objects) which can be represented by a corresponding scale. Furthermore, the preliminary experiments showed that the observers accepted a task to partition the chromaticity class into such subsets that, firstly, they were chromatically ordered; and secondly, the adjacent subsets were at an approximately equal perceptual distance from each other.[2] So, there is every reason to believe that each chromatic ordering has properties ensuring an interval scale for the corresponding component hue.

All the above gives grounds for the application of the theory of chromatically ordered structures for the object colours in a single-illuminant scene in the same way as was done for the light colours in Section 8.5.4. As the basic concepts of this theory are defined for an abstract set of stimuli, the main definitions (e.g., Definition 8.20) retain their meaning, and almost all the propositions remain

2 In this regard, it is appropriate to recall that the distinguishing feature of the Munsell collection is the supposed equality of the perceptual distance between neighbouring elements in it. Although many are rather skeptical about this claim, the very possibility of a colour atlas with such a property has never been questioned (as far as the authors are aware).

true when the notions of the light stimulus set and the colour-signal cone are replaced with the object stimulus set ($\mathcal{X}$) and the object-colour solid (S).

Specifically, generalising the results of the experiments on partial hue-matching mentioned above (Logvinenko & Beattie, 2011), we will assume that all the chromaticity classes, revealed in the partial hue-matching experiments conducted so far on colour objects in a single-illuminant scene, can be extended onto the whole set of the object stimuli, $\mathcal{X}$. The grey chromaticity class will be assumed to be neutral (in the sense of Definition 8.21) and to consist of the colour-object stimuli mapping to the object-colour-solid centre.[3] All the other (six) chromaticity classes will be assumed to be properly chromatically ordered.[4] As a result we get a chromatically ordered structure $\langle \mathcal{X}, H_0, H_1, \dots, H_m \rangle$, where $m = 6$ and H_0 stands for the grey chromaticity class; or in the alternative notation adopted in Section 8.5.4: $\langle \mathcal{X}, \succsim_1, \dots, \succsim_m \rangle$, where $m = 6$, and $\succsim_i$ is the chromatic ordering on the chromaticity class H_i extended onto the whole $\mathcal{X}$. $H_1, \dots, H_4$ will stand for the yellow, blue, green, and red, as in Section 8.5.1, and H_5, H_6 for black and white chromaticity classes respectively. Alternatively, the following self-evident notation will be also used: $H_Y, H_B, H_G, H_R, H_{Bk}, H_W$. The intersection of all the equivalence relations $\sim_i$ on $\mathcal{X}$ ($\sim_i$ being the symmetric part of $\succsim_i$) will also be (as in Section 8.5.4) called *chromatic equivalence* (denoted by $\approx$). The $\approx$-equivalence of two object stimuli x and y means that the observer cannot see any difference in any of the six component hues.

The experiments on partial hue-matching conducted on coloured objects in a single-illuminant scene testify for that, apart from the single neutral chromatic class, $C(H_0)$, there are six unitary chromatic classes, $C(H_Y), C(H_B), C(H_G), C(H_R), C(H_{Bk}), C(H_W)$; thus, six unitary colours[5]. Also, there are 12 binary chromatic classes ($C(H_Y, H_B)$, $C(H_G, H_R)$, $C(H_{Bk}, H_W)$ do not exist); and 8 ternary chromatic classes: $C(H_Y, H_G, H_{Bk})$, $C(H_Y, H_G, H_W)$, $C(H_Y, H_R, H_{Bk})$, $C(H_Y, H_R, H_W)$, $C(H_B, H_G, H_{Bk})$, $C(H_B, H_G, H_W)$, $C(H_B, H_R, H_{Bk})$, and $C(H_B, H_R, H_W)$. No quaternary object-colour was found so far. In other words, the degree of chromaticity for chromatically ordered structure $\langle \mathcal{X}, H_0, H_Y, H_B, H_G, H_R, H_{Bk}, H_W \rangle$ (see Definition 8.17) is 3.

It is found that chromaticity class H_Y overlaps with all the chromaticity classes except for H_B; and vice versa, H_B overlaps with all the chromaticity classes except for H_Y. Likewise, chromaticity classes H_G and H_R make non-empty intersection with every chromaticity class except with each other. Also this is the case for chromaticity classes H_{Bk} with H_W. This means that, in accord with Hering's anticipation, the three pairs of component hues ((Y, B), (G, R), and (Bk, W)) are not combined together in the colour of any object. In other words, these three pairs of chromaticity classes: (H_Y, H_B), (H_G, H_R), and (H_{Bk}, H_W), as well as the pairs of corresponding component hues: (Y, B), (G, R), and (Bk, W), are opponent to each other (according to the terminology introduced in Section 8.6.3).

Also, we will assume that there exist six chromatic scales $f_1, \dots, f_6$ on $\mathcal{X}$ (see Definition 8.23). In line with Definition 8.23, the sextuplet $(f_1(x), \dots, f_m(x))$ will be called the *chromaticness*[6]

---

3 Strictly speaking, the location of the point in the achromatic interval to which the members of the grey chromaticity classes map remains an open issue deserving more thorough investigation.

4 Recall that according to Definition 8.20 for chromaticity class H_i to be properly chromatically ordered means, in particular, there being a maximal element in H_i. We will assume that all the maximal elements are optimal object stimuli.

5 Recall that in this chapter 'colour' means 'object colour', unless otherwise indicated.

6 Recall that according to Definition 8.23 a chromatic scale is an interval scale. As in the case of light colours, we will assume every chromatic scale to take 1 on the maximal element in the corresponding chromaticity class, and zero on the objects outside it. In this case, the chromatic value is unique. So is the chromaticness of an object stimulus.

of object $x \in \mathcal{X}$, or simply *object chromaticness*.[7] As follows from the above, the chromaticness $(f_1(x), \dots, f_6(x))$ of any object x in $\mathcal{X}$ has maximum three non-zero components. The chromaticness of any neutral object (i.e., an object in H_0) is $0 \in \mathbf{R}^6$.

Two objects, x and y in $\mathcal{X}$ are chromatically equivalent when their chromaticnesses are equal, that is, $x \approx y \Leftrightarrow (f_1(x), \dots, f_6(x)) = (f_1(y), \dots, f_6(y))$. The quotient set $A/\approx$ will be referred to as the *chromaticity palette* for objects (denoted by *CP*). Each element of the chromaticity palette is uniquely represented by the chromaticness of its representatives. The chromaticity palette contains one element the chromaticness of which is zero ($0 \in \mathbf{R}^6$). This element will be referred to as the *neutral element*.[8] The chromaticity palette without the neutral element (i.e., $(A \backslash H_0)/\approx$) will be referred to as the *proper chromaticity palette* for objects.

Likewise in Section 8.5.4, define an equivalence relation (written as $\mathrel{\hat{=}}$) on the *CP* as follows: $(f_1(x), \dots, f_6(x)) \mathrel{\hat{=}} (f_1(y), \dots, f_6(y))$ if and only if there is such $k > 0$ that $(f_1(x), \dots, f_6(x)) = (kf_1(y), \dots, kf_6(y))$. The quotient set $CP/\mathrel{\hat{=}}$ will be referred to as the *hue palette* for objects, and the elements in $CP/\mathrel{\hat{=}}$ will be called *hues* for objects (in a single-illuminant scene). The class of $\mathrel{\hat{=}}$-equivalence containing the neutral element (with zero chromaticness) of *CP* comprises this element only. This class will be called the *improper hue*, the rest of hues being referred to as *proper hues*.

According to this definition of hue two object stimuli, x and y in $\mathcal{X}$, will match in hue if their chromaticnesses are proportional to each other, that is, there is such $k > 0$ that

$$(f_1(x), \dots, f_6(x)) = (kf_1(y), \dots, kf_6(y)). \tag{17.1}$$

Of great interest is to ascertain whether the results obtained with a traditional hue matching procedure will be in line with the prediction followed from (17.1). Likewise for lights, the notion of hue for objects appears to be also self-evident even for inexperienced observers. At any rate, they easy take up the task of hue matching and consistently accomplish it. If the results obtained by the hue matching technique will in line with those obtained by the hue scaling based on the object chromaticness (i.e., Eq. (17.1)), it will show that the observers' intuition behind the word 'hue' can be conceptualised as it is done here. In turn, it will justify the entire approach to object-colour perception presented in this chapter. Admittedly, one needs to conduct a systematic study into this issue. Until a study of this sort proves otherwise we will assume that these two ways of establishing the hue match (for object colours) are equivalent.

There being three pairs of opponent component hues allows introducing the following definition of opponent hue (similar to how it was done in the case of light colours). Consider an arbitrary proper hue determined by chromaticness $(f_1, \dots, f_6)$. Let $f_{i_1}, \dots, f_{i_m}$ differ from zero. Since only maximum three chromatic values can be non-zero, $m \leq 3$. Let us assume for certainty that $f_{i_1} = f^0_{i_1}, \dots, f_{i_m} = f^0_{i_m}$. Let $H_{j_1}, H_{j_2}, H_{j_3}$ be the chromaticity classes opponent to $H_{i_1}, H_{i_2}, H_{i_3}$ respectively. Consider a chromaticness where $f_{j_1} = kf^0_{i_1}, \dots, f_{j_m} = kf^0_{i_m}$ $(k > 0)$ and the rest of chromatic values are zero. We will say that the hue, determined by this chromaticness, is *perceptually opponent* to that determined by $(f_1, \dots, f_6)$. (Note that one can always choose k small enough to ensure the existence of an object colour with such the chromaticness.)

As for light colours, each proper hue for objects (i.e., a class of $\mathrel{\hat{=}}$-equivalence with non-zero chromaticness) can be naturally ordered if we define a binary relation $\gtrsim$ on it as follows. For any two $\mathrel{\hat{=}}$-equivalent chromaticnesses $(\xi_1, \dots, \xi_6)$ and $(\xi'_1, \dots, \xi'_6)$, that is, such that $(\xi_1, \dots, \xi_6) = (k\xi'_1, \dots, k\xi'_6)$ we will assume $(\xi_1, \dots, \xi_6) \gtrsim (\xi'_1, \dots, \xi'_6)$ if $k \geq 1$; and

7 Should not be confused with the chromaticness of lights introduced in Definition 8.23.
8 Naturally, the neutral element comprises the object stimuli that form the grey chromaticity class.

$(\xi_1', \dots, \xi_6') \gtrsim (\xi_1, \dots, \xi_6)$ if $0 < k \leq 1$. Clearly, $\gtrsim$ is a complete order. It will be called *purity*. The improper hue will be assumed to have zero purity. In other words, objects which appear grey (they map to the object-colour solid centre) will be thought of as having zero purity. If there is the maximal element in a class of $\doteqdot$-equivalence, it will be called a *full object colour*.

The object stimuli containing neither blackness nor whiteness (i.e., those not included in chromaticity classes H_{Bk} and H_W) make a circular Hering structure of the same kind as that formed by all the light stimuli (see Section 8.6.3). It follows that the set of object colours generated by these object stimuli (i.e., $\mathcal{X} \backslash (H_{Bk} \cup H_W)$) make the same manifold with boundary, topologically equivalent to a disc, as the chromaticity palette for the light-colours (see Section 8.6.1). Likewise, the subset of object colours evoked by the object stimuli that do not enter into chromaticity classes H_G and H_R (i.e., $\mathcal{X} \backslash (H_G \cup H_R)$) form the same circular Hering structure, and can be geometrically represented as a disc. The same is the case for the object colours produced by $\mathcal{X} \backslash (H_Y \cup H_B)$.

In fact, a subset of all the object stimuli containing an arbitrary fixed chromatic hue (i.e., neither black nor white), h, mixed with white in all possible proportions, and their opponent hues (of all possible purities) also form a circular Hering structure of the same type. Indeed, there will be two pairs of mutually opponent chromaticity classes: black and white, and the two classes induced by hue h and its opponent hue. This circular Hering structure can be also represented by a disc. For two different chromatic hues h_1 and h_2 the corresponding circular Hering structures will be represented by two discs intersecting over the segment of achromatic object colours[9] (i.e., the set of hues containing only black and white non-zero chromatic values, and grey). When h runs over all the possible chromatic hues (which can be, in turn, represented as a circle), the corresponding disc rotates, making a three-dimensional ball. Although being not quite rigorous, this line of reasoning shows that the chromatically ordered structure $\mathcal{H} = \langle \mathcal{X}, H_0, H_Y, H_B, H_G, H_R, H_{Bk}, H_W \rangle$ defined above can be considered as a smooth manifold with boundary, topologically homeomorphic to a ball in $\mathbf{R}^3$. (Though, this fact becomes practically obvious after adopting the foregoing, quite plausible assumption concerning the relationship between the chromatic equivalence and object metamerism.) For this reason $\mathcal{H}$ will be referred to as the *spherical Hering structure*.

As in the case of light colour, the question of the relationship between the chromatic equivalence, $\approx$, and object metamerism, $\sim$, arises. In this chapter, we will assume them to concur. It follows that each chromatic scale is a colour code. As a chromatic scale is by definition continuous (see Definition 8.23) the corresponding colour code is also to be continuous, which is in line with our intuition. Really, the closer to each other the colour signals to which the object stimuli map, the more similar their colours are. So, for each chromatic scale $f_i(x)$ $(i = 1, \dots, 6)$ there is such a unique continuous real-valued function c_i on the object-colour solid, $\Phi(\mathcal{X})$, that $f_i(x) = c_i(\Phi(x))$ for all $x \in \mathcal{X}$. (Its uniqueness stems from the uniqueness of the chromatic scale, due to the fact that its range is $[0; 1]$.)

Moreover, we will assume that there is the maximal element, i.e., the *full object colour*, in each proper class of $\doteqdot$-equivalence.[10] Therefore, there is a one-to-one correspondence between the set of proper hues for objects (in a single-illuminant scene) and the set of optimal object stimuli.

9 It should not be confused with the achromatic segment in the object-colour solid, which is a subset of the colour signals.

10 Generally speaking, the existence of full object colours follows from the continuity of the chromatic scales. However, in this context it is not as important.

17.1 Perceptual object-colour coordinates

In the same manner as for light colours (Eqs. (8.46) and (8.47)), it is possible to construct new, *opponent chromatic scales*, f_{YB} and f_{RG}, on $\mathcal{X}$ from chromatic scales f_Y, f_B, f_R, and f_G:

$$\forall x \in \mathcal{X} \qquad f_{YB}(x) = \begin{cases} f_Y(x), & \text{if } f_Y(x) \neq 0 \text{ (thus, } f_B(x) = 0), \\ -f_B(x), & \text{if } f_B(x) \neq 0 \text{ (thus, } f_Y(x) = 0), \\ 0, & \text{if } f_Y(x) = 0 \text{ and } f_B(x) = 0. \end{cases} \tag{17.2}$$

$$\forall x \in \mathcal{X} \qquad f_{RG}(x) = \begin{cases} f_R(x), & \text{if } f_R(x) \neq 0 \text{ (thus, } f_G(x) = 0), \\ -f_G(x), & \text{if } f_G(x) \neq 0 \text{ (thus, } f_R(x) = 0), \\ 0, & \text{if } f_R(x) = 0 \text{ and } f_G(x) = 0. \end{cases} \tag{17.3}$$

Similarly, one can define the opponent black-white chromatic scale, f_{WBk}, in terms of the chromatic scales f_W and f_{Bk}:

$$\forall x \in \mathcal{X} \qquad f_{WBk}(x) = \begin{cases} f_W(x), & \text{if } f_W(x) \neq 0 \text{ (thus, } f_{Bk}(x) = 0), \\ -f_{Bk}(x) & \text{if } f_Bk(x) \neq 0 \text{ (thus, } f_W(x) = 0), \\ 0 & \text{if } f_W(x) = 0 \text{ and } f_{Bk}(x) = 0. \end{cases} \tag{17.4}$$

As f_Y, f_B, f_R, f_G, f_{Bk}, and f_W are, in fact, the continuous chromatic codes, so are f_{YB}, f_{RG}, and f_{WBk}.

Following the tradition that has developed in the literature on colour, we will quite often refer to f_{WBk} as the *lightness* code. It is a purely terminological substitution. It does not mean that we believe, as many others do, that lightness is an independent perceptual attribute.[11] The lightness code is a synthetic and formal construction introduced for the sake of convenience.

Consider now a map $\Psi_{oc} : \mathcal{X} \to \mathbf{R}^3$, assigning to an object x in $\mathcal{X}$ vector $\left(f_{YB}(x), f_{RG}(x), f_{WBk}(x)\right)$. These map and vector will be referred to as the *perceptual object-colour map*, and the *perceptual object-colour coordinates* respectively (for objects in a single-illuminant scene). The $\mathbf{R}^3$ equipped with the coordinate axes corresponding to the opponent chromatic codes (17.2)–(17.4) will be referred to as the *perceptual object-colour space* (for objects in a single-illuminant scene). The image $\Psi_{oc}(\mathcal{X})$ will be called the *object-colour gamut* (in a single-illuminant scene).[12] The uniqueness of the chromatic scales entails the uniqueness of the object-colour gamut.

As the chromatic equivalence, $\approx$, is assumed to coincide with the object metamerism, the perceptual object-colour map is fully specified by its restriction to optimal object atlas $\mathcal{A}^3_{0.5}$, that is, by a map $\Psi_{oc}|_{\mathcal{A}^3_{0.5}} : \mathcal{A}^3_{0.5} \to \mathbf{R}^3$ that assigns an element in $\mathcal{A}^3_{0.5}$ its perceptual object-colour coordinates. Since the optimal object atlas, considered as the object-colour stimulus manifold, is homeomorphic to the unit ball B^3_1 in $\mathbf{R}^3$, and due to continuity of the perceptual object-colour map, the object-colour gamut, $\Psi_{oc}\left(\mathcal{A}^3_{0.5}\right)$, is also homeomorphic to the unit ball B^3_1. Moreover, it is the boundary of the object-colour stimulus manifold made by the optimal object stimuli, $Opt(\mathcal{X} \,|\, \Phi)$, that maps onto the boundary of the object-colour gamut, $\Psi_{oc}\left(\mathcal{A}^3_{0.5}\right)$. Note that given a point in the object-colour gamut $c \in \Psi_{oc}\left(\mathcal{A}^3_{0.5}\right)$, the ray from the origin through this point represents the object colours of the same hue as that of c. The opposite rays into which the origin decomposes the coordinate lines represent the opponent unitary colours. Therefore, the coordinate axes in the perceptual object-colour

11 'We refer to the perceived black, gray, or white value of a surface by the term *lightness*. Lightness is perceived reflectance.' (Gilchrist, 2006, p. 6).

12 Not to be confused with the light-colour gamut (Section 8.6.3).

space have clear perceptual meaning, unlike the coordinate axes in $\mathbf{R}^3$ into which the object-colour stimulus gamut is embedded (see Section 16.4.2), which have no particular interpretation and play a subsidiary role.

The trichromatic object-colour stimulus manifold, endowed with the perceptual object-colour coordinates, will be called the *colour-charted* object-colour stimulus manifold. A colour-charted object atlas, e.g., optimal object atlas $\mathcal{A}^3_{0.5}$, can be thought of as an *object-colour atlas* (for the given illumination). Colour-charting indicates for each element of the atlas (e.g., for each optimal metamer in $\mathcal{A}^3_{0.5}$) its object-colour chromaticness.

When the linear colour mechanisms, $\boldsymbol{\Phi}$, satisfy the weak regularity assumption (Definition 16.8), one can introduce $c\Delta\lambda\overline{\lambda}$-coordinates and consider the extended rectangular atlas, $\mathcal{AR}^3_{0.5}$, endowed with the perceptual object-colour coordinates, that will make it colour-charted. Colour-charting can be, formally, thought of as the map of the object-colour stimulus gamut, Γ_{Φ_p}, onto the object-colour gamut, $\Psi_{oc}(\mathcal{X})$. This map can be decomposed into the three opponent chromatic codes, f_{YB}, f_{RG}, and f_{WBk} ((17.2)–(17.4)), considered as functions of $c\Delta\lambda\overline{\lambda}$-coordinates: $f_{YB}(c, \Delta\lambda, \overline{\lambda})$, $f_{RG}(c, \Delta\lambda, \overline{\lambda})$, and $f_{WBk}(c, \Delta\lambda, \overline{\lambda})$.

The object-colour stimulus gamut, Γ_{Φ_p}, endowed with the perceptual object-colour coordinates will be called the *colour-charted* object-colour stimulus gamut. Identifying all the points in the colour-charted object-colour stimulus gamut differing only in purity, one can retract the colour-charted object-colour stimulus gamut (without its centre) onto its boundary (which is homeomorphic to a two-dimensional sphere S^2). Preimages of this retraction (i.e., the classes of $\mathrel{\overset{\circ}{=}}$-equivalence with non-zero chromaticness) will be referred to as the *object hue fibres*. As each class of $\mathrel{\overset{\circ}{=}}$-equivalence is a topological space (with the order topology induced by $\gtrsim$) homeomorphic to $(0, 1]$, the object hue fibres in the colour-charted object-colour stimulus gamut appear to be curved radii converging from every point in the colour-charted object-colour stimulus gamut boundary to the colour-charted object-colour stimulus gamut centre.

Each human observer has his/her own, individual opponent chromatic codes. The latter may differ even for observers with the same cone fundamentals. Such observers will have identical $c\Delta\lambda\overline{\lambda}$-coordinates but their opponent chromatic codes ((17.2)–(17.4)) might be different. It is only in experiment based on the hue-component scaling that the opponent chromatic codes can be identified. Since the rectangular spectral reflectance is impossible to implement, for the estimation of the value of the opponent chromatic codes at a particular triplet $(c, \Delta\lambda, \overline{\lambda})$ one can use any object stimulus with these $c\Delta\lambda\overline{\lambda}$-coordinates.

Practically, colour-charting can be done for only a finite sample of object stimuli. The choice of the sample is very important since the sample object stimuli must cover as much as possible the object-colour gamut, and do this as evenly as possible. As a first step, one can try the aforementioned colour charts, such as the Munsell book of colour and the Natural Colour System (NCS) colour collection which are commercially available and widely used in colour research (Section 16.1.1). Although, to the authors knowledge, colour charting has never been properly carried out for these colour charts,[13] some useful information can be obtained by comparing the $c\Delta\lambda\overline{\lambda}$-coordinates and the colour descriptors for the atlas elements.

13 Although the creators of the NCS colour collection claimed that it is based on the unique hues as suggested by Hering, the NCS colour notation is so confused and ambiguous that it is impossible to extract from it the chromaticness, as defined in this book. Not to mention that according to this notation a colour can contain blackness and whiteness at the same time, which contradicts the black-white opponency.

17.2 Perceptual correlates of $c\Delta\lambda\bar{\lambda}$-coordinates

In the theory of object colours, $c\Delta\lambda\bar{\lambda}$-coordinates play a role analogous to psychophysical colour coordinates (or cylindrical coordinates) in the theory of light colours (see Section 8.2). (Because of that, we will alternatively refer to $c\Delta\lambda\bar{\lambda}$-coordinates as the *psychophysical object-colour coordinates.*) In particular, the central wavelength $\bar{\lambda}$ appears to correlate with the hue of the object colours to the same extent as the variable θ in cylindrical coordinates with the hue of the light colours. At any rate, this is the case for hue as defined in the Munsell book of colour. Indeed, the $c\Delta\lambda\bar{\lambda}$-coordinates for the reflectance spectra of all the Munsell papers were evaluated for the CIE illuminant D_{65} using Smith & Pokorny's cone fundamentals (Logvinenko, 2009b). Figure 17.1 presents the $\Delta\lambda$ and $\bar{\lambda}$ coordinates of the papers with maximal Chroma (for each Value) from five pages of the Munsell book. (Such a representation as in Fig. 17.1 based only on two coordinates, $\Delta\lambda$ and $\bar{\lambda}$, will hereinafter be called the $\Delta\lambda\bar{\lambda}$ diagram.) The papers of the same Munsell Hue appear to have quite close central wavelengths provided they are of the same type.

Reparameterised spectral bandwidth, $\Delta\omega$, and central wavelength, $\bar{\omega}$, of all 1600 Munsell papers are depicted in Fig. 17.2. As one can see, the Munsell Hue order is, by and large, in line with the order in the central-wavelength axis. Likewise, the ordinal compatibility seems to take place between the spectral bandwidth and such Munsell descriptor as Value. The papers of the same Hue and larger Munsell Value are, as a rule, located higher in Fig. 17.2.

Though, it was hardly possible to expect a simple relationship between Munsell descriptors and the $c\Delta\lambda\bar{\lambda}$-coordinates. If we accept that the Munsell papers are perceptually equispaced (at any rate, that was Munsell's original intention), then note how unevenly (albeit quite smoothly) distributed the points representing them in Fig. 17.2. This unevenness is especially impressive when the Munsell papers are spherically represented as in Figs 17.3 and 17.4. The most part of the object-colour stimulus gamut turns out to be devoid of the Munsell papers. This unevenness is not a particular feature of the Munsell book of colour. Figure 17.5 shows that it takes place for the

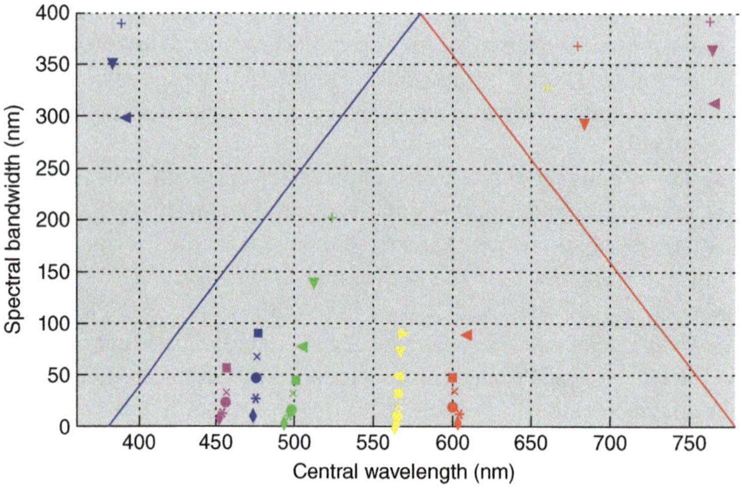

Figure 17.1 $\Delta\lambda$ and $\bar{\lambda}$ coordinates evaluated for Munsell papers of five hues: 10B, 10G, 10Y, 10R, and 10PB. The colour of the markers resembles the corresponding hue. The red and blue lines represent the main and opposite meridians, respectively. The triangle formed by these lines and the horizontal axis represents the reflectance spectra of type I. The points above the blue and red lines represent those of type II. Source: Logvinenko (2009b)/John Wiley and Sons.

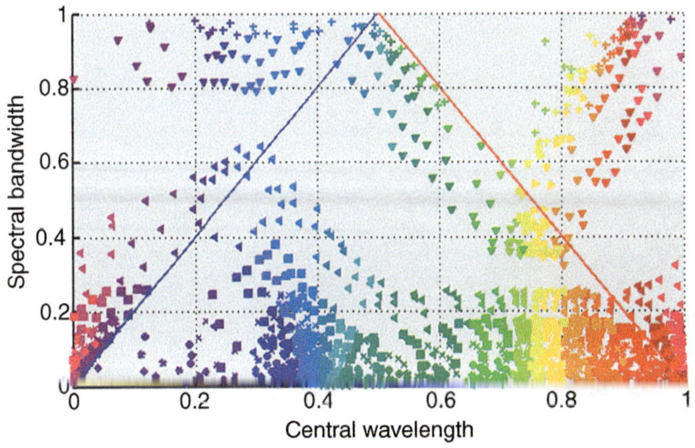

Figure 17.2 Reparameterised spectral bandwidth, $\Delta\omega$, and central wavelength, $\bar{\omega}$, (Eqs. (16.15) and (16.16)) of all 1600 Munsell papers. The papers of the same Munsell Hue are marked with symbols of the same colour. Papers of the same Munsell Value are represented by markers of the same shape. The red and blue lines represent the main and opposite meridians, respectively. Source: Logvinenko (2009b)/John Wiley and Sons.

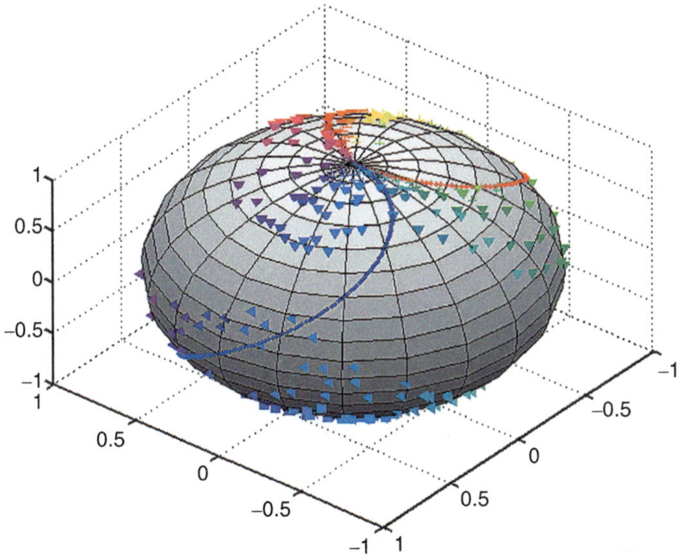

Figure 17.3 Spherical representation of Munsell papers by using $c\Delta\lambda\bar{\lambda}$-coordinates from Fig. 17.2 (top view). The red and blue lines represent the main and opposite meridians, respectively. Source: Logvinenko (2009b)/John Wiley and Sons.

NCS colour collection as well. It follows that the perceptual colour metric is quite different from that of the object-colour stimulus gamut.

Interestingly, it does not take a sophisticated experiment to notice that whiteness increases whereas blackness decreases with the spectral band. When the central wavelength is kept constant, and the spectral band gradually increases from zero, the hue, which is black for the zero spectral band, gets tinged with some chromatic component hues, losing its blackness. For some value of the spectral band the hue becomes chromatic (containing no blackness at all). Further increase

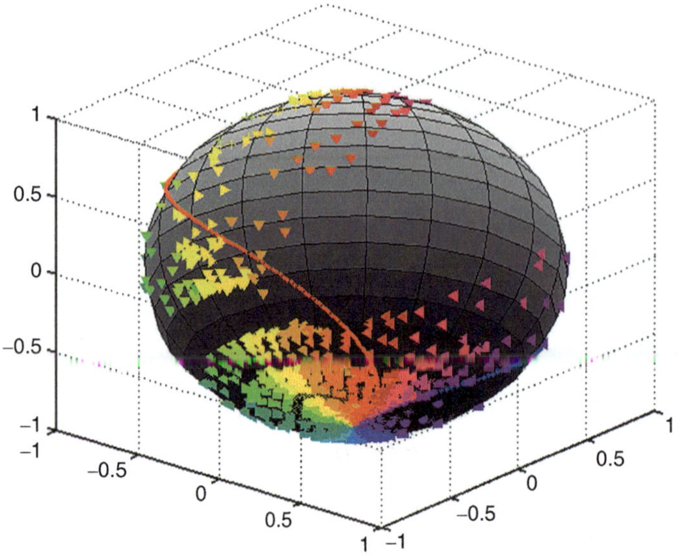

Figure 17.4 Spherical representation of Munsell papers by using $c\Delta\lambda\bar{\lambda}$-coordinates from Fig. 17.2 (bottom view). The red and blue lines represent the main and opposite meridians, respectively. Source: Logvinenko (2009b)/John Wiley and Sons.

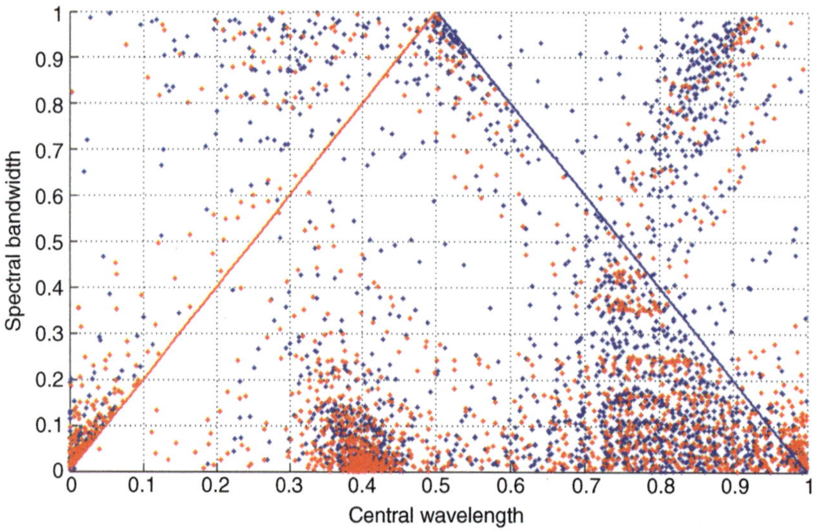

Figure 17.5 Reparameterised spectral bandwidth, $\Delta\omega$, and central wavelength, $\overline{\omega}$, for Munsell (blue dots) and NCS (red dots) papers. Source: Logvinenko (2009b)/John Wiley and Sons.

of the spectral band results in a tinge of whiteness. The latter becomes maximal at the maximal spectral band. Consequently, the north pole of the globe in Fig. 17.3 should have been depicted covered with a white cap, whose whiteness decreases in the direction of the equator. Likewise, the south pole in Fig. 17.4 should be covered with a black cap, with blackness gradually decreasing to the north.

17.3 Effect of illumination on object colour in a single-illuminant scene: Object-colour shift induced by illumination

The above-described experiments on partial hue-matching, resulting in the spherical Hering structure, were carried out under neutral illumination (Logvinenko, 2012; Logvinenko & Beattie, 2011; Logvinenko & Geithner, 2015). A remarkable fact is that the same spherical Hering structure was discovered in analogues experiments with chromatic illumination (Logvinenko & Beer, 2012).

Unexpectedness of these results is not so much that the chromatically ordered structure remains the same under chromatic illumination, but that the experiments did not bring about a trivial outcome under such conditions. Indeed, in the experiments with chromatic illumination the instruction given to the observers was the same as with neutral illumination: to decide whether two papers have a common hue. When the illuminant is, say, blue, then one could expect that there would be one large (blue) chromaticity class encompassing all the papers. The experimental results indicate that the observers just ignored (wholly on their own initiative) the common bluishness of lighting when answering the question concerning the availability of a common hue. It follows, in particular, that their decision was not based on the reflected light since the latter contained, with very few exceptions, a blue component. Curiously enough, these exceptions do only confirm the rule. For instance, as mentioned in the Outline (Chapter 1), a blue paper lit by a yellow light might reflect the neutral light. None of the observers participating in these experiments believed in this fact. In fact all they were practically unable to evaluate the chromaticity of the reflected light in the variegated scene illuminated by a chromatic light. Such discounting the illumination chromaticity in the experiments on partial hue-matching under chromatic illumination is key to understanding the structure of the object-colour manifold as presented in Section 19.3.

Thus, despite the paucity of available experimental data on the subject, it is safe to assume that everything said above concerning the spherical representation of the colour of objects lit by daylight remains valid for all admissible illuminants irrespective of their chromaticity. In particular, one can use the perceptual object-colour coordinates ((17.2)–(17.4)) in order to quantify object colours. Moreover, one can express these coordinates in terms of $c\Delta\lambda\bar{\lambda}$-coordinates.

At this point we are faced with the following problem of paramount importance. Can one consider the perceptual object-colour coordinates obtained under different illuminations to belong to the same perceptual object-colour space? Otherwise they are simply incommensurable. In other words, how are the spherical Hering structures for different illuminations related to one another, if they are connected at all? Note that even if these chromatically ordered structures are isomorphic, it does not mean that the component hues underlying them are phenomenologically identical. For example, anyone who has studied the colour of objects under different lighting conditions will probably agree that, say, blue in neutral and red light is experienced in different ways. On the other hand, the observers, even though yielding somewhat different chromaticity classes in the experiments on partial hue-matching under different illuminations, confidently used the same colour word name (e.g., blue, red, etc.) despite the difference in lighting conditions. However, one needs find more substantive arguments in favour of the connection between the chromatically ordered structures for different illuminations, than the use of common verbal categories. Quite the contrary, if they are found, these arguments could account for the use by observers of common colour categories under different lighting conditions.

As one such argument, one can point out the ability of observers to perform asymmetric (i.e., across-illuminant) colour matching. However, this argument is just as indirect as the use of common colour categories. It might always be argued that there can be different explanations for the across-illuminant colour matching. What determines this ability is a big problem waiting for solution. We will discuss this issue in more detail in Section 19.3.1.

In our opinion, direct evidence testifying that observers are able to put in correspondence the chromaticity classes obtained under different illuminations comes from an experiment on partial hue-matching employed objects simultaneously illuminated by two light sources of different chromaticity[14] (Logvinenko & Beer, 2012). Obtained without the use of any verbal categories, these data show not only which chromaticity class, produced under one illumination, corresponds to a chromaticity class, produced under another illumination, but they indicate that these two classes are, in fact, one and the same chromaticity class (Logvinenko & Beer, 2012). It follows that the perceptual object-colour spaces produced under different illuminations is one and the same space. Thus, the perceptual object-colour coordinates obtained under different illuminations can be treated as the same coordinates. Therefore, the perceptual object-colour coordinates can be used to compare object colours across different illumination conditions.

In general, there is every reason to expect the perceptual object-colour coordinates of an object to change when the lighting conditions change. Let Ψ_{oc} and Ψ'_{oc} be perceptual object-colour maps describing the same observer under illuminants p and p'; and vectors $\left(f_{YB}(x), f_{RG}(x), f_{WBk}(x)\right)$ and $\left(f'_{YB}(x), f'_{RG}(x), f'_{WBk}(x)\right)$ the corresponding perceptual object-colour coordinates of object x. The difference between the perceptual object-colour coordinates obtained under two different illuminations, p and p':

$$\Psi'_{oc}(x) - \Psi_{oc}(x) = \left(f'_{YB}(x), f'_{RG}(x), f'_{WBk}(x)\right) - \left(f_{YB}(x), f_{RG}(x), f_{WBk}(x)\right), \tag{17.5}$$

will be referred to as the *object-colour shift induced by the illuminant change* (from p to p').

Likewise the light-colour shift (Eq. (9.6)) in Chapter 9, the object-colour shift (17.5) can be decomposed into three coordinate shifts. In particular, the difference $f'_{YB}(x) - f_{YB}(x)$ represents the shift of the object colour along the yellow-blue axis in the perceptual object-colour space caused by the replacement of illuminant p with p'. The positive difference indicates either a yellow gain or a blue loss. The other two differences (i.e., $f'_{RG}(x) - f_{RG}(x)$ and $f'_{WBk}(x) - f_{WBk}(x)$) can be interpreted in a similar manner (i.e., as the red (respectively, white) gain or green (respectively, black) loss).

One reason for such an object-colour shift is the illuminant-induced object-colour stimulus shift discussed in Section 16.3. This shift reflects the fact that the illumination change results in the object to become metameric to a different rectangular metamer. Therefore, its $c\Delta\lambda\bar{\lambda}$-coordinates will be also different under the new illumination. In order to emphasise this point let us rewrite Eq. (17.5) as

$$\begin{aligned}\Psi'_{oc}(x) - \Psi_{oc}(x) = &\left(f'_{YB}\left(x'\left(c, \Delta\lambda, \bar{\lambda}\right)\right), f'_{RG}\left(x'\left(c, \Delta\lambda, \bar{\lambda}\right)\right), f'_{WBk}\left(x'\left(c, \Delta\lambda, \bar{\lambda}\right)\right)\right) \\ &- \left(f_{YB}\left(x\left(c, \Delta\lambda, \bar{\lambda}\right)\right), f_{RG}\left(x\left(c, \Delta\lambda, \bar{\lambda}\right)\right), f_{WBk}\left(x\left(c, \Delta\lambda, \bar{\lambda}\right)\right)\right),\end{aligned} \tag{17.6}$$

where $x\left(c, \Delta\lambda, \bar{\lambda}\right)$ and $x'\left(c, \Delta\lambda, \bar{\lambda}\right)$ are the $c\Delta\lambda\bar{\lambda}$-coordinates of object x under illuminants p and p' respectively.

The second reason is that colour-charting might also change under different illuminants. In other words, the perceptual object-colour map might be different for different illuminations. Unfortunately, there has been no study conducted on purpose concerning the effect of illumination on colour-charting, so our discussion will be limited to the available scarce indirect data on this issue. First of all, one has to point out that everything we know concerning this issue indicates that the perfect reflector (to be more exact, its implementable approximations) is perceived as unitary white

14 By analogy with asymmetric colour matching, one could call this method 'asymmetric partial hue-matching'.

no matter what illumination.[15] Likewise, the perfect absorber is always perceived as unitary black. Moreover, there is every indication that the achromatic interval in the object-colour solid is always assigned an achromatic colour (i.e., either black of some purity, or grey, or white of some purity).

However, there is some indication that colour-charting for the other elements of the object-colour stimulus gamut changes with illumination. It comes from experiment on asymmetric colour matching (Logvinenko & Tokunaga, 2011). It was found that the observers across-illuminant matches were not based on the equality of $c\Delta\lambda\bar{\lambda}$-coordinates (Logvinenko & Tokunaga, 2011). If one assumes that observers chose as a match the objects with equal perceptual object-colour coordinates, then the colour-charting constancy with respect to illumination will imply that the objects with the same $c\Delta\lambda\bar{\lambda}$-coordinates under both the illuminants should match each other. That this is not the case testifies that either the observers match is not based on the equality of perceptual object-colour coordinates (which is quite possible), or the colour-charting of the object-colour stimulus gamut is, generally, different for different illuminations.

Let us note in the end that since the illuminant-induced object-colour stimulus shift (i.e., the shift in $c\Delta\lambda\bar{\lambda}$-coordinates) can be theoretically evaluated, the experimental identification of opponent chromatic codes, f_{YB}, f_{RG}, and f_{WBk} for various illuminants is of paramount importance. Indeed, when f_{YB}, f_{RG}, and f_{WBk}, and f'_{YB}, f'_{RG} and f'_{WBk} are known (i.e., the object-colour stimulus gamut is colour-charted for illuminants p and p') the object-colour shift induced by the illuminant change (from p to p') can be theoretically predicted by using Eq. (17.6).

17.4 Object-colour perception by dichromats in a single-illuminant scene

Our approach to the dichromatic object-colour perception[16] is based on the assumptions analogous to those adopted in Section 8.4. In particular, we will assume that the perceptual structure of dichromatic hue palettes for objects is similar to that of the trichromatic one. Specifically, we will assume that it is entirely determined by the set of full object colours, which, in turn, are determined by the optimal object-colour stimuli (in the sense that the dichromatic object-colour solid (without its centre) can be represented as hue-fibre bundle). The dichromatic object-colour solid's centre will be assumed to be perceived as grey.[17] Moving from the dichromatic object-colour solid boundary point towards the centre along the corresponding hue-fibre is experienced as a change only in purity. Also, we will assume that dichromats perceive the interval connecting the north and south poles in the dichromatic object-colour solid the same way as trichromats, i.e., as a series of achromatic colours (from black through grey to white). The key issue is, therefore, how dichromats perceive their optimal object-colour stimuli.

The optimal stimuli for dichromatic colour vision can be derived from Eq. (14.11) if one drops off one summand. For example, omitting the last one, we will have the equation specifying the transition wavelengths of the optimal colour stimuli for protanopic colour vision:

$$\xi_1 p\left(\lambda; \lambda_S^{\max}\right) + \xi_2 p\left(\lambda; \lambda_M^{\max}\right) = 0. \tag{17.7}$$

15 Admittedly, it changes its colour appearance with illumination, however, this change is perceived as that in *lighting colour* rather than *material colour* (more on this distinction see in Chapter 19).
16 For more detail see Logvinenko (2014).
17 The presence of grey within the dichromatic object-colour solid follows from the continuity principle for object colours in the same way as for light colours. However, its location there is not predetermined, and must be established empirically.

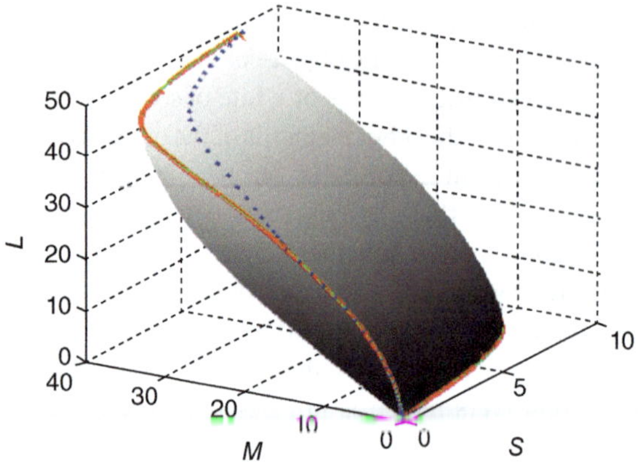

Figure 17.6 Object-colour solid with dichromatic orbits on its boundary surface. The coordinate axes are cone excitations (as in Fig. 15.5). Plankian radiator (6500 K) was used as an illuminant. The red, green, and blue dots mark the protanopic, deuteranopic, and tritanopic orbits, respectively. Note that although the deuteranopic and tritanopic orbits appear practically overlapping, they are, actually, not identical (see text). Source: Logvinenko (2014)/John Wiley and Sons.

Since Eq. (17.7) is equivalent to Eq. (14.11) where ξ_3 is assumed to be zero, the dichromatic optimal stimuli make a proper subset of the trichromatic object-colour optimal stimuli. The dichromatic optimal stimuli map to a closed contour on the trichromatic object-colour solid boundary surface that will be referred to as the *dichromatic orbit*[18] (Fig. 17.6).

When being projected parallel to the axis corresponding to the missing cone, the dichromatic orbit forms the boundary of the corresponding dichromatic object-colour solid. It should be emphasised that when the trichromatic object-colour solid collapses to the corresponding dichromatic object-colour solid, the dichromatic orbit does not merge with any other point of the trichromatic colour solid (simply because, being the optimal stimuli for dichromatic colour vision, they cannot be metameric to any other object.)

It follows that, for any colour signal (S', M') lying on the protanopic object-colour-solid boundary, there is only one such value L' that (S', M', L') belongs to the trichromatic object-colour solid. Namely, this is the (trichromatic) colour signal produced by the dichromatic optimal object-colour stimulus that gives rise to (S', M'). In other words, the missing component L' can be restored from the available two – S' and M' – provided that the trichromatic object-colour solid is known. Therefore, given the colour-charted trichromatic object-colour solid,[19] the trichromatic full object colour, which is also a dichromatic full object colour, can be restored from the colour signal even if one of its component is missing.

Thus, a trichromatic colour vision system that, for some reason, is blocked from the cone signals of one type is, in principle, able to (i) recognise the colour signals produced by the dichromatic object-colour optimal stimuli; (ii) restore the blocked component for them; and thus (iii) determine which trichromatic full object colour corresponds to which dichromatic object-colour optimal stimulus. In other words, such a system is capable, in principle, to perceive correctly a particular subset of the object-colour optimal stimuli, namely, those that serve as dichromatic object-colour optimal stimuli. Note that the latter is the only part of object-colour optimal stimuli that such a system can correctly perceive, despite the absence of one type of cones. For all other object-colour optimal stimuli, due to the presense of non-trivial confusion intervals, restoration of the missing cone signal is impossible.

---

18 As the optimal stimuli are solely determined by the cone-photopigment spectral sensitivities $g(\lambda; \lambda^{\max})$, they do not depend on either the prereceptor filters or the illuminant. Hence, although the dichromatic orbits depend on the illuminant, the dichromatic optimal stimuli (thus, the dichromatic object-colour palettes) do not.

19 In particular, this means that for each colour signal on the object-colour solid boundary, we know which full object-colour corresponds to it.

This gives reason to accept as a working hypothesis that the dichromatic palette of full object colours is a subset of the trichromatic palette of full object colours. Namely, these are those full object colours that correspond to the dichromatic object-colour optimal stimuli. If, as suggested above, the dichromatic hue palette for objects can be represented, similarly to the trichromatic, in the form of a hue bundle, determined by the full object colours, then for a complete description of the dichromatic object-colour perception, it is sufficient to (i) evaluate the dichromatic object-colour optimal stimuli; and (ii) identify the dichromatic hue fibres.

The first problem lends itself to a theoretical solution. Really, solving Eq. (17.7) for all possible pairs ξ_1 and ξ_2 permits us to calculate the protanopic optimal stimuli. Equation (17.7) is equivalent to the following:

$$p\left(\lambda;\lambda_M^{\max}\right)/p\left(\lambda;\lambda_S^{\max}\right)=k, \tag{17.8}$$

where k is an arbitrary real number. The graph of the function from the left side of Eq. (17.8) (referred to as the *M/S* cone response ratio curve) is plotted in Fig. 17.7 in semi-logarithmic coordinates. As can be seen, it intersects an arbitrary horizontal line at not more than two points. Therefore, the protanopic optimal colour stimuli are the reflectance step functions with no more than two transition wavelengths.

For Govardovskii's cone fundamentals (Definition 3.4) with the peak absorbance wavelength $\lambda_i^{\max}$ at 430, 530, and 560 nm, the *M/S* cone response ratio curve reaches its minimum, k_1, and maximum, k_2, at $\lambda_1 = 420$, and $\lambda_2 = 639$ nm respectively. Therefore, Eq. (17.8) has roots only when k is between k_1 and k_2. Let us denote $k_3 = p\left(\lambda_{\min};\lambda_S^{\max}\right)/p\left(\lambda_{\min};\lambda_M^{\max}\right)$ and $k_4 = p\left(\lambda_{\max};\lambda_S^{\max}\right)/p\left(\lambda_{\max};\lambda_M^{\max}\right)$ (as usually, $\lambda_{\min} = 380$ nm and $\lambda_{\max} = 780$ nm denote the ends of the visible spectrum). Denote also λ_3 and λ_4 the roots of Eq. (17.8) with $k = k_3$ and $k = k_4$ respectively, that lie within the visible spectrum interval. That is, consider such wavelengths λ_3 and λ_4 that $k_3 = p\left(\lambda_3;\lambda_S^{\max}\right)/p\left(\lambda_3;\lambda_M^{\max}\right)$ and $k_4 = p\left(\lambda_4;\lambda_S^{\max}\right)/p\left(\lambda_4;\lambda_M^{\max}\right)$; and $\lambda_3 \neq \lambda_{\min}$, $\lambda_4 \neq \lambda_{\max}$ (see Fig. 17.7). It was found that $\lambda_3 = 455$ nm, and $\lambda_4 = 583$ nm (Table 17.1). When

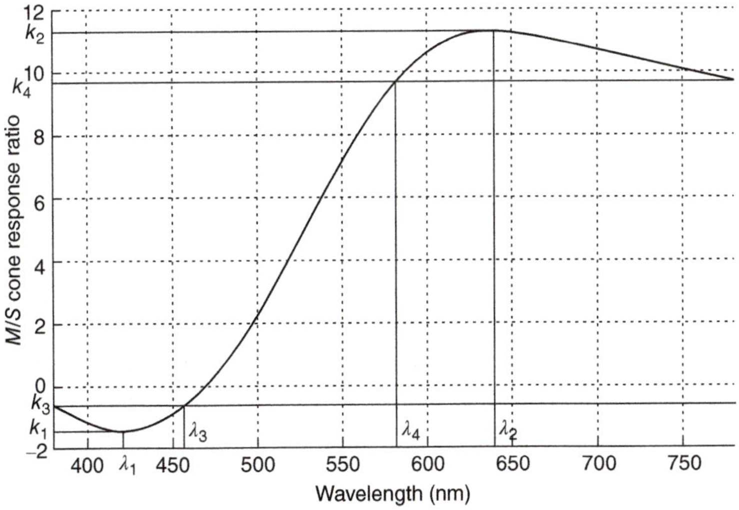

Figure 17.7 Log *M/S* cone response ratio curve for Govardovskii's cone fundamentals with $\lambda_S^{\max} = 430$ nm and $\lambda_M^{\max} = 530$ nm. The vertical axis represents the logarithm of the ratio of the responses of the *M*- and *S*-cones to monochromatic lights whose wavelength is plotted along the horizontal axis. Source: Logvinenko (2014)/John Wiley and Sons.

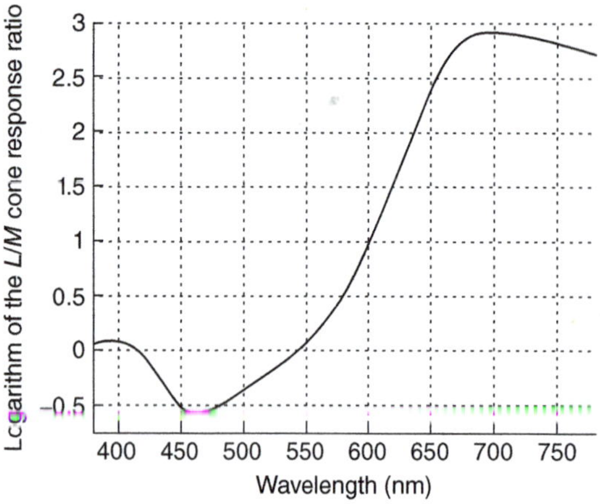

Figure 17.8 Log *L/M* cone response ratio curve for Govardovskii's cone fundamentals with $\lambda_M^{max} = 530$ nm and $\lambda_L^{max} = 560$ nm. The vertical axis represents the logarithm of the ratio of the responses of the *L*- and *M*-cones to monochromatic lights, the wavelength of which is plotted along the horizontal axis. Source: Logvinenko (2014)/John Wiley and Sons.

Table 17.1 Critical wavelengths (nm) specifying the optimal stimuli for protanopes and deuteranopes, as evaluated for Govardovskii's cone fundamentals with λ^{max} at 430, 530, and 560 nm.

	λ_1	λ_2	λ_3	λ_4
Protanope	420	639	455	583
Deuteranope	433	670	469	615

Source: Logvinenko (2014)/John Wiley and Sons.

either $k_1 < k < k_3$ or $k_4 < k < k_2$ Eq. (17.8) has exactly two different roots that lie between either λ_{min} and λ_3, or λ_4 and λ_{max} respectively. For $k_3 < k < k_4$ Eq. (17.8) has exactly one root in the interval between λ_3 and λ_4.

Hence, the protanopic optimal colour stimuli consist of the reflectance step functions with (i) no transition wavelength at all (that is, the perfect reflector, and the perfect absorber); (ii) one transition wavelength lying between λ_3 and λ_4; and (iii) two transition wavelengths lying between either λ_{min} and λ_3, or λ_4 and λ_{max}.

Similar analysis of the *L/S* cone response ratio curve (which looks similar to that in Fig. 17.7) yields the corresponding values of λ_1, λ_2, λ_3, and λ_4 for deuteranopes (Table 17.1). As these values are different for these two types of dichromacy, the optimal stimuli for protanopes and deuteranopes also differ.

As to the *L/M* cone response ratio curve, it has three local maxima within the visible spectrum interval – at 396, 464, and 701 nm (Fig. 17.8). As a result, it intersects some horizontal levels at three points. Therefore, the tritanopic optimal colour stimuli include the reflectance step functions with three (as well as all the fewer numbers) transition wavelengths. Specifically, apart from the perfect reflector and the perfect absorber, they include the reflectance step functions with (i) one transition wavelength between 552 and 665 nm; (ii) two transition wavelengths lying between either 408 and 549, or 665 and 780 nm; and (iii) three transition wavelengths, the first of which lies between 380 and 396 nm, the second between 396 and 408 nm, and the third between 549 and 552 nm.

Although the protanopic and deuteranopic orbits look overlapping in Fig. 17.6, in fact, the protanopic and deuteranopic optimal stimuli (thus, the protanopic and deuteranopic full object colours) differ significantly. The difference between them is easy to see when plotting their $\Delta\lambda$ and $\bar{\lambda}$ coordinates (as in Fig. 17.1). The dichromatic orbits in the $\Delta\lambda\bar{\lambda}$ diagram are depicted in

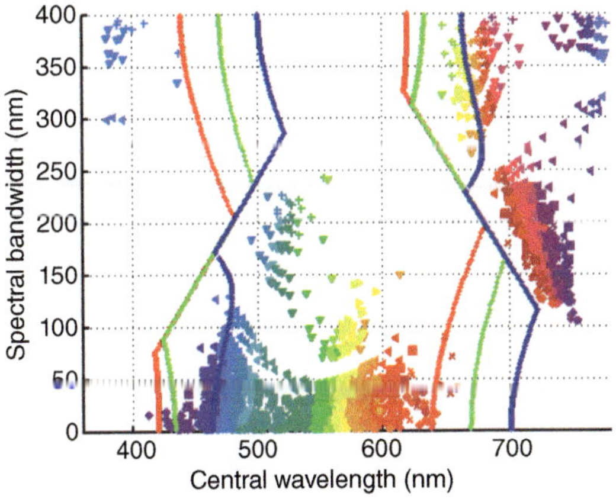

Figure 17.9 Dichromatic orbits on the $\Delta\lambda\bar{\lambda}$ diagram, along with Munsell papers. The protanopic orbit is marked in red, the deuteranopic orbit in green, and the tritanopic orbit in blue. Each of the three orbits has a rectilinear part over which it overlaps with the other two orbits. Therefore, although only the tritanopic (blue) orbit branches appear to be continuous, the branches of the protanopic and deuteranopic orbits also have no gaps. Papers of different Munsell values are marked with markers of different shapes. The contours of the constant Munsell value are vertically shifted relative to each other (the higher the Munsell value, the higher the position of the contour in the $\Delta\lambda\bar{\lambda}$ diagram). Source: Logvinenko (2014)/John Wiley and Sons.

Fig. 17.9. Although all three dichromatic orbits have a similar zigzagging form, they are, in fact, rather different.

Note that being a one-dimensional curve, each dichromatic orbit imposes a strong constraint on the combination of central wavelength and spectral bandwidth in dichromatic optimal stimuli. For instance, both protanopic and deuteranopic orbits contain at most two (or very rarely, three) dichromatic optimal stimuli with the same central wavelength. As the difference in spectral bandwidth is mainly experienced as the difference in whiteness/blackness, this means that most object-colour hues, available for protanopes and deuteranopes, are locked with one (or very rarely, two) predetermined value of either blackness or whiteness, if any. By and large, this is also the case for tritanopes.

It follows that dichromats cannot see an arbitrary hue with an arbitrary whiteness or blackness. In other words, in contrast with the trichromatic full object-colour palette represented by the colour sphere where one can approach the white pole moving along any meridian, moving along a dichromatic orbit one can approach the white pole only from exactly two meridians. This is also the case for the black pole.

As the perfect reflector does not exist in reality, we never see pure white, only a white tinged with a shade of some hue. As a results there are thousands of various shades of 'white'. It is easy to learn to distinguish between them despite the fact that, when seeing each of them for the first time, you are sure that it is pure white. For dichromats all this variety of whites will collapse into two groups: a 'cold' white and a 'warm' white. Likewise, there are two kinds of black in the dichromatic object-colour palette: 'cold' and 'warm'.

It must be said that, ideally, implementing the dichromatic optimal stimuli as real objects (surfaces) would be the best way to learn what colours the dichromats see. Unfortunately, this is impossible since there is no real surface with a rectangular spectral reflectance function. Yet, one can find a real surface the rectangular metamer of which (under given illumination) will be close to a particular dichromatic optimal stimulus in the $\Delta\lambda\bar{\lambda}$ diagram. In this context plotting the dichromatic chromaticity orbits in the $\Delta\lambda\bar{\lambda}$ diagram together with the rectangular metamers for the papers from Munsell book of colour, is quite instructive with regard to the colours that the dichromats can see (Fig. 17.9).

As pointed out above, there are large areas void of Munsell papers. Because of this only few Munsell papers happen to be in close vicinity to the dichromatic orbits. Figure 17.10 presents two samples of Munsell papers one of which comprises the Munsell papers closest to the left, and the other to the right branch of the tritanopic orbit in the $\Delta\lambda\bar{\lambda}$ diagram (Fig. 17.9). Please note that since these Munsell papers have different purity, the pallor of colour can be caused not only by the large spectral bandwidth, but also by low purity.

Although the colours in Fig. 17.10 are only a rough approximation to the tritanopic full object-colour palette, they unequivocally testify that the tritanopic full object colour palette consists of more than two hues. Moreover, it seems as if the shades of all the six component hues (red, green, blue, yellow, black, and white) are present in it. At any rate, when trichromatic observers are asked to assess the presence of component hues, as is usually done in experiments on hue scaling, their results support this observation (Logvinenko, 2021). However, in order to reliably identify the component hues in the sample in Fig. 17.10, the partial hue-matching technique (Section 8.5.1) should be used. If both trichromatic and tritanopic observers yield the same set of chromaticity classes, this will be convincing evidence in support of the approach outlined above. In turn, any significant difference between the results of trichromatic and tritanopic observers will indicate that the present approach, in its current form, should be reconsidered.

Unfortunately, the yellow and green component hues are present in the sample in Fig. 17.10 in such a reduced form, if any, that it creates a problem for inexperienced observers when they judge the presence of a common hue, as the partial hue-matching technique requires. Preliminary experiments show that even trichromatic observers sometimes yield one chromaticity class, comprising all yellowish and greenish chips, and sometimes two chromaticity classes – one for yellow and the other for green.[20] More experiments are to be done before we can get a conclusive result. Still, the analysis of the location of the dichromatic orbits in the $\Delta\lambda\bar{\lambda}$ diagram (Fig. 17.9) provides additional arguments in favour of the fact that the dichromatic hue palette for objects contain all the six component hues.

When inspecting Fig. 17.9 it is worth remembering that the proximity in the $\Delta\lambda\bar{\lambda}$ diagram is highly nonlinearly related to the subjective (perceptual) similarity, especially for central

Figure 17.10 Munsell papers, which tritanopes perceive in the same way as trichromats, provided that the reductionist hypothesis is correct. Source: Logvinenko (2014)/John Wiley and Sons.

20 True, when dealing with trichromats you can always replenish the sample with other chips so that all six chromaticity classes reveal themselves, and then see which chromaticity classes the chips from the sample will fall into.

wavelengths close to the ends of the visible spectrum interval. For instance, although the cluster of the Munsell papers in the right part of the $\Delta\lambda\bar{\lambda}$ diagram is quite far from the rectilinear fragment of the dichromatic orbits, subjectively this gap is hardly more than one Munsell step. Specifically, the bottom-right branches of all three dichromatic orbits fall in between the papers from Munsell pages 7.5R and 5R. Notably, it is these pages that most trichromatic observers choose paper from, when asked to pick out 'unique red' (Shamey et al., 2010; Logvinenko & Geithner, 2015). Therefore, all three types of dichromats have a hue in their colour palette that trichromats most likely describe as pure red (i.e., containing neither yellow nor blue).[21]

As mentioned above, these red hues in the dichromatic colour palettes are bound with only one (or two) spectral bandwidth values. As can be seen in Fig. 17.9, these values are rather low. It means that the dichromats will see red either in the pure form or being tinged with some amount of black, but never with white. This prediction is fully in line with verbal reports of dichromats (D. Jameson & Hurvich, 1978; Broackes, 2010a).

The upper-right branch of the tritanopic orbit overlaps with quite a few Munsell papers. In particular, Munsell papers 2.5Y9/4 and 2.5Y8.5/4 practically lie on the tritanopic orbit. Munsell Hue 2.5Y is what trichromatic observers usually choose as unique yellow (Shamey et al., 2010; Logvinenko & Geithner, 2015). Note that these papers have high Munsell Values. Thus, they contain a great deal of white. As the stimuli in the upper-right branch of the tritanopic orbit have large spectral bandwidth values, they are perceived as rather pale. Apart from these the upper-right branch of the tritanopic orbit consists of stimuli that trichromats perceive as yellow-green, yellow, and orange (Fig. 17.9). Therefore, it is safe to assume that the right half of the tritanopic orbit comprises the shades from yellow-green to red. The yellow-green and yellow hues are perceived by dichromats as very whitish (never blackish), the red hues always looking either pure or blackish (never whitish).

While no Munsell papers happen to lie near the upper-right branches of the protanopic and deuteranopic orbits, one can conjecture that they also contain pure yellow. Really, as can be seen in Fig. 17.9, being bent to the right, the loci of yellow hues must at some point cross both the protanopic and deuteranopic orbits. The difference between the protanopic and deuteranopic full object-colour palettes, on the one hand, and the tritanopic full object-colour palette, on the other, seems to be that the former contain more yellow-green and fewer orange and red shades.

Pure blue is also likely to be in the tritanopic full object-colour palette. Indeed, trichromatic observers usually choose unique blue from Munsell pages 10B or 2.5PB (Shamey et al., 2010; Logvinenko & Geithner, 2015). Munsell papers 10B7/6 and 2.5PB6/10 are located practically on the bottom-left branch of the tritanopic orbit.

The cluster of blue-green papers in the top-left corner of the $\Delta\lambda\bar{\lambda}$ diagram testifies that the loci of the pure blue is likely to be bent to the left, crossing somewhere the protanopic and deuteranopic orbits. Therefore, the protanopic and deuteranopic full object-colour palettes should contain pure blue. These palettes contain more purple-blue as compared to the tritanopic full object-colour palette.

Yet, all the dichromatic orbits lack pure green. Between the Munsell Hues 2.5 G and 5G, which trichromatic observers categorise as unique green (Shamey et al., 2010; Logvinenko & Geithner, 2015), and the left halves of the dichromatic branches there are a few Munsell Hues containing various shades of green-blue. Furthermore, shades of green are sparsely represented in the protanopic and deuteranopic full object-colour palettes. Specifically, blackish shades of green are completely

21 This hue cannot be named 'unitary red' because, as we can see below, it might be contaminated with some black. That is, it can be either unitary (red) or binary (blackish-red).

absent. Deuteranopes and protanopes might see only pure blue-green, and very whitish shades of yellow-green.

As for purple colours, tritanopes almost do not see them. Only blue-purple colours are available to them. Protanopes and deuteranopes see more purplish shades of blue-purple than tritanopes. Notably, all the dichromats see a purplish colour with large amount of black. Whitish shades of purple colours are not available in the dichromatic object-colour palettes.

To summarise, all six component hues are present in the dichromatic object-colour palettes. Moreover, five of the six component hues are experienced by dichromats in their pure form (i.e., as unitary colours or tinged with some black or white). It is only pure green that is absent in the dichromatic object-colour palettes.

The dichromatic object-colour palettes contain samples of each of 12 binary and 8 ternary hues, which are perceived by trichromats. Yet, in contrast to trichromatic colour vision, not all possible combinations of component hues occur in dichromatic vision. For example, the magnitude of the green component in the colours of the dichromatic object-colour palette is always rather low. Also, purple colours with moderate magnitude of the red component are absent in dichromatic colour vision.

Another important distinction of dichromatic colour vision is that, in the dichromatic object-colour palette, each combination of chromatic component colours is associated with the black (or white) component of only one (or two) particular magnitude, if any. In other words, the chromatic and achromatic components are not independent, they are locked together in dichromatic colour perception.

Hence, what makes the major difference between the dichromatic and trichromatic object colour palettes is neither the nomenclature of component hues (they are identical for both kinds of vision) nor the nomenclature of binary and ternary hues, but rather the range of magnitude that each component hue can take on when being combined into a compound object-colour.

As we can see, the dichromatic colour palettes are fundamentally different for object colours as compared to light colours. This is due to the fact that in the trichromatic light-colour palette there are only two non-confusable colours for each particular type of dichromacy, whereas there exists the entire continuum of non-confusable object colours, different for different types of dichromacy. In the object-colour-solid, the non-confusable colours are represented by a closed continuous meridian-like contour through the north and south poles. Hence, there is a fundamental difference between light-colour perception and object-colour perception in dichromats.

18

Object Metamer Mismatching

In view of (14.1), an illuminant change (i.e., replacing $p(\lambda)$ with $p'(\lambda)$) is, formally, equivalent to changing spectral sensitivity functions of the colour mechanisms. Thus, we consider the general situation when a set of colour mechanisms $\varphi_1, \ldots, \varphi_n$ is replaced by the other (linearly independent) set $\psi_1, \ldots, \psi_n$. The new set of colour mechanisms can be understood as a result of altering either the illuminant, or the colour mechanisms' spectral sensitivity (e.g., due to the replacement of the observer), or both. Respectively, a colour-signal map $\Phi = (\varphi_1, \ldots, \varphi_n)$ is replaced by $\Psi = (\psi_1, \ldots, \psi_n)$.

18.1 Metamer-mismatch regions

Metamer mismatching can be characterised by describing every metamer-mismatch region, that is, a volume, into which the former colour signal is spread as a result of the substitution of the colour mechanisms. More specifically, given $\Phi = (\varphi_1, \ldots, \varphi_n)$, $\Psi = (\psi_1, \ldots, \psi_n)$ and an object $x_0 \in \mathcal{X}$, the Ψ-image of the set of its metamers $M(\Psi, \Phi; x_0) := \Psi(\Phi^{-1}(\Phi(x_0)))$ will be referred to as the *metamer-mismatch region* associated with x_0.

Let us consider a colour-signal map $\Upsilon = (\Phi, \Psi) : \mathcal{X} \to \mathbf{R}^{2n}$,

$$\Upsilon(x) = (z; z') = (\langle x, \varphi_1\rangle, \ldots, \langle x, \varphi_n\rangle; \langle x, \psi_1\rangle, \ldots, \langle x, \psi_n\rangle),$$

and the corresponding object-colour solid $S_\Upsilon = \Upsilon(\mathcal{X})$, which is a compact convex subset in $\mathbf{R}^{2n}$. Clearly, the object-colour solid $S_\Phi = \Phi(\mathcal{X})$ is the z-projection of S_Υ,

$$S_\Phi = \{z \in \mathbf{R}^n : (z; z') \in S_\Upsilon, \ \ z' \in \mathbf{R}^n\}.$$

Furthermore, given a point $x_0 \in \mathcal{X}$ and its Φ-image $z_0 = \Phi(x_0)$, metamer-mismatch region $M(\Psi, \Phi; x_0) = \Psi(\Phi^{-1}(z_0))$ is exactly the cross-section of S_Υ, $M(\Psi, \Phi; x_0) = \{z' \in \mathbf{R}^n : (z_0; z') \in S_\Upsilon\}$. (Similarly, the object-colour solid $S_\Psi = \Psi(\mathcal{X})$ is the z'-projection of S_Υ and the metamer-mismatch region $M(\Phi, \Psi; x_0)$ is the cross-section of S_Υ, $M(\Phi, \Psi; x_0) = \{z \in \mathbf{R}^n : (z; z'_0) \in S_\Upsilon\}$, where $z'_0 = \Psi(x_0)$.) Note that in the view of Theorem 16.2, if $x_0 \in \mathcal{X}$ changes in such a way that $z_0 = \Phi(x_0)$ gets close to the boundary of the object-colour solid S_Φ, then the corresponding cross-section of $\Upsilon(\mathcal{X})$ (the metamer-mismatch region, $M(\Psi, \Phi; x_0)$) shrinks, degenerating to a point $z'_0 = \Psi(x_0)$ when z_0 reaches the boundary.

From this it follows that a similar approach can be used to evaluate the metamer-mismatch regions, which has been used in the description of the object-colour solid (Chapter 14). In particular, it suffices to describe the boundary of the metamer-mismatch region. Really, given a point

Foundations of Colour Science: From Colorimetry to Perception, First Edition. Alexander D. Logvinenko and Vladimir L. Levin.
© 2023 John Wiley & Sons Ltd. Published 2023 by John Wiley & Sons Ltd.

$z_0 \in int(S_\Phi)$, a pair $(z_0; z')$ is a boundary point of the object-colour solid S_Υ[1] if and only if $z' \in \partial M(\Psi, \Phi; x_0)$. Now, applying Theorem 14.2 gives us a measurable set $B \subseteq \Lambda$ with $\Upsilon(\chi_B) = (z_0; z')$ and $2n$ real numbers $\alpha_1, \dots, \alpha_n; \xi_1, \dots, \xi_n$ such that at least one of them is non-zero and

$$\sum_{i=1}^{n} \alpha_i \varphi_i(\lambda) \ + \ \sum_{i=1}^{n} \xi_i \psi_i(\lambda) \geq 0 \quad \text{a.e. on } B,$$

$$\sum_{i=1}^{n} \alpha_i \varphi_i(\lambda) \ + \ \sum_{i=1}^{n} \xi_i \psi_i(\lambda) \leq 0 \quad \text{a.e. on } \Lambda \setminus B.$$

That is, each $z' \in \partial M(\Psi, \Phi; x_0)$ is obtained as the Ψ-image of an indicator χ_B, where B satisfies the above inequalities for some coefficients $\alpha_1, \dots, \alpha_n; \xi_1, \dots, \xi_n$ with at least one of them being non-zero and the Φ-image of χ_B is z_0. We shall see that the *extreme points* of $M(\Psi, \Phi; x_0)$ are actually obtained using various $\alpha_1, \dots, \alpha_n; \xi_1, \dots, \xi_n$ with at least one non-zero ξ_i. True, these arguments are fairly informal and heuristic in nature, and they should be taken only as a prelude to the main result (Theorem 18.1), which we will begin to expound below.

Recall that $x_0 \in \mathcal{X}$ and $\Phi(x_0) \in int(\Phi(\mathcal{X}))$. We consider the set

$$M := \mathcal{X}_0 - x_0 = (\mathcal{X} - x_0) \cap E, \tag{18.1}$$

where E stands for $\ker(\Phi)$.

Let us show that the relative interior of M, that is, its interior in E, $int_E(M)$, is nonempty. Indeed, since $\Phi(x_0) \in int(\Phi(\mathcal{X}))$, there exists a point $x_* \in int(\mathcal{X})$ such that $\Phi(x_*) = \Phi(x_0)$. Now, since $x_* \in int(\mathcal{X})$, there is $\delta, 0 < \delta < 1$, such that $\delta \leq x_*(\lambda) \leq 1 - \delta$ a.e. on Λ, and since $\Phi(x_*) = \Phi(x_0)$, we have $x_* - x_0 \in M$. Moreover, for every $e \in E$ with $\|e\| < \delta$, we get $x_* + e \in \mathcal{X}_0$ hence $x_* - x_0 + e \in M$; therefore $x_* - x_0 \in int_E(M)$.

It can occur that the metamer-mismatch region for some x_0, $M(\Psi, \Phi; x_0) = \Psi(\mathcal{X}_0)$, is not of full dimension in $\mathbf{R}^n$. In this case, speaking of its interior or boundary, we always mean the relative interior and the relative boundary of $M(\Psi, \Phi; x_0)$, that is, its interior and boundary in the affine subspace Z in $\mathbf{R}^n$,

$$Z = \Psi(x_0) + Lin\big(\Psi(\mathcal{X}_0) - \Psi(x_0)\big) = \Psi(x_0) + Lin(\Psi(M)).$$

Similarly, the interior and the boundary of $\Psi(M)$ are understood with respect to the linear subspace $Lin(\Psi(M))$.

Lemma 18.1 *If z is an extreme point of $\Psi(\mathcal{X}_0)$, then there exists a point $x \in Ext(\mathcal{X}_0)$ such that $\Psi(x) = z$, and*

$$(x + \ker(\Psi)) \cap ri(\mathcal{X}_0) = \emptyset, \tag{18.2}$$

where $ri(\mathcal{X}_0) := x_0 + int_E(M)$.

Lemma 18.2 *If $z \in \Psi(x)$ where $x \in \mathcal{X}_0$, and (18.2) holds true, then z is a boundary point of $\Psi(\mathcal{X}_0)$.*

Lemmas 18.1 and 18.2 imply the following inclusions:

$$\partial\Psi(\mathcal{X}_0) \supseteq \{\Psi(x) : x \in Ext(\mathcal{X}_0), (x + \ker(\Psi)) \cap ri(\mathcal{X}_0) = \emptyset\} \supseteq Ext(\Psi(\mathcal{X}_0)).$$

Taking into account the Caratheodory theorem (e.g., see Holmes, 1975), the next two results follow immediately from Lemmas 18.1 and 18.2.

---

1 When $2n$ functions $\varphi_1, \dots, \varphi_n; \psi_1, \dots, \psi_n$ are not linearly independent, the object-colour solid S_Υ is not of full dimension in $\mathbf{R}^{2n}$. In this case by its boundary, we mean the relative one.

Corollary 18.1 *The metamer-mismatch region associated with x_0, $\Psi(\mathcal{X}_0)$, is the convex hull of the set $\{\Psi(x) : x \in Ext(\mathcal{X}_0), (x + \ker(\Phi)) \cap ri(\mathcal{X}_0) = \emptyset\}$.*

Corollary 18.2 *If the boundary of the metamer-mismatch region $\partial\Psi(\mathcal{X}_0)$ coincides with the set of its extreme points $Ext(\Psi(\mathcal{X}_0))$, then*

$$\partial\Psi(\mathcal{X}_0) = \{\Psi(x) : x \in Ext(\mathcal{X}_0), (x + \ker(\Psi)) \cap ri(\mathcal{X}_0) = \emptyset\}.$$

Lemma 18.3 *If $x \in Ext(\mathcal{X}_0)$, and $(x + \ker(\Psi)) \cap ri(\mathcal{X}_0) = \emptyset$, then there are real numbers $\xi_1, \dots, \xi_n$ not all equal to zero, such that the following equality holds*

$$\int_\Lambda f(\lambda)x(\lambda)\,d\lambda = \max\left\{\int_\Lambda f(\lambda)x'(\lambda)\,d\lambda : x' \in \mathcal{X}_0\right\}, \tag{18.3}$$

where

$$f(\lambda) := \sum_{i=1}^{n} \xi_i \psi_i(\lambda). \tag{18.4}$$

Theorem 18.1

(I) *If $z \in Ext(\Psi(\mathcal{X}_0))$, then there exist real numbers $\xi_1, \dots, \xi_n; \alpha_1, \dots, \alpha_n$ (not all of ξ_i being equal to zero), and a measurable set $B \subseteq \Lambda$ such that $x = \chi_B \in \mathcal{X}_0, z = \Psi(x)$, and for the function*

$$g(\lambda) = \sum_{i=1}^{n} \alpha_i \varphi_i(\lambda) + \sum_{i=1}^{n} \xi_i \psi_i(\lambda) \tag{18.5}$$

the following inequalities hold:

$$g(\lambda) \geq 0 \text{ a.e. on } B, \text{ and } g(\lambda) \leq 0 \text{ a.e. on } \Lambda \backslash B. \tag{18.6}$$

(II) *If $z \in \Psi(\mathcal{X}_0)$ and there exist real numbers $\xi_1, \dots, \xi_n; \alpha_1, \dots, \alpha_n$ (not all of ξ_i being equal to zero), and a measurable set $B \subseteq \Lambda$ such that $x = \chi_B \in \mathcal{X}_0, z = \Psi(x)$, and for the function (18.5) the inequalities (18.6) hold, then $z \in \partial(\Psi(\mathcal{X}_0))$.*

Theorem 18.1 along with Corollaries 18.1 and 18.2 provides the following procedure to evaluate the boundary of the metamer-mismatch region $\Psi(\mathcal{X}_0)$. Take two sets of real numbers $\xi_1, \dots, \xi_n$ (not all of ξ_i being equal to zero), and $\alpha_1, \dots, \alpha_n$. Construct a function $g(\lambda)$ as defined by (18.5). Consider sets $B(g) := \{\lambda : g(\lambda) > 0\}$ and $B_0(g) := \{\lambda : g(\lambda) = 0\}$. Then consider various sets $B = B(g) \cup B'$, where $B' \subseteq B_0(g)$. (If $mes(B_0(g)) = 0$, only $B = B(g)$ is to be considered.) If $\chi_B \in \mathcal{X}_0$, that is, $\Phi(x) = \Phi(x_0)$, then $\Psi(\chi_B)$ belongs to the boundary of the metamer-mismatch region $\Psi(\mathcal{X}_0)$. The points $x = \chi_B$ that can be obtained as a result of this procedure determine all $\Psi(\mathcal{X}_0)$ in the sense that $\Psi(\mathcal{X}_0)$ is the convex hull of their images $\Psi(x)$.

Let us consider the simplest illustrative example of a complete description of the metamer-mismatch region $\Psi(\mathcal{X}_0)$.

Example 18.1 *Suppose $n = 1$ so that function (18.5) takes the form*

$$g(\lambda) = \alpha\varphi(\lambda) + \xi\psi(\lambda),$$

where $\xi \neq 0$. Since $B(g) = B(\beta g)$ whenever $\beta > 0$, we can assume without loss of generality that $\xi = 1$ or $\xi = -1$. Let us consider α as a parameter. We have to examine two functions: $g_\alpha^+(\lambda) = \alpha\varphi(\lambda) + \psi(\lambda)$ and $g_\alpha^-(\lambda) = \alpha\varphi(\lambda) - \psi(\lambda)$. The corresponding sets $B(g)$ are as follows:

$$B(g_\alpha^+) = \left\{\lambda : \alpha > -\frac{\psi(\lambda)}{\varphi(\lambda)}\right\}, \quad B(g_\alpha^-) = \left\{\lambda : \alpha > \frac{\psi(\lambda)}{\varphi(\lambda)}\right\}.$$

Clearly $B(g_\alpha^+)$ is decreasing in α, $B(g_\alpha^+) = \emptyset$ for $\alpha \geq -\max\{\frac{\psi(\lambda)}{\varphi(\lambda)} : \lambda \in \Lambda\}$, and $B(g_\alpha^+) = \Lambda$ for $\alpha < -\min\{\frac{\psi(\lambda)}{\varphi(\lambda)} : \lambda \in \Lambda\}$. Similarly, $B(g_\alpha^-)$ is increasing in α, $B(g_\alpha^-) = \emptyset$ for $\alpha \leq \min\{\frac{\psi(\lambda)}{\varphi(\lambda)} : \lambda \in \Lambda\}$, and $B(g_\alpha^-) = \Lambda$ for $\alpha > \max\{\frac{\psi(\lambda)}{\varphi(\lambda)} : \lambda \in \Lambda\}$. Moreover, the function $\alpha \mapsto \Phi(\chi_{B(g_\alpha^+)}) = \int_{B(g_\alpha^+)} \varphi(\lambda) d\lambda$ is continuous and strictly decreasing on $S_1 = [-\max_\lambda \psi(\lambda)/\varphi(\lambda), -\min_\lambda \psi(\lambda)/\varphi(\lambda)]$ and takes all the values between 0 and $\Phi(\chi_\Lambda)$. Similarly, the function $\alpha \mapsto \Phi(\chi_{B(g_\alpha^-)}) = \int_{B(g_\alpha^-)} \varphi(\lambda) d\lambda$ is continuous and strictly increasing on $S_2 = [\min_\lambda \psi(\lambda)/\varphi(\lambda), \max_\lambda \psi(\lambda)/\varphi(\lambda)]$ and takes all the values between 0 and $\Phi(\chi_\Lambda)$. Then there exists a unique $\alpha = \alpha_1 \in S_1$ such that $\Phi(\chi_{B(g_{\alpha_1}^+)}) = \Phi(x_0)$ and a unique $\alpha = \alpha_2 \in S_2$ such that $\Phi(\chi_{B(g_{\alpha_2}^-)}) = \Phi(x_0)$. It follows then from Theorem 18.1 that $\Psi(\chi_{B(g_{\alpha_1}^+)}) = \int_{B(g_{\alpha_1}^+)} \psi(\lambda) d\lambda$ and $\Psi(\chi_{B(g_{\alpha_2}^-)}) = \int_{B(g_{\alpha_2}^-)} \psi(\lambda) d\lambda$ are boundary points of metamer-mismatch regions $\Psi(\mathcal{X}_0)$, and $\Psi(\mathcal{X}_0)$ is the segment joining these points.

When the spectral sensitivity functions φ_i and ψ_i are smooth enough, the spectral reflectances, mapping to the boundary of the metamer-mismatch region $\Psi(\mathcal{X}_0)$, (let us call them *sub-optimal*) are elementary step functions whose wavelength transitions are determined by zero-crossings of linear combination $g(\lambda)$ (18.5). If the system $\varphi_1, \ldots, \varphi_n, \psi_1, \ldots, \psi_n$ meets the Schrödinger–Maximov constraint, then the function $g(\lambda)$ has not more than $2n - 1$ zero-crossings. This is exactly what was claimed by Ohta and Wyszecki for the human colour vision ($n = 3$) – the number of these transitions is five (Ohta & Wyszecki, 1975). However, as we know now, the Schrödinger–Maximov constraint does not apply to the human visual system (see Section 14.4). Therefore, depending on the shape of the spectral sensitivity functions, the number of transitions for sub-optimal step functions may well exceed five. Furthermore, there are obvious limitations on the sub-optimal spectral reflectances following from that the optimal object stimuli cannot be the sub-optimal spectral reflectances at the same time. Indeed, since there is no metamerism on the object-colour solid boundary (Theorem 16.2), Theorem 18.1 is supposed to be applied only when $\Phi(x_0)$ is an internal point of the object-colour solid. If so, Φ-image of any optimal object stimulus (which belongs to the object-colour solid boundary) will differ from $\Phi(x_0)$, and should be taken out of consideration. For example, if the set of the optimal object stimuli comprises all the elementary step functions of order not more than 3, then the sub-optimal spectral reflectances must have at least three transitions.

If colour mechanisms $\varphi_1, \ldots, \varphi_n$ meet the Schrödinger–Maximov constraint and p is strictly positive on Λ, then a direct calculation shows that the mechanisms $\psi_1, \ldots, \psi_n$, where $\psi_i(\lambda) = p(\lambda)\varphi_i(\lambda), i = 1, \ldots, n$, meet the Schrödinger–Maximov constraint as well. The following simple example shows that the united system $\varphi_1, \ldots, \varphi_n, \psi_1, \ldots, \psi_n$ can fail to have that property.

Example 18.2 *Suppose there are $2n$ different points $\lambda_1, \ldots, \lambda_{2n} \in \Lambda$ such that $p(\lambda_1) = p(\lambda_2) = \cdots = p(\lambda_{2n}) = \alpha$, then the united system $\varphi_1, \ldots, \varphi_n, \psi_1, \ldots, \psi_n$ does not meet the Schrödinger–Maximov* *constraint. Indeed, we have* $\psi_i(\lambda_j) = \alpha\varphi_i(\lambda_j)$, $i = 1, \dots, n$, $j = 1, \dots, 2n$, *hence the matrix*

$$\begin{pmatrix} \varphi_1(\lambda_1) & \varphi_1(\lambda_2) & \dots & \varphi_1(\lambda_n) & \dots & \varphi_1(\lambda_{2n}) \\ \cdot & \cdot & \cdot & \cdot & \cdot & \cdot \\ \varphi_n(\lambda_1) & \varphi_n(\lambda_2) & \dots & \varphi_n(\lambda_n) & \dots & \varphi_n(\lambda_{2n}) \\ \psi_1(\lambda_1) & \psi_1(\lambda_2) & \dots & \psi_1(\lambda_n) & \dots & \psi_n(\lambda_{2n}) \\ \cdot & \cdot & \cdot & \cdot & \cdot & \cdot \\ \psi_n(\lambda_1) & \psi_n(\lambda_2) & \dots & \psi_n(\lambda_n) & \dots & \psi_n(\lambda_{2n}) \end{pmatrix}$$

is singular, that is, the Schrödinger–Maximov constraint fails.

18.2 Numerical evaluation of metamer-mismatch regions

We turn now to the problem of numerical evaluating metamer-mismatch regions in the case $n > 1$. The main idea is as follows. According to Theorem 18.1 we have to consider the function

$$g(\lambda) = g(\lambda; \alpha, \xi) = \sum_i \alpha_i\varphi_i(\lambda) + \sum_i \xi_i\psi_i(\lambda), \quad (\alpha, \xi) = (\alpha_1, \dots, \alpha_n, \xi_1, \dots, \xi_n)$$

and the set $B = B(\alpha, \xi) \subset \Lambda$ where that function is non-negative. Given a point z_0 in *int* $\Phi(\mathcal{X})$, we are interested in vectors $(\alpha, \xi) = (\alpha_1, \dots, \alpha_n, \xi_1, \dots, \xi_n) \in \mathbf{R}^{2n}$ for which not all ξ_i equal 0 and $\Phi(\chi_{B(\alpha,\xi)}) = z_0$. For any such (α, ξ), $\Psi(\chi_{B(\alpha,\xi)})$ is a boundary point of $\Psi(\mathcal{X}_0)$, and if the boundary of $\Psi(\mathcal{X}_0)$ coincides with the set of its extreme points (for the sake of simplicity, we shall accept that hypothesis), then every boundary point of $\Psi(\mathcal{X}_0)$ can be obtained in such a way. Furthermore, since $B(\theta\alpha, \theta\xi) = B(\alpha, \xi)$ whenever $\theta > 0$, we may assume that $|\alpha| \le 1, |\xi| \le 1$.

To make the issue more tractable, the following axiom will be adopted. This will considerably simplify the matter.

Axiom 18.1 $\Phi(\chi_{B(\alpha,\xi)})$ *and* $\Psi(\chi_{B(\alpha,\xi)})$ *continuously depend on* (α, ξ), *that is, all the integrals* $\int_{B(\alpha,\xi)}\varphi_i(\lambda)d\lambda, \int_{B(\alpha,\xi)}\psi_i(\lambda)d\lambda$ *are continuous functions of* (α, ξ).

If Axiom 18.1 is satisfied then for every $\varepsilon > 0$ one can find $\delta > 0$ such that

$$\left\|\Phi\left(\chi_{B(\alpha,\xi)}\right) - \Phi\left(\chi_{B(\alpha',\xi')}\right)\right\| < \varepsilon, \tag{18.7}$$

$$\left\|\Psi\left(\chi_{B(\alpha,\xi)}\right) - \Psi\left(\chi_{B(\alpha',\xi')}\right)\right\| < \varepsilon, \tag{18.8}$$

whenever $(\alpha, \xi), \left(\alpha', \xi'\right) \in [-1, 1]^{2n}$, $\left\|(\alpha, \xi) - \left(\alpha', \xi'\right)\right\| < \delta$ (here $\|\cdot\|$ stands for Euclidean norms in $\mathbf{R}^n$ and $\mathbf{R}^{2n}$ respectively). We now fix a large integer N and replace the $2n$-dimensional cube $[-1, 1]^{2n}$ with the grid T_N^{2n}, where

$$T_N := \left\{t_i = -1 + \frac{i-1}{N} : i = 1, \dots, 2N + 1\right\}.$$

We shall consider $2n$-dimensional vectors (α, ξ) whose components α_i and ξ_i belong to T_N. The total number of such vectors is $(2N + 1)^{2n}$ and the number of vectors for which not all ξ_i are 0 is $(2N+1)^{2n} - (2N+1)^n$. When $N = N(\varepsilon)$ is large enough, then for every $(\alpha, \xi) \in [-1, 1]^{2n}$ one can find $(\alpha', \xi') \in T_N^{2n}$ such that inequalities (18.7) and (18.8) are satisfied. Therefore, given ε, one can find for $N = N(\varepsilon)$ a node in the grid $(\alpha, \xi) \in T_N^{2n}$ such that $\Phi(\chi_{B(\alpha,\xi)})$ differs from z_0 at most by ε. Although $\chi_{B(\alpha,\xi)}$ is not, strictly speaking, included in $\mathcal{X}_0$, one can say that it 'almost' belongs to $\mathcal{X}_0$ and $\Psi(\chi_{B(\alpha,\xi)})$ 'almost' (up to ε) belongs to the boundary of $\Psi(\mathcal{X}_0)$. (Figuratively speaking, we will say that $\chi_{B(\alpha,\xi)}$ is 'almost' sub-optimal reflectance). On the other hand, if, for some $(\alpha, \xi) \in [-1, 1]^{2n}$, $\chi_{B(\alpha,\xi)}$ is an exact sub-optimal reflectance then, in virtue of (18.7) and (18.8), there is a close 'almost' sub-optimal reflectance which is a node of the grid T_N^{2n} and its Φ-image differs from z_0 less than by ε. Hence, there will be no loss in accuracy due to discretising the problem (and resorting to 'almost' sub-optimal reflectances).

The procedure of numerical evaluating the boundary $\partial(\Psi(\mathcal{X}_0))$ can be reduced then to generating 'almost' (up to ε) sub-optimal reflectances. Given $(\alpha, \xi) \in T_N^{2n}$, we form the function $\chi_{B(\alpha,\xi)}$ and calculate $\Phi(\chi_{B(\alpha,\xi)})$. If $\|\Phi(\chi_{B(\alpha,\xi)}) - z_0\| < \varepsilon$ then, up to ε, $\Psi(\chi_{B(\alpha,\xi)})$ is a boundary point of $\Psi(\mathcal{X}_0)$, and the set of all such $\Psi(\chi_{B(\alpha,\xi)})$ is accepted as the discrete approximation for the boundary $\partial(\Psi(\mathcal{X}_0))$. Notice that testing all $(\alpha, \xi) \in T_N^{2n}$ is not needed, since, for instance, any two sets of positive (α, ξ) give the same result: $B(\alpha, \xi) = \Lambda$ (we assume that all φ_i, ψ_i are non-negative).

A question arises: when is Axiom 18.1 satisfied? It is clear that the system of $2n$ functions, $\varphi_1, \dots, \varphi_n, \psi_1, \dots, \psi_n$ must be linearly independent since otherwise one can find $\alpha_1, \dots, \alpha_n, \xi_1, \dots, \xi_n$ (not all ξ_i equal to 0) such that $g(\lambda; \alpha, \xi) \equiv 0$ on Λ. Then $B(\alpha, \xi) = \Lambda$ but for close (α', ξ') the positivity set $B(\alpha', \xi')$ can differ in measure from Λ rather considerably, and in such a case $\Phi(\chi_{B(\alpha,\xi)})$, $\Psi(\chi_{B(\alpha,\xi)})$ are not continuous in (α, ξ). A similar argument shows that $\varphi_1, \dots, \varphi_n, \psi_1, \dots, \psi_n$ must be linearly independent on every segment $[a, b] \subset \Lambda$ and even on every subset of positive Lebesgue measure in Λ. This is a necessary condition for Axiom 18.1 to be satisfied. Further, for every (α, ξ) the null-set of the function $g(\cdot; \alpha, \xi)$, $\{\lambda : g(\cdot; \alpha, \xi) = 0\}$, must be Lebesgue negligible since otherwise it can occur that for any (α', ξ') close to (α, ξ) there exists a set of positive Lebesgue measure $M(\alpha', \xi') \subset \{\lambda : g(\cdot; \alpha, \xi) = 0\}$, on which the function $g(\cdot; \alpha', \xi')$ is negative. Certainly, this reasoning is not entirely strict but it could be formalised. Therefore, the condition that $mes\{\lambda : g(\lambda; \alpha, \xi) = 0\} = 0$ together with the above linear independence requirement proves to be almost necessary for the basic supposition to be satisfied. Let us give now a simple sufficient condition. It is as follows: all φ_i, ψ_i are continuous, for every (α, ξ) zeros of the function $g(\cdot; \alpha, \xi)$ compose a finite set, and all of them are continuous functions of (α, ξ). For this, in turn, it suffices the validity of two conditions as follows: (i) all φ_i, ψ_i are smooth, (ii) the system of $2n$ functions $\varphi_1, \dots, \varphi_n, \psi_1, \dots, \psi_n$ meet the Schrödinger–Maximov constraint (the number of zeros of $g(\cdot; \alpha, \xi)$ is then less $2n$), and for every (α, ξ) the implication

$$g(\lambda; \alpha, \xi) = 0 \Rightarrow \frac{dg(\lambda; \alpha, \xi)}{d\lambda} \neq 0 \tag{18.9}$$

holds true (the continuity of zeros of $g(\cdot; \alpha, \xi)$ in (α, ξ) follows then from (18.9) by the implicit function theorem). Of course, this is a rather restrictive sufficient condition which is far from a necessary one as it may be seen from Fig. 18.1.

Here, the middle curve (green line) is a graph of $g(\cdot; \alpha, \xi)$ whereas the upper (red line) and the lower (black line) curves are graphs of $g(\cdot; \alpha', \xi')$ where the corresponding (α', ξ') are close to (α, ξ). There are two roots for $g(\cdot; \alpha, \xi)$ (green line): at $\lambda_1 = 410.4$ nm and $\lambda_2 = 570$ nm. Condition (18.9) fails at the point λ_1. Indeed, when passing from (α, ξ) to close values (α', ξ') the first zero of g, λ_1, vanishes in one case (the upper curve) and becomes double, $\lambda_1' = 384$ nm and $\lambda_1'' = 438$ nm, in other (the lower curve). We have $B(\alpha, \xi) = [\lambda_{\min}, \lambda_2]$, $B(\alpha', \xi') = [\lambda_{\min}, \lambda_2'']$ for the upper curve, and $B(\alpha', \xi') = [\lambda_{\min}, \lambda_1'] \cup [\lambda_1'', \lambda_2']$ for the lower curve. Thus the Lebesgue measure of $B = B(\alpha, \xi)$

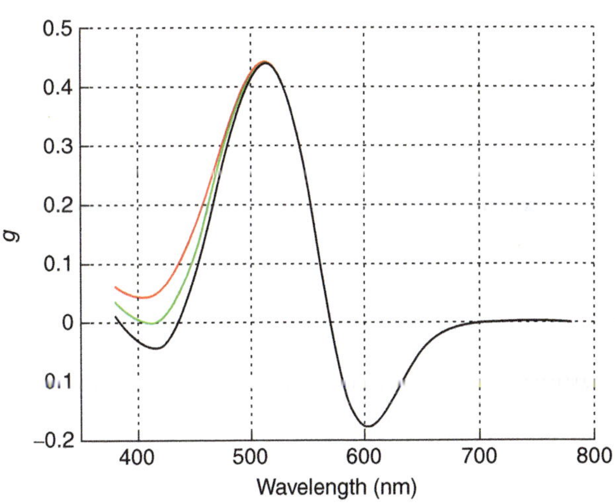

Figure 18.1 Example of a function $g(\lambda; \alpha, \xi)$ that does not satisfy Eq. (18.9) (see text).

changes continuously with changing (α, ξ); therefore $\Phi(\chi_{B(\alpha,\xi)})$ and $\Psi(\chi_{B(\alpha,\xi)})$ are continuous in (α, ξ), that is, the basic supposition is satisfied.

18.3 Indices of object metamer mismatching

By analogy with the light metamer-mismatch indices (Section 7.2), we introduce now similar quantitative characteristics for the object metamer mismatching caused by two sets of colour mechanisms, $\Phi = (\varphi_1, \dots, \varphi_n)$ and $\Psi = (\psi_1, \dots, \psi_n)$.

Proposition 18.1 *Suppose $x_0 \in \mathcal{X}$, $\Phi(x_0)$ is an interior point of the object-colour solid $S = \Phi(\mathcal{X})$, and $\mathcal{X}_0 = \{x \in \mathcal{X} : \Phi(x) = \Phi(x_0)\}$ stands for the set of stimuli metameric to x_0. Let*

$$\varepsilon := \sqrt{\sum_{i=1}^{n} \|\psi_i - \varphi_i\|_1^2}, \tag{18.10}$$

where $\|\cdot\|_1$ is defined in (A.4). For the sake of simplicity, assume that $\Phi(x_0) = \Psi(x_0) = z_0$[2]*. Then the metamer-mismatch region $\Psi(\mathcal{X}_0)$ is contained in the ε-neighbourhood of the colour signal $z_0 = \Phi(x_0)$:*

$$\|z_0 - z\| \leq \varepsilon$$

whenever $z \in \Psi(\mathcal{X}_0)$.

Thus, the less the $\mathbf{L}^1$-norms of the differences $\psi_i - \varphi_i$, the smaller the Ψ-spread $\Psi(\mathcal{X}_0)$ of the original signal $z_0 = \Phi(x_0)$.

Remark 18.1 *If $\varphi_i, \psi_i \in \mathbf{L}^\infty$ then a coarser estimate holds true as follows:*

$$\|z_0 - z\| \leq \varepsilon_1 := \sqrt{n} \max_{i=1,\dots,n} \|\psi_i - \varphi_i\|_1.$$

2 The required equality can always be achieved, without loss of generality, by appropriate rescaling of the colour mechanisms $\psi_1, \dots, \psi_n$, since this will not change the metamerism induced by them (see Section 18.4).

Moreover, when $\varphi_i, \psi_i \in \mathbf{L}^\infty$, then it may be replaced with a yet coarser one:

$$\|z_0 - z\| \leq \varepsilon_2 := \sqrt{n}(\lambda_{\max} - \lambda_{\min}) \max_{i=1,\dots,n} \|\psi_i - \varphi_i\|_\infty,$$

where $\|\cdot\|_\infty$ is defined in (A.5).

We consider now two modifications of colour mechanisms, $\Psi = (\psi_1, \dots, \psi_n)$ and $\Psi' = (\psi'_1, \dots, \psi'_n)$, and evaluate the proximity between the images $\Psi(\mathcal{X}_0)$ and $\Psi'(\mathcal{X}_0)$ using the Hausdorff metric

$$d_H(\Psi(\mathcal{X}_0), \Psi'(\mathcal{X}_0)) := \max\{ \max_{z \in \Psi(\mathcal{X}_0)} dist(z, \Psi'(\mathcal{X}_0)), \max_{z' \in \Psi'(\mathcal{X}_0)} dist(z', \Psi(\mathcal{X}_0))\},$$

where *dist* stands for the Euclidean distance. The next result follows straightforwardly from Proposition 18.1.

Corollary 18.3 *Given Ψ and Ψ' an estimate holds true as follows:*

$$d_H(\Psi(\mathcal{X}_0), \Psi'(\mathcal{X}_0)) \leq \varepsilon + \varepsilon',$$

where ε is given by (18.10) and

$$\varepsilon' = \sqrt{\sum_{i=1}^{n} \|\psi'_i - \varphi_i\|_1^2}. \tag{18.11}$$

In other words, any of two sets, $\Psi(\mathcal{X}_0)$ and $\Psi'(\mathcal{X}_0)$, is contained in the $(\varepsilon + \varepsilon')$-neighbourhood of the other.

In the following corollary, Φ and Ψ are replaced with Ψ and Ψ'. (As to the space $\mathbf{L}^2$, see Appendix A.1.)

Corollary 18.4 *Given $\Phi = (\varphi_1, \dots, \varphi_n)$, $\varphi_i \in \mathbf{L}^2$, $i = 1, \dots, n$, and two functions $p, p' \in \mathbf{L}^\infty$ such that $\frac{p}{p'} \in \mathbf{L}^2$, we consider mechanisms $\Psi = (\psi_1, \dots, \psi_n)$ and $\Psi' = (\psi'_1, \dots, \psi'_n)$ where $\psi_i(\lambda) = p(\lambda)\varphi_i(\lambda)$, $\psi'_i(\lambda) = p'(\lambda)\varphi_i(\lambda)$, $i = 1, \dots, n$. Suppose $x_0 \in \mathcal{X}$, $z_0 = \Psi(x_0)$ is an interior point of the object-colour solid $S = \Psi(\mathcal{X})$, and $\mathcal{X}_0 = \{x \in \mathcal{X} : \Psi(x) = \Psi(x_0)\}$ stands for the set of stimuli metameric to x_0 with respect to Ψ. The following estimation of the metamer-mismatch region $\Psi'(\mathcal{X}_0)$ holds then true:*

$$\|z_0 - z\| \leq \sqrt{\int_\Lambda |\frac{p(\lambda)}{p'(\lambda)} - 1|^2 d\lambda} \sqrt{\sum_{i=1}^{n} \int_\Lambda p'^2(\lambda)\varphi_i^2(\lambda) d\lambda} \tag{18.12}$$

whenever $z \in \Psi'(\mathcal{X}_0)$.

We now fix $\Psi = p\Phi$ and consider $x_0 \in \mathcal{X}$ for which $\Psi(x_0)$ is an interior point of $\Psi(\mathcal{X})$. Let $\mathcal{X}_0(x_0)$ denote the set of metamers of x_0 with respect to the colour-signal map Ψ. Given a different (altered) colour-signal map $\Psi' = p'\Phi$, we define the *first metamer-mismatch index for objects* (spectral reflectance functions) to be the quotient

$$I_1(x_0; \Psi') := \frac{diam\Psi'(\mathcal{X}_0(x_0))}{diam\Psi'(\mathcal{X})},$$

where *diam* stands for the diameter (the maximum distance between two points) of the corresponding set, i.e., the metamer-mismatch region $\Psi'(\mathcal{X}_0(x_0))$ and the object-colour solid $\Psi'(\mathcal{X})$ respectively.

It follows from this definition that $0 \leq I_1(x_0;\Psi') < 1$. Observe that if we take any positive number k and pass from p' to kp' (hence from Ψ' to $k\Psi'$), then both diameters are multiplied by k and their quotient does not change:

$$I_1(x_0;\Psi') = I_1(x_0;k\Psi'). \tag{18.13}$$

Taking into account (18.13), the next result is derived from Corollary 18.4.

Proposition 18.2 *Suppose p and p' are as in Corollary 18.4. An estimate holds then true as follows:*

$$I_1(x_0;\Psi') \leq 2\frac{\sqrt{\sum_{i=1}^{n} \|\psi_i'\|_2^2}}{\sqrt{\sum_{i=1}^{n} \|\psi_i'\|_1^2}}\sqrt{\int_\Lambda |\frac{p(\lambda)}{k_0 p'(\lambda)} - 1|^2 d\lambda}, \tag{18.14}$$

where k_0 is given by

$$k_0 = \frac{\int_\Lambda \frac{p^2(\lambda)}{p'^2(\lambda)} d\lambda}{\int_\Lambda \frac{p(\lambda)}{p'(\lambda)} d\lambda}. \tag{18.15}$$

We turn now to the following question. Given two objects, x_1 and x_2, we examine their images,

$$\Psi'(x_1) = \left(\int_\Lambda p'(\lambda)\varphi_1(\lambda)x_1(\lambda)d\lambda, \dots, \int_\Lambda p'(\lambda)\varphi_n(\lambda)x_1(\lambda)d\lambda \right),$$

$$\Psi'(x_2) = \left(\int_\Lambda p'(\lambda)\varphi_1(\lambda)x_2(\lambda)d\lambda, \dots, \int_\Lambda p'(\lambda)\varphi_n(\lambda)x_2(\lambda)d\lambda \right),$$

under various illuminants p'. If $p' \in \mathbf{L}^\infty$ and $\varphi_i \in \mathbf{L}^2, i = 1, \dots, n$, the next estimate follows easily from the Cauchy–Schwartz inequality (see Appendix A.1):

$$\|\Psi'(x_1) - \Psi'(x_2)\| \leq \|p'\|_\infty \left(\sum_{i=1}^{n} \|\varphi_i\|_2^2 \right)^{\frac{1}{2}} \|x_1 - x_2\|_2.$$

This estimate shows that the mean value of $\|\Psi'(x_1) - \Psi'(x_2)\| / \|p'\|_\infty$ over an arbitrary finite set of observed illuminants is small enough provided the $\mathbf{L}^2$-distance between the objects is small. This fact does not depend on whether x_1 and x_2 are metameric with respect to some illuminant p or not; therefore such a mean value cannot be accepted as a measure of metamerism.

As in the case of a similar index for light (Definition 7.2), a substantial drawback of the first metamer-mismatch index I_1 is that it is not affine invariant: if Ψ' and Ψ'' are two object-colour-signal maps representing the same altered metamerism relation, the values $I_1(x_0;\Psi')$ and $I_1(x_0;\Psi'')$ can differ. In this connection, it seems natural to introduce an invariant metamer-mismatch index as follows.

Definition 18.1 *Given $x_0 \in \mathcal{X}$ and two colour-signal maps Φ and Ψ on the object colour stimulus set $\mathcal{X}$, the volume of the metamer-mismatch region $M(\Psi, \Phi; x_0) := \Psi(\Phi^{-1}(\Phi(x_0)))$ divided by the volume of the object-colour solid $\Psi(\mathcal{X})$ will be referred to as the* n-dimensional metamer-mismatch index for objects,

$$I_n(x_0;\Psi) := \frac{Vol_n(M(\Psi, \Phi; x_0))}{Vol_n(\Psi(\mathcal{X}))}.$$

Since a ratio of two volumes is an affine invariant, this definition does not depend on the choice of the colour mechanisms $\Psi = (\psi_1, \dots, \psi_n)$ in terms of which the altered object-colour solid is determined.

18.4 Object-metamerism-preserving transformations of colour mechanisms

Of theoretical interest is the following question: what Ψ preserves metamerism? In other words, what change in the spectral sensitivity of colour mechanisms does not violate metamerism for any pair of objects?

Recall that the set of objects metameric to x_0 with respect to the colour mechanisms $\varphi_1, \dots, \varphi_n$ is denoted by $Met\left(\varphi_1, \dots, \varphi_n; x_0\right)$.

Proposition 18.3 *Suppose that two sets of colour mechanisms, $\varphi_1, \dots, \varphi_n$ and $\psi_1, \dots, \psi_n$, are linearly independent. If $0 < \delta \le x_0(\lambda) \le 1 - \delta$, then the following statements are equivalent:*

(i) $Met\left(\varphi_1, \dots, \varphi_n; x_0\right) = Met\left(\psi_1, \dots, \psi_n; x_0\right)$;
(ii) $Lin\left(\varphi_1, \dots, \varphi_n\right) = Lin\left(\psi_1, \dots, \psi_n\right)$.

Remark 18.2 *If any of the equivalent statements (i) or (ii) holds, then there exists a unique non-singular continuous linear operator $T : Lin\left(\varphi_1, \dots, \varphi_n\right) \to Lin\left(\psi_1, \dots, \psi_n\right)$ such that $T\varphi_i = \psi_i$, $i = 1, \dots, n$.*

So, when $x_0(\lambda)$ is an interior point in $\mathcal{X}$, its metamers are preserved only if $\psi_1, \dots, \psi_n$ belong to the linear subspace spanned over $\varphi_1, \dots, \varphi_n$. This may prove to be not the case when $x_0(\lambda)$ is a boundary point of $\mathcal{X}$ (see Example 18.3).

Given colour mechanisms $\varphi_1, \dots, \varphi_n$ and an object-colour stimulus $x_0 \in \mathcal{X}$, we define, by analogy with Section 6.2, a vector subspace in $\mathbf{L}^1$, $L := Lin\left(\varphi_1, \dots, \varphi_n\right)$, its annihilator $L^0 = \left\{y \in \mathbf{L}^\infty : \left\langle \varphi_i, y \right\rangle = 0,\ i = 1, \dots, n\right\}$, and the set

$$H_{x_0} := \left\{y \in L^0 : y = -x_0 + x,\ x \in \mathcal{X}\right\} = \left\{y \in L^0 : -x_0 \le y \le 1 - x_0\right\}.$$

The next simple result is an analogue of Lemma 6.2.

Lemma 18.4 *The following statements are equivalent:*

(i) $Met\left(\varphi_1, \dots, \varphi_n; x_0\right) \subseteq Met\left(\psi; x_0\right)$;
(ii) $\psi \in H^0_{x_0}$.

Example 18.3 *Let $\varphi_1 \equiv 1, \varphi_2 = \chi_{[\lambda_{\min}, \lambda_0]}$, and $x_0 = \chi_{[\lambda_0, \lambda_{\max}]}$, where $\lambda_0 = \frac{1}{2}\left(\lambda_{\min} + \lambda_{\max}\right)$. One has $L^0 = \left\{y \in \mathbf{L}^\infty : \int_\Lambda y(\lambda)\, d\lambda = 0,\quad \int_{\lambda_{\min}}^{\lambda_0} y(\lambda)\, d\lambda = 0\right\}$, $H_{x_0} = \left\{y \in L^0 : -\chi_{[\lambda_0, \lambda_{\max}]} \le y \le \chi_{[\lambda_{\min}, \lambda_0]}\right\}$. Consider any $y \in H_{x_0}$. Then $0 \le y(\lambda) \le 1$ for $\lambda \in \left[\lambda_{\min}, \lambda_0\right]$ along with $\int_{\lambda_{\min}}^{\lambda_0} y(\lambda)\, d\lambda = 0$ yields $y(\lambda) = 0$ a.e. on that segment. Furthermore, $-1 \le y(\lambda) \le 0$ for $\lambda \in \left[\lambda_0, \lambda_{\max}\right]$ along with $\int_\Lambda y(\lambda)\, d\lambda = 0$ and $y(\lambda) = 0$ on the first segment yields $y(\lambda) = 0$ on $\left[\lambda_0, \lambda_{\max}\right]$. Thus $y = 0$ and $H_{x_0} = \{0\}$. Therefore, $H^0_{x_0} = \mathbf{L}^1$, and $Met(\varphi_1, \varphi_2; x_0) = \left\{x_0\right\}$. It follows that for any $\varphi_3, \dots, \varphi_n$; $\psi_3, \dots, \psi_n \in \mathbf{L}^1_+$, $Met\left(\varphi_1, \varphi_2, \varphi_3, \dots, \varphi_n; x_0\right) = Met\left(\varphi_1, \varphi_2, \psi_3, \dots, \psi_n; x_0\right) = \left\{x_0\right\}$.*

It follows from Proposition 18.3 that, as in the case of metamerism of lights, metamerism for *all* objects will be preserved only by a linear transformation within the subspace $Lin(\varphi_1, \dots, \varphi_n)$. Likewise, any invertible transformation of the colour signal will also retain metamerism. Therefore, metamerism of objects is preserved by only those transformations which can be represented as $G \circ T$, where T is a linear operator defined in Remark 18.2, and $G : \mathbf{R}^n_+ \to \mathbf{R}^n_+$ is a one-to-one map.

As follows from the foregoing proposition, changing the illuminant will preserve metamerism for all objects only when it does not affect the form of its spectral composition. In other words, it is only variations in the illuminant intensity that can retain all the metamers unaltered.

Proposition 18.4 *Given linearly independent colour mechanisms $\varphi_1, \dots, \varphi_n$, suppose that $\psi_i(\lambda) = \varphi_i(\lambda)\, p(\lambda)$ $(i = 1, \dots, n)$, where $\varphi_1, \dots, \varphi_n \in \mathbf{L}^\infty_+$, $p \in \mathbf{L}^1_+$. Then the following two statements are equivalent:*

(i) *colour mechanisms $\psi_1, \dots, \psi_n$ belong to $Lin(\varphi_1, \dots, \varphi_n)$;*
(ii) *$p(\lambda) \equiv Const.$*

19

Object-Colour Perception in a Multiple-Illuminant Scene

Multiple-illuminant scene implies there being at least two different light sources illuminating the reflecting objects in the scene. The simplest example is a bipartite field made of two different adjacent papers, $x_1(\lambda)$ and $x_2(\lambda)$, each of which lit by its own light: $p_1(\lambda)$ and $p_2(\lambda)$, respectively. When $x_1(\lambda) = x_2(\lambda)$ the physical border between the fields is constituted by the difference between the illuminants: $p_1(\lambda)$ and $p_2(\lambda)$. In this case, we will talk about *illumination border*. When $p_1(\lambda) = p_2(\lambda)$ the physical border between the fields is constituted by the difference between the spectral reflectance functions $x_1(\lambda)$ and $x_2(\lambda)$. In this case, we will talk about *reflectance border*. A bipartite field with the illumination border is an experimental tools to study light colours (Part I). A bipartite field with the reflectance border is an experimental tools to study object colours in a single-illuminant scene. Both are just particular cases of the more general situation, considered in this chapter, when the border between the two fields is produced by the difference in both spectral reflectance and illumination.

The reflected lights (i.e., $x_1(\lambda)\,p_1(\lambda)$ and $x_2(\lambda)\,p_2(\lambda)$) can be either metameric, or not metameric. When the reflected lights are metameric, visual experience of two kinds is possible. First, an observer sees no border between the two halves of the bipartite field. In this case, we say that the colour of the objects (i.e., the two papers making the bipartite field) is identical; or more formally, the pairs $(x_1(\lambda), p_1(\lambda))$ and $(x_2(\lambda), p_2(\lambda))$ are *colour equivalent*. Being physically juxtaposed is not essential in this definition of *object/light colour equivalence*. When the papers are separated from each other they will be referred to as colour equivalent if when imagined as abutting via mental translation there would still be no border apparent between them. Needless to say, reflecting objects, lit so as to make up colour equivalent pairs, have, by definition, the same colour (i.e., object-colour).

Second, despite the metamerism of the reflected lights, the observer may happen to see clearly the colour border. In vision literature, there is a number of demonstrations proving that this can be the case. For example, Tokunaga & Logvinenko (2010c) implemented Koffka's idea (Koffka, 1935) by presenting a yellow Munsell chip lit by such blue light that the reflected light was practically metameric to that reflected by the blue Munsell chip lit by the yellow light. Moreover, both the reflected lights were metameric to the light reflected by the grey Munsell chip lit by the neutral light. Still, the blue chip was clearly perceived as blue, the yellow chip as yellow, and the grey chip as grey (see also Fig. 19.1 and Logvinenko (2005a)). Hence, metamerism is necessary but not sufficient condition for object/light colour equivalence.

Two object/light pairs which are not colour equivalent, will be referred to as *colour different*. Colour equivalent object/light pairs can be made colour different by either varying their the objects' spectral reflectance or the lights' spectral power distribution, or both. It is important to emphasise that, usually, the colour difference induced by varying spectral reflectance appears qualitatively different from the colour difference induced by varying lights.

Foundations of Colour Science: From Colorimetry to Perception, First Edition. Alexander D. Logvinenko and Vladimir L. Levin.
© 2023 John Wiley & Sons Ltd. Published 2023 by John Wiley & Sons Ltd.

Figure 19.1 Although this is just a sample of computer graphics, it illustrates well how different colours can be invoked by the same colour signals. Despite the apparent difference between the colour of the blue tiles at the top of the left cube and the yellow tiles at the top of the right cube, these tiles are actually identical. Source: Purves & Lotto (2003).

Consider, for example, the experimental display in Fig. 19.2 used by Logvinenko & Tokunaga (2011). It comprises two identical sets of Munsell chips one of which was illuminated by a yellow, and the other by a blue light. As a matter of fact, the colour difference between any identical (but differently illuminated) chips is qualitatively different from the colour difference between any chips within each of the two frames of equal illumination (Tokunaga & Logvinenko, 2010a,b,c). It is to conceptualise this qualitative difference that the two notions – *material colour* and *lighting colour* – have been introduced (Logvinenko, 2013b). Specifically, the colour difference between chips within a field of equal illumination is mainly the difference in material colour, the colour difference between identical chips in the fields of different illumination being basically the difference in lighting colour.[1]

The existence of the colour difference of two types entails the dimensionality of the set of object colours is likely to be higher than three. This conjecture has been supported by the results of an experiment conducted by Tokunaga & Logvinenko (2010a) using the multidimensional scaling (MDS).

1 The difference between material and lighting components of object colour is so self-evident that some believe it can be discerned even in pictorial representations on colour monitors, such as in Fig. 19.1 (Craven & Foster, 1992; Foster et al., 2001). Yet, we find it hardly appropriate to use colour monitors to simulate complex multilight scenes so as to study object colours, especially, the difference between the lighting and material components. The fact is that the lighting/material duality of object colours is found to disappear in pictorial representations (Logvinenko et al., 2008). How this difference can, still, be rendered on pictures is a separate question. However, this concerns pictorial rather than natural perception.

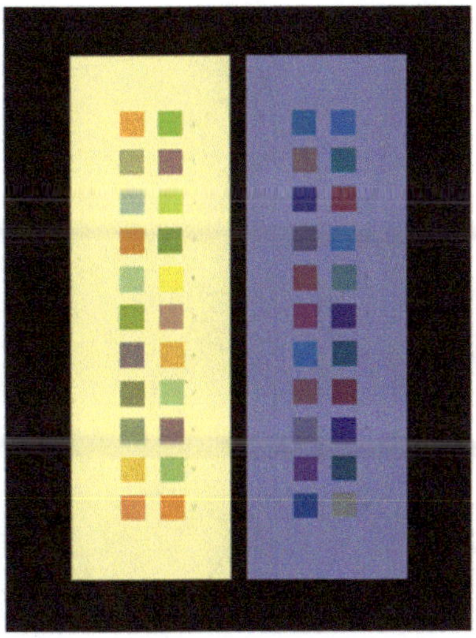

Figure 19.2 Experimental display used by Logvinenko & Tokunaga (2011) to explore simultaneous colour constancy. Source: Logvinenko & Tokunaga (2011)/Brill.

Their experimental display was rather similar to that in Fig. 19.2 except for only seven Munsell papers were employed. Six different lights were used to illuminate the papers. The chromaticity coordinates of the lights reflected by all the 35 object/light pairs are given in Fig. 19.3. Observers were asked to rate the dissimilarity between all the pairs of the 35 object/light pairs. The MDS output configuration is presented in Fig. 19.4. Each marker represents an object/light pair. The distances between the markers represent the averaged (across observers and repetitions) dissimilarities between the corresponding object/light pairs. Specifically, the distances are believed to be in the same order as the dissimilarities. It is worth mentioning that the observers were not given any specification for what constitutes dissimilarity. This was left for them to determine themselves. No one experienced any problems with making dissimilarity judgements. As the object/light pairs were identical with respect to all physical dimensions apart from the spectral reflectance of the papers and the spectral power distribution of the lights, it is safe to believe that the dissimilarity in this experiment can be considered as an operational counterpart of the notion of colour difference as defined above.

There are two features of the MDS output configuration in Fig. 19.4 worth to be pointed out in the present context. First, there are some object/light pairs which map to a cluster of points so close to each other in the chromaticity diagram that one can count them either practically metameric or differing mainly in luminance. Yet, the markers corresponding to these object/light pairs are rather far apart in Fig. 19.4. It means that the colour difference of these object/light pairs is not based on the chromatic difference between the reflected lights. This corroborates the claim that metamerism does not imply object/light colour equivalence.

Second, the MDS output configuration seems to have the shape of a torus deformed by the unavoidable measurement errors. Given a paper lit by some light, the marker corresponding to it will revolve around a small circle of the torus when it is the illumination that varies. Going through the papers lit by the same light will make the marker revolve around the large circle of the torus. As varying the paper spectral reflectance results in mainly the material colour difference, and varying the light spectral composition results in mainly the lighting colour difference, it follows from Fig. 19.4 that the material and lighting dimensions are transversal to each other. Specifically, the material hue varies along the large circle of the torus, the lighting hue varying along the smaller circle.

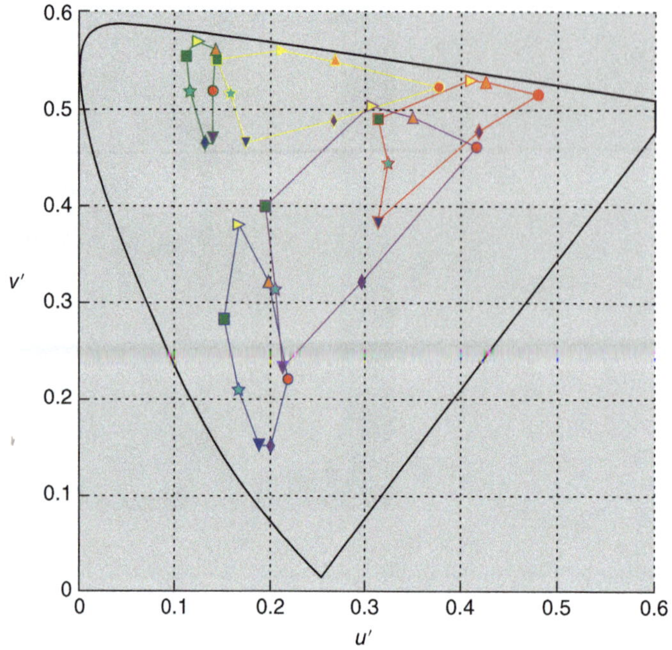

Figure 19.3 Lights reflected by Munsell papers under various illuminations, as presented in the CIE 1976 uniform chromaticity diagram. The shape and face colour of the markers indicate the colour of the paper. (Circle – 5R4/14, triangle (up) – 5YR7/12, triangle (pointing right) – 5Y8/12, square – 5G6/10, pentagram – 10BG5/8, triangle (pointing down) – 5PB5/12, diamond – 10P5/12.) The edge colour of the markers indicates the colour of the illumination. The Munsell papers lit by the same light are connected with a line whose colour indicates the colour of the illumination. Source: Tokunaga and Logvinenko (2010a)/The Optical Society of America.

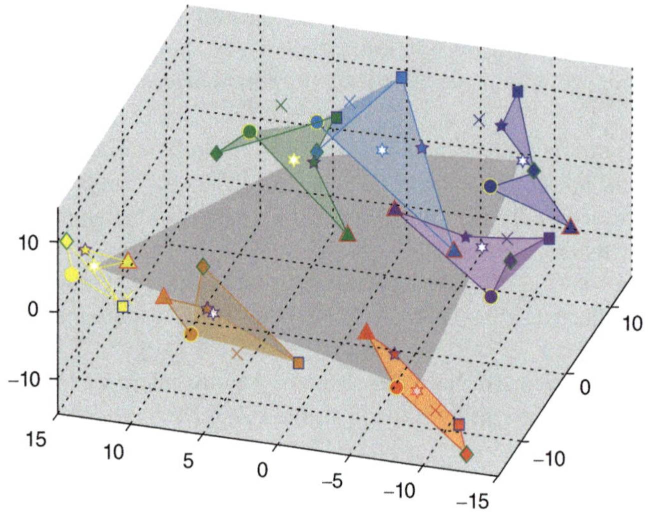

Figure 19.4 Output configuration produced by the non-metric MDS algorithm. Each marker represents a Munsell paper illuminated with a particular light. The colour notation is the same as in Fig. 8.17. Source: Tokunaga and Logvinenko (2010a)/The Optical Society of America.

One might ask: what is the difference between the notion of material and lighting colour dimensions on the one hand, and the notion of surface and apparent illumination colours, as widely present in literature, on the other? Above all, the lighting colour is a perceptual attribute of the object whereas the apparent illumination colour is a perceptual attribute of the illumination. For instance, being in shadow, the lighting colour of an object might become darker despite of clear awareness of the apparent illumination colour being the same for the whole scene. Also, the notions of surface colour and apparent illumination colour implicitly imply the doctrine of intrinsic colours (R. Shepard, 1992) which seems to be so natural that it is usually taken for granted (Maloney, 2003). According to this doctrine, there is on the one hand, the apparent surface colour which, being a perceptual attribute of the surface, is entirely determined by the surface spectral reflectance function; and on the other, the apparent illumination colour which, being a perceptual attribute of the illumination, is entirely determined by the illuminant spectral radiant power distribution (D. I. A. MacLeod, 2003; Mausfeld, 2003). Note that the doctrine of intrinsic colours implies the separability of two sorts. First, it implies the perceptual separability (i.e., the surface colour is experienced as qualitatively different and independent from the apparent illumination colour), which seems quite plausible for the material and lighting colours. Second, it implies the separability with respect to the physical characteristics of object and illumination. However, as will be shown below, the separability of both sorts does not take place. This means that the material and lighting colours cannot be defined separately and independently; one as a perceptual characteristic of only the object, and the other as a perceptual characteristic of only the light. Both are perceptual characteristics of the object/light pair.

19.1 Object/light colour equivalence and its inseparability

In this section, we define object colour as a class of equivalence on the set of object/light pairs. Consider a set of spectral reflectance functions, $\mathcal{X}$ (in this chapter it will be referred to as the *object set*), a set of spectral power distributions, $\mathcal{P}$ (referred to as the *light set*), and their Cartesian product, $\mathcal{X}\times\mathcal{P}$, that is, the set of 'object/light' pairs (x, p) which will be referred to as the *object/light stimulus set*. (For brevity the elements of $\mathcal{X}$ and $\mathcal{P}$ will be called as *objects* and *lights* respectively.) We will assume the object/light colour equivalence to be an equivalence relation on the Cartesian product $\mathcal{X}\times\mathcal{P}$ (denoted as $\sim$). Its relation to metamerism, $\approx$ (defined on $\mathcal{P}$), will be formalised as follows. For any lights p_1 and p_2 in $\mathcal{P}$, and any objects x_1 and x_2 in $\mathcal{X}$

$$\left(x_1, p_1\right) \sim \left(x_2, p_2\right) \Rightarrow x_1 p_1 \approx x_2 p_2, \tag{19.1}$$

where x_1p_1 and x_2p_2 stand for the lights reflected from the objects x_1 and x_2 illuminated by the lights p_1 and p_2 respectively.

The quotient set $(\mathcal{X}\times\mathcal{P})/\sim$ (i.e., the set of the classes of colour equivalent object/light pairs) will be referred to as the *object/light colour stimulus set*. A notation C will be used for the object/light colour stimulus set; that is, $C =: (\mathcal{X}\times\mathcal{P})/\sim$. Formally, C coincides with the set of object colours (in the same way as the colour stimulus space (see Definition 2.5) coincides with the set of light colours). However, we will make a distinction between the set of object colours and the object/light stimulus set because, as in the case of light colours, the former has a richer internal structure that cannot be derived from the latter alone.

In the view of object metamer mismatching, condition (19.1) implies that the object/light colour equivalence cannot be factored into an equivalence defined solely on the set of objects, $\mathcal{X}$, and an equivalence defined solely on the set of lights, $\mathcal{P}$. Indeed, assume that there exists

(i) an equivalence relation $\sim_x$ on $\mathcal{X}$; and (ii) an equivalence relation $\sim_p$ on $\mathcal{P}$ such that for any $x_1, x_2 \in \mathcal{X}$, and $p_1, p_2 \in \mathcal{P}$

$$(x_1, p_1) \sim (x_2, p_2) \Leftrightarrow (x_1 \sim_x x_2) \text{ and } (p_1 \sim_p p_2). \tag{19.2}$$

In this case, if for some light $p \in \mathcal{P}$ the object/light pairs (x_1, p) and (x_2, p) are colour equivalent, i.e., $(x_1, p) \sim (x_2, p)$, then for any other light $p' \in \mathcal{P}$ the object/light pairs (x_1, p') and (x_2, p') should be colour equivalent too. Thus, the reflected lights x_1p and x_2p must be metameric for any incident light p. However, this is impossible because two objects reflecting metameric lights under one illuminant may reflect non-metameric lights under another. It follows that the object/light colour equivalence cannot be reduced to a colour equivalence on objects and a colour equivalence on lights that are independent of each other. In other words, the object/light colour equivalence is *not separable*.

The object/light colour equivalence inseparability unequivocally testifies that object colour is a property of object/light pairs, not the reflecting objects themselves, thereby undermining the notion of the intrinsic colour of an object which assumes that colour is a property of objects alone (R. Shepard, 1992; Maloney, 2003). Indeed, assuming the existence of intrinsic colours (no matter how they are called: (i) apparent illumination and object colours, or (ii) material and lighting colours) proves to be incompatible with another, also very natural assumption, namely, that the object/light colour equivalence implies the metamerism of the reflecting lights (19.1). So, of these two incompatible assumptions one should be abandoned. Since, to the authors' knowledge, there is no evidence that pairs of object/light producing different triplets of cone excitations can be experienced as colour equivalent, it is the notion of intrinsic colour that has to be sacrificed. For this reason, we find terminology for object colour such as 'perceived reflectance' (Arend, 1994; Gilchrist, 2006; Kingdom, 2011) somewhat misleading since, as we can see, object colour is not a subjective representation of the object reflectivity.[2]

It must be said that if you incline to believe in intrinsic colours there is no need for the concept of object/light colour equivalence since the notion of intrinsic colours makes it trivial and unnecessary. Not surprisingly, almost nothing is know about its properties. Something, however, can be gleaned from the fragmentary information on so-called simultaneous colour constancy when real reflecting objects illuminated by several lights were employed. Still, despite the shortage of available empirical data, this concept (i.e., object/light colour equivalence) is worth being considered in more detail before proceeding with the development of the theoretical framework for studying object colour in a multiple-illuminant scene.

19.2 Object/light atlas

It would be useful to have a subset in $\mathcal{X} \times \mathcal{P}$ which could represent the object/light colour stimulus set C (i.e., the set, $(\mathcal{X} \times \mathcal{P})/\sim$, of the classes of colour equivalent object/light pairs) just as a full object atlas (Definition 16.4) represents the object-colour stimulus set in a single illuminant scene (Definition 16.1). This can be done by picking out exactly one object/light pair in each colour equivalent class. The following definition is a natural generalisation of the notion of object atlas.

---

2 One can draw an analogy between, on the one hand, intrinsic object and illumination colours, and on the other, Newton's 'absolute space' and 'absolute time'. Much like the space-time inseparability undermines the existence of the absolute space and absolute time, the material-lighting inseparability undermines the existence of intrinsic object and illumination colours. Paraphrasing Einstein stating that 'space and time are modes in which we think, not conditions in which we exist'; one can say that material and lighting colours are modes in which we perceive objects, not the intrinsic properties of such objects.

Definition 19.1 *Product set $\mathcal{A}_x\times\mathcal{A}_p$ ($\mathcal{A}_x \subset \mathcal{X}$, $\mathcal{A}_p \subset \mathcal{P}$) such that any different object/light pairs (a_x, a_p) and (a'_x, a'_p) in $\mathcal{A}_x\times\mathcal{A}_p$ ($a_x, a'_x \in \mathcal{A}_x; a_p, a'_p \in \mathcal{A}_p$) are not colour equivalent, i.e., $(a_x, a_p) \nsim (a'_x, a'_p)$, will be called* object/light atlas. *When each class in $(\mathcal{X}\times\mathcal{P})/\sim$ contains a pair from $\mathcal{A}_x\times\mathcal{A}_p$, the object/light atlas $\mathcal{A}_x\times\mathcal{A}_p$ is called* full. *Multiple sets $\mathcal{A}_x$ and $\mathcal{A}_p$ will be referred to as the* object atlas *and the* light atlas *respectively.*

Note that in this definition we use the same term (object atlas) as in Section 16.2 (e.g., Definitions 16.4 and 16.5) despite that in Definition 19.1 the object atlas is defined in terms of the object/light colour equivalence whereas in Definitions 16.4 and 16.5 via metamerism (of the reflected lights). This is justified by the fact that if $\mathcal{A}_x\times\mathcal{A}_p$ in Definition 19.1 is the full object/light atlas, then $\mathcal{A}_x$ will be a full illuminant-invariant object atlas in the sense of Definition 16.5 provided that $\mathcal{P} = \mathcal{A}_p$. Indeed, the existence of $a_x \in \mathcal{A}_x$ colour equivalent to $x \in \mathcal{X}$ under light $p \in \mathcal{P}$ implies the reflected lights xp and a_xp to be metameric because of condition (19.1).

The converse is certainly not true because the metamerism of reflecting objects does not entail their colour equivalence. As pointed out above, metamerism guarantees colour equivalence neither in multiple-illuminant scenes (see Section 20.1 and Chapter 21) nor even in a single-illuminant scene.

In the absence of experimental data on the object/light colour equivalence, the question of which sets of objects can serve as an object atlas $\mathcal{A}_x$ remains open, not to mention that it depends on the choice of $\mathcal{X}\times\mathcal{P}$. For example, full illuminant-invariant object atlases (in particular, the optimal object atlas $\mathcal{A}_{0.5}$ (Definition 16.6) considered in Section 16.3 as the object-colour stimulus manifold) seem to be plausible candidates to the multiple set $\mathcal{A}_x$ (i.e., the object atlas) in an object/light atlas (Definition 19.1). However, they should, generally, be tested in experiment for any particular $\mathcal{X}\times\mathcal{P}$. Meanwhile, for the theoretical purpose, some of these atlases will be considered assuming that $\mathcal{X}\times\mathcal{P}$ is chosen so that these can serve as an object atlas $\mathcal{A}_x$.

The situation is similar with the light atlas $\mathcal{A}_p$. In particular, the three-parameter set of lights $\mathcal{A}$ (Eq. (5.4)) called the colour stimulus manifold in Section 5.1 would deserve taking consideration as a candidate to $\mathcal{A}_p$. Unfortunately, $\mathcal{A}$ as defined by Eq. (5.4) involves monochromatic lights that are not included in $\mathcal{P}$. Note, however, that a monochromatic light $\delta(\lambda - \mu)$ can be approximated by a light with Gaussian spectral power distribution of the form of

$$\left(\sqrt{2\pi}\sigma\right)^{-1} \exp\left(-((\lambda - \mu)/\sigma)^2\right), \tag{19.3}$$

letting σ approach zero. Therefore, the set of lights of the form (5.4) where monochromatic lights are replaced with Gaussians (19.3) with sufficiently small σ (written as $\mathcal{G}_\sigma$) may well serve as a light atlas. It will be called a *pseudo-monochromatic light atlas*. Evidently, the object/light atlas $\mathcal{A}_x\times\mathcal{G}_\sigma$ cannot be full for the whole $\mathcal{X}\times\mathcal{P}$, no matter how small σ. Still, fullness can be achieved by excluding from $\mathcal{P}$ the lights with an extremely narrow-band spectral power distribution. In other words, $\mathcal{A}_x\times\mathcal{G}_\sigma$ can happen to be a full object/light atlas for $\mathcal{X}\times\mathcal{P}_\sigma$, where $\mathcal{P}_\sigma$ is a proper subset of $\mathcal{P}$. For sufficiently small σ, the difference between $\mathcal{P}_\sigma$ and $\mathcal{P}$ can be made negligible.[3]

19.3 Object/light colour stimulus manifold

19.3.1 Asymmetric colour matching

Asymmetric colour matching implies asking observers to set up colour matches between differently illuminated objects (as in Fig. 19.2). In such a situation, a genuine match is extremely unlikely

3 Further, whenever using the object/light atlas $\mathcal{A}_x\times\mathcal{G}_\sigma$, we will always restrict our consideration to such a subset of lights, $\mathcal{P}_\sigma$, that $\mathcal{A}_x\times\mathcal{G}_\sigma$ is a full object/light atlas for $\mathcal{X}\times\mathcal{P}_\sigma$.

since it can occur only when the object/light colour equivalence takes place. So, not surprisingly a number of researchers reported a proper asymmetric colour match to be impossible to achieve (Foster, 2003). 'At this match point, however, the test and the match surfaces looked different, and the observers felt as if further adjustments of the match surface should produce a better correspondence. Yet turning any of the knobs or combinations of knobs only increased the perceptual difference' (Brainard et al., 1997, p. 2098). Therefore, one has to admit that the instruction on asymmetric colour match is rather ambiguous and confusing since, because of the difference in lighting colour, a proper colour match (i.e., the object/light colour equivalence) cannot, generally, be established. Nevertheless, in spite of that, the observers will usually accept such an instruction and put in correspondence the colours of differently illuminated objects. As these matches cannot obviously be based on the object/light colour equivalence (i.e., they cannot be colour matches in the proper sense'), the following three questions arise. Why do observers agree to undertake the asymmetric colour match at all? What underlies the asymmetric colour match? Are asymmetric colour matches systematic?

A tentative answer to the first question can be as follows. Since an illumination border (as in Fig. 19.2) introduces unavoidable colour difference between the objects under different lights, quite possibly observers attempt to minimise the dissimilarity between these objects when making the asymmetric (across illuminant) object-colour match. With this hypothesis in mind, a few researchers (Logvinenko & Maloney, 2006; Logvinenko & Tokunaga, 2011) explicitly asked their observers to find a paper in one lighting frame that they saw as subjectively least dissimilar to a given paper in another lighting frame. (When both arrays are equally lit, the least-dissimilar match becomes an exact colour match.) The asymmetric colour matches, obtained using this technique, were rather consistent. The individual differences between observers for least-dissimilar matching were not found to be larger than for symmetric colour matching.

As to the second question, recall that, as revealed by the method of partial hue-matching, the observers produce the same spherical Hering structure, regardless of the illuminant chromaticity (Logvinenko & Beer, 2012). Even more intriguing were the results of the experiments with two different illuminants (Logvinenko & Beer, 2012). In these experiments, the observers had to decide whether an object under one chromatic illumination had a hue in common with an object under the other chromatic illumination. The remarkable fact established is that the chromaticity classes derived in the experiments with illuminants of different chromaticities practically coincided with those obtained under the neutral illuminant (Logvinenko & Beer, 2012). Moreover, the very fact that the observers accept the partial-hue-matching task (in this experimental condition) and yield systematic and consistent results proves that observers are able to establish a correspondence between the chromaticity classes (thus, component hues) under one illumination with the corresponding chromaticity classes (thus, component hues) under the other illumination (Logvinenko & Beer, 2012). Hence, it is safe to generalise this fundamental empirical fact as that they are able to establish a correspondence between the spherical Hering structures induced under different illuminations. In particular, they are likely to be able to judge whether the chromaticness of two objects lit by different lights is same or different. So, we come to the following hypothetical answer to the second question. Two objects lit by different lights make an asymmetric match if they have identical object chromaticness (see Eq. (17.1)).

It should be noted that this hypothesis should not be understood as the claim that observers are able to evaluate in number the chromaticness. Having such a perceptual attribute as chromaticness (which can be, in principle, measured quantitatively by an experimenter), and the observer ability to directly judge it in number, are two different issues. Just as an issue of existence of the sensation magnitude, and the observer ability to directly judge it in number, are two different issues. Fechner and Stevens were in agreement on the former but in strong disagreement on the latter.

All what this hypothesis claims is that, looking at the display like in Fig. 19.2, an observer sees all the papers different in colour, the papers lit by the same light are arranged in a structure which can be likened to a sphere. Due to that, the observer is capable to establish which papers have the same disposition in this structure. How accurately s/he can perform this, is a separate issue and requires a special investigation.

At the same time, it is necessary to emphasise the unitary nature of object colours. It is the phenomenological unitarity that is meant here. That the object colours in multi-illuminant scenes are dually structured (with respect to lighting and material perceptual attributes) in no way contradicts this phenomenological unitarity. In view of the importance of this issue, dwell on it in somewhat more detail, and consider a few examples. A set of boxes containing a chocolate and a cookie, both of different types, has a dual structure, but it is not unitary. Nor is unitary a set of things made of different metals and having different shapes. On the other hand, the set of tones together with these tones shifted down in pitch by a constant interval will be unitary provided that this interval large enough to make sure that all the tones are different. At the same time we have here the same tone structure represented twice due to the transposition. This is a good analogy to what occurs when we perceive the display like in Fig. 19.2: the same perceptual structure implemented in the different pieces of the sensory fabric.

As for the third question, we note that the asymmetric colour matches based on least-dissimilarity were found to meet the properties of symmetry and transitivity (Logvinenko & Tokunaga, 2011). (Transitivity of the least-dissimilar colour matching means that if for an object x_1 illuminated by a light p_1 the least-dissimilar match is an object x_2 illuminated by a light p_2, and an object x_3 illuminated by a light p_3 is judged as least-dissimilar to the object x_2 illuminated by the light p_2, then the object x_3 illuminated by the light p_3 is also judged as least-dissimilar to the object x_1 illuminated by the light p_1.) Although it was confirmed only for matches based on least-dissimilarity,[4] we will assume that transitivity holds for asymmetric colour matching in general.

As the asymmetric colour matching is obviously reflexive, its symmetry and transitivity allows conceptualising it as an equivalence relation on the object/light stimulus set $\mathcal{X}\times\mathcal{P}$. Let us designate the asymmetric-colour-matching relation as $\backsim_m$. In particular, when the asymmetric colour matching is based on least-dissimilarity, $(x_1, p_1) \backsim_m (x_2, p_2)$ means that the object x_1 under the light p_1 looks least dissimilar (amongst all the other objects under this light) to the object x_2 under the light p_2. Also, the object/light colour equivalence will be assumed to be compatible with the asymmetric-colour-matching relation (i.e., the object/light colour equivalence is a congruence with respect to the asymmetric-colour-matching relation). Therefore, the asymmetric-colour-matching relation can be thought to be defined on the object/light colour stimulus set C too. As there is a one-to-one correspondence between the object/light colour stimulus set and any full object/light atlas, the asymmetric-colour-matching relation $\backsim_m$ can be thought of as being defined on (full) object/light atlases as well. For the sake of convenience we will state below some important properties of the asymmetric-colour-matching relation $\backsim_m$ in terms of a full object/light atlas.[5]

This asymmetric-colour-matching relation $\backsim_m$ partitions the object/light colour stimulus set C into classes of $\backsim_m$equivalence. The $\backsim_m$equivalence class to which the colour stimulus (a_x, a_p) belongs will be denoted by $[(a_x, a_p)]_{\backsim_m}$. The quotient set $C/\backsim_m$ (i.e., the set of $\backsim_m$equivalence classes) will be denoted by C_m.

Although the assumption that the asymmetric colour match under equal illumination reduces to the exact match seems obvious, it will be formalised as the following axiom.

4 In fact, we are simply not aware of any other attempts to test the transitivity of asymmetric colour matching.
5 Since in this chapter and further we mainly deal with full object/light atlases, we will omit the adjective 'full' hereinafter for the sake of brevity, unless otherwise stated.

Axiom 19.1 *Given an object/light atlas $\mathcal{A}_x \times \mathcal{A}_p$, and the asymmetric-colour-matching relation $\backsim_m$ on it, for any two pairs $(a'_x, a'_p) \in \mathcal{A}_x \times \mathcal{A}_p$ and $(a''_x, a''_p) \in \mathcal{A}_x \times \mathcal{A}_p$*

$$\text{if } (a'_x, a'_p) \backsim_m (a''_x, a''_p), \text{ and } a'_p = a''_p, \text{ then } a'_x = a''_x.$$

It follows from Axiom 19.1 that in the object/light atlas $\mathcal{A}_x \times \mathcal{A}_p$, the asymmetric colour match is unique for a fixed pair of lights. Indeed, given $(a_x, a_p) \backsim_m (a'_x, a'_p)$, assume that there is another a''_x such that $(a_x, a_p) \backsim_m (a''_x, a'_p)$. As the asymmetric-colour-matching relation $\backsim_m$ is transitive, we have $(a''_x, a'_p) \backsim_m (a'_x, a'_p)$. Axiom 19.1 entails that $a'_x = a''_x$. This will be summarised as follows.

Proposition 19.1 *Consider three pairs (a_x, a_p), (a'_x, a'_p), (a''_x, a'_p) in an object/light atlas $\mathcal{A}_x \times \mathcal{A}_p$, where the asymmetric-colour-matching relation $\backsim_m$ is defined. If $(a_x, a_p) \backsim_m (a'_x, a'_p)$ and $(a_x, a_p) \backsim_m (a''_x, a'_p)$, then $a'_x = a''_x$.*

We will assume that the asymmetric colour match exists for every light. More formally, we will assume that the following axiom holds.

Axiom 19.2 *Given an object/light atlas $\mathcal{A}_x \times \mathcal{A}_p$, for any $\backsim_m$equivalence class, i.e., for any $\mathfrak{m} \in C_m$, and for any element a_p of the light atlas $\mathcal{A}_p$ there is an element a_x in the object atlas $\mathcal{A}_x$ such that the pair (a_x, a_p) belongs to the $\backsim_m$equivalence class $\mathfrak{m}$.*

It follows that one can establish a one-to-one correspondence between the light atlas $\mathcal{A}_p$ and any class of $\backsim_m$equivalence. Indeed, consider an arbitrary $\mathfrak{m} \in C_m$, and define the map $\mathcal{A}_p \to \mathfrak{m}$ that assigns to an element a_p of the light atlas $\mathcal{A}_p$ the pair in $\mathfrak{m}$ with the light component a_p. Axiom 19.2 and Proposition 19.1 secure that such pair exists and that it is unique.

19.3.2 Material colour

The asymmetric-colour-matching relation indicates the existence of a particular type of object-colour resemblance which object/light pairs share being not colour equivalent. Above, based on introspective experience, the perceptual attribute of object colour underlying this resemblance has been called material colour. In other words, it has been suggested that asymmetric colour match is based on equality in terms of the material colour. Now we will use the $\backsim_m$equivalence to conceptualise the term 'material colour'. Specifically, when two object/light pairs belong to the same $\backsim_m$equivalence class we will say that they have, by definition, the same *material colour*.

The set of $\backsim_m$equivalence classes C_m will be referred to as the *material colour set* and its elements *material colours*. The canonical map $C \to C_m$ assigning each object/light colour stimulus (thus, object colour) the $\backsim_m$equivalence class to which it belongs (that is, a material colour) will be called the *canonical material colour map*. Using the canonical map to assign to each object/light pair its object colour (i.e., $(\mathcal{X} \times \mathcal{P}) \to (\mathcal{X} \times \mathcal{P})/\sim$), one can extend the canonical material colour map to the object/light set:

$$M : (\mathcal{X} \times \mathcal{P}) \to C_m. \tag{19.4}$$

The map M will be referred to as the *material colour map*. It assigns to each object/light pair a material colour. Ideally,[6] equality $M(x_1, p_1) = M(x_2, p_2)$ is required for asymmetric (i.e., across-illuminant) colour matching between the object/light pairs (x_1, p_1) and (x_2, p_2).

---

6 In reality, an asymmetric colour match, as any human performance, is influenced by many factors, which leads to random fluctuations in the observer's results.

Given an object/light atlas $\mathcal{A}_x \times \mathcal{A}_p$, one can also establish a one-to-one correspondence between the object atlas $\mathcal{A}_x$ and the set of $\backsim_m$equivalence classes C_m. Let us fix an arbitrary element a_p in the light atlas $\mathcal{A}_p$. Define the map

$$M_{a_p} : C_m \to \mathcal{A}_x \tag{19.5}$$

which assigns to each $\backsim_m$equivalence class $\mathfrak{m}$ an element $a_x \in \mathcal{A}_x$ such that $(a_x, a_p) \in \mathfrak{m}$, that is, $M_{a_p}(\mathfrak{m}) = a_x$. Axiom 19.2 guarantees that such an element a_x exists for any $\mathfrak{m} \in C_m$. Proposition 19.1 guarantees that a_x is unique. Therefore, the map M_{a_p} is bijective. Therefore, $\mathcal{A}_x$ represents C_m. In other words, the set of $\backsim_m$equivalence classes, C_m, has the same structure as an object atlas. If the optimal object atlas $\mathcal{A}_{0.5}$ can serve as $\mathcal{A}_x$ then C_m can be represented as a 3D ball.[7]

The way the object atlas $\mathcal{A}_x$ represents the material colour set depends on the choice of the illuminant. It seems natural to choose some reference illuminant, for example, the illuminant with flat spectrum, to represent the material colour set. Let $a_p \in \mathcal{A}_p$ will be such a reference illuminant. Then, for each material colour, $\mathfrak{m} \in C_m$, the map M_{a_p} defined in (19.5) will determine in the object atlas $\mathcal{A}_x$ its element $a_x = M_{a_p}(\mathfrak{m})$. This element will represent the material colour $\mathfrak{m}$. This means that the object a_x under the reference illuminant a_p will have the material colour $\mathfrak{m}$. In other words, if one wishes to ascertain what colour appearance the material colour $\mathfrak{m}$ has under the reference illuminant a_p, one has to observe the object $a_x = M_{a_p}(\mathfrak{m})$ under the illuminant a_p. Alternatively, one can observe any object/light pair (x, p) colour equivalent to (a_x, a_p) (i.e., such that $(x, p) \sim (a_x, a_p)$).

19.3.3 Lighting colour

There is every indication that the object colours belonging to the same $\backsim_m$equivalence class have the same structure as that revealed when studying light colours in Part I, and the perceptual difference between the object colours in the same $\backsim_m$equivalence class can be described in terms of the same perceptual attributes as for light colours (e.g., hue, brightness, and saturation). At any rate, it is the case for a whitish surface illuminated by various chromatic lights. As mentioned before, such a display is one possible experimental implementation of the colorimetric paradigm. Moreover, the surface does not have to be white. It seems as if many other surfaces will also do.

At first glance, everyday experience indicates that we are quite good at recognising which patches cast on different objects are produced by the same light source. For example, if the display like in Fig. 19.2 is produced by two identical sets of light sources illuminating the bipartite field made by two rectangular sheets of paper with different reflectances, there seems no problem to find out the identical patches in both the halves. From this we can conclude that we are likely to be good at relating the object colours belonging to different $\backsim_m$equivalence classes. However, it is not quite clear how good we are at relating the object colours belonging to different $\backsim_m$equivalence classes in more complicated scenes. For instance, given a variegated surface patchy illuminated, can an observer determine which bits of the surface are equally illuminated? Will observers be capable of making the asymmetric match between patches of light cast on different objects? In other words, are human observers capable of an asymmetric across-object colour match? Unfortunately, there is not enough experimental data to answer these questions.

In this section, we will look into the consequences of assuming that an asymmetric across-object colour match is possible just as an asymmetric across-illuminant colour match. Moreover, we will assume that such an across-object match induces an equivalence relation (written as $\backsim_l$) on the object/light colour stimulus set (and thereby on any object/light atlas, and also on the object/light

7 As pointed out above, $\mathcal{A}_{0.5}$ can always serve as $\mathcal{A}_x$ but, possibly, for a somewhat smaller set $\mathcal{X}' \subset \mathcal{X}$.

set). The quotient set $C/\backsim_l$ (i.e., the set of classes of $\backsim_l$equivalence) will be denoted by C_l. For the sake of theoretical analysis, it will be also assumed that axioms analogous to Axioms 19.1 and 19.2 hold true.

Axiom 19.3 *Given an object/light atlas $\mathcal{A}_x \times \mathcal{A}_p$, and the $\backsim_l$equivalence on it, for any two pairs $(a'_x, a'_p) \in \mathcal{A}_x \times \mathcal{A}_p$ and $(a''_x, a''_p) \in \mathcal{A}_x \times \mathcal{A}_p$*

$$\text{if } (a'_x, a'_p) \backsim_l (a''_x, a''_p), \text{ and } a'_x = a''_x, \text{ then } a'_p = a''_p.$$

Axiom 19.4 *Given an object/light atlas $\mathcal{A}_x \times \mathcal{A}_p$, for any $\backsim_l$equivalence class, i.e., for any $\mathfrak{l} \in C_l$, and for any element a_x of the object atlas $\mathcal{A}_x$ there is an element a_p in the light atlas $\mathcal{A}_p$ such that the pair (a_x, a_p) belongs to the $\backsim_l$equivalence class $\mathfrak{l}$.*

As in the case of material colour, Axiom 19.3 entails a proposition similar to Proposition 19.1.

Proposition 19.2 *Consider three pairs (a_x, a_p), (a'_x, a'_p), (a'_x, a''_p) in an object/light atlas $\mathcal{A}_x \times \mathcal{A}_p$, where the asymmetric-colour-matching relation $\backsim_l$ is defined. If $(a_x, a_p) \backsim_l (a'_x, a'_p)$ and $(a_x, a_p) \backsim_l (a'_x, a''_p)$, then $a'_p = a''_p$.*

Similarly, there is a one-to-one correspondence between the object atlas $\mathcal{A}_x$ and any class of $\backsim_l$equivalence. For an arbitrary $\mathfrak{l} \in C_l$, define the map $\mathcal{A}_x \to \mathfrak{l}$ that assigns to an element a_x of the object atlas $\mathcal{A}_x$ the pair in $\mathfrak{l}$ with the object component a_x. Existence and uniqueness of such a pair follows from Axioms 19.3 and 19.4.

The set of $\backsim_l$equivalence classes C_l will be referred to as the *lighting colour set*, its elements being called *lighting colours*; and the canonical map $C \to C_l$ the *canonical lighting colour map*. The canonical lighting colour map assigns each object/light colour stimulus (thus, object colour) the $\backsim_l$equivalence class to which it belongs, that is, the lighting colour. As in the case of the canonical material colour map, the canonical lighting colour map can be extended to the object/light set: $L : (\mathcal{X} \times \mathcal{P}) \to C_l$. The map L will be referred to as the *lighting colour map*. It assigns to each object/light pair a lighting colour.

Likewise, one can choose some reference object, for example, the perfect reflector, to represent the lighting colour set. Let $a_x \in \mathcal{A}_x$ be such a reference reflectance. Then the map L_{a_x} defined in (19.6) will determine for each lighting colour a light a_p which will represent this lighting colour. In other words, in order to find out the appearance this lighting colour produces on the object a_x, one has to observe the light a_p falling onto the object a_x. Alternatively, one can observe any object/light pair (x, p) colour equivalent to (a_x, a_p) (i.e., such that $(x, p) \sim (a_x, a_p)$).

Finally, given an object/light atlas $\mathcal{A}_x \times \mathcal{A}_p$, one can also establish a one-to-one correspondence between the light atlas $\mathcal{A}_p$ and the lighting colour set C_l. Let us fix an arbitrary element a_x in the object atlas $\mathcal{A}_x$ and define the map

$$L_{a_x} : C_l \to \mathcal{A}_p \tag{19.6}$$

which assigns to an $\backsim_l$equivalence class $\mathfrak{l}$ a pair $(a_x, a_p) \in \mathcal{A}_x \times \mathcal{A}_p$ such that $(a_x, a_p) \in \mathfrak{l}$, that is, $L_{a_x}(\mathfrak{l}) = a_p$. Axiom 19.4 guarantees that such element a_p exists for any $\mathfrak{l} \in C_l$. Proposition 19.2 guarantees that a_p is unique. Therefore, the map L_{a_x} is bijective. Therefore, the set of $\backsim_l$equivalence classes has the same structure as a light atlas. In the case of the pseudo-monochromatic light atlas (19.3), the lighting colour set can be represented as a 3D cone (without a vertex).

19.3.4 Object/light colour stimulus manifold. Material and lighting components of object/light colour stimulus manifold. Material- and lighting-colour coordinates

Of great interest is the situation when each class of $\backsim_l$equivalence intersects with one and only one class of $\backsim_m$equivalence.[8] In this case, each object/light pair (x, p) is assigned a unique pair $(\mathfrak{m}, \mathfrak{l})$ where $\mathfrak{m} \in C_m$, and $\mathfrak{l} \in C_l$. (The lighting and material colours, $\mathfrak{m}$ and $\mathfrak{l}$, will be referred to as the *material* and *lighting components* of the object colour of the object/light pair (x, p).) Therefore, a map $\mathcal{X} \times \mathcal{P} \to C_m \times C_l$ is defined which assigns each object/light pair its material and lighting components.

Given a reference light, a_{p_r}, and a reference object, a_{x_r}, the material and lighting components can be represented by the elements of the object, $\mathcal{A}_x$, and light, $\mathcal{A}_p$ atlases using the maps $M_{a_{p_r}}$ and $L_{a_{x_r}}$, defined in (19.5) and (19.6). In other words, a map $R : \mathcal{X} \times \mathcal{P} \to \mathcal{A}_x \times \mathcal{A}_p$ is determined which assigns each object/light pair (x, p) a pair $\left(a_x, a_p\right)$ where $a_x = M_{a_{p_r}}(M(x, p))$, and $a_p = L_{a_{x_r}}(L(x, p))$. Note that the pair $\left(a_x, a_p\right)$ is not colour equivalent to the pair (x, p), that is, $\left(a_x, a_p\right) \nsim (x, p)$. Being illuminated by the light a_p, the element a_x of the object atlas will not look the same as the object x lit by the light p. Still, the element of object atlas a_x illuminated by the reference illuminant a_{p_r}, will have the same material colour as the object x lit by light p. Also, falling onto the reference object a_{x_r}, the element a_p of the light atlas will have the same lighting colour as the object x lit by light p.

Note the difference in the range of these two maps. While the map R actually amounts to two maps with three-dimensional ranges, the range of the colour map $\mathcal{X} \times \mathcal{P} \to \mathcal{A}_x \times \mathcal{A}_p$ is six-dimensional.

In practice, the material and lighting colour components can be independently identified in an experiment on asymmetric colour matching, in particular, using the least-dissimilarity criterion. Specifically, the material colour component of the object/light pair (x, p) can be found by fixing the reference illumination and choosing an element of the object atlas $\mathcal{A}_x$ least-dissimilar to (x, p). In order to establish the lighting colour component of (x, p), one has to find an element of the light atlas $\mathcal{A}_p$ that, being cast onto the reference object, will produce an object colour least-dissimilar to (x, p). Both the experiments will be of the three-knob type.

It is worth mentioning that it is, probably, the material and lighting colour components that correspond to what was meant by those authors who admitted that, apart from the colour of an object, we also always see the colour of its illumination (Helmholtz, 1867; Mausfeld, 2003). Thus, one needs in total six numbers to quantify both (Mausfeld, 1998; D. I. A. MacLeod, 2003). Note, however, that although thinking separately, as these authors appear to do, of the colour of an object and of the colour of its illumination seems intuitively plausible, this intuition has a physicalist root. The major difference between such a physicalist view of colour and the view presented here is that the material colour component is not a characteristic of the object's reflectivity independent of its illumination. As shown below (Section 19.4), the material colour component of an object, generally, changes with its illumination.

Nevertheless, the set of colours rendered by the object atlas $\mathcal{A}_x$ in a single-illuminant scene (i.e., being lit by a single (reference) light) can be thought of as a good approximation to the material colour set C_m. True, in a single-illuminant scene the colour of an object varies not only in its material component. One cannot exclude a possibility of variation in the lighting component too. However, firstly, as mentioned early, the range of the lighting variation in a single-illuminant scene is hardly to be significant, if any. Secondly, if the observers are able to abstract from differences in lighting component when the chromaticity difference between the illuminants was as large as in Logvinenko & Beer's experiment, they can definitely do so in a single-illuminant scene too.

8 Although this assumption seems natural and entirely plausible, it requires experimental verification.

As a matter of fact, the conceptual framework for colour perception in a single-illuminant scene has been developed so thoroughly in Section 17 just because of the anticipation of its subsequent use as a concrete implementation of the material-colour-set. The whole point to pay so much attention to single-illuminant scenes (that do not occur in the real world) was that they could be a good approximation to a scene where it is the material colour that varies, the lighting colour remaining fixed.

Denote by $C^{\mathfrak{l}}$ a subset of the object colours in C with an equal lighting component, $\mathfrak{l}$, that is, $C^{\mathfrak{l}} = C_m \times \mathfrak{l}$. From the foregoing, it is safe to assume a spherical Hering structure $\langle \mathcal{X}, H_0, H_Y, H_B, H_G, H_R, H_{Bk}, H_W \rangle$ with three pairs of mutually opponent chromaticity classes to exist on $C^{\mathfrak{l}}$. Moreover, one can also assume that there are the three material opponent chromatic scales, $f^{\mathfrak{l}}_{M-YB}$, $f^{\mathfrak{l}}_{M-RG}$, and $f^{\mathfrak{l}}_{WBk}$ on $C^{\mathfrak{l}}$. Given an object colour $(\mathfrak{m}, \mathfrak{l})$ in $C^{\mathfrak{l}}$, a triplet $\left(f^{\mathfrak{l}}_{M-YB}(\mathfrak{m}), f^{\mathfrak{l}}_{M-RG}(\mathfrak{m}), f^{\mathfrak{l}}_{WBk}(\mathfrak{m})\right)$ will be referred to as its $\mathfrak{l}$-*material-colour coordinates* in $C^{\mathfrak{l}}$. Assigning the $\mathfrak{l}$-material-colour coordinates to each element of $C^{\mathfrak{l}}$ makes it $\mathfrak{l}$-*colour-charted*.[9]

Consider the $\mathbf{R}^3$ equipped with the coordinate axes corresponding to the material opponent chromatic scales $f^{\mathfrak{l}}_{M-YB}$, $f^{\mathfrak{l}}_{M-RG}$, and $f^{\mathfrak{l}}_{WBk}$. A map $MC^{l} : C^{\mathfrak{l}} \to \mathbf{R}^3$, assigning to an object colour $(\mathfrak{m}, \mathfrak{l})$ its $\mathfrak{l}$-material-colour coordinates $(f^{\mathfrak{l}}_{M-YB}(\mathfrak{m}), f^{\mathfrak{l}}_{M-RG}(\mathfrak{m}), f^{\mathfrak{l}}_{WBk}(\mathfrak{m}))$ determines a volume $MC^{l}\left(C^{\mathfrak{l}}\right)$ in $\mathbf{R}^3$ that will be called the $\mathfrak{l}$-*material-colour gamut*. It is topologically equivalent to a ball in $\mathbf{R}^3$.

All the notions introduced in Section 17 to characterise object colours (in a single-illuminant scene) can be reformulated so as to characterise material colours in $C^{\mathfrak{l}}$. Specifically, one can talk of $\mathfrak{l}$-*material chromaticness*, $\mathfrak{l}$-*material hue*, $\mathfrak{l}$-*material purity* and so on.

When the extended rectangular atlas, $\mathcal{AR}^3_{0.5}$ is used as the object atlas, $\mathcal{A}_x$, it follows from Eq. (19.5) that one can use $c\Delta\lambda\bar{\lambda}$-coordinates in order to represent a material colour $\mathfrak{m}$ in $C^{\mathfrak{l}}$. In this case, its $\mathfrak{l}$-material-colour coordinates, $\left(f^{\mathfrak{l}}_{M-YB}(\mathfrak{m}), f^{\mathfrak{l}}_{M-RG}(\mathfrak{m}), f^{\mathfrak{l}}_{WBk}(\mathfrak{m})\right)$, can be expressed as functions of $c\Delta\lambda\bar{\lambda}$-coordinates: $f^{\mathfrak{l}}_{M-YB}\left(c(\mathfrak{m}), \Delta\lambda(\mathfrak{m}), \bar{\lambda}(\mathfrak{m})\right)$, $f^{\mathfrak{l}}_{M-RG}\left(c(\mathfrak{m}), \Delta\lambda(\mathfrak{m}), \bar{\lambda}(\mathfrak{m})\right)$ and $f^{\mathfrak{l}}_{M-WBk}$ $\left(c(\mathfrak{m}), \Delta\lambda(\mathfrak{m}), \bar{\lambda}(\mathfrak{m})\right)$.

Likewise, the light atlas $\mathcal{A}_p$ the lights of which fall on the reference object can be considered as a practical implementation of a subset of the object colours in C with an equal material component. Given a material colour $\mathfrak{m}$, denote $C^{\mathfrak{m}} = \mathfrak{m} \times C_l$. If the perfect reflector (or its realistic implementation) is taken as the reference object, we will get, practically, a good example of the colorimetric paradigm. Therefore, there is every reason to assume on $C^{\mathfrak{m}}$ the circular Hering structure $\langle \mathcal{X}, H_0, H_Y, H_B, H_G, H_R \rangle$ with two mutually opponent pairs of chromaticity classes (as described in Section 8.6.3). Furthermore, the two lighting opponent chromatic scales, $f^{\mathfrak{m}}_{L-YB}$, $f^{\mathfrak{m}}_{L-RG}$, and one lighting non-opponent chromatic scale $f^{\mathfrak{m}}_{L-Br}$ will be assumed to be defined on $C^{\mathfrak{m}}$. The triplet $(f^{\mathfrak{m}}_{L-YB}(\mathfrak{l}), f^{\mathfrak{m}}_{L-RG}(\mathfrak{l}), f^{\mathfrak{m}}_{L-Br}(\mathfrak{l}))$ will be referred to as the $\mathfrak{m}$-*lighting-colour coordinates* of object colour $(\mathfrak{m}, \mathfrak{l})$ in $C^{\mathfrak{m}}$. The same way as it is done in Section 8.5, one can introduce the notions of $\mathfrak{m}$-*lighting chromaticness*, $\mathfrak{m}$-*lighting brightness*, $\mathfrak{m}$-*lighting hue* and $\mathfrak{m}$-*lighting purity*. Endowing $C^{\mathfrak{m}}$ with $\mathfrak{m}$-lighting-colour coordinates makes it $\mathfrak{m}$-*colour-charted*.

When a pseudo-monochromatic light atlas (see Eq. (19.3)) is used as $\mathcal{A}_p$, $\mathfrak{m}$-lighting chromaticness and $\mathfrak{m}$-lighting brightness $\left(f^{\mathfrak{m}}_{L-YB}(\mathfrak{l}), f^{\mathfrak{m}}_{L-RG}(\mathfrak{l}), f^{\mathfrak{m}}_{L-Br}(\mathfrak{l})\right)$ can be written as functions of three variables: $(f^{\mathfrak{m}}_{L-YB}(k(\mathfrak{l}), \sigma(\mathfrak{l}), \mu(\mathfrak{l})), f^{\mathfrak{m}}_{L-RG}(k(\mathfrak{l}), \sigma(\mathfrak{l}), \mu(\mathfrak{l})), f^{\mathfrak{m}}_{L-Br}(k(\mathfrak{l}), \sigma(\mathfrak{l}), \mu(\mathfrak{l})))$, where k stands for a multiplication coefficient in front of the Gaussian in Eq. (19.3).

Given $\mathbf{R}^3$ with the coordinate axes corresponding to the lighting opponent chromatic scales $f^{\mathfrak{m}}_{L-YB}$, $f^{\mathfrak{m}}_{L-RG}$, and the lighting brightness scale $f^{\mathfrak{m}}_{L-Br}$, a map $LC^{\mathfrak{m}} : C^{\mathfrak{m}} \to \mathbf{R}^3$, assigning to object

9 Despite the fact that $\mathfrak{l}$ denotes a lighting component, it is material-colour charting that is meant here.

colour $(\mathfrak{m}, \mathfrak{l})$ its $\mathfrak{m}$-lighting-colour coordinates $(f^{\mathfrak{m}}_{L-YB}(\mathfrak{l}), f^{\mathfrak{m}}_{L-RG}(\mathfrak{l}), f^{\mathfrak{m}}_{L-Br}(\mathfrak{l}))$ determines a volume $LC^{\mathfrak{m}}(C^{\mathfrak{m}})$ in $\mathbf{R}^3$ that will be called the *$\mathfrak{m}$-lighting-colour gamut.*

Everything that has been established for the shape of the light-colour gamut in Section 8.6.3 is also valid with respect to the shape of the $\mathfrak{m}$-lighting-colour gamut. We only note here that the absence of the vertex in $LC^{\mathfrak{m}}(C^{\mathfrak{m}})$ is even more obvious than in the light-colour gamut. Really, without light there is no object-colour perception. One cannot assume the otherwise, even for the sake of theoretical abstraction and completeness. Therefore, the shape of the $\mathfrak{m}$-lighting-colour gamut $LC^{\mathfrak{m}}(C^{\mathfrak{m}})$ can be thought of as an irregular cylinder, narrowing down and up, and remaining open at the top and bottom. Formally, it is topologically equivalent to a circular cone, without the vertex, in $\mathbf{R}^3$.

An interesting question is whether the material opponent chromatic scales defined for one lighting colour are related somehow to the material opponent chromatic scales defined for a different lighting colour. We have already argued above that the chromaticity classes (thus, the component hues underlying them as well) obtained for different lighting conditions can be put in correspondence. However, it does not follow from this that it is possible to construct a common (for these lighting conditions) chromatic scale. An important prerequisite for the existence of such a scale is the ability of observers to judge how the chromatic values for different lighting conditions are related. For example, is the amount of white which the perfect-reflector colour contains equal or different under bright and dim light? Although the answer can only be given experimentally, direct observation suggests that the perfect-reflector whiteness in the light of the moon seems to be smaller than in the light of the sun. So one can assume that we are capable of comparing the material chromatic values of an object when their lighting colour is different. In other words, we assume that we are able to use the same material chromatic scale regardless the differences in the lighting component. (Note that so far we have left aside this question, without prejudging any answer. For example, the material chromatic value of any full unitary object-colour has been assumed to be 1 no matter what its lighting component is. Our ability to see the difference between them (due to different lighting components) does not mean that there is a common yard-stick by means of which one can measure this difference.)

If the material component hues are commensurable under different lighting conditions, one can assume the existence of three material opponent chromatic scales, f_{M-YB}, f_{M-RG}, and f_{WBk} on the whole $C_m \times C_l$. Each of these is a function of both variables $\mathfrak{m}$ and $\mathfrak{l}$: $f_{M-YB}(\mathfrak{m}, \mathfrak{l})$, $f_{M-RG}(\mathfrak{m}, \mathfrak{l})$, and $f_{M-WBk}(\mathfrak{m}, \mathfrak{l})$. (The difference with the previous case is that, being an interval scale, each of the previous material opponent chromatic scales can be rescaled independently from the others, for example, setting its maximum chromatic value to be 1.) Likewise, if the lighting component hues are commensurable for different material colours, one can introduce three common lighting chromatic scales, $f_{L-YB}(\mathfrak{m}, \mathfrak{l})$, $f_{L-RG}(\mathfrak{m}, \mathfrak{l})$, and $f_{L-Br}(\mathfrak{m}, \mathfrak{l})$ on $C_m \times C_l$.

Given a map $\mathcal{X} \times \mathcal{P} \to C_m \times C_l$ assigning material and lighting colours $\mathfrak{m}$ and $\mathfrak{l}$ to an object/light pair (x, p), the sextuplet $(f_{M-YB}(\mathfrak{m}(x,p), \mathfrak{l}(x,p)),\ f_{M-RG}(\mathfrak{m}(x,p), \mathfrak{l}(x,p)),\ f_{M-WBk}(\mathfrak{m}(x,p), \mathfrak{l}(x,p)),\ f_{L-YB}(\mathfrak{m}(x,p), \mathfrak{l}(x,p)),\ f_{L-RG}(\mathfrak{m}(x,p), \mathfrak{l}(x,p)),\ f_{L-Br}(\mathfrak{m}(x,p), \mathfrak{l}(x,p)))$ will be called the *colour coordinates* of (x, p).

Consider the $\mathbf{R}^6$ equipped with the coordinate axes corresponding to the chromatic scales f_{M-YB}, f_{M-RG}, f_{WBk}, f_{L-YB}, f_{L-RG}, and f_{L-Br}, which will be referred to as the *object/light colour space.* A map $CC : \mathcal{X} \times \mathcal{P} \to C_m \times C_l \to \mathbf{R}^6$, assigning to an object/light pair (x, p) its colour coordinates determines a volume $CC(C_m \times C_l)$ in $\mathbf{R}^6$ that will be called the *object/light colour gamut.* The restriction of this map on the object/light colour stimulus set C (i.e., the set, $(\mathcal{X} \times \mathcal{P})/\sim$) makes the latter *colour-charted.* Likewise, the restriction of CC on an object/light atlas, in particular, on $\mathcal{A}_x \times \mathcal{A}_p$, makes this atlas colour-charted.

Topologically the object-colour gamut is equivalent to $\mathcal{A}_x \times \mathcal{A}_p$ which is homeomorphic to the Cartesian product of the *3D* ball and the *3D* circular cone (without a vertex). Therefore, it can be considered as a six-dimensional topological manifold (referred to as the *object/light colour manifold).* As a *3D* circular cone without a vertex can be thought as the Cartesian product of a *2D* ball (i.e., a disc) and the real line, the object/light colour manifold can be geometrically represented as the Cartesian product of the *3D* ball, the *2D* ball, and the line. The Cartesian product of the *3D* and *2D* balls is known to be homeomorphic to a *5D* ball. Therefore, the object/light colour manifold can be represented as the Cartesian product of the *5D* ball and the line.

Recall, however, that normalising the chromatic scales to have minimum and maximal values of zero and unity, respectively, uniquely defines the object/light colour gamut. In this case, the object-colour gamut cannot be represented as the Cartesian product of a single object-colour gamut and a single lighting-colour gamut. For each value of the lighting brightness the cross section through it has the shape of an irregular *5D* volume topologically equivalent to the *5D* ball. Its exact shape can be established only by experiment. In particular, there is every indication that the object-colour gamut shrinks when the lighting brightness gets dimmer. Specifically, as mentioned above, Logvinenko & Maloney (2006) found that the perceptual dissimilarity between the blackest and whitest papers they employed diminished when their illumination reduced. This corresponds to introspective observations: in shadow, all colours look less pronounced. The same light grey paper, when in shadow, appears as if it contains less whiteness than when it is in highlight.

19.4 Material colour shift induced by illumination change. Implication for the problem of 'colour constancy'

The material colour map M (see Eq. (19.4)) assigns to two object/light pairs (x, p_1) and (x, p_2) with the same object component, generally speaking, different material colours: $M(x, p_1) \neq M(x, p_2)$. This phenomenon will be referred to as the *material colour shift* induced by illumination change. Quantitatively, it can be expressed as the difference in I-material-colour coordinates for (x, p_1) and (x, p_2). The material colour shift should be distinguished from the illuminant-induced object-colour stimulus shift (i.e., a change of the object component a_x of the colour stimulus (a_x, a_p) for an object/light pair (x, p) induced by a change in the light p) (see Section 16.3) since the latter depends on the choice of the object/light atlas whereas the former does not. In other words, the material colour shift is a genuine colour phenomenon independent of the colour representation.

The material colour shift of an object x caused by switching the illuminant p to the illuminant p' can be revealed as follows. Given an object/light atlas $\mathcal{A}_x \times \mathcal{A}_p$, establish a pair $(a_x, a_p) \in \mathcal{A}_x \times \mathcal{A}_p$ such that $(a_x, a_p) \sim (x, p)$. For simplicity consider $a_p \in \mathcal{A}_p$ as a reference illuminant. Therefore, the element of the object atlas, $a_x \in \mathcal{A}_x$, represents the material colour $M(x, p)$ of the object x under the reference illumination. Then, let a'_x be the element of the object atlas that represents the material colour, $M(x, p')$, of the object x under some other illumination p' (i.e., $a'_x = M_{a_p}(M(x, p')))$. The difference between a_x and a'_x characterises the illuminant-induced object-colour stimulus shift induced by switching from p to p'. An apparent difference between a_x and a'_x under the reference illuminant a_p gives an idea of the material colour shift of an object x induced by a switch from p to p'. If the atlas $\mathcal{A}_x \times \mathcal{A}_p$ is colour-charted, the material colour shift can be characterised by the difference in the material components of the colour coordinates of (a_x, a_p) and (a'_x, a_p), i.e., $(f_{M-YB}(a_x, a_p), f_{M-RG}(a_x, a_p), f_{WBk}(a_x, a_p))$ and $(f_{M-YB}(a'_x, a_p), f_{M-RG}(a'_x, a_p), f_{WBk}(a'_x, a_p))$.

The change of the material colour caused by an illuminant creates a problem for manufacturers of light sources. Indeed, a natural requirement for an artificial light source is that, ideally, it does not change the material colour of the object compared to what it would be in daylight (Schanda, 2007b). As any alteration of light source generally results in some change in both the lighting and material colours, this requirement can be reformulated as producing a minimal change in the material colour with respect to daylight, assuming daylight is considered to be the reference illuminant. As suggested elsewhere (Logvinenko, 2009a), an average material-colour shift can be taken as a quantitative index of colour rendering for one illuminant with respect to another (reference) illuminant. The smaller the average material-colour shift, the better the colour rendering.

Distinguishing between the material and lighting components of an object-colour helps to unravel one of perhaps the most confusing problems in the science of colour: the so-called colour constancy. Although the colour constancy issue has been the focus of colour scientists for more than a century (Katz, 1999; Pokorny et al., 1991; Jameson & Hurvich, 1989; Brainard, 2004; Shevell & Kingdom, 2008; Foster, 2011; Brainard & Radonjic, 2014), it still remains rather vague. In fact, neither the very phenomenon of colour constancy with respect illumination, nor the problems arising from it, have ever been clearly articulated. Admitting that when illumination changes, the object colour appearance changes in one aspect, and it remains the same in the others, the investigators described this as 'relative constancy' (Brainard, 2009), 'relational constancy' (Foster & Nascimento, 1994), or 'qualitative constancy' (Trigt, 2007) of object colour with respect to illumination. However, it has never been specified what exactly in the object colour changes and what remains constant when the illumination changes.

The conceptual framework put forth in this chapter allows a new look at the old problem. When the illumination on object x, switches from p to p' the colour appearance will remain completely constant only if the object/light pair (x, p) is colour equivalent to the object/light pair (x, p'). Otherwise, (that is, when $(x, p) \nsim (x, p')$) the object x will appear different under these two illuminations. It might happen, however, that while $(x, p) \nsim (x, p')$ the material colour of the object x will be the same under both illuminations, that is, $M(x, p) = M(x, p')$. In this case, one can say that *material-colour constancy* (with respect to illumination) takes place. Constancy of material colour means no material-colour shift. It seems natural to refer to the case when non-zero material colour shift occurs (that is, when $M(x, p) \neq M(x, p')$) as *material-colour inconstancy*. As the material-colour shift is, generally, different for different objects, material-colour inconstancy of various degrees should be expected for different objects under the same illuminant. Such dependence of colour inconstancy on the object involved has indeed been reported (Foster, 2011; Logvinenko & Tokunaga, 2011).

It should be noted that because of metamer mismatching, different objects, x_1 and x_2, having the same material colour under the reference illuminant, p, (i.e., $M(x_1, p) = M(x_2, p)$), will, generally, have different material colours under another illuminant, p', (i.e., $M(x_1, p') \neq M(x_2, p')$). It follows that the objects x_1 and x_2 will exhibit material colour inconstancy of different degree. In particular, it might happen that after switching to the illuminant p', one of two colour equivalent (under the illuminant p) objects, say x_1, will show material colour constancy (i.e., $M(x_1, p) = M(x_1, p')$) whereas the other, x_2, material colour inconstancy (i.e., $M(x_1, p) \neq M(x_1, p')$). Therefore, the material colour constancy is a property of an object rather than a colour. It follows that it would be misleading to claim that some colours are less constant than the others (Kulikowski & Vaitkevicius, 1997). It is objects, not colours, that might exhibit more or less material colour inconstancy.

20

Object-Colour Indeterminacy

20.1 Trade-off between object and light components

Given an object/light atlas $\mathcal{A}_x \times \mathcal{A}_p$ (with respect to linear colour mechanisms $\Phi = (\varphi_1(\lambda), \varphi_2(\lambda), \varphi_3(\lambda))$), consider an object/light pair $(x, p) \in \mathcal{X} \times \mathcal{P}$. Subject to the conditions set out in the previous chapter (e.g., Φ satisfies the weak regularity assumption, p being admissible with respect to Φ), there exists a unique pair $(a_x, a_p) \in \mathcal{A}_x \times \mathcal{A}_p$ colour equivalent to (x, p). As follows from condition (19.1) the light reflected from the object x (i.e., xp) is metameric to the light reflected from the object a_x (i.e., $a_x a_p$). Note that there are many other pairs (a'_x, a'_p) in $\mathcal{A}_x \times \mathcal{A}_p$ such that the light $a'_x a'_p$ is metameric to xp. In other words, the following equation

$$\int_{\lambda_{\min}}^{\lambda_{\max}} a_x(\lambda)\, a_p(\lambda)\, \varphi_i(\lambda)\, d\lambda = z_i, \quad i = 1, 2, 3; \tag{20.1}$$

where $z_i = \varphi_i(xp) = \int_{\lambda_{\min}}^{\lambda_{\max}} x(\lambda)\, p(\lambda)\, \varphi_i(\lambda)\, d\lambda$, does not determine uniquely the pair (a_x, a_p).

In fact, for a fixed colour signal (z_1, z_2, z_3) equations (20.1) determine a whole manifold of object/light stimuli (a_x, a_p) such that $a_x a_p$ is metameric to xp. We will refer to it as the *object-light invariance manifold*. When $\mathcal{A}_x$ and $\mathcal{A}_p$ are three-dimensional manifolds the object-light invariance manifold is, generally, three-dimensional too. When the object/light atlas $\mathcal{A}_x \times \mathcal{A}_p$ is colour-charted, the object-light invariance manifold will be called the *material-lighting colour invariance manifold*.

For any light component a_p there exists at most one object component a_x so that (a_x, a_p) meets Eq. (20.1) provided that colour signal (z_1, z_2, z_3) is fixed; and vice versa, fixing the object component a_x uniquely determines the light component a_p in Eq. (20.1). The relationship between a_x and a_p that meet Eq. (20.1) will be referred to as the *invariant relationship between object and light components*. When a_x and a_p are in this relationship (i.e., meet Eq. (20.1)) we will say that they are compatible with each other (with regard to this relationship). In the colour-charted object/light atlas, compatible a_x and a_p define the material and lighting colours, respectively, which will be also called compatible with each other (with regard to the material-lighting colour invariance manifold).

The invariant relationship between object and light components can be revealed, to the first approximation, by means of the implicit function theorem the way it has been done in Section 16.4.3. Specifically, Eq. (16.21) can be used to ascertain how light component a_p affects object component a_x provided that the extended rectangular atlas, $\mathcal{AR}^3_{0.5}$, serves as an object atlas.

Consider a particular object component $a^0_x(\lambda)$, and a light component compatible with it (with respect to Eq. (20.1)), which we denote by $p^0(\lambda)$ to make the notation the same as in Section 16.4.3. Equation (16.21) predicts how much the coordinates $\left(c^0, \Delta\lambda^0, \overline{\lambda}^0\right)$ of $a^0_x(\lambda)$ will change as a result of a small change of the light component $p^0(\lambda)$: $p(\lambda) = p^0(\lambda) + \Delta p(\lambda)$. As shown in Section 16.4.3,

Foundations of Colour Science: From Colorimetry to Perception, First Edition. Alexander D. Logvinenko and Vladimir L. Levin.
© 2023 John Wiley & Sons Ltd. Published 2023 by John Wiley & Sons Ltd.

the partial derivative $\frac{\partial F}{\partial v}(u^0, v^0)$ appearing in Eq. 16.21 is a linear map $DF : \mathbf{L}^1 \to \mathbf{R}^3$ defined by the three linear functionals DF_i:

$$DF_i(q(\lambda)) = c \int_{\overline{\lambda}^0 - 0.5\Delta\lambda^0}^{\overline{\lambda}^0 + 0.5\Delta\lambda^0} q(\lambda)\, \varphi_i(\lambda)\, d\lambda + \int_{\lambda_{\min}}^{\lambda_{\max}} (0.5\,(1-c))\, q(\lambda)\, \varphi_i(\lambda)\, d\lambda, \quad (i = 1, 2, 3). \tag{20.2}$$

Applying the 3×3 matrix $-\left(\frac{\partial F}{\partial u}(u^0, v^0)\right)^{-1}$ to the vector $DF(\Delta p(\lambda)) = \big(DF_1(\Delta p(\lambda)), DF_2(\Delta p(\lambda)), DF_3(\Delta p(\lambda))\big)$ one gets the alteration of the object-component coordinates $\left(c^0, \Delta\lambda^0, \overline{\lambda}^0\right)$ induced by the light-component increment $\Delta p(\lambda)$.

Note that $DF(q)$ amounts to $\Phi(a_x^0 q)$. In other words, $DF(q)$ is just the colour signal of the light reflected by $a_x^0(\lambda)$ when it is illuminated by $q(\lambda)$. Since $\frac{\partial F}{\partial u}(u^0, v^0)$ coincides with the Jacobian of $\Phi_p|_{AR^3_{0.5}}$ determined by Eq. (16.13), $-\left(\frac{\partial F}{\partial u}(u^0, v^0)\right)^{-1}$ times $DF(\Delta p(\lambda))$ yields such coordinates $(c, \Delta\lambda, \overline{\lambda})$ that define the object component which under illumination $p^0(\lambda)$ produces the colour-signal vector which is the additive inverse of the colour-signal vector $\Phi(a_x^0 \Delta p)$ (i.e., $-\Phi(a_x^0 \Delta p)$).

Though, such complementarity directly follows from Eq. (16.21) after it is rewritten as

$$\frac{\partial F}{\partial u}(u^0, v^0) \circ du = -\left(\frac{\partial F}{\partial v}(u^0, v^0) \circ dv\right). \tag{20.3}$$

Here $\frac{\partial F}{\partial v}(u^0, v^0) \circ dv$ is the colour signal induced by the light-component increment dv, and $\left(\frac{\partial F}{\partial u}(u^0, v^0)\right) \circ du$ is the colour signal induced by the object-component increment du.

So, a change in light component a_p in Eq. (20.1) requires a change of the object component a_x (so as to meet Eq. (20.1)) which induces the same alteration of the colour signal as that induced by the change of a_p only in the opposite direction.

When a light component a_p is from a three-parameter set (light atlas) $\mathcal{A}_p$, v will be a triplet of parameters, and $\frac{\partial F}{\partial v}(u^0, v^0)$ will be a 3×3 matrix. If it is invertible Eq. (20.3) can be resolved with respect to dv:

$$dv = -\left(\frac{\partial F}{\partial v}(u^0, v^0)\right)^{-1} \circ \frac{\partial F}{\partial u}(u^0, v^0) \circ du. \tag{20.4}$$

It shows how light component a_p covaries with object component a_x in Eq. (20.1).

In order to illustrate the trade-off between object and light components a_x and a_p in Eq. (20.1) within a larger range, consider the object atlas comprising the Gaussian spectral reflectance functions and the light atlas also comprising the Gaussian spectral power distributions as suggested by Logvinenko (2013b).[1] Specifically, consider a three-parameter set, $\mathcal{G}_x$, of spectral reflectance functions, $g_m(\lambda; k_m, \theta_m, \mu_m)$, which are defined as follows. Denote as usual $\Lambda = \lambda_{\max} - \lambda_{\min}$. When $\mu_m \le (\lambda_{\max} + \lambda_{\min})/2$ we have: for $\lambda_{\min} \le \lambda \le \mu_m + \Lambda/2$

$$g_m(\lambda; k_m, \theta_m, \mu_m) = k_m \exp\left[-\theta_m(\lambda - \mu_m)^2\right]; \tag{20.5}$$

and for $\mu_m + \Lambda/2 < \lambda \le \lambda_{\max}$

$$g_m(\lambda; k_m, \theta_m, \mu_m) = k_m \exp\left[-\theta_m(\lambda - \mu_m - \Lambda)^2\right]. \tag{20.6}$$

1 Gaussians were also previously used to model objects and lights by some other authors (Nikolaev, 1985; D. I. A. MacLeod & Golz, 2003).

When $\mu_m > (\lambda_{\max} + \lambda_{\min})/2$ we have: for $\lambda_{\min} \le \lambda \le \mu_m - \Lambda/2$

$$g_m(\lambda; k_m, \theta_m, \mu_m) = k_m \exp\left[-\theta_m(\lambda - \mu_m + \Lambda)^2\right] \tag{20.7}$$

and for $\mu_m - \Lambda/2 < \lambda \le \lambda_{\max}$

$$g_m(\lambda; k_m, \theta_m, \mu_m) = k_m \exp\left[-\theta_m(\lambda - \mu_m)^2\right]. \tag{20.8}$$

For $0 \le k_m \le 1$, $\lambda_{\min} \le \mu_m \le \lambda_{\max}$, and positive θ_m Eqs. (20.5–20.6) and (20.7–20.8) define a Gaussian on the visible spectrum circle with its maximum (equal to k_m) being at the wavelength μ_m.

Consider also a three parameter set, $\mathcal{G}_p$, of spectral power distribution functions of the same form, that is, when $\mu_l \le (\lambda_{\max} + \lambda_{\min})/2$

$$g_l(\lambda; k_l, \theta_l, \mu_l) = k_l \exp\left[-\theta_l(\lambda - \mu_l)^2\right] \tag{20.9}$$

if $\lambda_{\min} \le \lambda \le \mu_l + \Lambda/2$; and

$$g_l(\lambda; k_l, \theta_l, \mu_l) = k_l \exp\left[-\theta_l(\lambda - \mu_l - \Lambda)^2\right], \tag{20.10}$$

if $\mu_l + \Lambda/2 < \lambda \le \lambda_{\max}$. Likewise, when $\mu_l > (\lambda_{\max} + \lambda_{\min})/2$

$$g_l(\lambda; k_l, \theta_l, \mu_l) = k_l \exp\left[-\theta_l(\lambda - \mu_l + \Lambda)^2\right] \tag{20.11}$$

if $\lambda_{\min} \le \lambda \le \mu_l - \Lambda/2$; and if $\mu_l - \Lambda/2 < \lambda \le \lambda_{\max}$

$$g_l(\lambda; k_l, \theta_l, \mu_l) = k_l \exp\left[-\theta_l(\lambda - \mu_l)^2\right]. \tag{20.12}$$

Although $\mathcal{G}_x \times \mathcal{G}_p$ is not a full object/light atlas, it represents quite a part of the object/light colour stimulus set C.

Despite that there is no simple relationship between the parameters k_m, θ_m, and μ_m (respectively, k_l, θ_l, and μ_l) and the perceptual attributes of the colour of the object $g_m(\lambda; k_m, \theta_m, \mu_m)$ (respectively, the light $g_l(\lambda; k_l, \theta_l, \mu_l)$), these parameters allow fairly simple interpretation. Specifically, the coefficient k_l can be thought of as the main determinant of the brightness of light with the Gaussian spectral power distribution (20.9). The coefficient k_m is a stimulus correlate of the lightness (i.e., blackness/whiteness) of the object with the Gaussian spectral reflectance function (20.5) – the smaller k_m, the blacker the object's colour. The peak wavelength μ_m (respectively, μ_l) can be considered, to the first approximation, as the stimulus correlate of hue.

Assume now that $\mu_m, \mu_l \le (\lambda_{\max} + \lambda_{\min})/2$, and $\lambda_{\min} \le \lambda \le \max(\mu_m, \mu_l) + \Lambda/2$. In this case, for each i, equations analogous to (20.1) will take the form

$$\int_{\lambda_{\min}}^{\lambda_{\max}} k_m k_l \exp\left[-\theta_m(\lambda - \mu_m)^2\right] \exp\left[-\theta_l(\lambda - \mu_l)^2\right] s_i(\lambda)\, d\lambda = \varphi_i(xp). \tag{20.13}$$

It follows that any pair of k'_m and k'_l such that $0 \le k'_m \le 1$, $0 \le k'_l$, and $k'_m k'_l = k_m k_l$ will satisfy Eq. (20.13).

Equation (20.13) also bound parameters θ_m and θ_l, and μ_m and μ_l. Indeed, they can be transformed as

$$A \int_{\lambda_{\min}}^{\lambda_{\max}} \exp\left[-\left(\theta_m + \theta_l\right)\left(\lambda - \frac{\theta_m \mu_m + \theta_l \mu_l}{\theta_m + \theta_l}\right)^2\right] s_i(\lambda)\, d\lambda = \varphi_i(xp), \tag{20.14}$$

where

$$A = k_m k_l \exp\left[-\left(\theta_m \mu_m^2 + \theta_l \mu_l^2\right) + \frac{\left(\theta_m \mu_m + \theta_l \mu_l\right)^2}{\theta_m + \theta_l}\right].$$

One can show that for each triplet (k_m, θ_m, μ_m) there is not more than one triplet (k_l, θ_l, μ_l) such that the sextuplet $(k_m, \theta_m, \mu_m, k_l, \theta_l, \mu_l)$ meets Eq. (20.14); and vice versa, given a triplet (k_l, θ_l, μ_l) there is not more than one triplet (k_m, θ_m, μ_m) such that the sextuplet $(k_m, \theta_m, \mu_m, k_l, \theta_l, \mu_l)$ meets Eq. (20.14).

To look into the relationship between the object and light components, (k_m, θ_m, μ_m) and (k_l, θ_l, μ_l), satisfying (20.14), let us hold the shape of the Gaussian under the integral sign in Eq. (20.14) constant, i.e., let us assume that

$$\theta_m + \theta_l = \theta, \text{ and } \frac{\theta_m \mu_m + \theta_l \mu_l}{\theta_m + \theta_l} = \mu, \tag{20.15}$$

where μ and θ are some constants (determining the shape of the Gaussian under the integral sign). In this case, the left-hand-side terms will be proportional to the corresponding right-hand-side terms in (20.14). Then for a fixed chromaticity of the reflected-light (i.e., when

$$\varphi_i(xp)/\left(\varphi_1(xp) + \varphi_2(xp) + \varphi_3(xp)\right) = c_i \quad (i = 1, 2, 3),$$

where c_i are constants) the trade-off between object and light components is described by Eq. (20.15), which can be rewritten as

$$\theta_m = \theta - \theta_l, \tag{20.16}$$

$$\mu_m = \frac{\theta}{\theta - \theta_l}\mu - \theta_l \mu_l. \tag{20.17}$$

It follows from Eq. (20.16) that the parameters θ_m and θ_l are bound by a simple linear relation. The parameters μ_m and μ_l are also linearly related to each other (20.17). When θ_m and θ_l are kept constant, μ_l decreases when μ_m increases, and vice versa.

Let us complete the last two equations with that bounding k_m and k_l:

$$k_m k_l = k, \tag{20.18}$$

where k is the parameter determining the amplitude of the Gaussian under the integral sign.

Equations (20.16–20.18) specify how the corresponding parameters of the object and light components $g_m(\lambda; k_m, \theta_m, \mu_m)$ and $g_l(\lambda; k_l, \theta_l, \mu_l)$ interrelate when they are in the invariant relationship with each other. We see the form of reciprocity similar to that following from Eqs. (16.21) and (20.4). Specifically, if the illumination changes so that μ_l increases, then μ_m decreases, and vice versa (provided that θ_m and θ_l are kept constant).

Since parameters k_m and k_l can be treated as the main determinants for lightness and lighting-brightness, respectively, Eq. (20.18) allows to expect the lightness and lighting-brightness to reciprocally covary for such object and light components $g_m(\lambda; k_m, \theta_m, \mu_m)$ and $g_l(\lambda; k_l, \theta_l, \mu_l)$ that are in the invariant relationship with each other. Likewise, as the parameters μ_m and μ_l can be considered, at least to a first approximation, as the stimulus correlates of the material and lighting hue respectively, then Eq. (20.17) indicates that there is a reciprocal relationship between the material and lighting hues for those object and light components $g_m(\lambda; k_m, \theta_m, \mu_m)$ and $g_l(\lambda; k_l, \theta_l, \mu_l)$ which are in the invariant relationship with each other. In Section (20.2) we present some evidence supporting this conjecture.

20.2 Trade-off between material and lighting colours

The invariance relationships between the object and light components of light/object atlases quantify, generally speaking, the intuitively clear physical fact that the same light coming to the eye can be produced by, on the one hand, say, a blue surface (i.e., appearing blue under daylight) illuminated by a yellow light, and on the other hand, a yellow surface (i.e., appearing yellow under daylight) illuminated by a blue light. (An illustrative example of such a situation is given in Fig. 19.1.) Just as a retinal image of the same size can be produced by a long stick at a short distance from the eye and a short stick at a longer distance. In order to produce the same retinal size, α, the physical size, S, and distance, D, have to be in the following relation:

$$\frac{S}{D} = 2\arctan\left(\frac{\alpha}{2}\right), \tag{20.19}$$

or for relatively small α:

$$\frac{S}{D} \cong \alpha. \tag{20.20}$$

Note that Eq. (20.20) is purely a physical relation. It tells us about perception of size and distance nothing more than a relationship between the distal and proximal stimuli (pertinent to size-distance perception). Yet, it was a great achievement when it was discovered that a similar relationship holds between the apparent counterparts of S and D, on the one hand, and the retinal size α, on the other:

$$\frac{\tilde{S}}{\tilde{D}} \propto \alpha, \tag{20.21}$$

where $\tilde{S}$ and $\tilde{D}$ stand for the apparent size and distance, respectively, $\propto$ meaning equality up to a proportionality constant. Equation (20.21) is known as the *apparent size-distance invariance hypothesis* (Kilpatrick & Ittelson, 1953; Ittelson, 1960).

This relationship is supported by the celebrated Ames demonstrations (Ittelson, 1952) which show that, for a fixed retinal size, when the apparent distance is illusory misperceived, the corresponding misperception of the apparent distance takes place, and vice versa, in accordance with Eq. (20.21). As a matter of fact, Eq. (20.21) has been long known as the so-called Emmert law for after-images (Brown, 1965).

The greatness of this discovery is that it showed a new consistency pattern in perception. Instead of the conventional 'stimulus-percept' determination, Eq. (20.21) claims the existence of the 'percept-percept' coupling.[2] The latter means that one perceptual attribute may affect another perceptual attribute (even if the proximal stimulation remains constant). Using the word 'affect' we try to refrain from the term 'determine'. It would be, perhaps, wrong to say that one perceptual attribute determines another since it is generally believed that perceptual attributes are determined by stimuli.[3] Interestingly, even in 'deterministic terminology', Eq. (20.21) implies that, instead of the apparent size, the retinal size 'determines' the ratio of the apparent size and distance. Equation (20.21) can be considered as a normative rule imposed on such the perceptual attributes as apparent size and distance. Specifically, it is a rule prescribing which pairs of $\tilde{S}$ and $\tilde{D}$ are admissible for a given α.

A similar relationship was claimed to hold between the apparent form of a flat object and its apparent slant for a fixed retinal form. For example, having a square form on the retina, we can

2 The importance of such 'percept-percept' coupling for the theory of perception was emphasised by Epstein (1982).
3 Though, from a philosophical point of view, the statement that a material event can determine a mental event (that is, one that does not belong to the material world, by definition) sounds even more controversial.

perceive a vertical square or a slanted rectangle. Likewise, the retinal form determines a family of admissible pairs of 'apparent form' and 'apparent slant'. It is known as the *apparent form-slant invariance hypothesis* (Ittelson, 1960; Epstein & Park, 1963). Unfortunately, we find it difficult, if not impossible, to express this in a similar way as it is done in Eq. (20.21) for the apparent size and distance, because it is unclear how the apparent slant is represented in perception. Gibson believed that there was no perceptual counterpart of the physical slant, insisting that it is a 'form-at-slant' as a whole that is perceived.[4] However, as we will see below (Section 20.2.3) one can point out a special perceptual attribute (shading) which plays a role of the perceptual counterpart of the physical slant.

It turns out that the material and lighting colours are bound by some invariant relationship, similar to how the apparent size and the apparent distance are connected, as well as the apparent form and slant. Furthermore, just as in the case with the latter, we come to this conclusion not only because of the existence of the invariant relationship between object and light components. Of course, since each element of an object/light atlas uniquely represents an object colour, it follows that the object colours represented by the elements of an object/light atlas, which are in this invariant relationship, are similarly related. But besides this, there is abundant evidence – visual phenomena and demonstrations – testifying for that the same colour signal can bring about only particular combinations of material and lighting colours. That is, between material and lighting colours there is a specific trade-off. In what follows we consider these phenomena and demonstrations, starting from the achromatic case for the sake of simplicity.

20.2.1 Invariant relationship between lightness and lighting brightness

For an experimental study of the relationship between lightness and lighting brightness, it would be ideal to have such a subset of objects and lights that each pair of object/light in it brings about a hueless object colour; that is, the material and lighting colours of which would be achromatic. However, this is hard to achieve due to the metamer mismatching. The metamer-mismatch volume computed for the centre of the object-colour solid, caused by the shift from illuminant D65 to A, was found to be quite large (Logvinenko, Funt, & Godau, 2015). Even without the colour-charted object-colour solid at hand, it indicates that much of this metamer-mismatch volume will not be experienced as hueless, as far as the material colour is concerned. Interestingly, the magnitude of metamer mismatching remains significant even when both the illuminants, inducing it, appear achromatic (Logvinenko, Funt, & Godau, 2015). As follows from Proposition 18.4, metamer mismatching can be eliminated only if the spectral power distributions of the illuminants, $p_1(\lambda)$ and $p_2(\lambda)$, are proportional to each other, i.e., $p_1(\lambda) = kp_2(\lambda)$, where $k > 0$.

Alternatively, one can consider only objects with flat spectral reflectance function, for which the illuminant-induced object-colour stimulus shift is always zero. Although such objects are obviously impossible to physically implement, every established colour collection (such as the Munsell book of colour) includes a series of papers, the spectral reflectance function of which is approximately flat. Each of these papers can be specified by a single number, namely, the average reflectance coefficient, r (which can also be interpreted as the surface albedo).

Under daylight both the material colour and the lighting colour of these papers appear neutral (hueless), the lightness and lighting brightness being the only perceptual attributes of them. Such an achromatic scene is convenient to begin studying the relationship between lightness and lighting brightness. In fact, any illuminant will be suitable, provided that it ensures the achromaticity of the material and lighting colours.

---

4 In fact, Gibson used the phrase 'shape-at-slant' but we reserve the term 'shape' for another notion.

So, we come up with the following particular case of the idealised experimental paradigm expressed by Eq. (II.4). We have a matte, flat, spatially homogeneous surface illuminated by light coming at every point of the surface from one direction. The spectral reflectance function of the surface is assumed to be flat, with r being its reflectance coefficient. Given the spectral power distribution of the illuminant in the form of $kI_0(\lambda)$, where k denotes a variable factor, the luminance of the light entering the eye, Lum, can be expressed as

$$Lum \propto rk. \tag{20.22}$$

Therefore, as long as Lum is kept constant, the colour signal produced by these object/light pairs remains the same. (Note that when the luminance is fixed at some level, Eq. (20.22) becomes an analogue of Eq. (20.18) for achromatic scenes considered here.) Increasing (respectively, decreasing) r results in an increase (respectively, a decrease) of lightness.[5] Likewise, increasing (respectively, decreasing) k will result in an increase (respectively, a decrease) of the lighting brightness.[6]

About lightness perception

It so happened that lightness perception has formed as a discipline on its own, little associated with colour science in general. This cannot but come as a surprise because of at least two things. First, an idea of creating a science about only two colours (black and white) seems as bizarre as an idea of the science about two pitches or two notes. Second, lightness perceptionists intend to study black and white only in their pure form, not contaminated by any other hue. If such an undertaking is feasible at all, then only for ideal objects with a flat spectral reflection function. For any real objects, including the aforementioned achromatic Munsell chips, their reflectivity cannot be characterised by a single number. Specifically, the albedo of real objects depends on the illumination (Logvinenko, Funt, & Godau, 2015), thus it cannot be considered as a possible stimulus for their lightness.

Equation (20.22) is of the same kind as Eq. (20.19). It relates the distal stimuli for lightness and lighting brightness with the proximal stimulus, i.e., luminance.[7] More interestingly, Eq. (20.22) turns out to have its perceptual counterpart, just like the Eq. (20.19) has its perceptual counterpart in the form of Eq. (20.21). Historically, the first mention of such a possibility is known in literature as the *albedo hypothesis*, which, in our terminology, states that, under a fixed luminance of the reflected light, Lum, lightness, L_{WBk}, reciprocally varies with lighting brightness, B_l:

$$L_{WBk}B_l \propto Lum. \tag{20.23}$$

Albedo hypothesis

Although Helmholtz also claimed something similar (Helmholtz, 1867), Koffka (1935, p. 244) was probably the first to explicitly point out that in conditions where proximal stimulation is constant, any change in lightness is accompanied with a reciprocal change in apparent illumination[8] (for a review see Beck, 1972; Gilchrist, 2006, Chapter 8).

---

5 That is, either an increase (respectively, a decrease) of whiteness, or a decrease (respectively, an increase) of blackness.

6 In some publication on this subject, the lighting brightness in such a paradigm was referred to as the surface-brightness in order to emphasise that it is a perceptual attribute of object colour rather than light colour (Logvinenko & Maloney, 2006; Logvinenko et al., 2008).

7 As a caveat, note that albedo can be considered as a stimulus correlate of lightness only within this particular experimental paradigm, designed to study only achromatic object colours. It would be a mistake to attribute this role to albedo beyond this paradigm. In general, it is hardly possible to find a physical parameter that can play the role of a stimulus for lightness, because, as shown below, the concept of a stimulus itself does not seem to apply to object colours.

8 As mentioned before, the early authors used the term ‘apparent illumination’ to refer to what we believe is the lighting colour.

Despite that the albedo hypothesis seems quite feasible, its repeated experimental tests yielded somewhat controversial results. While some investigators claimed to have found evidence to support the albedo hypothesis (Kozaki & Noguchi, 1976; Noguchi & Kozaki, 1985; Logvinenko & Menshikova, 1994), others failed to verify it (R. B. MacLeod, 1940; Beck, 1961; Oyama, 1968; Kozaki, 1973; Rutherford & Brainard, 2002).

It should be pointed out that in all these experiments, along with lightness, the latter group of researchers measured the apparent illumination understood as the ambient illumination which is, actually, a characteristic of the illuminant rather than the illuminated objects. Hence, it is hardly surprising that the subjective estimates of the illuminant or ambient light turned out to be poorly correlated with lightness.

Surely, for a fixed light reaching the eye from an object, one might perceive many combinations of its lightness and lighting brightness. Suffice it to recall the aforementioned Gelb effect (Gelb, 1929), or its more sophisticated version the staircase Gelb effect (Gilchrist, 2017), which demonstrate that black object lit by an intense light can be perceived as white. In this case, the lighting colour is 'mistakenly' perceived as being dimmer than under the veridical perception. As a result, the material colour turns from black to white. Yet, the relationship (20.23) as it stands cannot hold true due to incorrect predictions arising from it. Indeed, it implies a linear relationship between lightness and luminance when lighting brightness is kept constant. However, as is well established, lightness increases with luminance more slowly than a linear function (Wyszecki & Stiles, 1982, pp. 493–499); which is quite understandable, since lightness has a natural maximum (whiteness of the perfect reflector), whereas luminance can theoretically increase infinitely. Still, as argued elsewhere (Logvinenko, 1997), Eq. (20.23) lends itself to the following elaboration so as to avoid this contradiction:

$$L_{WBk}B_l \propto g\,(Lum), \tag{20.24}$$

where g is some monotonically increasing function (with negative acceleration). We will refer to (20.24) as the *lightness/lighting-brightness invariance hypothesis*. It is, of course, only a hypothesis that still needs experimental verification. It is worth noting that in the present context, it is not the specific form of Eq. (20.24) that matters, but the very fact that the lightness of the material colour is perceptually coupled with the brightness of the lighting colour, reciprocally covarying with each other.

Apart from the Gelb effect, there are many other demonstrations testifying in favour of such an invariant relationship between lightness and lighting brightness. An essential feature of these demonstrations is the alteration of the lighting brightness without changing the luminance. For example, in some experiments, the manipulation of the apparent spatial layout was used to alter the lighting colour (Gilchrist, 1977; Logvinenko & Menshikova, 1994). Rearranging depth cues one can make an object be illusorily[9] perceived at a different distance where the illumination is different. It turns out that such an apparent depth shift results in a change in lightness. A white surface, after having been apparently moved to a brighter area, looks blackish (Logvinenko, 2005a). Figure 20.1 depicts the experimental display used by Logvinenko (2005a). Vertical strips cut out of white paper are presented in front of a background made of the same paper painted black. Illumination of the strips and the background were independently set to reflect equal amounts of light. Despite their equiluminance the strips appeared white and the background appeared black. That is, the identical colour signals brought about two different pairs of material and lighting colours: white material colour with dim lighting colour (the strips), and black material colour with bright lighting colour (the background).

---

9 The word 'illusorily' should not be misunderstood in the sense that there was something wrong with perception. It just means that the perception was not veridical. See more about the concept of visual illusion in Gregory (1997) and Todorović (2020).

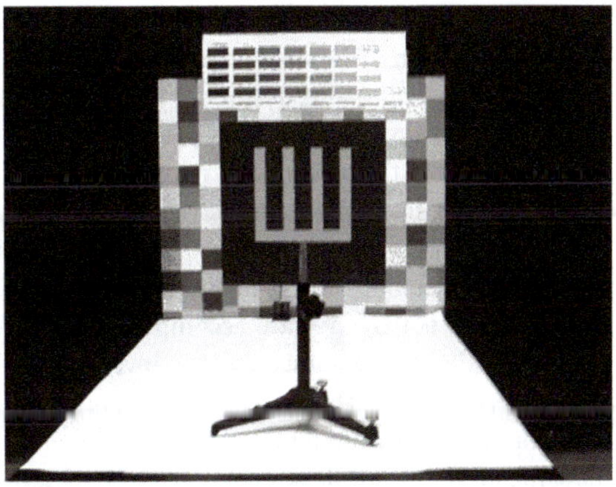

Figure 20.1 Photo of the experimental display, taken under a single illuminant. In the experiment, the illumination was arranged so that there was no luminance border between the strips and the background. Source: Logvinenko (2005a).

When one looks through the strips and focuses on the background, a wallpaper phenomenon (Woodworth, 1938; Logvinenko & Belopolsky, 1994; Logvinenko et al., 2001) can be experienced. Namely, the strips can be perceived to be located in the same plane as the background. In this case, apart from changing their apparent location, the strips also change in appearance. They now appear to have nearly the same blackness as the background.

Presumably, the illusory depth shift into a higher illuminated plane induces a shift in lighting brightness which, in turn, causes a change of the strips' lightness. Indeed, when the strips are perceived to be coplanar with the background they appear to be under the same level of illumination as the background. In other words, their lighting brightness is now the same as that of the background. Since the strips are equiluminant with the background and their lighting colour is identical to that of the background, they have to acquire the same material colour as the background provided that the relationship (20.24) holds true.

Another example of manipulation with lighting colour by illusory change of the apparent spatial layout is shown in Fig 20.2. Looking at a paper cone on a desktop (Fig. 20.2a) through a pseudoscope[10] induces an inversion of apparent depth: that is, it makes the cone look like a conical hole in

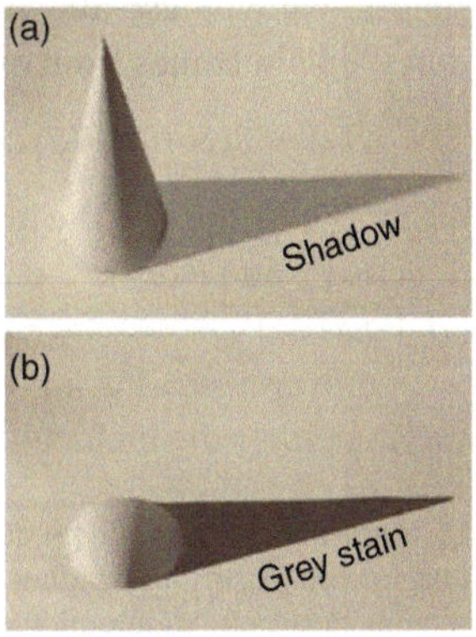

Figure 20.2 Sketch of the stimulus display with a paper cone (see text). Source: Logvinenko (2009c)/Color Society of Australia.

10 A pseudoscope is an optical device used to reverse the sign of binocular disparity. A prismatic pseudoscope (based on a pair of Dove prisms), as that used in the present study, reverses from left to right the image each eye receives when the actual scene is viewed. As a result, binocular disparity and thus apparent depth relations are reversed as well. The object, which is normally perceived as being in front, when viewing through the pseudoscope appears to be behind; concavities look convexities, and vice versa.

the desktop (Fig. 20.2b).[11] The apparent depth reversal of the cone in Fig. 20.2 also leads to an apparent transformation of the shadow cast by the cone. Specifically, it causes the shadow, which the now 'invisible' cone casts on the desktop (Fig. 20.2b), to change its appearance and look like a pigmented area, a sort of stain (Logvinenko & Menshikova, 1994). In other words, under pseudoscopic vision a border between the shadowed area and the other part of the desktop is no longer perceived as a lighting border. Instead, it is perceived as a material border, that is, the shadowed area is perceived as a material insert of a different colour. Before the depth reversal the shadowed area looks to be of white material colour – the same white as the rest of the desktop. The difference between these two types of white is that they are accompanied by different lighting colours: the shadowed area has the dimmer lighting colour. After the reversal the shadowed area turns into a black stain (like an ink spot). Now it looks black, not white. In other words, the depth reversal causes a transformation of the lighting edge into a material edge. Therefore, after the depth reversal the lighting colour is the same for the whole desktop whereas the material colour is different: the black area surrounded by white. Such a perceptual transformation is in line with the relationship (20.24).

20.2.2 Invariant relationship between lightness, lighting brightness and shading brightness

That both lightness and lighting brightness can be essentially affected by rearranging the apparent spatial layout of a scene without any physical changes in it leads to the thought that the lightness is constrained not only by the lighting brightness but by the apparent spatial layout as well. To address this issue, one needs to take into account the spatial variable. Specifically, Eq. (20.22) should be expanded as:

$$Lum(\overline{\alpha}) \propto r(\overline{\alpha})k(\overline{\alpha}), \tag{20.25}$$

where $\overline{\alpha}$ denotes a spatial coordinate as defined in Section 13.3.2, and $Lum(\overline{\alpha})$ stands for the luminance pattern.[12]

In the foregoing section, the spatial layout of the scene was assumed to be the simplest. Namely, the surface was assumed to be flat and its orientation with respect to the illumination direction to be fixed. Now we abandon these restrictions and will consider smooth curved surfaces of various shapes. We will also allow the light flux to vary across space (keeping, however, the assumption that the light comes from one direction). In this case, Eq. (20.25) takes the following form:

$$Lum(\overline{\alpha}) \propto r(\overline{\alpha})F(\overline{\alpha})\cos\big(\theta_i(\overline{\alpha})\big), \tag{20.26}$$

where $F(\overline{\alpha})$ stands for the luminous flux falling at $\overline{\alpha}$ per unit area (perpendicular to the illumination direction), and $\theta_i(\overline{\alpha})$ is the angle made by the illumination direction with the surface normal at $\overline{\alpha}$ (cf. Eq. (II.4)).

As one can see, for a given reflectance pattern $r(\overline{\alpha})$, the luminance pattern can be altered by varying either the incident illumination (i.e., the illumination pattern[13] $F(\overline{\alpha})$) or the shape and/or

11 It must be said that the pseudoscopic transformation of the cone's apparent shape is not the result of thinking, imagination or any other higher cognitive process. Neither is it subject to deliberate conscious control – it occurs spontaneously and unintentionally (as Necker's cube reversal). Undoubtedly, it has a perceptual nature. Under pseudoscopic viewing you see a solid object and you do not doubt that it is not an illusion or a hallucination. In other words, a pseudoscopic image is felt to be as real as a perceptual image under normal viewing. This allows us to believe that by studying pseudoscopic vision one can gain knowledge about normal perception.

12 Recall that, according to the definition in Section 13.3.3, luminance pattern (13.20) is a particular case of the colour-signal-code pattern (13.19) when $k_S = 0$ and $k_M = k_L$.

13 Although it was defined at the beginning of Part II in terms of the radiant flux, we will continue to use the concept of the illumination pattern, even when the radiant flux is replaced by a luminous flux.

orientation of the surface (or both). Interestingly, we are capable to perceptually distinguish these two situations. For instance, we are able to distinguish when the lighting brightness is reduced due to decreasing the ambient illumination and when because the surface is slanted away from the direction of the prevailing illumination. We usually use the words 'cast shadow' and 'attached shadow' to refer to the first and second cases, respectively.

It has been shown recently that these terms (cast and attached shadows) verbally reflect two different perceptual attributes that are locked into the reciprocal relationship (Logvinenko, 2015a). Specifically, it was found that the cast and attached shadows could not cancel out each other. When the decrease of luminance, caused by turning the surface away from the direction of the prevailing illumination, was made up for by the corresponding increase in illumination (i.e., luminous flux) so as to maintain the luminance constant, the observers continued to perceive overlapping cast and attached shadows as clearly different from each other. For example, the same white surface looked either being brightly illuminated in shadow or being dimly illuminated in highlight despite that in both cases it sent to the eye the light of the same luminance.

So, allowing the surface slant to vary permits to reveal that the lighting brightness bifurcates into two perceptual attributes, for one of which we will leave the old term 'lighting brightness', while the other will be called *shading brightness*. Or generalising, one can say that the lighting colour bifurcates into the lighting colour as such and the *shading colour*.[14]

If the reciprocity between the lighting brightness and shading brightness found by Logvinenko (2015a) is confirmed by further experiments, then Eq. (20.24) can be generalised as

$$L_{WBk}(\overline{\alpha})B_l(\overline{\alpha})Sh(\overline{\alpha}) \propto g\,(Lum(\overline{\alpha})), \tag{20.27}$$

where $L_{WBk}(\overline{\alpha})$, $B_l(\overline{\alpha})$, and $Sh(\overline{\alpha})$ are the lightness (i.e., whiteness, grey or blackness), the lighting brightness and the shading brightness at the point in the visual field with the spatial coordinate $\overline{\alpha}$.

We will call $L_{WBk}(\overline{\alpha})$, $B_l(\overline{\alpha})$, and $Sh(\overline{\alpha})$ the *lightness image*, the *lighting-brightness image* and the *shading-brightness image*, respectively. In the general case (i.e., when consideration is not restricted to only lightness, lighting- and shading-brightnesses), we will use the terms 'material-colour image', 'lighting-colour image', and 'shading image'.

Recall that only objects with geometrico-chromatically separable reflectivity (see Eq. (II.2)) are taken into consideration throughout the book, because otherwise the concept of a spectral reflectance function, as a function of one variable λ, does not make sense. For such objects, as pointed out earlier, the effect of their shape and orientation can be viewed as just scaling their spectral reflectance function. Therefore, the shading image is reduced to the shading-brightness image. In other words, the shading image for these objects is supposed to be achromatic.

20.2.3 Shading as a sensory basis of shape

By and large, there is a consensus on that under certain conditions (e.g., the lack of spatial variations in the surface spectral reflectance and illumination) the surface spatial relief, that is, the surface shape can be inferred from the luminance pattern, $Lum(\overline{\alpha})$. A number of computational algorithms have been proposed for inferring 'shape from shading'[15] (Horn, 1989; Zhang et al., 1999). Giving

14 It is because of this that the notion of a shading pattern was introduced as a special case of a lighting pattern (that is, when a lighting pattern is formed solely due to the shape and/or orientation of the surface). The experiments by Logvinenko (2015a) allow to suggest that a lighting pattern created by only an illumination pattern is perceived quantitatively differently from the identical shading pattern, even they both bring about the same luminance pattern.

15 Note that in literature on 'shape from shading', shading means what is here called the luminance pattern, i.e., $Lum(\overline{\alpha})$.

credit to the ingenuity and efficiency of some of these algorithms, one cannot help noting that their authors do not even raise the question of how the shape is represented in human perception. Is there any sensory basis for the surface shape? Or the term 'apparent shape' is just a figure of speech? The use of the term 'infer' hints that the answer is negative, that is, the 'inferred shape' is just as extrasensory as thought. However, in such a case, the shape-from-shading algorithms are nothing more than exercises in computational mathematics that have nothing to do with human perception.

We suggest that shading, as described by Logvinenko (2015a), plays the role of such a sensory basis for shape. To be more exact, it is shading-brightness image, $Sh(\overline{\alpha})$, that 'portraits' shape in our perception. In a sense, shading is an apparent shape.

In this regard, it is appropriate to recall Gibson who believed that perceiving the surface shape can be thought of as picking up the invariants of the gradient of its texture (Gibson, 1979). However, we are as good at perceiving the shape of textureless surfaces too. We believe that the gradient of the shading-brightness image renders the surface shape along with the texture gradient, the importance of which for shape perception was emphasised by Gibson.

The following pseudoscopic demonstration (Logvinenko, 2009c) shows that the lighting image can completely convert into a shading-image giving rise to an apparent shape which was not perceived before the transformation. An observer looks at a vertical background, covered with brown paper, through a hole made in a screen (Fig. 20.3a). Viewing through the pseudoscope is supposed to reverse the apparent relative spatial position of the screen and the background (more specifically, the circular area of the background – the only part of the background which can be seen by an observer). Therefore, what is expected to be perceived through the pseudoscope is a brown disc suspended in front of the screen (Fig. 20.3b). Evidently, it would be a bizarre percept: a material object, like a disc cut out of solid cardboard, hanging in the air without any visible support. At any rate, even though it occasionally comes to existence, such a perceptual image is never stable and tends to turn to more plausible percepts which, generally, differ for different observers. The image most frequently encountered in this situation was as follows.

As the illumination of the background was not completely homogeneous, at some point, residual cast shadows on the disc began to be perceived as attached shadows, that is, as deformations of the spatial relief of the disc. Specifically, the highlight in the centre of the disc brought about the perceptual image of three-dimensional stalactite attached to the screen and facing the observer. It looked as solid and 'material' as all the other things in the environment. Appearing like a clay icicle

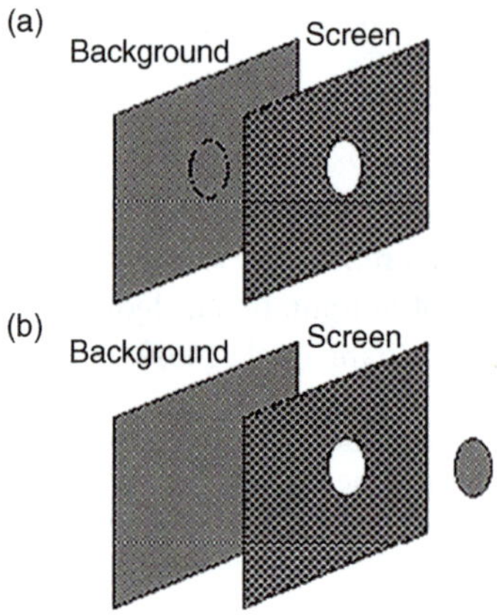

Figure 20.3 Sketch of the display used to demonstrate the trade-off between lighting-brightness and shading-brightness. Source: Logvinenko (2009c)/Color Society of Australia.

homogeneously lit, it did not give the impression of an illusion or an 'ethereal ghost'. Its exact shape depended on the residual highlight pattern.

So, we observe a trade-off between lighting-brightness and shading-brightness (thus, spatial relief) akin to the trade-off that occurs between lighting-brightness and lightness in the previous demonstration (Fig 20.2). Without taking into account this trade off it is almost impossible to understand why the pseudoscopic viewing leads to the apparent stalactite coming to existence virtually from nothing. Unlike the pseudoscopic transformation of the cone in Fig 20.2, where the convex cone turns into a similar concave cone, the apparent stalactite in this demonstration literally emerges from the flat surface for no apparent reason.

20.2.4 Invariant relationship between material-colour image and lighting-colour image in the chromatic domain

The trade-off between material and lighting colours occurs, of course, not only in the achromatic, but also in the chromatic domain. The data in its favour are abundantly present in colour vision literature. However, these are not systemised, and scattered in various, sometimes difficult to access sources. The situation is also aggravated by the fact that these data were published under different headings (e.g., chromatic induction, colour shadows), the relevance of which to the issue under consideration is not always easy to see. Still, the very existence of a trade-off between the material and lighting colours in the chromatic domain can be demonstrated using the same experimental set-up as in Fig. 20.1.

Indeed, when the background illumination in Fig. 20.1 varied in chromaticity the colour of the strips, perceived as being coplanar with the background, also varied. Specifically, under the red background illumination the strips looked greenish, under the green illumination reddish. In other words, the apparent material colour of the grey strips apparently shifted in the direction opposite to the lighting colour of the background (with which the strips are perceived to be coplanar).

It should be noted that although the described transformation of the material and lighting colours of the strips (due to changing the apparent spatial layout) is in line with the trade-off between object and light components (Section 20.1), certainly, one can hardly expect that the likely invariant relationship between the material- and lighting-colour images in general can be expressed as simply as in Eq. (20.27). In order for this relationship to be formulated in quantitative terms, more experimental research is needed.

20.3 Object-colour indeterminacy in variegated scenes. Impact of articulation

After all these demonstrations convincingly testifying for the existence of the trade-off between material and lighting colours have been given above, one cannot refrain from the following naturally arising question. Why does one need, strictly speaking, resort to such tricks (as the pseudoscopic vision and the wall-paper phenomenon) in order to prove that this trade-off really exists? Why do we never come across it in natural, everyday vision?

By the way, an analogous question is also pertinent to the trade-off between the apparent size and distance. Though, in the latter case the answer could be that, thanks to the numerous cues for distance available in natural scenes, the unique, veridical perception of distance is ensured, which entails (due to the trade-off between the apparent size and distance) the unique, veridical perception of size.

Interestingly, such a line of reasoning is quite applicable to one of the three images, present in Eq. (20.27), namely, shading image. In natural scenes, there is plenty of information about the shape of the surfaces constituting these scenes (e.g., the texture gradients, the importance of which was underlined by Gibson). This information can essentially constraint (if not uniquely determine) the shading image, and thus, reduce the object-colour indeterminacy in general. However, even though the shape (thus, the shading image too) is perceived completely veridically, there remains the indeterminacy caused by the trade-off between the material-colour image and the lighting-colour image. Furthermore, this indeterminacy cannot be reduced, say, by extracting the available physical information about the illumination to ensure the 'veridical' lighting-colour image because the concept of veridicality has no relevance to colour.

Natural scenes usually contain a large number of fragments with different reflectance illuminated by a multiple light source. In such a scene, each spatial point defines three equations as Eq. (20.1). If the resultant material- and lighting-colour images brought about by the scene are constrained only by these systems of simultaneous equations, then our perception will seem completely mysterious. Indeed, the cluster of the colour signals produced by the scene is, in theory, compatible with any set of compatible (with regard to the material-lighting colour invariance manifold) pairs of material and lighting colours. Even in a single-illuminant scene all points can, in principle, be assigned different lighting colours. However, in reality, it never happens. Our experience shows that we almost never see more illuminants than in reality. As repeatedly pointed out before, the variability of the lighting colour in a single-illuminant scene occurs within a rather small range, if any. One might conjecture (e.g., see Bergström, 1994) that there is a built-in 'perceptual rule' according to which the lighting-colour variations in a scene should be as minimal as possible. We will call this the *minimum lighting-variation rule.* It assumes that the visual system tries to find a solution to a system of simultaneous equation (20.1) with as small the number of lighting colours as possible.

Adopting the minimum lighting-variation rule makes the problem of recovering the object colour from the colour signals more tractable. Consider, for example, a single-illuminant variegated scene. The colour-signal cluster[16] produced by such a scene is always compatible with a cluster of material colours paired with some, one and the same lighting colour. Moreover, when the number of reflectances in the scene is large, the colour indeterminacy is considerably reduced. Scenes with many different objects are sometimes referred to as 'articulated' (Maloney & Schirillo, 2002). Due to the lack of better, we will use the term 'articulation' to loosely characterise the number of different objects (and/or different lights) in a scene. The more different objects (respectively lights) in a scene, the more it is articulated with respect to objects (respectively lights).

In order to ascertain the role of articulation, consider a colour-signal surface coinciding with the whole object-colour-solid boundary. First of all, note that in this case it can be restored from its tangent bundle, that is, from the derivatives of the colour-signal array (13.10). Really, the gradient encoding, suggested to take place in V1 in Section 13.4, allows to evaluate the colour-signal surface up to an unknown affine transformation. Unlike the linear part of this transformation (which is not a principal problem), the additive term (in the affine transformation) is, generally, a problem. Note, however, that the additive term should not translate the colour-signal

16 The colour-signal cluster can be thought of as a finite sample of the colour-signal surface as defined in Section 13.4. The latter is defined for one particular ambient optic array, and assumed to be continuous. In the present context, we will treat colour-signal cluster as any sample of the colour signals contained in the object-colour solid, including the extreme case when it coincides with the object-colour solid.

surface outside the object-colour-solid boundary.[17] Therefore, when the colour-signal surface coincides with the object-colour-solid boundary, the additive term must be zero. In other words, the colour-signal-surface indeterminacy arising from the gradient encoding reduces to only an unknown linear transformation which can be neglected in the present context.

Assuming that the material colour of the perfect reflector is unique white, one can get its lighting colour. (Recall that given a colour signal, if a lighting colour is singled out in the material-lighting colour invariance manifold determined by this colour signal, then the corresponding material colour is uniquely determined.) Then if we assume, for simplicity, that all the material colours in the scene are coupled with the same lighting colour (just ignoring possible minor changes in the lighting colour), we can find the material colour corresponding to every colour signal in the colour-signal cluster.

The similar situation takes also place in the case of a finite colour-signal cluster provided that it is sufficiently dense and, by and large, uniformly samples the object-colour solid. Indeed, the more the colour-signal cluster is scattered, the smaller the range of its possible translations within the object-colour solid, thus, the less the indeterminacy in question is.

Since the bottom apex of the object-colour solid practically coincides with the bottom tip of the colour-signal cone (see Proposition 14.2), its shape does not depend on the illuminant. If the number of colour signals (in the colour-signal cluster) is large enough to identify the bottom apex (of the object-colour solid) by its specific shape, one can estimate the location of the object-colour-solid vertex (corresponding to the perfect absorber) in the colour-signal cluster. Then, this can be used to anchor the colour-signal cluster.

Due to symmetry, the very top of the object-colour solid has the same shape as the very bottom. Therefore, one can find the part of the colour-signal cluster corresponding to the upper apex of the object-colour solid, and then to estimate the location of the object-colour-solid vertex corresponding to the perfect reflector.[18] Under the same assumptions as above (namely, the perfect reflector is white, and there is only one lighting colour), one can assign the material colour to the entire cluster of colour signals. Of course, such an assignment is not unique since an estimation of the location of the upper object-colour-solid vertex results from an approximate (perhaps, statistical) evaluation procedure. However, when the number of points in a colour-signal cluster increases, this uncertainty decreases.

Indeed, for any particular lighting colour $\mathfrak{l}$, and the n colour signals, either there is a cluster of the n compatible material colours (i.e., those which can be represented by a cluster of corresponding n points in the $\mathfrak{l}$-material-colour gamut), or not. When such a material-colour cluster exists, we will say that the lighting colour $\mathfrak{l}$ is compatible with the colour-signal cluster. In order for the lighting colour $\mathfrak{l}$ to be compatible with a colour-signal cluster, $\mathfrak{l}$ has to belong to the material-lighting colour invariance manifold defined by each colour signal in the colour-signal cluster under consideration. Clearly, the wider the colour-signal cluster is (in particular, the closer its extreme points to the object-colour-solid boundary), the harder it is to fulfil this condition, therefore, the smaller the range of compatible lighting colours for this colour-signal cluster. In turn, the smaller the range of compatible lighting colours, the less significant the material-colour indeterminacy in the scene.

---

17 Here we assume that the visual system 'knows' the shape of its own object-colour solid.

18 The colour signal produced by the perfect reflector will be the colour signal of the illuminant. Note that unlike the illuminant-estimation algorithm that simply assigns white to the colour signal with maximum luminance (i.e., assume that this colour signal corresponds to the perfect reflector), we suggest using additional information regarding the shape of the colour-signal cluster. (For various illuminant-estimation algorithms, see, e.g., Hordley (2006)).

When the illumination also varies in the scene it is rather unlikely that there is a single lighting colour compatible with the colour-signal cluster. Indeed, note that not for every set of colour signals one can resolve the system of simultaneous Eq. (20.1) with respect to a_x for a fixed lighting component a_p. A necessary condition for this is that the colour-signal cluster can be inscribed in some object-colour solid. Moreover, if there are a few object-colour solids encompassing the colour-signal cluster then there are as many different lighting components for each of which there exists a solution (with the single lighting colour) for Eq. (20.1) as the number of these object-colour solids. When this condition is not satisfied the scene can be parsed into frames within each of which the lighting colour is perceived approximately constant.[19]

Since natural scenes are highly articulated in terms of both objects and lights, the object-colour variations in such scenes, caused by the colour indeterminacy in question, are so reduced that go unnoticed in everyday life.

20.4 Implication for measuring object colour

As in colorimetry, measuring the colour of an object/light pair (x, p) (from the object/light stimulus set $(\mathcal{X}\times\mathcal{P})$) means finding which class of the object/light colour equivalence, $\sim$, it belongs to; that is, ascertaining which object/light colour stimulus in $C = (\mathcal{X}\times\mathcal{P})/\sim$ contains (x, p). When C is represented by an object/light atlas, $\mathcal{A}_x\times\mathcal{A}_p$, measuring the colour of (x, p) reduces to finding an element $(a_x, a_p) \in \mathcal{A}_x\times\mathcal{A}_p$ colour equivalent to (x, p). If $\mathcal{A}_x\times\mathcal{A}_p$ is colour-charted then, being expressed in terms of the colour coordinates, the colour of (x, p) can be represented as a point in the object/light colour gamut (Section 19.3.4).

Since all the object/light atlases considered above are hard (or impossible) to be physically realised, for a practical implementation of the object/light colour measurement one has to use some real objects and lights, the combination of which can serve as a suitable approximation to the object/light atlas. The spectral properties of these objects and lights and their number are not so important if they are colour-charted, and cover more or less evenly the object/light colour gamut. Much more important is the next problem which does not exist in colorimetry.

It turns out that it is impossible to employ the colorimetric experimental paradigm with the bipartite field (Fig. I.1) to measure object colours. At first glance, it seems that to do so one needs to place an object x lit by light p in one half of the field, and an object a_x lit by light a_p in the second half. Then varying both a_x and a_p, one could, in principle, find the desired pair (a_x, a_p) colour equivalent to (x, p). However, this will not work out. The problem is that, as a matter of fact, such a scene is highly indeterminate, as far as object colour is concerned.

Really, a scene consisting of a single spatially homogeneous object, lit by a single light source, will not be perceived as an object at all. Instead, it will lead to a special visual experience called *Ganzfeld*[20] which some see as a sort of hallucination. At least two objects and/or two lights in a scene are needed to avoid the Ganzfeld effect. Although the bipartite field as described above allows preventing the Ganzfeld effect, nevertheless it remains a great deal indeterminate. There is an

---

19 In some experimental conditions, one can observe an opaque object perceived as self-luminous object. As Gilchrist (2006) notes, this might happen, for example, when white is assigned to an object that does not reflect light of maximum luminance in the scene. This can be thought of as a kind of perceptual solution to a problem that has no solution.

20 The Ganzfeld effect usually occurs when the entire field of view is unstructured (i.e., uniform). It shows that spatial modulation is just as important as temporal one for normal perception.

infinite number of various pairs of material and lighting colours which are compatible with the pair of colour signals created by such a scene. Furthermore, any material colour (in combination with the appropriate (i.e., compatible) lighting colour) is admissible in this scene. For instance, when a scene is achromatic, an object can, in principle, assume any achromatic colour, from black through grey to white. The Gelb effect and the staircase Gelb effect show that under appropriate circumstances these admissible combinations can be carried out in our perception, even extreme ones. Hence, the trade-off between material and lighting colours leads to the fact that the object colour of the bipartite scene is really indeterminate and ambiguous, and the range of this ambiguity is almost unlimited.

Yet, as pointed out above, the colour indeterminacy is considerably reduced in scenes with many different objects (i.e., in articulated scenes). It follows that measuring object colours should be carried out by using an experimental set-up as articulated as possible. In particular, when finding an element (a_x, a_p) in the object/light atlas $\mathcal{A}_x \times \mathcal{A}_p$ colour equivalent to a given object/light pair, (x, p), it is desirable to present simultaneously as many objects from $\mathcal{A}_x$ as possible to ensure the best possible articulation in terms of objects. For example, when the Munsell collection is used, all the Munsell chips are best presented simultaneously.[21] Since object/light colour equivalence implies metamerism of the reflected lights, the choice of objects predetermines their illumination. Specifically, the illumination should be arranged so as to achieve the equality of the colour signals from all the objects. That is, for every a_x, one should chose a_p so that light $a_x a_p$ is metameric to light xp. When a three-parameter set of lights serves a light atlas $\mathcal{A}_p$, there is at most one element a_p in $\mathcal{A}_p$ that provides such equality of the colour signals. Moreover, this element a_p can be calculated explicitly. Therefore, during establishing object/light colour equivalence, each object a_x must be presented under its own illumination, which can be derived from the condition of equality of the colour signals. This will result in a high degree of articulation, not only in terms of objects, but also in terms of lights, which in turn will ensure a low degree of object-colour indeterminacy. All that remains for the observer in this experiment is to chose an element (a_x, a_p), which is colour equivalent to (x, p); that is, one that looks exactly like (x, p).

Such a device consisting of a sample from the object atlas $\mathcal{A}_x$, the elements of which are illuminated by lights from the light atlas $\mathcal{A}_p$ so as to maintain all colour signals equal, can be called an *object colorimeter*. It can be used to find an element $(a_x, a_p) \in \mathcal{A}_x \times \mathcal{A}_p$, which is colour equivalent to any given object/light pair (x, p). This means that, using this tool, one can establish a genuine match (i.e., the equality of both the material and lighting components). When $\mathcal{A}_x \times \mathcal{A}_p$ is colour-charted, such an object colorimeter will provide measuring the object colour of (x, p) in terms of colour coordinates (as defined in Section 19.3.4). Interestingly, such an object colorimeter frees observers from adjusting the stimulation in order to make a match. All necessary settings (of objects illumination) can be made in advance[22] (in automatic mode). The observer is supposed to simply select the pair (a_x, a_p) that creates a match. This can be seen as an important advantage of the object colorimeter over the classic colorimeter for measuring light colours. An object colorimeter can be used to measure the colour of objects not only in normal viewing conditions but also under such that cause its apparent change, as in the case of coloured shadows.

---

21 The Munsell colour atlas can be used as a practical implementation of the object atlas only if none of the employed illuminants can make some Munsell chips metameric.

22 True, this can be done only if the fundamental colour mechanisms of the observer are known in advance. Yet, it all depends on the requirements for the measurement accuracy. If, for instance, they are such that they make it possible to neglect individual differences, then one can use Govardovskii's cone fundamentals (Eq. (3.18)) or one of the CIE standardised sets of cone fundamentals (see Section 3.3).

Coloured shadows

Although each of us has encountered with coloured shadows at least once in our lives, it is Benjamin Thompson who is commonly cited as the first to publish a special study on this phenomenon. Shining a beam of daylight and candlelight on a wooden rod, he watched the shadows cast on a sheet of white paper. There were two shadows. The paper under one of the shadows was illuminated only by candlelight, and under the other only by daylight. The first shadow looked yellow, the second surprisingly blue. 'That the shadow corresponding to the beam of daylight, which is illuminated by the yellow light of a candle, should be of a yellowish hue, is not surprising; but why is the shadow corresponding to the light of the candle, and which is illuminated by no other light than the apparently white light of heavens, *blue*? I at first thought that it might arise from the blueness of the sky; but finding that the broad daylight, reflected from the roof of a neighbouring house covered with the whitest new-fallen snow, produced the same blue colour, and if possible of a still more beautiful tint, I was obliged to abandon that opinion,' (Thompson, 1876, pp. 53–54).

The conspicuous complementarity of shadow colours leads to the conjecture that it is simply a particular case of colour induction, namely, chromatic simultaneous contrast. To make sure that this is not the case, it is enough to replicate the sheet of paper with shadows on the screen of a colour monitor. Really, a rectangular uniform pattern (imitating the sheet of paper) with two vertical strips over it (imitating the shadows) appear completely different, even if the colorimetric coordinates (i.e., colour signals) of the rectangle and strips are identical to those of the sheet of paper and shadows on it. Firstly, the rectangular pattern on the screen looks quite yellowish whereas the sheet of paper looks white with a barely visible (for experienced observers) tint of yellow illumination, if any. Secondly, the chromatic simultaneous contrast is hardly observed on the colour-monitor screen. Specifically, the vertical strips have practically the same colour as that they have when displayed in isolation on the screen. In particular, the strip imitating daylight shadow looks almost neutral (possibly with a slight bluish tint). As Thompson emphasised, the shadow 'corresponding to the light of the candle – and consequently illuminated by the light of the heavens – was of the most beautiful *blue* that it is possible to imagine.' (Thompson, 1876, p. 51) Hence, any model of chromatic simultaneous contrast based on colorimetric coordinates cannot successfully predict both coloured shadows and their (colorimetrically equivalent) imitation on a colour-monitor screen.

The point, of course, is not the terminology. If you like, you can call coloured shadows a kind of chromatic simultaneous contrast. But then you need to understand how the shadow simulation differs from the original. What is so essential in the original setup that is lost in the monitor simulation? The lack of what on the monitor screen leads to the loss (or a very significant reduction) of the phenomenon in question (chromatic simultaneous contrast, if you like)? This lost factor is likely to be the key to a correct understanding of the essence of this phenomenon, and without which its explanation is impossible.

In fact, the answer is quite obvious. With a colour-monitor screen we can simulate the light (its colorimetric characteristics) entering the eye, but we cannot simulate shadows as such. When watching the material sheet of paper lit by real light sources, we perceive object colours. The very term 'coloured shadow' (implying lighting colour) testifies for this. As mentioned before, what we see on a monitor screen cannot be referred to as an object colour. True, it is not a light colour either, but it gravitates more towards the latter rather than the former. Thus, according to the terminology in the previous chapter, the words 'coloured shadows' mean that the lighting colour of the two shadowed areas is dramatically different: yellow and blue. The problem that struck Benjamin Thompson can be stated as follows. Why does the lighting colour of the white paper lit by daylight, which is hueless when there is no other illuminant, becomes blue in the presence of candlelight? Why does slightly yellowish surrounding lighting affect so much the lighting colour of the shadowed area hitherto perceived neutral? Why is this effect not observed (at least not that spectacularly) when simulated on a monitor screen?

To appreciate the profoundness of this problem it is worth noting that the lighting colour of the white sheet surrounding the two shadowed areas is hardly perceived yellow. Being a mixture of daylight and candlelight, the paper illumination tinged with yellow. Nevertheless, the lighting colour of the sheet illuminated by this mixture is perceived as to be nearly neutral. Even if it contains a yellow tint, its magnitude is very small. Strangely enough, Benjamin Thompson did not consider it necessary to report what the lighting colour of the white paper was like. However, in our opinion, this circumstance is of decisive importance for understanding the phenomenon under consideration.

As a matter of fact, in natural scenes, the lighting colour induced by the prevailing illumination tends to lose saturation, appearing sometimes completely hueless. Usually, such discoloration goes unnoticed. For instance, those who are not involved in colour measurement are often surprised to learn how yellow candlelight is. When the lighting colour corresponding to the prevailing illumination is neutral, one could call this discoloration 'lighting-colour discounting', if only the term 'illumination discounting' was not already widely used in the vision literature.[23] We suspect that some kind of lighting-colour discounting took place in Thompson's demonstration. At any rate, we clearly observed it when reproducing his demonstration.

Now assume that the lighting colour of the white sheet is perceived completely hueless. Let us also assume that the border around the shadowed areas is perceived as a lighting-colour border (i.e., an 'illumination border'). Then, the colour-signal contrast at the border of the shadowed area lit by only daylight will signal a shift of the lighting colour at this border towards blue. Likewise, the colour-signal contrast at the border of the shadowed area lit by only candlelight will indicate a shift of the lighting colour at this border towards yellow. If the lighting colour of the white paper is tinged with some amount of yellow, then the saturation of the 'blue' shadow will be reduced, and the saturation of the 'yellow' shadow will increase. Depending on which lighting colour is assigned to the white paper, the saturation of both shadows can vary in the reciprocal manner.

To test this account of coloured shadows, one needs to measure the lighting colour of all three areas of the sheet of paper (i.e., the two shadows and the rest of the sheet), which can be done by means of a colour-charted object colorimeter. With such a colorimeter at hand, one can measure not only the lighting colours but also the material colours of shadowed areas. Besides, measuring the material colours in the scene is even necessary, since the possibility of changing it cannot be ruled out. Specifically, the white paper under the blue shadow might take on a slightly blueish material-colour tint. If this happens, the blue lighting-colour of the shadowed area will lose some of its saturation due to the trade-off between the lighting and material colour components. In general, any explanation of coloured shadows should be preceded by a careful measurement of them. Before trying to explain why we see this way and not otherwise, we must first find out what we actually see there. At this stage, a colour-charted object colorimeter can become an invaluable tool.

23 Illumination discounting is usually associated with light adaptation in photoreceptors. The tendency, which we tentatively call lighting-colour discounting, has nothing to do with light adaptation. This is the consequence of the fundamental indeterminacy of the object colour in general and its lighting component in particular. We believe that this can be thought as a sort of another built-in perceptual rule, according to which the lighting colour corresponding to the prevailing illumination in a scene tends to appear hueless.

21

On Perception in General: An Outline of an Alternative Approach

One lesson to draw from the previous Chapter (20) is that colour cannot be studied in isolation from other perceptual attributes, not last of which being, shape. Another one is that the stimulus-response approach is not suitable for studying colour since, strictly speaking, there is no stimulus for object colours. Curiously enough, the developers of computer programs that simulate human vision realise this very well. For instance, Barrow & Tenenbaum (1978) see an important problem for vision in recovering from the luminance pattern in Eq. (20.30) the three multipliers in its right side, which they call the *intrinsic images*.[1] Since, obviously, this equation cannot be uniquely solved with respect to these intrinsic images, they suggest to make use of some natural constraints on the intrinsic images. Admitting the enormous complexity of the problem, they are led by the belief that after all, humans manage somehow to succeed in it.[2] At first glance, the lightness image, the lighting-brightness image, and the shading-brightness image, the existence of which we tried to argue for above, corroborate such belief. Really, are they not solutions that the visual system has come up with to deal with Eq. (20.30)?

This question is not easy to answer. It is far too involved for simple answers. On the one hand, yes, in a sense, the lightness image, the lighting-brightness image and the shading-brightness image can be thought of as the perceptual counterparts of the reflectance pattern, the illumination pattern and the shading pattern respectively. On the other hand, firstly, this can be the case only with respect to the particular, idealised experimental paradigm for which the latter concepts have been defined (the geometrico-chromatically separable reflectivity, matte objects, collimated illuminant, etc.). Secondly, even within this paradigm, for the same reasons that the material colour of an object cannot be thought of as a perceptual counterpart of its spectral reflectance,[3] the lightness image cannot be considered as a perceptual solution for the reflectance pattern $r(\overline{\alpha})$ in Eq. (20.30). Likewise, the lighting-brightness image cannot be considered as a perceptual solution for illumination pattern because lighting colour is not a perceptual characteristic of illumination.

Thus, even if the material-colour image, the lighting-colour image, and the shading image result from the solution of some equation, this equation will most likely resemble Eq. (20.31) rather than Eq. (20.30). In reality, as argued above, the material-colour image, the lighting-colour image, and

1 For more information on intrinsic images in the context of colour perception, see, e.g., Arend (1994).
2 'Attempts to construct computer models for the interpretation of arbitrary scenes have resulted in such poor performance, limited range of abilities, and inflexibility that, were it not for the human existence proof, we might have been tempted long ago to conclude that high-performance, general-purpose vision is impossible.' Barrow & Tenenbaum, 1978, p. 3).
3 One of these reasons is metamer mismatching, which in the achromatic case is even greater compared to the trichromatic case (Logvinenko, Funt, & Godau, 2015). True, it can be ignored if consideration is limited to only objects with flat spectral reflectance functions. However, such idealised objects do not exist in the real world. For real objects, metamer mismatching always takes place.

Foundations of Colour Science: From Colorimetry to Perception, First Edition. Alexander D. Logvinenko and Vladimir L. Levin.
© 2023 John Wiley & Sons Ltd. Published 2023 by John Wiley & Sons Ltd.

the shading image are definitely linked by certain relationships. The exact nature of this relationship has yet to be established by way of experiment. For now, we only note that, as stated above, this relationship cannot be a mirror replica of the corresponding physical relationship (i.e., Eq. (20.30)), which is actually Eq. (20.31). Whatever its ultimate form, this relationship can be interpreted as an equation in which the material-colour image, the lighting-colour image, and the shading image serve unknowns.

Perceptual equations

It is important to point out that Eq. (20.31) is an equation with respect to perceptual rather than physical unknowns. Note that Eq. (20.25) can be also interpreted as a perceptual equation. Another example is the apparent depth distance hypothesis (Foley, 1967):

$$\frac{\Delta\tilde{D}}{\tilde{D}} \propto \eta, \tag{21.1}$$

where $\Delta\tilde{D}$ and $\tilde{D}$ are apparent depth and distance respectively; and η is retinal disparity.

Expressions (20.25) and (21.1) can be treated as perceptual equations where the attributes of the colour-signal array (i.e., visual angle, binocular parallax, and the like) play the role of known variables; and the perceptual attributes (i.e., apparent size, distance, depth, lightness, lighting-brightness, and shading) are unknowns to be determined. Admittedly, none of these equations, nor even the whole system of equations they comprise, can be solved uniquely. This is obvious from the fact that the number of unknowns is greater than the number of knows. Yet, some other perceptual equations have also been found, which include these and other perceptual attributes, in particular, the apparent form, the apparent slant, and the apparent speed (Epstein & Park, 1963; Epstein, 1982). The result is that the same perceptual attributes may be involved in a variety of perceptual equations. Consequently, if we consider the system of perceptual equations as a whole, the number of perceptual equations may exceed the number of unknowns (i.e., the number of perceptual attributes).[4]

Introducing the notion of perceptual equation can be thought of as a move away from the psychophysical approach. Yet it is still overburdened with psychophysical associations. Indeed, a perceptual equation implies an input (the colour-signal array) and an output (the material-colour image, the lighting-colour image, and the shading image). Although a unique solution is not supposed for these perceptual equations, the ears of the stimulus-response paradigm are clearly visible here. Nevertheless, we find the perceptual-equation metaphor useful since it clarifies some important distinctions between the psychophysical approach and its possible alternatives, one of which is outlined in the most general terms in Section 21.2.

Outside of this idealised paradigm, things get even worse. In this case, it is not entirely clear what in the outer world can even correspond to the reflectance, lighting and shading patterns, not to mention the material, lighting and shading images. Really, for non-Lambertian objects the effect of the geometric BRDF factor (i.e., $R_s\left(\omega_i, \omega_r, \mathbf{x}\right)$ in Eq. (II.2)) is confounded with the effect of the object shape and orientation, and thereby the reflectance, lighting and shading patterns cannot be defined as simply as for Lambertian objects. Strictly speaking, these patterns cannot be defined for non-Lambertian objects at all. For instance, as mentioned earlier, the geometrico-chromatic separability does not hold for real objects. It follows that for such objects the very concept of the spectral reflectance function is not applicable, thus, for such objects, the reflectance pattern (i.e., one of the three intrinsic images) simply does not exist. Therefore the problem is not how the visual system recovers the intrinsic images, and, in particular, the reflectance pattern. The visual system cannot

4 When a system of perceptual equations becomes overdetermined (hence inconsistent), its approximate solution may still be possible.

recover what does not exist. The problem is why the visual system produces such outputs as the material colour image and the lighting colour image if there is no clear stimulus that can correlate to them. Briefly, what is the purpose for which colour vision has evolved and been retained in evolution?

21.1 What is colour for?

According to the most common point of view, colour vision is a way to conduct an approximate spectral analysis of light. As this view implicitly presumes light colour, it does not embrace object colour (unless we revive the old fallacy that the colour of an object is the colour of reflected light). In addition, a three-component spectral analysis is much too crude. As mentioned before, the manatis shrimps are equipped with 16 types of photoreceptors (Marshall & Oberwinkler, 1999). This points out that, if the objective were the spectral analysis of the light entering the eye, humans could, in principle, perform it much better.

Another answer to the eponymous question of this section is that colour could possibly be for finding fruits among foliage. 'Primate colour vision has been shaped by the need to find coloured fruits amongst foliage, and the fruits themselves have evolved to be salient to primates and so secure dissemination of their seeds.' (Regan et al., 2001) For instance, the cherries in the left part of Fig. 21.1 are visible only due to the chromaticity difference between the pixels, the luminance of all the corresponding pixels in both halves being equal. Hence, an ability to see colour (thus, discriminate between the pixels) allows us to detect the cherries against the foliage, which otherwise would remain invisible. Paying tribute to the wit of this hypothesis, it should be noted, however, that scenes such as Fig. 21.1 are extremely unlikely in nature. It takes some effort to produce a fully equiluminant picture as in the left part of Fig. 21.1 even if you have access to Photoshop.[5] After all, it is hard to believe that such a complex mechanism as colour vision has evolved for dealing with a problem which can be solved in the natural environment by other readily available means. There must have been another, more fundamental, reason for colour vision to come into existence.

Figure 21.1 The image on the right was created so that the luminance of every pixel was equal to that of the corresponding pixel in the image on the left.

5 Besides, what is equiluminant for one observer may not be for another.

We are deeply convinced that we cannot understand what colour vision is for until a more general question is answered: what do we need perception in general for?

As a matter of fact, colour textbooks have been mainly written by the scientists with a physicalistic mindset. Their physicalism reveals itself in the fact that they, explicitly or implicitly, consider the eye a physical device, even if imperfect, for measuring the physical properties of light and objects reflecting (or transmitting) light. Recall, for instance, likening the eye to the spectroradiometer, as mentioned above. However, what could be the benefit to an animal of being aware of the spectrum of light that lands in its eye? And not just being aware, but having a core understanding of this spectrum? And then on top of that, to comprehend it in such a specific form as colour? True, from a physical point of view, the eye processes electromagnetic radiation. But so do plants, performing photosynthesis. Either way, we can hardly understand the essence of these processes without realising the wider context in which each of them is included.

Everyone would, probably, agree that perception is an aspect of animal behaviour. Notably, it is this aspect that provides animals with the information necessary for the implementation of behaviour in their environment. Yet, disagreement seems inevitable if one tries to specify what is meant exactly by 'animal behaviour', 'animal environment', and 'information'. So as not to get bogged down in fruitless theoretical debates about what 'animal behaviour' is, let us confine ourselves to such particular but important forms of it as locomotions and manipulations. As these take place in the environment, for them to occur one needs some information about the environment. It is clear that the way you understand what 'the environment' is determines what information is necessary for behaviour in this environment.

A physicalist understands 'the environment' as the physical world, or some part of it. The physical world means the world as described in textbooks on physics. Accordingly, information about the environment defined in physical terms cannot be anything other than physical information. Moreover, in this case, it seems quite logical to understand perception as an optical inquiry into the physical world, with only the help of such poor optical instrument as the eye. However, in this case, the physicalist is absolutely unable to tell us why perception brings about colour, which does not exist in the physical world.

The distinction between the environment and the physical world is a pivotal point for Gibson's ecological approach to perception (Gibson, 1979). Without being able to go into all the details, we note only the following here. For terrestrial animals, which humans are, their environment is constituted by 'a *medium*, *substances*, and the *surfaces* that separate them' (Gibson, 1979, p. 16). From a physical point of view, substances are mostly solid, the medium being liquid or gaseous. Water and atmosphere are the major examples of media. Media are usually transparent or nearly transparent, while substances are opaque to light. The boundaries between the medium and substances constitute surfaces. A surface separating solid substances from the medium is usually opaque. A surface separating water from the atmosphere is transparent to some degree.

An important (from a behavioural point of view) distinction between media and substances is that the first afford locomotion through them to the animal, whereas the second do not. Consequently, opaque surfaces serve as obstacles; therefore, they should be avoided during locomotion. Thus, it is vitally important for animals to distinguish between transparent and opaque surfaces in the terrestrial environment. Being unable to distinguish between these may cost life.

One of the most important characteristics of an opaque surface is its shape. It is needless to stress how important it is for the animal to have reliable information about the shape of the surfaces that make up the spatial layout of the environment. As successful shape-from-shading algorithms have shown, the shape of a surface can be restored from gradients of only one cone-fundamental pattern (see Eq. (13.11)), for example, the luminance pattern, provided that the surface has a spatially uniform spectral reflectivity and is illuminated by a single remote light source. However, spatial variations in illumination and spatial variations in the surface spectral reflectivity also affect cone-fundamental patterns. Hence, the contribution made by shape should be isolated and perceptually represented in its own way so as to perceptually represent the shape itself. This can be done only if the surface-spectral-reflectivity contribution, and the contribution made by the spectral power distribution of the ambient illumination, are somehow taken into account.

Note, however, that given this objective, it is not necessary to accurately estimate the surface spectral reflectivity or the illumination spectral power distribution. These estimates are only needed to facilitate the evaluation of surface shape. On the contrary, shape evaluation should be done as accurately as possible, as it is essential for locomotion and manipulation. Any rough evaluation of the contributions made by the surface spectral reflectivity and the illuminant spectral power distribution will suffice if it secures accurate shape estimation. Therefore, approximate estimation of surface spectral reflectivity and illumination spectral power distribution plays an ancillary role in the process of estimation of the shape. In this context, colour perception is, in a sense, a by-product of shape perception.

Of course, surface spectral reflectivity may have ecological validity, so even its rough estimation can be of use from an ecological point of view. It can indicate the chemical composition of the substances behind these surfaces. Therefore, such estimates can provide some ecologically useful information regarding the substances, such as edibility or poisonousness. For instance, redness may indicate fruit ripeness. However, this does not mean at all, either that a surface colour is a perceptual estimate of the surface spectral reflectivity or even that such an estimate is the objective of colour perception.

Summing up, that estimating the surface spectral reflectivity cannot be one of the goals of perception follows from the fact that it is only for Lambertian surfaces that their capacity to reflect light can be characterised by a function of a single variable – a wavelength. As repeatedly emphasised above, in the general case, surface reflectivity varies with both the directions of the incident and the reflected light. As a result, in addition to wavelength, spectral surface reflectance depends on two directions: the incident and the outgoing. Hence, estimation of surface spectral reflectance cannot be the goal of perception simply because the latter does not exist for real surfaces.[6] This means that both the proximal and distal stimuli do not exist for object colour. Hence, the psychophysical approach treating perception as a response to the stimulus turns out to be inadequate. At least, this approach does not work in the case of colour perception of the real environment, as opposed to the simplified worlds (e.g., the 'flat patchy world') which have so far been the favourite object of research among colour scientists. A possible alternative to the psychophysical approach to perception will be discussed in Section 21.2.

6 Even for those idealised objects for which it exists, the accuracy of its perceptual estimation is seriously undermined by metamer mismatching (see in Section 19.1 our arguments against the notion of intrinsic colours). In other words, the degree of metamer mismatching shows how inexact the estimate could be. Note, however, that this inexactness does not undermine the aforementioned ecological informativeness of this estimate. Does it really matter what shade of red the apple is, to be sure that the fruit is ripe?

21.2 The need for a new approach to perception. A linguistic metaphor

As stated above, perception provides animals with information about their environment. To be more precise, perception represents to animals their environment, opening up for them the field of possible activity. Furthermore, perception also represents the animal itself in its environment. In particular, as Gibson convincingly showed, the locomotion of animals is represented in their perception along with other movements and events in the environment. Gibson argued that motion and self-motion intertwine with one another in perception. Generally, perception is inseparable from behaviour, being one of its most important aspects.

As Gibson has repeatedly stressed, the environment of an animal depends on its way of living. In the same part of the physical world, there may be many different animals with very different environments. Since the environment is not the same as the physical world, information about it cannot be derived directly from data provided by any physical device, including the animal sense organs. The latter, by the way, are rather similar for many different animals whose environments may differ very much. For example, the eyes of many species are rather similar from an anatomo-physiological point of view. So are their outputs. Nevertheless, being equipped with similar eyes, the animals obtain with these eyes information about their own environments which are very different. (Compare, for instance, the environment of a fish, bird, and terrestrial animal such as a cat or dog.)

Helmholtz compared sensory data with 'hieroglyphs' that were to be further deciphered. It must be noted that Helmholtz's metaphor might be well justified, but only partially. On the one hand, this metaphor fairly enough points out the non-physical nature of the relationship between the sense-organ output and the environment. That is quite true that the senses can only signal, indicate the environment. However, on the other hand, it implicitly exaggerates the sense-organ outputs in the process of perception, giving the impression that perception is simply the result of inferring from data provided by the senses. As Gibson argued, the senses should be considered only as part of the perceptual systems,[7] and perception itself is a process of actively picking-up of information about the environment that unfolds over time. We will return to this important issue later.

Now note that if we agree that the sense organs provide us with just hieroglyphs (i.e., signs or symbols) for later use, to derive information about the environment, then we must admit that perception is inherently dual, there being two different levels at which it occurs. On the one hand, perception is based on the impartial data provided by the sense organs; on the other, it provides the representation of the environment in its immanent form. In order to reflect this duality, we will use the terms 'sensory fabric' and 'perceptual content' (Logvinenko, 1981, 1985, 1993). Certainly, such a distinction is hardly novel. For generations, scientists have used a dual system of terms to describe visual experience (for review, see e.g., Hatfield & Epstein, 1979): 'sensation and perception' (Reid, 1785/1969), 'sensory core and perceptual context' (Titchener, 1902), 'primary image and perceptual image' (Helmholtz, 1867), 'the visual field and the visual world' (Gibson, 1950). Despite the terminological diversity, all these authors impart more or less the same meaning into these pairs of terms. In fact, such a terminological dualism arises from the distinction between proximal and distal stimulation, which they all adhered to. Specifically, they understood the first (sensory) term as a subjective representation of the proximal stimulus, and the second (perceptual) term as an internal representation of the external (physical) world. Almost all of them (except Gibson, being the most well-known exception) shared Helmholtz's view that perception is inferred from sensations.

7 For more on this idea, see Gibson (1966).

Our understanding of sensory fabric is, by and large, in line with this tradition – sensory fabric can be, roughly, thought of as a subjective representation of the ambient optic array. However, an important distinction between perceptual content and its precursors is that perceptual content refers to the environment rather than the distal physical objects.

Specifically, we believe that the sensory fabric is, in a sense, directly determined by the ambient optic array pre-processed by the sense organs. In other words, the 'ambient optic array – sensory fabric' connection is a psychophysical one. As we saw in Section 13.4, the ambient optic array evokes the colour-signal array which is then encoded as the colour-signal surface (to be more exact, its tangent bundle) in the primary visual cortex. Both the colour-signal array and the colour-signal surface can be thought of, by and large, as a neural substratum of sensory fabric.

Yet, it would be an oversimplification to consider sensory fabric as simply a 'mental copy' of its neural substratum, not to mention the ambient optic array. For instance, as is well known, natural (such as a 'blind spot') and acquired scotomas present in the retina are absent in the sensory fabric of the perceived image. They are believed to be 'filled in' somehow as a result of the processing of the retinal image by the higher mechanisms in the visual system. There are also another visual phenomena, for explanation of which authors have resorted to the hypothetical process of 'filling-in' (Yarbus, 1965/1967; Pessoa, Thompson, & Noe, 1998). One can believe or not in the existence of such a process, but the phenomena themselves unequivocally testify against the idea that sensory fabric is a mental replica of the proximal stimulation, or its neural substratum, whichever of the two.

Still, introspective analysis of perceptual images corroborates the idea that there is something in them that depends only on the ambient optic array. Let us recall the celebrated reversible figures, for example, the Necker cube, which became widely known thanks to Gestalt psychologists. They show that rather different perceived content can be combined with the same sensory fabric. The observation of pseudoscopical transformations leads to the same conclusion. One of the most compelling features of these transformations is that the more dramatically the scene changes perceptually, the stronger the impression that it remains, in a sense, the same. What remains unchanged is, in our opinion, the sensory fabric, while the perceptual content undergoes the transformations. Really, since the pseudoscope does not change the structure of the ambient optic array, and thus, the colour-signal array, their representation in the visual image, whatever this may be, should remain unchanged. Note that pseudoscopic evidence is more appropriate here, since real objects, rather than pictures (which are reversible figures), undergo the transformations in question.

Actually, in order to be convinced that perceptual content is underpinned by sensory fabric which is quite different from it, there is no need in a pseudoscope. Thorough introspective analysis of natural scenes brings about a remarkably different perceptual experience in comparison with what we usually get without such an analytical set. For a detailed description of this experience, we refer to E. Mach (1959) and J. Gibson (1950). Gibson coined the term 'perception of the visual field' to distinguish this sort of visual experience from the ordinary one, which he called 'perception of the visual world'. It should be noted, however, that the sensory fabric of the percept does not enter awareness in everyday life – a special attitude is needed to gain access to it. The reason for this is quite obvious – there is no need to experience it in everyday life. Normally we perceive things.

A natural question arises then. In terms of which attributes can sensory fabric be described? In analytic psychology of sensations, this problem is known as the problem of sensation attributes. It is believed that the attributes of sensations are those ultimate elements into which a percept can be broken down by thorough analytical introspection.

In this regard, it is interesting to recall a criterion that was proposed by Helmholtz (1867) to distinguish between the primary image and the perceptual image. Specifically, he pointed out that there is some sensory content in consciousness which could not be altered by any individual experience. So, such persistence may be taken as a distinctive feature of the sensory fabric in comparison with the perceptual content. Indeed, we believe that the perceptual content can be altered by perceptual adaptation, training, and learning, but that the sensory fabric is hardly susceptible to such alteration.

It is worth mentioning also that a similar problem was discussed in the context of iconic memory research. Recall that it was found that only a few sensory attributes could be retained in iconic memory, namely, brightness, hue, shape, and location – i.e., solely those which were traditionally referred to as the attributes of sensation (Boring, 1942). Some researchers believed that iconic memory reflected the receptor stage of processing information in the visual system (e.g., Sakitt, 1975). Although this problem – i.e., how to distinguish sensory fabric attributes from perceptual content attributes – is far too involved to allow a simple answer,[8] it seems that it is only sensory fabric (as distinguished from the perceptual content) that can be fully retained in iconic memory.

We will return to this problem later. At this moment, we are pointing out that, for example, the variables on the right hand side of perceptual equations (e.g., Eqs. (20.25), (21.1), (20.27)) are relevant to sensory fabric (at least, such as we are using this term), whereas those on the left are attributes of perceptual content. Note only that we do not claim that the variables on the right hand side of perceptual equations are the attributes of sensory fabric. These are the attributes of the ambient optic array or its neural counterpart, e.g., the colour-signal array. We believe that the latter should have some subjective representation. Its form will be discussed later.[9]

Having acknowledged that perception is a two-tier entity (whatever the terminology), it is important to specify how these tiers are related to each other. By all accounts, perceptions are inferred from sensations. This stance just reflects the more general belief that the knowledge of the external (physical) world, contained in the percept, can be only inferred from the proximal stimulation. Really, as a matter of fact, a huge amount of information about the external world is missing in the proximal stimulation. Once again, we face here the same physicalist metaphor. A subject equipped with poor physical measuring instruments (i.e., the sense organs) has to resort to the power of inference using past experience, logic, and other means from the repertoire of a scientist so as to infer the physical properties of the fragment of the external (physical) world, which producing the proximal stimulation in question. However, if what is to be inferred is the environment, and not the physical world, then the inference must be of a different kind, not the one used in physical research, if the term 'inference' can be used at all in this context.

As already mentioned, sensory fabric can only play the role of a sign, pointing out (i.e., signalling) what fragment of the environment the animal is facing. Thus, perceptual content can be inferred from sensory fabric in much the same way that linguistic meaning is inferred from the sounds of speech. This linguistic analogy turns out to be more profound than it might seem at first glance. Indeed, one of the distinctive features of human language is the existence of the plane of expression and the plane of content. As follows from the above, there is every indication that sensory fabric acts as the former, and perceptual content as the latter. In other words, sensory fabric and perceptual content play the role of the signifier and signified, respectively.

However, there being a semiotic connection between sensory fabric and perceptual content does not suffice to justify the linguistic analogy. As well known, speech generation and speech

8 This problem leads us to a more general and more complex problem, namely, the problem of the mental representation of perceptual attributes, their awareness, and the extent to which these attributes are accessible for introspection.

9 Actually, these equations should be rewritten in terms of the attributes of the sensory fabric rather than the proximal stimulation (the ambient optical array). However, it requires a deeper and more detailed knowledge of sensory fabric than that which is currently available.

perception are underpinned by the hierarchical structure that is the human language, which comprises a number of layers arranged one above the other. Elements of one level, each consisting of elements of the immediate lower level, are combined into respective elements of the immediate upper level. For instance, phonemes (units of the bottom level) are combined into morphemes, which, in turn, are gathered into larger units lexemes (e.g., words). Importantly, both the level of expression and the level of content are, in turn, hierarchically structured in this way. According to the linguistic terminology, the lower level serves as the 'expression' (i.e., form) level for the higher level, considered as the 'content' level.

In fact, psychologists have long known that perceptual content is organised into a hierarchical structure. For instance, as stated before, apparent shape, exists in perception in the form of shading. It is important to stress that apparent shape and shading belongs to different levels of perceptual image. The level of shading serves as the expression plane with respect to the apparent shape which belong the content plane (with respect to the shading level). In turn, the level of apparent shape serves as the expression level for apparent motion. Indeed, apparent motion is perceived in the form of a change in the apparent spatial layout. Furthermore, the level of apparent motion lends itself as the expression level for apparent causality. Michotte (1946/1963) reported that when two moving light patches are spatially and temporally contiguous, and the speed of the first patch exceeds that of the second, a unique visual impression might emerge at a certain ratio of their speeds. Namely, one cannot help seeing the first object 'launch' the second.

However, the linguistic analogy (i.e., the comparison of visual perception with speech perception) will remain superficial unless we clarify what is meant by visual language.[10] An important feature of the human language is the existence of the two types of rules: paradigmatic and syntagmatic. In our opinion, perceptual equations can be considered as an example of such paradigmatic rules. As a matter of fact, perceptual equations not only relate perceptual content with sensory fabric (more precisely, some of their attributes), but they also impose some constraints on the attributes of perceptual content. Specifically, they implicitly show in what combinations the attributes of perceptual content can form part of the perceptual image, and in which they cannot. For instance, Eq. (20.27) can be considered as a rule, i.e., prescriptive, for which pairs of material lightness and lighting brightness can be combined for a given sensory fabric. The most convincing arguments for the availability of such rules in visual perception are provided in abundance by pseudoscopic vision. So, let us consider from this point of view some of the most persuasive pseudoscopic phenomena that corroborate the existence of perceptual paradigmatic rules; considering both the phenomena themselves, and also considering them in the context of the linguistic analogy in general.

It must be said that perceptual transformations under pseudoscopic vision raise a few fundamental questions concerning the very nature of the perceptual process that are very difficult to answer by established perceptual theories. The first puzzling thing concerning pseudoscopic vision is the question of why depth reversal does not occur instantaneously and automatically. As a matter of fact, it always takes some time for the depth reversal to happen.[11] Why, then, do pseudoscopic transformations carry on even after depth reversal has occured? Next: why are some pseudoscopic

10 Although the words 'visual language' can be found in prior literature (e.g., Cavanagh, 2021), they were used as a metaphor, evoking some associations with speech, rather than language.

11 During the first few seconds (sometimes minutes – it depends how experienced an observer is), the scene viewed through a pseudoscope is perceived in normal (not inverted) relief. Only after this period does depth reversal occur. It always occurs gradually. It starts from flattening the existent normal depth relief (convex parts of the scene decrease in height, and concave parts become progressively more and more shallow). At some point the scene looks flat (without depth at all). Then depth appears again, but in reverse order, becoming progressively stronger and stronger, until it has become of the same magnitude as before, but of the opposite sign.

images more stable than others? Then: why are some visuals, such as human faces, impossible to transform at all?

The key to understanding pseudoscopic transformation is that it is a form of resolving a perceptual conflict between sensory information about the spatial relief of a surface, from binocular disparity on the one hand, and from shading and some other sources (e.g., monocular motion parallax) on the other. To understand why it takes that specific form, recall that the visual process begins with some event in the environment which induces some disturbance of the ambient optic array according to the laws of ecological optics (Gibson, 1979). This perturbation of the proximal stimulation causes a corresponding perturbation of the sensory fabric (a sensory event) according to psychophysical laws. This sensory event is then converted into a perceptual event.

It is important to point out that the process of converting sensory fabric into perceptual content is not passive decoding or deciphering sensory fabric, for which there is numerous evidence, amongst which maybe the most convincing is that the same sensory fabric is compatible with multiple perceptual content. To be precise, the same sensory fabric can be coupled with multiple perceptual content giving rise to different percepts. Such extensive 'polysemy' of sensory fabric manifests itself very vividly in pseudoscopic vision. Moreover, experiments with pseudoscopic vision reveal that vision is an active process of building a perceptual image, the essence of which is that the visual system tries various versions of perceptual content compatible with a given sensory fabric. We will refer to these as *perceptual hypotheses*.[12]

As stated above, a remarkable feature of pseudoscopic vision, which makes it an indispensable source of information about the very process of perception, is that the sensory fabric which comes into existence under pseudoscopic viewing is self-contradictory. The latter means that there exists no perceptual content ensuring a self-consistent resultant percept. This is the gist of the perceptual conflict under pseudoscopic vision. So, being unable to produce non-contradictory holistic perceptual hypotheses, the visual system starts producing local ones trying to master local fragments of the sensory fabric. As a result the visual scene comes apart. Various local perceptual hypotheses replace each other in different parts of the visual field, and we can experience such a kaleidoscopic dynamic for as long as we look through the pseudoscope, since it may stop only when a holistic self-consistent perceptual image is established, but that is impossible. Since the process of perception cannot end in any way, it continues to unfold over time again and again, allowing us to 'glance in' at this process, which is normally hidden from our introspection.

We illustrate the above with the following demonstration. Consider a display consisting of two objects on a table: one is easily pseudoscopically reversible (e.g., a paper cone), and the other is not reversible at all (e.g., a statuette with a complicated relief). Through a pseudoscope you will see the statuette next to a conical hole in the tabletop. In other words, one half of the display (with the statuette) will be perceived veridically, and the other (with the cone) – with an inverted relief. Such perceptual image will be quite stable, despite the fact that it is still controversial. However, the perceptual image becomes very unstable if we lay, say, a pencil on the tops of the cone and the statuette. The end of the pencil next to the statuette is always seen above the tabletop, and the end of the pencil that touches the tip of the cone is always visible under the tabletop. The perception of the middle of the pencil is not stable and changes all the time. For instance, one half of the pencil can be seen as fallen down through the table, and the other half as remaining over the table, both parallel to the tabletop. In this case, the pencil is seen as broken. However, if you start to scrutinise

12 A view of perception as a process of generating and verifying hypotheses was put forward by Richard Gregory. 'Perceptions may be compared with hypotheses in science. The methods of acquiring scientific knowledge provide a working paradigm for investigating processes of perception.' (Gregory, 1980, p. 181) However, as follows from this passage, being completely physicalistic, his approach is vulnerable to the criticism outlined above.

the site of the break, this version of the image will give way to another. Specifically, the pencil may look unbroken and straight but tilted. One half of it seems to be immersed in the table, the second sticking up from the table at an angle. Interestingly, the area of the tabletop (around the cone), in which the pencil appears to be partly immersed, is no longer perceived opaque. Instead, it looks like an insertion made of amber or some similar transparent material.[13] Hence, the same fragment of sensory fabric gives rise to either the material colour of an opaque object or the colour of a transparent medium.

As another example, consider the variety of perceptual hypotheses which come about when a background is viewed pseudoscopically through a circular hole in the screen (Fig. 20.3). What is perceived in the first moments of pseudoscopic viewing is a perceptual hypothesis corresponding to the real spatial arrangement of the surfaces – the background is seen to be behind the screen. However, such a hypothesis cannot be sustained for long since there is a powerful cue (binocular disparity) which testifies against it. The second hypothesis is the reversed counterpart of the first, that is, a circular area of background hanging up in the air in front the screen. However, it is a very rare and very unstable perceptual hypothesis. The third perceptual hypothesis is a stalactite looking like a clay pot. The fourth perceptual hypothesis is a transparent stalactite looking like an amber pot. And finally, the fifth hypothesis is a formless stalactite. It should be noted that in all these cases, you feel that you see different real things, and at the same time, in a sense, you see the one same thing. It is in order to qualify such a dual experience that the distinction between sensory fabric and perceptual content has been introduced.[14] In these terms, what is seen is the same sensory fabric coupled with different perceptual contents. Specifically, this can give rise to the same material colour coupled with different combinations of attached and cast shadows. As a result, different shape reliefs come to existence, despite the fact that nothing changes in the sensory fabric.

However, we do not see under pseudoscopic transformations everything which we could, in principle, see. As a matter of fact, objects with complicated reliefs cannot be easily and stably perceived in reversed form. There seem to be certain constraints on perceptual hypothesis. Notably, these constraints have a perceptual rather than a physical nature. Some perceptual hypotheses which are compatible with the sensory fabric from a physical point of view, are highly unlikely and are never seen to appear. For example, as mentioned above, a human face never appears reversed in depth, although a mask shaped like a human face can (though not without effort) be transformed. Probably, such an image as a concave live human face does not exist in the perceptual vocabulary or there exists some rule in the visual language that forbids such percept.

So, one might suggest that the process of generating and testing perceptual hypotheses is governed by special rules (which we will refer to as 'perceptual rules'). Arguably, they play the same role in the visual process as grammatical rules of natural language do in speech perception. Some of these rules determine compatibility of sensory fabric with perceptual hypotheses. These can be formally expressed as certain relationships between attributes of perceptual content and those of the sensory fabric (e.g., the perceptual equations above). Other seem to apply to different items of perceptual content.

For instance, looking through the pseudoscope at a cup of tea from above leads to an image of an amber cover over the inverted cup. If you gently shake the cup, the previously liquid tea, subsequently transformed into the solid amber, undergoes a further transformation: it begins to look like a gelatinous cover, which trembles due to the shaking. However, when the strength of

13 The tabletop was made of light brown lacquered pine.

14 It is worth noting that pseudoscopic demonstrations provide an opportunity to see the difference between sensory fabric and perceptual content with quite literally your own eyes.

shaking become incompatible with the gelatinous consistency, the inverted percept breaks down, and you see the cup in the normal relief, same as without a pseudoscope.

Along with the demonstrations described above, this demonstration also testifies that, in trying to resolve a perceptual conflict, the visual system can alter the material properties of the image forming the subject of a perceptual hypothesis: a solid object can turn into liquid, and vice versa. Once again we see that the same sensory fabric can give rise to an opaque object, a transparent solid object, and a transparent liquid. In other words, the material that we see changes despite the fact that the object's form, its sensory fabric, remains unchanged. Specifically, the consistency of the material changes. It goes from complete rigidity, through thick viscosity, to semifluidness.[15] It is important to note that we do *see* this consistence. It is a perceptual experience, not the result of thinking.[16]

A change in material property is accompanied by a change of colour. The colour of liquid tea is somewhat different from the colour of a solid piece of amber, which tea turns into after the pseudoscopic reversal of the apparent spatial relief. A similar difference is observed when the cone is perceived in the inverse relief. Specifically, a weak shade of blue appears on the surface of the apparent conical hole despite that the cone is made from a snow-white sheet of paper. We believe that the bluish shade is caused by the illuminant. The colour of illumination is not perfectly neutral, it contains a tint of blue. However, its purity is so low that it is practically invisible. As shown elsewhere, lighting colours appear less pronounced than material colours even if the light entering the eye is the same (Logvinenko, 2005b). One might suggest that the blue tint does not exceed this threshold when in the lighting-colour mode, but it does exceed it when in the material-colour mode.

In this regard, it is worth recalling that the main colour conversion that accompanies the material change is a change in the colour palette. It is not only that the object colours are subjectively experienced in a different way than the light colours. As established in previous chapters, the important fact is that they form different chromatic structures. Although no systematic study has yet been carried out, our preliminary experiments indicate that the colour palette of a transparent medium is different from both the object-colour palette and the light-colour palette. Despite the fact that, from a mathematical point of view, there is no difference between the spectral transmittance function of the filter and the spectral reflectance function of the object, the partial hue-matching experiment (with coloured filters) brought about only four component hues, the same as for light colours. Whilst the lack of white seems understandable (the perfect transmitter does not reveal itself at all for the eye), the lack of black comes as a surprise. True, since a filter transmitting no light at all ceases to be a filter, the very term 'black filter' would be an oxymoron. On the other hand, at first glance, the colour of many filters appears somewhat blackish. Still, no black component hue has been found as yet. Nevertheless, it would be wrong to suggest that the colour manifold for medium colours is similar to that for light colours. Our intuition is that the colour manifold for medium colours can be thought of as a sphere with the poles removed.

Thus, the same fragment of the sensory fabric can signify very different elements of the perceptual content, in particular, different colours. The signified colour, in turn, signifies the corresponding material property. One can, of course, say that one and the same fragment of the sensory fabric can

---

15 This raises a question concerning hypothetical specific psychophysical mechanisms that determine material properties in our perception (see e.g., Fleming, 2014, 2017). If they existed, they would had to continue to maintain the same output until the proximal stimulation changed. Since the proximal stimulation in the described demonstration did not change, it is difficult to reconcile the dramatic changes in material properties observed in this case with the existence of these hypothetical mechanisms.

16 Knowing that tea is a liquid does not prevent us from seeing it as a piece of solid amber after the pseudoscopic inversion.

take on different 'perceptual meanings'. However, the point, surely, is not in the terminology, but in the importance of fact claimed above, i.e., that the elements of perceptual content of one type can serve as the expression level for the elements of the perceptual content of another type, which play the role of the content level for the former.

So, a perceptual image is not just a set of primitives, such as colour attributes, material attributes, and such others as are only yet to be described. It is a more complex hierarchical entity. Furthermore, the most important property of perception has not yet even been mentioned at all: it is structured not only in space, but also in time. The core point of Gibson's ecological approach is that perception is a process that is carried out in time and accompanied by motion. He even substitutes the classical 'point of observation' by a notion of the 'path of observation' in order to emphasise the dynamic nature of perception. It should be recognised that this temporal aspect is at odds with the very term 'perceptual image', which implies a static picture, or a series of pictures ('snapshots') that replace each other in time. After all, 'perceptual image' turns out to be a metaphor as well. Adhering to the linguistic analogy in general, we prefer to replace this misleading (in our opinion), metaphor (i.e., 'visual image') with another one: the *visual utterance*.

If our conjecture that visual utterances are mediated by the underlying visual language is justified, then the whole strategy of studying visual perception should be radically revised. Recall that, in linguistics, language, speech, generation, and perception of utterances are the four different subject matters, with priority being given to language. Really, if we observe a person speaking an unknown language, but are unaware of the very existence of such language, it would hardly be a good idea to begin with psycholinguistics (i.e., a discipline concerned with the mechanisms by which the language is processed in the brain). Oddly enough, in our view, this is exactly what currently happens in the studying of perception. Indeed, within the framework of our linguistic analogy, almost the entire modern science of perception can be likened to phonetics, the main concern being with the analysis of the primitives (i.e., the visual analogue of phonemes or their distinctive features). Moreover, many visual scientists are inspired by a hope to find and describe the primitives of visual perception, and the neural mechanisms responsible for processing them.

However, a phoneme cannot be specified as a particular sound. After Ferdinand de Saussure's work (Saussure de, 1916/1983), hardly anyone doubts that phonemes exist as a formal system of differential elements, and cannot be understood (and/or specified) outside of this system. Being parts of the system, phonemes are defined by their relations to one another within this system. The differential nature of phonemes explains why even the absence of sound (for instance, in the end of a word) can play the role of a phoneme. The study of phonemes is hardly possible without taking into account the fact that they form a relational structure within the language.

Likewise, the search for the primitives of visual perception (i.e., the visual phonemes) is unlikely to be successful without first recognising the existence of the underlying visual language, which forms the structure that makes them primitives. Thus, the study of the visual language seems to us to be the key problem, which, if left unsolved, will seriously hinder further progress in visual perception.

22

Epilogue

The main achievements in colour science are rightly associated with colorimetry. The success of colorimetry is primarily due to the fact that it rests on a solid theoretical foundation. The latter allowed the development of advanced colour measurement techniques used in many applications. Beyond colorimetry, the importance of a theoretical foundation is often underestimated. Yet, experimental research differs from empirical research in that behind the first lies a theoretical model of the phenomenon being studied, and in the second, such model is missing, and is replaced by everyday ideas about this phenomenon. For instance, examining rates of salivation can be an routine empirical study of salivary glands, but can also become an ingenuous experiment on the physiological activity of the cerebral cortex, provided that you keep the concept of conditional reflex in mind. Nevertheless, it is even worse when an inadequate conceptual framework is used, as is exemplified by the above-mentioned attempts to look into colour perception from a colorimetric standpoint.

The inadequacy of the colorimetric paradigm for studying colour perception is due to the unsuitability of the stimulus-response paradigm – that underlies colorimetry – for colour perception. Indeed, as argued above, there is neither a distal nor proximal stimulus for object-colour. Moreover, object-colour cannot be considered a response for several reasons. It is important to understand that we are not talking about terminological purity, but rather about the fact that the experimenter is not quite in control of object-colour as an experimental dependent variable.

Specifically, there being no stimulus for object-colours implies that, unlike light-colours, manipulating the physical variables of the experimental display does not secure full control over object-colours. In particular, keeping the spectral reflectivity of an object stable does not guarantee that its colour will not change. As pseudoscopic demonstrations show, because of object-colour indeterminacy, even maintaining stability in both the spectral reflectivity of an object and the spectral characteristics of its illumination does not guarantee this. For natural scenes, the possible effect of object-colour indeterminacy is mitigated by the fact that such scenes are usually highly articulated as far as both lights and objects are concerned. As noticed in Section 20.3, the more objects and/or light sources are involved simultaneously, the less the effect of the trade-off between material and lighting colours. In those colour constancy experiments where highly articulated displays were used, almost perfect constancy was usually recorded (Maloney & Schirillo, 2002). On the other hand, a display consisting of a single object lit by a single light source (as that used by Gelb) is prone to producing a rather unstable percept. In this regard, the attempt to repeat Logvinenko & Maloney's experiment (Logvinenko & Maloney, 2006), as undertaken by Madigan & Brainard, who 'slightly' modified the method (Madigan & Brainard, 2014) is quite demonstrative. Specifically, the stimulus objects were not presented all at once (as Logvinenko & Maloney did),

Foundations of Colour Science: From Colorimetry to Perception, First Edition. Alexander D. Logvinenko and Vladimir L. Levin.
© 2023 John Wiley & Sons Ltd. Published 2023 by John Wiley & Sons Ltd.

but one at a time, thereby reducing the scene articulation to a minimum. Not surprisingly, the experiment brought about completely different results.

Actually, the situation with controlling object-colour as an experimental dependent variable is even more nuanced because of the hierarchical structure of the perceptual image mentioned above. When we focus on the higher levels, the content of the lower levels can slip away from our consciousness, and vice versa. Undoubtedly, everyone is familiar with this situation in speech perception. When, for example, listening to someone speak, if we pay too much attention to their accent, we may lose the thread of the conversation. And, vice versa, if you were completely carried away by the content of a dispute, you might not remember in exactly which words the main arguments were presented.[1] Something akin to this can take place in visual perception as well. For instance, after a prolonged adaptation to inversion, observers in experiments with inverting glasses reported that before they were asked by the experimenter whether their vision was still upside down, they thought their vision was upright. However, afterwards they were not so sure.

It is worth mentioning that inverted vision is, in general, quite pertinent to the issues discussed in this context. Although Stratton (1896) and later Kohler (1964) reported that perceptual adaptation to an optically inverted visual field took place in their prolonged experiments with inverting goggles, other researchers claimed that they could not reproduce their achievement (Peterson & Peterson, 1938; Snyder & Pronko, 1952; Linden et al., 1999). Interestingly, there was no disagreement on visual-motor adaptation, which, as reported by all experimenters, did occur after several days of prolonged adaptation. In particular, Kohler reported that his participants could ride a bicycle through the streets in such goggles. How could such highly coordinated motor skills be possible without perceptual adaptation? This discrepancy looks mysterious, especially given the lack of 'any change in the retinotopy of early visual areas observed in the functional magnetic resonance images' obtained in a more recent study (Linden et al., 1999, p. 480). Still, everything falls into place if one assumes that the long-term exposure to the inverted visual field results in the generation of upright perceptual content without any changes in the sensory fabric, which remains inverted (Logvinenko, 1981). In a sense, observers acquired a second (hybrid) visual language. This explains why, after removing the inverting spectacles, no aftereffects were observed (Logvinenko, 1974; Logvinenko & Zhedunova, 1980). Indeed, the acquisition of a foreign language usually does not lead to the loss of the native language. This hypothesis is also consistent with the fact that there is no need to permanently wear inverting spectacles in order to achieve perceptual adaptation (Logvinenko & Zhedunova, 1981). Just as, in order to master a foreign language, we do not need to fully and permanently immerse ourselves in that new language's environment.

Thus, due to the multi-layered structure of perceptual images, it is very difficult, if not impossible, to obtain a direct, unequivocal answer to such a seemingly simple question as 'what do you see?' At first glance, when studying colour, one can get round this problem with the usual instructions for observers to be more attentive, concentrate, and accurately describe what colour they see. However, what possible answer can be expected from an observer if she or he looks, as in the aforementioned case, at a blue surface illuminated with such a yellow light, that the reflected light is metameric to daylight? From our experience, we may suggest that most inexperienced (in colour) observers would, probably, not even notice the colour of the lighting at all. After we drew their attention to it, they might realise that the colour of the lighting was yellow. (By the way, artists and photographers are, usually, able to evaluate the lighting colour, some even better than the experimenters.) Note, however, that the question: 'what colour do you see?' ceases being simple after this happens. Not to

1 During scientific meetings it has happened that the author of this book could not later remember what language was used by him to raise a controversial question, even while the idea being expressed was still clear in his memory.

mention that almost all of our observers refused to believe that the reflected light was neutral. Admittedly, after some training in introspection, they might change their minds – but that will make the question concerning the colour they see even more difficult to answer.[2]

It follows that the stimulus-response paradigm is entirely extraneous to colour perception, not only because of the lack of stimulus for colour, but also because there is nothing in perception that lends itself as a definitive, sharp, and unique response. Interestingly, the rejection of the 'stimulus-response' paradigm raises serious doubts about the prospects of searching for the elusive 'neural mechanisms' of colour vision, in which so many scientists are currently involved. True, the functioning of the eye and brain is compatible with the stimulus-response paradigm. However, if the perception cannot be placed in the Procrustean bed of the stimulus-response paradigm, then it is very unlikely that there are such neural units that the perception is simply a mirror reflection of the activity of them, as the single-neuron doctrine suggests (Barlow, 1972). Hence, the question of the neurophysiological substratum of colour vision requires some clarification. Namely, what is it the substratum, that will be searched for? Assuming that the eye and the brain process some information, we face the following question: what sort of information is processed there? Or, in Marr's words, what does the brain compute? If our hypothesis about the visual language is not without reason, then a neurophysiologist trying to decipher the neural code can be compared to a spy, trying to hack into the mailbox of a computer, who not only does not know the language in which these emails are written, but does not even suspect the existence of this language.

If we abandon the doctrine of the direct nature of perception and suppose that it is mediated by visual language, the opportunity opens up to take a fresh look at the perception of pictures and their connection with natural vision. As repeatedly emphasised earlier, the perception of pictures differs from the perception of natural scenes. Why, then, do the pictures resemble the physical objects they represent? In what sense does the picture depict its physical prototype?

An analysis of the following sentence may shed light on this issue. It was made by a linguist Henry Gleason, Jr., from non-existent words, but in a grammatically proper way: 'The iggle squiggs trazed wombly in the harlish hoop'. Since all the words' roots in this sentence are meaningless, the phrase has no meaning at the lexical level. Nevertheless, as all affixes are real and used grammatically correctly, they ensure that the phrase provides some meaning at the morphological level. So, the grammatical structure of the phrase yields enough information to get an idea of some action, but with unknown entities.

If the visual language exists, with its grammar and syntax, then pictures can be thought of as an analogue of such a phrase, which does not refer to any physical object, but nonetheless conveys some semantic information through its grammatical and syntactic structure.

2 Besides, as described by Mach (1959), Gibson (1950), and some others (e.g., Logvinenko, 1981), intense peering into the visual scene can result in a special kind of visual experience – the perception of visual field in Gibson's terminology.

References

Abney, W. de W. (1910). On the change in hue of spectrum colours by dilution with white light. *Proceedings of the Royal Society (London)*, A83, 120–127.

Abramov, I., & Gordon, J. (1994). Color appearance: on seeing red – or yellow, or green, or blue. *Annual Review of Psychology*, 45, 451–485.

Adelson, E. H. (1993). Perceptual organization and the judgment of brightness. *Science*, 262, 2042–2044.

Adelson, E. H., & Pentland, A. P. (1996). The perception of shading and reflectance. In D. C. Knill & W. Richards (Eds.), *Perception as Bayesian inference* (pp. 409–423). New York, NY, USA: Cambridge University Press.

Alpern, M., Kitahara, K., & Krantz, D. H. (1983). Perception of colour in unilateral tritanopia. *Journal of Physiology, London*, 335, 683–697.

Arend, L. (1994). Surface colors, illumination, and surface geometry: intrinsic-image models of human color perception. In A. Gilchrist (Ed.), *Lightness, brightness, and transparency* (pp. 159–213). Hillsdale, New Jersey: Lawrence Erlbaum Associates, Inc.

Barlow, H. B. (1972). Single units and sensation: a neuron doctrine for perceptual psychology? *Perception*, 1, 371–394.

Barrow, H. G., & Tenenbaum, J. M. (1978). Recovering intrinsic scene characteristics from images. In A. R. Hanson, E. M. A. Riseman (Eds.) *Computer vision systems* (pp. 3–26). New York e.a.: Academic Press.

Bauml, K., & Wandell, B. A. (1996). Color appearance of mixture gratings. *Vision Research*, 36, 2849–2864.

Baylor, D. A., Fuortes, M. G., & O'Bryan, P. M. (1971). Receptive fields of cones in the retina of the turtle. *Journal of Physiology*, 214, 265–294.

Beck, J. (1961). Judgments of surface illumination and lightness. *Journal of Experimental Psychology*, 61, 368–375.

Beck, J. (1972). *Surface color perception*. Ithaca and London: Cornell University Press.

Bergström, S. S. (1994). Color constancy: arguments for a vector model for the perception of illumininination, color, and depth. In A. Gilchrist (Ed.), *Lightness, brightness, and transparency* (pp. 257–286). Hillsdale, New Jersey: Lawrence Erlbaum Associates, Inc.

Birch, J. (2001). *Diagnosis of defective colour vision* (2nd ed.). Oxford e.a.: Butterworth Heinemann.

Blackwell, H. R. (1963). Neural theories of simple visual discriminations. *Journal of the Optical Society of America*, 53, 129–160.

Blackwell, H. R. (1972). Luminance difference thresholds. In D. Jameson & L. Hurvich (Eds.), *Handbook of sensory physiology: Vol. VII/4. Visual psychophysics* (pp. 78–101). Berlin e.a.: Springer-Verlag.

Foundations of Colour Science: From Colorimetry to Perception, First Edition. Alexander D. Logvinenko and Vladimir L. Levin.
© 2023 John Wiley & Sons Ltd. Published 2023 by John Wiley & Sons Ltd.

Bloomfield, S. A., & Völgyi, B. (2014). Mind the gap: the functional roles of neuronal gap junctions in the retina. In J. S. Werner & L. M. Chalupa (Eds.), *The new visual neurosciences* (pp. 93–105). Cambridge, MA, London, England: MIT Press.

Bongard, M. M., Smirnov, M. S., & Friedrich, L. (1958). The four-dimensional colour space of the extrafovel retinal area of the human eye. In *Visual problems of colour, vol. 6* (pp. 325–330). London: N. P. L. Symposium No. 8, Her Majesty's Stationery Office.

Boring, E. G. (1942). *Sensation and perception in the history of experimental psychology*. New York: Appleton-Century-Crofts, Inc.

Boring, E. G. (1946). Mind and mechanism. *The American Journal of Physiology*, 59, 173–192.

Bowmaker, J. K. (1991). Photoreceptors, photopigments and oil droplet. In P. Gouras (Ed.), *Vision and visual dysfunction. Vol. 6. The perception of colour* (pp. 108–127). Houndmills and London: The Macmillan Press.

Bowmaker, J. K., Dartnall, H. J. A., & Mollon, J. D. (1980). Microspectrophotometric demonstration of four classes of photoreceptors in an old world primate, *Macaca fascicularis*. *Journal of Physiology*, 298, 131–143.

Boynton, R. M. (1996). History and current status of a physiologically based system of photometry and colorimetry. *Journal of the Optical Society of America A*, 13, 1609–1621.

Boynton, R. M., & Gordon, J. (1965). Bezold-Brucke hue shift measured by color-naming technique. *Journal of the Optical Society of America*, 55, 78–86.

Boynton, R. M., & Kaiser, P. K. (1968). Vision: the additivity law made to work for heterochromatic photometry with bipartite fields. *Science*, 116, 366–368.

Boynton, R. M., Ikeda, M., & Stiles, W. (1964). Interaction among chromatic mechanisms as inferred from positive and negative increment thresholds. *Vision Research*, 4, 87–117.

Brainard, D. H. (2004). Color constancy. In L. Chalupa & J. Werner (Eds.), *The visual neurosciences* (pp. 948–961). Cambridge, MA: MIT Press.

Brainard, D. H. (2009). Color constancy. In B. Goldstein (Ed.), *The Sage encyclopedia of perception* (pp. 253–257). Los Angeles, CA: SAGE Publications.

Brainard, D. H. (2015). Color and the cone mosaic. *Annual Review of Vision Science*, 1, 519–546.

Brainard, D. H., & Radonjic, A. (2014). Color constancy. In J. S. Werner & L. M. Chalupa (Eds.), *The new visual neurosciences* (pp. 545–556). Cambridge, MA, London, England: MIT Press.

Brainard, D. H., & Stockman, A. (2010). Colorimetry. Chap. 10 in M. Bass (Ed.), *OSA Handbook of optics: Vol. III. Vision and vision optics* (3rd ed., pp. 10.1–10.56). New York e.a.: McGraw-Hill, Inc.

Brainard, D. H., Brunt, W. A., & Speigle, J. M. (1997). Color constancy in the nearly natural image. I. Asymmetric matches. *Journal of the Optical Society of America A*, 14, 2091–2110.

Bridges, D. S., & Mehta, G. B. (1995). *Representations of preference orderings. Lecture notes in economics and mathematical systems*. Berlin, Heidelberg, New York: Springer-Verlag.

Brindley, G. S. (1960). *Physiology of the retina and the visual pathway*. London: Edward Arnold Ltd.

Broackes, J. (2010a). Unilateral colour vision defects and the dimensions of dichromat experience. *Ophthalmic and Physiological Optics*, 30, 672–684.

Broackes, J. (2010b). What do the color-blind see? In J. Cohen & M. Matthen (Eds.), *Color ontology and color science* (pp. 291–405). Cambridge, MA: The MIT Press.

Brown, J. L. (1965). Afterimages. In C. H. Graham (Ed.), *Vision and visual perception* (pp. 479–503). New York, London, Sydney: John Wiley & Sons, Inc.

Brown, J. L., & Mueller, C. G. (1965). Brightness discrimination and brightness contrast. In C. H. Graham (Ed.), *Vision and visual perception* (pp. 208–250). New York, London, Sydney: John Wiley & Sons, Inc.

Buchsbaum, G., & Gottschalk, A. (1983). Trichromacy, opponent colours coding and optimum colour information transmission in the retina. *Proceedings of the Royal Society of London. Series B, Biological Sciences*, 220, 89–113.

Buck, S. L. (2004). Rod-cone interactions in human vision. In L. Chalupa & J. Werner (Eds.), *The visual neurosciences* (pp. 863–878). Cambridge, MA: MIT Press.

Buck, S. L. (2014). The interaction of rod and cone signals: pathways and psychophysics. In J. S. Werner & L. M. Chalupa (Eds.), *The new visual neurosciences* (pp. 485–497). Cambridge, MA, London, England: MIT Press.

Burns, M. E., & Pugh, E. N., Jr. (2014). Visual transduction by rod and cone photoreceptors. In J. S. Werner & L. M. Chalupa (Eds.), *The new visual neurosciences* (pp. 7–18). Cambridge, MA, London, England: MIT Press.

Burns, S. A., Elsner, A. E., Pokorny, J., & Smith, V. C. (1984). The Abney effect: chromaticity coordinates of unique and other constant hues. *Vision Research*, 24, 479–489.

Bush, R. B., Galanter, E., & Luce, R. D. (1963). Characterization and classification of choice experiments. In R. D. Luce, R. R. Bush, & E. Galanter (Eds.), *Handbook of mathematical psychology: Vol. I.* (pp. 77–102). New York and London: John Wiley & Sons, Inc.

Byrne, A., & Hilbert, D. R. (2010). How do things look to the color-blind? In J. Cohen & M. Matthen (Eds.), *Color ontology and color science* (pp. 259–290). Cambridge, MA: The MIT Press.

Callaway, E. M. (2014). Cell types and local circuits in primary visual cortex of the macaque monkey. In J. S. Werner & L. M. Chalupa (Eds.), *The new visual neurosciences* (pp. 353–366). Cambridge, MA, London, England: MIT Press.

Campbell, S. L., & Meyer, C. D., Jr. (1979). *Generalised inverses of linear transformations*. London, San Francisco, CA, Melbourne: Pitman.

Cavanagh, P. (2021). The language of vision. *Perception*, 50, 195–215.

Cavonius, C. R., & Estevez, O. (1975). Contrast sensitivity of individual colour mechanisms of human vision. *Journal of Physiology*, 248, 649–662.

Chichilnisky, E. J., & Wandell, B. A. (1995). Photoreceptor sensitivity changes explain color appearance shifts induced by large uniform backgrounds in dichoptic matching. *Vision Research*, 35, 239–254.

Cicerone, C. M., Krantz, D. H., & Larimer, J. (1975). Opponent-process additivity - III. Effect of moderate chromatic adaptation. *Vision Research*, 15, 1125–1135.

CIE. (2006). *Fundamental Chromaticity Diagram with Physiological Axes – Part 1. CIE: 170–171* (Tech. Rep.). Vienna: Central Bureau of the Commission Internationale de l'Èclairage.

CIE. (2015). *Fundamental Chromaticity Diagram with Physiological Axes – Part 2: Spectral Luminous Efficiency Functions and Chromaticity Diagrams. CIE: 170–171* (Tech. Rep.). Vienna: Central Bureau of the Commission Internationale de l'Èclairage.

Cohen, J. B. (2001). *Visual color and color mixture. The fundamental color space*. Urbana and Chicago: University of Illinois Press.

Cohen, J. B., & Kappauf, W. E. (1982). Metameric color stimuli, fundamental metamers, and Wyszecki's metameric blacks. *American Journal of Psychology*, 95, 537–564.

Cohen, J. B., & Kappauf, W. E. (1985). Color mixture and fundamental metamers: theory, algebra, geometry, application. *American Journal of Psychology*, 98, 171–259.

Cohn, T., & Lasley, D. (1975). Spatial summation of foveal increments and decrements. *Vision Research*, 15, 389–399.

Cole, G. R., Hine, T., & McIlhagga, W. (1993). Detection mechanisms in L-, M- and S-cone contrast space. *Journal of the Optical Society of America A*, 10, 38–51.

Cole, G. R., Hine, T., & McIlhagga, W. (1994). Estimation of linear detection mechanisms for stimuli of medium spatial frequency. *Vision Research*, 34, 1267–1278.

Cole, G. R., Stromeyer, C. F., & Kronauer, R. E. (1990). Visual interactions with luminance and chromatic stimuli. *Journal of the Optical Society of America A*, 7, 128–140.

Craven, B. J., & Foster, D. H. (1992). An operational approach to colour constancy. *Vision Research*, 32, 1359–1366.

Curcio, C. A., Allen, K. A., Sloan, K. R., Lerea, C. L., Hurley, J. B., Klock, I. B., et al. (1991). Distribution and morphology of human cone photoreceptors stained with anti-blue opsin. *Journal of Comparative Neurology*, 312, 610–624.

Dartnall, H. J. A. (1953). The interpretation of spectral sensitivity curves. *British Medical Bulletin*, 9, 24–30.

Dartnall, H. J. A., Bowmaker, J. K., & Mollon, J. D. (1983). Human visual pigments: microspectrophotometric results from the eyes of seven persons. *Proceedings of the Royal Society of London. Series B*, 220, 115–130.

Day, M. M. (1958). *Normed linear spaces*. Berlin, Gottingen, Heidelberg: Springer-Verlag.

Debreu, G. (1954). Representation of a preference ordering by a numerical function. In R. M. Thrall, C. H. Coombs, & R. L. Davis (Eds.), *Decision processes* (pp. 159–165). New York: John Wiley & Sons, Inc.

Debreu, G. (1964). Continuity properties of Paretian utility. *International Economic Review*, 5, 285–293.

Dennett, D. C. (1991). *Consciousness explained*. Boston, MA: Little, Brown and Company.

De Valois, R., & De Valois, K. (1988). *Spatial vision*. New York: Oxford University Press.

De Valois, R. L., & De Valois, K. K. (1993). A multi-stage color model. *Vision Research*, 33, 1053–1065.

DeVries, S. H., Qi, X., Smith, R., Makous, W., & Sterling, P. (2002). Electrical coupling between mammalian cones. *Current Biology*, 12, 1900–1907.

Dubois, E. (2009). *The structure and properties of color spaces and the representation of color images*. Morgan & Claypool Publishers. (doi: doi: 10.2200/S00224ED1V01Y200910IVM011).

Dunford, N., & Schwartz, J. T. (1958). *Linear operators. Part I: General theory*. New York, London: Interscience Publishers.

Ejima, Y., & Takahashi, S. (1985). Interaction between short- and longer-wavelength cones in hue cancellation codes: nonlinearities of hue cancellation as a function of stimulus intensity. *Vision Research*, 25, 1911–1922.

Epstein, W. (1982). Percept-percept couplings. *Perception*, 11, 75–83.

Epstein, W., & Park, J. N. (1963). Shape constancy: functional relationships and theoretical formulations. *Psychological Bulletin*, 60, 265–288.

Eskew, R. T., Jr. (2008). Chromatic detection and discrimination. In T. D. Albright & R. H. Masland (Eds.), *The senses: a comprehensive reference* (Vol. 2, pp. 101–117). San Diego, CA: Academic Press.

Eskew, R. T., Jr. (2009). Higher order color mechanisms: A critical review. *Vision Research*, 49, 2686–2704.

Eskew, R. T., Jr., McLellan, J. S., & Giulianini, F. (1999). Chromatic detection and discrimination. In K. R. Gegenfurtner & L. T. Sharpe (Eds.), *Color vision: from genes to perception* (pp. 345–368). Cambridge: Cambridge University Press.

Evans, R. M. (1974). *The perception of color*. New York: John Wiley & Sons, Inc.

Fairchild, M. D. (2005). *Color appearance models* (2nd ed.). Chichester, UK: Wiley.

Fairman, H. S. (1991). Recommended terminology for matrix R and metamerism. *Color Research & Application*, 16, 337–341.

Falmagne, J.-C. (1985). *Elements of psychophysical theory*. New York: Oxford University Press.

Fechner, G. T. (1860). *Elemente der Psychophysik [Elements of psychophysics]*. Leipzig: Breitkopf und Härtel.

Feynman, R. P., Leighton, R. B., & Sands, M. L. (1963). *The Feynman lectures on physics* (Vol. 1). Reading, MA: Addison-Wesley Publishing Company.

Fishburn, P. C. (1970). *Utility theory for decision making*. New York: John Wiley and Sons.

Fleming, R. W. (2014). Visual perception of materials and their properties. *Vision Research*, 94, 62–75.

Fleming, R. W. (2017). Material perception. *Annual Reviews of Vision Science*, 3, 365–388.

Foley, J. M. (1967). Binocular disparity and perceived relative distance: an examination of two hypotheses. *Vision Research*, 7, 655–670.

Foster, D. H. (2003). Does colour constancy exist? *Trends in Cognitive Science*, 7, 439–443.

Foster, D. H. (2011). Color constancy. *Vision Research*, 51, 674–700.

Foster, D. H., & Nascimento, S. M. C. (1994). Relational colour constancy from invariant cone-excitation ratios. *Proceedings of the Royal Society of London. Series B*, 257, 115–121.

Foster, D. H., Nascimento, S. M. C., Amano, K., Arend, L., Linnell, K. J., Nieves, J. L., Plet, S., Foster, J. S. (2001). Parallel detection of violations of color constancy. *Proceedings of the National Academy of Sciences of the United States of America*, 98, 8151–8156.

Foster, D. H., Nascimento, S. M. C., Craven, B. J., Linnell, K. J., Cornelissen, F. W., & Brenner, E. (1997). Four issues concerning colour constancy and relational colour constancy. *Vision Research*, 37, 1341–1345.

Gantmacher, F. R. (2004). *Teoriya matrits [The theory of matrices]* (5th ed.). Moscow: Physmathlit.

Gegenfurtner, K. R. (2003). Cortical mechanisms of colour vision. *Nature Reviews Neuroscience*, 4, 563–572.

Gelb, A. (1929). Die 'Farbenkonstanz' der Sehdinge. In A. Bethe, G. v. Bergmann, G. Embden, and A. Ellinger (Eds.), *Handbuch der normalen und pathologischen Physiologie*: Bd. 12 (pp. 594–678). Berlin: Springer-Verlag.

Georgeson, M. A., May, K. A., Freeman, T. C. A., & Hesse, G. S. (2007). From filters to features: scale-space analysis of edge and blur coding in human vision. *Journal of Vision*, 7(13):7, 1–21.

Gescheider, G. (1997). *Psychophysics: the fundamentals* (3rd ed.). New Jersey, USA: Lawrence Erlbaum Associates.

Gibson, J. (1950). *The perception of the visual world*. Boston, MA: Houghton Mifflin.

Gibson, J. (1966). *The senses considered as perceptual systems*. Boston, MA: Houghton Mifflin.

Gibson, J. (1979). *The ecological approach to visual perception*. Boston, MA: Houghton Mifflin.

Gilchrist, A. (1977). Perceived lightness depends on perceived spatial arrangement. *Science*, 195, 185–187.

Gilchrist, A. (2006). *Seeing black and white*. New York: Oxford University Press.

Gilchrist, A. (2017). The staircase Gelb illusion. In A. G. Shapiro & D. Todorović (Eds.), *The Oxford compendium of visual illusions* (pp. 367–372). New York: Oxford Scholarship Online.

Goethe, J. W. von. (1970). *Theory of colours* (C. L. Eastlake, Trans.). Cambridge, MA and London, England: The MIT Press. (Original work published 1810; Introduction by Deane B. Judd).

Gordon, J., & Abramov, I. (1988). Scaling procedures for specifying color appearance. *Color Research & Application*, 13, 146–152.

Gordon, J., Abramov, I., & Chan, H. (1994). Describing color appearance: hue and saturation scaling. *Perception & Psychophysics*, 56, 27–41.

Govardovskii, V. I., Fyhrquist, N., Reuter, T., Kuzmin, D. G., & Donner, K. (2000). In search of the visual pigment template. *Visual Neuroscience*, 17, 509–528.

Graham, N. (1989). *Visual pattern analysers*. Oxford: Oxford University Press.

Grassmann, H. (1853). Zur Theorie der Farbenmischung. *Annalen der Physik und Chemie* (J. C. Poggendorff's Annalen), 89, 69–84.

Green, D. G. (1968). The contrast sensitivity of the colour mechanisms of the human eye. *Journal of Physiology*, 196, 415–429.

Green, D. M., & Swets, J. A. (1966). *Signal detection theory and psychophysics*. New York: John Wiley & Sons Ltd.

Gregory, R. L. (1980). Perceptions as hypotheses. *Philosophical Transactions of the Royal Society of London B: Biological Sciences*, 290, 181–197.

Gregory, R. L. (1997). Visual illusions classified. *Trends in Cognitive Sciences*, 1, 190–194.

Guild, J. (1931). The colorimetric properties of the spectrum. *Philosophical Transactions of the Royal Society of London*, A230, 149–187.

Guilford, J. P. (1956). *Psychometrics methods* (2nd ed.). New York, Toronto, London: McGraw-Hill Book Company Inc.

Guth, S. L. (1991). Model for color vision and light adaptation. *Journal of the Optical Society of America A*, 8, 976–993.

Hardin, C. (1988). *Color for philosophers. Unweaving the rainbow*. Cambridge, MA: Hackett.

Hartridge, H. (1945). The change from trichromatic to dichromatic vision in the human retina. *Nature*, 155, 657–662.

Hatfield, G. (2003). Objectivity and subjectivity revisited: colour as a psychobiological property. In R. Mausfeld & D. Heyer (Eds.), *Colour perception: mind and the physical world* (pp. 187–202). Oxford, New York: Oxford University Press.

Hatfield, G. C., & Epstein, W. (1979). The sensory core and the medieval foundations of early modern perceptual theory. *Industry and Science Information*, 70, 363–384.

von Helmholtz, H. (1867). *Handbuch der physiologischen Optik*. Leipzig: Voss.

von Helmholtz, H. (1970). The sensations of vision. In D. L. MacAdam (Ed.), *Sources of color sciences* (pp. 84–100). Cambridge, MA and London, England: The MIT Press. (Reprinted from *Handbuch der physiologischen Optik*, 1867, Hamburg: Voss).

Hering, E. (1964). *Outlines of a theory of the light sense* (L. M. Hurvich & D. Jameson, Trans.). Cambridge, MA: Harvard University Press. (Original work published 1920).

Hess, R. F., Mullen, K. T., & Zrenner, E. (1989). Human photopic vision with only short wavelength cones: post-receptoral properties. *Journal of Physiology*, 417, 151–172.

Hilbert, D. R. (1987). *Color and color perception: a study in anthropocentric realism*. Stanford: Center for the Study of Language and Information.

Hirsch, M. W. (Ed.). (1976). *Differential topology*. New York: Springer.

Holmes, R. B. (1975). *Geometric functional analysis and its applications*. New York, Heidelberg, Berlin: Springer-Verlag.

Holmgren, F. (1881). How do the colour-blind see the different colours. *Proceedings of the Royal Society of London*, 44, 35–43.

Hordley, S. D. (2006). Scene illuminant estimation: past, present, and future. *Color Research & Application*, 31, 303–314.

Horn, B. K. P. (1989). Obtaining shape from shading information. In B. K. P. Horn & M. J. Brooks (Eds.), *Shape from shading* (pp. 121–171). Cambridge, MA and London, England: The MIT Press.

Hubel, D. H., & Wiesel, T. N. (1962). Receptive fields, binocular interaction and functional architecture in the cat's visual cortex. *Journal of Physiology, London*, 160, 106–154.

Hung, P. C., & Berns, R. S. (1995). Determination of constant hue loci for a CRT gamut and their predictions using color appearance spaces. *Color Research & Application*, 20, 285–295.

Hunt, R. W. G. (1977). The specifications of colour appearance. I. Concepts and terms. *Color Research & Application*, 2, 55–68.

Hunt, R. W. G. (1998). *Measuring colour* (3rd ed.). Chichester, England: Fountain Press.

Hurlbert, A. C. (1998). Computational models of color constancy. In V. Walsh & J. J. Kulikowski (Eds.), *Perceptual constancy: why things look as they do* (pp. 283–322). Cambridge, UK; New York, USA: Cambridge University Press.

Hurvich, L. M., & Jameson, D. (1955). Some quantitative aspects of an opponent-colors theory. II. Brightness, saturation, and hue in normal and dichromatic vision. *Journal of the Optical Society of America*, 45 (8), 602–616.

Hurvich, L. M., & Jameson, D. (1957). An opponent-process theory of color vision. *Psychological Review*, 64, 384–404.

Ioffe, A. D., & Tikhomirov, V. M. (1974/1979). Theory of extremal problems, (Russian) Series in Nonlinear Analysis and Its Applications, Izdat. "Nauka", Moscow, 1974. English translation by Karol Makowski: Theory of extremal problems. Studies in Mathematics and its Applications, v. 6 North-Holland Publishing Co., Amsterdam-New York, 1979.

Ittelson, W. H. (1952). *The Ames demonstrations in perception: a guide to their construction and use.* Princeton, NJ: Princeton University Press.

Ittelson, W. H. (1960). *Visual space perception.* New York: Springer Publishing Company, Inc.

Jameson, D., & Hurvich, L. (1972). Color adaptaion: sensitivity, contrast, after-images. In D. Jameson & L. Hurvich (Eds.), *Handbook of sensory physiology: Vol. VII/4. Visual psychophysics* (pp. 568–581). Berlin e.a.: Springer-Verlag.

Jameson, D., & Hurvich, L. M. (1978). Dichromatic color language: 'reds' and 'greens' don't look alike but their colors do. *Sensory Processes*, 2, 146–155.

Jameson, D., & Hurvich, L. M. (1989). Essay concerning color constancy. *Annual Reviews of Psychology*, 40, 1–22.

Jameson, K. A., Highnote, S. M., & Wasserman, L. M. (2001). Richer color experience in observers with multiple photopigment opsin genes. *Psychonomic Bulletin and Review*, 8, 244–261.

Johnson, E. N., Hawken, M. J., & Shapley, R. (2001). The spatial transformation of color in the primary visual cortex of the macaque monkey. *Nature Neuroscience*, 4, 409–416.

Johnson, E. N., Hawken, M. J., & Shapley, R. (2004). Cone inputs in macaque primary visual cortex. *Journal of Neurophysiology*, 91, 2501–2514.

Johnson, E. N., Hawken, M. J., & Shapley, R. (2008). The orientation selectivity of color-responsive neurons in macaque V1. *Journal of Neuroscience*, 28 (32), 8096–8106.

Jordan, G., & Mollon, J. D. (1997). Adaptation of colour vision to sunlight. *Nature*, 386, 135–136.

Jordan, G., Deeb, S. S., Bosten, J. M., & Mollon, J. D. (2010). The dimensionality of color vision in carriers of anomalous trichromacy. *Journal of Vision*, 10 (8):12, 1–19.

Judd, D. B. (1948). Color perceptions of deuteranopic and protanopic observers. *Journal of Research of the National Bureau of Standards*, 41, 247–271.

Judd, D. B. (1951). *Colorimetry and artificial daylight* (Technical Committee No. 7. Report of Secretariat United States Commission). Stockholm: International Commission on Illumination, Twelfth Session. (pp. 1–60).

Judd, D. B., & Wyszecki, G. (1975). *Color in business, science, and industry* (3rd ed.). New York, Chichester, Brisbane, Toronto: John Wiley & Sons, Inc.

Judd, D. B., MacAdam, D. L., Wyszecki, G., Budde, H. W., Condit, H. R., Henderson, S. T., & Simonds, J. L. (1964). Spectral distribution of typical daylight as a function of correlated color temperature. *Journal of the Optical Society of America*, 54, 1031–1040.

Kaiser, P. K. (1971). Minimally distinct border as a preferred psychophysical criterion in visual heterochromatic photometry. *Journal of the Optical Society of America*, 61, 966–971.

Kaiser, P. K., & Boynton, R. M. (1996). *Human color vision* (2nd ed.). Washington, DC: Optical Society of America.

Kantorovich, L. V., & Akilov, G. P. (1982). *Functional analysis* (2nd ed.). Oxford e.a.: Pergamon Press.

Kaplan, E. (2014). The M, P, and K pathways of the primate visual system revisited. In J. S. Werner & L. M. Chalupa (Eds.), *The new visual neurosciences* (pp. 215–226). Cambridge, MA, London, England: MIT Press.

Katz, D. (1999). *The world of colour*. London: Routledge. (Original work published 1911).

Kelly, D. H. (1974). Spatio-temporal frequency characteristics of color-vision mechanisms. *Journal of the Optical Society of America*, 64, 983–990.

Kelly, D. H., & Burbeck, C. A. (1984). Critical problems in spatial vision. *Critical Reviews in Biomedical Engineering*, 10, 125–177.

Kelly, D. H., & Savoie, R. E. (1978). Theory of flicker and transient responses. III. An essential nonlinearity. *Journal of the Optical Society of America*, 68, 1481–1490.

Kilpatrick, F. P., & Ittelson, W. H. (1953). The size-distance invariance hypothesis. *Psychological Review*, 60, 223–231.

Kingdom, F. A. A. (2011). Lightness, brightness and transparency: a quarter century of new ideas, captivating demonstrations and unrelenting controversy. *Vision Research*, 51, 652–673.

Knoblauch, K., & Maloney, L. T. (1996). Testing the indeterminancy of linear color mechanisms from color discrimination data. *Vision Research*, 36, 295–306.

Koenderink, J. J. (1987). Color atlas theory. *Journal of the Optical Society of America A*, 4, 1314–1321.

Koenderink, J. J. (2010). *Color for the sciences*. Cambridge, MA, London, England: The MIT Press.

Koenderink, J. J., & van Doorn, A. J. (1987). Representation of local geometry in the visual system. *Biological Cybernetics*, 55, 367–375.

Koenderink, J. J., & van Doorn, A. J. (2003). Perspectives on colour space. In R. Mausfeld & D. Heyer (Eds.), *Colour perception: mind and the physical world* (pp. 1–56). Oxford, New York: Oxford University Press.

Koffka, K. (1935). *Principles of gestalt psychology*. New York: Harcourt Brace & Co.

Kohler, I. (1964). *The formation and transformation of the perceptual world*. New York: International Universities Press.

Kolmogorov, A. N., & Fomin, S. V. (1999). *Elements of the theory of functions and functional analysis*. Mineola, New York: Dover Publications, Inc.

König, A., & Dieterici, C. (1892). Die Grundempfindungen in normalen und anomalen Farbensystemen und ihre Intensitätsverteilung im Spektrum. *Zeitschrift für Psychologie*, 4, 241–347.

Kozaki, A. (1973). Perception of lightness and brightness of achromatic surface color and impression of illumination. *Japanese Psychological Research*, 15, 194–203.

Kozaki, A., & Noguchi, K. (1976). The relationship between perceived surface lightness and perceived illumination: a manifestation of perceptual scission. *Psychological Research*, 39, 1–16.

Kramer, R. H. (2014). Horizontal cells: lateral interactions at the first synapse in the retina. In J. S. Werner & L. M. Chalupa (Eds.), *The new visual neurosciences* (pp. 63–74). Cambridge, MA, London, England: MIT Press.

Kranda, K., & King-Smith, P. E. (1979). Detection of coloured stimuli by independent linear systems. *Vision Research*, 19, 733–746.

Krantz, D. H. (1971). Integration of just-noticeable differences. *Journal of Mathematical Psychology*, 8, 591–599.

Krantz, D. H. (1975a). Color measurement and color theory: I. Representation theorem for Grassmann structures. *Journal of Mathematical Psychology*, 12, 283–303.

Krantz, D. H. (1975b). Color measurement and color theory: II. Opponent-colors theory. *Journal of Mathematical Psychology*, 12, 304–327.

Krantz, D. H., Luce, R. D., Suppes, P., & Tversky, A. (Eds.). (1971). *Foundations of measurement: Vol. 1. Additive and polynomial representations*. New York and London: Academic Press.

Krauskopf, J. (1963). Effect of retinal image stabilization on the appearance of heterochromatic targets. *Journal of the Optical Society of America*, 53, 741–744.

Krauskopf, J. (1980). Discrimination and detection of changes in luminance. *Vision Research*, 20, 671–677.

Krauskopf, J., Williams, D. R., & Heeley, D. W. (1982). Cardinal directions of color space. *Vision Research*, 22, 1123–1131.

Kries, J. (1905). Die Gesichtsempfindungen. In W. Nagel (Ed.), *Handbuch der Physiologie des Menschen: Vol. 3* (pp. 109–282). Braunschweig: Vieweg.

Kuehni, R. G. (2003). *Color space and its divisions: color order from antiquity to the present*. Hoboken, NJ: John Wiley & Sons, Inc.

Kuehni, R. G., & Schwarz, A. (2008). *Color ordered: a survey of color order systems from antiquity to the present*. Oxford; New York: Oxford University Press, Inc.

Kulikowski, J. J., & King-Smith, P. E. (1973). Spatial arrangement of line, edge and grating detectors revealed by subthreshold summation. *Vision Research*, 13, 1455–1478.

Kulikowski, J. J., & Vaitkevicius, H. (1997). Colour constancy as a function of hue. *Acta Psychologica*, 97, 25–35.

Lamb, T. (1995). Photoreceptor spectral sensitivities: common shape in the long-wavelength region. *Vision Research*, 35, 3083–3091.

Lamb, T. (1999). Photopigments and the biophysics of transduction in cone photoreceptors. In K. R. Gegenfurtner & L. T. Sharpe (Eds.), *Color vision: from genes to perception* (pp. 89–101). Cambridge: Cambridge University Press.

Larimer, J., Krantz, D. H., & Cicerone, C. M. (1975a). Opponent process additivity - I. Red/green equilibria. *Vision Research*, 14, 1127–1140.

Larimer, J., Krantz, D. H., & Cicerone, C. M. (1975b). Opponent process additivity - II. Yellow/blue equilibria and nonlinear models. *Vision Research*, 15, 723–731.

Lennie, P., Pokorny, J., & Smith, V. C. (1993). Luminance. *Journal of the Optical Society of America A*, 10, 1283–1293.

Levin, V. L. (1983). Measurable utility theorems for closed and lexicographic preference relations. *Soviet Mathematics – Doklady*, 27, 639–643.

Levin, V. L. (1985). *Vipuklyi analiz v prostranstvah izmerimyh funkcii i ego primenenie v matematike i ekonomike [Convex analysis in spaces of measurable functions and its applications in mathematics and economics]*. Moscow: Nauka.

Levin, V. L. (1999). Dual representations of convex sets and Gâteaux differentiability spaces. *Set-Valued Analysis*, 7, 133–157.

Lindeberg, T. (2013). A computational theory of visual receptive fields. *Biological Cybernetics*, 107, 589–635.

Linden, D. E., Kallenbach, U., Heinecke, A., Singer, W., & Goebel, R. (1999). The myth of upright vision. A psychophysical and functional imaging study of adaptation to inverting spectacles. *Perception*, 28, 469–481.

Locke, J. (1975). An essay concerning human understanding P. H. Nidditch (Ed.). (The Clarendon Edition of the Works of John Locke). Oxford: Oxford University Press. (Original work published 1689).

Logvinenko, A. D. (1974). Visual image and inverted vision. *Voprosi Psychologii*, Issue 5, 19–28. (in Russian).

Logvinenko, A. D. (1981). *Zritel'noe vospriyatie prostranstva [Visual space perception]*. Moscow: Moscow University Press.

Logvinenko, A. D. (1985). *Chuvstvennaya osnova vospriyatiya prostranstva [Sensory basis of space perception]*. Moscow: Moscow University Press.

Logvinenko, A. D. (1990). Nonlinear analysis of spatial vision using first- and second-order Volterra transfer functions measurement. *Vision Research*, 30 (12), 2031–2057.

Logvinenko, A. D. (1993). Perceptual equations: implications for computer vision. *The Irish Journal of Psychology*, 14, 330–342.

Logvinenko, A. D. (1996a). Convexity of the threshold body implies a peak detector. *Spatial Vision*, 10, 105–123.

Logvinenko, A. D. (1996b). On cardinal directions in spatial pattern space and falsifying multi channel detection models. *Spatial Vision*, 10, 189–200.

Logvinenko, A. D. (1997). Invariant relationship between achromatic colour, apparent illumination, and shape of surface: implications for the colour perception theories. In C. M. Dickinson, I. J. Murray, & D. Carden (Eds.), *John Dalton's colour vision legacy* (pp. 591–597). London: Taylor & Francis.

Logvinenko, A. D. (1998). On derivation of spectral sensitivities of the human cones from trichromatic colour matching functions. *Vision Research*, 38, 3207–3211.

Logvinenko, A. D. (1999). Lightness induction revisited. *Perception*, 28, 803–816.

Logvinenko, A. D. (2001). High-spatial-frequency tritanopia: S-filling-in or S-filtering-out? *Perception*, 30, 223–232.

Logvinenko, A. D. (2003a). A fair test of the effect of a shadow-incompatible luminance gradient on the simultaneous lightness contrast. *Perception*, 32, 717–720.

Logvinenko, A. D. (2003b). Method of quadratic approximation: a new approach to identification of analysers and channels in human vision. *Journal of Mathematical Psychology*, 47, 495–506.

Logvinenko, A. D. (2005a). Does luminance contrast determine lightness? *Spatial Vision*, 18, 337–346.

Logvinenko, A. D. (2005b). On achromatic colour appearance. In J. L. Nieves & J. Hernndez-Andrs (Eds.), *AIC Colour 05. The 10th congress of the international colour association.* 8–13 May 2005, Granada, Spain, Part 1 (pp. 639–642). Granada, Spain: Comit Espaol del Color.

Logvinenko, A. D. (2006). Does colour matching follow Grassmann's laws? In *Symposium on 75 Years of the CIE Standard Colorimetric Observer* (pp. 133–138). Vienna, Austria: CIE Publication x030:2006, Commission Internationale de l'Eclairage.

Logvinenko, A. D. (2009a). A new colour rendering index. In D. Smith, P. Green-Armytage, M. A. Pope, & N. Harkness (Eds.), *Proceedings of the 11th congress of the international colour association (AIC 2009). CD*. Sydney: Colour Society of Australia.

Logvinenko, A. D. (2009b). An object-color space. *Journal of Vision*, 9 (11): 5, 1–23.

Logvinenko, A. D. (2009c). Pseudoscopic colour illusions. In D. Smith, P. Green-Armytage, M. A. Pope, & N. Harkness (Eds.), *Proceedings of the 11th congress of the international colour association (AIC 2009). CD*. Sydney: Colour Society of Australia.

Logvinenko, A. D. (2012). A theory of unique hues and colour categories in the human colour vision. *Color Research & Application*, 37, 109–116.

Logvinenko, A. D. (2013a). Individual differences in human colour vision as derived from Stiles & Burch 10° colour matching functions. *Color Research & Application*, 38, 96–109.

Logvinenko, A. D. (2013b). Object-colour manifold. *International Journal of Computer Vision*, 101, 143–160.

Logvinenko, A. D. (2014). On the colours dichromats see. *Color Research & Application*, 39, 112–124.

Logvinenko, A. D. (2015a). The achromatic object-colour manifold is three-dimensional. *Perception*, 44, 243–268.

Logvinenko, A. D. (2015b). The colour cone. *Journal of the Optical Society of America A*, 32, 314–322.

Logvinenko, A. D. (2015c). The geometric structure of color. *Journal of Vision*, 15 (1): 16, 1–9.

Logvinenko, A. D. (2021). On the dichromatic object colour palette. *Vision Research*, 178, 112–116.

Logvinenko, A. D., & Beattie, L. L. (2006). Partial colour matching: a new method to evaluate colour appearance and derive unique hues. *3rd European Conference on Colour in Graphics, Imaging, and Vision Proceedings*, 262–265.

Logvinenko, A. D., & Beattie, L. L. (2011). Partial hue-matching. *Journal of Vision*, 11 (8): 6, 1–16.

Logvinenko, A. D., & Beer, A. (2012). Color constancy investigated via partial hue-matching. *Journal of Vision*, 12 (4): 17, 1–20.

Logvinenko, A. D., & Belopolsky, V. I. (1994). Convergence as a cue for distance. *Perception*, 23, 207–217.

Logvinenko, A. D., & Geithner, C. (2015). Unique hues as revealed by unique-hue selecting versus partial hue-matching. *Attention, Perception & Psychophysics*, 77 (3), 883–894.

Logvinenko, A. D., & Hutchinson, S. J. (2005). One blue colour channel or two? *Perception*, 34, 921–925.

Logvinenko, A. D., & Hutchinson, S. J. (2006). An effect of luminance contrast on high-spatial-frequency tritanopia. *Perception*, 35, 171–184.

Logvinenko, A. D., & Hutchinson, S. J. (2007). Evidence for the existence of colour mechanisms producing unique hues as derived from a colour illusion based on spatio-chromatic interactions. *Vision Research*, 47, 1315–1334.

Logvinenko, A. D., & Kane, J. (2003). Luminance gradient can break background-independent lightness constancy. *Perception*, 32, 263–268.

Logvinenko, A. D., & Kane, J. (2004). Hering's and Helmholtz's types of simultaneous lightness contrast. *Journal of Vision*, 4 (12), 1102–1110.

Logvinenko, A. D., & Lu, T. (2008). Unique hue isochromes in the equi-illuminant plane. Paper presented at the *Annual Meeting of the Vision Science Society*, Naples, Florida, USA.

Logvinenko, A. D., & Maloney, L. T. (2006). The proximity structure of achromatic surface colors and the impossibility of asymmetric lightness matching. *Perception & Psychophysics*, 68, 76–83.

Logvinenko, A. D., & Menshikova, G. Y. (1994). Trade-off between achromatic colour and perceived illumination as revealed by the use of pseudoscopic inversion of apparent depth. *Perception*, 23, 1007–1023.

Logvinenko, A. D., & Tokunaga, R. (2011). Colour constancy as measured by least dissimilar matching. *Seeing and Perceiving*, 24, 407–452.

Logvinenko, A. D., & Zhedunova, L. G. (1980). Adaptation to inverted vision. *Voprosi Psychologii*, Issue 6, 97–108. (in Russian).

Logvinenko, A. D., & Zhedunova, L. G. (1981). Adaptation to inverted vision: no need to constantly wear inverting glasses. *Voprosi Psychologii*, Issue 6, 83–92. (in Russian).

Logvinenko, A. D., Byth, W., & Vityaev, E. E. (2004). In search of an elusive hard threshold: a test of observer's ability to order sub-threshold stimuli. *Vision Research*, 44, 287–296.

Logvinenko, A. D., Epelboim, J., & Steinman, R. M. (2001). The role of vergence in the perception of distance: a fair test of Bishop Berkeley's claim. *Spatial Vision*, 15, 77–97.

Logvinenko, A. D., Funt, B., & Godau, C. (2015). How metamer mismatching decreases as the number of colour mechanisms increases with implications for colour and lightness constancy. *Vision Research*, 113, 65–70.

Logvinenko, A. D., Funt, B. V., Mirzaei, H., & Tokunaga, R. (2015). Rethinking colour constancy. *PLoS ONE*, 10 (9), e0135029. doi: 10.1371/journal.pone.0135029. PMID: 26356217; PMCID: PMC4565710.

Logvinenko, A. D., Kane, J., & Ross, D. A. (2002). Is lightness induction a pictorial illusion? *Perception*, 31, 73–82.

Logvinenko, A. D., Petrini, K., & Maloney, L. T. (2008). A scaling analysis of the snake lightness illusion. *Perception & Psychophysics*, 70, 828–840.

Logvinenko, A. D., Tyurin, Y. N., & Sawey, M. (2002). Equi-discrimination contours around an achromatic colour are convex but not balanced. *Journal of Mathematical Psychology*, 46, 486–509.

Lomonosov, M. V. (1757). *Oratio de origine lucis*, Petropoli, Typis Academiae Scientiarum.

Luce, R. D., & Green, D. M. (1974). Detection, discrimination, and recognition. In C. E. Carterette & M. P. Friedman (Eds.), *Handbook of perception. vol. II. Psychophysical judgment and measurement* (pp. 299–342). New York, San Francisco, CA, London: Academic Press.

MacAdam, D. L. (1942). Visual sensitivities to color differences in daylight. *Journal of the Optical Society of America*, 32, 247–274.

MacAdam, D. L. (1955). Color measurement and whiteness. *TAPPI*, 38, 78–87.

Mach, E. (1959). *The analysis of sensations*. New York: Dover Publications, Inc.

MacLeod, R. B. (1940). Brightness-constancy in unrecognized shadows. *Journal of Experimental Psychology*, 27, 1–22.

MacLeod, D. I. A. (2003). New dimensions in color perception. *Trends in Cognitive Sciences*, 7, 97–99.

MacLeod, D. I. A. (2010). Into the neural maze. In J. Cohen & M. Matthen (Eds.), *Color ontology and color science* (pp. 151–178). Cambridge, MA: The MIT Press.

MacLeod, D. I. A., & Boynton, R. M. (1979). Chromaticity diagram showing cone excitation by stimuli of equal luminance. *Journal of the Optical Society of America*, 69, 1183–1186.

MacLeod, D. I. A., & Golz, J. (2003). A computational analysis of colour constancy. In R. Mausfeld & D. Heyer (Eds.), *Colour perception: mind and the physical world* (pp. 205–242). Oxford, New York: Oxford University Press.

Madigan, S. C., & Brainard, D. H. (2014). Scaling measurements of the effect of surface slant on perceived lightness. *i-Perception*, 5, 53–72.

Maloney, L. T. (1999). Physics-based approaches to modeling surface color perception. In K. R. Gegenfurtner & L. T. Sharpe (Eds.), *Color vision: from genes to perception* (pp. 387–416). Cambridge, UK: Cambridge University Press.

Maloney, L. T. (2003). Surface color perception and environmental constraints. In R. Mausfeld & D. Heyer (Eds.), *Colour perception: mind and the physical world* (pp. 279–300). Oxford, New York: Oxford University Press.

Maloney, L. T., & Schirillo, J. A. (2002). Color constancy, lightness constancy, and the articulation hypothesis. *Perception*, 31, 135–139.

Marmarelis, P. Z., & Marmarelis, V. Z. (1978). *Analysis of physiological systems: The white-noise approach*. New York and London: Plenum Press.

Marr, D. C. (1982). *Vision: a computational investigation into the human representation and processing of visual information*. San Francisco: W. H. Freeman and Company.

Marshall, J., & Oberwinkler, J. (1999). The colourful world of the mantis shrimp. *Nature*, 401, 873–874.

Masland, R. H. (2012). The neuronal organization of the retina. *Neuron*, 76, 266–280.

Mausfeld, R. (1998). Color perception: from Grassmann codes to a dual code for object and illumination colors. In W. G. K. Backhaus, R. Kliegl, & J. S. Werner (Eds.), *Color vision: perspectives from different disciplines* (pp. 219–250). Berlin; New York: Walter de Gruyter.

Mausfeld, R. (2003). 'Colour' as part of the format of different perceptual primitives: the dual coding of colour. In R. Mausfeld & D. Heyer (Eds.), *Colour perception: mind and the physical world* (pp. 381–430). Oxford, New York: Oxford University Press.

Maximov, V. V. (1984). *Transformatsii tsveta pri izmenenii osvescheniya [Transformations of colour under illumination changes]*. Moscow: Nauka.

Maxwell, J. C. (1855). Experiments on colour, as perceived by the eye, with remarks on colour-blindness. *Transactions of the Royal Society of Edinburgh*, 21, 275–298.

Maxwell, J. C. (1860). On the theory of the compound colours and the relations of the colours of the spectrum. *Philosophical Transactions of the Royal Society*, 150, 57–84.

Maxwell, J. C. (1970). On color vision. In D. L. MacAdam (Ed.), *Sources of color sciences* (pp. 75–83). Cambridge, MA and London, England: The MIT Press. (Reprinted from *Proceedings of the Royal Institution of Great Britan*, 1872, 6, 260–271).

McKeefry, D. J., Murray, I. J., & Kulikowski, J. J. (2001). Red-green and blue-yellow mechanisms are matched in sensitivity for temporal and spatial modulation. *Vision Research*, 41, 245–255.

Michotte, A. (1963). *The perception of causality* (T. R. Miles & E. Miles, Trans.). London: Methuen. (Original work published 1946).

Miller, D. T., & Roorda, A. (2010). Adaptive optics in retinal microscopy and vision. Chap. 15 in M. Bass (Ed.), *OSA Handbook of optics: Vol. III. Vision and vision optics* (3rd ed., pp. 15.1–15.30). New York e.a.: McGraw-Hill, Inc.

Mollon, J. D. (1982). A taxonomy of tritanopias. *Documenta Ophthalmologica Proceedings Series*, 33, 87–101.

Mollon, J. D. (2003). The origins of modern color science. In S. K. Shevell (Ed.), *The science of color* (2nd ed., pp. 1–39). Amsterdam e.a.: Elsevier.

Montag, E. D. (1994). Surface color naming in dichromats. *Vision Research*, 34, 2137–2151.

Montag, E. D., & Boynton, R. M. (1987). Rod influence in dichromatic surface color perception. *Vision Research*, 27, 2153–2162.

Mullen, K. T. (1985). The contrast sensitivity of human colour vision to red-green and blue-yellow chromatic gratings. *Journal of Physiology*, 359, 381–400.

Munsell, A. H. (1915). *Atlas of the Munsell color system*. Malden, MA: Wadsworth-Holland.

Neitz, J., & Neitz, M. (2011). The genetics of normal and defective color vision. *Vision Research*, 51, 633–651.

Neitz, M., Neitz, J., & Jacobs, G. H. (1993). More than three different cone pigments among people with normal color vision. *Vision Research*, 33, 117–122.

Neugebauer, H. E. J. (1956). Quality factor for filters whose spectral transmittances are different from color mixture curves, and its application to color photography. *Journal of the Optical Society of America*, 46, 821–824.

Newton, I. (1730). *Optiks, or a treatise of the reflections, refractions, inflections and colours of light*. London: William Innys.

Nicodemus, F. E., Richmond, J. C., Hsia, J. J., Ginsberg, I. W., & Limperis, T. (1977). *Geometrical considerations and nomenclature for reflectance*. Washington, D.C.: Institute for Basic Standards, National Bureau of Standards.

Nida-Rümelin, M., & Suarez, J. (2009). Reddish green: a challenge for modal claims about phenomenal structure. *Philosophy and Phenomenological Research*, 78, 346–391.

Nikolaev, P. P. (1985). A colour constancy model with continuous spectral functions. *Biophysics*, 30, 112–117.

Noguchi, K., & Kozaki, A. (1985). Perceptual scission of surface-lightness and illumination: an examination of the Gelb effect. *Psychological Research*, 47, 19–25.

Noorlander, C., & Koenderink, J. J. (1983). Spatial and temporal discrimination ellipsoids in color space. *Journal of the Optical Society of America*, 73, 1533–1543.

Ohta, N. (1975). Generating metameric object colors. *Journal of the Optical Society of America*, 65, 1081–1082.

Ohta, N., & Robertson, A. R. (2005). *Colorimetry: fundamentals and applications*. Chichester, England; Hoboken, NJ, USA: John Wiley & Sons, Ltd.

Ohta, N., & Wyszecki, G. (1975). Theoretical chromaticity-mismatch limits of metamers viewed under different illuminants. *Journal of the Optical Society of America*, 65, 327–333.

Ohta, N., & Wyszecki, G. (1976a). Color changes caused by specified changes in the illuminant. *Color Research & Application*, 1, 17–21.

Ohta, N., & Wyszecki, G. (1976b). Designing illuminants that render given objects in prescribed colors. *Journal of the Optical Society of America*, 66, 269–275.

Oyama, T. (1968). Stimulus determinants of brightness constancy and the perception of illumination. *Japanese Psychological Research*, 10, 146–155.

Packer, O., & Williams, D. R. (2003). Light, the retinal image, and photoreceptors. In S. K. Shevell (Ed.), *The science of color* (2nd ed., pp. 41–102). Amsterdam e.a.: Elsevier.

Palmer, G. (1777). *Theory of colours and vision*. London: S. Leacroft.

Patel, A. S., & Jones, R. W. (1968). Increment and decrement visual thresholds. *Journal of the Optical Society of America*, 58, 696–699.

Pessoa, L., Thompson, E., & Noë, A. (1998). Finding out about filling-in: a guide to perceptual completion for visual science and the philosophy of perception. *Behavioral and Brain Sciences*, 21, 723–748.

Peterson, J., & Peterson, J. K. (1938). Does practice with inverting lenses make vision normal? *Psychological Monograph*, 50, 12–37.

Phelps, R. R. (1966). *Lectures on Choquet's theorem*. Princeton, NJ: D. Van Nostrand Company, Inc.

Phelps, R. R. (1993). *Convex functions, monotone operators and differentiability*. Vol. 1364 of Lecture Notes Math., second edn, New York: Springer-Verlag.

Poirson, A. B., & Wandell, B. A. (1993). Appearance of colored patterns: pattern-color separability. *Journal of the Optical Society of America A*, 10, 2458–2470.

Poirson, A. B., & Wandell, B. A. (1996). Pattern-color separable pathways predict sensitivity to simple colored patterns. *Vision Research*, 36, 515–526.

Poirson, A. B., Wandell, B. A., Varner, D. C., & Brainard, D. H. (1990). Surface characterizations of color thresholds. *Journal of the Optical Society of America A*, 7, 783–789.

Pokorny, J., Lutze, M., Cao, D., & Zele, A. J. (2006). The color of night: surface color perception under dim illuminations. *Visual Neuroscience*, 23, 525–530.

Pokorny, J., Shevell, S. K., & Smith, V. C. (1991). Colour appearance and colour constancy. In P. Gouras (Ed.), *Vision & visual dysfunction: Vol. 6. The perception of colour* (pp. 43–61). London, UK: Macmillam.

Pridmore, R. W. (2007). Effects of luminance, wavelength and purity on the color attributes: brief review with new data and perspectives. *Color Research & Application*, 32, 208–222.

Purdy, D. M. (1931). Spectral hue as a function of intensity. *American Journal of Psychology*, 43, 541–559.

Purves, D., & Lotto, R. B. (2003). *Why we see what we do: an empirical theory of vision*. Sunderland, MA: Sinauer Associates, Inc.

Regan, B., Julliot, C., Simmen, B., Vienot, F., Charles-Dominique, P., & Mollon, J. (2001). Fruits, foliage and the evolution of primate colour vision. *Philosophical Transactions of the Royal Society B*, 356, 229–283.

Reid, T. (1969). *Essay on the intellectual powers of man*. Cambridge, MA: M.I.T Press. (Original work published 1785).

Reinhard, E., Khan, E. A., Akyuz, A. O., & Johnson, G. (2008). *Color imaging. Fundamentals and applications*. Wellesley, MA: A. K. Peters, Ltd.

Revonsuo, A., & Newman, J. (1999). Binding and consciousness. *Consciousness and Cognition*, 8, 123–127.

Rieke, F. (2008). Seeing in the dark: retinal processing and absolute visual threshold. In T. D. Albright & R. H. Masland (Eds.), *The senses: a comprehensive reference* (Vol. 1, pp. 393–412). San Diego, CA: Academic Press.

Riggs, L. A. (1965). Light as a stimulus for vision. In C. H. Graham (Ed.), *Vision and visual perception* (pp. 1–38). New York, London, Sydney: John Wiley & Sons, Inc.

Roberts, F. S. (1973). Tolerance geometry. *Notre Dame Journal of Formal Logic*, 14, 68–76.

Robson, J. G. (1980). Neural images: the physiological basis of spatial vision. In C. S. Harris (Ed.), *Visual coding and adaptability* (pp. 177–214). Hillsdale, New Jersey: Lawrence Erlbaum.

Rockafellar, R. (1970). *Convex analysis*. Princeton, New Jersey: Princeton University Press.

Roorda, A. (2011). Adaptive optics for studying visual function: a comprehensive review. *Journal of Vision*, 11 (5): 6, 1–21.

Rushton, W. A. H. (1972). Visual pigments in man. In H. J. A. Dartnall (Ed.), *Handbook of sensory physiology: Vol. VII/1. Photochemistry of vision* (pp. 364–394). Berlin e.a.: Springer-Verlag.

Rutherford, M., & Brainard, D. (2002). Lightness constancy: a direct test of the illumination-estimation hypothesis. *Psychological Science*, 13, 142–149.

Sakitt, B. (1975). Locus of short-term visual storage. *Science*, 190, 1318–1319.

Sankeralli, M. J., & Mullen, K. T. (1996). Estimation of the L-, M-, and S-cone weights of the postreceptoral detection mechanisms. *Journal of the Optical Society of America A*, 13, 906–915.

de Saussure, F. (1983). *Course in general linguistics* (C. Bally & A. Sechehaye, Eds. & R. Harris, Trans.). La Salle, Illinois: Open Court (Original work published 1916).

Schanda, J. (Ed.). (2007a). *Colorimetry: Understanding the CIE System*. New York: John Wiley & Sons, Inc.

Schanda, J. (2007b). Color rendering of light sources. In J. Schanda (Ed.), *Colorimetry*: Understanding the CIE System (pp. 207–217). New York: John Wiley & Sons, Inc.

Scheibner, H. M. O., & Boynton, R. M. (1968). Residual red-green discrimination in dichromats. *Journal of the Optical Society of America*, 58, 1151–1158.

Schmitt, F. J. M. (1976). A method for the treatment of metamerism in colorimetry. *Journal of the Optical Society of America*, 66, 601–608.

Schnapf, J. L., & Baylor, D. A. (1987). How photoreceptor cells respond to light. *Scientific American*, 256, 40–47.

Schnapf, J. L., & Schneeweis, D. M. (1999). Electrophysiology of cone photoreceptors in the primate retina. In K. R. Gegenfurtner & L. T. Sharpe (Eds.), *Color vision: from genes to perception* (pp. 103–112). Cambridge: Cambridge University Press.

Schreider, J. A. (1975). *Equality, resemblance and order*. Moscow: Mir.

Schrödinger, E. (1920). Theorie der Pigmente von größter Leuchtkraft. *Annalen der Physik*, 62, 603–622. (English translation in Erwin Schrödinger's Color Theory: Translated with Modern Commentary. Keith K. Niall, Editor (pp. 115–131). NY: Springer, 2017).

Schrödinger, E. (1970). Outline of a theory of color measurement for daylight vision. In D. L. MacAdam (Ed.), *Sources of color sciences* (pp. 134–182). Cambridge, MA, and London, England: The MIT Press. (Reprinted from *Annalen der Physik*, 1920, 63, 397–447, 481–520)

Sekiguchi, N., Williams, D. R., & Brainard, D. H. (1993). Efficiency in detection of isoluminant and isochromatic interference fringes. *Journal of the Optical Society of America A*, 10, 2118–2133.

Shafer, S. A. (1985). Using color to separate reflection components. *Color Research & Application*, 10, 210–218.

Shamey, R., Sedito, M. G., & Kuehni, R. G. (2010). Comparison of unique hue stimuli determined by two different methods using Munsell color chips. *Color Research & Application*, 35, 419–424.

Shapley, R. (2019). Physiology of color vision in primates. In *Oxford research encyclopedia of neuroscience*. Oxford University Press. (DOI:10.1093/ACREFORE/9780190264086.013.74).

Shapley, R., & Enroth-Cugell, C. (1984). Visual adaptation and retinal gain controls. *Progress in Retinal Research*, 3, 263–346.

Shapley, R., & Hawken, M. J. (2002). Neural mechanisms for color perception in the primary visual cortex. *Current Opinion in Neurobiology*, 12, 426–432.

Shapley, R., & Hawken, M. J. (2011). Color in the cortex: single- and double-opponent cells. *Vision Research*, 51, 701–717.

Shapley, R., Hawken, M. J., & Johnson, E. (2014). Color in the primary visual cortex. In J. S. Werner & L. M. Chalupa (Eds.), *The new visual neurosciences* (pp. 569–586). Cambridge, MA, London, England: MIT Press.

Sharpe, L. T., Stockman, A., Jagla, W., & Jägle, H. (2005). A luminous efficiency function, $V^*(\lambda)$, for daylight adaptation. *Journal of Vision*, 5 (11), 948–968.

Shepard, R. (1992). The perceptual organization of colors: an adaptation to regularities of the terrestrial world? In J. H. Barkow, L. Cosmides, & J. Tooby (Eds.), *The adapted mind: evolutionary psychology and the generation of culture* (pp. 495–531). New York: Oxford University Press.

Shepard, T. G., Swanson, E. A., McCarthy, C. L., & Eskew, R. T., Jr. (2016). A model of selective masking in chromatic detection. *Journal of Vision*, 16 (9): 3, 1–17.

Shepherd, A. J. (1997). A vector model of colour contrast in a cone-excitation colour space. *Perception*, 26, 455–470.

Shevell, S. K., & Kingdom, F. A. A. (2008). Color in complex scenes. *Annual Review of Psychology*, 59, 143–166.

Shoemaker, S. (1982). The inverted spectrum. *The Journal of Philosophy*, 79, 357–381.

Short, A. D. (1966). Decremental and incremental visual thresholds. *Journal of Physiology, London*, 185, 646–654.

Shreider, Y. A. (1970). Tolerance spaces. *Kibernetika*, 6 (2), 124–128.

Sloan, L. L., & Wollach, L. (1948). A case of unilateral deuteranopia. *Journal of the Optical Society of America*, 38, 502–509.

Smith, V. C., & Pokorny, J. (1975). Spectral sensitivity of the foveal cone photopigments between 400 and 500 nm. *Vision Research*, 15, 161–171.

Smith, V. C., & Pokorny, J. (1977). Large-field trichromacy in protanopes and deuteranopes. *Journal of the Optical Society of America*, 67, 213–220.

Smith, V. C., & Pokorny, J. (1996). The design and use of a cone chromaticity space: a tutorial. *Color Research & Application*, 21, 375–383.

Smith, V. C., & Pokorny, J. (2003). Color matching and color discrimination. In S. K. Shevell (Ed.), *The science of color* (2nd ed., pp. 103–148). Amsterdam e.a.: Elsevier.

Snyder, F. W., & Pronko, N. H. (1952). *Vision with spatial inversion*. Wichita, KS: University of Wichita Press.

Speranskaya, N. I. (1959). Determination of spectrum color co-ordinates for twenty-seven normal observers. *Optics and Spectroscopy*, 7, 424–428.

Stevens, S. S. (1957). On the psychophysical law. *Psychological Review*, 64, 153–181.

Stevens, S. S. (1961). To honor Fechner and repeal his law. *Science*, 133, 80–86.

Stiles, W. S., & Burch, J. M. (1955). Interim report to the Commission Internationale de l'Éclairage, Zurich, 1955, on the National Physical Laboratory's investigation of colour-matching (1955). *Optica Acta*, 2, 168–181.

Stiles, W. S., & Burch, J. M. (1959). N.P.L. colour-matching investigation: final report (1958). *Optica Acta*, 6, 1–26.

Stockman, A. (2019). Cone fundamentals and CIE standards. *Current Opinion in Behavioral Sciences*, 30, 87–93.

Stockman, A., & Brainard, D. H. (2010). Color vision mechanisms. Chap. 11 in M. Bass (Ed.), *OSA Handbook of optics: Vol. III. Vision and vision optics* (3rd ed., pp. 11.1–11.104). New York e.a.: McGraw-Hill, Inc.

Stockman, A., & Brainard, D. H. (2015). Fundamentals of color vision I: Color processing in the eye. In A. J. Elliot, M. D. Fairchild, & A. Franklin (Eds.), *Handbook of color psychology* (pp. 27–69). Cambridge: Cambridge University Press.

Stockman, A., & Sharpe, L. T. (1999). Cone spectral sensitivities and color matching. In K. R. Gegenfurtner & L. T. Sharpe (Eds.), *Color vision: from genes to perception* (pp. 53–87). Cambridge: Cambridge University Press.

Stockman, A., & Sharpe, L. T. (2000). The spectral sensitivities of the middle- and long-wavelength-sensitive cones derived from measurements in observers of known genotype. *Vision Research*, 40, 1711–1737.

Stratton, G. M. (1896). Some preliminary experiments on vision without inversion of the retinal image. *Psychological Review*, 3, 611–617.

Suppes, P., Krantz, D. H., Luce, R. D., & Tversky, A. (Eds.). (1989). *Foundations of measurement: Vol. 2. Geometrical, threshold, and probabilistic representations*. San Diego e. a.: Academic Press.

Takahama, K., & Nayatani, Y. (1972). New method for generating metameric stimuli of object colors. *Journal of the Optical Society of America*, 62, 1516–1520.

Teller, D. Y. (1984). Linking propositions. *Vision Research*, 24, 1233–1246.

Thompson, B. (1876). *The complete works of Count Rumford. Published by The American academy of arts and sciences* (Vol. 5). London: Macmillan and Co.

Tikhonov, A. N., & Arsenin, V. Ya. (1977). *Solutions of ill-posed problems*. New York: Winston.

Titchener, E. B. (1902). *A textbook of psychology*. New York: Macmillan.

Todorović, D. (2020). What are visual illusions? *Perception*, 49, 1128–1199.

Tokunaga, R., & Logvinenko, A. D. (2010a). Hue manifold. *Journal of the Optical Society of America A*, 27, 2551–2557.

Tokunaga, R., & Logvinenko, A. D. (2010b). Material and lighting dimensions of object colour. *Vision Research*, 50, 1740–1747.

Tokunaga, R., & Logvinenko, A. D. (2010c). Material and lighting hues of object colour. *Ophthalmic and Physiological Optics*, 30, 611–617.

Tootell, R. B. H., Nelissen, K., Vanduffel,W., & Orban, G. A. (2004). Search for color 'center(s)' in macaque visual cortex. *Cerebral Cortex*, 14, 353–363.

Treutwein, B. (1995). Adaptive psychophysical procedures. *Vision Research*, 35, 2503–2522.

Trezona, P. W. (1973). The tetrahcromatic colour match as a colorimetric technique. *Vision Research*, 13, 9–25.

Trezona, P. W. (1976). Aspects of peripheral colour vision. In E. B. Streiff (Ed.), *Modern problems in ophthalmology*, vol. 27 (pp. 52–70). Basel: S. Karger.

van Trigt, C. (1994). Metameric blacks and estimating reflectance. *Journal of the Optical Society of America A*, 11, 1003–1024.

van Trigt, C. (2007). Linear models in color constancy theory. *Journal of the Optical Society of America A*, 24, 2684–2691.

Tsukamoto, Y., Masarachia, P., Schein, S. J., & Sterling, P. (1992). Gap junctions between the pedicles of macaque foveal cones. *Vision Research*, 32, 1809–1815.

Valberg, A. (2001). Unique hues: an old problem for a new generation. *Vision Research*, 41, 1645–1657.

Valberg, A. (2005). *Light, vision, color*. Chichester, England: John Wiley & Sons, Ltd.

Valberg, A., Lange-Malecki, B., & Seim, T. (1991). Colour changes as a function of luminance contrast. *Perception*, 20, 655–668.

Valeton, J. M., & van Norren, D. (1983). Light adaptation of primate cones: an analysis based on extracellular data. *Vision Research*, 23, 1539–1547.

Viénot, F., Brettel, H., Ott, L., M'Barek, A. B., & Mollon, J. D. (1995). What do colour-blind people see? *Nature*, 376, 127–128.

Volbrecht, V. J., & Kliegl, R. (1998). The perception of blackness: an historical and contemporary review. In W. G. K. Backhaus, R. Kliegl, & J. S. Werner (Eds.), *Color vision: perspectives from different disciplines* (pp. 187–206). Berlin; New York: Walter de Gruyter.

Vora, P. L., & Trussell, H. J. (1993). Measure of goodness of a set of color scanning filters. *Journal of the Optical Society of America A*, 10, 1499–1508.

Vos, J. J. (1978). Colorimetric and photometric properties of a 2^0 fundamental observer. *Color Research & Application*, 3, 125–128.

Wagner, G., & Boynton, R. M. (1972). Comparison of four methods of heterochromatic photometry. *Journal of the Optical Society of America*, 62, 1508–1515.

Walraven, J. (1976). Discounting the background – the missing link in the explanation of chromatic induction. *Vision Research*, 16, 289–295.

Wandell, B. A. (1995). *Foundations of vision*. Massachusetts: Sinauer Associates, Inc.

Watson, A. B. (1986). Temporal sensitivity. In K. R. Boff, L. Kaufman, & J. P. Thomas (Eds.), *Handbook of perception and human perfomance: Vol. 1. Sensory processes and perception. Chapter 6* (pp. 6.1–6.43). New York: John Wiley & Sons, Inc.

Watson, J. B. (1919). *Psychology from the standpoint of a behaviorist*. Philadelphia, PA: J. B. Lippincott.

Webster, M. A. (2015). Individual differences in color vision. In A. J. Elliot, M. D. Fairchild, & A. Franklin (Eds.), *Handbook of color psychology* (pp. 197–215). Cambridge: Cambridge University Press.

Webster, M. A., & MacLeod, D. I. A. (1988). Factors underlying individual differences in the color matches of normal observers. *Journal of the Optical Society of America A*, 5, 1722–1735.

Werner, J. S., & Wooten, B. R. (1979). Opponent chromatic mechanisms: relation to photopigments and hue naming. *Journal of the Optical Society of America*, 69, 422–434.

Whittle, P. (1994). The psychophysics of contrast brightness. In A. Gilchrist (Ed.), *Lightness, brightness, and transparency* (pp. 35–110). Hillsdale, New Jersey: Lawrence Erlbaum Associates, Inc.

Whittle, P. (2003). Contrast colours. In R. Mausfeld & D. Heyer (Eds.), *Colour perception: mind and the physical world* (pp. 115–138). Oxford, New York: Oxford University Press.

Wiesel, T. N., & Hubel, D. H. (1966). Spatial and chromatic interactions in the lateral geniculate body of the rhesus monkey. *Journal of Neurophysiology*, 29, 1115–1156.

Williams, D. R., & Roorda, A. (1999). The trichromatic cone mosaic in the human eye. In K. R. Gegenfurtner & L. T. Sharpe (Eds.), *Color vision: from genes to perception* (pp. 113–122). Cambridge: Cambridge University Press.

Williams, D. R., MacLeod, D. I. A., & Hayhoe, M. M. (1981). Foveal tritanopia. *Vision Research*, 21, 1341–1356.

Willmer, E. N. (1944). Colour of small objects. *Nature*, 153, 774–775.

Wilson, H. R., & Wilkinson, F. (1997). Evolving concepts of spatial channels in vision: from independence to nonlinear interactions. *Perception*, 26, 939–960.
Wittgenstein, L. (1977). *Remarks on colour* (G. E. M. Anscombe, Ed. & L. L. McAlister & M. Schättle, Trans.). Oxford: Basil Blackwell.
Woodworth, R. S. (1938). *Experimental psychology*. New York: H. Holt and Company.
Worthey, J. A. (1988). Calculation of metameric reflectances. *Color Research & Application*, 13, 76–84.
Wright, W. D. (1928–29). A re-determination of the trichromatic coefficients of the spectral colours. *Transactions of the Optical Society*, 30, 141–164.
Wright, W. D. (1946). *Researches on normal and defective colour vision*. London, UK: Henry Kimpton.
Wright, W. D. (1996). The origins of the CIE system. In P. K. Kaiser & R. M. Boynton (Eds.), *Human color vision* (pp. 534–543). Washington, DC: Optical Society of America.
Wu, S., Amari, S.-I., & Nakahara, H. (2002). Population coding and decoding in a neural field: a computational study. *Neural Computation*, 14, 999–1026.
Wuerger, S., & Xiao, K. (2014). Color vision, opponent theory. In M. R. Luo (Ed.), *Encyclopedia of color science and technology* (pp. 1–6). Berlin, Heidelberg: Springer-Verlag.
Wyszecki, G. (1986). Color appearance. In K. R. Boff, L. Kaufman, & J. P. Thomas (Eds.), *Handbook of perception and human performance: Vol. 1. Sensory processes and perception*. Chapter 9 (pp. 9.1–9.57). New York: John Wiley & Sons, Inc.
Wyszecki, G., & Stiles, W. S. (1982). *Color science: concepts and methods, quantitative data and formulae* (2nd ed.). New York: John Wiley & Sons, Inc.
Yarbus, A. L. (1967). *Eye movements and vision* (B. Haigh, Trans.). New York: Plenum Press. (Original work published 1965).
Young, T. (1802). The Bakerian Lecture. On the theory of light and colours. *Philosophical Transactions of the Royal Society of London*, 92, 12–48.
Young, R. A. (1987). The Gaussian derivative model for spatial vision: I. Retinal mechanisms. *Spatial Vision*, 2, 273–293.
Zaidi, Q. (1998). Identification of illuminant and object colors: heuristic-based algorithms. *Journal of the Optical Society of America A*, 15, 1767–1776.
Zeeman, E. C. (1962). The topology of the brain and visual perception. In M. K. Fort (Ed.), *The topology of 3-manifolds* (pp. 240–256). Englewood Cliffs, New Jersey: Prentice-Hall.
Zeki, S. M. (1973). Colour coding in rhesus monkey prestriate cortex. *Brain Research*, 53, 422–427.
Zhang, R., Tsai, P.-S., Cryer, J. E., & Shah, M. (1999). Shape from shading: a survey. *IEEE Transactions on Pattern Analysis and Machine Intelligence*, 21, 690–706.

Appendix A

Some Auxiliary Facts from Functional Analysis

A.1 Banach spaces of measures and functions, and stimulus spaces

The mathematical models of lights, stimuli, and colour mechanisms used in this book are elements of some Banach spaces of (finite) measures and functions. In this appendix, we remind some concepts of functional analysis related to these Banach spaces. More detailed information with complete proofs may be found in Dunford & Schwartz (1958) and Holmes (1975).

A positive Borel measure σ on a compact set Λ in $\mathbf{R}^1$ (or on an abstract metrisable compact space Λ) is a countably additive non-negative real-valued function on the Borel σ-algebra in Λ. Here the countable additivity means

$$\sigma\left(\bigcup_{i=1}^{\infty} B_i\right) = \sum_{i=1}^{\infty} \sigma(B_i),$$

where B_i are any disjoint Borel sets in Λ. The support of such σ, $spt(\sigma)$, is defined to be the minimum closed subset of full measure in Λ.

The Dirac measure (delta function) at λ, δ_λ, is defined by

$$\delta_\lambda(B) = \begin{cases} 1, \text{ if } \lambda \in B, \\ 0, \text{ if } \lambda \notin B. \end{cases}$$

A general (signed) Borel measure is a real-valued set function that can be represented as a difference of two positive Borel measures. For every Borel measure σ, there exists such a representation $\sigma = \sigma_+ - \sigma_-$, which is minimal in the sense that if $\sigma = \sigma_1 - \sigma_2$, where σ_1, σ_2 are positive Borel measures, then $\sigma_1 \geq \sigma_+$, $\sigma_2 \geq \sigma_-$. The representation $\sigma = \sigma_+ - \sigma_-$ is called Jordan's decomposition of σ. Moreover, the space of Borel measures on Λ is a vector lattice, and $\sigma_+ = \sigma \vee 0$, $\sigma_- = (-\sigma) \vee 0$, $|\sigma| = \sigma \vee (-\sigma) = \sigma_+ + \sigma_-$.

Given a visible segment of spectrum $\Lambda = [\lambda_{\min}, \lambda_{\max}]$, we consider two variants of a stimulus space, (1) and (2), as follows.

(1) The Banach space (in fact, the Banach lattice) of Borel measures σ on Λ with the standard norm (a total variation)

$$\|\sigma\| = |\sigma|(\Lambda), \tag{A.1}$$

where $|\sigma| = \sigma_+ + \sigma_-$, the positive measures σ_+ and σ_- are elements of Jordan's decomposition of σ.

By the Riecz representation theorem (Holmes, 1975), the space of Borel measures on Λ is exactly the dual Banach space $\mathbf{C}^* = \mathbf{C}^*(\Lambda)$, where $\mathbf{C} = \mathbf{C}(\Lambda)$ is the Banach space of continuous real-valued

Foundations of Colour Science: From Colorimetry to Perception, First Edition. Alexander D. Logvinenko and Vladimir L. Levin.
© 2023 John Wiley & Sons Ltd. Published 2023 by John Wiley & Sons Ltd.

functions on Λ equipped with the uniform norm

$$\|\varphi\|_{\mathbf{C}} := \max_{\lambda\in\Lambda} |\varphi(\lambda)|, \quad \varphi \in \mathbf{C}. \tag{A.2}$$

Therefore, $\|\sigma\| = \sup\left\{\langle\varphi,\sigma\rangle : \varphi \in \mathbf{C}, \|\varphi\|_{\mathbf{C}} \le 1\right\}$, where

$$\langle\varphi,\sigma\rangle := \int_\Lambda \varphi(\lambda)\,\sigma(d\lambda), \quad \varphi \in \mathbf{C},\ \sigma \in \mathbf{C}^*. \tag{A.3}$$

Furthermore, since Λ is a segment of the real line, $\int_\Lambda \varphi(\lambda)\,\sigma(d\lambda)$ proves to be the Stilties integral $\int_\Lambda \varphi(\lambda)\,dg_\sigma(\lambda)$, where g_σ is a function of bounded variation given by $g_\sigma(\lambda) = \sigma\left(\left[\lambda_{\min}, \lambda\right]\right)$. In particular, for the Dirac measure $\sigma = \delta_{\lambda_0}$ the corresponding g_σ is given by

$$g_\sigma(\lambda) = \begin{cases} 0, \text{if } \lambda < \lambda_0, \\ 1, \text{if } \lambda \ge \lambda_0, \end{cases}$$

and we get

$$\int_\Lambda \varphi(\lambda)\,\delta_{\lambda_0}(d\lambda) = \int_\Lambda \varphi(\lambda)\,dg_\sigma(\lambda) = \varphi\left(\lambda_0\right).$$

(2) The Banach space $\mathbf{L}^1 = \mathbf{L}^1(\Lambda)$ of Lebesgue integrable real-valued functions on Λ with the norm

$$\|y\|_1 := \int_\Lambda |y(\lambda)|\,d\lambda, \quad y \in \mathbf{L}^1. \tag{A.4}$$

(To be more precise, elements of $\mathbf{L}^1$ are functions defined up to their values on Lebesgue negligible sets, that is, elements of $\mathbf{L}^1$ are the Lebesgue equivalence classes of Lebesgue integrable functions rather than the functions themselves.)

Each $y \in \mathbf{L}^1$ may be identified with a Lebesgue absolutely continuous Borel measure on Λ, $\sigma(d\lambda) = y(\lambda)d\lambda$, and for such σ one has $|\sigma|(d\lambda) = |y(\lambda)|d\lambda$ and $\|\sigma\| = |\sigma|(\Lambda) = \|y\|_1$. Thus, the map $\mathbf{L}^1 \ni y \to y(\lambda)d\lambda \in \mathbf{C}^*$ is an isometric linear embedding of $\mathbf{L}^1$ into $\mathbf{C}^*$.

The dual Banach space $\mathbf{L}^{1*}$ is the space $\mathbf{L}^\infty = \mathbf{L}^\infty(\Lambda)$ of (the Lebesgue equivalence classes of) essentially bounded Lebesgue measurable real-valued functions on Λ with the norm

$$\|\varphi\|_\infty = \operatorname*{ess\,sup}_{\lambda\in\Lambda} |\varphi(\lambda)| := \inf\{\alpha > 0 : |\varphi(\lambda)| \le \alpha \ \text{ a.e.}\}, \quad \varphi \in \mathbf{L}^\infty, \tag{A.5}$$

so that $\varphi y \in \mathbf{L}^1$ and

$$\langle y,\varphi\rangle = \int_\Lambda y(\lambda)\,\varphi(\lambda)d\lambda \le \|y\|_1\|\varphi\|_\infty,$$

whenever $\varphi \in \mathbf{L}^\infty, y \in \mathbf{L}^1$. Thus, elements of $\mathbf{L}^1$ are weak* continuous linear functionals on $\mathbf{L}^\infty$. The space $\mathbf{L}^1$ is not reflexive, that is, $\mathbf{L}^1$ is a proper subspace in $\mathbf{L}^{\infty*} = \left(\mathbf{L}^{1*}\right)^*$, and on $\mathbf{L}^\infty$ there exist continuous linear functionals different from elements of $\mathbf{L}^1$. Notice that $\mathbf{C}(\Lambda)$ is a closed vector subspace in $\mathbf{L}^\infty(\Lambda)$, and for $\varphi \in \mathbf{C}(\Lambda)$ two norms as given by (A.2) and (A.5) are equal: $\|\varphi\|_{\mathbf{C}} = \|\varphi\|_\infty$.

In addition to $\mathbf{L}^1$ and $\mathbf{L}^\infty$, we make use of the Hilbert space $\mathbf{L}^2$, whose elements are (Lebesgue equivalent classes of) Lebesgue measurable real-valued functions on Λ with finite norm

$$\|\varphi\|_2 := \left(\int_\Lambda |\varphi^2(\lambda)|d\lambda\right)^{\frac{1}{2}}, \quad \varphi \in \mathbf{L}^2.$$

The natural embeddings hold true: $\mathbf{L}^\infty \subset \mathbf{L}^2 \subset \mathbf{L}^1$, and for any $\varphi \in \mathbf{L}^\infty$ one has $\|\varphi\|_1 \le \sqrt{mes(\Lambda)}\|\varphi\|_2 \le mes(\Lambda)\|\varphi\|_\infty$. For any two functions $\varphi, \psi \in \mathbf{L}^2$ their product $\varphi\psi$ belongs to $\mathbf{L}^1$ and satisfies the Cauchy–Schwartz inequality $\|\varphi\psi\|_1 \le \|\varphi\|_2\|\psi\|_2$, hence, any $\psi \in \mathbf{L}^2$ may be considered as a continuous linear functional on $\mathbf{L}^2$:

$$\langle \varphi, \psi \rangle = \int_\Lambda \varphi(\lambda)\psi(\lambda)d\lambda, \quad \varphi \in \mathbf{L}^2.$$

Moreover, every continuous linear functional on $\mathbf{L}^2$ has this form, that is, the space $\mathbf{L}^2$ coincides with its dual: $\mathbf{L}^{2*} = \mathbf{L}^2$.

In dual Banach spaces such as $\mathbf{C}^*$ and $\mathbf{L}^\infty = \mathbf{L}^{1*}$, an important role is played by the weak* *topology*. Given a dual Banach space E^*, the sets

$$V_{e_1,\dots,e_n} := \{e^* \in E^* : \langle e_k, e^* \rangle < 1, \; k = 1, \dots, n\},$$

where $\{e_1, \dots, e_n\}$ runs over various finite subsets in E, constitute a base of weak* neighbourhoods of the origin. A subset G in E^* is called weak* *open* if, for every $e^* \in G$, $G - e^*$ contains one of neighbourhoods $V_{e_1,\dots,e_n}$. The weak* topology is metrisable if and only if E is finite-dimensional, in which case it coincides with the norm topology. Every weak* closed set is norm closed, but the converse is not true when E is infinite-dimensional. A net (generalized sequence) e^*_ν in E^* converges weakly* to a point $e^* \in E^*$ if and only if the net of real numbers $\langle e, e^*_\nu - e^* \rangle$ converges to 0 whenever $e \in E$. A linear functional on E^*, $l(e^*)$, is continuous in weak* topology (weak* *continuous*) if and only if it is of the form $l(e^*) = \langle e, e^* \rangle$ where e is an element of E.

In Part I, lights are represented by elements of the convex cone $\mathbf{C}^*_+ = \mathbf{C}^*_+(\Lambda)$ in $\mathbf{C}^* = \mathbf{C}^*(\Lambda)$, which consists of positive measures, while in Part II by elements of the convex cone $\mathbf{L}^1_+ = \mathbf{L}^1_+(\Lambda)$ in $\mathbf{L}^1 = \mathbf{L}^1(\Lambda)$ consisting of non-negative functions. Reflecting objects (in Part II) are represented by elements of the cone $\mathbf{L}^\infty_+ = \mathbf{L}^\infty_+(\Lambda)$ of non-negative functions in $\mathbf{L}^\infty = \mathbf{L}^\infty(\Lambda)$.

A.2 Convex analysis

Here we remind briefly some known facts from convex analysis that are used throughout this book. For details see Rockafellar (1970) and Phelps (1993).[1]

Let E be a Hausdorff locally convex space and E^* its dual. Given a subset M in E, its *polar* in E^* is defined as the set

$$M^0 := \{e^* \in E^* : \langle e, e^* \rangle \le 1 \; \forall e \in M\}. \tag{A.6}$$

Similarly, the polar in E of a subset $M' \subset E^*$ is defined as follows:

$$M'^0 := \{e \in E : \langle e, e^* \rangle \le 1 \; \forall e^* \in M'\}.$$

If M is a linear subspace in E, then its polar becomes the annihilator subspace: $M^0 = \{e^* \in E^* : \langle e, e^* \rangle = 0 \, \forall e \in M\}$, and if M is a convex cone, then its polar M^0 becomes the dual (polar) cone $\{e^* \in E^* : \langle e, e^* \rangle \le 0 \, \forall e \in M\}$. Similarly, the polar in E of a linear subspace $M' \subset E^*$ is its annihilator, and if M' is a convex cone, then its polar is the dual cone in E: $M'^0 = \{e \in E : \langle e, e^* \rangle \le 0 \; \forall e^* \in M'\}$. For every nonempty subset, M in E its *bipolar* M^{00} is defined to be the polar of its polar: $M^{00} = (M^0)^0 = \{e \in E : \langle e, e^* \rangle \le 1 \; \forall e^* \in M^0\}$. The *bipolar theorem* says that the bipolar of a nonempty

1 Ioffe & Tikhomirov (1974/1979) and Holmes (1975) might also be helpful.

subset $M \subset E$ is exactly the closed convex hull of the set $M \cup \{0\}$ (see e.g., Holmes, 1975, p. 68); therefore, the equivalence holds true as follows:

$$[M \text{ is closed and convex, and } 0 \in M] \Leftrightarrow M = M^{00}. \tag{A.7}$$

A subset B in E is said to be *absorbing* if for every $e \in E$ there is $\alpha > 0$ such that $\beta e \in B$ whenever $0 \le \beta < \alpha$. A subset B in E is called a *body* if it is closed, convex, and absorbing.[2] Given a body B, its *gauge* or *Minkowski function* is the following function $\mu(\cdot; B)$ on E:

$$\mu(e; B) := \inf \left\{ \alpha > 0 : \frac{e}{\alpha} \in B \right\}. \tag{A.8}$$

This function is non-negative, lower semi-continuous (lsc), and *sublinear* that is, *semi-additive* $(\mu(e^1 + e^2; B) \le \mu(e^1; B) + \mu(e^2; B)\ \ \forall e^1, e^2 \in E)$ and *positively homogeneous* $(\mu(\alpha e; B) = \alpha\mu(e; B)$ $\forall \alpha \ge 0, e \in E)$[3], and B is its sublevel set,

$$B = \{e \in E : \mu(e; B) \le 1\}. \tag{A.9}$$

Moreover, for any lsc sublinear function $\mu : E \to \mathbf{R}_+$ the set $B_\mu = \{e \in E : \mu(e) \le 1\}$ is a body whose gauge is μ, $\mu = \mu(\cdot; B_\mu)$. Therefore, every non-negative lsc sublinear function is referred to as a *gauge*.

The simplest examples of a body and its gauge in a Banach space are the unite ball and the norm respectively.

Note that the gauge of a body B is continuous if and only if 0 is an interior point of B. If E is a Banach (or any barreled) space, then every body B in it is a neighbourhood of 0, hence every gauge on a Banach space is continuous. In general case, a gauge need not be continuous. A simplest counter-example: the dual Banach space E^* endowed with the weak* topology and the standard dual norm in it.[4] Clearly, the dual norm is weak* lsc but is not weak* continuous. It is the gauge of the unite ball in E^*, which is not a weak* neighbourhood of the origin.

Given a subset M' in E^*, the function $s(\cdot; M')$ on E,

$$s(e; M') := \sup_{e^* \in M'} \langle e, e^* \rangle, \tag{A.10}$$

is called the *support* function of M'. It follows from (A.7)–(A.10) that for every body B the equality holds

$$\mu(e; B) = s(e; B^0), \quad e \in E. \tag{A.11}$$

Let E and E_1 be Banach spaces, B_1 a body in E_1, and $\Phi : E \to E_1$ a continuous linear operator. Then the polar B_1^0 is weak* compact, and the following equality holds:

$$(\Phi^{-1}(B_1))^0 = \Phi^*(B_1^0). \tag{A.12}$$

Here, inclusion $\Phi^*(B_1^0) \subseteq (\Phi^{-1}(B_1))^0$ is obvious, while the converse inclusion is established via the separation theorem.

---

2 In the literature on locally convex spaces, the term *barrel* is used instead of *body*, and a space in which every barrel is a neighbourhood of the origin is named *barreled*. Such are, in particular, Banach spaces.

3 Here, lower semi-continuity of $\mu(\cdot; B)$ is equivalent to closedness of B, positive homogeneity of $\mu(\cdot; B)$ is equivalent to absorptivity of B, and taking into account positive homogeneity convexity of $\mu(\cdot; B)$ is equivalent to its semi-additivity and to convexity of B.

4 In particular one can take the standard norm (total variation) in the space $\mathbf{C}^*$ of Borel measures endowed with the weak* topology.

Let us list some properties of polars.

(i) A polar M^0 of any set $M \subset E$ is convex and weakly* closed.
(ii) If M is a neighbourhood of the origin in E then its polar M^0 is a weak* compact subset in E^*.
(iii) Let $\{B_i\}$ be a family of non-empty sets in E, then

$$\left(\bigcup_i B_i\right)^0 = \bigcap_i B_i^0; \tag{A.13}$$

and if every B_i contains 0, then

$$\left(\bigcap_i B_i\right)^0 = cl\ co \bigcup_i B_i^0, \tag{A.14}$$

where *cl co* stands for the closed convex hull.

Further information on dual (polar) representation of convex sets can be found in Levin (1999).

A function $f : E \to \mathbf{R}^1 \cup \{+\infty\}$ with non-empty *effective domain* $dom(f) := \{e \in E : f(e) < +\infty\}$ is called *proper*. A proper function is called *convex* if its *epigraph* $epi(f) := \{(e, \alpha) : f(e) \le \alpha\}$ is a convex set in $E \times \mathbf{R}^1$. A sublinear function is convex.

Given a proper convex function f, its *subdifferential* at a point $e_0 \in dom(f)$ is the set of continuous linear functionals

$$\partial f(e_0) := \{e^* \in E^* : \langle e - e_0, e^* \rangle \le f(e) - f(e_0) \quad \forall e \in E\}. \tag{A.15}$$

It is a weak* closed convex set in E^*, and if f is continuous at e_0 then $\partial f(e_0)$ is weak* compact. If f is lsc on E then $\partial f(e_0)$ is non-empty whenever $e_0 \in dom(f)$. Elements of the subdifferential are named *subgradients.*

If $f = \mu(\cdot; B)$ is a gauge of a body B, then

$$\partial \mu(0; B) = B^0 \tag{A.16}$$

and

$$\partial \mu(e_0; B) = \{e^* \in B^0 : \langle e_0, e^* \rangle = \mu(e_0; B)\}, \quad e_0 \in E. \tag{A.17}$$

Let $B \subset E$ be a closed convex set with nonempty interior, and e_0 its boundary point ($e_0 \in bdry(B)$). A functional $e^* \in E^*$ and a hyperplane $H = \{e \in E : \langle e, e^* \rangle = \alpha\}$ ($\alpha = \langle e_0, e^* \rangle$) in E determined by it are said to be *supporting* B at e_0 if $\langle e, e^* \rangle \le \langle e_0, e^* \rangle$ for all $e \in B$. The set $N(e_0|B)$ of all such functionals is a weak* closed cone in E^*. It follows from (A.17) that if B is a body then the subdifferential of the gauge $\mu(\cdot; B)$ at a *boundary* point ($e_0 \in bdry(B)$) comprises functionals from the polar of the body supporting B and equal 1 at e_0:

$$\partial \mu(e_0; B) = \{e^* \in B^0 : \langle e_0, e^* \rangle = 1\} = \{e^* \in N(e_0|B) : \langle e_0, e^* \rangle = 1\}. \tag{A.18}$$

Sometimes lsc proper functions are referred to as *closed*. Given a closed convex function f, its *directional derivative* at a point $e_0 \in dom(f)$ is defined to be a sublinear function on E

$$e \mapsto f'(e_0; e) := \lim_{t \downarrow 0} \frac{f(e_0 + te) - f(e_0)}{t} = \inf_{t>0} \frac{f(e_0 + te) - f(e_0)}{t}, \tag{A.19}$$

and the subdifferential of f at $e_0 \in dom(f)$ equals the subdifferential of $f'(e_0; \cdot)$ at 0:

$$\partial f(e_0) = \partial f'(e_0; 0). \tag{A.20}$$

Appendix B

Proofs

Proof of Proposition 2.3: I. If $\varphi_1, \dots, \varphi_n$ are linearly independent, then, by Lemma 2.1, there exist n points in A, $a_1, \dots, a_n$, such that $\det(\varphi_i(a_j) \neq 0$. Suppose that $\sum_1^n t_j a_j \sim \sum_1^n u_j a_j$ for some positive numbers t_j, u_j, $j = 1, \dots, n$; then $\sum_1^n (t_j - u_j)\varphi_i(a_j) = 0$, $i = 1, \dots, n$, and since $\det(\varphi_i(a_j) \neq 0$, we get $t_j = u_j$, $j = 1, \dots, n$. Moreover, for any system of $m > n$ lights, $b_1, \dots, b_m$, there exists a non-trivial solution to linear equations

$$\sum_{j=1}^{m} x_j \varphi_i(b_j) = 0, \quad i = 1, \dots, n,$$

which implies the n-dimensionality of the Grassmann structure.

If now $\varphi_1, \dots, \varphi_n$ are not linearly independent, then, again by Lemma 2.1, $\det(\varphi_i(a_j) = 0$ whenever $a_1, \dots, a_n \in A$. Therefore, for any $a_1, \dots, a_n$ in A, there exists a non-trivial solution $(x_1, \dots, x_n)$ to linear equations

$$\sum_{j=1}^{n} x_j \varphi_i(a_j) = 0, \quad i = 1, \dots, n,$$

hence $\sum_1^n x_j a_j \sim 0$, which implies that the law of n-chromacy cannot be true.

The second assertion of I is obvious.

II. Note that equivalence relations $\sim_\Phi$ and $\sim_{\Phi'}$ are determined by the spaces $Lin(\varphi_1, \dots, \varphi_n)$ and $Lin(\varphi'_1, \dots, \varphi'_m)$ rather than the corresponding functionals, φ_i and φ'_j, as such. Therefore, $Lin(\varphi_1, \dots, \varphi_n) = Lin(\varphi'_1, \dots, \varphi'_m)$ implies $\sim_\Phi = \sim_{\Phi'}$.

Furthermore, it follows from I that the Grassmann structure $(A, \sim_\Phi)$ is r-dimensional where r is the rank (= the maximum number of linear independent functionals) of the system $(\varphi_1, \dots, \varphi_n)$. Therefore, if $\sim_\Phi = \sim_{\Phi'}$, then both the spaces $Lin(\varphi_1, \dots, \varphi_n)$ and $Lin(\varphi'_1, \dots, \varphi'_m)$ are r-dimensional; hence there are subsystems of $(\varphi_1, \dots, \varphi_n)$ and $(\varphi'_1, \dots, \varphi'_m)$ consisting of r functionals (to simplify the notation, we will assume that these functionals are $\varphi_1, \dots, \varphi_r$ and $\varphi'_1, \dots, \varphi'_r$ respectively), such that $Lin(\varphi_1, \dots, \varphi_n) = Lin(\varphi_1, \dots, \varphi_r)$ and $Lin(\varphi'_1, \dots, \varphi'_m) = Lin(\varphi'_1, \dots, \varphi'_r)$. Now, applying the Krantz Theorem yields us the existence of a nonsingular $r \times r$ matrix (t_{ij}) such that $\varphi'_i = \sum_{j=1}^r t_{ij}\varphi_j$, $i = 1, \dots, r$; hence $Lin(\varphi_1, \dots, \varphi_r) = Lin(\varphi'_1, \dots, \varphi'_r)$, which completes the proof. □

Proof of Lemma 2.1: If $\det(\varphi_i(a_j)) \neq 0$, then all φ_i, $i = 1, \dots, n$, are linearly independent. This is obvious. Suppose now that $\varphi_1, \dots, \varphi_n$ are linearly independent. The linear independence of φ_1, φ_2 means $\varphi_1 \neq \alpha\varphi_2$ whenever $\alpha \in \mathbf{R}$, which implies the existence of two points $a_1, a_2 \in A$ such

Foundations of Colour Science: From Colorimetry to Perception, First Edition. Alexander D. Logvinenko and Vladimir L. Levin.
© 2023 John Wiley & Sons Ltd. Published 2023 by John Wiley & Sons Ltd.

that the two-dimensional vectors $(\varphi_1(a_1), \varphi_1(a_2))$ and $(\varphi_2(a_1), \varphi_2(a_2))$ are linearly independent, that is,

$$\det\begin{pmatrix} \varphi_1(a_1) & \varphi_1(a_2) \\ \varphi_2(a_1) & \varphi_2(a_2) \end{pmatrix} \neq 0.$$

Then there is the unique solution α, β to the system

$$\varphi_3(a_1) = \alpha\varphi_1(a_1) + \beta\varphi_2(a_1), \tag{B.1}$$

$$\varphi_3(a_1) = \alpha\varphi_1(a_1) + \beta\varphi_2(a_1), \tag{B.2}$$

and as $\varphi_3 \neq \alpha\varphi_1 + \beta\varphi_2$, there exists a_3 such that

$$\varphi_3(a_3) \neq \alpha\varphi_1(a_3) + \beta\varphi_2(a_3). \tag{B.3}$$

Now, evaluating (with help of Cramer's rule) α and β from (B.1) and (B.2), we see that inequality (B.3) is rewritten as $\det(\varphi_i(a_j)) \neq 0$. Thus, for $n = 3$ the proof is completed. The general result is established by a similar argument, based on applying induction on the number of functions and Cramer's rule. □

Proof of Corollary 2.1: It suffices to verify that one new point a_{m+1} can be added to $a_1, \dots, a_m$ with preserving linear independence. Suppose the contrary. Then for each $a \in A \setminus \{a_1, \dots, a_m\}$ the system of $m + 1$ vectors consisting of $(\varphi_1(a_j), \dots, \varphi_n(a_j))$, $j = 1, \dots, m$, and $(\varphi_1(a), \dots, \varphi_n(a))$ proves to be linearly dependent hence $(\varphi_1(a), \dots, \varphi_n(a))$ is represented as a linear combination of m vectors $(\varphi_1(a_j), \dots, \varphi_n(a_j))$, $j = 1, \dots, m$. It follows from Lemma 2.1 that there are $a'_1, \dots, a'_n \in A$ such that the vectors $(\varphi_1(a'_j), \dots, \varphi_n(a'_j))$, $j = 1, \dots, n$, are linearly independent. On the other hand, each of them can be represented as a linear combination of $m < n$ vectors $(\varphi_1(a_j), \dots, \varphi_n(a_j))$, $j = 1, \dots, m$; a contradiction. □

Proof of Proposition 2.6: (i) ⇒ (ii) Since Φ satisfies the statement of the Krantz Theorem, $E(\sim) = \Phi^{-1}(0)$ is a vector subspace of co-dimension n in E. Therefore, its annihilator $E(\sim)^{\#}$ in the space of all (not necessarily continuous) linear functionals on E, is the n-dimensional vector subspace $Lin(\varphi_1, \dots, \varphi_n)$ spanned over $\varphi_1, \dots, \varphi_n$. The annihilator of $E(\sim)$ in the dual space E^* consisting of *continuous* linear functionals on E, $E(\sim)^0 := \{\varphi \in E^* : \langle e, \varphi\rangle = 0\ \forall e \in E(\sim)\}$ cannot be broader than $E(\sim)^{\#}$; therefore, $dim\ E(\sim)^0 \leq n$ and $E(\sim)^0 = Lin(\psi_1, \dots, \psi_m)$ where $\psi_1, \dots, \psi_m \in E^*$, $m \leq n$. Furthermore, as $E(\sim)$ is closed, it follows from the bipolar theorem (e.g., see Corollary I.6.3 in Day, 1958) that $E(\sim)$ coincides with the annihilator of $E(\sim)^0$ in E,

$$E(\sim) = \{e \in E : \langle e, \varphi\rangle = 0 \quad \forall \varphi \in E(\sim)^0\},$$

or, equivalently, $E(\sim) = \{e : \langle e, \psi_i\rangle = 0, i = 1, \dots, m\}$. In other words, for any $a_1, a_2 \in A$, $a_1 \sim a_2$ if and only if $\langle a_1, \psi_i\rangle = \langle a_2, \psi_i\rangle$, $i = 1, \dots, m$. Now applying Proposition 2.3 yields $m = n$ and $\varphi_1, \dots, \varphi_n \in Lin(\psi_1, \dots, \psi_n) \subset E^*$; hence Φ is continuous.

(ii) ⇒ (i) Given a net (generalised sequence) e_ν in $E(\sim)$ which converges to an element $e \in E$, we have to show that $e \in E(\sim)$. Take an operator $\Phi = (\varphi_1, \dots, \varphi_n)$ satisfying the statement of the Krantz Theorem. Since it is continuous, we get $\langle e, \varphi_i\rangle = \lim_\nu \langle e_\nu, \varphi_i\rangle$, $i = 1, \dots, n$, and as $e_\nu \in E(\sim)$ and $\varphi_i \in E^*$, $i = 1, \dots, n$, we get $\langle e, \varphi_i\rangle = \lim_\nu \langle e_\nu, \varphi_i\rangle = 0$, $i = 1, \dots, n$; therefore, $e \in E(\sim)$.

Consider now the case where E is $\mathbf{C}^*$ equipped with the weak* topology, and $A = \mathbf{C}^*_+$.

(iii) ⇒ (i) Although the weak* topology in $\mathbf{C}^*$ is not metrisable, its restriction to every closed ball with the centre at 0 is. This follows from the fact that $\mathbf{C}$ is separable. Therefore, the implication follows from the Krein–Schmulyan theorem (see Theorem V.5.7 in Dunford & Schwartz, 1958, or

Theorem II.5.5 in Day, 1958), stating that a vector subspace M in the dual Banach space F^* (in our case, $F = \mathbf{C}$ and $M = E(\sim)$) is weak* closed if and only if its intersection with the unit ball in F^* is so.

(i) $\Rightarrow$ (iii) Obvious. □

Proof of Proposition 3.1: It follows from Proposition 2.6 that the Grassmann structure under consideration is represented by m linearly independent weak* continuous linear functionals on $\mathbf{C}^*$, and as any such a functional is of the form (2.11), there exist m linearly independent continuous functions on Λ, $\varphi_1, \dots, \varphi_m$, such that for any $\sigma_1, \sigma_2 \in \mathbf{C}^*_+$,

$$\sigma_1 \sim \sigma_2 \Leftrightarrow \langle \varphi_i, \sigma_1 \rangle = \langle \varphi_i, \sigma_2 \rangle, \quad i = 1, \dots, m. \tag{B.4}$$

Let A denote the cone of non-negative linear combinations of delta functions. Since $\varphi_1, \dots, \varphi_m$ are linearly independent functions on Λ, we derive from (B.4) that the metamerism $\sim$ determines on A a m-dimensional Grassmann structure. On the other hand, from (3.1) it follows that it is n-dimensional. Therefore, $m = n$, and there exists a non-singular $n \times n$ matrix (s_{ij}) such that, for every $\lambda \in \Lambda$, $\langle \varphi_i, \delta_\lambda \rangle = \sum_{j=1}^n s_{ij} \langle t_j, \delta_\lambda \rangle$, $i = 1, \dots, n$. But this means $Lin(\varphi_1, \dots, \varphi_n) = Lin(t_1, \dots, t_n)$, and the proof is concluded by applying Proposition 2.3. □

Proof of Theorem 3.1: (i) $\Rightarrow$ (ii) We define the colour mechanisms f_{t_i} as follows. For every $\sigma \in \mathbf{C}^*_+$,

$$\mathsf{f}_{t_i}(\sigma) := \langle t_i, \sigma \rangle = \int_\Lambda t_i(\lambda)\sigma(d\lambda), \quad i = 1, \dots, n.$$

Clearly, $\mathsf{f}_{t_i}(\delta_\lambda) = t_i(\lambda)$, $i = 1, \dots, n$. We have to show that for any $\sigma_1, \sigma_2 \in \mathbf{C}^*_+$, $\sigma_1 \sim \sigma_2 \Leftrightarrow \langle t_i, \sigma_1 \rangle = \langle t_i, \sigma_2 \rangle$, $i = 1, \dots, n$. Denote by V the vector subspace in $\mathbf{C}^*$ that consists of linear combinations of delta functions, and by V_+, the convex cone in it, consisting of non-negative combinations. Since $t_1, \dots, t_n \in \mathbf{C}$, relation (3.1) along with the axioms of n-dimensional Grassmann structure implies that, for every $\sigma \in V_+$,

$$\sigma \sim \sum_{i=1}^n \langle t_i, \sigma \rangle \delta_{\lambda_i} \tag{B.5}$$

and that the right hand side of (B.5) is the unique element of $Lin(\delta_{\lambda_1}, \dots, \delta_{\lambda_n})$ metameric to σ.

Taking into account that the linear combination in the right hand side of (3.1) is determined uniquely for every λ and substituting $\lambda = \lambda_j$, we get $t_i(\lambda_i) = 1$ and $t_i(\lambda_j) = 0$ if $i \neq j$. Now, transitivity of $\sim$ along with the uniqueness of the right hand side of (B.5) implies the equivalence

$$\sigma_1 \sim \sigma_2 \Leftrightarrow \langle t_i, \sigma_1 \rangle = \langle t_i, \sigma_2 \rangle, \quad i = 1, \dots, n \tag{B.6}$$

whenever $\sigma_1, \sigma_2 \in V_+$.

Now let σ_1 and σ_2 be two arbitrary positive measures. It follows from Proposition 2.6 that there are n linearly independent functions $\varphi_i \in \mathbf{C}$, $i = 1, \dots, n$, such that $\sigma_1 \sim \sigma_2$ if and only if $\langle \varphi_i, \sigma_1 - \sigma_2 \rangle = 0$. Since V is weakly* dense in $\mathbf{C}^*$ (this follows, for example, from Lemma I.5.2 in Day, 1958), one can find a net (generalised sequence) σ_ν in V such that $\sigma_\nu \to \sigma_1 - \sigma_2$, where $\to$ means the weak* convergence in $\mathbf{C}^*$. However, in general, the equalities $\langle \varphi_i, \sigma_\nu \rangle = 0$, $i = 1, \dots, n$, need not hold true. We try to change σ_ν in such a way that the changed measure $\sigma'_\nu \in V$ satisfy these equalities with preserving the weak* convergence $\sigma'_\nu \to \sigma_1 - \sigma_2$. To this end, we apply Lemma 2.1 to linearly independent functions on Λ, $\varphi_1, \dots, \varphi_n$, and find n points $\lambda'_j \in \Lambda$, $j = 1, \dots, n$, such that $\det(\varphi_i(\lambda'_j)) \neq 0$.

It follows that there exists the unique solution $x_1^\nu, \dots, x_n^\nu$ to a system of n linear equations:

$$\sum_{j=1}^{n} x_j \varphi_i(\lambda_j') = \langle \varphi_i, \sigma_\nu \rangle, \quad i = 1, \dots, n, \tag{B.7}$$

and since $\langle \varphi_i, \sigma_\nu \rangle \to \langle \varphi_i, \sigma_1 - \sigma_2 \rangle = 0$, we get

$$\lim_\nu x_j^\nu = 0, \quad j = 1, \dots, n. \tag{B.8}$$

The required measure σ_ν' is now defined as follows:

$$\sigma_\nu' := \sigma_\nu + x_1^\nu \delta_{\lambda_1'} + \cdots + x_n^\nu \delta_{\lambda_n'}.$$

Taking into account that $x_1^\nu, \dots, x_n^\nu$ is a solution to (B.7), we get $\langle \varphi_i, \sigma_\nu' \rangle = 0$, $i = 1, \dots, n$, that is, $\sigma_{\nu 1}' \sim \sigma_{\nu 2}'$, where $\sigma_{\nu 1}'$ and $\sigma_{\nu 2}'$ are respectively the positive and the negative parts of σ_ν'. Since $\sigma_{\nu 1}'$ and $\sigma_{\nu 2}'$ are elements of V_+, we can apply (B.6), which results in $\langle t_i, \sigma_\nu' \rangle = 0$, $i = 1, \dots, n$. Also, we derive from (B.8) that $\sigma_\nu' \to \sigma_1 - \sigma_2$; therefore,

$$\langle t_i, \sigma_1 - \sigma_2 \rangle = \lim_\nu \langle t_i, \sigma_\nu \rangle = \lim_\nu \langle t_i, \sigma_\nu' \rangle = 0, \quad i = 1, \dots, n.$$

We have thus proved that $\sigma_1 \sim \sigma_2$ implies $\langle t_i, \sigma_1 \rangle = \langle t_i, \sigma_2 \rangle$, $i = 1, \dots, n$.

Similarly, if $\langle t_i, \sigma_1 \rangle = \langle t_i, \sigma_2 \rangle$, $i = 1, \dots, n$, then applying Lemma 2.1 to functions t_i, taking into account equivalence (B.6), and arguing as above with the measure

$$\sigma_\nu'' := \sigma_\nu + y_1^\nu \delta_{\lambda_1'} + \cdots + y_n^\nu \delta_{\lambda_n'},$$

instead of σ_ν', where $y_1^\nu, \dots, y_n^\nu$ is the unique solution to

$$\sum_{j=1}^{n} y_j t_i(\lambda_j'') = \langle t_i, \sigma_\nu \rangle, \quad i = 1, \dots, n,$$

we see that $\langle \varphi_i, \sigma_1 \rangle = \langle \varphi_i, \sigma_2 \rangle$, $i = 1, \dots, n$, that is, $\sigma_1 \sim \sigma_2$.

(ii)⇒(i) According to Proposition 2.6, $\sigma_1 \sim \sigma_2$ if and only if $\langle \varphi_i, \sigma_1 \rangle = \langle \varphi_i, \sigma_2 \rangle$, $i = 1, \dots, n$, where all φ_i are some continuous functions on Λ. Furthermore, from Proposition 2.3 it follows that $Lin(\mathrm{f}_{t_1}, \dots, \mathrm{f}_{t_n}) = Lin(\varphi_1, \dots, \varphi_n)$. Therefore, each functional f_{t_i} is generated by a continuous function which is a linear combination of $\varphi_1, \dots, \varphi_n$, and as for every $\lambda \in \Lambda$, one has $\mathrm{f}_{t_i}(\delta_\lambda) = t_i(\lambda)$, that continuous function is necessarily t_i. □

Proof of Proposition 3.3: Given a $\sigma = \sum_{k=1}^{m} \alpha_k \delta_{\lambda_k'} \in A$, we derive from (II′) that

$$\delta_{\lambda_k'} \sim t_1(\lambda_k')\delta_{\lambda_1} + t_2(\lambda_k')\delta_{\lambda_2} + t_3(\lambda_k')\delta_{\lambda_3},$$

and (3.11) holds with $\lambda = \lambda_k'$, $k = 1, \dots, m$. Now, since $\langle A, \sim \rangle$ is a Grassmann structure, we get

$$\sigma \sim \left(\sum_{k=1}^{m} \alpha_k t_1(\lambda_k') \right) \delta_{\lambda_1} + \left(\sum_{k=1}^{m} \alpha_k t_2(\lambda_k') \right) \delta_{\lambda_2} + \left(\sum_{k=1}^{m} \alpha_k t_3(\lambda_k') \right) \delta_{\lambda_3},$$

that is, (3.9) holds with

$$t_i(\sigma) = \sum_{k=1}^{m} \alpha_k t_i(\lambda_k'), \quad i = 1, 2, 3.$$

Furthermore, for any two different $i, j \in \{1, 2, 3\}$ one has

$$t_i(\sigma) + t_j(\sigma) = \sum_{k=1}^{m} \alpha_k \left(t_i(\lambda_k') + t_j(\lambda_k') \right),$$

and (3.12) follows then from (3.11). □

Proof of Proposition 3.4: Let $\mathbf{a}_i \in E/E(\sim)$, $i = 1, \ldots, n$, be linear independent colour stimuli from Definition 3.3. Then $\mathbf{a}_1, \ldots, \mathbf{a}_n$ is a basis in the colour stimulus space $E/E(\sim)$. Take $\mathbf{a}_1^*, \ldots, \mathbf{a}_n^*$ to be the biorthogonal system in the algebraic dual space[1] $(E/E(\sim))^\star$ and consider linear functionals $\varphi_1, \ldots, \varphi_n$ on the light stimulus space E such that $\varphi_i = \pi^*\left(\mathbf{a}_i^*\right)$ where $\pi^* : (E/E(\sim))^\star \to E^\star$ is the conjugate colour stimulus map. Being linear independent, $\mathbf{a}_1^*, \ldots, \mathbf{a}_n^*$ form a basis in the n-dimensional space $(E/E(\sim))^\star$, and their linear independence implies linear independence of $\varphi_1, \ldots, \varphi_n$. Since $\left\langle e, \varphi_j \right\rangle = \left\langle \pi(e), \mathbf{a}_j^* \right\rangle$, we get a chain of equivalences

$$\left\langle e, \varphi_j \right\rangle = 0, j = 1, \ldots, n \Leftrightarrow \pi(e) = 0 \Leftrightarrow e \sim 0,$$

hence, whenever $a, b \in A$,

$$a \sim b \Leftrightarrow \varphi_j(a) = \varphi_j(b), j = 1, \ldots, n, \tag{B.9}$$

that is, $\Phi = \left(\varphi_1, \ldots, \varphi_n\right)$ is a homomorphism of $\langle A, \sim\rangle$ into $\mathbf{R}^n$. Moreover, for every i, one has

$$\left\langle e, \varphi_j \right\rangle = 0, j \neq i \Leftrightarrow \langle \pi(e), \mathbf{a}_j^* \rangle = 0, j \neq i \Leftrightarrow \pi(e) \in L_i \Leftrightarrow e \sim_i 0,$$

hence,

$$a \sim_i b \Leftrightarrow \varphi_j(a) = \varphi_j(b) \text{ for } j \neq i, \tag{B.10}$$

that is, $\Phi_i = \left(\varphi_1, \ldots, \varphi_{i-1}, \varphi_{i+1}, \ldots, \varphi_n\right)$ is a homomorphism of $\left\langle A, \sim_i \right\rangle$ into $\mathbf{R}^{n-1}$ for each i.

If there are another linear independent functionals $\varphi_1', \ldots, \varphi_n'$ on E satisfying (B.9) and (B.10) then $E(\sim_i) = \{e \in E : \Phi_i'(e) = 0\}, i = 1, \ldots, n$, where $\Phi_1' = (\varphi_2', \ldots, \varphi_n')$, $\Phi_i' = (\varphi_1', \ldots, \varphi_{i-1}', \varphi_{i+1}', \ldots, \varphi_n')$ for $1 < i < n$, and $\Phi_n' = (\varphi_1', \ldots, \varphi_{n-1}')$. Now, as any linear functional belonging to the annihilator of $E(\sim_i)$ is a linear combination of φ_k, $k \neq i$, there are real numbers α_{jk}^i, $j \neq i, k \neq i$, such that

$$\varphi_j' = \sum_{k \neq i} \alpha_{jk}^i \varphi_k. \tag{B.11}$$

It follows from (B.11) that for any $i_1 \neq i_2$

$$\sum_{k \neq i_1} \alpha_{jk}^{i_1} \varphi_k = \sum_{k \neq i_2} \alpha_{jk}^{i_2} \varphi_k \quad \forall j \notin \{i_1, i_2\}. \tag{B.12}$$

Taking into account linear independence of $\varphi_1, \ldots, \varphi_n$, (B.12) implies

$$\alpha_{jk}^{i_1} = \alpha_{jk}^{i_2} \text{ for } i_1 \neq i_2,\ j \notin \{i_1, i_2\},\ k \notin \{i_1, i_2\}, \tag{B.13}$$

$$\alpha_{ji_2}^{i_1} = 0, \alpha_{ji_1}^{i_2} = 0 \text{ for } i_1 \neq i_2,\ j \notin \{i_1, i_2\}. \tag{B.14}$$

Now, one derives from (B.14) (by taking $i_1 = i, i_2 = k$) that $\alpha_{jk}^i = 0$ whenever $i \neq j$, $i \neq k$, $j \neq k$, and from (B.13) it follows that α_{jj}^i does not depend on i. Thus, (B.11) is rewritten as $\varphi_j' = \alpha_j \varphi_j$ where $\alpha_j := \alpha_{jj}^i$. Finally, all α_j are non-zero because each of φ_j' is non-zero. □

Proof of Theorem 4.1: Since Φ is linear and weak* continuous, and P is convex and weak* compact, the Φ-image of P, $\Phi(P)$, is a convex compact subset in $\mathbf{R}^n$. By the Caratheodory theorem (e.g., see Holmes, 1975), every $z \in \Phi(P)$ is represented as a convex combination of at most $n + 1$ extreme points of $\Phi(P)$; therefore, $\Phi(P)$ is the convex hull (closing is not needed) of $Ext(\Phi(P))$.

Let us show that $Ext(\Phi(P)) \subseteq \Phi(\Delta)$. Suppose $\sigma_0 \in P$ and $z = \Phi\left(\sigma_0\right) \in Ext(\Phi(P))$. The set $M := \{\sigma \in P : \Phi(\sigma) = z\}$ is convex, bounded and weak* compact. By the Krein–Milman theorem (Holmes, 1975), the set $Ext(M)$ is then nonempty. Take any $\sigma \in Ext(M)$ and show that $\sigma \in \Delta$.

1 The algebraic dual space to a vector space is the space of all linear functionals on it.

Suppose $\sigma \notin \Delta$. Since $\Delta = Ext(P)$, there exist $\sigma_1, \sigma_2 \in P, \sigma_1 \neq \sigma_2$, such that $\sigma = \frac{1}{2}(\sigma_1 + \sigma_2)$. But then $\Phi(\sigma) = z = \frac{1}{2}(z_1 + z_2)$ where $z_1 = \Phi(\sigma_1), z_2 = \Phi(\sigma_2)$. Since z is an extreme point of $\Phi(P)$, we get $z_1 = z_2 = z$; therefore, $\sigma_1, \sigma_2 \in M$, which contradicts to $\sigma \in Ext(M)$. □

Proof of Proposition 4.4: Suppose $c \in \mathbf{R}^3$ is the unique vector such that $H = \{z \in \mathbf{R}^3 : \langle c, z\rangle = 1\}$ hence $g_H(z) = z/\langle c, z\rangle$ whenever $z \in \Phi(\mathbf{C}_+^*)\backslash\{0\}$.

(i) ⇔ (ii) Clearly (ii) is equivalent to the equality $\{\Phi(\delta_\lambda) : \lambda_1 \leq \lambda \leq \lambda_2\} = I(\Phi(\delta_{\lambda_1}), \Phi(\delta_{\lambda_2}))$; therefore, it suffices to show that

$$g_H(I(\Phi(\delta_{\lambda_1}), \Phi(\delta_{\lambda_2}))) = I(g_H(\Phi(\delta_{\lambda_1})), g_H(\Phi(\delta_{\lambda_2}))). \tag{B.15}$$

Suppose $x \in g_H(I(\Phi(\delta_{\lambda_1}), \Phi(\delta_{\lambda_2}))$, then there is $0 \leq \alpha \leq 1$ such that

$$x = g_H(\alpha\Phi(\delta_{\lambda_1}) + (1-\alpha)\Phi(\delta_{\lambda_2})) = \frac{\alpha\Phi(\delta_{\lambda_1}) + (1-\alpha)\Phi(\delta_{\lambda_2})}{\langle c, \alpha\Phi(\delta_{\lambda_1}) + (1-\alpha)\Phi(\delta_{\lambda_2})\rangle}$$

$$= \beta g_H(\Phi(\delta_{\lambda_1})) + (1-\beta)g_H(\Phi(\delta_{\lambda_2})),$$

where

$$\beta = \frac{\alpha\langle c, \Phi(\delta_{\lambda_1})\rangle}{\alpha\langle c, \Phi(\delta_{\lambda_1})\rangle + (1-\alpha)\langle c, \Phi(\delta_{\lambda_2})\rangle},$$

and as $0 \leq \beta \leq 1$, we obtain $x \in I(g_H(\Phi(\delta_{\lambda_1})), g_H(\Phi(\delta_{\lambda_2})))$. On the other hand, if $x \in I(g_H(\Phi(\delta_{\lambda_1})), g_H(\Phi(\delta_{\lambda_2})))$ then there is $0 \leq \beta \leq 1$ such that $x = \beta g_H(\Phi(\delta_{\lambda_1})) + (1-\beta)g_H(\Phi(\delta_{\lambda_2}))$, which implies $x = g_H(\alpha\Phi(\delta_{\lambda_1}) + (1-\alpha)\Phi(\delta_{\lambda_2}))$ where

$$\alpha = \frac{\beta\langle c, \Phi(\delta_{\lambda_2})\rangle}{(1-\beta)\langle c, \Phi(\delta_{\lambda_1})\rangle + \beta\langle c, \Phi(\delta_{\lambda_2})\rangle}$$

and as $0 \leq \alpha \leq 1$, we have $x \in g_H(I(\Phi(\delta_{\lambda_1}), \Phi(\delta_{\lambda_2})))$, and equality (B.15) is thus completely established.

(i) ⇒ (iii) Given λ', $\lambda_1 < \lambda' < \lambda_2$, there is $0 < \alpha < 1$ such that $g_H(\Phi(\delta_{\lambda'})) = \alpha g_H(\Phi(\delta_{\lambda_1})) + (1-\alpha)g_H(\Phi(\delta_{\lambda_2}))$; therefore,

$$\frac{\Phi(\delta_{\lambda'})}{\langle c, \Phi(\delta_{\lambda'})\rangle} = \alpha\frac{\Phi(\delta_{\lambda_1})}{\langle c, \Phi(\delta_{\lambda_1})\rangle} + (1-\alpha)\frac{\Phi(\delta_{\lambda_2})}{\langle c, \Phi(\delta_{\lambda_2})\rangle}.$$

Now, as Φ is a linear operator, the latter is rewritten as

$$\Phi(\delta_{\lambda'}) = \Phi\left(t_1\delta_{\lambda_1} + t_2\delta_{\lambda_2}\right),$$

where $t_1 = \alpha\langle c, \Phi(\delta_{\lambda'})\rangle/\langle c, \Phi(\delta_{\lambda_1})\rangle > 0$, $t_2 = (1-\alpha)\langle c, \Phi(\delta_{\lambda'})\rangle/\langle c, \Phi(\delta_{\lambda_2})\rangle > 0$, and the result follows.

(iii) ⇒ (i) It follows from the colour-signal-cone-boundary hypothesis that the subset $D'(H; \lambda_1, \lambda_2)$ in H bounded by the piece of the curve $\{g_H(\delta_\lambda) : \lambda_1 \leq \lambda \leq \lambda_2\}$ and the linear segment $I(g_H(\Phi(\delta_{\lambda_1})), g_H(\Phi(\delta_{\lambda_2})))$ is convex.[2] Now, as $\Phi(\delta_{\lambda'}) = t_1\Phi(\delta_{\lambda_1}) + t_2\Phi(\delta_{\lambda_2}))$, one derives from here

$$g_H(\Phi(\delta_{\lambda'})) = \alpha g_H(\Phi(\delta_{\lambda_1})) + (1-\alpha)g_H(\Phi(\delta_{\lambda_2})),$$

2 This subset is the intersection of $D'(H)$ with the half-space determined by the straight line through the segment $I(g_H(\Phi(\delta_{\lambda_1}), g_H(\Phi(\delta_{\lambda_2})))$.

where

$$\alpha = \frac{t_1\langle c, \Phi(\delta_{\lambda_1})\rangle}{t_1\langle c, \Phi(\delta_{\lambda_1})\rangle + t_2\langle c, \Phi(\delta_{\lambda_2})\rangle},$$

and as $0 < \alpha < 1$, $g_H(\Phi(\delta_{\lambda'}))$ is an interior point of the linear segment $I(g_H(\Phi(\delta_{\lambda_1})), g_H(\Phi(\delta_{\lambda_2})))$. Now, taking into account the convexity of the set $D'(H; \lambda_1, \lambda_2)$, this implies that $\lambda_1 < \lambda' < \lambda_2$ and the whole set $\{g_H(\Phi(\delta_\lambda)) : \lambda_1 \le \lambda \le \lambda_2\}$ coincides with the segment $I(g_H(\Phi(\delta_{\lambda_1})), g_H(\Phi(\delta_{\lambda_2})))$. □

Proof of Theorem 5.1: First show that $\Phi_{|\mathcal{A}}$ is one-to-one. If there are two lights in $\mathcal{A}$, $\sigma_1 = p_1\sigma_1' + q_1\sigma_0$ and $\sigma_2 = p_2\sigma_2' + q_2\sigma_0$, with the same image $\Phi(\sigma_1) = \Phi(\sigma_2)$, then $(q_2 - q_1)\Phi(\sigma_0) = p_1\Phi(\sigma_1') - p_2\Phi(\sigma_2')$. If now $\sigma_1 \ne \sigma_2$ then we have one of the following two cases:

1. $q_1 \ne q_2$ (for definiteness, we will assume $q_1 < q_2$);
2. $q_1 = q_2$ but $p_1\sigma_1' \ne p_2\sigma_2'$.

In case 1, we have to consider the following four sub-cases: (i) both σ_1' and σ_2' are in $\Delta(\Lambda')$; (ii) $\sigma_1' \in \Delta(\Lambda')$, $\sigma_2' \notin \Delta(\Lambda')$; (iii) $\sigma_1' \notin \Delta(\Lambda')$, $\sigma_2' \in \Delta(\Lambda')$; (iv) $\sigma_1' \notin \Delta(\Lambda')$ and $\sigma_2' \notin \Delta(\Lambda')$.

In sub-case (i), one has $(q_2 - q_1)\sigma_0 \sim p_1\delta_{\lambda_1} - p_2\delta_{\lambda_2}$ where $\lambda_1, \lambda_2 \in \Lambda'$, $\lambda_1 \ne \lambda_2$. But this is impossible. Indeed, if both p_1 and p_2 are strictly positive, we get a contradiction with condition (ii) of Definition 5.1, and if $p_1 > 0$ and $p_2 = 0$ then $\Phi(\sigma_0) = t\Phi(\delta_{\lambda_1})$ where $t = p_1/(q_2 - q_1) > 0$, and as $\Phi(\sigma_0)$ is an interior point of $\Phi(A)$, this contradicts the colour-signal-cone-boundary hypothesis; finally, the situation where $p_2 > 0$ and $p_1 = 0$ is also impossible because then $(q_2 - q_1)\Phi(\sigma_0) = -p_2\Phi(\delta_{\lambda_2})$, i.e., $(q_2 - q_1)\langle\varphi_i, \sigma_0\rangle = -p_2\varphi_i(\lambda_2)$, $i = 1, 2, 3$, and as $\sum_1^3 \varphi_i(\lambda_2) > 0$, we obtain a contradiction with $\Phi(\sigma_0) \in \Phi(A)$.

In sub-case (ii), one has $(q_2 - q_1)\sigma_0 \sim p_1\delta_{\lambda_1} - p_2(\alpha\delta_{\lambda'_{\min}} + (1-\alpha)\delta_{\lambda'_{\max}})$ where $\lambda_1 \in \Lambda'$, $0 < \alpha < 1$; therefore, $\Phi(\sigma_0) = t_1\Phi(\delta_{\lambda_1}) - t_2\Phi(\delta_{\lambda'_{\min}}) - t_3\Phi(\delta_{\lambda'_{\max}})$ where all t_i are non-negative, and $t_2 = 0$ if and only if $t_3 = 0$. If $t_2 = t_3 = 0$, this contradicts the colour-signal-cone-boundary hypothesis, and if $t_2 \ne 0, t_3 \ne 0$, we obtain a contradiction with condition (ii) of Definition 5.1 (the case that $\lambda_1 = \lambda'_{\min}$, $t_1 > t_2$ or $\lambda_1 = \lambda'_{\max}$, $t_1 > t_3$) or with condition (iii) of Definition 5.1 (the case that $\lambda_1 \notin \{\lambda'_{\min}, \lambda'_{\max}\}$, $t_1 \ne 0$) or with hypothesis (ii) of the theorem (the case that $t_1 = 0$ or $\lambda_1 = \lambda'_{\min}$ and $t_1 \le t_2$ or $\lambda_1 = \lambda'_{\max}$ and $t_1 \le t_3$).

In sub-case (iii), one has $(q_2 - q_1)\sigma_0 \sim p_1(\alpha\delta_{\lambda'_{\min}} + (1-\alpha)\delta_{\lambda'_{\max}}) - p_2\delta_{\lambda_2}$ where $\lambda_2 \in \Lambda'$, $0 < \alpha < 1$; therefore, $\Phi(\sigma_0) = t_1\Phi(\delta_{\lambda'_{\min}}) + t_2\Phi(\delta_{\lambda'_{\max}}) - t_3\Phi(\delta_{\lambda_2})$, where all t_i are non-negative, and $t_1 = 0$ if and only if $t_2 = 0$. If $t_1 = t_2 = 0$, this contradicts hypothesis (ii) of the theorem, and if $t_1 \ne 0$, $t_2 \ne 0$, and $t_3 = 0$, we obtain a contradiction with the colour-signal-cone-boundary hypothesis. It remains to consider the situation where all t_i are strictly positive. Thus, suppose $t_1 > 0$, $t_2 > 0$, $t_3 > 0$ and take the measure $\sigma = t_1\delta_{\lambda'_{\min}} + t_2\delta_{\lambda'_{\max}} + t_3\delta_{\lambda_2}$. We obtain $\frac{1}{2}\Phi(\sigma) + \frac{1}{2}\Phi(\sigma_0) = t_1\Phi(\delta_{\lambda'_{\min}}) + t_2\Phi(\delta_{\lambda'_{\max}}) \in \partial\Phi(A)$, but this is impossible: the middle of the segment connecting a point $\Phi(\sigma)$ of the convex cone $\Phi(A)$ with its interior point $\Phi(\sigma_0)$ cannot be a boundary point of $\Phi(A)$.

Finally, in sub-case (iv), $(q_2 - q_1)\sigma_0 \sim u_1\delta_{\lambda'_{\min}} + u_2\delta_{\lambda'_{\max}}$ where $u_1, u_2 \in \mathbf{R}$. But this contradicts the colour-signal-cone-boundary hypothesis (when both u_1 and u_2 are non-negative) or hypothesis (ii) of the theorem (when both u_1 and u_2 are non-positive) or condition (ii) of Definition 5.1 (when the signs of u_1 and u_2 are different).

Thus, we have established that in all the cases no $\sigma_1, \sigma_2 \in \mathcal{A}$ such that $\sigma_1 \ne \sigma_2$ but $\Phi(\sigma_1) = \Phi(\sigma_2)$ can exist, that is, the map $\Phi_{|\mathcal{A}}$ is one-to-one. We will show now that $\Phi_{|\mathcal{A}}$ maps $\mathcal{A}$ onto the whole of the colour-signal cone $\Phi(A)$. Let us take a two-dimensional plane $H \in \mathcal{H}(\Phi)$ (i.e., H is a plane in $\mathbf{R}^3$ such that $0 \notin H$, H has common points with the colour-signal cone $\Phi(A)$, and no ray of the colour-signal cone is parallel to H; see Chapter 4). Let us consider the conical envelope of $\Phi(\Delta'(\Lambda'))$. Its intersection with H is a closed curve $g_H(\Phi(\Delta'(\Lambda')))$ bounding a closed domain $D(H)$

in H (see Chapter 4 for details). Since the colour-signal-cone-boundary hypothesis is supposed to hold, the set $D(H)$ is convex and coincides with the intersection of H with the colour-signal cone (i.e., $D(H) = H \cap \Phi(\mathcal{A})$). Then for every $\sigma \in \mathcal{A}$, $\sigma \neq 0$, there is a unique $t = t(\sigma) > 0$ such that $t(\sigma)\Phi(\sigma) = g_H(\Phi(\sigma)) \in D(H)$. (Here, $t(\sigma) = 1/\langle c, \Phi(\sigma)\rangle$ where $c = (c_1, c_2, c_3) \in \mathbf{R}^3$ is the unique vector such that $H = \{z \in \mathbf{R}^3 : \langle c, z\rangle = 1\}$.) Since $D(H)$ is convex, the segments joining $g_H(\Phi(\sigma_0))$ with the points $g_H(\Phi(\sigma')) \in g_H(\Phi(\Delta'(\Lambda')))$ cover the whole $D(H)$. Then there are $\sigma' \in \Delta'(\Lambda')$ and $0 \leq \alpha \leq 1$ such that $t(\sigma)\Phi(\sigma) = \alpha t(\sigma_0)\Phi(\sigma_0) + (1-\alpha)t(\sigma')\Phi(\sigma')$; therefore, $\Phi(\sigma) = \Phi(\sigma_*)$ where

$$\sigma_* = \frac{\alpha t(\sigma_0)}{t(\sigma)}\sigma_0 + \frac{(1-\alpha)t(\sigma')}{t(\sigma)}\sigma' \in \mathcal{A},$$

and the surjectivity of $\Phi_{|\mathcal{A}}$ is thus established.

It remains to verify continuity of $\Phi_{|\mathcal{A}}$ and $\Phi^{-1}_{|\mathcal{A}}$. Continuity of $\Phi_{|\mathcal{A}}$ follows immediately from the definition of the weak* topology in $\mathbf{C}^*$ if one takes into account that $\varphi_i \in \mathbf{C}(\Lambda)$, $i = 1, 2, 3$. As the light stimulus cone $\Phi(\mathcal{A})$ is closed in $\mathbf{R}^n$, the continuity of $\Phi^{-1}_{|\mathcal{A}}$ will follow if we show that the Φ-image of every closed set $\mathbf{F} \subset \mathcal{A}$ is closed in $\mathbf{R}^n$. Thus, given a sequence $z^k = (z_1^k, z_2^k, z_3^k) \in \Phi(\mathbf{F})$ tending to $z = (z_1, z_2, z_3) \in \mathbf{R}^3$, we have to show that $z \in \Phi(\mathbf{F})$. Since $\Phi_{|\mathcal{A}}$ is one-to-one, there are uniquely determined measures $\sigma_k = p_k\sigma'_k + q_k\sigma_0 \in \mathbf{F} \subset \mathcal{A}$ such that $z^k = \Phi_{|\mathcal{A}}(\sigma_k)$, i.e., $z_i^k = p_k\langle\varphi_i, \sigma'_k\rangle + q_k\langle\varphi_i, \sigma_0\rangle$ $(i = 1, 2, 3)$, $k = 1, 2, \ldots$. Being convergent, the sequence (z^k) is bounded: there is a number $C > 0$ such that $\|z^k\|_1 = |z_1^k| + |z_2^k| + |z_3^k| \leq C$ for all k. Since all three φ_i are continuous functions on a compact set Λ and their sum is strictly positive, there is a number $\varepsilon > 0$ such that $\varphi_1(\lambda) + \varphi_2(\lambda) + \varphi_3(\lambda) \geq \varepsilon$ for all $\lambda \in \Lambda$. Note that for every $\sigma' \in \Delta'(\Lambda')$ one has $\|\sigma'\| = \langle \mathbf{1}_\Lambda, \sigma'\rangle = 1$. Now, as all p_k and q_k are non-negative, we obtain $C \geq |z_1^k| + |z_2^k| + |z_3^k| = z_1^k + z_2^k + z_3^k \geq p_k\langle\varphi_1 + \varphi_2 + \varphi_3, \sigma'_k\rangle \geq p_k\varepsilon\langle\mathbf{1}_\Lambda, \sigma'_k\rangle = p_k\varepsilon$ and, similarly, $C \geq z_1^k + z_2^k + z_3^k \geq q_k\langle\varphi_1 + \varphi_2 + \varphi_3, \sigma_0\rangle \geq q_k\varepsilon\langle\mathbf{1}_\Lambda, \sigma_0\rangle = q_k\varepsilon\|\sigma_0\|$; therefore,

$$0 \leq p_k \leq C/\varepsilon, \quad 0 \leq q_k \leq C/(\varepsilon\|\sigma_0\|).$$

Taking into account these inequalities and the weak* compactness of the set $\Delta'(\Lambda')$, we can chose (weakly*) convergent subsequences $\sigma'_{k_n} \to \sigma' \in \Delta'(\Lambda')$, $p_{k_n} \to p \geq 0$, and $q_{k_n} \to q \geq 0$. Then the sequence σ_{k_n} converges weakly* to the measure $\sigma = p\sigma' + q\sigma_0$; therefore, $\sigma \in \mathbf{F}$, and as

$$z = \lim_{n\to\infty} z_{k_n} = \lim_{n\to\infty} \Phi(\sigma_{k_n}) = \Phi(\sigma),$$

the result follows. □

Proof of Lemma 5.1: Since all φ_i are weak* continuous linear functionals on $\mathbf{C}^*$, the set $\{b \in \mathbf{C}^*_+ : b \sim a\} = \{b : \langle\varphi_i, b\rangle = \langle\varphi_i, a\rangle,\ i = 1, \ldots, n\}$ is *weak** closed and convex. Furthermore, as $\sum_1^n \varphi_i(\lambda)$ is a continuous function on a compact set Λ, it follows from (5.12) that there exists a positive number ε such that

$$\sum_1^n \varphi_i(\lambda) \geq \varepsilon \text{ for all } \lambda \in \Lambda. \tag{B.16}$$

Now, for every b, $b \sim a$, one has $\varepsilon\|b\| = \varepsilon\langle 1_\Lambda, b\rangle \leq \langle\sum_1^n \varphi_i, b\rangle = \langle\sum_1^n \varphi_i, a\rangle$; then $\|b\| \leq \frac{1}{\varepsilon}\langle\sum_1^n \varphi_i, a\rangle$; that is, $\{b : b \sim a\}$ is norm bounded hence *weak** compact. □

Proof of Theorem 6.1: I. Suppose σ is an extreme point of $Met(\varphi_1, \ldots, \varphi_n; \sigma_0)$. Let us show that its support $F = spt(\sigma)$ can contain no more than n points. Indeed, if not, then the space $\mathbf{C}(F)$ is either infinite-dimensional or m-dimensional with $m > n$. In the first case, F is an infinite set; in the second, it consists of m points ($m > n$). In both cases, there exists such a function $\varphi_0 \in \mathbf{C}(F)$ that $\varphi_0 \neq 0$, and

$$\int_F \varphi_i(\lambda)\,\varphi_0(\lambda)\,\sigma(d\lambda) = 0, \qquad i = 1, \dots, n. \tag{B.17}$$

Indeed, every φ_i determines a linear functional, $\varphi_i \otimes \sigma$ on $\mathbf{C}(F)$:

$$\langle \varphi, \varphi_i \otimes \sigma \rangle_F := \int_F \varphi_i(\lambda)\,\varphi(\lambda)\,\sigma(d\lambda), \ \ \varphi \in \mathbf{C}(F),$$

and since the dimension of $\mathbf{C}(F)$ is more than n, the linear subspace

$$\{\varphi \in \mathbf{C}(F) : \langle \varphi, \varphi_i \otimes \sigma \rangle_F = 0, \ i = 1, \dots, n\}$$

is non-trivial. So, there is a non-zero $\varphi_0 \in \mathbf{C}(F)$ satisfying (B.17), and we assume without loss of generality that $|\varphi_0| < 1$. Applying the Urysohn–Tietse theorem (e.g., see Dunford & Schwartz, 1958, Theorem I.5.3), we extend φ_0 with preserving this inequality and continuity to the whole segment Λ so that

$$\langle \varphi_i, \varphi_0 \otimes \sigma \rangle := \int_F \varphi_i(\lambda)\,\varphi_0(\lambda)\,\sigma(d\lambda) = 0, \quad i = 1, \dots, n.$$

Here, the measure $\varphi_0 \otimes \sigma$ is defined by

$$\langle \varphi, \varphi_0 \otimes \sigma \rangle = \int_\Lambda \varphi(\lambda)\,\varphi_0(\lambda)\,\sigma(d\lambda), \qquad \varphi \in \mathbf{C}.$$

(Since $spt(\sigma) = F$, the value of the above integral does not depend on the behaviour of φ outside F.)

We have $\sigma = 1/2(\sigma_1 + \sigma_2)$ where $\sigma_1 = \sigma + \varphi_0 \otimes \sigma$, and $\sigma_2 = \sigma - \varphi_0 \otimes \sigma$. Both measures, σ_1 and σ_2, are positive because $|\varphi_0| < 1$, and as $\langle \varphi_i, \sigma \pm \varphi_0 \otimes \sigma \rangle = \langle \varphi_i, \sigma \rangle = \langle \varphi_i, \sigma_0 \rangle$, $i \in \overline{1, n}$, we have $\sigma_1, \sigma_2 \in Met(\varphi_1, \dots, \varphi_n; \sigma_0)$. Furthermore, $\sigma_1 \neq \sigma$ and $\sigma_2 \neq \sigma$ since $\varphi_0 \neq 0$ on the support of σ. Thus, contrary to our supposition, σ is not an extreme point of $Met(\varphi_1, \dots, \varphi_n; \sigma_0)$. This contradiction shows that $F = spt(\sigma)$ contains at most n points. Hence, $\sigma = \sum_1^n \alpha_i \delta_{\lambda_i}$ where $\alpha_i \geq 0$, $\lambda_i \in \Lambda$, $i = 1, \dots, n$. To complete the proof note that system (6.6) is satisfied because σ is metameric to σ_0.

II. Suppose $\sigma = \frac{1}{2}(\sigma_1 + \sigma_2)$ where $\sigma_1, \sigma_2 \in Met(\varphi_1, \dots, \varphi_n; \sigma_0)$. Since the supports of these two measures cannot be larger than the support of σ and $spt(\sigma) \subseteq \{\lambda_1, \dots, \lambda_n\}$, one has

$$\sigma_j = \sum_{i=1}^n \alpha_i^j \delta_{\lambda_i}, \ j = 1, 2.$$

Moreover, as $\sigma_1, \sigma_2 \in Met(\varphi_1, \dots, \varphi_n; \sigma_0)$, we get

$$\sum_{i=1}^n \alpha_i^j \varphi_k(\lambda_i) = \langle \varphi_k, \sigma_0 \rangle, \quad k = 1, \dots, n; \ j = 1, 2. \tag{B.18}$$

Note that, taking into account (6.6) and (6.7), $x_1 = \alpha_1, \dots, x_n = \alpha_n$ is the unique solution to a system of linear equations

$$\sum_{i=1}^n x_i \varphi_k(\lambda_i) = \langle \varphi_k, \sigma_0 \rangle, \quad k = 1, \dots, n.$$

This, along with (B.18), implies $\sigma_1 = \sigma_2$, therefore, σ is an extreme point of $Met(\varphi_1, \dots, \varphi_n; \sigma_0)$. □

Proof of Theorem 6.2: Suppose $Met(\varphi_1, \dots, \varphi_n; \sigma_0)$ contains a measure σ which is different from σ_0. According with Lemma 5.1, the object metamer set $Met(\varphi_1, \dots, \varphi_n; \sigma_0)$ is convex and *weak** compact; therefore, one can take σ to be its extreme point. Theorem 6.1 yields then $\sigma = \sum_{i=1}^{n} \alpha_i \delta_{\lambda_i}$ where $\alpha_i \geq 0$, $\sum_{i=1}^{n} \alpha_i \varphi_k(\lambda_i) = \langle \varphi_k, \sigma_0 \rangle$, $k = 1, \dots, n$. That is, $\Phi(\sigma_0) = \sum_{i=1}^{n} \alpha_i \Phi(\delta_{\lambda_i})$, and all α_i are non-negative. Furthermore, since $\sigma_0 = \alpha\delta_\lambda$ and since we accept that two monochromatic lights with different wavelengths cannot be metameric, at least two of the coefficients α_i must be strictly positive. It follows then from Proposition 4.2(a) that $g_H(\Phi(\sigma_0))$ cannot be an extreme point of $H \cap \Phi(\mathbf{C}_+^*)$. Now, as $\Phi(\sigma_0) \in H$, one gets $\Phi(\sigma_0) = g_H(\Phi(\sigma_0))$; therefore, in contradiction with the assumption of the theorem, $\Phi(\sigma_0)$ is not an extreme point of $H \cap \Phi(\mathbf{C}_+^*)$. □

Proof of Lemma 6.2: (i) ⇒ (ii) Consider any $\mu = -\sigma_0 + \sigma$ from H_{σ_0}. Since $\langle \varphi_i, \mu \rangle = 0$, we have $\sigma \in Met(\varphi_1, \dots, \varphi_n; \sigma_0)$. It follows from (i) that $\sigma \in Met(\psi; \sigma_0)$. Therefore, $\langle \psi, \mu \rangle = \langle \psi, -\sigma_0 + \sigma \rangle = 0$, that is, $\psi \in H_{\sigma_0}^0$.

(ii) ⇒ (i) Take any $\sigma \in Met(\varphi_1, \dots, \varphi_n; \sigma_0)$. Then $\sigma \in \mathbf{C}_+^*$ and $\mu = -\sigma_0 + \sigma \in L^0$. Therefore, $\mu \in H_{\sigma_0}$, and as $\psi \in H_{\sigma_0}^0$, we obtain $\langle \psi, \mu \rangle = \langle \psi, -\sigma_0 + \sigma \rangle = 0$, that is, $\sigma \in Met(\psi; \sigma_0)$. □

Proof of Proposition 6.1: Let us take any $\sigma \in Met(\varphi_1, \dots, \varphi_n; \delta_{\lambda_0})$. Since $\chi_\Lambda \in L$, we have $\langle \chi_\Lambda, \sigma \rangle = \langle \chi_\Lambda, \delta_{\lambda_0} \rangle = 1$; therefore, σ is a probability measure. Similarly, $\langle \varphi, \sigma \rangle = \langle \varphi, \delta_{\lambda_0} \rangle = \varphi(\lambda_0)$; and since $\varphi(\lambda) < \varphi(\lambda_0)$ for $\lambda \neq \lambda_0$, it follows that $spt(\sigma) = \{\lambda_0\}$. Since the only probability measure σ with $spt(\sigma) = \{\lambda_0\}$ is δ_{λ_0}, we get $H_{\delta_{\lambda_0}} = \{0\}$ and $H_{\delta_{\lambda_0}}^0 = \mathbf{C}$, and (6.12) follows. □

Proof of Proposition 6.2: (i)⇒(ii) is obvious.
(ii) ⇒ (i) Suppose (i) fails. Then there is $\psi \in Lin(\psi_1, \dots, \psi_m) \backslash Lin(\varphi_1, \dots, \varphi_n)$ (or $\psi \in Lin(\varphi_1, \dots, \varphi_n) \backslash Lin(\psi_1, \dots, \psi_m)$). It follows from the separation theorem that a measure $\mu \in \mathbf{C}^*$ exists such that $\langle \varphi_i, \mu \rangle = 0$, $i = 1, \dots, n$, and $\langle \psi, \mu \rangle \neq 0$. Let us take its Jordan decomposition $\mu = \mu_+ - \mu_-$ and consider $\sigma_0 = \mu_-$, $\sigma = \mu_+$. We have $\sigma_0, \sigma \in \mathbf{C}_+^*$, $\sigma \in Met(\varphi_1, \dots, \varphi_n; \sigma_0)$, and $\sigma \notin Met(\psi; \sigma_0)$, hence $\sigma \notin Met(\psi_1, \dots, \psi_m; \sigma_0)$.

The equivalence of (i) and (ii) is thus established, and the rest of the proposition follows easily from it. □

Proof of Proposition 7.1: The first statement (convexity and compactness of $M(\Psi, \Phi; \sigma_0)$) follows directly from Lemma 6.1 if one takes into account equality (7.5) along with linearity and continuity of Ψ. Convexity of $Msm(\approx, \sim; \sigma_0)$ follows from (7.6) if one takes into account that $Met(\varphi_1, \dots, \varphi_n; \sigma_0)$ is convex and Ψ is a linear operator. Furthermore, $Met(\varphi_1, \dots, \varphi_n; \sigma_0)$ is weak* compact by Lemma 6.1, and as the operator Ψ is weak* continuous, it follows from (7.6) that $Msm(\approx, \sim; \sigma_0)$ is weak* closed. The weak* compactness of $Msm(\approx, \sim; \sigma_0)$ will then follow if we show that it is norm bounded.

Since $\sum_1^n \psi_i(\lambda)$ is a continuous function on a compact set Λ, it follows from (7.8) that there exists a positive number ε such that $\sum_1^n \psi_i(\lambda) \geq \varepsilon$ for all $\lambda \in \Lambda$. Now, for every $\sigma \in Msm(\approx, \sim; \sigma_0)$, one has $\varepsilon \|\sigma\| \leq \langle \sum_1^n \psi_i, \sigma \rangle$, and as $\langle \psi_i, \sigma \rangle = \langle \psi_i, \sigma' \rangle$, $i = 1, \dots, n$, where σ' is some element of $Met(\varphi_1, \dots, \varphi_n; \sigma_0)$, we get

$$\begin{aligned} \varepsilon \|\sigma\| &\leq \sum_1^n \langle \psi_i, \sigma \rangle = \sum_1^n \langle \psi_i, \sigma' \rangle \\ &\leq \| \sum_1^n \psi_i \|_{\mathbf{C}} \max \{ \|\sigma'\| \ : \ \sigma' \in Met(\varphi_1, \dots, \varphi_n; \sigma_0) \}. \end{aligned} \tag{B.19}$$

It follows from (B.19) that $Msm(\approx, \sim; \sigma_0)$ is norm bounded hence *weak** compact. □

Proof of Theorem 7.1: I. Consider the set $N_z = \Psi^{-1}(z) \cap \mathbf{C}_+^*$. Clearly, it is convex and weak* closed. Further, being weak* closed and norm bounded (the latter follows from (7.8)), N_z is weak* compact. Then, by the Krein–Milman theorem, it has extreme points. Take σ to be one of them.

Suppose σ is not an extreme point of $Msm(\approx, \sim; \sigma_0) = \Psi^{-1}(M(\Psi, \Phi; \sigma_0))$. Then there exist two different lights σ_1 and σ_2 in $Msm(\approx, \sim; \sigma_0)$ such that at least one of them does not belong to N_z and $\sigma = \frac{1}{2}(\sigma_1 + \sigma_2)$ (both the lights cannot be in N_z because σ is an extreme point of N_z), which implies $z = \frac{1}{2}(z^1 + z^2)$ where $z^i = \Psi(\sigma_i) \in M(\Psi, \Phi; \sigma_0),\ i = 1, 2$. Since $\sigma_i \notin N_z$ for at least one i, the corresponding $z^i \neq z$; then $z^i \neq z$ for both $i = 1, 2$, and therefore, z cannot be an extreme point of $M(\Psi, \Phi; \sigma_0)$, a contradiction.

II. As follows from (7.3), for every light $\sigma \in Msm(\approx, \sim; \sigma_0)$ there is a light $\sigma' \in Met(\sim; \sigma_0)$ such that $\sigma \in Met(\approx; \sigma')$. Then $\sigma' \in Met(\approx; \sigma)$ hence $\sigma' \in M(\sigma, \sigma_0)$. Since $M(\sigma, \sigma_0) = Met(\sim; \sigma_0) \cap Met(\approx; \sigma)$, it follows from Lemma 6.1 that $M(\sigma, \sigma_0)$ is convex and weak* compact. By the Krein–Milman theorem, the set of its extreme points, $Ext(M(\sigma, \sigma_0))$, is then non-empty; therefore, we can take σ' to be one of them. Let us verify that σ is an extreme point of $Met(\approx, \sigma')$. Indeed, if not, then there exist $\sigma_1, \sigma_2 \in Met(\approx; \sigma')$ such that $\sigma_1 \neq \sigma_2$ and $\sigma = \frac{1}{2}(\sigma_1 + \sigma_2)$. But then $\sigma_i \in Msm(\approx, \sim; \sigma_0)$, $i = 1, 2$, which contradicts the assumption that σ is an extreme point of $Msm(\approx, \sim; \sigma_0)$. The first statement of II is fully proved.

Furthermore, taking into account that $\sigma' \in Ext(M(\sigma, \sigma_0))$, the same argument as used in the proof of Theorem 6.1(I) shows that the support of σ' comprises at most $2n$ points, which implies representation (7.10). Now, the inclusions $\sigma' \in Met(\sim; \sigma_0)$ and $\sigma' \in Met(\approx; \sigma)$ are rewritten as (7.11) and (7.12) respectively. Finally, as follows from Theorem 6.1(I), $\sigma \in Ext(Met(\approx; \sigma'))$ amounts to (7.9). □

Proof of Proposition 7.2: Since $\Psi(Met(\varphi_1, \dots, \varphi_n; \sigma_0))$ is a compact subset in $\mathbf{R}^n$, there are two points in it, $z^1 = \Psi(\sigma_1)$ and $z^2 = \Psi(\sigma_2)$ (where $\sigma_1, \sigma_2 \in (Met(\varphi_1, \dots, \varphi_n; \sigma_0))$, such that

$$diam\Psi(Met(\varphi_1, \dots, \varphi_n; \sigma_0)) = \|z^1 - z^2\|.$$

Now, taking into account that $\Phi(\sigma_1) = \Phi(\sigma_2) = z_0$, where $z_0 = \Phi(\sigma_0)$, and applying estimate (7.17), we get

$$\begin{aligned} \|z^1 - z^2\| &\leq \|z^1 - z_0\| + \|z^2 - z_0\| \leq (\|\sigma_1\| + \|\sigma_2\|)\sqrt{\sum_{i=1}^{n} \|\psi_i - \varphi_i\|_{\mathbf{C}}^2} \\ &\leq 2\max\{\|\sigma\| \ : \ \sigma \in Met(\varphi_1, \dots, \varphi_n; \sigma_0)\}\sqrt{\sum_{i=1}^{n} \|\psi_i - \varphi_i\|_{\mathbf{C}}^2}, \end{aligned}$$

and the result follows. □

Proof of Proposition 8.7: As $\langle \mathbb{S}, \mathcal{H}, \vartheta \rangle$ is simple, for each i $\vartheta^{-1}(h_i) \neq \emptyset$, thus $H_i \neq \emptyset$. Each H_i is a τ-class. Indeed, fix arbitrarily an element $x \in \vartheta^{-1}(h_i)$ and consider a τ-class K containing x. Then x must partially match in hue any element in K. It follows that $K \subseteq H_i$, and as K is a τ-class, we conclude that $K = H_i$.

Let us show now that $\mathbf{H} = \{H_1, \dots, H_m\}$ (resp., $\mathbf{H} = \{H_0, H_1, \dots, H_m\}$) is a basis. If $x\tau y$ then there is a component hue $h_i \in \vartheta(x) \cap \vartheta(y)$, thus $x, y \in H_i$. Conversely, if $x, y \in H_i$ then $h_i \in \vartheta(x) \cap \vartheta(y)$, thus $x\tau y$.

Let $\mathcal{K}$ be another basis in $\langle \mathbb{S}, \mathcal{H}, \vartheta \rangle$. There must be a τ-class K in $\mathcal{K}$ which contains $\vartheta^{-1}(h_i)$. It can be only H_i. Therefore, $\mathcal{K}$ includes $\mathbf{H}$. However, the latter is a basis itself. Hence, $\mathcal{K}$ coincides with $\mathbf{H}$. □

Proof of Proposition 10.1: It follows directly from Definition 10.1 that $\sim_\varsigma$ is an equivalence. Furthermore, if $a\varsigma b$ and $a' \sim_\varsigma a$, then $b \in \varsigma(a)$ and $\varsigma(a') = \varsigma(a)$; hence $b \in \varsigma(a')$ that is, $a'\varsigma b$. Suppose now that $\sim$ is an equivalence on A such that $a\varsigma b$ implies $a'\varsigma b$ whenever $a' \sim a$. If $a, b \in A$ and

$a \sim b$, then for every $c \in \varsigma(a)$ one has $c \in \varsigma(b)$ and, similarly, for every $c \in \varsigma(b)$ one has $c \in \varsigma(a)$; that is, $a \sim_\varsigma b$. □

Proof of Proposition 10.2: It follows immediately from Definition 10.3 that v_ς is a tolerance and the implication $\Leftarrow$ in (10.7) holds true. Suppose now that $x v_\varsigma y$ and take any $a \in \pi^{-1}(x)$, $b \in \pi^{-1}(y)$. Then there exist $a' \in \pi^{-1}(x)$, $b' \in \pi^{-1}(y)$ such that $a' \varsigma b'$, and since $\pi(a) = \pi(a') = x$, $\pi(b) = \pi(b') = y$, we get $a \sim_\varsigma a'$, $b \sim_\varsigma b'$. Now applying twice Proposition 10.1 yields $a \varsigma b$, and the implication $\Rightarrow$ in (10.7) is thus established. □

Proof of Corollary 10.1: If $x, y \in A(\sim_\varsigma)$ and $v_\varsigma(x) = v_\varsigma(y)$, then it follows from Proposition 10.2 that for any $a \in \pi^{-1}(x)$, $b \in \pi^{-1}(y)$,

$$\varsigma(a) = \pi^{-1}(v_\varsigma(x)) = \pi^{-1}(v_\varsigma(y)) = \varsigma(b),$$

that is, $a \sim_\varsigma b$, which may be rewritten as $x = \pi(a) = \pi(b) = y$. □

Proof of Proposition 10.3: If $a \sim b$ then $\pi(a) = \pi(b)$, and we get $\varsigma(a) = \pi^{-1}(v(\pi(a))) = \pi^{-1}(v(\pi(b))) = \varsigma(b)$, that is, $a \sim_\varsigma b$.

Suppose v is simple. If $a \sim_\varsigma b$ then $\varsigma(a) = \varsigma(b)$. It follows from (10.9) that in such a case $v(\pi(a)) = v(\pi(b))$, and since v is simple, this implies $\pi(a) = \pi(b)$, that is, $a \sim b$.

Suppose v is not simple. Then there are x, y in $A(\sim_\varsigma)$, $x \neq y$ such that $v(x) = v(y)$. If we assume that $\sim_\varsigma$ coincides with $\sim$, then (10.9) will imply $\varsigma(a) = \pi^{-1}(v(\pi(a)))$, $a \in A$. Therefore, for any $a \in \pi^{-1}(x)$, $b \in \pi^{-1}(y)$ we have $\varsigma(a) = \varsigma(b)$, that is, $a \sim b$. This implies $\pi(a) = \pi(b)$, hence $x = y$; a contradiction. □

Proof of Proposition 10.4: According to Corollary 10.2, $\sim'$ coincides with $\sim_\Phi$, and as $\approx_\Phi$ is the restriction of $\sim_\Phi$ to A, the first statement of the Proposition follows. The equivalence (i) $\Leftrightarrow$ (ii) is then obvious; and (ii) $\Leftrightarrow$ (iii) follows from Corollary 10.2. □

Proof of Proposition 10.5: We have $v'(x) = \{z \in \mathbf{R}^n : d(x, z) \leq \varepsilon\}$ for every $x \in \mathbf{R}^n$ and $v(x) = \{z \in M : d(x, z) \leq \varepsilon\}$ for every $x \in M$. Then, as follows from Definition 10.4, the validity of (10.11) whenever $x \neq y$ means precisely that the corresponding tolerance (v' or v) is simple. □

Proof of Corollary 10.4: Given points $x, y \in \mathbf{R}^n$, $x \neq y$, we consider the ray $\{t(y - x) : t \geq 0\}$ passing through $y - x$. Since g infinitely increases along the ray, the number $t_{\max} = \max\{t > 0 : g(t(y - x)) \leq \varepsilon\}$ is well-defined and strictly positive. Let us take $z = x + t_{\max}(x - y)$. We get $x - z = t_{\max}(y - x)$, $y - z = (t_{\max} + 1)(y - x)$, hence $g(x - z) \leq \varepsilon$, $g(y - z) > \varepsilon$. Therefore, condition (10.11) is satisfied, and the result follows from Proposition 10.5. □

Proof of Proposition 10.6: Given $x = (x_1, \ldots, x_n)$ and $y = (y_1, \ldots, y_n)$ with $x \neq y$, we have to show that $v'(x) \neq v'(y)$. Since $x \neq y$, one has $x_{i_*} \neq y_{i_*}$ for at least one index i_*. Take an arbitrary set of m indices, $J = \{i_1, \ldots, i_m\}$, such that i_* is one of them. Since $i_* \in J$ and $x_{i_*} \neq y_{i_*}$, there exists a vector $z = (z_1, \ldots, z_n)$ such that $|z_i - x_i| \leq \varepsilon$ for $i \in J$ and $|z_i - y_i| > \varepsilon$ for $i \notin J \backslash \{i_*\}$. (For example, one can take $z_i = x_i$ for $i \in J \backslash \{i_*\}$, $z_i = y_i + \varepsilon + 1$ for $i \notin J$ and choose z_{i_*} from the condition that $|z_{i_*} - x_{i_*}| \leq \varepsilon$ and $|z_{i_*} - y_{i_*}| > \varepsilon$.) Since the number of indices i, for which $|z_i - x_i| \leq \varepsilon$, is at least m while the number of indices i, for which $|z_i - y_i| \leq \varepsilon$, is at most $m - 1$, we have $z \in v'(x)$ but $z \notin v'(y)$; that is, $v'(x) \neq v'(y)$. □

Proof of Theorem 10.1: Given $x, y \in M$, we have to show that $v(x) = v(y)$ implies $x = y$. One has $v(x) = \{z \in M : g(x - z) \leq \varepsilon\} = (x + U(\varepsilon)) \cap M$ where $U(\varepsilon) = \{u \in \mathbf{R}^n : g(u) \leq \varepsilon\}$ and, similarly, $v(y) = (y + U(\varepsilon)) \cap M$. Suppose that $v(x) = v(y)$, consider a point $z \in bdry(v(x)) \cap int\, M$ (here, bdry($v(x)$) stands for the boundary on $v(x)$), and let z^* be a subgradient[3] of g at z. Then $u = z - x$ is a boundary point of $U(\varepsilon)$ (hence $g(u) = g(z-x) = g(x-z) = \varepsilon$), and z^* is a supporting functional of $x + U(\varepsilon)$ at z, that is, $z^* \neq 0$ and

$$\langle z', z^* \rangle \leq \langle z, z^* \rangle \quad \text{whenever } z' \in x + U(\varepsilon). \tag{B.20}$$

We derive from (B.20) that the functional z^* is supporting to $U(\varepsilon)$ at u:

$$\langle u', z^* \rangle \leq \langle u, z^* \rangle \quad \text{whenever } u' \in U(\varepsilon).$$

A similar argument shows that $v = z - y$ is a boundary point of $U(\varepsilon)$ (hence $g(v) = g(z-y) = g(y-z) = \varepsilon$), and the functional z^* is supporting to $U(\varepsilon)$ at v:

$$\langle u', z^* \rangle \leq \langle v, z^* \rangle \quad \text{whenever } u' \in U(\varepsilon).$$

Now, since $U(\varepsilon) = \{u' \in \mathbf{R}^n : g(u') \leq \varepsilon\}$ and z^* is supporting to $U(\varepsilon)$ both at u and at v, we derive from hypothesis (c) that u and v are linearly dependent. Furthermore, $g(u) = g(v) = \varepsilon$ implies $u \neq 0$, $v \neq 0$. Therefore, there exists a number $\alpha > 0$ such that $u = \alpha v$ or $u = \alpha(-v)$. Since g is a convex function satisfying (a) and (b), and $g(v) = g(-v) = \varepsilon$, it follows that the function h on $\mathbf{R}_+$, $h(t) = g(tv) = g(t(-v))$, is convex and non-decreasing hence strictly increasing for $t \geq t_0$, where $t_0 = \max\{t \geq 0 : g(tv) = 0\}$. Now, $g(v) = \varepsilon > 0$ implies $t_0 < 1$, while $g(u) = \varepsilon$ implies $t_0 < \alpha$, and taking into account that $g(u) = \varepsilon$, $h(1) = g(v) = \varepsilon$, and h is strictly increasing on $\{t : t \geq t_0\}$, we get $\alpha = 1$. Thus, $u = v$ or $u = -v$. If $u = v$, then $x = y$. If $u = -v$, then $x - y = 2v$ hence $g(x-y) = h(2) > h(1) = \varepsilon$, but in such a case $y \notin v(x)$, and as always $y \in v(y)$, we obtain a contradiction with our supposition. Thus, we have completely proved that $v(x) = v(y)$ implies $x = y$, that is, v is simple. □

Proof of Lemma 10.1: Since $\Phi(A) \subset \mathbf{R}^n_+$, we get $\|z\|_1 = z_1 + \cdots + z_n$ for all $z = (z_1, \ldots, z_n) \in \Phi(A)$; therefore, $\|x+y+z\| + \|x+y+z'\| = \|x+z\| + \|x+z'\| + 2\|y\| \geq \|x+z\| + \|x+z'\|$ whenever $x, y, z, z' \in \Phi(A)$, and we get

$$d(x+y+z, x+y+z') = \frac{\|z-z'\|}{\|x+y+z\| + \|x+y+z'\| + 1}$$

$$\leq \frac{\|z-z'\|}{\|x+z\| + \|x+z'\| + 1} = d(x+z, x+z'),$$

which completes the proof. □

Proof of Lemma 10.2: One has

$$v(x) = \{y \in \Phi(A) : \|x-y\| \leq \varepsilon(\|x\|_1 + \|y\|_1 + 1)\},$$

and as $\Phi(A) \subset \mathbf{R}^n_+$, we get $\|x\|_1 + \|y\|_1 + 1 = \langle x+y, x_0^* \rangle + 1$, where

$$\langle z, x_0^* \rangle := z_1 + \cdots + z_n \quad \forall z = (z_1, \ldots, z_n) \in \mathbf{R}^n,$$

and the convexity of $v(x)$ follows. The inclusion $v(x) \supset B_\varepsilon(x) \cap \Phi(A)$ is an obvious consequence of the inequality $d(x,y) \leq \|x-y\|$, $x, y \in \mathbf{R}^n_+$. □

Proof of Theorem 10.2: Let $x = (x_1, \ldots, x_n), y = (y_1, \ldots, y_n) \in \Phi(A)$, and $v(x) = v(y)$. Then $d(x, y) \leq \varepsilon$ since otherwise $y \notin v(x)$, $y \in v(y)$. One has

3 Recall (see Section A.2 in Appendix A) that a subgradient of a convex function g at a point z is a linear functional z^* on $\mathbf{R}^n$ such that $\langle z' - z, z^* \rangle \leq g(z') - g(z)$ whenever $z' \in \mathbf{R}^n$. The set $\partial g(z)$ of all such functionals is called the subdifferential of g at z.

$$v(x) = \{z = (z_1, \dots, z_n) \in \Phi(A) : \|x - z\| \le \varepsilon(x_1 + \cdots + x_n + z_1 + \cdots + z_n + 1)\},$$

$$v(y) = \{z = (z_1, \dots, z_n) \in \Phi(A) : \|y - z\| \le \varepsilon(y_1 + \cdots + y_n + z_1 + \cdots + z_n + 1)\}.$$

Suppose that $x \neq y$, consider the line $L_0 = \{x + t(y - x) : t \in \mathbf{R}\}$ passing through x and y, and find a point $z_* \in B_\varepsilon(x) \cap int\,\Phi(A)$ such that the ray $R(z_*) := \{\alpha z_* : \alpha \ge 0\}$ has no common points with L_0, $R(z_*) \cap L_0 = \emptyset$. Since $v(x) \supset B_\varepsilon(x) \cap \Phi(A)$ (see Lemma 10.2), $z_* \in v(x)$, hence $d(x, z_*) \le \varepsilon$.

Consider on $\mathbf{R}_+$ a function h, $h(t) = d(x, tz_*)$. The inequality $d(x, z_*) \le \varepsilon$ is then rewritten as $h(1) \le \varepsilon$. Note now that since $\sqrt{n}\|z\| \ge \|z\|_1$ for each $z \in \mathbf{R}^n$, we get

$$\liminf_{t\to\infty} h(t) = \liminf_{t\to\infty} d(x, tz_*) \ge \liminf_{t\to\infty} \frac{1}{\sqrt{n}} \frac{\|tz_* - x\|_1}{\|tz_*\|_1 + \|x\|_1 + 1}$$
$$\ge \lim_{t\to\infty} \frac{1}{\sqrt{n}} \frac{t\|z_*\|_1 - \|x\|_1}{t\|z_*\|_1 + \|x\|_1 + 1} = \frac{1}{\sqrt{n}} > \varepsilon, \tag{B.21}$$

and as h is a continuous function on $\mathbf{R}_+$, there exist $t \ge 1$ such that $h(t) = d(x, tz_*) = \varepsilon$. Let t_* be the maximum t with this property:

$$t_* = \max\{t \ge 1 : d(x, tz_*) = \varepsilon\}.$$

Then $z_0 := t_* z_* \in v(x) = v(y)$, therefore $d(y, z_0) \le \varepsilon$. Let us show that $d(y, z_0) = \varepsilon$. Indeed, if $d(y, z_0) < \varepsilon$, then $d(y, (1+\delta)z_0) < \varepsilon$ where $\delta > 0$ is large enough, but in such a case the definition of t_* along with (B.21) implies $d(x, (1+\delta)z_0) = d(x, (1+\delta)t_* z_*) = h((1+\delta)t_*) > \varepsilon$, therefore, $z_0 \in v(y)$ but $z_0 \notin v(x)$; a contradiction. Thus, $z_0 \in int\,\Phi(A)$ and $d(x, z_0) = d(y, z_0) = \varepsilon$, that is, $z_0 \in S(x) \cap \mathrm{int}\,\Phi(A) = S(y) \cap \mathrm{int}\,\Phi(A)$ where

$$S(x) = \{z \in \mathbf{R}^n : \|x - z\| = \varepsilon(x_1 + \cdots + x_n + z_1 + \cdots + z_n + 1)\}, \tag{B.22}$$

$$S(y) = \{z \in \mathbf{R}^n : \|y - z\| = \varepsilon(y_1 + \cdots + y_n + z_1 + \cdots + z_n + 1)\}. \tag{B.23}$$

Since $S(x)$ and $S(y)$ are smooth surfaces, differentiating the equalities in (B.22) and (B.23) at $z = z_0$ with respect to z_i, $i = 1, \dots, n$, yields

$$\frac{x - z_0}{\|x - z_0\|} = \frac{y - z_0}{\|y - z_0\|} = \varepsilon(1, \dots, 1). \tag{B.24}$$

It follows from (B.24) that z_0 belongs to L_0 but this contradicts the initial choice of z_*, and the obtained contradiction proves that $x = y$. □

Proof of Proposition 11.1: Suppose $\varphi(e) \neq 0$ for some $e \in E(\sim_\varsigma)$. Since $E(\sim_\varsigma)$ is a vector subspace, we can assume that $\varphi(e) > 0$. Now, since ς' is a regular extension of ς, $E(\sim_\varsigma) \subset B'_{a^0}$; therefore, $a^0 + Ne \in \varsigma'(a^0)$ whenever $N > 0$. Then $\varphi(a^0) + N\varphi(e) = \varphi(a^0 + Ne) \le \varphi(a)$ but the latter is impossible when N is large enough. □

Proof of Proposition 11.2: This is a particular case of the separating hyperplane theorem: since $int(\varsigma'(a^0))$ is an open convex set and $e \notin \mathrm{int}(\varsigma'(a^0))$, there is a functional $\varphi \in E^*$ separating e from $int(\varsigma'(a^0))$.[4] □

Proof of Proposition 11.3: One has $\varsigma'(a^0) = \Phi^{-1}(v'(x^0))$, where v' is a tolerance on $\mathbf{R}^n$ given by

$$xv'y \Leftrightarrow d(x, y) \le \varepsilon \quad (x, y \in \mathbf{R}^n).$$

4 In other words, φ is a supporting functional of $\varsigma'(a^0)$ at e; see Section A.2 in Appendix A.

Since condition (iii) means $\nu'(x^0) = \nu(x^0)$, we get $\varsigma'(a^0) = \Phi^{-1}(\nu(x^0))$. Now, taking into account that Φ is a continuous linear map, (i) implies that $\varsigma'(a^0)$ is closed and convex, and (ii) implies $a^0 \in int(\varsigma'(a^0))$.

Further, a chain of implications

$$\begin{aligned} e \in E(\sim_\varsigma) &\Leftrightarrow \Phi(e) = 0 \Leftrightarrow \Phi(a^0 + e) = \Phi(a^0) \\ &\Rightarrow \Phi(a^0 + e) \in \nu'(x^0) \Leftrightarrow e \in \varsigma'(a^0) - a^0 = B'_{a^0} \end{aligned}$$

shows that ς' is a regular extension of ς.

To prove (11.9), observe that $e \in \varsigma'(a^0)$ means $\Phi(e) \in \nu'(x^0)$, and since $\nu'(x^0) = \nu(x^0) \subset \Phi(A)$ (see (11.8)), there exists $a \in \varsigma(a^0)$ such that $\Phi(a) = \Phi(e)$. Then $e - a \in E(\sim_\varsigma)$, and we get $e = a + (e - a) \in \varsigma(a^0) + E(\sim_\varsigma)$. The inclusion $\varsigma'(a^0) \subseteq \varsigma(a^0) + E(\sim_\varsigma)$ is thus established.

In order to establish the converse inclusion, notice that $a \in \varsigma(a^0)$ means $d(x^0, \Phi(a)) \leq \varepsilon$, and $e \in E(\sim_\varsigma)$ means $\Phi(e) = 0$. Then $\Phi(a + e) = \Phi(a)$ hence $d(x^0, \Phi(a + e)) \leq \varepsilon$, that is, $a + e \in \varsigma'(a^0)$, and the result follows. □

Proof of Corollary 11.2: It follows from (11.9) that there is an element $a \in A$ such that $e - a \in E(\sim_\varsigma)$. Since, the interior of the closed convex set $\varsigma'(a^0)$ is nonempty, it follows from Proposition 11.1 and Corollary 11.1 that $\varphi \in E(\sim_\varsigma)^0$ and φ supports $\varsigma'(a^0)$ at a. Then $a \in bdry(\varsigma'(a^0))$, and the result follows. □

Proof of Theorem 11.1: (i) By Theorem 10.1, ν is simple. Then, by Proposition 10.10, the Grassmann structure $\langle A, \sim\rangle$ is induced by the colour matching structure $\langle A, \varsigma\rangle$, and ς' is an extension of ς. Since for every $e \in E(\sim)$, $g(\Phi(a^0 - (a^0 + e))) = g(-\Phi(e)) = g(0) = 0 < \varepsilon$, we get $e \in B'_{a^0}$, that is, ς' is a regular extension of ς. Furthermore, as $e \in B'_{a^0}$ means $g(\Phi(e)) \leq \varepsilon$ and g is a convex function on $\mathbf{R}^n$ satisfying $g(x) > 0$ for any $x \neq 0$, one has $\lim_{\alpha \to +\infty} g(\Phi(\alpha e)) = \lim_{\alpha \to +\infty} g(\alpha \Phi(e)) = +\infty$ whenever $e \notin E(\sim)$. Therefore, $\alpha e \in B'_{a^0}$ for all $\alpha > 0$ implies $\Phi(e) = 0$ that is, $e \in E(\sim)$. Thus, we have proved that ς' is a strongly regular extension of ς.

(ii) Since $x^0 = \Phi(a^0)$ is an interior point of the colour-signal cone $\Phi(A)$, there is a number $r > 0$ such that the ball $B_r(x^0) = \{x \in \mathbf{R}^n : \|x - x^0\| \leq r\}$ is contained in $\Phi(A)$. Let us show that in such a case one can find an $\varepsilon(x^0) > 0$ such that

$$\{x \in \mathbf{R}^n : g(x^0 - x) \leq \varepsilon(x^0)\} \subset B_r(x^0). \tag{B.25}$$

Indeed, if (B.25) fails, then there is a sequence (x^k) such that $g(x^0 - x^k) \leq \frac{1}{k}$ and $\|x^0 - x^k\| > r$. Then, taking into account that g is convex, we get for $y^k = \frac{r}{\|x^0 - x^k\|}(x^0 - x^k)$, $\|y^k\| = r$ and $g(y^k) \leq \frac{r}{\|x^0 - x^k\|} g(x^0 - x^k) < g(x^0 - x^k) \leq \frac{1}{k}$. We will assume, by passing if needed to a subsequence, that y^k tends to some y. Then $\|y\| = \lim_{k\to\infty} \|y^k\| = r$ and $g(y) = \lim_{k\to\infty} g(y^k) = 0$; therefore, $y = 0$, a contradiction. Thus, (B.25) holds true, hence equality (11.8) is satisfied. Then for any $0 < \varepsilon \leq \varepsilon(x^0)$, all suppositions of Proposition 11.3 are satisfied, and applying the proposition shows that $\varsigma'(a^0)$ is closed convex and has the nonempty interior. Now, applying Corollary 11.1 completes the proof. □

Proof of Theorem 11.2: i) By Theorem 10.2, ν is simple. The same argument as used in the proof of Theorem 11.1, (i), shows that the Grassmann structure $\langle A, \sim\rangle$ is induced by the colour matching structure $\langle A, \varsigma\rangle$, and ς' is an extension of ς. Let us show that ς' is a strongly regular extension of ς. Since for every $e \in E(\sim)$,

$$d(\Phi(a^0), \Phi(a^0 + e)) = \frac{\|\Phi(e)\|}{2\|\Phi(a^0)\|_{1,\alpha} + 1} = 0,$$

ς' is a regular extension of ς. Moreover, ς' is a strongly regular extension of ς, since if $Ne \in B'_{a^0}$ and $e \notin E(\sim)$, then

$$\frac{N\|\Phi(e)\|}{2\|\Phi(a^0)\|_{1,\alpha} + N\|\Phi(e)\|_{1,\alpha} + 1} \leq \frac{N\|\Phi(e)\|}{\|\Phi(a^0)\|_{1,\alpha} + \|\Phi(a^0 + Ne)\|_{1,\alpha} + 1} \leq \varepsilon,$$

and tending N to ∞ leads to a contradictory chain of inequalities

$$\frac{1}{\|\alpha\|} = \min_{x \neq 0} \frac{\|x\|}{\|x\|_{1,\alpha}} \leq \frac{\|\Phi(e)\|}{\|\Phi(e)\|_{1,\alpha}} \leq \varepsilon.$$

(ii) Since $x^0 = \Phi(a^0)$ is an interior point of $\Phi(A)$, there is a number $r > 0$ such that the ball $B_r(x^0) = \{x \in \mathbf{R}^n \ ; \ \|x - x^0\| \leq r\}$ is contained in $\Phi(A)$. Let us show that in such a case one can find an $0 < \varepsilon(x^0) < 1/\|\alpha\|$ such that

$$\{x \in \mathbf{R}^n : d(x^0, x) \leq \varepsilon(x^0)\} \subset B_r(x^0). \tag{B.26}$$

Indeed, if (B.26) fails, then there is a sequence (x^k) such that $d(x^0, x^k) \leq \frac{1}{k}$ and $\|x^0 - x^k\| > r$. If the sequence (x^k) is bounded, then we can assume, by passing if needed to a subsequence, that it tends to some x, and as $\|x^0 - x\| = \lim_{k\to\infty} \|x^0 - x^k\| \geq r$, it follows that $x \neq x^0$, hence $d(x^0, x) > 0$. On the other hand, $d(x^0, x) = \lim_{k\to\infty} d(x^0, x^k) = 0$; a contradiction. Suppose now that (x^k) is unbounded, and, by passing if needed to a subsequence, we can assume that $\lim_{k\to\infty} \|x^k\| = \infty$. Then

$$0 = \lim_{k\to\infty} d(x^0, x^k) \geq \lim_{k\to\infty} \frac{\|x^k\| - \|x^0\|}{\|x^0\|_{1,\alpha} + \|x^k\|_{1,\alpha} + 1} = \frac{1}{\|\alpha\|};$$

again a contradiction. Thus, (B.26) holds, and as $B_r(x^0) \subset \Phi(A)$, equality (11.8) is satisfied. Taking into account (11.8) along with Lemma 10.2 and Remark 10.12, we derive from Proposition 11.3 that the set $\varsigma'(a^0)$ is closed convex and has the nonempty interior. Then applying Corollary 11.1 completes the proof. □

Proof of Lemma 11.1: (i) ⇔ (ii) Since measures $\sigma_1, \dots, \sigma_N$ are linearly independent elements of $\mathbf{C}^*$, the signed measure $\sum_1^N \alpha_i \sigma_i$ is non-zero if and only if at least one of the coefficients α_i is non-zero. Similarly, since functions $\varphi_1, \dots, \varphi_N$ are linearly independent elements of $\mathbf{C}$, the function $\sum_1^N \beta_i \varphi_i$ is non-zero if and only if at least one of the coefficients β_i is non-zero. It follows that (i) and (ii) are equivalent and mean one and the same thing, namely the existence for any $(\alpha_1, \dots, \alpha_N) \neq (0, \dots, 0)$ of an element $\varphi \in E_N^*$ such that $\langle \varphi, \sum_1^N \alpha_i \sigma_i \rangle \neq 0$ and the existence for any $(\beta_1, \dots, \beta_N) \neq (0, \dots, 0)$ of an element $e \in E_N$ such that $\langle \sum_1^N \beta_i \varphi_i, e \rangle \neq 0$.

(ii) ⇔ (iv) Obvious.

(iii) ⇒ (i) Obvious.

(iv) ⇒ (iii) Fix $e_j = \sigma_j$ and define $\varphi'_i := \sum_{j=1}^N x_{ij} \varphi_j$, $i = 1, \dots, N$, where $(x_{ij})_{ij}$ is the inverse matrix of $(\langle \varphi_i, \sigma_j \rangle)_{ij}$. We get

$$\langle \varphi'_i, \sigma_j \rangle = \sum_{k=1}^N x_{ik} \langle \varphi_k, \sigma_j \rangle = \delta_{ij}, \quad i, j \in \{1, \dots, N\},$$

where δ_{ij} equals 1 for $i = j$ and 0 for $i \neq j$.[5] □

5 A similar argument shows that $\langle \varphi_i, e_j \rangle = \delta_{ij}$ for $e_j = \sum_{k=1}^N x_{kj} \sigma_k$. Thus, in (iii) we can take $e_j = \sigma_j$ and select $\varphi'_i \in E_N^*$ or take $\varphi'_i = \varphi_i$ and select $e_j \in E_N$ in such a way that the biorthogonality condition $\langle \varphi'_i, e_j \rangle = \delta_{ij}$ is satisfied.

Proof of Proposition 11.4:

1. It follows from the separating hyperplane theorem that

$$\varsigma'(a^0) = \bigcap_{e',\varphi} \{e \in E : \langle \varphi, e \rangle \leq \langle \varphi, e' \rangle\},$$

where the intersection is taken over all $e' \in bdry(\varsigma'(a^0))$ and all $\varphi \in \mathbf{C}$ supporting $\varsigma'(a^0)$ at e'. Now, applying Corollary 11.2 completes the proof.
2. If $a \in bdry(\varsigma'(a^0))$, then, by the separating hyperplane theorem, there is a functional $\varphi \in \mathbf{C}$ supporting $\varsigma'(a^0)$ at a, which implies $e = a - a^0 \in Q$ and $t_e = 1$.
3. This follows from (11.4) if one takes into account that $a^0 + te$ belongs to A and does not belong to $\varsigma(a^0)$ when $t - t_e > 0$ is small enough. □

Proof of Proposition 11.5: (i) ⇒ (ii) By supposition (I), ς' is a regular extension of ς, therefore, $E(\sim_\varsigma) \subset B'_{a^0}$ (see Definition 11.4). Now, since the functional φ determines the tangential (hence supporting) hyperplane of B'_{a^0} at e_0, one has $\langle \varphi, e \rangle = 0$ for all $e \in E(\sim_\varsigma)$, that is, $\varphi \in E(\sim_\varsigma)^0$, and as it is accepted that $E(\sim_\varsigma)^0 \subset E_N^*$, one gets $\varphi \in E_N^*$. Furthermore, we can assume, by passing if needed from φ to $-\varphi$, that φ is a supporting functional of B'_{a^0} at e_0:

$$\langle \varphi, e \rangle \leq \langle \varphi, e_0 \rangle \quad \forall e \in B'_{a^0},$$

and since e_0 is a tangency point of B'_{a^0}, it follows that there is a number $\alpha > 0$ such that $\alpha\varphi = \nabla\mu(e_0; B'_{a^0})$. In order to prove that e_0 is a tangency point of $B'_{a^0} \cap E_N$ in E_N, it suffices to show that the subdifferential $\partial\mu(e_0; B'_{a^0} \cap E_N)$ consists of a sole functional on E_N which is then $\nabla\mu(e_0; B'_{a^0} \cap E_N)$. To this end, we consider the linear operator $T : E_N \to E = \mathbf{C}^*$ which is the natural (identical) imbedding. The conjugate operator $T^* : E^* = \mathbf{C} \to E_N^*$ maps then any functional $\varphi' \in E^*$ into its restriction to E_N, $T^*\varphi' = \varphi'|E_N$. Now, taking into account the equality $\mu(e; B'_{a^0} \cap E_N) = \mu(Te; B'_{a^0})$, $e \in E_N$, and applying the well-known formula of convex analysis[6]

$$\partial\mu(e_0; B'_{a^0} \cap E_N) = T^*\partial\mu(Te_0; B'_{a^0}),$$

we see that $\partial\mu(e_0; B'_{a^0} \cap E_N)$ comprises the sole functional $\nabla\mu(e_0; B'_{a^0})|E_N$.

(ii)⇒(i) Since e_0 is a tangency point of $B'_{a^0} \cap E_N$ in E_N and for any functional $\varphi' \in \partial\mu(e_0; B'_{a^0})$ one has $\varphi'|E_N \in \partial\mu(e_0; B'_{a^0} \cap E_N)$, it follows that $\varphi'|E_N = \nabla\mu(e_0; B'_{a^0} \cap E_N)$. We will assume without loss of generality (by passing, if needed from φ to $\alpha\varphi$ or $-\alpha\varphi$ where $\alpha > 0$) that $\varphi|E_N \in \partial\mu(e_0; B'_{a^0} \cap E_N)$ hence $\varphi|E_N = \nabla\mu(e_0; B'_{a^0} \cap E_N)$. Since φ' is supporting to B'_{a^0}, it belongs to $E(\sim_\varsigma)^0$ hence to E_N^*. We have thus two functionals from E_N^*, φ and φ', whose restrictions to E_N coincide. One has $\langle e, \varphi - \varphi' \rangle = 0$ whenever $e \in E_N$, and as the examined bilinear form on $E_N \times E_N^*$ is non-degenerate, this implies $\varphi' = \varphi$. We have thus proved that the subdifferential $\partial\mu(e_0; B'_{a^0})$ consists of the sole functional φ, and the result follows. □

Proof of Proposition 12.1: We have to show that the polar of the colour preprocessor M is precisely the threshold body B'_{a^0}:

$$M^0 := \{a \in E : \varphi_e(a) \leq 1 \quad \forall \varphi_e \in M\} = B'_{a^0}. \tag{B.27}$$

Since each colour mechanism $\varphi_e \in M$ is most sensitive for e (hence supporting to the threshold body at its boundary point $e \in C \subset bdry(B'_{a^0})$; see Definition 11.5), one has $\varphi_e(a) \leq \varphi_e(e) = 1$ for all $a \in B'_{a^0}$, that is, $B'_{a^0} \subseteq M^0$.

Let us show that $M^0 \subseteq B'_{a^0}$. Thus, let a be an arbitrary element of M^0. We assume that $a \notin E(\sim)$, otherwise the inclusion $a \in B'_{a^0}$ is obvious. Then $e = \alpha a$ is a boundary point of B'_{a^0} for some $\alpha > 0$. If

6 It follows from the Hahn–Banach theorem (see, e.g., Holmes, 1975).

$e \in C$, then $\varphi_e \in M$, and as $a \in M^0$, $\varphi_e(a) \le 1$. At the same time one has $\alpha\varphi_e(a) = \varphi_e(e) = 1$, hence $\varphi_e(a) = \frac{1}{\alpha}$. Then $\frac{1}{\alpha} \le 1$ and $a = \frac{1}{\alpha} e \in B'_{a^0}$.

It remains to consider the case where $e = \alpha a \notin C$. Since C is dense in the threshold body boundary, in this case there exists a net (generalised sequence) $(e_\nu) \subset C$ that tends (*weakly**) to e. Therefore, the net(a_ν), $a_\nu := \frac{1}{\alpha} e_\nu$, tends to $\frac{1}{\alpha} e = a$. Now, as $a_\nu \in B'_{a^0}$ (this is shown in the preceding paragraph) and B'_{a^0} is *weak** closed, $a = \lim_\nu a_\nu \in B'_{a^0}$, and equality (B.27) is thus completely established. □

Proof of Proposition 12.2: Since $\mathbf{d}$ is a detection function and e_0 is a threshold stimulus, one has $B'_{a^0} = L_\varepsilon[\mathbf{d}]$ and $\mathbf{d}(e_0) = \varepsilon$. Hence,

$$\mathbf{d}(e) \le \mathbf{d}(e_0) = \varepsilon \quad \forall e \in B'_{a^0}. \tag{B.28}$$

If now $\varphi \in \partial \mathbf{d}^\varepsilon(e_0)$ then for all $e \in B'_{a^0}$ $\varphi(e) - \varphi(e_0) \le \mathbf{d}(e) - \mathbf{d}(e_0) \le 0$, and the functional φ (hence $\phi := \frac{\varphi}{\varphi(e_0)}$) is thus supporting the threshold body at e_0. Moreover, $\phi(e) = \frac{\varphi(e)}{\varphi(e_0)} \le 1 = \phi(e_0)$ whenever $e \in B'_{a^0}$, that is, $\frac{\varphi}{\varphi(e_0)} \in ms\big(e_0|(B'_{a^0})^0\big)$. □

Proof of Theorem 14.1: (i)⇒(ii) Suppose the interior of S is empty. Taking into account that $0 \in S$, this means that $Z = Lin(S)$ is a proper vector subspace in $\mathbf{R}^n$. Therefore, there is a non-zero vector $c = (c_1, \dots, c_n) \in \mathbf{R}^n$ such that $\langle z, c\rangle = 0$ for all $z \in Z$. Now fix an arbitrary $\lambda_0 \in \Lambda$ and take in Λ any sequence of closed segments Δ_k such that $\lambda_0 \in \Delta_k$ $(k = 1, 2, \dots)$ and $\lim_{k\to\infty} |\Delta_k| = 0$, where $|\Delta_k|$ stands for the length of Δ_k. Since

$$\begin{aligned} z^k &:= \frac{1}{|\Delta_k|}\left(\int_{\Delta_k} \varphi_1(\lambda)\, d\lambda, \dots, \int_{\Delta_k} \varphi_n(\lambda)\, d\lambda\right) \\ &= \frac{1}{|\Delta_k|}\Big(\langle \chi_{\Delta_k}, \varphi_1\rangle, \dots, \langle \chi_{\Delta_k}, \varphi_n\rangle\Big) \in Z \end{aligned}$$

and all φ_i are continuous, we get

$$z := (\varphi_1(\lambda_0), \dots, \varphi_n(\lambda_0)) = \lim_{k\to\infty} z^k \in Z;$$

therefore, $\langle z, c\rangle = 0$, that is, $\sum_{i=1}^n c_i\varphi_i(\lambda_0) = 0$. Taking into account that λ_0 was taken arbitrarily from Λ, we conclude that, in contradiction with the assumption, $\varphi_1, \dots, \varphi_n$ are linearly dependent.

(ii)⇒(i) Suppose $\varphi_1, \dots, \varphi_n$ are linearly dependent, hence there is a non-zero vector $c = (c_1, \dots, c_n) \in \mathbf{R}^n$ such that $\sum_{i=1}^n c_i\varphi_i(\lambda) = 0$ whenever $\lambda \in \Lambda$. Then, for every $z = \Phi(x) \in S$,

$$\langle z, c\rangle = \sum_{i=1}^n c_i z_i = \sum_{i=1}^n c_i \int_\Lambda x(\lambda)\varphi_i(\lambda)\, d\lambda = \int_\Lambda x(\lambda)\sum_{i=1}^n c_i\varphi_i(\lambda)\, d\lambda = 0,$$

that is, S lies in the hyperplane $H = \{z : \langle z, c\rangle = 0\}$, in contradiction to (ii). □

Proof of Proposition 14.2: Recall (see Section A.1) that any element $x \in \mathcal{X}$ is identified with the measure $x(\lambda)d\lambda \in \mathbf{C}_+^*$, and as $\Phi(\mathbf{C}_+^*)$ is a closed convex cone (see Corollary 4.1), we get *cl con* $\Phi(\mathcal{X}) \subseteq \Phi(\mathbf{C}_+^*)$. In order to complete the proof, it remains to check that *con* $\Phi(\mathcal{X})$ is dense in $\Phi(\mathbf{C}_+^*)$. Since $\varphi_1, \dots, \varphi_n \in \mathbf{C}$, Φ is a weak* continuous linear operator $\mathbf{C}^* \to \mathbf{R}^n$; therefore, the density of *con* $\Phi(\mathcal{X}) = \Phi($*con* $\mathcal{X})$ in $\Phi(\mathbf{C}_+^*)$ will follow if we show that *con* $\mathcal{X}$ is weakly* dense in $\mathbf{C}_+^*$. Suppose *con* $\mathcal{X}$ is not weakly* dense in $\mathbf{C}_+^*$, that is, there is a measure $\sigma \in \mathbf{C}_+^*$ which does not belong to the weak* closure of *con* $\mathcal{X}$ (written as $w^* -$ *cl con* $\mathcal{X}$). Let $(\mathbf{C}^*, w^*)$ denote the locally convex linear topological space $\mathbf{C}^*$ equipped with the weak* topology. We have in $(\mathbf{C}^*, w^*)$ a closed convex cone $w^* - cl\ con\ \mathcal{X}$ and a point $\sigma \notin w^* - cl\ con\ \mathcal{X}$. By the separation theorem (see Holmes, 1975), there exists a continuous linear functional on $(\mathbf{C}^*, w^*)$, that is, a function $\varphi \in \mathbf{C}$, such that

$$\int_\Lambda \varphi(\lambda)\ \sigma(d\lambda) > \sup \left\{ \int_\Lambda \varphi(\lambda)x(\lambda)\ d\lambda\ :\ x \in w^* - cl\ con\ \mathcal{X} \right\}. \tag{B.29}$$

Since $w^* - cl\ con\ \mathcal{X}$ is a convex cone, the right-hand side of (B.29) equals 0, and we get

$$\int_\Lambda \varphi(\lambda)\ \sigma(d\lambda) > 0 \geq \int_\Lambda \varphi(\lambda)x(\lambda)\ d\lambda \tag{B.30}$$

whenever $x \in con\ \mathcal{X}$. It follows from the right-hand side of (B.30) that $\varphi(\lambda) \leq 0$. Indeed, if $\varphi(\lambda_0) > 0$ for some $\lambda_0 \in \Lambda$ then, by continuity of φ, there is a segment $\Lambda_0 \subset \Lambda$ such that $\varphi(\lambda) > 0$ for all $\lambda \in \Lambda_0$. Then the indicator of Λ_0, $x = \chi_{\Lambda_0}$ belongs to $\mathcal{X}$, and we get $\int_\Lambda \varphi(\lambda)x(\lambda)\ d\lambda = \int_{\Lambda_0} \varphi(\lambda)x(\lambda)\ d\lambda > 0$. This contradicts the right-hand side of (B.30), so $\varphi(\lambda) \leq 0$, but this, in turn, contradicts the left-hand side of (B.30). Thus $con\ \mathcal{X}$ is weak* closed in $\mathbf{C}_+^*$, and the proof is completed. □

Proof of Lemma 14.1: Let us suppose $\Phi(x_0) \in \partial S = \partial\Phi(\mathcal{X})$ and show that $(x_0 + \ker(\Phi)) \cap int(\mathcal{X}) = \emptyset$. Indeed, otherwise there should be an element x in this set, and since Φ is a continuous (even more, *weak** continuous) linear map of $\mathcal{X}$ onto S, $\Phi(x)$ has to belong to $int(S)$. On the other hand, since $x - x_0 \in \ker(\Phi)$ and Φ is a linear operator, one has $\Phi(x) = \Phi(x_0) + \Phi(x - x_0) = \Phi(x_0) \in \partial S$, a contradiction. □

Proof of Theorem 14.2: Suppose that $x_0 = \chi_B$ is an optimal object stimulus, that is, $\Phi(x_0) \in \partial S = \partial\Phi(\mathcal{X})$. Then, by Lemma 14.1, $(x_0 + \ker(\Phi)) \cap int(\mathcal{X}) = \emptyset$. By the separation theorem (e.g., see Dunford & Schwartz, 1958, Chapter V, 2, Th. 8), there exists a continuous linear functional $f \neq 0$ on the space $\mathbf{L}^\infty$ separating the convex set $x_0 + \ker(\Phi)$ and the open convex set $int(\mathcal{X})$, so that

$$\langle x_0 + y, f \rangle \geq \alpha \geq \langle x, f \rangle \tag{B.31}$$

whenever $y \in \ker(\Phi)$, $x \in \mathcal{X}$.

The functional f can, generally speaking, be an element of the space $(\mathbf{L}^\infty)^*$ which is broader than $\mathbf{L}^1$. However, in this particular case since $\ker(\Phi)$ is a vector subspace, the left hand side inequality in (B.31) implies $\langle y, f \rangle = 0$ for all $y \in \ker(\Phi)$, that is, $\ker(\Phi)$ is contained in the hyperplane $\{y : \langle y, f \rangle = 0\}$. As well-known (e.g., see Holmes, 1975), in this case f can be represented as a linear combination of functionals φ_i, $i = 1, \ldots, n$. In other words, there are real numbers $\xi_1, \ldots, \xi_n$, such that

$$f = \sum_{i=1}^{n} \xi_i \varphi_i. \tag{B.32}$$

Therefore, $f \in \mathbf{L}^1$, and as $f \neq 0$, we have $(\xi_1, \ldots, \xi_n) \neq (0, \ldots, 0)$. It also follows from (B.31) that

$$\langle x_0, f \rangle \geq \langle x, f \rangle \text{ for all } x \in \mathcal{X}. \tag{B.33}$$

Let us show now that $f(\lambda) \geq 0$ a.e. on B, and $f(\lambda) \leq 0$ a.e. on $\Lambda \backslash B$.

If there is a set of positive measure $B' \subset B$ such that $f(\lambda) < 0$ on B', then $\langle x_0, f \rangle = \langle \chi_B, f \rangle = \int_B f(\lambda)\,d\lambda < \int_{B\backslash B'} f(\lambda)\,d\lambda = \langle \chi_{B\backslash B'}, f \rangle$, that contradicts (B.33) for $x = \chi_{B\backslash B'}$.

Likewise, if there is a set of positive measure $A \subset \Lambda \backslash B$ such that $f(\lambda) > 0$ on A, then $\langle x_0, f \rangle = \int_B f(\lambda)\,d\lambda < \int_B f(\lambda)\,d\lambda + \int_A f(\lambda)\,d\lambda = \int_{A\cup B} f(\lambda)\,d\lambda = \langle \chi_{A\cup B}, f \rangle$, that contradicts (B.33) for $x = \chi_{A\cup B}$.

So, we have proved that for an optimal object stimulus $x_0 = \chi_B$ there are real numbers $\xi_1, \ldots, \xi_n$ not all of which are equal to zero and such that inequalities (14.5) and (14.6) hold true.

Suppose now that for some real numbers $\xi_1, \ldots, \xi_n$ not all equal to zero (14.5) and (14.6) hold true. Let us consider f as defined by (B.32). According to Lemma 14.1, optimality of

$x_0 = \chi_B$ will be established if we show that $(x_0 + \ker(\Phi)) \cap int(\mathcal{X}) = \emptyset$. Suppose that there is an $x \in (x_0 + \ker(\Phi)) \cap int(\mathcal{X})$. Then $x = x_0 + y$ where $y \in \ker(\Phi)$ and $0 < \delta \le x(\lambda) \le 1 - \delta$. It follows from (B.32) that $\langle y, f\rangle = 0$. Since $x = x_0 + y$, we have $\langle x, f\rangle = \langle x_0 + y, f\rangle = \langle x_0, f\rangle = \int_B f(\lambda)\,d\lambda$. On the other hand, $0 < x(\lambda) \le 1 - \delta$, hence $\langle x, f\rangle = \int_\Lambda f(\lambda)x(\lambda)\,d\lambda \le \int_B f(\lambda)x(\lambda)\,d\lambda \le (1-\delta)\int_B f(\lambda)\,d\lambda = (1-\delta)\langle x_0, f\rangle < \langle x_0, f\rangle$. We obtain a contradiction, which completes the proof. □

Proof of Proposition 14.3: Since the Schrödinger–Maximov constraint holds true, for any two different λ_1, λ_2 in Λ there exists a non-zero vector $(\xi_1, \xi_2, \xi_3) \in \mathbf{R}^3$ such that λ_1 and λ_2 are the only solutions to

$$\xi_1\varphi_1(\lambda) + \xi_2\varphi_2(\lambda) + \xi_3\varphi_3(\lambda) = 0. \tag{B.34}$$

Fix any λ_1, λ_2 satisfying $\lambda_{\min} \le \lambda_1 < \lambda_2 \le \lambda_{\max}$. If we could find $(\xi_1, \xi_2, \xi_3) \in \mathbf{R}^3$ such that λ_1, λ_2 satisfy (B.34) and

$$\xi_1\frac{d\varphi_1(\lambda_j)}{d\lambda} + \xi_2\frac{d\varphi_2(\lambda_j)}{d\lambda} + \xi_3\frac{d\varphi_3(\lambda_j)}{d\lambda} \ne 0, \quad j = 1, 2, \tag{B.35}$$

then the graph of the function

$$\xi_1\varphi_1(\lambda) + \xi_2\varphi_2(\lambda) + \xi_3\varphi_3(\lambda) \tag{B.36}$$

will intersect the abscissa axis in both points, λ_1 and λ_2 (in such a case, the signs of two values in (B.35) prove to be opposite); therefore, function (B.36) will be positive in the interior points of the segment $[\lambda_1, \lambda_2]$ and negative outside it or vice versa. Replacing (ξ_1, ξ_2, ξ_3) with $(-\xi_1, -\xi_2, -\xi_3)$, one can see that in both cases $x_2^{\lambda_1,\lambda_2}$ and $1 - x_2^{\lambda_1,\lambda_2}$ are optimal object stimuli. Thus, the result will follow if we show that there is a non-zero vector $(\xi_1, \xi_2, \xi_3) \in \mathbf{R}^3$ such that λ_1 and λ_2 satisfy (B.34) and (B.35).

To this end, consider a system of three linear equations:

$$\begin{aligned}
&\xi_1\varphi_1(\lambda_1) + \xi_2\varphi_2(\lambda_1) + \xi_3\varphi_3(\lambda_1) = 0,\\
&\xi_1\varphi_1(\lambda_2) + \xi_2\varphi_2(\lambda_2) + \xi_3\varphi_3(\lambda_2) = 0,\\
&\xi_1\frac{d\varphi_1}{d\lambda}(\lambda_1) + \xi_2\frac{d\varphi_2}{d\lambda}(\lambda_1) + \xi_3\frac{d\varphi_3}{d\lambda}(\lambda_1) = 1.
\end{aligned}$$

By virtue of the regularity assumption, this system has a unique solution ξ_1, ξ_2, ξ_3. Moreover, it follows from the third of these equations that function (B.36) increases in a neighbourhood of λ_1, and as it equals 0 only in two points, λ_1 and λ_2, it is strictly positive between these points. Therefore, its derivative at λ_2 cannot be positive:

$$\xi_1\frac{d\varphi_1}{d\lambda}(\lambda_2) + \xi_2\frac{d\varphi_2}{d\lambda}(\lambda_2) + \xi_3\frac{d\varphi_3}{d\lambda}(\lambda_2) \le 0. \tag{B.37}$$

If (B.37) is satisfied as the strict inequality, then (B.35) holds true, and the result follows.

Suppose now that (B.37) holds with the equality sign, and applying again the regularity assumption, find a unique solution ξ_1', ξ_2', ξ_3' to the system:

$$\begin{array}{ccccccc}
\xi_1'\varphi_1(\lambda_1) & + & \xi_2'\varphi_2(\lambda_1) & + & \xi_3'\varphi_3(\lambda_1) & = & 0,\\
\xi_1'\varphi_1(\lambda_2) & + & \xi_2'\varphi_2(\lambda_2) & + & \xi_3'\varphi_3(\lambda_2) & = & 0,\\
\xi_1'\frac{d\varphi_1}{d\lambda}(\lambda_2) & + & \xi_2'\frac{d\varphi_2}{d\lambda}(\lambda_2) & + & \xi_3'\frac{d\varphi_3}{d\lambda}(\lambda_2) & = & -1.
\end{array}$$

Arguing as before, we get

$$\xi_1'\frac{d\varphi_1}{d\lambda}(\lambda_1) + \xi_2'\frac{d\varphi_2}{d\lambda}(\lambda_1) + \xi_3'\frac{d\varphi_3}{d\lambda}(\lambda_1) \ge 0. \tag{B.38}$$

If (B.38) is satisfied as the strict inequality, then (B.35) holds true, and the result follows. If (B.38) is satisfied as the equality, then (B.34) and (B.35) are satisfied for the vector $(\xi_1 + \xi_1', \xi_2 + \xi_2', \xi_3 + \xi_3') \in \mathbf{R}^3$, and the result follows as well. □

Proof of Proposition 15.1: By the definition of the equator (see Definition 15.1), $z \in Eqt(S)$ if and only if there exists a non-zero vector $(\xi_1, \dots, \xi_n) \in \mathbf{R}^n$ satisfying (15.5) and such that $z \in H(\xi_1, \dots, \xi_n; c(\xi_1, \dots, \xi_n)) \cap S$. The result will follow if we show that the latter, in turn, is equivalent to the validity of the representation $z = \Phi'(\chi_B)$ where B satisfies

$$\{\lambda : \xi_1\varphi_1'(\lambda) + \cdots + \xi_n\varphi_n'(\lambda) > 0\} \subset B \subseteq \{\lambda : \xi_1\varphi_1'(\lambda) + \cdots + \xi_n\varphi_n'(\lambda) \geq 0\}. \quad \text{(B.39)}$$

Let us define a function $g(\lambda) = \xi_1\varphi_1'(\lambda) + \cdots + \xi_n\varphi_n'(\lambda)$. If B satisfies (B.39) then for every $x \in \mathcal{X}$ the inequality holds: $\int_B g(\lambda)d\lambda \geq \int_\Lambda x(\lambda)g(\lambda)d\lambda$. If now $z = \Phi'(\chi_B)$ then $z \in S$, and for every $z' = \Phi'(x) \in S$ we have

$$\xi_1 z_1 + \cdots + \xi_n z_n = \int_B g(\lambda)d\lambda \geq \int_\Lambda x(\lambda)g(\lambda)d\lambda = \xi_1 z_1' + \cdots + \xi_n z_n';$$

Therefore, $\int_B g(\lambda)d\lambda = c(\xi_1, \dots, \xi_n)$ and $z \in H(\xi_1, \dots, \xi_n; c(\xi_1, \dots, \xi_n))$. On the other hand, any $z \in S$ has a representation $z = \Phi'(\chi_{B'})$ (if $z \in \partial S$, it follows from Theorem 14.2 that such a set B' is unique up to Lebesgue equivalence, but we make no use of this fact); so if $z \in H(\xi_1, \dots, \xi_n; c(\xi_1, \dots, \xi_n)) \cap S$, we take such a representation and show that, up to Lebesgue equivalence, B' satisfies (B.39). Indeed, if not, then at least one of two sets $B_1 = \{\lambda \in B' : g(\lambda) < 0\}$ and $B_2 = \{\lambda \in \Lambda \setminus B' : g(\lambda) > 0\}$ is of positive Lebesgue measure. Therefore,

$$c(\xi_1, \dots, \xi_n) = \xi_1 z_1 + \cdots + \xi_n z_n = \int_{B'} g(\lambda)d\lambda < \int_{(B'\setminus B_1)\cup B_2} g(\lambda)d\lambda$$

$$= \xi_1 \int_{(B'\setminus B_1)\cup B_2} \varphi_1'(\lambda)d\lambda + \cdots + \xi_n \int_{(B'\setminus B_1)\cup B_2} \varphi_n'(\lambda)d\lambda,$$

and as $\left(\int_{(B'\setminus B_1)\cup B_2} \varphi_1'(\lambda)d\lambda, \dots, \int_{(B'\setminus B_1)\cup B_2} \varphi_n'(\lambda)d\lambda\right) = \Phi'(\chi_{(B'\setminus B_1)\cup B_2}) \in S$, this contradicts the definition of $c(\xi_1, \dots, \xi_n)$. □

Proof of Theorem 16.1: If $x = \chi_B \in \mathcal{X}_0$, then $x \in Ext(\mathcal{X}_0)$. This is obvious. Let us prove the opposite. Suppose $x \in \mathcal{X}_0$ is not an indicator. Then there is a set B_1 such that $mes(B_1) > 0$ and $0 < \delta \leq x(\lambda) \leq 1 - \delta$ for $\lambda \in B_1$. Since $mes(B_1) > 0$, the space $\mathbf{L}^\infty(B_1)$ is infinite-dimensional. Therefore, there exists a non-zero function $x_* \in \mathbf{L}^\infty(B_1)$ satisfying

$$\int_{B_1} \varphi_i(\lambda)x_*(\lambda)\,d\lambda = 0, \quad i = 1, \dots, n.$$

We normalise it so that $\|x_*\|_\infty \leq \delta$, that is, $-\delta \leq x_*(\lambda) \leq \delta$ almost everywhere on B_1. Let us extend x_* to the whole Λ by setting $x_*(\lambda) = 0$ for $\lambda \in \Lambda\setminus B_1$. It is obvious that $\Phi(x_*) = 0$. In view of linearity of Φ, both $x_1 = x + x_*$ and $x_2 = x - x_*$ belong to $\mathcal{X}_0$. So, we have $x_1, x_2 \in \mathcal{X}_0$, $x_1 \neq x_2$, and $x = \frac{1}{2}(x_1 + x_2)$. Thus, x is not an extreme point of $\mathcal{X}_0$. □

Proof of Theorem 16.2: By Theorem 14.2, there exists non-zero vector $(\xi_1, \dots, \xi_n) \in \mathbf{R}^n$ such that, for the function

$$f(\lambda) = \sum_{i=1}^{n} \xi_i\varphi_i(\lambda)$$

the following inequalities hold

$$f(\lambda) \geq 0 \text{ a.e. on } B, \tag{B.40}$$

$$f(\lambda) \leq 0 \text{ a.e. on } \Lambda \backslash B. \tag{B.41}$$

Suppose $x \in Met(\varphi_1, \dots, \varphi_n; x_0)$, hence $\Phi(x) = \Phi(x_0)$. Then

$$\langle x, f \rangle = \sum_{i=1}^{n} \xi_i \langle x, \varphi_i \rangle = \sum_{i=1}^{n} \xi_i \langle x_0, \varphi_i \rangle = \langle x_0, f \rangle. \tag{B.42}$$

On the other hand,

$$\langle x, f \rangle = \int_B x(\lambda) f(\lambda)\, d\lambda + \int_{\Lambda \backslash B} x(\lambda) f(\lambda)\, d\lambda. \tag{B.43}$$

Since $x \in \mathcal{X}$, one has $0 \leq x(\lambda) \leq 1$. Taking into account (B.40) and (B.41) we get

$$\int_B x(\lambda) f(\lambda)\, d\lambda \leq \int_B f(\lambda)\, d\lambda = \langle x_0, f \rangle, \tag{B.44}$$

$$\int_{\Lambda \backslash B} x(\lambda) f(\lambda)\, d\lambda \leq 0. \tag{B.45}$$

Now it follows from (B.42)–(B.45) that $x(\lambda) = 1$ a.e. on B and $x(\lambda) = 0$ a.e. on $\Lambda \backslash B$, that is, $x = x_0$. □

Proof of Corollary 16.1: First show that $Opt(\mathcal{X})$ is *weak** closed in $\mathbf{L}^\infty = \mathbf{L}^{1*}$. Indeed, since $\mathbf{L}^1$ is separable, the restriction of the *weak** topology to every bounded set in $\mathbf{L}^\infty$, particularly to $\mathcal{X}$, is metrisable. Therefore, it suffices to verify that the limit $x \in \mathcal{X}$ of every *weak** convergent sequence $(x_n) \subset Opt(\mathcal{X})$ lies in $Opt(\mathcal{X})$. Since x_n converges *weakly** to x and Φ is *weak** continuous as an operator $\mathbf{L}^\infty \to \mathbf{R}^n$, $\Phi(x_n)$ converges to $\Phi(x)$ in $\mathbf{R}^n$. Now, as all $\Phi(x_n)$ lie in ∂S and ∂S is closed in $\mathbf{R}^n$, we get $\Phi(x) \in \partial S$ that is, $x \in Opt(\mathcal{X})$. The *weak** closedness of $Opt(\mathcal{X})$ is thus established. Therefore, $Opt(\mathcal{X})$ is *weak** compact. Now it remains to notice that, being a one-to-one continuous map of a compact Hausdorf space, $\Phi : Opt(\mathcal{X}) \to \partial S$ is a homeomorphism. □

Proof of Proposition 16.1: Since $\mathcal{A}$ is a *weak** compact atlas, Φ' and Φ'' are *weak** continuous, it follows that $\Phi'(\mathcal{A})$ and $\Phi''(\mathcal{A})$ are compact subsets in $\mathbf{R}^n$ and $\rho_{\mathcal{A}} = \Phi'' \circ \Phi'|_{\mathcal{A}}^{-1} : \Phi'(\mathcal{A}) \to \Phi''(\mathcal{A})$ is a continuous one-to-one map hence a homeomorphism. □

Proof of Lemma 18.1: Consider the set

$$\mathcal{X}_z := \Psi^{-1}(z) \cap \mathcal{X}_0 = \{x \in \mathcal{X}_0 : \Psi(x) = z\}.$$

This set is convex and *weak** compact. Therefore, by the Krein–Milman theorem, $Ext(\mathcal{X}_z) \neq \emptyset$. Let us choose an arbitrary $x \in Ext(\mathcal{X}_z)$ and show that x is an extreme point of $\mathcal{X}_0$. Indeed, if not, then there exist $x_1, x_2 \in \mathcal{X}_0$, $x_1 \neq x_2$, such that $x = \frac{1}{2}(x_1 + x_2)$. We have $z = \Psi(x) = \frac{1}{2}(z^1 + z^2)$, where $z^1 = \Psi(x_1)$ and $z^2 = \Psi(x_2)$. Since $z^1, z^2 \in \Psi(\mathcal{X}_0)$ and $z \in Ext(\Phi(\mathcal{X}_0))$, one has $z^1 = z^2 = z$. The latter entails $x_1, x_2 \in \mathcal{X}_z$, which contradicts the assumption that $x \in Ext(\mathcal{X}_z)$. Thus, we conclude that $x \in Ext(\mathcal{X}_0)$.

It remains to verify (18.2). If it fails then there is $x_1 \in (x + \ker(\Psi)) \cap ri(\mathcal{X}_0) = (x + \ker(\Psi)) \cap (x_0 + int_E(M))$. This implies $\Psi(x_1 - x_0) \in int(\Psi(M))$ hence $\Psi(x_1) \in \Psi(x_0) + int(\Psi(M)) = int(\Psi(\mathcal{X}_0))$. However, the latter cannot be true because $\Psi(x_1) = \Psi(x) = z \in Ext(\Psi(\mathcal{X}_0))$. □

Proof of Lemma 18.2: Suppose z is not a boundary point of $\Psi(\mathcal{X}_0)$, that is, $z \in int(\Psi(\mathcal{X}_0)) = \Psi(x_0) + int(\Psi(M))$. Then there exists a point $x_1 \in x_0 + int_E(M) = ri(\mathcal{X}_0)$ such that $z = \Psi(x_1)$.

But then $x_1 - x \in \ker(\Psi)$, which, in turn, implies $x_1 \in (x + \ker(\Psi)) \cap ri(\mathcal{X}_0)$ in contradiction to (18.2). □

Proof of Lemma 18.3: We have $x - x_0 \in E$ and $(x - x_0 + \ker(\Psi)) \cap int_E(M) = \emptyset$; then by the separation theorem, there is a linear functional $e^* \in E^*, e^* \neq 0$, separating two convex sets in E, $x - x_0 + (\ker(\Psi) \cap E)$ and $int_E(M)$, so that

$$\langle x - x_0 + e, e^* \rangle \geq \alpha > \langle y, e^* \rangle \text{ for all } e \in \ker(\Psi) \cap E,\ y \in int_E(M);$$

therefore,

$$\langle x - x_0 + e, e^* \rangle \geq \alpha \geq \langle y, e^* \rangle \text{ for all } e \in \ker(\Psi) \cap E,\ y \in M. \tag{B.46}$$

The left hand side inequality implies

$$\langle e, e^* \rangle = 0 \text{ for all } e \in \ker(\Psi) \cap E, \tag{B.47}$$

and the right hand side inequality implies

$$\langle y, e^* \rangle \leq \alpha \text{ for all } y \in M. \tag{B.48}$$

Let us consider the subspace $E_1 := E + \ker(\Psi)$ in $\mathbf{L}^\infty$ and extend e^* to a linear functional e_1^* on E_1 by the formula

$$\langle e + x, e_1^* \rangle := \langle e, e^* \rangle \text{ for all } e \in E, x \in \ker(\Psi). \tag{B.49}$$

So extended functional e_1^* is well-defined, since if $e + x = e' + x', e' \in E, x' \in \ker(\Psi)$, then $e - e' = x' - x \in \ker(\Psi) \cap E$, and applying (B.47) yields $\langle e - e', e_1^* \rangle = \langle e - e', e^* \rangle = 0$ hence $\langle e + x, e_1^* \rangle = \langle e' + x', e_1^* \rangle$. Furthermore, it follows from (B.47) and (B.49) that

$$\langle x, e_1^* \rangle = 0 \text{ for all } x \in \ker(\Psi). \tag{B.50}$$

Now we extend e_1^* to a linear functional f on the whole of $\mathbf{L}^\infty$ as follows. We fix some (arbitrary) Hamel's base (= a maximum linear independent system of vectors) $\mathcal{B}(E_1)$ in E_1 and find, by applying Theorem I.1.1 in Day (1958), a Hamel's base $\mathcal{B}(\mathbf{L}^\infty)$ in $\mathbf{L}^\infty$ that contains $\mathcal{B}(E_1)$. Let $\mathcal{B}(\mathbf{L}^\infty) = \{x_\gamma : \gamma \in \Gamma\}$ where Γ is an infinite set of indices, and let

$$\Gamma_1 := \{\gamma \in \Gamma : x_\gamma \in \mathcal{B}(E_1)\}.$$

Every $x \in \mathbf{L}^\infty$ is represented uniquely in form $x = \sum_{\gamma \in \Gamma(x)} \alpha_\gamma x_\gamma$, where $\Gamma(x)$ is a finite subset of Γ. We define

$$\langle x, f \rangle = \sum_{\gamma \in \Gamma(x) \cap \Gamma_1} \alpha_\gamma \langle x_\gamma, e_1^* \rangle.$$

Then

$$\langle x, f \rangle = \langle x, e_1^* \rangle \text{ for } x \in E_1, \tag{B.51}$$

so that f is actually an extension of e_1^* to $\mathbf{L}^\infty$. A priori, the linear functional f need not be continuous; however, in fact, it proves to be even weak* continuous. Indeed, it follows from (B.51) and (B.50) that $\langle x, f \rangle = 0$ for all $x \in \ker(\Psi)$; therefore, by Lemma V.3.10 in Dunford & Schwartz (1958) (see also Holmes, 1975), f is a linear combination of $\psi_1, \ldots, \psi_n$, $f = \sum_{i=1}^n \xi_i \psi_i$ and hence $f \in \mathbf{L}^1$. Since $f \neq 0$, not all of ξ_i are equal to zero. Now, for $e = 0$ (B.46) yields

$$\int_\Lambda f(\lambda) x(\lambda)\, d\lambda \geq \int_\Lambda f(\lambda) x'(\lambda)\, d\lambda \text{ for all } x' = x_0 + y \in \mathcal{X}_0,$$

and the result follows. □

Proof of Theorem 18.1: I. It follows from Lemmas 18.1 and 18.3 that there exist (i) an indicator $x = \chi_B \in Ext(\mathcal{X}_0)$, and (ii) n real numbers $\xi_1, \dots, \xi_n$, not all of which are equal to zero, such that for the function f as defined by (18.5) the equality (18.3) holds. Let us consider an infinite linear program having the form of maximising the linear functional on $\mathbf{L}^\infty$,

$$\langle x', f \rangle = \int_\Lambda f(\lambda) x'(\lambda)\, d\lambda,$$

subject to the following constraints:

$$0 \le x'(\lambda) \le 1$$

$$\int_\Lambda \varphi_1(\lambda) x'(\lambda)\, d\lambda = r_1,$$

$$\dots$$

$$\int_\Lambda \varphi_n(\lambda) x'(\lambda)\, d\lambda = r_n,$$

where $r_i = \int_\Lambda \varphi_i(\lambda) x_0(\lambda)\, d\lambda,\ i = 1, \dots, n$.

The equality (18.3) means that the maximum value of the linear program is attained at $x = \chi_B$. Recall that $\Phi(x_0) \in int(\Phi(\mathcal{X}))$. It follows then from the definition of $\mathcal{X}_0$ that there exists $x_* \in ri(\mathcal{X}_0) = x_0 + int_E(M) \subset int(\mathcal{X})$ such that $\Phi(x_*) = \Phi(x_0)$. This means that x_* satisfies all the constraints of the linear program and the first (inequality) constraint is satisfied strictly in the sense that $\delta \le x_*(\lambda) \le 1 - \delta$ a.e. on Λ where $0 < \delta < 1$. As follows from a generalised Kuhn–Tucker theorem (see Levin, 1985, p. 69), in such a case there exist real numbers $\alpha_1, \dots, \alpha_n$ such that the maximum of the functional $g = f + \sum_{i=1}^{n} \alpha_i \varphi_i$ on the set $\mathcal{X}$ equals the maximum value of the linear program, and it is attained at the same element $x = \chi_B$. But this means exactly (18.6). Indeed, if there is a set of positive measure $B' \subset B$ such that $g(\lambda) < 0$ on B', then

$$\langle \chi_B, g \rangle = \int_B g(\lambda)\, d\lambda < \int_{B \setminus B'} g(\lambda)\, d\lambda = \langle \chi_{B \setminus B'}, g \rangle,$$

in contradiction to the maximality of g on $\mathcal{X}$. Similarly, if there is a set of positive measure $A \subset \Lambda \setminus B$ such that $g(\lambda) > 0$ on A, then

$$\langle \chi_B, g \rangle = \int_B g(\lambda)\, d\lambda < \int_{A \cup B} g(\lambda)\, d\lambda = \langle \chi_{A \cup B}, g \rangle,$$

again, in contradiction to the maximality of g on $\mathcal{X}$.

II. If we show that (18.2) holds true, the result will follow from Lemma 18.2. Suppose then that (18.2) fails, that is, one can find an element $x_* \in (x + \ker(\Psi)) \cap ri(\mathcal{X}_0)$. Then one has $x_* = x + e$, where $\Psi(e) = 0$, and hence

$$\Psi(x_*) = \Psi(x). \tag{B.52}$$

At the same time, both x and x_* belong to $\mathcal{X}_0$, thus

$$\Phi(x_*) = \Phi(x) = \Phi(x_0). \tag{B.53}$$

Now, from (B.52) and (B.53) we derive

$$\langle x_*, g \rangle = \langle x, g \rangle. \tag{B.54}$$

On the other hand, since $x_* \in ri(\mathcal{X}_0) \subseteq int(\mathcal{X})$, there is a number δ, $0 < \delta < 1$, such that $\delta \le x_*(\lambda) \le 1 - \delta$ almost everywhere, so we have got

$$\langle x_*, g\rangle \leq (1-\delta)\int_\Lambda g(\lambda)\,d\lambda \leq (1-\delta)\int_B g(\lambda)\,d\lambda$$
$$= (1-\delta)\langle \chi_B, g\rangle = (1-\delta)\langle x, g\rangle < \langle x, g\rangle$$

in contradiction to (B.54). □

Proof of Proposition 18.1: Taking into account that $\Phi(x) = z_0$ and $\|x\|_\infty \leq 1$ whenever $x \in \mathcal{X}_0$, we get for every $z = \Psi(x) \in \Psi(\mathcal{X}_0)$

$$\|z_0 - z\| = \sqrt{\sum_{i=1}^{n}\langle x, \psi_i - \varphi_i\rangle^2} \leq \sqrt{\sum_{i=1}^{n}\|\psi_i - \varphi_i\|_1^2}$$
$$\leq \sqrt{n(\max_{i=1,\ldots,n}\|\psi_i - \varphi_i\|_1)^2} = \varepsilon.$$

□

Proof of Corollary 18.3: Given arbitrary points $z = \Psi(x) \in \Psi(\mathcal{X}_0)$, $z' = \Psi'(x') \in \Psi'(\mathcal{X}_0)$, we have

$$\|z - z'\| \leq \|z - z_0\| + \|z_0 - z'\| \leq \varepsilon + \varepsilon',$$

and the result follows. □

Proof of Corollary 18.4: Proposition 18.1 for Φ, Ψ replaced with Ψ, Ψ' implies the estimation

$$\|z_0 - z\| \leq \sqrt{\sum_{i=1}^{n}\|\psi_i' - \psi_i\|_1^2} \tag{B.55}$$

whenever $z \in \Psi'(\mathcal{X}_0)$. Further, as

$$\|\psi_i' - \psi_i\|_1^2 = \left(\int_\Lambda |p'(\lambda) - p(\lambda)|\varphi_i(\lambda)d\lambda\right)^2 \tag{B.56}$$

$$= \left(\int_\Lambda |1 - \frac{p(\lambda)}{p'(\lambda)}|p'(\lambda)\varphi_i(\lambda)d\lambda\right)^2, \quad i = 1, \ldots, n, \tag{B.57}$$

applying the Cauchy–Schwartz inequality (see Section A.1) yields

$$\left(\int_\Lambda |1 - \frac{p(\lambda)}{p'(\lambda)}|p'(\lambda)\varphi_i(\lambda)d\lambda\right)^2 \leq \int_\Lambda |1 - \frac{p(\lambda)}{p'(\lambda)}|^2 d\lambda \int_\Lambda p'^2(\lambda)\varphi_i^2(\lambda)d\lambda, \quad i = 1, \ldots, n, \tag{B.58}$$

and the required estimation results from (B.55)–(B.58). □

Proof of Proposition 18.2: We first compute the diameter of the object-colour solid, $diam\Psi'(\mathcal{X})$. To this end, we consider any two points in $\Psi'(\mathcal{X})$, $z^1 = \Psi'(x_1)$ and $z^2 = \Psi'(x_2)$, and compute the Euclidean distance between them. Taking into account that $p'(\lambda)\varphi_i(\lambda) \geq 0$, we get

$$\|z^1 - z^2\| = \left(\sum_{i=1}^{n}\left(\int_\Lambda p'(\lambda)\varphi_i(\lambda)(x_1(\lambda) - x_2(\lambda))d\lambda\right)^2\right)^{\frac{1}{2}}$$
$$\leq \left(\sum_{i=1}^{n}\left(\int_\Lambda p'(\lambda)\varphi_i(\lambda)|x_1(\lambda) - x_2(\lambda)|d\lambda\right)^2\right)^{\frac{1}{2}}$$
$$\leq \left(\sum_{i=1}^{n}\left(\int_\Lambda p'(\lambda)\varphi_i(\lambda)d\lambda\right)^2\right)^{\frac{1}{2}} = \|\Psi'(1) - \Psi'(0)\|;$$

therefore

$$diam\Psi'(\mathcal{X}) = \|\Psi'(1) - \Psi'(0)\| = \left(\sum_{i=1}^{n}\left(\int_{\Lambda} p'(\lambda)\varphi_i(\lambda)d\lambda\right)^2\right)^{\frac{1}{2}} = \sqrt{\sum_{i=1}^{n}\|\psi_i'\|_1^2}. \tag{B.59}$$

Let us evaluate $diam\Psi'(\mathcal{X}_0(x_0))$. Given two arbitrary points $z^1, z^2 \in \Psi'(\mathcal{X}_0(x_0))$, we have $\|z^1 - z^2\| \leq \|z^1 - z_0\| + \|z_0 - z^2\|$, where $z_0 = \Psi(x_0)$. Then $\|z^1 - z^2\| \leq 2\max_{z \in \Psi'(\mathcal{X}_0(x_0))}\|z_0 - z\|$, and applying Corollary 18.4 yields

$$diam\Psi'(\mathcal{X}_0(x_0)) = \max_{z^1, z^2 \in \Psi'(\mathcal{X}_0(x_0))}\|z^1 - z^2\|$$
$$\leq 2\sqrt{\int_{\Lambda}\left|\frac{p(\lambda)}{p'(\lambda)} - 1\right|^2 d\lambda}\sqrt{\sum_{i=1}^{n}\int_{\Lambda} p'^2(\lambda)\varphi_i^2(\lambda)d\lambda},$$

which can be rewritten as

$$diam\Psi'(\mathcal{X}_0(x_0)) \leq 2\sqrt{\int_{\Lambda}\left|\frac{p(\lambda)}{p'(\lambda)} - 1\right|^2 d\lambda}\sqrt{\sum_{i=1}^{n}\|\psi_i'\|_2^2}. \tag{B.60}$$

Now, taking into account equality (18.13) and applying (B.59) and (B.60) we get

$$I_1(x_0; \Psi') \leq 2\sqrt{\int_{\Lambda}\left|\frac{p(\lambda)}{kp'(\lambda)} - 1\right|^2 d\lambda}\frac{\sqrt{\sum_{i=1}^{n}\|\psi_i'\|_2^2}}{\sqrt{\sum_{i=1}^{n}\|\psi_i'\|_1^2}},$$

and as this holds true for any $k > 0$, we can take minimum in k. It remains to show that this minimum is attained at $k = k_0$ as given by (18.15). But this minimum is the $\mathbf{L}^2$-distance of the element 1_Λ from the line through $f = \frac{p}{p'}$. Indeed, $\int_\Lambda |\frac{p(\lambda)}{kp'(\lambda)} - 1|^2 d\lambda = \|\alpha f - 1_\Lambda\|_2$ where $\alpha = \frac{1}{k}$; therefore, the optimal α is determined from the condition that $1_\Lambda - \alpha f$ is orthogonal to f, which gives value (18.15) for k. □

Proof of Proposition 18.3: (i) ⇒ (ii) As above, let us define $L := Lin\{\varphi_1, \dots, \varphi_n\}$, $L^0 := \{y \in \mathbf{L}^\infty : \langle\varphi_i, y\rangle = 0,\ i = 1, \dots, n\}$, $H_{x_0} := -x_0 + Met(\varphi_1, \dots, \varphi_n; x_0)$, and $H_{x_0}^0 := \{\psi \in \mathbf{L}^1 : \langle\psi, y\rangle = 0\ \forall y \in H_{x_0}\}$. It follows from (i) that $\psi_1, \dots, \psi_n \in H_{x_0}^0$. Therefore, if we show that $H_{x_0}^0 = L$ the statement (ii) will be proved. Fix any $y \in L^0$ and choose $\varepsilon > 0$ such that $0 \leq x_0 + \varepsilon y \leq 1$. We have $x_0 + \varepsilon y \in Met(\varphi_1, \dots, \varphi_n; x_0)$, so $\varepsilon y \in H_{x_0}$. Since y is an arbitrary element of L^0, and $H_{x_0}^0 \subseteq L^0$, we get $Lin\{H_{x_0}\} = L^0$. Hence $H_{x_0}^0 = \left(Lin\{H_{x_0}\}\right)^0 = (L^0)^0 = L$.

(ii) ⇒ (i) is obvious. □

Proof of Lemma 18.4: It is similar to that of Lemma 6.2. □

Proof of Proposition 18.4: (i) ⇒ (ii) Statement (i) is equivalent to that there is a $n \times n$ matrix $A = (a_{ij})$ such that for each $\lambda \in \Lambda$:

$$p(\lambda)\,\varphi_1(\lambda) = a_{11}\varphi_1(\lambda) + \cdots + a_{1n}\varphi_n(\lambda),$$
$$\dots$$
$$p(\lambda)\,\varphi_n(\lambda) = a_{n1}\varphi_1(\lambda) + \cdots + a_{nn}\varphi_n(\lambda).$$

Let us show that $A - p(\lambda)I$, where I is the identity matrix, is a zero matrix. Indeed, for every $\lambda \in \Lambda$ it maps the vector $(\varphi_1(\lambda), \dots, \varphi_n(\lambda)) \in \mathbf{R}^n$ in the null-vector.

Consider in $\mathbf{R}^n$ a vector subspace E generated by such vectors, so that $E = Lin\{e(\lambda) = (\varphi_1(\lambda), \dots, \varphi_n(\lambda)) : \lambda \in \Lambda\}$. It suffices to verify that $E = \mathbf{R}^n$. If $E \neq \mathbf{R}^n$ then there exists a vector $b = \left(b_1, \dots, b_n\right) \in \mathbf{R}^n, b \neq 0$, such that $\langle e(\lambda), b\rangle = \sum_{i=1}^n \varphi_i(\lambda)\, b_i = 0$ for all $\lambda \in \Lambda$. But this contradicts our hypothesis that $\varphi_1, \dots, \varphi_n$ are linearly independent. Thus $E = \mathbf{R}^n$ hence $A - p(\lambda)I$ is a zero matrix. Then $a_{ij} = 0$ for $i \neq j$, and $a_{ii} = p(\lambda)$ for each $i = 1, \dots, n$ and every $\lambda \in \Lambda$; therefore, $p(\lambda) = a_{11} = \dots = a_{nn}$ is constant.

(ii)⇒(i) is obvious. □

Index

Note: Page numbers in bold refer to tables, *italic* refer to figures, 'n' refer to notes.

a

absorbance 4n12, 14, 16, 46, 51, 83, 83n8, 83n9, 84, 85, 102, 109, 110, 130, 153, 155–157, 157n7, 158, *159*, 161–164, 180, 181, 213, 214, 362, 403
absorptance 4n12, 13, 83, 84, *85*, 86, *88*, 92, 104, 105, 109, *153*, 162, 180, *181*, 317
adaptation
- adaptation state 4n13, 5n16, 37, 39, 43, 44, 302, 319–321, 326, 327n12
- adaptation to inverted vision 474
- light adaptation 18, 18n37, 37, 39, 41, 44, 47, 141, 316, 317, 319–321, 326, 327, 334, 371n3, 457n23

albedo 352, 444–446
annihilator, of a vector subset 139, 278, 283, 286–287, 499
array
- ambient optic array 39, 57, 328, 352, 452n16, 465, 466, 468
- colour-signal array 41, 43–45, 328–330, 333, 333n24, 333n28, 334, 348, 352, 360, 465, 466
- colour-stimulus array 328–330, 333
- light-stimulus array 39–41, 328–333, 333n24, 335, 346, 347, 349, 354

assimilation, chromatic 6
assumption, regularity 46, 48, 363–365, 381, 385–387, 395, 439, 522
atlas
- light atlas 52, 54–56, 426–427, 427n3, 429, 429n5, 430–436, 439–441, 444, 454–455
  - pseudo-monochromatic light atlas 427, 432, 434
- light colour atlas 15, 16
- object atlas 47, 48, 50–52, 54, 55, 377–381, 383, 385–387, 394, 395, 426–427, 430–434, 436, 439, 440, 443, 455n21
  - colour-charted object atlas 51, 395
  - full object atlas 50, 378, 379, 426
  - illuminant-invariant object atlas 378, 427
  - optimal object atlas 48, 379–381, 385, 386, 394, 395, 427, 431
  - rectangular object atlas 48, 50, 381, 383
    - extended rectangular atlas 48, 386, 395, 434, 439
- object-colour atlas 378, 395
- object/light atlas 52, 54, 56, 426–427, 427n3, 429, 429n5, 430–432, 435, 436, 439, 441, 444, 454, 455
  - light component of object/light atlas 52, 56, 443
  - object component of object/light atlas 52

b

background 34, 37, 44, 45, 57, 59, 60, 201, 202, 252, 275, 276, 280, 280n1, *281*, 282, 283, 285, 287–291, 302, 303n4, 304–306, 320, 325–327, 327n12, 328, *345*, 357, 446, *447*, 450, 451, 469
- remote background 37, *281*, 302

basis, of partial colour matching structure 220–223

Foundations of Colour Science: From Colorimetry to Perception, First Edition. Alexander D. Logvinenko and Vladimir L. Levin.
© 2023 John Wiley & Sons Ltd. Published 2023 by John Wiley & Sons Ltd.

brightness 2n4, 3, 3n7, 7n19, 9, 9n25, 11n28, 19, 20, 24, 24n41, 25, 26, 28, 45, 56, 57, 101n5, 187–212, 214, 238, 238n40, 238n41, 239–243, 249, 249n54, 251, 252, 254, 258, 259, 259n11, 260, 261, 325, 328n13, 431, 434, 436, 441, 444–449, 466
  lighting brightness 56, 57, 202, 436, 442, 444–451, 459–460, 467
  $\mathfrak{m}$-lighting brightness 434
  shading brightness 57, 448–450, *450*, 451, 459
brightness-robustness index 261

C

catch, quantum 4, 4n13, 5, 5n16, 8, 13, 38, 39, 82, 317, 320, 321
cell
  colour-luminance double-opponent cell 334, 341
  L − M ganglion cell 322–326
  $M + L$-double-opponent cell 344
  $M - L$-double-opponent cell 341–344
  M − L ganglion cell 322–326
  simple cortical cell 334–341
    principal direction of simple cortical cell 335
  $S - (M + L)$-double-opponent cell 343–345, 348
chromaticity 21–24, 49, 50, 54, 55, 222–224, *224*, 225, *225*, 226–230, 233–234, 236, 237, 237n39, 238–239, 239n45, 239n46, 240, 241, 243, 249, 253–255, 255n5, 390–393, 399, 400, 406, 406n20, 428, 434, 435
  chromaticity gamut 101, 101n6, 102, *102*, 109, 186
  chromaticity palette 23, 24, 234, 237, 237n39, 238, 239, 241, 392, 393
    neutral element in chromaticity palette 234, 392
    proper chromaticity palette 23, 24, 234, 237, 237n39, 241, 392
  chromaticity plane 101, 101n6, 110, 155, *156*, 159, 182, 186, 363
  chromaticity space 207
  chromaticity structure 23, 24, 230, 233, 234
    proper chromaticity structure 23
chromaticness 23–25, 28, 55, 231, 233–235, 235n33, 238, 240, 241, 243, 248, 252, 254, 255n6, 349, 391, 391n6, 392, 392n7, 395, 395n13, 428
  chromaticness gamut 25, 240, 241, 243, 248
    iso-bright chromaticness gamut 243
  chromaticness of light 231, 392n7
  chromaticness plane 240, 241
  $\mathfrak{l}$-material chromaticness 434
  $\mathfrak{m}$-lighting chromaticness 434
  object chromaticness 392, 428
chromatic value 231, 233, 236, 237, 241, 244–247, 392, 393, 401n6, 435
CIE 1931 Standard Colorimetric Observer *76*, *77*, 78, *91*, *180*, 181
CIE 1964 Supplementary Standard Colorimetric Observer 77, *77*, 165, 166, *167*
circle, of visible spectrum 381, 441
class
  chromatic class 222, 223, 227, 235–238, 240, 255, 391
  chromaticity class 21–24, 49, 50, 54, 55, 222–224, *224*, 225, *225*, 226, 227–230, 234, 236, 238, 239, 239n45, 239n46, 240, 243, 249, 253–255, 255n5, 390–393, 399, 400, 406, 406n20, 428, 434, 435
    binary chromaticity class 235–237, 391
    neutral chromaticity class 24, 222, 234, 238, 239, 391
    opponent chromaticity classes 239, 243, 393, 434
    singular chromaticity class 222
    unitary chromaticity class 222, 223, 225, 227, 235, 236, 240, 391
  component hue class 220–223
  τ-class 219–221, 237, 513
  τ-preclass 219, 220
closure
  spectral colour-signal closure 114, 211
  spectral colour-signal conical closure 114
  spectral light stimulus closure 114
  spectral light stimulus conical closure 114, 211, 234, 242, 242n49
code
  brightness code 195–203, 242, 258, 259

colour code 26, 27, 43, 50, 64–72, 187, 195, 232, 242, 243, 245, 249–251, 323, 324, 393
Grassmann colour code 64–72, 243, 245
continuous Grassmann colour code 70–72
linear colour code 69, 71
non-trivial colour code 69
ordinal chromatic colour code 232
colour-signal code 43, 69, 232, 242–245, 251, 331, 332, 333n24, 334, 335, 342, 343, 345–348, 348n12
$M-L$-colour-signal code 342
ordinal chromatic colour-signal code 252
lightness code 394
luminance code 198, 334
opponent chromatic code 26, 27, 242–244, 247, 248, 258, 321–326, 394, 395, 401
colorimetric paradigm 5n16, 6, 19, 26, 54, 187, 243, 252, 258n9, 289, 431, 434, 473
colorimetry xi, xii, xiii, 5–9, 11, 13–16, 19, 28, 29, 29n46, 31, 31n48, 32, 36, 41, 74, 74n2, *76*, 187, 187n2, 204, 238n42, 248n53, 253, 320, 349, 454, 473
colour
achromatic colour 2, 86, 401, 455
binary colour 22, 23, 28, 228, 228n29, 235n33, 237, 241, 255n6
chromatic colour 2, 250, 251
full colour 23, 24, 47, 233–237, 240, 241, 242n49, 243, 247, 255
full-colour contour 241, 243
iso-bright full-colour contour 243
light colour xii, 5–7, 7n21, 8, 8n24, 9, 10n27, 11n28, 15, 16, 19–22, 24–28, 45–56, 58–72, 62n4, 73–141, 143–252, 217n22, 253–261, 263–355, 357–408, 391n6, 392–394, 394n12, 396, 400, 401n17, 408, 409–419, 421–437, 439–457, 445n6, 455, 456, 459–471, 473–475, 527
complementary light colours 116n7, 241
full light-colour 233n32
opponent light colours 241
lighting colour 8, 53, 54, 54n70, 55–58, 401n15, 422, 423, 425, 426n2, 428, 431–436, 439, 443–455, 457, 457n23, 459–461, 470, 473
material colour 8, 53, 54, 54n70, 55–58, 401n15, 422, 423, 430–437, 434n9, 444, 446–449, 451–453, 455, 457, 459–461, 469, 470
medium colour 5, 7, 470
neutral colour 235, 238, 280
object colour xii, 7, 7n21, 8, 21, 45–56, 58, 62n4, 200, 215, 217n22, 349, 351–355, 357–419, 421–437, 439–457, 459–471, 473–475, 527
full object colour 51, 393, 393n10, 401, 402, 402n19, 403–407
lighting component of object-colour 54, 422n1, 433–437
material component of object-colour 422n1, 434
related colour 5, 7, 8
shading colour 354, 449
spectral colour 66, 80, 138, 148, 204
simple spectral colour 138, 148
unitary colour 22, 23, 236, 237, 241, 349, 391, 394, 408
unrelated colour 5, 8
colour category 2, 187, 226–229, 399
colour chart 50, 51, 378, 380, 395, 400–402, 434, 434n9, 435, 439, 444, 454, 455, 457
colour constancy 8, 52, 53, 202, *423*, 426, 436–437, 473
material-colour constancy 437
colour detection 36–38, 59, 252, 280, *281*, 301–316
colour detector 36, 285, 308, 313, 315
colour inconstancy 437
material-colour inconstancy 437
colour pyramid 191
colour signal 11, 14–17, 19–21, 26, 28, 30, 43, 45, 50, 67, 69, 70, 93–115, 137, 138, 151, 159, 194, 195, 203, 210–214, 229, 230, 232, 242–245, 250–251, 257, 258, 265, 272–271, 288, 328, 331, 332, 333n24, 334, 335, 342, 343, 345–348, 348n12, 357, 359, 390, 391, 453, 509, 510, 517
additive colour signal 70, 70n14, 141, 348
coloured shadows 455–457

condition
  concordance condition 23, 236, 237
  independence condition 89
  monotonicity condition 275, 276
  two-concomitant-hue condition 236, 237
cone
  colour-signal cone 14–17, 19–21, 26, 28, 30, 45, 50, 67, 69, 70, 93–115, 137, 138, 151, 159, 194, 195, 203, 210–214, 229, 230, 232, 242, 243, 250, 251, 257, 258, 265, 272–274, 288, 328, 348, 357, 359, 390, 391, 453, 509, 510, 517
    protanopic colour-signal cone 212
  colour stimulus cone 10, 11, 14, 15, 66, 67, 81, 94, 111, 112, 115, 203, 205, 218, 229, 268, 271–273, 274n10, 276, 277, 348
  light stimulus cone 9, 10, 14, 16, 26, 61–65, 68–70, 72, 93, 94, 113, 114n4, 117, 143, 150, 192–195, 203, 205, 229, 230, 238, 248, 255, 267, 268, 271, 272, 274, 277, 280n1, 301, 328, 510
cone fundamental 12, 13, 13n32, 14n33, 17, 19, 26, 27, 30, 31, 35–38, 40, 40n59, 41, 41n60, 43, 46–48, 51, 86, 86n11, 86n12, *88*, 89–95, 100, *101*, *103*, 104, *104*, 107, 121, 123, 125, 152, 163–167, *167*, 196, 198, 213, 243, 247, 286, 315, 316, 322, 329–333, 333n25, 333n28, 334, 348, 361, 363, 365, 368, 376, 376n2, 384, 395, 396, 403, *403*, 404, *404*, 455n22, 463
  Govardovskii's cone fundamentals 46–48, 51, 86, *88*, 92, 94, 104, 123, 125, 198, 210, 213, 363, 365, 368, 376, 384, 403, *403*, 404, *404*
coordinates
  chromaticity coordinates 200, 207n12, *366*, *374*, 423
  colorimetric coordinates 11, 15, 70n14, *76*, 208, 456
  light-colour coordinates 260
    cylindrical coordinates 11n28, 20, 119, 206, 208, 242, 396
    n-colour stimulus coordinates 277, 303, 304
    perceptual light-colour coordinates 25, 26, 28, 210, 240–242, 255–258, 261
    psychophysical colour coordinates 11n28, 20, 202–210, 396
    $qp\lambda$-coordinates 115–117, 121–128
      achromatic component of $qp\lambda$-coordinates 116
      chromatic component of $qp\lambda$-coordinates 116
  lighting-colour coordinates 433–436
    $\mathfrak{m}$-lighting-colour coordinates 433–436
  material-colour coordinates 434, 436
    $\mathfrak{l}$-material-colour coordinates 434, 436
  object-colour coordinates 50, 51, 54, 55, 394–396, 399–401
    $c\Delta\lambda\bar{\lambda}$-coordinates 48–50, 383, 386, 387, 395, 396, *397*, *398*, 399–401, 434
    perceptual object-colour coordinates 50, 51, 54, 55, 394–395, 399–401
    psychophysical object-colour coordinates 396
curve
  constant hue curve 204, 204n9
  spectral colour-signal curve 94, *94*, *95*, 96, 105, 106, 112, 114, 115, 138, 211, 213, 360–362
    protanopic spectral colour-signal curve *212*
  spectral curve 14, 94, 109, 153, *154*, 211, 243, 255
  spectral light stimulus curve 114

d

derivative
  directional derivative 336–343, 339n34, 501
deuteranope 51, 85, 86, 92, 210n16, 214, 215, 404, 405, 408
dichromat 5n14, 11, 12, 20, 21, 50, 51, 76, 78, 87, 88, 131, 210–215, 401–408
dichromatic orbit 50, 51, 402, *402*, 402n18, 404, 405, *405*, 407
difference, just noticeable 199
dispersion of light-colour 28, 257–260
  brightness midpoint of light-colour dispersion 259
  brightness-range of light-colour dispersion 259
  light-colour dispersion region 28, 257, 259
  relative brightness range of light-colour dispersion 260

relative RG-range of light-colour dispersion 259
relative YB-range of light-colour dispersion 259
RG-midpoint of light-colour dispersion 259
RG-range of light-colour dispersion 259
YB-midpoint of light-colour dispersion 259
YB-range of light-colour dispersion 259

e

effect
- Abney effect 204, 241n47
- Bezold-Brücke effect 205, 208, 254
- Helmholtz-Kohlrausch effect 196, 200

energy
- radiant energy 4, 4n12, 6, 7, 18, 61, 63

environment
- admissible environment 6, 7
- test environment 6

equation
- Naka-Rushton equation 318
- perceptual equations 58, 460, 460n4, 466, 467

equivalence
- chromatic equivalence 23, 230, 231, 233, 238, 239, 241, 391, 393, 394
- light-colour equivalence 24–26, 52, 239, 242, 242n48, 249, 249n54, 250, 251, 258, 421, 423, 425–429, 454, 455
- matching equivalence 29, 268, 269, 271, 273
- object/light colour equivalence 52, 421, 423–429, 454, 455
  - inseparability of object/light colour equivalence 52, 426
- psychophysical equivalence 35, 303, 303n3

f

fabric, sensory 7n19, 57, 328n13, 429, 464–470, 466n9, 469n14, 474

fibre, *see also* fibre bundle
- object hue fibre 395

fibre bundle 19, 21, 24, 202–208, 215, 237, 237n39, 348n43
- brightness fibre bundle 202–208
- hue fibre bundle 19, 24, 207, 215, 237, 237n39, 401
- purity fibre bundle 207

field, receptive 321–323, 325, 330n19, 331, 333, 335, 338

frequency, spatial 215, 344

function
- bidirectional reflectance distribution function (BRDF) 352, 353
- characteristic function 359, *see also* indicator
- colour matching function xii, 6n18, 12, 13, 13n32, 19, 26, 31, 31n48, 34, 46, 74–82, 86, 87, 89, 90n14, 91, *91*, 92, 95, 98, 103, 120, *121*, 167–179, 181, 197, 198, 277n13, 279, 316, 361
  - CIE 1931 colour matching functions 76, 78, 79, 90, 90n14, 95, 103
  - CIE XYZ 2015 colour matching functions 78, 92
- component hue function 218, 219
- detection function 34–37, 294, 302–305, 307, 309–313, 315, 520
  - *n*-colour stimulus coordinate representation of detection function 303, 304
  - psychophysically equivalent detection functions 35, 36, 303
- discrimination function 34, 302, 315
- generalised function 9
- material spectral reflectance function 352, 353
- Minkowski function 35, 291, 292, 302, 500
- psychometric function 282, 301, 302, 307
- psychophysical function 3, 199
- radiant spectral density function 63, 63n5, *116*, 130, 135, 166, 179, 328, 355, 357, 379, 398
- sensitivity function 13, 33, 35, 39, 42, 71, 71n15, 76, 81, 93, 103, 109, 118, 135, 136, 152, *163*, 164, 178, 179, 280–282, 285, 303, 305, 308–312, 318, 323, 324, 331–336, 338, 342, *343*, 344, 346, 357, 360–362, 371, 409, 412
- spectral luminous efficiency function 76, 78, 197
- spectral reflectance function 17, 45–48, 51, 352, 353, 355, 357, 359, 361, 375, 405, 416, 421, 440, 441, 444, 445, 449, 459n3, 460, 470
- step function 46, *131*, 361, 363–364, 403, 404, 412

function (*contd.*)
elementary step function 48, 361–365, 367, 368, 371–373, 380, 381, 383, 412
degree of elementary step function 361, 362
type of elementary step function 361

g

gain control 42, 319, 333n27, 334
gamut
colour-signal gamut 70
light-colour gamut 55, 240–243, 243n50, 248, 254, 254n2, 255, 255n6, 256, 256n8, 257–259, 394n12, 435
perceptual light-colour gamut 25, 28
lighting-colour gamut 435, 436
m-lighting-colour gamut 435
material-colour gamut 434, 453
ɩ-material-colour gamut 434, 453
object-colour gamut 50, 51, 55, 386, 394, 395, 436
object-colour stimulus gamut 48, 50, 385–386, 395–397, 401
colour-charted object-colour stimulus gamut 395
object/light colour gamut 435, 436, 454
Ganzfeld 357n1, 454, 454n20
gauge 173, 282, 285, 291, 302, 303, 309–311, 500, 501
Govardovskii's template 83, 86, 92, 107, 125, 128, 132, 386
gradient 43, 43n62, 43n63, 44, 45, 57, 291, 293n20, 296, 297, 312, 313, 329, 330n18, 334, 339–346, 348, 450, 452, 453, 463
gradient encoding 43, 43n63, 44, 340–345, 348, 452, 453
directional gradient-encoding 43, 340–345, 348

h

homeomorphism
colour-graduation homeomorphism 242, 257
colour-signal-conversion homeomorphism 257
colour-stimulus-conversion homeomorphism 257
hue 2, 2n4, 3, 5, 6, 6n17, 9, 11n28, 15, 15n34, 16, 19–21, 21n38, 21n39, 22, 23, 23n40, 24, 24n42, 25, 27, 28, 49, 101n5, 187, 188, *188*, 189–191, 191n6, 202–241, 243–250, 252, 255, 256, 280, 322, 328n13, 344, 347, 380, 390–397, 399–401, 403–408, 423, 428, 431, 434, 435, 441, 442, 445, 456, 466, 470, 513
achromatic hue 217
binary hue 218, 224
component hue xii, 21, 21n38, 22, 23, 23n40, 24, 24n42, 25, 27, 28, 38, 49, 51, 55, 216–222, 223n26, 225–231, 235, 235n36, 236, 237, 237n39, 238, 238n42, 239–241, 243, 244, *244*, 245, 246, *246*, 247, *247*, 249, 249n54, 390–392, 397, 399, 406, 408, 428, 435, 470, 513
component-hue value 218, 220
concomitant component hues 22, 24
opponent component hues 49, 243, 392
dichromatic hue 51, 212–214, 401, 403, 406
improper hue 23, 24, 216, 233, 234, 392, 393
ɩ-material hue 434
m-lighting hue 434
non-spectral hue 188, *188*, 216, 216n18
opponent hues 49, 241, 392, 393
p-nary hue 218
proper hue 23, 233–235, 237, 392, 393
unique hue 21, 205n10, 217, 217n22, 223, 228, *245*, 255, 322, 395n13
unitary hue 205n10, 218, 226, 245
hue palette 21, 28, 51, 188n3, 211, 214, 215, 229, 233, 235, 236, 392, 401, 403, 406
proper hue palette 233
hyperplane, supporting 369, 519
hypothesis
albedo hypothesis 445, 446
apparent form-slant invariance hypothesis 444
apparent size-distance invariance hypothesis 443
colour-signal-cone-boundary hypothesis 96–109, 113–115, 138, 203, 508–510
strict colour-signal-cone-boundary hypothesis 138

strong colour-signal-cone-boundary hypothesis 96–105, 109, 114, 138
cone-fundamentals hypothesis 86, 361
weak cone-fundamentals hypothesis 86
cone-photopigment quantum catch hypothesis 82, 86
weak cone-photopigment-quantum-catch hypothesis 82, 86
directional gradient-encoding hypothesis 43, 341–343, 345, 348
flat patchy world hypothesis 354
fundamental-colour-mechanisms hypothesis 82
weak fundamental-colour-mechanisms hypothesis 82
Grassmann-opponent-chromatic-code hypothesis 243–248
lightness/lighting-brightness invariance hypothesis 446
metameric matching hypothesis 73, 86, 96, 153n1, 220, 259
perceptual hypotheses 468–470
reductionist hypothesis 20, 21, 50, 51, 211, 214, 215, *406*
rod intrusion hypothesis 78

illuminant 17, 45–55, 116, *116*, 130, 331, 355, 355n5, 357, 360, 361, 371, 371n3, 375–377, 379, 380, 383, 385, 386–388, 399–401, *402*, 402n18, 409, 417, 419, 421, 425, 426, 428, 431, 433, 436, 437, 444–446, *447*, 453, 453n18, 455n21, 456, 459, 463, 470
admissible illuminant 386, 399
illuminant D65 116, *116*, 396, 444
image
cone-photoreceptor image 41, 43, 330, 332, 333
intrinsic image 459, 459n1, 460
lighting image 354
lighting-brightness image 449, 459
lightness image 449, 459
material image 354
neural image 40, 42, 43, 330–334
shading image 354
shading-brightness image 449, 450, 459
index
light metamerism index 141
metamer-mismatch index 148, 154, 155, 186, 416, 417
chromaticity metamer-mismatch index 155–159, *157*, *158*, *160*, *169*, 180, 186
linear metamer-mismatch index 153, *161*, *169*
metamer-mismatch index for objects 416, 417
first metamer-mismatch index for objects 416
n-dimensional metamer-mismatch index for objects 417
n-dimensional metamer-mismatch index 150, 151
alternative *n*-dimensional metamer-mismatch index 151
one-dimensional metamer-mismatch index 149
volumetric metamer-mismatch index 153, 156, *156*, 157, 160, *168*, 169, 171, 173, 175–178, 186
indicator 315, 359, 360, 362, 363, 364n5, 370, 376, 381, 381n6, 385, 410, 521, 523, 526
intensity, spectral 351
interval
dichromatic confusion interval 20, 210, 212
effective visible spectrum interval 109, 114n5
purple interval 101, *103*, 110, 211, 216, 216n18, 243

j

junctions, gap 320, 333

k

Krantz Theorem 67–72, 75, 89, 503, 504

l

language, visual 467, 467n10, 469, 471, 474, 475
law
Abney's law 196, 197
Fechner Law 200
Grassmann's laws xi, xii, 10, 12, 18, 28, 29, 29n46, 64, 67, 70, 71, 73, 194, 196, 232, 251, 263, 267, 279

law (*contd.*)
  Weber-Fechner Law 200
  Weber's Law 199, 276
level, of discriminability 280–282, 302, 307, 310
light xi–xiii, 1–141, 143–261, 263–355, 375, 377, 389, 389n1, 390–394, 396, 399–400, *403*, *404*, 408, 415, 417, 419, 421–423, *424*, 425–437, 439–450, *450*, 451–457, 459–463, 467, 469n13, 470, 473, 474, 475, 497, 499, 503, 507, 509, 512, 513
  colour independent lights 65, 66n7, 67
  complementary lights 114, 203, 204
  improper light 62
  metameric lights xiii, 17, 26, 26n44, 47, 141, 266, 375, 375n1, 426
  nil light 62, 62n4, 67, 67n8, 136, 136n2, 199
line
  neutral line 19, 205, 207, 214, 330
  purple line 100, *101*, *103*
locus, spectral 101, *101–103*, 102, 104, 105, 109, 110, 155, *155*, 363
luminance 40–45, 56–58, 155, 160, *161*, *162*, 186, 190, 196, 198–201, 205, 263, 305, 307n8, 322, 325, 326, 334–341, 344, 350, 423, 445, 446, *447*, 448, 449, 449n14, 453n18, 454n19, 459, 461, *461*, 463

m

manifold
  colour stimulus manifold 15, 16, 19, 20, 24, 47, 55, 111–134, 148, 205–208, 234, 242, 248–252, 257, 258, 328, 375, 427
    colour-graduated colour-stimulus manifold 242, 257, 258
  light-colour manifold 24, 25, 54n70, 234–252, 254–256
  material-lighting colour invariance manifold 439, 452, 453
  object-colour stimulus manifold 47, 48, 49, 375–408, 427
    colour-charted object-colour stimulus manifold 395
    trichromatic regular object-colour stimulus manifold 381–385
  object/light colour manifold 54, 55, 436
  object-light invariance manifold 439
map
  colour map 16, 25n43, 149n3, 281, 360, 430, 432, 433, 436
  colour-signal map 67, 67n9, 68, 70, 71, 73–75, 79, 93, 94, 96, 111, 113, 114, 117, 137, 143, 144, 148–151, 203, 232, 242, 250, 257, 265, 272–274, 278, 283–285, 288, 303, 309, 328, 329n15, 348, 358, 359, 361, 365, 376–379, 383, 409, 416, 417
  colour stimulus map xi, xii, 66, 111, 116–120, 121, 123, 268, 277, 278, 301, 507
    conjugate colour stimulus map 277, 278, 507
    non-linear colour stimulus map 116–121, 123
    null-space of colour stimulus map 277, 278, 281, 301, 359
  convex map 314, 315
    subdifferential of convex map 314
  light-colour map 25, 26, 240–242, 256, 258, 258n10
    anchored light-colour map 240
  lighting colour map 432
    canonical lighting colour map 432
  material colour map 430, 432, 436
    canonical material colour map 430, 432
  non-linear object-colour stimulus map 380–381, 385, 387
  perceptual object-colour map 394, 400
match 5, 53, 64, 125, 133, 138, 188, 206, 220, 263
  brightness match 189, 190, 191, 240
    heterochromatic brightness match 19, 190, 200, 238, 240
  colour match 5n16, 9, 13, 17, 18, 27, 29, 30, 36, 53, *59*, 64, 78, 79, 82, 117, 125, 168, 190, 194, 210, 249, 249n54, 251, 252, 266, 271, 287, 306, 326, 327, 427, 428, 429, 430, 431
    asymmetric colour match 53, 326–328, 400n14, 401, 427–430, 432, 433
  hue match 188, 203, 206n11, 392
    component hue match 219
matching 29, 143, 210, 216, 263, 264, 271
  colour matching xi, 9, 10, *59*, 64, 74–81, *91*, *92*, 120–134, 167–179, 188, 263–278, 279–300

asymmetric colour matching 326–328, 400n14, 401, 427–430, 429n4, 432, 433
partial colour matching 21n39, 219–223, 226–228
hue matching 19, 191, 203, 205, 206, 206n11, 219, 226, 232, 233, 392
partial hue-matching 217–228, 232, 234–236, 238, 239, 255, 256, 390, 391, 399, 400, 400n14, 406, 428, 470
metameric matching 10, 17, 29n46, 30, 73–92, 143n1, 210, 249, 279
matrix, pseudo-inverse 134, 298, 298n27
mechanism, colour 11–14, 17, 30, 32, 34, 36, 38, 61–72, 93, 139–141, 306, 317–360
cardinal colour mechanisms 35, 37
fundamental colour mechanisms 12–14, 17, 18, 25, 26, 30, 31, 31n48, 32–35, 38, 39, 46, 47, 81–93, 132–135, 138, 148, 196, 249, 250, 250n57, 258, 264, 265, 285, 286, 294–300, 304–307, 306n6, 307, 376, 381, 455n22
most sensitive colour mechanism 36, 285, 308, 310–315
metamer 10, 66, 115, 141
optimal metamer 48–50, 379, 381, 385, 387, 395
chromatic contrast of optimal metamer 379
spectral waveform of optimal metamer 379
quasi-metamer 277
rectangular metamer 48, 50, 51, 381, 383, 384, 386, 400, 405, 406
central wavelength of rectangular metamer 48, 382
chromatic contrast of rectangular metamer 48, 383
imaginary rectangular metamer 48, 386
spectral bandwidth of rectangular metamer 48, 382
metamerism xiii, 10, 17, 64, 73, 117, 143, 150, 317, 327, 375, 375n1, 421
degree of metamerism 17, 18, 141
fundamental metamerism 12, 13, 25, 38, 40n59, 249–252
light metamerism 17, 47, 111, 135–141, 377
matching metamerism 29, 29n46, 30–34, 263–279, 283, 288, 294, 301, 304, 305
null-space of metamerism 65
object metamerism 17, 47, 51, 375–378, 380, 393, 394
mismatching, of metamers 14, 17–20, 27, 28, 51–53, 143–186, 208–210, 253, 257, 260, 277, 409–419, 425, 437, 444, 460n3, 463n6
chromaticity metamer-mismatch area 154, *155*, *156*, 158, *159*, *164–167*, 166, 179, 180, *180–182*, *185*
metamer-mismatch region 143–149, 151–186, 257, 259, 409–417
metamer-mismatch volume 152, 153, *154*, 155, 156, 161, 162, *163*, *164*, 165, *166*, 168, 172, 179, 182, *185*, 260, 444
model
colorimetric model of colour matching 13, 35, 82, 263, 264, 306
detection model 35, 302, 303, 303n3, 304, 307–308, 310, 311, 313, 316
convex detection model 313
criterion-independent detection model 303
detector-summator model 304
multi-channel detection model 307–308, 310
peak-detector model 304, 306, 308–315
p-norm colour detection model 304, 306, 306n6, 307, 309, 316
polar peak-detector model 308
sublinear detection model 308
multi-channel sublinear detection model 310–311
Lambertian model 353
stochastic model of colour matching 267
monochromat 76
blue-cone monochromat 85, 92
rod monochromat 85

O

object-colour solid xii, 45–48, 50, 357–364, 365, 369, 370, 371, 376, 377, 380, 383, 384, 386, 390, 391, 393, 401, *402*, 408–410, 412, 415, 416, 417, 418, 444, 452, 452n16, 453, 454, 527
black pole of object-colour solid 365–367
centre of object-colour solid 48, 391, 393
equator of object-colour solid 369–374

object-colour solid (*contd.*)
regular object-colour solid 46, 47, 365, 366, *366*, 367, *368*, *369*, *372*, *373*, 385, 386
trichromatic regular object-colour solid 365–384
meridians of trichromatic regular object-colour solid 365–369
spherical representation of object-colour solid 384
white pole of object-colour solid 358, 366
opponency
cone opponency 322, 323
spatial opponency 322
order
cyclic order 235–237
hue cyclic order 235–237
linear order 191, 204, 237
purity order 20, 204, 206, 233, 241
saturation order 241
weak order 191, 228, 229
ordering
achromatic ordering 192–195, 230, 232, 238, 239, 249
brightness ordering 191, 192, 195, 238, 240, 249
chromatic ordering 22, 23, 49, 228–234, 238, 239, 249, 254, 390, 391
proper chromatic ordering 230
continuous achromatic ordering 193–195, 249
purity ordering 204

p

paradigm, colorimetric experimental 5, 6, 8, 9, 389, 454
parallelism, psycho-physiological 44
pattern
colour signal code pattern 40, 43–44, 58, 332–334, 342, 343, 345–348, 448
luminance pattern 42, 45, 57, 58, 334–341, 350, 448, 448n12, 449, 449n14, 449n15, 459, 463
$(M+L)$ pattern 41, 43, *see also* luminance pattern
$(M-L)$ pattern 41, 342
$(S-(M+L))$ pattern 41
cone excitation pattern 41–43, 333, 348
cone fundamental pattern 40–42, 332–334, 463
illumination pattern 354, 448, 448n13, 449n14, 459
lighting pattern 354, 449n14
reflectance pattern 354, 448, 459, 460
shading pattern 354, 449n14, 459, 460
perception
perceptual adaptation 466, 474
perceptual conflict 468, 470
perceptual content 464–469, 469n11, 470, 471, 474
perceptual rules 452, 457n23, 469
photon 4, 5, 13, 264, 320
photoreceptor 4, 5, 16, 18, 20, 39, 40, 41, 41n60, 42, 44, 46, 75, 78, 83, 84, 85, 111, 147, 197, 250, 251, 264, 272, 286, 306, 316–320, 321, 322, 326, 327, 330, 331, 333, 333n26, 345, 348, 351, 361, 371, 371n3, 457n23, 461
cone photoreceptor 5, 5n14, 5n16, 11n30, 12–14, 16, 27, 37–39, 41, 43, 50, 75, 76, 78, 82, 85, 86, 210, 264, 317–323, 326, 330–334
rod photoreceptor 78
phototransduction, visual 4, 13
Planck's constant 4, 130
plane, visual 332, 333n28, 335n30, 337, 345, 345n39
point
extreme point 17, 18, 94, 94n2, 98, 135, 137, 144–146, 149, 152, 359, 376, 410, 411, 413, 453, 507–509, 511–513, 523, 524
tangency point 214, 289, 289n11, 291–293, 293n19, 293n21, 294, 519
polar 35–37, 237, 285, 308–312, 499–501, 519
preorder 191–195, 204, 228–230, 232
complete preorder 191–195, 228, 229, 232
continuous preorder 194, 195
order dense preorder 192, 193
total preorder 191
preprocessor
colour preprocessor 35, 308, 309, 311, 313, 519
primaries 22, 26, 42, 65, 73, 75, 76, 80, *121*, 348, 464–466
missing primaries 89

monochromatic primaries 74, 75, *76*, 79–81, *121*, 251
rinciple
principle of continuity 71, 212
retinotopy principle 330
univariance principle 4, 13
roperty, of strict convexity 106, 107, 112, 138, 203
rotanope 51, 85, 92, 131, 210n16, 214, 215, 404, **404**, 405, 408
seudoscope 447, 447n10, 450, 465, 467n11, 468–470
urity 20, 23, 28, 47, 51, 131, 188, 188n4, 189, 189n5, 190, 191, 202–212, 214, 228n29, 232–234, 238n42, 241, 247, 249n54, 252, 380, 393, 395, 401, 406, 434, 470, 473
colorimetric purity 434
l-material purity 434
m-lighting purity 16, 131

adiation, electromagnetic 3, 4, 462
adiator, black-body 388
egion, of matching criterion 265, 265n5, 266, 267
relation
achromatic matching relation
local achromatic matching relation 239
brightness matching relation 192, 239
colour matching relation 29, 267, 268, 271, 274, 276, 279, 283, 288, 294, 301
partial colour matching relation 226, 227
stochastic colour matching relation 267, 280
tolerance extension of colour matching relation 282, 283
regular tolerance extension of colour matching relation 282–285
strongly regular tolerance extension of colour matching relation 282–285
hue matching relation 19
partial hue matching relation 218, 226
indiscriminate relation 30–32, 34, 268, 272, 274, 276, 276n11, 283, 285, 288, 301–316
relationship
invariant relationship between object and light components 439, 444
response-intensity relationship 318, 319
RG-robustness index 261
robustness, of light to medium transmittance change 119, 209, 210

S

saturation 2n4, 9, 11n28, 16, 19, 20, 101n5, 187, *188*, 188n4, 189n5, 202, 204, 232–234, 241, *246*, 247, *247*, *248*, 328n13, 431, 457
scale
achromatic scale 20, 191–202, 230, 231
interval achromatic scale 231
ordinal achromatic scale 192, 230
brightness scale 20, 24, 26, 191, 196, 203, 207, 240, 242, 249, 258, 434
interval brightness scale 196
ordinal brightness scale 192–196
chromatic scale 23–26, 49, 230–232, 234, 236, 238, 240–243, 246, 246n52, 254, 259n11, 392, 391n6, 393, 394, 435, 436
interval chromatic scale 23
opponent chromatic scale 50, 240–243, 259n11, 394, 434, 435
ordinal chromatic scale 230, 232
scaling
component-hue scaling 249, 249n54
Fechnerian scaling 198–200, 204
hue scaling 19, 21, 23, 23n40, 27, 228, 236n38, 246, 247, 249, 249n54, 392, 406
Schrödinger-Maximov constraint 362, 363, 412–414, 522
segment
achromatic segment 358, 369–372, 374, 380, 383, 384, 393n9
complementary colour-signal segment 114
complementary light stimulus segment 114, 203, 204
sensitivity, spectral 12, 13, 18, 37, 38, 39, 42, 43, 46, 71n15, 76, 81, 84, 86, *90*, *92*, 92, 93, 93n1, 103, 109, 111, 113, 118, 135, 136, 137, 152, 153n5, 161, 178, 179, 250, 272, 286, 318, 323, 351, 357, 360, 361, 362, 402n18, 409, 412, 418

sensitivity, spectral (*contd.*)
  cone spectral sensitivity 12n31, 13, 13n32, 33, 84, 86, *88*, *90–92*, 100, *165*, 316, 348n42
separability
  Debreu order separability 193
  geometrico-chromatical separability of reflectivity 352
  spatio-chromatic separability of neural cells 39, 323, 324, 331–333, 341, 346
set
  colour matching set 33, 34, 271, 279, 280–285, 287, 288, 291–294
    extended colour matching set 282–285, 287, 288, 291, 291n15
  colour stimulus set 10, 10n27, 11, 219–221, 229, 268, 271
  constant hue set 204
  indiscriminable set 30–33, 33n53, 34, 35, 35n54, 36, 138, 272, 301, 306, 306n6, 307, 315, 316
  lighting colour set 432
  light set 425
  material colour set 430, 431, 433, 434
  metamer set 10, 10n27, 14–18, 51, 66, 67, 135–138, 141, 143, 257, 317, 378
  object-colour stimulus set 47, 375, 380, 426
  object/light colour stimulus set 52, 425, 426, 429, 431, 435, 441
  object/light stimulus set 52, 425, 429, 454
  object metamer set 47, 49, 376, 511
  object set 425
  set of colour mechanisms preserving metamerism 140, 418–419
  threshold set 280–283, 285, 291, 294, 296n24, 301–303, 305–311
    balanced threshold set 305
    extended threshold set 281n3, 282, 283, 285, 291, 291n16, 294, 296n24, 301–303, 308–310
shift
  average brightness-shift 260
  average RG-shift 260
  average YB-shift 260
  central-wavelength shift 387
  chromatic-contrast shift 387
  colour stimulus shift 20, 117–119, 128, 208
    colour stimulus shift induced by medium transmittance change 117
    relative maximal colour stimulus shift 118
  equivalent (filter-induced) peak-wavelength shift 129
  illuminant-induced object-colour shift 50, 400, 401
  light-colour shift 28, 258–260, 400
  material colour shift 436–437
  maximal β-shift 209
  maximal brightness-shift 260
  maximal λ-shift 209
  maximal p-shift 208, 209
  maximal q-shift 208, 209
  maximal RG-shift 260
  maximal θ-shift 209
  maximal YB-shift 260
  minimal brightness-shift 261
  minimal RG-shift 261
  minimal YB-shift 261
  object-colour stimulus shift 49, 50, 381, 386–388, 400, 401, 436, 444
    illuminant-induced object-colour stimulus shift 381, 387, 388, 400, 401, 436, 444
  psychophysical colour shift 119, 208, 208n14, 209, 258
  Purkinje shift 197
  relative maximal β-shift 209
  relative maximal λ-shift 209
  relative maximal p-shift 209
  relative maximal q-shift 209
  relative maximal θ-shift 209
  spectral-bandwidth shift 387
simultaneous contrast 6, 201, 456
  simultaneous brightness contrast 200, *201*
  simultaneous chromatic contrast 385
space
  colour-signal space 14, 20, 25, 30, 34, 35, 46, 67, 67n9, 70, 79, 94, *95*, 100, 115, 143, 148, 152, 153, 210, 213, 244, 245, 250, 251, 260, 265, 274n10, 316, 322, 330, 360, 371, 379
  colour stimulus space 10, 11, 61–72, 111, 273, 274n10, 277, 330, 425, 507
  cone excitation space 14n33, 38, 94, *94*, *95*, *211*, *212*, 263

light stimulus space 9–13, 33, 34, 66, 71, 136, 141, 186n2, 271, 277, 278, 282, 507
linear colour mechanism space 278, 279, 283, 285–294, 371
object/light colour space 435
perceptual light-colour space 25, 28, 240, 243, 256, 256n7, 258–260
spectrum
absorptance spectrum 13, 83, 84, 86, 92, 317
absorptivity spectrum 83, 83n9
action spectrum 84
electromagnetic spectrum 3, 4
photopigment absorption spectrum 4, 12n31
radiant spectrum 4–9, 59, 63
visible spectrum xii, 4, 5, 5n14, 9, 14, 46, 48, 63, *90*, 95, 101, 105, 110, 152, 153, 168, 179, 196, 197, 243, 245, 355, 359, 376, 381, 384, 403, 404, 407, 441
effective visible spectrum 14, 15, 17, 109, 110, 114n5, 123n11, 243, 363
stimulus
colour stimulus 10, 11, 14, 42, 61–72, 74, 111–134, 187, 203, 204, 248–252, 258, 317–321, 375
hueless colour stimulus 19
imaginary colour stimulus 66
neutral colour stimulus 238, 280
light stimulus 35, 36, 38, 39, 138, 281, 328–330
imaginary light stimulus 280n1
object-colour stimulus 47, 48, 375, 380, 402, 418, 426
object stimulus 47, 48, 49, 357, 367, 377, 378, 379, 380, 381, 383, 387, 388, 391, 395
complementary object stimuli 379
imaginary object stimulus 386
optimal object stimulus 51, 359–361, 363, 370, 378, 379, 381, 382, 384, 412, 521
strictly optimal object stimulus 370
sub-optimal object stimulus 412
structure
achromatically ordered structure 192–194
chromatically ordered structure 22, 23, 49, 54, 192–194, 229–232, 234, 237, 237n39, 238, 249, 253, 254, 390, 391, 393, 399
chromatic structure 22–27, 215–234, 253, 256n7, 470
circular chromatic structure 24–26, 237, 239, 239n45
m-chromatic structure 21, 22, 222, 227, 229, 236
circular Hering structure 24–26, 54, 55, 239–248, 254–256, 393, 434
colour matching structure 29, 30, 34, 38, 263–301, 517
component hue structure 218–223, 227, 228
complete component hue structure 218, 221
degree of chromaticity of component hue structure 218
simple component hue structure 220–222
Grassmann structure 10, 11, 14, 27, 29, 30, 64–70, 73–92, 125, 128, 135, 231, 242–243, 279–300
achromatically ordered Grassmann structure 194
chromatically ordered Grassmann structure 231, 232, 234, 242, 249, 249n55
continuous Grassmann structure 11, 12, 70–75, 112–114, 117, 120, 135, 203, 265, 284, 285
induced Grassmann structure 30, 32–34, 112, 271, 276, 277, 279, 282–284
n-pigment Grassmann-Govardovskii structure 92
n-pigment Grassmann structure 84
3-pigment Grassmann structure 46, 152, 362
regular Grassmann structure 67, 81
weak* continuous Grassmann structure 73–75, 112–114, 117, 120, 135, 203, 265, 284, 285
Hering-Grassmann structure 26, 28, 242–243, 242n48, 249, 250, 257, 258n10, 260n12
partial colour matching structure 219–223
m-chromatic partial colour matching structure 222
partial hue-matching structure 218, 220–222
spherical Hering structure 49, 50, 54, 55, 393, 399, 428, 434
subset, equi-bright 19, 192, 203, 204, 206, 207
subspace, of linear colour mechanisms 30, 31, 33, 34, 36

surface
  colour-signal surface 329, 330, 343–350, 452, 452n16, 453, 465
    tangent bundle of colour-signal surface 345–350, 452, 465
  equi-discrimination surface 33–37, 281, 316
  isochromatic surface 244, 245
  spectral colour-signal conical surface 96, *97*, 205
  spectral surface 14, 463
  threshold surface 292, 292n18, 293, 293n19, 294, 296, 297, 299, 300, 307, 308

t

target 59, 280, *281*, 281, 282, 304, 316, 327
  subthreshold target 282
  superthreshold target 281
technique
  forced-choice technique 267, 267n7, 280
  hue-cancelation technique 245
temperature, colour 130–132, 388
tetrachromat 253
threshold 35, 198, 199, 200, 218, 243n50, 252, 263, 280–282, 283, 285, 291, 301, 302, 347
  absolute threshold 199, 200
  difference threshold 199
tolerance 120, 218–220, 227, 228, 267, 267n8, 268–276, 279, 282–300, 311, 314, 316
  simple tolerance 227, 269–275, 284
  tolerance class 120, 219
  tolerance space 120, 219
transformation, of light-colour gamut 256, 258
trichromat 5n14, 11, 12, 20, 21, 51, 75, 76, 78, 86-88, 105, 157n8, 131, 210, 210n15, 211, 213, 214, 215, 250, 251n59, 253, 256, 401, 406, 406n20, 407, 408
  deuteranomalous trichromat 86, 157n8, 250
  protanomalous trichromat 86, 131, 250, 405, 406, 408
tritanope 51, 85, 86, 214, 215, 405, 406, 408
tritanopia 215
  high-spatial-frequency tritanopia 215
  small-field tritanopia 215
  tritanopia of the central foveola 215

u

utterance, visual 471

v

value
  chromatic value of light 245–247, 435
  value, of component-hue 218, 220
vision
  dichromatic vision 11, 215, 253, 408
  inverted vision 474
  mochromatic vision 5, 76
  photopic vision 75, 197
  pseudoscopic vision 348, 448, 451, 467, 468
  scotopic vision 75, 197
  tetrachromatic vision 210
  trichromatic vision 15, 125, 234, 250, 253

w

wave, electromagnetic 3, 4, 5, 216, 351
  monochromatic electromagnetic wave 5
wavelength 3, 4, 15, 16, 63, *76*, 84, *85*, *87*, *112*, 107, 109, *110*, 116, 118, 130–132, 136, 153, 158, 180, 315, 355, 359, 360, *367*, *404*, **404**
  central wavelength 48, 49, 382, 384–387, 396–398, 405
  complementary wavelength 6, 15, 116, 119
  dominant wavelength 7, 15, 16, 115, 116, 119, 121–123, 125, 131, 204

y

YB-robustness index 261

Printed in the USA/Agawam, MA
September 16, 2026

910661.193